Fundamentals of noise and vibration analysis for engineers

Fundamentals of noise and vibration analysis for engineers

M.P.NORTON
Department of Mechanical Engineering, University of Western Australia

CAMBRIDGE
UNIVERSITY PRESS

Published by the Press Syndicate of the University of Cambridge
The Pitt Building, Trumpington Street, Cambridge CB2 1RP
40 West 20th Street, New York, NY 10011–4211, USA
10 Stamford Road, Oakleigh, Melbourne 3166, Australia

First published 1989
Reprinted 1994

British Library cataloguing in publication data
Norton, M.P. (Michael Peter), 1951-
Fundamentals of noise and vibration analysis for engineers
1. Mechanical vibration. Analysis
I. Title
620.3

Library of Congress cataloguing in publication data
Norton, M.P. (Michael Peter), 1951-
Fundamentals of noise and vibration analysis for engineers / M.P. Norton
p. cm.
Includes bibliographies and index.
ISBN 0 521 34148 5 - ISBN 0 521 34941 9 (paperback)
1. Noise. 2. Vibration. I. Title.
TA365.N67 1989
620.2'3 - dc19 88-34385

ISBN 0 521 34148 5 hard covers
ISBN 0 521 34941 9 paperback

Transferred to digital printing 2001

To

my parents

my wife Christine

and

my daughters Rebecca and Jessica

Contents

Preface

The study of noise and vibration and the interactions between the two is now fast becoming an integral part of mechanical engineering courses at various universities and institutes of technology around the world. There are many undergraduate text books available on the subject of mechanical vibrations, and there are also a relatively large number of books available on applied noise control. There are also several text books available on fundamental acoustics and its physical principles. The books on mechanical vibrations are inevitably only concerned with the details of vibration theory and do not cover the relationships between noise and vibration. The books on applied noise control are primarily designed for the practitioner and not for the engineering student. The books on fundamental acoustics generally concentrate on physical acoustics rather than on engineering noise and vibration and are therefore not particularly well suited to the needs of engineers. There are also several excellent specialist texts available on structural vibrations, noise radiation and the interactions between the two. These texts do not, however, cover the overall area of engineering noise and vibration, and are generally aimed at the postgraduate research student or the practitioner. There are also a few specialist reference handbooks available on shock and vibration and noise control – these books are also aimed at the practitioner rather than the engineering student.

The main purpose of this book is to attempt to provide the engineering student with a unified approach to the fundamentals of engineering noise and vibration analysis and control. Thus, the main feature of the book is the bringing of noise and vibration together within a single volume instead of treating each topic in isolation. Also, particular emphasis is placed on the interactions beween sound waves and solid structures, this being an important aspect of engineering noise and vibration. The book is primarily designed for undergraduate students who are in the latter stages of their engineering course. It is also well suited to the postgraduate student who is in the initial stages of a research project on engineering noise and vibration and to the practitioner, both of whom might wish to obtain an overview and/or a revision of the fundamentals of the subject.

This book is divided into eight chapters. Each of these chapters is summarised in the introductory comments. Because of the wide scope of the contents, each chapter has its own nomenclature list and its own detailed reference list. A selection of problems relating to each chapter is also provided at the end of the book together with solutions. Each of the chapters has evolved from lecture material presented by the author to (i) undergraduate mechanical engineering students at the University of Western Australia, (ii) postgraduate mechanical engineering students at the University of Western Australia, and (iii) practising engineers in industry in the form of short specialist courses. The complete text can be presented in approximately seventy-two lectures, each of about forty-five minutes duration. Suggestions for subdividing the text into different units are presented in the introductory comments.

The author hopes that this book will be of some use to those who choose to purchase it, and will be pleased and grateful to hear from readers who identify some of the errors and/or misprints that will undoubtedly be present in the text. Suggestions for modifications and/or additions to the text will also be gratefully received.

M. P. Norton

Acknowledgements

This book would not have eventuated had it not been for several people who have played an important role at various stages in my career to date. Whilst these people have, in the main, not had any direct input into the preparation of this book, their contributions to the formulation of my thoughts and ideas over the years have been invaluable to say the least. Special acknowledgements are due to M. K. Bull, D. A. Bies, R. E. Luxton, J. M. Pickles and J. R. Dyer of the University of Adelaide, D. C. Gibson of CSIRO Division of Construction and Engineering, F. J. Fahy of the Institute of Sound and Vibration Research, and B. J. Stone of the University of Western Australia.

Acknowledgements are also due to several of my colleagues and my postgraduate students at the University of Western Australia. The help given to me by J. Soria, P. R. Keswick, W. K. Chiu, and L. O. Kirkham is gratefully acknowledged. Also, the support provided by S. Mitton and his staff at Cambridge University Press is gratefully acknowledged.

Last, but not least, special acknowledgements are due to my family: my parents for encouraging me to pursue an academic career; my wife Christine for enduring the very long hours that I have had to work during the gestation period of this book; and my young daughters, Rebecca and Jessica.

Introductory comments

A significant amount of applied technology pertaining to noise and vibration analysis and control has emerged over the last twenty years or so. It would be an impossible task to attempt to cover all this material in a text book aimed at providing the reader with a fundamental basis for noise and vibration analysis. This book is therefore only concerned with some of the more important fundamental considerations required for a systematic approach to engineering noise and vibration analysis and control, the main emphasis being the industrial environment. Thus, this book is specifically concerned with the fundamentals of noise and vibration analysis for mechanical engineers, structural engineers, mining engineers, production engineers, maintenance engineers, etc. It embodies eight self-contained chapters, each of which is summarised here.

The first chapter, on mechanical vibrations, is a review of some fundamentals. This part of the book assumes no previous knowledge of vibration theory. A large part of what is presented in this chapter is covered very well in existing text books. The main difference is the emphasis on the wave–mode duality, and the reader is encouraged to think in terms of both waves and modes of vibration. As such, the introductory comments relate to both lumped parameter models and continuous system models. The sections on the dynamics of a single oscillator, forced vibrations with random excitation and multiple oscillator are presented using the traditional 'mechanical vibrations' approach. The section on continuous systems utilises both the traditional 'mechanical vibrations' approach and the wave impedance approach. It is in this section that the wave–mode duality first becomes apparent. The wave impedance approach is particularly useful for identifying energy flow characteristics in structural components and for estimating energy transmission and reflection at boundaries. The contents of chapter 1 are best suited to a second year or a third year course unit (based on a total course length of four years) on mechanical vibrations.

The second chapter, on sound waves, is a review of some fundamentals of physical acoustics. Like the first chapter, this chapter assumes no previous working knowledge of acoustics. Sections are included on a classical analysis of the homogeneous wave equation, fundamental sound source models and the inhomogeneous wave equation associated with aerodynamic sound, with particular attention being given to Lighthill's acoustic analogy and the Powell–Howe theory of vortex sound. The distinction between the homogeneous and the inhomogeneous acoustic wave equations is continually emphasised. The chapter also includes a discussion on how reflecting surfaces can affect the sound power characteristics of sound sources – this important practical point is often overlooked. The contents of chapter 2 are best suited to a third year or a fourth year course unit on fundamental acoustics.

The third chapter complements chapters 1 and 2, and is about the interactions between sound waves and solid structures. It is very important for engineers to come to grips with this chapter, and it is the most important fundamental chapter in the book. Wave–mode duality concepts are utilised regularly in this chapter. The chapter includes discussions on the fundamentals of fluid–structure interactions, radiation ratio concepts, sound transmission through panels, the effects of fluid loading, and impact noise processes. The contents of chapter 3 are best suited to a third year or a fourth year course unit. The optimum procedure would be to combine chapters 2 and 3 into a single course unit.

The fourth chapter is a fairly basic chapter on noise and vibration measurements and control procedures. A large part of the contents of chapter 4 is readily available in the noise and vibration control handbook literature with three exceptions: firstly, constant power, constant volume and constant pressure sound source concepts are discussed in relation to the effects of rigid, reflecting boundaries on the sound power characteristics of these sound sources; secondly, the economic issues in noise and vibration control are discussed; and, thirdly, sound intensity techniques for sound power measurement and noise source identification are introduced. The contents of chapter 4 are best suited to a fourth year course unit on engineering noise and vibration control. By the very nature of the wide range of noise and vibration control procedures, several topics have had to be omitted from the chapter. Some of these topics include mufflers, outdoor sound propagation, community noise, air conditioning noise, transmission lines and filters, psychological effects, etc.

The fifth chapter is about the analysis of noise and vibration signals. It includes discussions on deterministic and random signals, signal analysis techniques, analogue and digital signal analysis procedures, random and bias errors, aliasing, windowing, and measurement noise errors. The contents of chapter 5 are best suited to a fourth year unit on engineering noise and vibration noise control, and are best combined with chapters 4 and 8 for the purposes of a course unit.

The sixth and seventh chapters involve specialist topics which are more suited to postgraduate courses. Chapter 6 is about the usage of statistical energy analysis procedures for noise and vibration analysis. This include energy flow relationships, modal densities, internal loss factors, coupling loss factors, non-conservative coupling, the estimation of sound radiation from coupled structures, and relationships between dynamic stress and strain and structural vibration levels. Chapter 7 is about flow-induced noise and vibrations in pipelines. This includes the sound field inside a cylindrical shell, the response of a cylindrical shell to internal flow, coincidence, and other pipe flow noise sources. These two chapters can be included either as optional course units in the final year of an undergraduate course, or as additional reading material for the course unit based on chapters, 4, 5 and 8.

The eighth chapter is a largely qualitative description of noise and vibration as a diagnostic tool (i.e. source identification and fault detection). Magnitude and time domain signal analysis techniques, frequency domain signal analysis techniques, cepstrum analysis techniques, sound intensity analysis techniques, and other advanced signal analysis techniques are described here. The chapter also includes five specific practical test cases. The contents of chapter 8 are best suited to a fourth year unit on engineering noise and vibration noise control, and are best combined with chapters 4 and 5 for the purposes of a course unit.

Based upon the preceding comments, the following subdivision of the text is recommended for the purposes of constructing course units.

(1) 2nd year unit	mechanical vibrations ($\sim$14 hrs) chapter 1 (sections 1.1–1.8)
(2) 3rd year unit	waves in structures and fluids ($\sim$14 hrs) chapter 1 (section 1.9), chapter 2 (sections 2.1, 2.2)
(3) 3rd or 4th year unit	structure–sound interactions ($\sim$18 hrs) chapter 2 (sections 2.3, 2.4), chapter 3
(4) 4th year unit*	engineering noise control ($\sim$18 hrs) chapters 4, 5, 8
(5) optional specialist units and/or additional reading	statistical energy analysis and pipe flow noise ($\sim$8 hrs) chapters 6, 7.

* Chapters 2 and 3 should be a prerequisite for the engineering noise control unit.

1

Mechanical vibrations: a review of some fundamentals

1.1 Introduction

Noise and vibration are often treated separately in the study of dynamics, and it is sometimes forgotten that the two are inter-related – i.e. they simply relate to the transfer of molecular motional energy in different media (generally fluids and solids respectively). It is the intention of this book to bring noise and vibration together within a single volume instead of treating each topic in isolation. Central to this is the concept of wave–mode duality; it is generally convenient for engineers to think of noise in terms of waves and to think of vibration in terms of modes. A fundamental understanding of noise, vibration and interactions between the two therefore requires one to be able to think in terms of waves and also in terms of modes of vibration.

This chapter reviews the fundamentals of vibrating mechanical systems with reference to both wave and mode concepts since the dynamics of mechanical vibrations can be studied in terms of either. Vibration deals (as does noise) with the oscillatory behaviour of bodies. For this oscillatory motion to exist, a body must possess inertia and elasticity. Inertia permits an element within the body to transfer momentum to adjacent elements and is related to density. Elasticity is the property that exerts a force on a displaced element, tending to return it to its equilibrium position. (Noise therefore relates to oscillatory motion in fluids whilst vibration relates to oscillatory motion in solids.)

Oscillating systems can be treated as being either linear or non-linear. For a linear system, there is a direct relationship between cause and effect and the principle of superposition holds – i.e., if the force input doubles, the output response doubles. The relationship between cause and effect is no longer proportional for a non-linear system. Here, the system properties depend upon the dependent variables, e.g. the stiffness of a non-linear structure depends upon its displacement.

In this book, only linear oscillating systems which are described by linear differential equations will be considered. Linear system analysis adequately explains the behaviour of oscillatory systems provided that the amplitudes of the oscillations are very small relative to the system's physical dimensions. In each case, the system (possessing inertia and elasticity) is initially or continuously excited in the presence of external forces which tend to return it to its undisturbed position. Noise levels of up to about 140 dB (~25 m from a jet aircraft at take off) are produced by linear pressure fluctuations. Most engineering and industrial type noise sources (which are generally less than 140 dB) and the associated mechanical vibrations can therefore be assumed to behave in a linear manner. Some typical examples are the noise and vibration characteristics of industrial machinery, noise and vibration generated from high speed gas flows in pipelines, and noise and vibration in motor vehicles.

The vibrations of linear systems fall into two categories – free and forced. Free vibrations occur when a system vibrates in the absence of any externally applied forces (i.e. the externally applied force is removed and the system vibrates under the action of internal forces). A finite system undergoing free vibrations will vibrate in one or more of a series of specific patterns: for instance, consider the elementary case of a stretched string which is struck at a chosen point. Each of these specific vibration patterns is called a mode shape and it vibrates at a constant frequency, which is called a natural frequency. These natural frequencies are properties of the finite system itself and are related to its mass and stiffness (inertia and elasticity). It is interesting to note that if a system were infinite it would be able to vibrate freely at any frequency (this point is relevant to the propagation of sound waves). Forced vibrations, on the other hand, take place under the excitation of external forces. These excitation forces may be classified as being (i) harmonic, (ii) periodic, (iii) non-periodic (pulse or transient), or (iv) stochastic (random). Forced vibrations occur at the excitation frequencies, and it is important to note that these frequencies are arbitrary and therefore independent of the natural frequencies of the system. The phenomenon of resonance is encountered when a natural frequency of the system coincides with one of the exciting frequencies. The concepts of natural frequencies, modes of vibration, forced vibrations and resonance will be dealt with later on in this chapter, both from an elastic continuum viewpoint and from a macroscopic viewpoint.

The concept of damping is also very important in the study of noise and vibration. Energy within a system is dissipated by friction, heat losses and other resistances, and any damped free vibration will therefore diminish with time. Steady-state forced vibrations can be maintained at a specific vibrational amplitude because the required energy is supplied by some external excitation force. At resonance, it is only the damping within a system which limits vibrational amplitudes. Both solids and fluids possess damping, and the response of a practical system (for example, a built-up plate or shell structure) to a sound field is dependent upon both structural damping and acoustic radiation damping. The concepts of structural damping will be introduced in this chapter and discussed in more detail in chapter 6 together with acoustic radiation damping.

A macroscopic (modal) analysis of the dynamics of any finite system requires an understanding of the concept of degrees of freedom. The degrees of freedom of a system are defined as the minimum number of independent co-ordinates required to describe its motion completely. An independent particle in space will have three degrees of freedom, a finite rigid body will have six degrees of freedom (three position components and three angles specifying its orientation), and a continuous elastic body will have an infinite number of degrees of freedom (three for each point in the body). There is also a one to one relationship between the number of degrees of freedom and the natural frequencies (or modes of vibration) of a system – a system with p degrees of freedom will have p natural frequencies and p modes of vibration. Plates, shell and acoustic volumes, for instance, have many thousands of degrees of freedom (and therefore natural frequencies/modes of vibration) within the audible frequency range. As far as mechanical vibrations of structures (shafts, machine tools, etc.) are concerned, certain parts of the structures can often be assumed to be rigid, and the system can therefore be reduced to one which is dynamically equivalent to one with a finite number of degrees of freedom. Many mechanical vibration problems can thus be reduced to systems with one or two degrees of freedom.

An engineering description of the time response of vibrating systems can be obtained by solving linear differential equations based upon mathematical models of various equivalent systems. When a finite-number-of-degrees-of-freedom model is used, the system is referred to as a lumped-parameter system. Here, the real system is approximated by a series of rigid masses, springs and dampers. When an infinite-number-of-degrees-of-freedom model is used, the system is referred to as a continuous or a distributed-parameter system. The differential equation governing the motion of the structure is still the same as for the lumped-parameter system except that the mass, damping and stiffness distributions are now continuous and a wave-type solution to the equations can therefore be obtained. This wave–mode duality which is central to the study of noise and vibration will be discussed in some detail at the end of this chapter.

1.2 Introductory wave motion concepts – an elastic continuum viewpoint

A wave motion can be described as a phenomenon by which a particle is disturbed such that it collides with adjacent particles and imparts momentum to them. After collision, the particles oscillate about their equilibrium positions without advancing in any particular direction, i.e. there is no nett transport of the particles in the medium. The disturbance, however, propagates through the medium at a speed which is characteristic of the medium, the kinematics of the disturbance, and any external body forces on the medium. Wave motion can be described by using either molecular or particulate models. The molecular model is complex and cumbersome, and the particulate model is the preference for noise and vibration analysis. A particle is a volume element which is large enough to contain millions of molecules such that it is considered to be a continuous medium, yet small enough such that its thermodynamic and acoustic variables are constant. Solids can store energy in shear and compression, hence several types of waves are possible, i.e. compressional (longitudinal) waves flexural (transverse or bending) waves, shear waves and torsional waves. Fluids, on the other hand, can only store energy in compression. Wave motion is simply a balance between potential and kinetic energies, with the potential energy being stored in different forms for different wave-types. Compressional waves store potential energy in longitudinal strain, and flexural waves store it in bending strain.

Some elementary examples of wave motion are the propagation of sound in the atmosphere due to a source such as blast noise from a quarry, bending motions in a metal plate (such as a machine cover) which is mechanically excited, and ripples in a moving stream of water due to a pebble being thrown into it. In the case of the sound radiation associated with the blasting process at the quarry, the waves that are generated would travel both upwind and downwind. Likewise, the ripples in the stream would also travel upstream and downstream. In both these examples the disturbances propagate away from the source without being reflected. For the case of the finite metal plate, a series of standing waves would be established because of wave reflection at the boundaries. In each of the three examples there is, however, no nett transport of mass particles in the medium.

It is important to note at this stage that it is mathematically convenient to model the more general time-varying wave motions that are encountered in real life in terms of summations of numerous single frequency (harmonic) waves. The discussions in this book will therefore relate to such models. The properties of the main types of wave motions encountered in fluids and solids are now summarised. Firstly, there are two different velocities associated with each type of harmonic wave motion. They are: (i) the velocity at which the disturbance propagates through the medium (this velocity is characteristic of the properties of the medium, the kinematics of the disturbance, and any external body forces on the medium), and (ii) the velocity of the oscillating mass particles in the medium (this particle velocity is a measure of the amplitude of the disturbance which produces the oscillation, and relates to the vibration or sound pressure level that is measured). These two types of velocities which are associated with harmonic waves are illustrated in Figure 1.1 for the case of compressional and flexural wave motions on an arbitrary free surface. For the compressional (longitudinal) wave, there are alternate regions of expansion and compression of the mass particles, and the particle and wave velocities are in the same direction. The propagation of sound waves in air and longitudinal waves in bars is typical of such waves. For the flexural (transverse or bending) wave, the particle velocity is perpendicular to the direction of wave propagation. The bending motion of strings, beams, plates and shells is typical of this type of wave motion. It will be shown later on (in chapter 3) that bending waves are the only type of structural waves that contribute directly to noise radiation and transmission through structures (e.g. aircraft fuselages). The main reason for this is that the particle velocity (and structural displacement) is perpendicular to the direction of wave propagation, as illustrated in Figure 1.1(b). This produces an effective disturbance of the adjacent fluid particles and results in an effective exchange of energy between the structure and the fluid. It will also be shown in chapter 3 that the bending wave velocity varies with frequency whereas other types of wave velocities (compressional, torsional, etc.) do not.

Fig. 1.1. Illustration of wave and particle velocities.

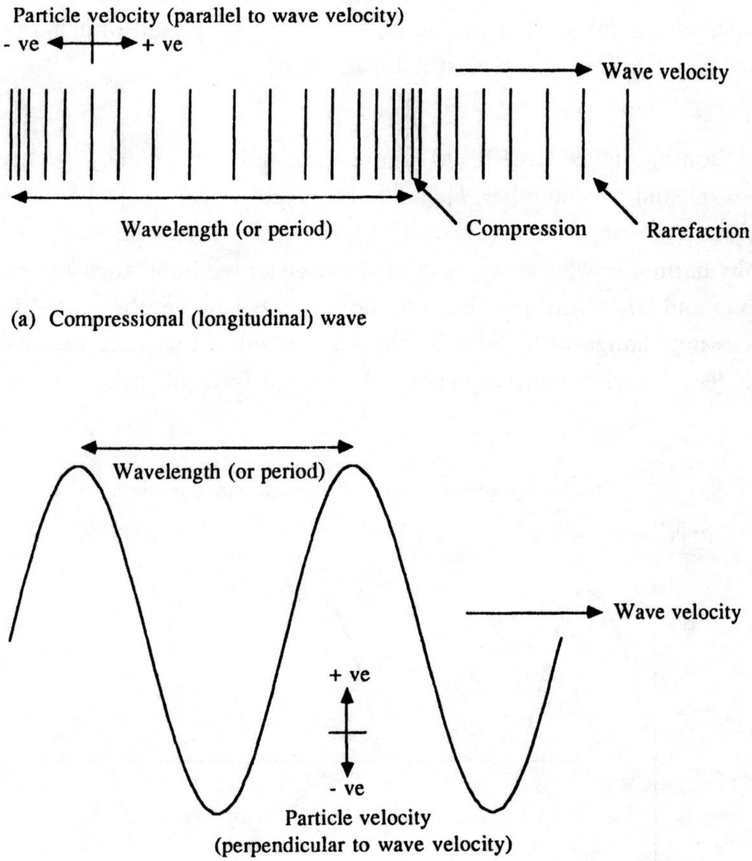

(a) Compressional (longitudinal) wave

(b) Flexural (transverse or bending) wave

Any wave motion can be represented as a function of time, of space or of both. Time variations in a harmonic wave motion can be represented by the radian (circular) frequency ω. This parameter represents the phase change per unit increase of time, and

$$\omega = 2\pi/T, \tag{1.1}$$

where T is the temporal period of the wave motion. This relationship is illustrated in Figure 1.2. The phase of a wave (at a given point in time) is simply the time shift relative to its initial position. Spatial variations in such a wave motion are represented by the phase change per unit increase of distance. This parameter is called the wavenumber, k, where

$$k = \omega/c, \tag{1.2}$$

and c is the wave velocity (the velocity at which the disturbance propagates through the medium). This wave velocity is also sometimes called the phase velocity of the wave – it is the ratio of the phase change per unit increase of time to the phase change per unit increase of distance. Now, the spatial period of a harmonic wave motion is described by its wavelegth, λ, such that

$$k = 2\pi/\lambda. \tag{1.3}$$

This relationship is illustrated in Figure 1.3, and the analogy between radian frequency, ω, and wavenumber, k, can be observed.

If the wave velocity, c, of an arbitrary time-varying wave motion (a summation of numerous harmonic waves) is constant for a given medium, then the relationship between ω and k is linear and therefore non-dispersive – i.e. the spatial form of the wave does not change with time. On the other hand, if the wave velocity, c, is not constant (i.e. it varies with frequency), the spatial form of the wave changes with

Fig. 1.2. Time variations for a simple wave motion.

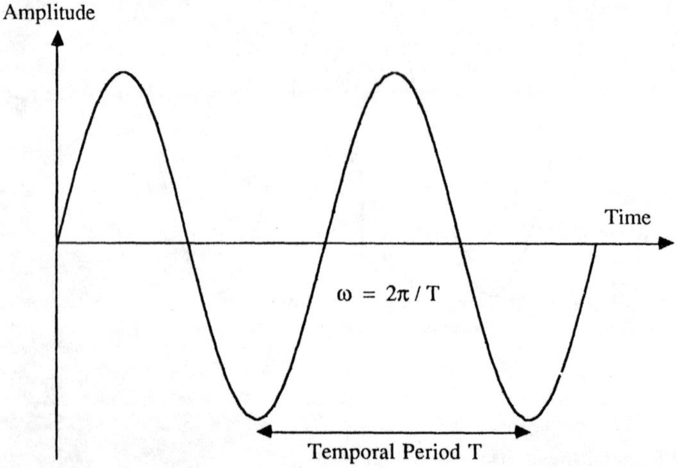

time and is therefore dispersive. It is a relatively straightforward exercise to show that a single frequency wave is non-dispersive but that a combination of several waves of different freqencies is dispersive if they each propagate at different wave velocities. Dispersion relationships are very important in discussing the interactions between different types of wave motions (e.g. interactions between sound waves and structural waves). When a wave is non-dispersive, the wave velocity, c, is constant and therefore $\partial\omega/\partial k$ (the gradient of equation 1.2) is also constant. When a wave is dispersive, both the wave velocity, c, and the gradient of the corresponding dispersion relationships are variables. This is illustrated in Figure 1.4. The gradient of the

Fig. 1.3. Spatial variations for a simple wave motion.

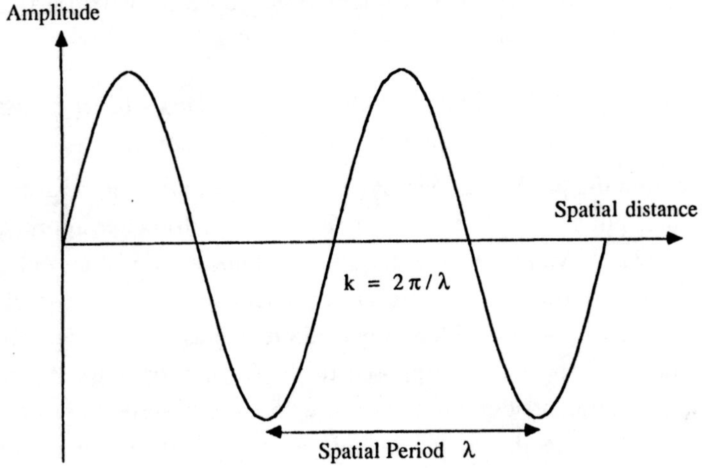

Fig. 1.4. Linear and non-linear dispersion relationships.

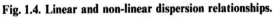

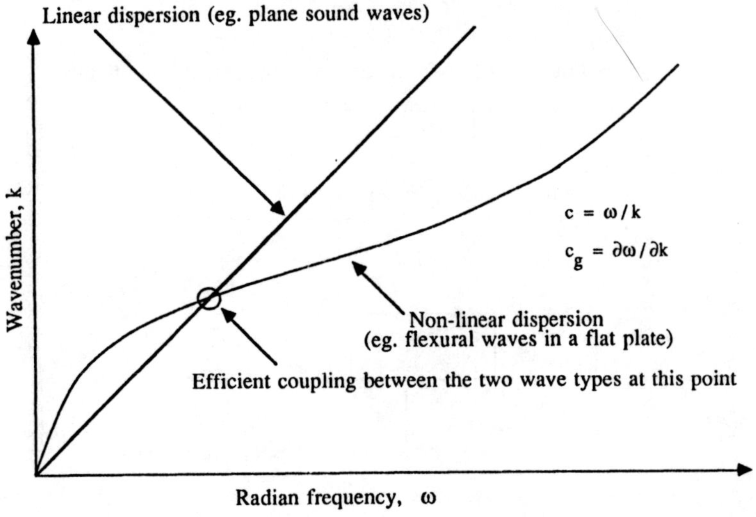

dispersion relationship is termed the group velocity,

$$c_g = \partial\omega/\partial k, \tag{1.4}$$

and it quantifies the speed at which energy is transported by the dispersive wave. It is the velocity at which an amplitude function which is impressed upon a carrier wave packet (a time-varying wave motion which can be represented as a summation of numerous harmonic waves) travels, and it is of great physical importance. Plane sound waves and compressional waves in solids are typical examples of non-dispersive waves, and flexural waves in solids are typical examples of dispersive waves. If the dispersion relationship of any two types of wave motions intersect, they then have the same frequency, wavenumber, wavelength and wavespeed. This condition (termed 'coincidence') allows for very efficient interactions between the two wave-types, and it will be discussed in some detail in chapters 3 and 7.

1.3 Introductory multiple, discrete, mass–spring–damper oscillator concepts – a macroscopic viewpoint

When considering the mechanical vibrations of machine elements and structures one generally utilises either the lumped or the distributed parameter approach to study the normal modes of vibration of the system. Engineers are often only concerned with the estimation of the first few natural frequencies of a large variety of structures, and the macroscopic approach with multiple, discrete, mass–spring–damper oscillators is therefore more appropriate (as opposed to the wave approach). When modelling the vibrational characteristics of a structure via the macroscopic approach, the elements that constitute the model include a mass, a spring, a damper and an excitation. The elementary, one-degree-of-freedom, lumped-parameter oscillator model is illustrated in Figure 1.5.

Fig. 1.5. One-degree-of-freedom, lumped-parameter oscillator.

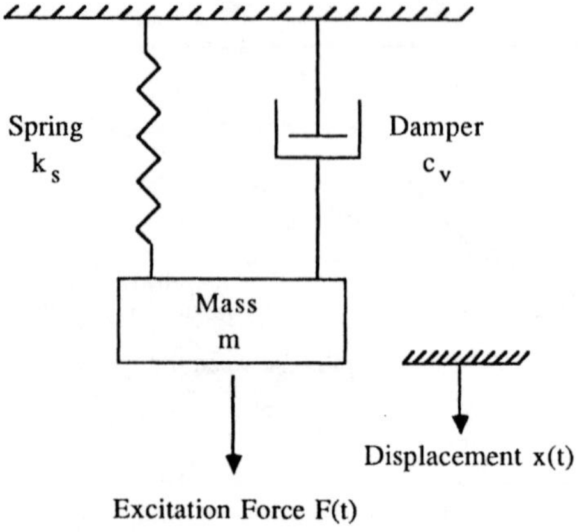

The excitation force provides the system with energy which is subsequently stored by the mass and the spring, and dissipated in the damper. The mass, m, is modelled as a rigid body and it gains or loses kinetic energy. The spring (with a stiffness k_s) is assumed to have a negligible mass, and it possesses elasticity. A spring force exists when there is a relative displacement between its ends, and the work done in compressing or extending the spring is converted into potential energy – i.e. the strain energy is stored in the spring. The spring stiffness, k_s, has units of force per unit deflection. The damper (with a viscous-damping coefficient c_v) has neither mass nor stiffness, and a damping force will be produced when there is relative motion between its ends. The damper is non-conservative because it dissipates energy. Various types of damping models are available, and viscous damping (i.e. the damping force is proportional to velocity) is the most commonly used model. The viscous-damping coefficient, c_v, has units of force per unit velocity. Other damping models include coulomb (or dry-friction) damping, hysteretic damping, and velocity-squared damping. Fluid dynamic drag on bodies, for example, approximates to velocity-squared damping (the exact value of the exponent depends on several other variables).

The idealised elements that make up the one-degree-of-freedom system form an elementary macroscopic model of a vibrating system. In general, the models are somewhat more complex and involve multiple, discrete, mass–spring–damper oscillators. In addition, the masses of the various spring components often have to be accounted for (for instance, a coil spring possesses both mass and stiffness). The low frequency vibration characteristics of a large number of continuous systems can be approximated by a finite number of lumped parameters. The human body can be approximated as a linear, lumped-parameter system for the analysis of low frequency (<200 Hz) shock and vibration effects. A simplified multiple, discrete, mass–spring–damper model of a human body standing on a vibrating platform is illustrated in Figure 1.6. The natural frequencies of various parts of the human body can be estimated from such a model, and the subsequent effects of external shock and vibration can therefore be analysed.

The concepts of multiple, discrete, mass–spring–damper models can be extended to analyse the vibrations of continuous systems (i.e. systems with an infinite number of degrees of freedom, natural frequencies, and modes of vibration) at higher frequencies by re-modelling the structure in terms of continuous or distributed elements. Mathematically, the problem is usually first set up in terms of the wave equation and subsequently generalised as an eigenvalue problem in terms of modal mass, stiffness and damping. The total response is thus a summation of the modal responses over the frequency range of interest.

It should be noted that the generally accepted convention in most of the literature is the symbol c for both the wave (phase) velocity and the viscous-damping coefficient, and the symbol k for both the wavenumber and the spring stiffness. To avoid this conflicting use of symbols, the symbol c will denote the wave (phase) velocity, the symbol c_v, the viscous-damping coefficient, the symbol k the wavenumber, and the symbol k_s, the spring stiffness.

1.4 Introductory concepts on natural frequencies, modes of vibration, forced vibrations and resonance

Natural frequencies, modes of vibration, forced vibrations and resonance can be described both from an elastic continuum and a macroscopic viewpoint. The existence of natural frequencies and modes of vibration relates to the fact that all real physical systems are bounded in space. A mode of vibration (and the natural frequency associated with it) on a taut, fixed string can be interpreted as being composed of two waves of equal amplitude and wavelength travelling in opposite directions

Fig. 1.6. A simplified, multiple, discrete mass–spring–damper model of a human body standing on a vibrating platform.

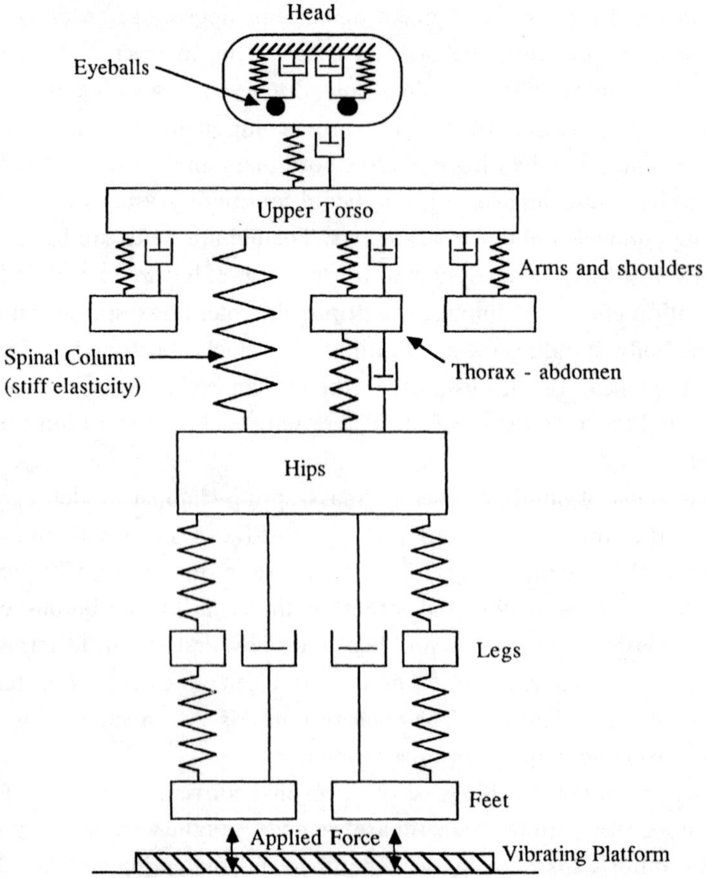

between the two bounded ends. Alternatively, it can be interpreted as being a standing wave, i.e. the string oscillates with a spatially varying amplitude within the confines of a specific stationary waveform. The first interpretation of a mode of vibration relates to the wave model, and the second to the macroscopic model. Both describe the same physical motion and are mathematically equivalent – this will be illustrated in section 1.9.

The concepts discussed above can be be illustrated by means of a simple example. Let us consider a piece of string which is stretched and clamped at its ends, as illustrated in Figure 1.7(a). The string is plucked at some arbitrary point and allowed to vibrate freely. At the instant that the string is plucked, a travelling wave is generated in each direction (i.e. towards each clamped end of the string). It is important to

Fig. 1.7. Schematic illustration of travelling and standing waves for a stretched string.

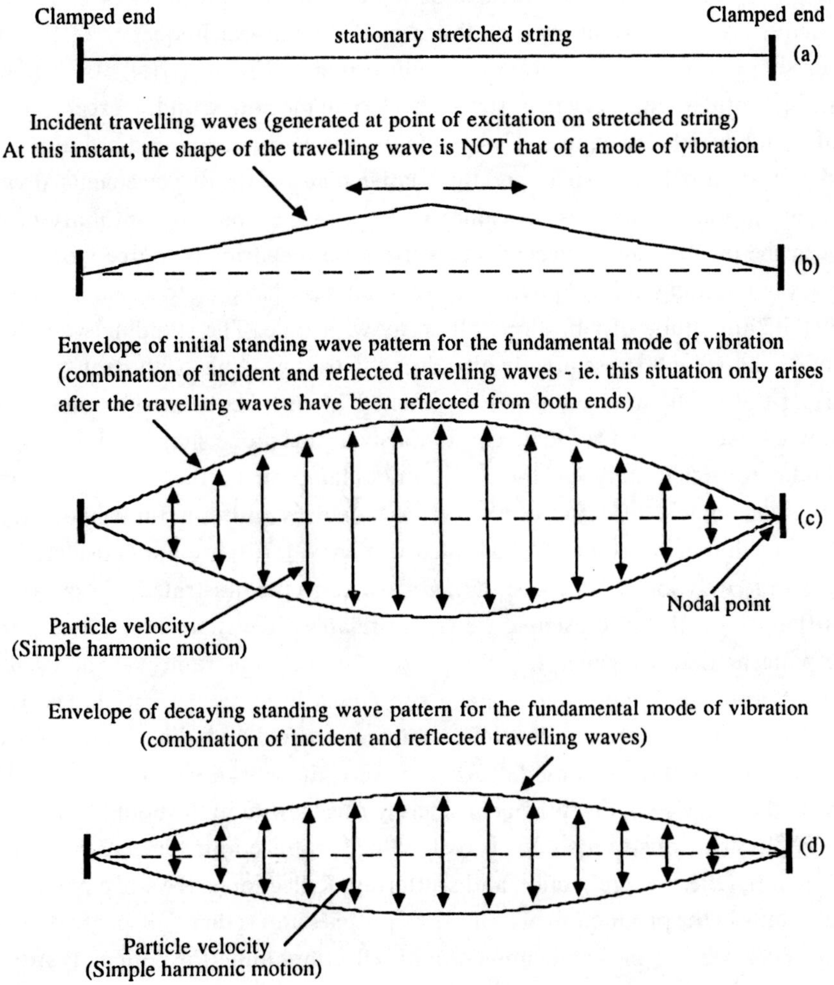

recognise that, at this instant, the shape of the travelling wave is not that of a mode of vibration (Figure 1.7b) since a standing wave pattern has yet to be established. The travelling waves move along the string until they meet the clamped ends, at which point they are reflected. After these initial reflections (one from each clamped end) there is a further short time interval during which time the total motion along the stretched string is the resultant effect of the incident waves and the reflected waves which have yet to reach the starting point. During this time, the standing wave pattern has still not yet been established. Once the reflected waves meet, a situation arises where there is a combination of waves of equal amplitudes travelling in opposite directions. This gives rise to a stationary vibration with a spatially dependent amplitude. The standing wave pattern (mode of vibration) is thus established and the wave propagation process keeps on repeating itself. Depending on how the string is excited, different modes of vibration will be excited – the fundamental mode is most easily observed and this is assumed to be the case for the purposes of this example. The frequency of the resulting standing wave is a natural frequency of the string, and its shape is a mode of vibration, as illustrated in Figure 1.7(c) – each point on the string vibrates transversely in simple harmonic motion, with the exception of the nodal points which are at rest. The points of zero and maximum amplitude for the standing wave are fixed in space and the relative phase of the displacements at various points along the string takes on values of 0 or π – i.e. there is continuity of phase between the incident and reflected waves. Because the string is in free vibration and it possesses damping (all real physical systems possess damping to a smaller or greater extent), its amplitude of vibration will decay with time. The standing wave pattern (or mode shape) will subsequently also decay with time. This is illustrated in Figure 1.7(d). If the string were continuously excited at this frequency by some external harmonic force, it would resonate continuously at this mode shape and its amplitude would be restricted only by the amount of damping in the string – resonance occurs when some external forced vibration coincides with a natural frequency.

The travelling wave concepts illustrated in Figure 1.7(b) are not considered in the macroscopic viewpoint, but the standing wave concepts illustrated in Figures 1.7(c) and (d) are – i.e. the macroscopic viewpoint relates directly to the various standing wave patterns that are generated due to the physical constraints on the oscillating system. Lumped-parameter models can subsequently be used to study the various natural frequencies, forced vibrations and resonances, and the vibrations can be analysed in terms of normal modes. Alternatively, the vibrations can be studied from the wave motion viewpoint. When considering this viewpoint, it should be recognised that the physical constraints upon a real system produce four types of wave motion – diffraction, reflection, refraction and scattering. Reflection is the wave phenomenon which results in the production of natural frequencies and is therefore of great practical significance. Because of the phenomenon of reflection, finite (or bounded) structures can only vibrate freely at specific natural frequencies. Infinite (unbounded) structures on the other hand, can vibrate freely at any frequency. This point is particularly important when considering the interactions of sound fields (waves in an unbounded fluid medium) and bounded structures.

1.5 The dynamics of a single oscillator – a convenient model

In this section, oscillatory motion is mathematically described for the simplest of cases – a macroscopic, single-degree-of-freedom, mass–spring oscillator. The single oscillator is a classical problem, and it is covered in great detail in a variety of texts on mechanical vibrations. Some of the more important results relating to free and forced vibrations of the single oscillator will be presented in this section, and most of these results will be used repeatedly throughout this text.

1.5.1 Undamped free vibrations

It can be intuitively recognised that the motion of a rigid mass, m, on the end of a massless spring with a stiffness k_s, as illustrated in Figure 1.8(a), will be of an oscillatory nature.

The simple oscillatory system has one degree of freedom and its motion can be described by a single co-ordinate, x. The spring is stretched by an amount δ_{static} due to the mass, m, and this stretched position is defined as the equilibrium position for the system. All dynamic motion is subsequently about this equilibrium position. A simple force balance during equilibrium shows that

$$mg = k_s\delta_{static}, \tag{1.5}$$

where g is the gravitational acceleration constant. If the mass is now displaced below its equilibrium position (as illustrated in Figure 1.8b) and released, its equation of

Fig. 1.8. Free-body diagrams for undamped free vibrations of a single oscillator.

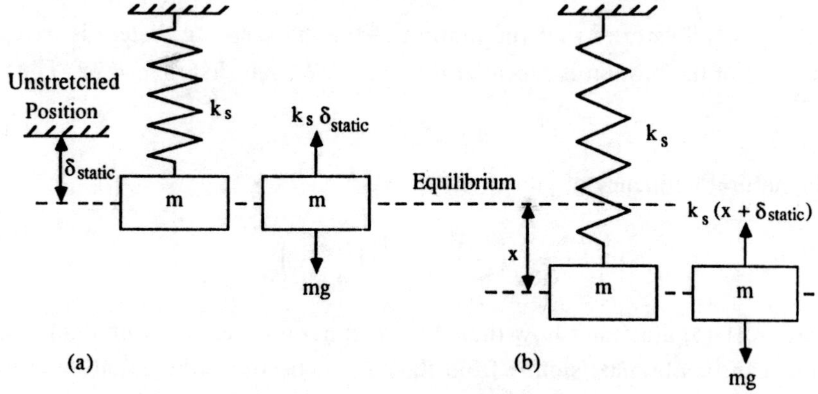

motion can be obtained from Newton's second law. Hence,

$$m\ddot{x} = mg - k_s x - k_s \delta_{static}, \tag{1.6}$$

and therefore

$$m\ddot{x} + k_s x = 0. \tag{1.7}$$

Note that $\ddot{x} = \partial^2 x/\partial t^2$ and $\dot{x} = \partial x/\partial t$, etc.

If a constant, $\omega_n = (k_s/m)^{1/2}$, is now defined, then

$$\ddot{x} + \omega_n^2 x = 0. \tag{1.8}$$

Equation (1.8) is an important homogeneous, second-order, linear, differential equation with a solution of the form $x(t) = A \sin \beta t + B \cos \beta t$. Differentiation and substitution into equation (1.8) readily show that this is a solution if $\beta = \omega_n$. The complete general solution is thus

$$x(t) = A \sin \omega_n t + B \cos \omega_n t, \tag{1.9}$$

where A and B are arbitrary constants (evaluated from the initial conditions) and ω_n is the radian (circular) frequency at which the system oscillates. It is, in fact, the natural frequency of the mass–spring system.

If the mass has an initial displacement x_0 and an initial velocity v_0 at time $t = 0$, then equation (1.9) becomes

$$x(t) = (v_0/\omega_n) \sin \omega_n t + x_0 \cos \omega_n t, \tag{1.10}$$

Now, if $x_0 = X \sin \psi$ and $v_0/\omega_n = X \cos \psi$, such that X is the amplitude of the motion and ψ is the initial phase angle, then

$$X = \{x_0^2 + (v_0/\omega_n)^2\}^{1/2}, \tag{1.11}$$

$$\tan \psi = (\omega_n x_0)/v_0, \tag{1.12}$$

and

$$x(t) = X \sin (\omega_n t + \psi). \tag{1.13}$$

Equation (1.13) illustrates that the motion of the mass–spring system is harmonic, i.e. the cycle of the motion is repeated in time $t = T$ such that $\omega_n T = 2\pi$. Thus,

$$T = 2\pi/\omega_n = 2\pi(m/k_s)^{1/2}, \tag{1.14}$$

and the natural frequency of vibration, f_n, is

$$f_n = \frac{1}{2\pi}\left(\frac{k_s}{m}\right)^{1/2} = \frac{1}{2\pi}\left(\frac{g}{\delta_{static}}\right)^{1/2}. \tag{1.15}$$

Equation (1.15) illustrates how the undamped natural frequency of a mass–spring oscillator can be obtained simply from its static deflection. The equation is widely used in practice to estimate the fundamental vertical natural frequency during vibration isolation calculations for various types of machines mounted on springs.

Equation (1.8) can also be solved by using complex algebra, and it is instructive to demonstrate this at this point. Complex algebra will be used later on in this book both for noise and vibration analyses. The solution to equation (1.8), using complex algebra, is

$$\mathbf{x}(t) = \mathbf{A}\,e^{i\omega_n t} + \mathbf{B}\,e^{i\omega_n t}. \tag{1.16}$$

It should be noted here that complex quantities are presented in **bold** type in this book. The complex constants **A** and **B** are complex conjugates and can be obtained from the initial conditions (initial displacement x_0 and initial velocity v_0 at time $t = 0$). It can easily be shown that

$$\mathbf{A} = \{x_0 - i(v_0/\omega_n)\}/2, \tag{1.17a}$$

and

$$\mathbf{B} = \{x_0 + i(v_0/\omega_n)\}/2. \tag{1.17b}$$

Substitution of **A** and **B** into equation (1.16) yields

$$\mathbf{x}(t) = (v_0/\omega_n)\sin\omega_n t + x_0\cos\omega_n t = x(t), \tag{1.18}$$

which is in fact equation (1.10). Both the initial conditions are real, therefore the solution also has to be real. When using complex algebra one therefore only needs to concern oneself with the real part of the complex solution.

It should be noted that if $\mathbf{z} = x + iy$ is a complex number, then $x = (\mathbf{z} + \mathbf{z}^*)/2 = \mathrm{Re}(\mathbf{z})$ where Re stands for the real part of the complex quantity **z**, and * represents the complex conjugate.

1.5.2 Energy concepts

The equation of motion (equation 1.8) can also be obtained from energy concepts. Energy is conserved for the particular case of free undamped vibrations since there are no excitation or damping forces present. This energy therefore is the sum of the kinetic energy of the mass and the potential energy of the spring. If damping were introduced, an energy dissipation function would have to be included. For free undamped vibrations,

$$T + U = \text{constant}, \tag{1.19a}$$

and

$$\mathrm{d}(T + U)/\mathrm{d}t = 0, \tag{1.19b}$$

where T is the kinetic energy and U is the potential energy.

The kinetic energy of the mass is established by the amount of work done on the mass in moving it over a specified distance. Hence,

$$T = \int_0^x m\frac{dv}{dt}\,dx = \int_0^v m\frac{dx}{dt}\,dv = \int_0^v mv\,dv,$$

and thus

$$T = \tfrac{1}{2}mv^2 = \tfrac{1}{2}m\dot{x}^2. \tag{1.20}$$

The potential energy of the spring is associated with its stiffness – i.e.

$$U = \int_0^x k_s x\,dx = \tfrac{1}{2}k_s x^2. \tag{1.21}$$

By substituting equations (1.20) and (1.21) into equation (1.19), and defining $\omega_n = (k_s/m)^{1/2}$ as before, the equation of motion (equation 1.8) can be obtained.

When the mass is at its maximum displacement, x_{max}, it is instantaneously at rest and therefore has no kinetic energy. Since there is conservation of energy, the total energy is therefore now equal to the maximum potential energy. Alternatively, when $x = 0$ (i.e. the mass passes through its equilibrium position), the system has no potential energy. The kinetic energy is a maximum at this point and the mass has maximum velocity. Thus,

$$T_{max} = U_{max}. \tag{1.22}$$

Later on in this book, the spatial and time averages of a variety of different signals, rather than the instantaneous values, will be considered. For instance, one might be concerned with the average vibration on a machine cover. Here, one would have to take vibration measurements at numerous locations and over some specified time interval to subsequently obtain an averaged vibration level. Average values are particularly useful when using energy concepts to solve noise and vibration problems. For an arbitrary signal, $x(t)$, the mean-square value over a time period, T, is

$$\langle x^2 \rangle = \frac{1}{T}\int_0^T x^2(t)\,dt, \tag{1.23}$$

and the root-mean-square value, x_{rms}, is

$$x_{rms} = \langle x^2 \rangle^{1/2}. \tag{1.24}$$

If the signal, $x(t)$, is harmonic (i.e. $x(t) = X\sin(\omega t + \psi)$), then

$$x_{rms} = \langle x^2 \rangle^{1/2} = \frac{X}{\sqrt{2}}. \tag{1.25}$$

Equation (1.25) relates to a time-averaged signal at a single point in space and its phase, ψ, has been averaged out. The spatial average is subsequently obtained by an arithmetic average of a number of point measurements. It is represented in this book by $\overline{}$. A space and time-averaged signal is thus represented as $\langle\overline{}\rangle$.

Following on from the above discussion, it can be shown that the average kinetic and potential energies are equal. The time-averaged kinetic energy of a vibrating mass is

$$\langle T\rangle = \frac{m\langle v^2\rangle}{2} = \frac{mV^2}{4}, \tag{1.26}$$

where v is its velocity and V is the velocity amplitude. (Note that $\langle v^2\rangle = V^2/2$.) Now, the time-averaged potential energy of the spring is

$$\langle U\rangle = \frac{k_s\langle x^2\rangle}{2} = \frac{k_s(V^2/\omega_n^2)}{4} = \frac{mV^2}{4}, \tag{1.27}$$

since $\omega_n = (k_s/m)^{1/2}$, and $\langle x^2\rangle = \langle v^2\rangle/\omega_n^2$. Thus,

$$\langle T\rangle = \langle U\rangle, \tag{1.28}$$

and the total energy of the undamped system is $mV^2/2$, or $m\langle v^2\rangle$.

1.5.3 Free vibrations with viscous damping

All real systems exhibit damping – energy is lost and the vibration decays with time when the excitation is removed. The exact description of the damping force associated with energy dissipation is difficult – it could be a function of displacement, velocity, stress or some other factors. In general, damping can be modelled by incorporating an arbitrary function, $a(t)$, which decreases with time and a constant, γ, which is related to the amount of damping into the equation of motion for a mass–spring system (equation 1.13). Thus,

$$x(t) = a(t)X \sin(\gamma\omega_n t + \psi). \tag{1.29}$$

The viscous-damping model, which is proportional to the first power of velocity, is commonly used in engineering to model the vibrational characteristics of real systems. With viscous damping, it can be shown that the function $a(t)$ is an exponential such that

$$x(t) = X_T\,\mathrm{e}^{-\beta t} \sin(\gamma\omega_n t + \psi), \tag{1.30}$$

and the resulting motion lies between two exponentials $a(t) = \pm X_T\,\mathrm{e}^{-\beta t}$, as illustrated in Figure 1.9. X_T is the amplitude of the transient, damped, oscillatory motion. The viscous-damping force is represented by

$$F_v = -c_v\dot{x}, \tag{1.31}$$

where c_v is the viscous-damping coefficient. Symbolically, it is designated by a dashpot (Figure 1.5). For free vibrations with damping, the equation of motion now becomes (from Newton's second law)

$$m\ddot{x} + c_v\dot{x} + k_s x = 0. \tag{1.32}$$

This is a homogeneous second-order differential equation with a solution of the form $x = A\,e^{st}$, where A and s are constants. Substitution of this solution into equation (1.32) yields

$$ms^2 + c_v s + k_s = 0. \tag{1.33}$$

This is the characteristic equation of the system and it has two roots:

$$s_{1,2} = \frac{1}{2m}\{-c_v \pm (c_v^2 - 4mk_s)^{1/2}\}. \tag{1.34}$$

Hence,

$$x(t) = B_1\,e^{s_1 t} + B_2\,e^{s_2 t}, \tag{1.35}$$

where B_1 and B_2 are arbitrary constants which are evaluated from the initial conditions. The following terms are now defined:

$$\text{(i) } \omega_n^2 = k_s/m; \qquad \text{(ii) } c_v/m = 2\zeta\omega_n; \qquad \text{(iii) } \zeta = c_v/(4mk_s)^{1/2},$$

where ω_n is the natural frequency (as defined previously), and ζ is the ratio of the viscous-damping coefficient to a critical viscous-damping coefficient. The critical viscous-damping coefficient is the value of c_v which reduces the radical to zero in equation (1.34), i.e. $c_{vc} = (4mk_s)^{1/2}$, and therefore $\zeta = c_v/c_{vc}$. Equations (1.32)–(1.34) can now be re-expressed as

$$\ddot{x} + 2\zeta\omega_n\dot{x} + \omega_n^2 x = 0, \tag{1.36}$$

$$s^2 + 2\zeta\omega_n s + \omega_n^2 = 0, \tag{1.37}$$

$$s_{1,2} = -\zeta\omega_n \pm \omega_n(\zeta^2 - 1)^{1/2}. \tag{1.38}$$

Three cases of interest arise. They are (i) $\zeta > 1$, (ii) $\zeta < 1$, and (iii) $\zeta = 1$.

Fig. 1.9. Schematic illustration of an exponentially damped, sinusoidal motion.

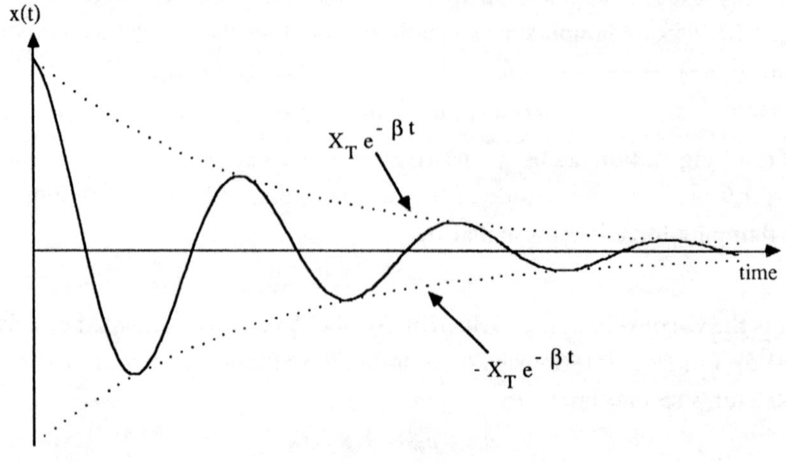

(i) $\zeta > 1$

Here, the roots $s_{1,2}$ are real, distinct and negative since $(\zeta^2 - 1)^{1/2} < \zeta$, and the motion is overdamped. The general solution (equation 1.35) becomes

$$x(t) = B_1\, e^{\{-\zeta + (\zeta^2 - 1)^{1/2}\}\omega_n t} + B_2\, e^{\{-\zeta - (\zeta^2 - 1)^{1/2}\}\omega_n t}, \tag{1.39}$$

and the overdamped motion is not oscillatory, irrespective of the initial conditions. Because the roots are negative, the motion diminishes with increasing time and is aperiodic, as illustrated in Figure 1.10. It is useful to note that, for initial conditions x_0 and v_0, the constants B_1 and B_2 are

$$B_1 = \frac{x_0\omega_n\{\zeta + (\zeta^2 - 1)^{1/2}\} + v_0}{2\omega_n(\zeta^2 - 1)^{1/2}},$$

and

$$B_2 = \frac{-x_0\omega_n\{\zeta - (\zeta^2 - 1)^{1/2}\} - v_0}{2\omega_n(\zeta^2 - 1)^{1/2}}.$$

(ii) $\zeta < 1$

Here, the roots are complex conjugates, the motion is underdamped and the general solution (equation 1.35) becomes

$$x(t) = e^{-\zeta\omega_n t}\{B_1\, e^{i(1-\zeta^2)^{1/2}\omega_n t} + B_2\, e^{-i(1-\zeta^2)^{1/2}\omega_n t}\}$$

$$= X_T\, e^{-\zeta\omega_n t} \sin\{(1 - \zeta^2)^{1/2}\omega_n t + \psi\}. \tag{1.40}$$

The underdamped motion is oscillatory (cf. equation 1.30) with a diminishing amplitude, and the radian frequency of the damped oscillation is

$$\omega_d = \omega_n(1 - \zeta^2)^{1/2} = \omega_n\gamma. \tag{1.41}$$

The underdamped oscillatory motion (commonly referred to as damped oscillatory motion) is illustrated in Figure 1.11. The amplitude, X_T, of the motion and the initial

Fig. 1.10. Aperiodic, overdamped, viscous-damped motion ($\zeta > 1.0$).

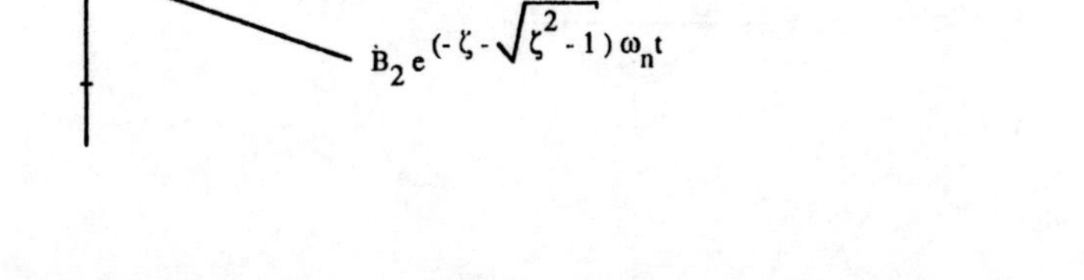

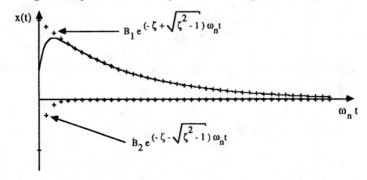

phase angle ψ can be obtained from the initial displacement, x_0, of the mass and its initial velocity, v_0, and they are, respectively,

$$X_T = \frac{\{(x_0\omega_d)^2 + (v_0 + \zeta\omega_n x_0)^2\}^{1/2}}{\omega_d}, \tag{1.42a}$$

and

$$\psi = \tan^{-1}\frac{x_0\omega_d}{v_0 + \zeta\omega_n x_0}. \tag{1.42b}$$

(iii) $\zeta = 1$

Here, both roots are equal to $-\omega_n$, and the system is described as being critically damped. Physically, it represents a transition between the oscillatory and the aperiodic damped motions. The general solution (equation 1.35) becomes

$$x(t) = (B_1 + B_2)\,e^{-\omega_n t}. \tag{1.43}$$

Because of the repeated roots, an additional term of the form $t\,e^{-\omega_n t}$ is required to retain the necessary number of arbitrary constants to satisfy both the initial conditions. Thus, the general solution becomes

$$x(t) = (B_3 + B_4 t)\,e^{-\omega_n t}, \tag{1.44}$$

where B_3 and B_4 are constants which can be evaluated from the initial conditions. For initial conditions x_0 and v_0, the constants are

$$B_3 = x_0,$$

and

$$B_4 = v_0 + \omega_n x_0.$$

Fig. 1.11. Oscillatory, underdamped, viscous-damped motion ($\zeta < 1.0$).

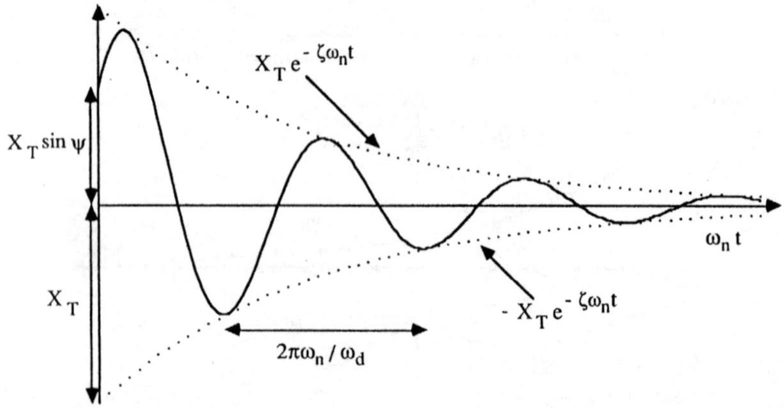

Critically damped motion is the limit of aperiodic motion and the motion returns to rest in the shortest possible time without oscillation. This is illustrated in Figure 1.12. The property of critical damping of forcing the system to return to rest in the shortest possible time is a useful one, and it has many practical applications. For instance, the moving parts of many electrical instruments are critically damped.

Some useful general observations can now be made about damped free vibrations. They are:

(i) $x(t)$ oscillates only if the system is underdamped ($\zeta < 1$);

(ii) ω_d is always less than ω_n;

(iii) the motion $x(t)$ will eventually decay regardless of the initial conditions;

(iv) the frequency ω_d and the rate of the exponential decay in amplitude are properties of the system and are therefore independent of the initial conditions;

(v) for $\zeta < 1$, the amplitude of the damped oscillator is $X_T e^{-\beta t}$, where $\beta = \zeta\omega_n$.

The parameter β is related to the decay time (or time constant) of the damped oscillator – the time that is required for the amplitude to decrease to 1/e of its initial value. The decay time is

$$\tau = 1/\beta = 1/\zeta\omega_n. \tag{1.45}$$

If $\beta < \omega_n$ the motion is underdamped and oscillatory; if $\beta > \omega_n$ the motion is aperiodic; and if $\beta = \omega_n$, the motion is critically damped and aperiodic. The case when $\beta < \omega_n$ (i.e. $\zeta < 1$) is generally of most interest in noise and vibration analysis.

Fig. 1.12. Aperiodic, critically damped, viscous-damped motion ($\zeta = 1$).

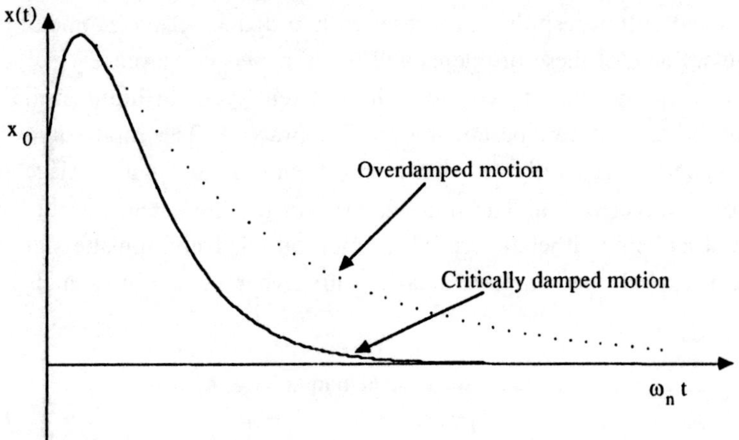

The equation for underdamped oscillatory motion (equation 1.40) can also be expressed as a complex number. It is the imaginary part of the complex solution

$$\mathbf{x}(t) = \mathbf{X}_\mathrm{T}\, \mathrm{e}^{-\beta t}\, \mathrm{e}^{\mathrm{i}\omega_\mathrm{d} t}, \tag{1.46}$$

where $\mathbf{X}_\mathrm{T} = X_\mathrm{T}\, \mathrm{e}^{\mathrm{i}\psi}$. The imaginary part of the solution is used here because equation (1.40) is a sine function. If it were a cosine function the real part of the complex solution would have been used. Equation (1.46) can be rewritten as

$$\mathbf{x}(t) = \mathbf{X}_\mathrm{T}\, \mathrm{e}^{\mathrm{i}(\omega_\mathrm{d} + \mathrm{i}\beta)t} = \mathbf{X}_\mathrm{T}\, \mathrm{e}^{\mathrm{i}\boldsymbol{\omega}_\mathrm{d} t}, \tag{1.47}$$

where $\boldsymbol{\omega}_\mathbf{d} = \omega_\mathrm{d} + \mathrm{i}\beta$ is the complex damped radian frequency. The complex damped radian frequency thus contains information about both the damped natural frequency of the system and its decay time.

The equation of motion for damped free vibrations (equation 1.32) can also be obtained from energy concepts by incorporating an energy dissipation function into the energy balance equation. Hence,

$$\mathrm{d}(T + U)/\mathrm{d}t = -\Pi, \tag{1.48}$$

where Π is power (the negative sign indicates that power is being removed from the system). Power is force $\times$ velocity, and the power dissipated from a system with viscous damping is

$$\Pi = F_\mathrm{v}\dot{x} = c_\mathrm{v}\dot{x}^2. \tag{1.49}$$

1.5.4 Forced vibrations: some general comments

So far, only the free vibrations of systems have been discussed. A linear system vibrating under the continuous application of an input excitation is now considered. This is illustrated schematically in Figure 1.13. In general, there can be many input excitations and output responses, together with feedback between some of the inputs and outputs. Some of these problems will be discussed in chapter 5.

It is useful at this stage to consider the different types of input excitations and output responses that can be encountered in practice. The input or output of a vibration system is generally either a force of some kind, or a displacement, or a velocity, or an acceleration. The time histories of the input and output signals can be classified as being either deterministic or random. Deterministic signals can be expressed by explicit mathematical relationships, whereas random signals have to be

Fig. 1.13. A single input–output linear system.

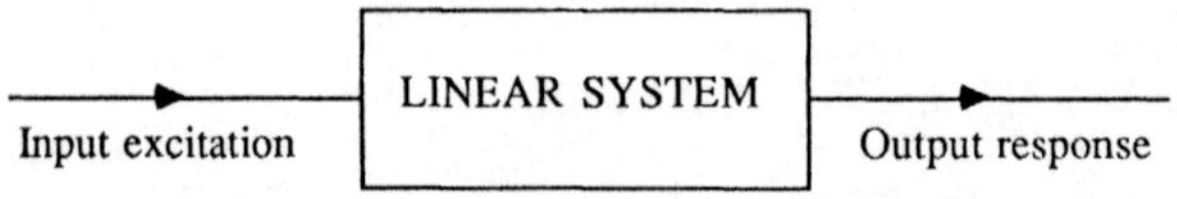

described in terms of probability statements and statistical averages. Typical examples of deterministic signals are those from electrical motors, rotating machinery and pumps. In these examples, a few specific frequencies generally dominate the signal. Some typical random signals include acoustical pressures generated by turbulence, high speed gas flows in pipeline systems, and the response of a motor vehicle travelling over a rough road surface. Here, the frequency content of the signals is dependent upon statistical parameters. Figure 1.14 is a handy flow-chart which illustrates the different types of input and output signals (temperatures, pressures, forces, displacements, velocities, accelerations, etc.) that can be encountered in practice. Therefore, the chart is not limited to only noise and vibration problems. It is worth reminding the reader at this stage that, in addition to all these various types of input excitation and output response functions, a system's response itself can, in principle, be either linear or non-linear. As mentioned in the introduction, only linear systems will be considered in this book.

Fig. 1.14. Flow-chart illustrating the different types of input and output signals.

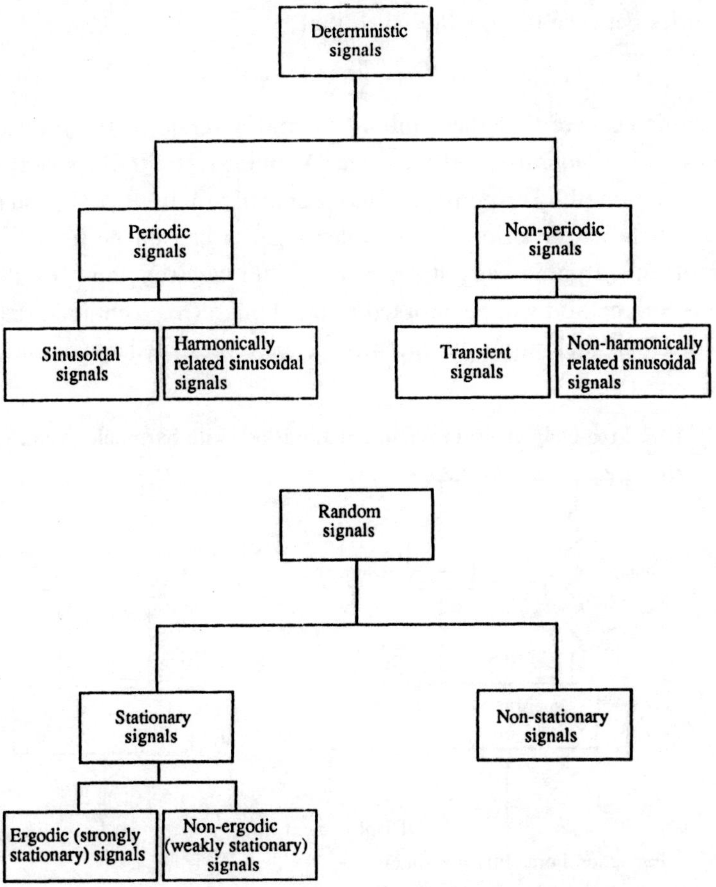

1.5.5 Forced vibrations with harmonic excitation

Now consider a viscous-damped, spring–mass system excited by a harmonic (sinusoidal) force, $F(t) = F \sin \omega t$, as illustrated in Figure 1.15. As mentioned in the previous sub-section, both the input and output to a system can be one of a range of functions (force, displacement, pressure, etc.). In this sub-section, an input force and an output displacement shall be considered initially. The differential equation of motion can be readily obtained by applying Newton's second law to the body. It is

$$m\ddot{x} + c_v\dot{x} + k_s x = F \sin \omega t. \tag{1.50}$$

This is a second-order, linear, differential equation with constant coefficients. The general solution is the sum of the complementary function ($F \sin \omega t = 0$) and the particular integral. The complementary function is just the damped, free, oscillator. This part of the general solution decays with time, leaving only the particular solution to the particular integral. This part of the general solution (the particular solution) is a steady-state, harmonic, oscillation at the forced excitation frequency. The output displacement response, $x(t)$, lags the input force excitation, $F(t)$, by a phase angle, ϕ, which varies between 0° and 180° such that

$$x(t) = X \sin (\omega t - \phi). \tag{1.51}$$

It should be noted here that the symbols X_T and ψ relate to the transient part of the general solution (equation 1.40), whereas X and ϕ relate to the steady-state part. The general solution (total response) is thus the sum of equations (1.40) and (1.51).

Phasors, Laplace transforms and complex algebra can all be used to study the behaviour of an output, steady-state, response for a given input excitation. The complex algebra method will be adopted in this book. This technique requires both the input force and the output displacement to be represented as complex numbers.

Fig. 1.15. Free-body diagram for forced vibrations with harmonic excitation.

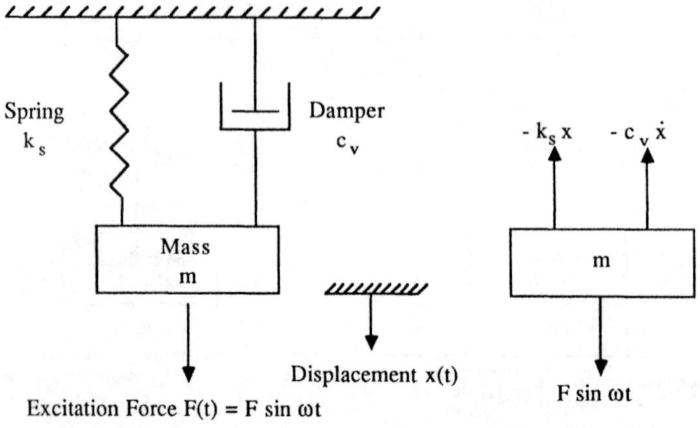

Since the forcing function is a sine term, the imaginary part will be used – if it were a cosine, the real part would have been used. Thus,

$$F \sin \omega t = \text{Im}\,[\mathbf{F}\, e^{i\omega t}], \tag{1.52a}$$

and

$$X \sin (\omega t - \phi) = \text{Im}\,[\mathbf{X}\, e^{i\omega t}], \tag{1.52b}$$

where **F** is the complex amplitude of $F(t)$ and **X** is the complex amplitude of $x(t)$, i.e.

$$\mathbf{F} = F\, e^{-i0} = F, \tag{1.53a}$$

and

$$\mathbf{X} = X\, e^{-i\phi}. \tag{1.53b}$$

The output displacement is thus

$$x(t) = X \sin (\omega t - \phi) = \text{Im}\,[X\, e^{i(\omega t - \phi)}] = \text{Im}\,[\mathbf{X}\, e^{i\omega t}]. \tag{1.54}$$

The complex displacement, **X**, contains information about both the amplitude and phase of the signal. By replacing $x(t)$ by $\mathbf{X}\, e^{i\omega t}$ and $F \sin \omega t$ by $\mathbf{F}\, e^{i\omega t}$ in the equation of motion (equation 1.50), with the clear understanding that finally only the imaginary part of the solution is relevant, one gets

$$-m\omega^2 \mathbf{X}\, e^{i\omega t} + ic_v\omega \mathbf{X}\, e^{i\omega t} + k_s \mathbf{X}\, e^{i\omega t} = \mathbf{F}\, e^{i\omega t}. \tag{1.55}$$

Several important comments can be made in relation to equation (1.55). They are:

(i) the displacement lags the excitation force by a phase angle ϕ, which varies between 0° and 180°;

(ii) the spring force is opposite in direction to the displacement;

(iii) the damping force lags the displacement by 90° and is opposite in direction to the velocity;

(iv) the inertia force is in phase with the displacement and opposite in direction to the acceleration.

Solving for **X** yields

$$\mathbf{X} = \frac{\mathbf{F}}{\{k_s - m\omega^2 + ic_v\omega\}}. \tag{1.56}$$

The output displacement amplitude, X, is obtained by multiplying equation (1.56) by its complex conjugate. Hence,

$$X = \frac{F}{\{(k_s - m\omega^2)^2 + (c_v\omega)^2\}^{1/2}}. \tag{1.57}$$

The phase angle, ϕ, is obtained by replacing $\mathbf{X}$ by $X\,e^{-i\phi}$ and $\mathbf{F}$ by $F\,e^{-i0} = F$ in equation (1.56) and equating the imaginary parts of the solution to zero. Hence,

$$\phi = \tan^{-1} \frac{c_v \omega}{k_s - m\omega^2}. \tag{1.58}$$

Equations (1.57) and (1.58) represent the steady-state solution. They can be non-dimensionalised by defining $X_0 = F/k_s$ as the zero frequency (D.C.) deflection of the spring–mass–damper system under the action of a steady force, F. In addition, $\omega_n = (k_s/m)^{1/2}$; $\zeta = c_v/c_{vc}$; $c_{vc} = 2m\omega_n$ as before. With these substitutions,

$$\frac{X}{X_0} = \frac{1}{[\{1 - (\omega/\omega_n)^2\}^2 + \{2\zeta\omega/\omega_n\}^2]^{1/2}}, \tag{1.59}$$

and

$$\phi = \tan^{-1} \frac{2\zeta\omega/\omega_n}{1 - (\omega/\omega_n)^2}. \tag{1.60}$$

Equations (1.59) and (1.60) are plotted in Figures 1.16(a) and (b), respectively. The main observation is that the damping ratio, ζ, has a significant influence on the amplitude and phase angle in regions where $\omega \approx \omega_n$. The magnification factor (i.e. the amplitude displacement ratio), X/X_0, can be greater than or less than unity depending on the damping ratio, ζ, and the frequency ratio, ω/ω_n. The phase angle, ϕ, is simply a time shift ($t = \phi/\omega$) of the output displacement, $x(t)$, relative to the force excitation, $F(t)$. It varies from 0° to 180° and is a function of both ζ and ω/ω_n. It is useful to note that, when $\omega = \omega_n$, $\phi = 90°$. This condition is generally referred to as phase resonance.

The general solution for the motion of the mass–spring–damper system is, as mentioned earlier, the sum of the complementary function (transient solution, i.e. equation 1.40) and the particular integral (steady-state solution, i.e. equation 1.51). It is therefore

$$x(t) = X_T\,e^{-\zeta\omega_n t} \sin(\omega_d t + \psi) + X \sin(\omega t - \phi). \tag{1.61}$$

The transient part of the solution always decays with time and one is generally only concerned with the steady-state part of the solution. There are some exceptions to this rule, and a typical example involves the initial response of rotating machinery during start-up. Here, one is concerned about the initial transient response before the steady condition is attained.

It can be shown that the steady-state amplitude, X, is a maximum when

$$\frac{\omega}{\omega_n} = (1 - 2\zeta^2)^{1/2}. \tag{1.62}$$

The maximum value of X is

$$X_r = \frac{X_0}{2\zeta(1-\zeta^2)^{1/2}}, \tag{1.63}$$

and the corresponding phase angle at $X = X_r$ is

$$\phi = \tan^{-1}\frac{(1-2\zeta^2)^{1/2}}{\zeta}. \tag{1.64}$$

This condition is called amplitude resonance. In general, it is different from phase resonance ($\phi = 90°$). If $\zeta > 1/\sqrt{2}$, the maximum value of X would occur at $\omega = 0$; i.e. it would be due to the zero frequency deflection of the mass–spring–damper. This is illustrated in Figure 1.16(a).

Fig. 1.16. (a) Magnification factor for a one-degree-of-freedom, mass–spring–damper system; (b) phase angle for a one-degree-of-freedom, mass–spring–damper system.

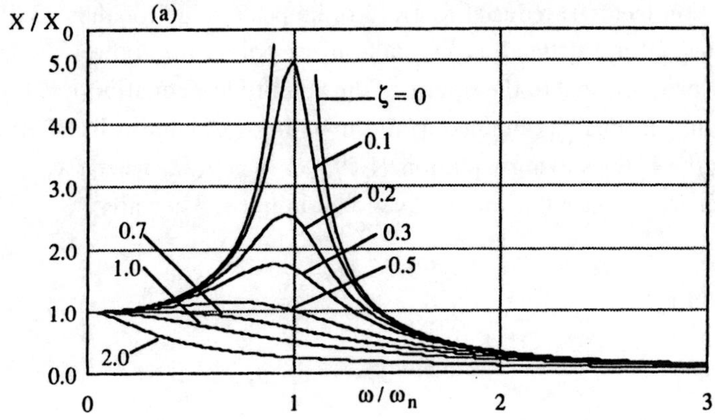

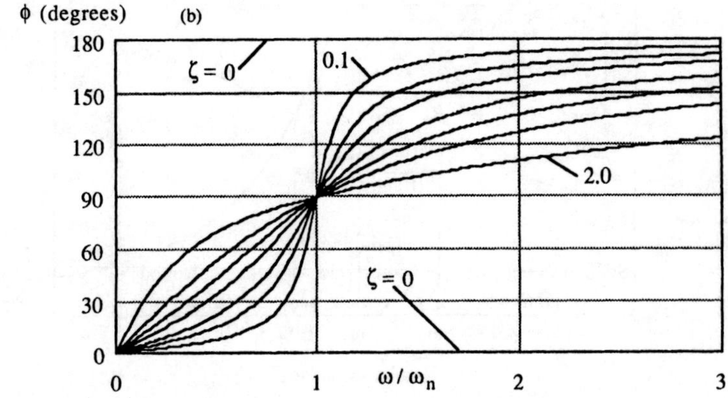

For most practical situations, however, ζ is small (<0.05) and

$$X_r = \frac{X_0}{2\zeta}\left(1 + \frac{\zeta^2}{2}\right) \approx \frac{X_0}{2\zeta}. \tag{1.65}$$

The corresponding phase angle is

$$\phi \approx \tan^{-1}\frac{1}{\zeta}. \tag{1.66}$$

For these cases of small damping, amplitude resonance and phase resonance are assumed to be equal, i.e. $\phi \approx 90°$, and therefore $\omega \approx \omega_n$. The magnification factor at resonance is thus $\sim 1/2\zeta$ and it is called the Q factor or the quality factor, i.e.

$$\frac{X_r}{X_0} = \frac{1}{2\zeta} = Q. \tag{1.67}$$

The quality factor is described physically as a measure of the sharpness of the response at resonance and is a measure of the system's damping. The points where the magnification factor is reduced to $1/\sqrt{2}$ of its peak value (or the -3 dB points, i.e. $20\log_{10}(1/\sqrt{2})$) are defined as the half-power points (the power dissipated by the damper is proportional to the square of the amplitude – equation 1.49). The damping in a system can thus be obtained from the half-power bandwidth. This is illustrated in Figure 1.17. By solving equation (1.59) for $X_{max}/\sqrt{2}$, where $X_{max} = X_r/X_0$, the half-power frequencies (ω_1 and ω_2) can be obtained. They are

$$\omega_{1,2} = (1 \pm \zeta)\omega_n, \tag{1.68}$$

and therefore

$$Q = \frac{1}{2\zeta} = \frac{\omega}{\omega_2 - \omega_1}. \tag{1.69}$$

Fig. 1.17. Half-power bandwidth and half-power points for a linear oscillator.

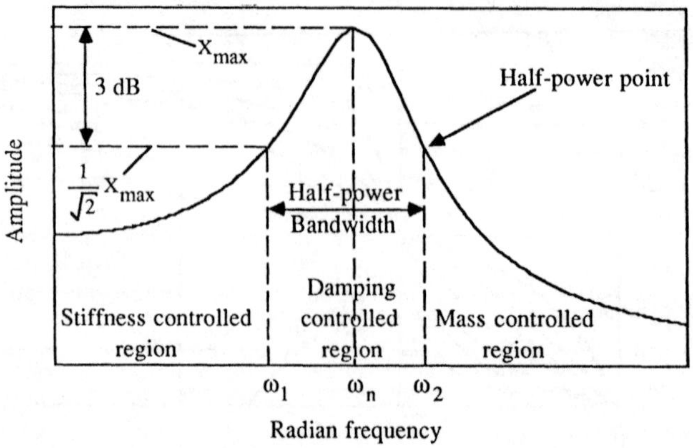

The Q factor is also related to the decay time τ (see equation 1.45) such that

$$Q = \frac{\omega_n \tau}{2}. \tag{1.70}$$

So far in this sub-section, solutions have been sought for the output steady-state displacement, X. The complex ratio of the output displacement to the input force, $\mathbf{X}/\mathbf{F}$, (i.e. equation 1.56) is a frequency response function and it is commonly referred to as a receptance. There are a range of different force–response relationships that are of general engineering interest. The more commonly used ones are presented in Table 1.1. In many applications in noise and vibration, the impedance (force/velocity; $\mathbf{F}/\mathbf{V}$), and the mobility (velocity/force; $\mathbf{V}/\mathbf{F}$), are often of interest. Expressions similar to equation (1.56) can be readily obtained by solving the equation of motion. It is a relatively straightforward exercise to show that the mechanical impedance, F/V, of the mass–spring–damper system in Figure 1.15 is

$$\frac{\mathbf{F}}{\mathbf{V}} = \mathbf{Z_m} = c_v + \mathrm{i}(m\omega - k_s/\omega). \tag{1.71}$$

The real part of the impedance is called the mechanical resistance, and the imaginary part is called the mechanical reactance ($m\omega$ is the mass or the mechanical inertance term, and ω/k_s is the mechanical compliance term). If the mechanical resistance term is dominant, the system's response is damping controlled; if the mechanical inertance term is dominant, the system's response is mass controlled; if the mechanical compliance term is dominant, the system's response is stiffness controlled. In noise and vibration control it is often important to identify which of the three (mass, stiffness or damping) dominates.

Frequency response functions such as impedance and mobility are important tools and will be used throughout this book. For a known input, the knowledge of the frequency response function of a system allows for the estimation of the output response. In complex systems, impedance and mobility concepts are very useful for analysing vibrational energy and power flow. They are used extensively in the dynamic analysis of structures and can be applied to either lumped-parameter, oscillator models or wave-motion models.

Table 1.1. *Different types of frequency response functions.*

Displacement/force	Receptance
Force/displacement	Dynamic stiffness
Velocity/force	Mobility
Force/velocity	Impedance
Acceleration/force	Inertance
Force/acceleration	Apparent mass

The instantaneous power developed by a force $F(t) = F \sin \omega t$ producing a displacement $x(t) = X \sin(\omega t - \phi)$ on the system in Figure 1.15 is

$$\Pi = F(t)\frac{\mathrm{d}x}{\mathrm{d}t} = \omega XF \sin \omega t \cos(\omega t - \phi)$$

$$= \tfrac{1}{2}\omega XF\{\sin \phi + \sin(2\omega t + \phi)\}. \tag{1.72}$$

The first term in the brackets is a constant and it represents the steady flow of work per unit time. The second term represents the fluctuating component of power. It averages to zero over any time interval which is a multiple of the period. The time-averaged power is thus

$$\langle \Pi \rangle = \tfrac{1}{2}\omega XF \sin \phi. \tag{1.73}$$

Now, from equation (1.71),

$$Z_\mathrm{m} = |\mathbf{Z}_\mathrm{m}| = \{c_\mathrm{v}^2 + (m\omega - k_\mathrm{s}/\omega)^2\}^{1/2}, \tag{1.74}$$

and

$$\sin \phi = \frac{c_\mathrm{v}}{Z_\mathrm{m}}. \tag{1.75}$$

Thus, the time-averaged power is

$$\langle \Pi \rangle = \frac{\omega XFc_\mathrm{v}}{2Z_\mathrm{m}} = \frac{VFc_\mathrm{v}}{2}\frac{V}{F} = \tfrac{1}{2}V^2 c_\mathrm{v}, \tag{1.76}$$

where $V = |\mathbf{V}|$.

The time-averaged power can also be obtained by using complex numbers, and it is instructive to obtain it this way at this stage. It is

$$\langle \Pi \rangle = \frac{1}{T}\int_0^T \Pi(t)\,\mathrm{d}t. \tag{1.77}$$

Hence,

$$\langle \Pi \rangle = \frac{\omega}{2\pi}\int_0^{2\pi/\omega} \mathrm{Re}\,[\mathbf{F}\,\mathrm{e}^{\mathrm{i}\omega t}]\,\mathrm{Re}\,[\mathbf{V}\,\mathrm{e}^{\mathrm{i}\omega t}] = \tfrac{1}{2}\,\mathrm{Re}\,[\mathbf{F}\mathbf{V}^*]. \tag{1.78}$$

It should be noted that the real parts of force and velocity are used here. The reason for this is explained in section 1.7.

$\langle \Pi \rangle$ can now be represented in terms of the mechanical impedance, $\mathbf{Z}_\mathrm{m}$, where

$$\langle \Pi \rangle = \tfrac{1}{2}|\mathbf{F}|^2\,\mathrm{Re}\,[\mathbf{Z}_\mathrm{m}^{-1}] = \tfrac{1}{2}|\mathbf{V}|^2\,\mathrm{Re}\,[\mathbf{Z}_\mathrm{m}] = \tfrac{1}{2}V^2 c_\mathrm{v}. \tag{1.79}$$

Equations (1.76) and (1.79) represent the time-averaged power delivered to the oscillator by the force. During steady-state oscillations this has to equal the power dissipated by the damper. The maximum power delivered (and dissipated) occurs when $\sin\phi = 1$, i.e. at resonance.

1.5.6 Equivalent viscous-damping concepts – damping in real systems

Damping exists in all real systems and very rarely is it viscous – viscous damping only exists when the velocity between two lubricated surfaces is sufficiently low such that laminar flow conditions exist. Many different types of damping can exist in practice, and the most commonly encountered include structural (hysteretic) damping, coulomb (dry-friction) damping, and velocity-squared (aerodynamic drag) damping. Because most mechanical systems are essentially lightly damped (i.e. the effect of damping is insignificant except near a resonance), it is possible to obtain approximate models of non-viscous damping in terms of equivalent viscous dampers. This subsequently allows for the continued usage of the simple vibration models, based upon viscous damping, developed in the last sub-section.

In proceeding to develop the concept of an equivalent viscous damper, one first needs to evaluate the energy dissipated per cycle by the damping force. The criteria for equivalence between the actual damping mechanism and viscous damping are (i) equal energy dissipation per cycle of vibration, and (ii) similar harmonic relative displacements.

For viscous damping, the energy dissipated per cycle by the damping force is

$$U_{\rm d} = \int_0^T c_{\rm v}\dot{x}\frac{{\rm d}x}{{\rm d}t}\,{\rm d}t = c_{\rm v}X^2\omega^2\int_0^{2\pi/\omega}\cos^2(\omega t - \phi)\,{\rm d}t = \pi c_{\rm v}\omega X^2. \tag{1.80}$$

The equivalent viscous damping can subsequently be determined from the equation

$$U_{\rm d} = \pi c_{\rm v_{eq}}\omega X^2, \tag{1.81}$$

where $U_{\rm d}$ has to be evaluated for the particular type of damping force.

The principles of equivalent viscous damping can best be illustrated by means of an example. The form of damping that is most relevant to engineering noise and vibration control is structural damping, and this will now be considered. When structural materials such as steel or aluminium are cyclically stressed, energy is dissipated within the material. A hysteresis loop is formed, hence the commonly used term 'hysteretic damping'. Experimental observations clearly show that the energy dissipated per cycle of stress is proportional to the square of the strain amplitude. The constant of proportionality is generally only valid over specific ranges of frequency and temperature – i.e. there will be different constants of proportionality over different frequency and temperature ranges. Hence, for a given frequency and temperature range,

$$U_{\rm d} = \alpha X^2, \tag{1.82}$$

where X is the displacement amplitude. This can now be equated to equation (1.81) to obtain the equivalent viscous-damping coefficient, and thus

$$c_{\mathrm{veq}} = \frac{\alpha}{\pi\omega}. \tag{1.83}$$

The complex differential equation of motion for a one-degree-of-freedom system is therefore

$$-m\omega^2\mathbf{X}\,\mathrm{e}^{\mathrm{i}\omega t} + \mathrm{i}\frac{\alpha}{\pi\omega}\,\omega\mathbf{X}\,\mathrm{e}^{\mathrm{i}\omega t} + k_s\mathbf{X}\,\mathrm{e}^{\mathrm{i}\omega t} = \mathbf{F}\,\mathrm{e}^{\mathrm{i}\omega t}. \tag{1.84}$$

Equation (1.84) can be re-written as

$$-m\omega^2\mathbf{X} + k_s(1 + \mathrm{i}\eta)\mathbf{X} = \mathbf{F}, \tag{1.85}$$

where

$$\eta = \frac{\alpha}{\pi k_s} \tag{1.86}$$

is the structural loss factor and

$$\mathbf{k}_s = k_s(1 + \mathrm{i}\eta) \tag{1.87}$$

is the complex stiffness.

The structural loss factor, η, is an important parameter which is extensively used in structural dynamics. It will be discussed in some detail in chapter 6. The analysis in sub-section 1.5.5 for the magnification factor and the phase angle can now be repeated, and it can be readily shown that for structural damping

$$\frac{X}{X_0} = \frac{1}{[\{1 - (\omega/\omega_n)^2\}^2 + \eta^2]^{1/2}}, \tag{1.88}$$

and

$$\phi = \tan^{-1}\frac{\eta}{1 - (\omega/\omega_n)^2}. \tag{1.89}$$

For a viscous-damped system, $X/X_0 = 1/2\zeta$ at resonance. Hence

$$\eta = 2\zeta = \frac{1}{Q}, \tag{1.90}$$

i.e. the structural loss factor is twice the viscous damping ratio and inversely proportional to the quality factor.

The two other most commonly encountered forms of non-viscous damping are coulomb (dry-friction) and velocity-squared damping. The analyses for obtaining the equivalent viscous-damping coefficients are available in most fundamental texts on mechanical vibrations (see reference list at the end of this chapter) and therefore only the results will be presented here. The equivalent viscous-damping coefficient for coulomb damping is

$$c_{\mathrm{veq}} = \frac{4\mu F_{\mathrm{N}}}{\pi\omega X}, \tag{1.91}$$

where μ is the coefficient of friction and F_{N} is the normal force. The equivalent viscous-damping coefficient for velocity-squared (aerodynamic) damping is

$$c_{\mathrm{veq}} = \frac{8}{3\pi} C_{\mathrm{F}}\omega X, \tag{1.92}$$

where C_{F} is a constant which is related to the drag coefficient, C_{D}, the exposed surface area, A, of the body, and the density, ρ, of the fluid in which it is immersed (i.e. $C_{\mathrm{F}} = \rho C_{\mathrm{D}} A/2$). It should be noted that both of these types of damping are non-linear, i.e. they are functions of the amplitude of the vibration.

1.5.7 Forced vibrations with periodic excitation

Harmonically related periodic signals are often encountered in forces in machinery, and the vibration models developed in the previous sub-sections therefore need to be generalised. Periodic signals are deterministic and can thus be expressed by explicit mathematical relationships – i.e. they can be developed into a Fourier series.

A function, $F(t)$, is periodic if $F(t) = F(t+T)$ where $T = 2\pi/\omega$. The Fourier series expansion of $F(t)$ is

$$F(t) = \frac{a_0}{2} + \sum_{n=1,2}^{\infty} (a_n \cos n\omega t + b_n \sin n\omega t), \tag{1.93}$$

where

$$a_0 = \frac{2}{T}\int_0^T F(t)\,\mathrm{d}t, \tag{1.94a}$$

$$a_n = \frac{2}{T}\int_0^T F(t)\cos n\omega t\,\mathrm{d}t, \tag{1.94b}$$

and

$$b_n = \frac{2}{T}\int_0^T F(t)\sin n\omega t\,\mathrm{d}t. \tag{1.94c}$$

Thus, for a periodic force, $F(t)$, applied to a one-degree-of-freedom system, the equation of motion is

$$m\ddot{x} + c_{\mathrm{v}}\dot{x} + k_{\mathrm{s}}x = \frac{a_0}{2} + \sum_{n=1,2}^{\infty} (a_n \cos n\omega t + b_n \sin n\omega t). \tag{1.95}$$

The coefficients a_0, a_n, and b_n are the Fourier coefficients, and the periodic force $F(t)$ is now expressed as a Fourier series. The steady-state response to each harmonic component is thus calculated separately and the total response obtained by linear superposition. It is

$$x(t) = \frac{a_0}{2k_s} + \sum_{n=1,2}^{\infty} \frac{(a_n/k_s)}{[\{1 - n^2(\omega/\omega_n)^2\}^2 + \{2\zeta n\omega/\omega_n\}^2]^{1/2}} \cos(n\omega t - \phi_n)$$
$$+ \sum_{n=1,2}^{\infty} \frac{(b_n/k_s)}{[\{1 - n^2(\omega/\omega_n)^2\}^2 + \{2\zeta n\omega/\omega_n\}^2]^{1/2}} \sin(n\omega t - \phi_n). \tag{1.96}$$

The phase angle, ϕ_n, is given by

$$\phi_n = \tan^{-1} \frac{2\zeta n\omega/\omega_n}{1 - n^2(\omega/\omega_n)^2}. \tag{1.97}$$

The first term in equation (1.96) is a static term and the terms within the summation signs are the contributions of the various harmonically related terms. Each individual term is similar to equation (1.59) and its corresponding phase is similar to equation (1.60).

1.5.8 Forced vibrations with transient excitation

When the forcing function, $F(t)$, is non-periodic it cannot be represented by a Fourier series, and other forms of solution have to be utilised. Several techniques are available for the solution of the equations of motion, including Fourier transforms, Laplace transforms, and the convolution integral. The convolution integral procedure will be adopted here initially and it will subsequently lead to the usage of the Fourier transform.

An arbitrary, non-periodic, forcing function, $F(t)$, can be approximated by a series of pulses of short duration, $\Delta\tau$, as illustrated in Figure 1.18. The convolution integral procedure for the estimation of the output response involves the linear summation of the product of each of the pulses in the input signal with a suitable pulse response function which is associated with the system.

To understand the behaviour of this pulse response function, the concept of an 'impulse response' needs to be introduced. Consider a rectangular pulse of unit area, as illustrated in Figure 1.19. A unit impulse is obtained by letting the time duration, T, of the pulse approach zero whilst maintaining the unit area. In the limit, a unit impulse, $\delta(t)$, of infinite height and zero width is produced. The unit impulse concept is an important one in noise and vibration studies and, as will be seen later on in this book, it has significant practical applications in noise and vibration.

A unit impulse has two main properties. They are:

$$\text{(i) } \delta(t) = 0 \text{ for } t \neq 0, \qquad \text{and (ii) } \int_{-\infty}^{\infty} \delta(t)\,\mathrm{d}t = 1. \tag{1.98}$$

If the unit impulse occurs at time $t = \tau$ instead of at time $t = 0$, then it is translated along the time axis by an amount τ, and

$$\int_{-\infty}^{\infty} \delta(t - \tau)\,\mathrm{d}t = 1. \tag{1.99}$$

This is illustrated in Figure 1.20. The function $\delta(t)$ is also known as the Dirac delta function.

Fig. 1.18. The approximation of an arbitrary function, *F*(*t*), by a series of pulses of short duration, Δτ.

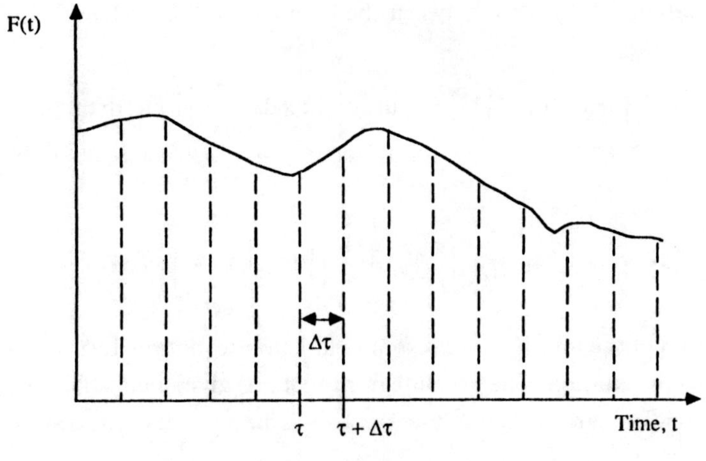

Fig. 1.19. A rectangular pulse of unit area.

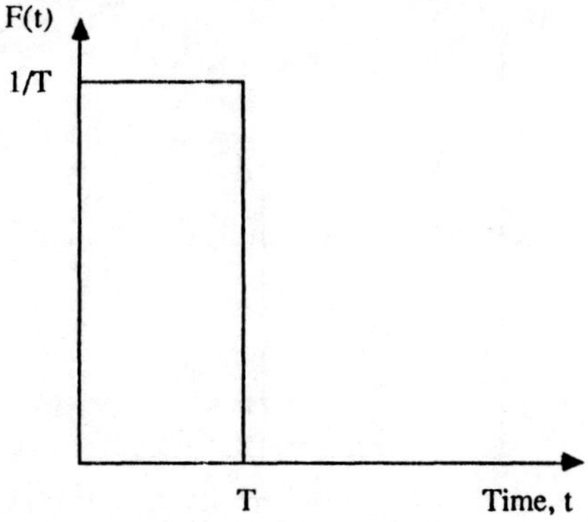

The excitation of a one-degree-of-freedom, mass-spring-damper, system with a unit impulse is now considered. The solution to this problem will subsequently lead to a general solution for an arbitrary transient force input. For an input force excitation $F(t) = \delta(t)$, the equation of motion is

$$m\ddot{x} + c_v\dot{x} + k_s x = \delta(t). \tag{1.100}$$

The system is at rest before the unit impulse is applied, and is motion immediately before and after the impulse is applied needs to be carefully considered. The unit impulse, $\delta(t)$, is applied at time $t = 0$ (by its definition). The small time interval immediately before $t = 0$ is $t = 0^-$, and the small time interval immediately after $t = 0$ is $t = 0^+$. Hence, the unit impulse is completed at $t \geqslant 0^+$. Because the system is at rest before the unit impulse is applied the initial displacement and velocity conditions at 0^- are

$$x(0^-) = \dot{x}(0^-) = 0. \tag{1.101}$$

The initial displacement condition at 0^+ is obtained by integrating the equation of motion (equation 1.100) twice between the limits 0^- and 0^+. Hence,

$$\iint_{0^-}^{0^+} m\ddot{x}\,\mathrm{d}t\,\mathrm{d}t + \iint_{0^-}^{0^+} c_v\dot{x}\,\mathrm{d}t\,\mathrm{d}t + \iint_{0^-}^{0^+} k_s x\,\mathrm{d}t\,\mathrm{d}t = \iint_{0^-}^{0^+} \delta(t)\,\mathrm{d}t\,\mathrm{d}t. \tag{1.102}$$

This leads to

$$m\{x(0^+) - x(0^-)\} + \int_{0^-}^{0^+} c_v x\,\mathrm{d}t + \iint_{0^-}^{0^+} k_s x\,\mathrm{d}t\,\mathrm{d}t = \iint_{0^-}^{0^+} \delta(t)\,\mathrm{d}t\,\mathrm{d}t. \tag{1.103}$$

The properties of the unit impulse response function (equation 1.98) are now used to solve the above equation. The first integration of $\delta(t)$ gives a constant which is unity, and the second integration gives a zero. This is because the integration of a finite

Fig. 1.20. Unit impulse at $t = 0$ and $t = \tau$.

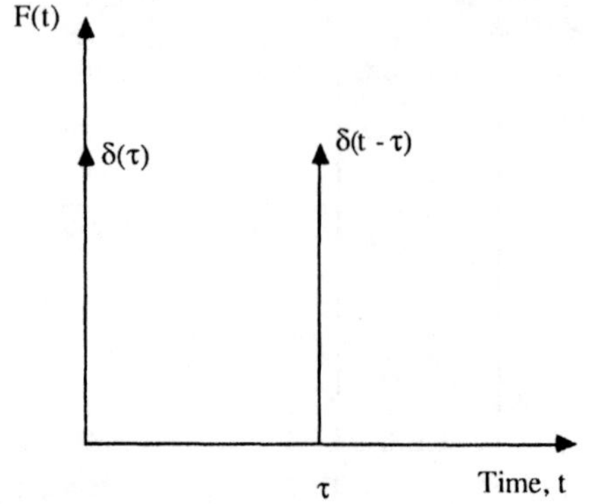

quantity over an infinitesimal interval is zero (only the integration of an infinite quantity over an infinitesimal interval produces a finite result). This argument can also be extended to the integrals on the left hand side of equation (1.103). Because the motion, $x(t)$, is finite, the integrals reduce to zero. Hence the initial displacement condition $x(0^+)$ is zero since $x(0^-)$ is also zero.

The initial velocity condition at 0^+ is obtained by integrating the equation of motion (equation 1.100) once between the limits 0^- and 0^+. Here,

$$m\{\dot{x}(0^+)-\dot{x}(0^-)\}+c_v\{x(0^+)-x(0^-)\}+\int_{0^-}^{0^+} k_s x\,dt=\int_{0^-}^{0^+}\delta(t)\,dt. \quad (1.104)$$

Using similar arguments to those in the preceding paragraph, it can be shown that all the terms on the left hand side of equation (1.104) are zero with the exception of the initial velocity condition at 0^+. The unit impulse function on the right hand side integrates to unity. Hence, at time $t \geqslant 0^+$, the equation of motion of the system is simply

$$m\ddot{x}+c_v\dot{x}+k_s x=0, \quad (1.105)$$

with the initial conditions

$$x(0^+)=0, \qquad \text{and } \dot{x}(0^+)=1/m. \quad (1.106)$$

When the input to the mass–spring–damper sytem is a unit impulse, its response, $x(t)$, is commonly referred to as its unit impulse response. The unit impulse response function is generally referred to in the literature as $h(t)$ – i.e. the symbol $x(t)$ is replaced by the symbol $h(t)$. For the above initial conditions, the solution to equation (1.105) can be readily obtained from equations (1.40) and (1.42). The phase angle ψ is zero since the initial displacement condition is zero and the unit impulse response function, $h(t)$, is

$$h(t)=\frac{1}{(1-\zeta^2)^{1/2}\omega_n m}\,e^{-\zeta\omega_n t}\sin\{(1-\zeta^2)^{1/2}\omega_n t\}. \quad (1.107)$$

If the unit impulse occurs at $t=\tau$ instead of at $t=0$ then the unit impulse response is delayed by the amount of time τ, and

$$h(t-\tau)=\frac{1}{(1-\zeta^2)^{1/2}\omega_n m}\,e^{-\zeta\omega_n(t-\tau)}\sin\{(1-\zeta^2)^{1/2}\omega_n(t-\tau)\}. \quad (1.108)$$

If the magnitude of the impulse is F instead of unity then the initial velocity condition at 0^+ is F/m instead of $1/m$, and the output response, $x(t)$, is

$$x(t)=Fh(t-\tau). \quad (1.109)$$

For the case of the arbitrary, non-periodic, forcing function $F(t)$, which was approximated by a series of pulses of short duration, $\Delta\tau$, as illustrated in Figure 1.18, the magnitude of each pulse is defined by the pulse area $F(t)\Delta\tau$. The system response to each individual pulse is given by equation (1.109) and is just the product of the unit impulse response and the pulse magnitude – i.e. $h(t-\tau)F(\tau)\Delta\tau$. The total response due to the arbitrary, non-periodic, forcing function $F(t)$ is the linear superposition of the system's response to each individual pulse. In the limit as $\Delta\tau$ approaches zero,

$$x(t) = \int_0^t F(\tau)h(t-\tau)\,\mathrm{d}\tau. \tag{1.110}$$

Equation (1.110) is the convolution integral. It states that the output response of a linear system to an arbitrary input is the convolution of the unit impulse response with the input signal.

It should be pointed out here that in this sub-section an input force and an output displacement have been used to illustrate the principles of the unit impulse response function and the convolution integral. In general, either the input or the output can be a force, a displacement, a velocity or an acceleration and thus there can be a range of different types of unit impulse response functions. There is an analogy between these different types of unit impulse response functions and the different types of frequency response functions discussed in sub-section 1.5.5. It will be shown in the next section that the unit impulse response function is in fact simply the inverse Fourier transform of the frequency response function.

1.6 Forced vibrations with random excitation

The response of a one-degree-of-freedom system to periodic and non-periodic signals has been summarised in section 1.5. In the case of periodic excitation, the input signal is resolved into its individual frequency components and the resultant output signal obtained by summation. For the case of non-periodic, transient, excitation the same principles are applied except that the summations are replaced by an integral because the individual frequency components are no longer discrete – i.e. they are continuously distributed. The signal is, however, still deterministic and can therefore be expressed by an explicit mathematical relationship.

Quite often, in noise and vibration analysis, the input signal to some system cannot be described by an explicit mathematical relationship. It is random in nature (i.e. the time history of the signal is neither periodic nor transient but is continuous and does not repeat itself) and has to be described in terms of probability statements and statistical averages – this class of vibrations is termed random vibrations. Also, if the input to a system is random, its output vibrations will also be random. Some typical examples of random vibrations are the turbulent flow over an aircraft body; the response of ships to ocean waves; the effects of internal flow disturbances (e.g. bends valves or orifice plates) on the vibration response and the sound radiation from pipes with internal fluid flow (e.g. nuclear reactors, heat exchangers and gas pipelines); the response of the suspension systems of road vehicles to rough roads; and, sound fields generated by jet engine exhausts. A time history of a typical random signal containing numerous frequency components is illustrated in Figure 1.21.

Four types of statistical functions are used to describe random signals. They are:

(i) mean-square values and the variance – they provide information about the amplitude of the signal;

(ii) probability distributions – they provide information about the statistical properties of the signal in the amplitude domain;

(iii) correlation functions – they provide information about the statistical properties of the signal in the time domain;

(iv) spectral density functions – they provide information about the statistical properties of the signal in the frequency domain.

An individual time history of a random signal is called a sample record, and a collection of several such records constitutes an ensemble average of a random, or a stochastic, process. A random process is ergodic (or strictly stationary) if all the probability distributions associated with it are time-invariant – i.e. all the probability distributions taken along any single sample record are the same as along a different sample record of the same process. It is weakly stationary if only its first and second order probability distributions are invariant with time. A random process is non-stationary when its probability distributions are not stationary with respect to a change of the time scale – i.e. they vary with time. It is important to remember that, when using the terms stationary and non-stationary, it is the probability distributions that are being referred to, and not the process itself.

Most random physical phenomena that are of interest to engineers can be approximated as being stationary – if a signal is very long compared with the period of the lowest frequency component of interest, it is approximately stationary. Therefore, only the random vibrations of stationary signals will be presented in this section. Random vibrations of non-stationary signals are discussed in the specialist

Fig. 1.21. A time history of a typical random signal.

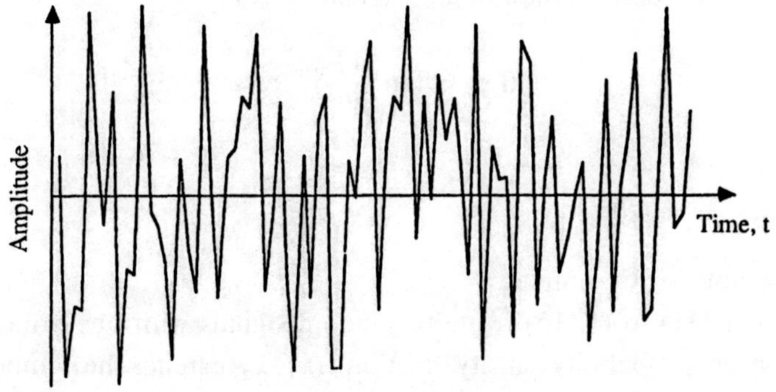

literature (see reference list at the end of this chapter). The discussions will also be limited at this stage to the specific case of the single oscillator – random vibrations of continuous systems will be discussed in section 1.9. Throughout this section, therefore, a linear system with a single input and a single output, as illustrated in Figure 1.13, will be considered. The input will be assumed to be a random signal, $x(t)$, and the output will be defined as $y(t)$. The system will be modelled as a single-degree-of-freedom, mass–spring–damper.

1.6.1 Probability functions

The expected or mean value of a function $x(t)$ is given by

$$E[x(t)] = \frac{1}{T}\int_0^T x(t)\,dt = \int_{-\infty}^{\infty} xp(x)\,dx, \tag{1.111}$$

where $p(x)$ is the probability density function. It specifies the probability, $p(x)\,dx$, that a random variable lies in the range x to $x + dx$. For a stationary random process, $E[x(t)] = E[x]$. This is because a stationary random process is time-invariant. It is sometimes referred to as the first statistical moment. The second statistical moment, or the mean-square value, $E[x^2]$, is the average value of x^2 and is given by

$$E[x^2] = \frac{1}{T}\int_0^T x^2\,dt = \int_{-\infty}^{\infty} x^2 p(x)\,dx. \tag{1.112}$$

The time integrals in equations (1.111) and (1.112) are approximations with the assumption that, for sufficiently large T, all values within the range 0 to T are equally probable. Hence, $dt/T \sim p(x)\,dx$.

The positive square root of $E[x^2]$ is the root-mean-square value of the signal. The standard deviation of $x(t)$, denoted by σ, and the variance, σ^2, are defined by

$$\sigma^2 = E[x^2] - \{E[x]\}^2. \tag{1.113}$$

Quite often, in modern signal analysis, the data is recorded digitally. Here, the mean value, the mean-square value and the variance can be obtained directly from the digital time history data of the random signal, $x(t)$. Thus

$$E[x] = \lim_{N\to\infty} \frac{1}{N}\sum_{i=1}^{N} x_i(t), \tag{1.114}$$

$$E[x^2] = \lim_{N\to\infty} \frac{1}{N}\sum_{i=1}^{N} x_i^2(t), \tag{1.115}$$

and N is the number of samples.

Equations (1.111) to (1.115) relate to random signals with one random variable. The second-order probability density function $p(x_1, x_2)$ extends the number of random variables from one to two. Also, if the two signals are statistically independent, then $p(x_1, x_2) = p(x_1)p(x_2)$. These concepts can subsequently be extended to multiple random variables, and the equations for the mean and mean-square values suitably modified. The discussions in this section will be limited to single random variables – i.e. single input and single output systems.

1.6.2 Correlation functions

The auto-correlation function for a random signal, $x(t)$, provides information about the degree of dependence of the value of x at some time t on its value at some other time $t+\tau$. For a stationary random signal, the auto-correlation depends upon the time separation, τ, and is independent of absolute time. It is defined as

$$R_{xx}(\tau) = E[x(t)x(t+\tau)] = \lim_{T\to\infty} \frac{1}{T}\int_0^T x(t)x(t+\tau)\,\mathrm{d}t. \tag{1.116}$$

Note that $p(x)\,\mathrm{d}x$ has been replaced by $\mathrm{d}t/T$ in the above equation – i.e. for sufficiently large T, all values of the random signal, $x(t)$, are equally likely within the range 0 to T.

A correlation coefficient (a normalised correlation function), $\rho_{xx}(\tau)$, can now be defined as

$$\rho_{xx}(\tau) = \frac{E[\{x(t_1) - m_x\}\{x(t_2) - m_x\}]}{\sigma_x \sigma_x} = \frac{R_{xx}(\tau) - m_x^2}{\sigma_x^2}, \tag{1.117}$$

where m_x is the mean value of the signal. When $\tau \to 0$, $\rho_{xx} \to 1$ and when $\tau \to \infty$, $\rho_{xx} \to 0$. The auto-correlation function is an even function, it does not contain any phase information, and its maximum value always occurs at $\tau = 0$. For periodic signals, $R_{xx}(\tau)$ is always periodic, and for random signals it always decays to zero for large values of τ. It is therefore a useful tool for identifying deterministic signals which would otherwise be masked in a random background. A typical auto-correlation signal is illustrated in Figure 1.22.

The cross-correlation function between two different stationary random signals (e.g. the input, $x(t)$, and the output, $y(t)$) is defined as

$$R_{xy}(\tau) = E[x(t)y(t+\tau)] = \lim_{T\to\infty} \frac{1}{T}\int_0^T x(t)y(t+\tau)\,\mathrm{d}t. \tag{1.118}$$

The cross-correlation function indicates the similarity between two signals as a function of the time shift, τ. Unlike the auto-correlation function, it is not an even function, and $R_{xy}(\tau) = R_{yx}(-\tau)$. It has many applications in noise and vibration, including the detection of time delays between signals, transmission path delays in room acoustics, airborne noise analysis, noise source identification, radar and sonar applications. A typical cross-correlation function is illustrated in Figure 1.23.

Auto- and cross-correlation functions and their applications will be discussed in detail in chapter 5. They are introduced in this chapter because of their relevance to the random excitation of single oscillators.

1.6.3 Spectral density functions

The spectral density function is the Fourier transform of the correlation coefficient. A general Fourier transform pair, $\mathbf{X}(\omega)$ and $x(t)$ is defined as

Fig. 1.22. A typical auto-correlation function for a stationary random signal.

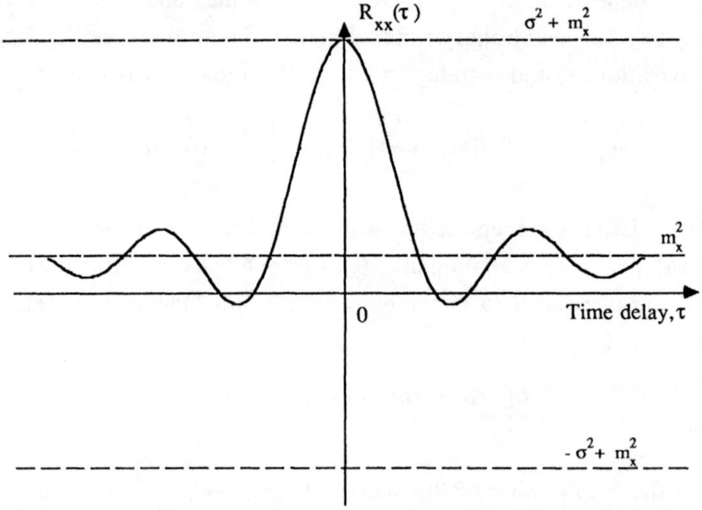

Fig. 1.23. A typical cross-correlation function for a stationary random signal.

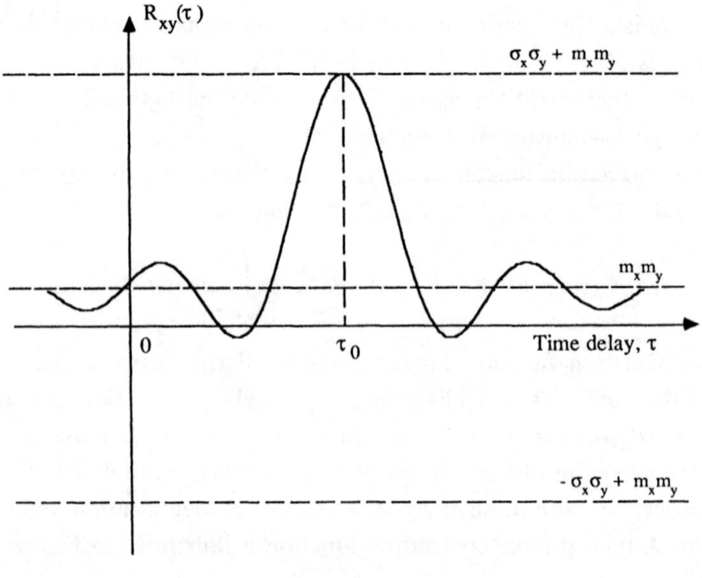

$$\mathbf{X}(\omega) = \frac{1}{2\pi} \int_{-\infty}^{\infty} x(t)\, \mathrm{e}^{-\mathrm{i}\omega t}\, \mathrm{d}t, \tag{1.119a}$$

and

$$x(t) = \int_{-\infty}^{\infty} \mathbf{X}(\omega)\, \mathrm{e}^{\mathrm{i}\omega t}\, \mathrm{d}\omega. \tag{1.119b}$$

$\mathbf{X}(\omega)$ is the Fourier transform of $x(t)$ and it is a complex quantity. Classical Fourier analysis also introduces the condition that

$$\int_{-\infty}^{\infty} |x(t)|\, \mathrm{d}t < \infty,$$

i.e. classical theory is valid for functions which are absolutely integrable and decay to zero when $|t| \to \infty$. Stationary random signals do not decay to zero with time. This problem is overcome by Fourier analysing the correlation function instead (the correlation function of a random signal decays to zero with increasing τ). It is important to note that the frequency content of the stationary random signal is not lost in the process.

The Fourier transform of $R_{xx}(\tau)$ and its inverse are thus given by

$$\mathbf{S}_{\mathbf{xx}}(\omega) = S_{xx}(\omega) = \frac{1}{2\pi} \int_{-\infty}^{\infty} R_{xx}(\tau)\, \mathrm{e}^{-\mathrm{i}\omega\tau}\, \mathrm{d}\tau, \tag{1.120a}$$

and

$$R_{xx}(\tau) = \int_{-\infty}^{\infty} S_{xx}(\omega)\, \mathrm{e}^{\mathrm{i}\omega\tau}\, \mathrm{d}\omega. \tag{1.120b}$$

$S_{xx}(\omega)$ is the auto-spectral density of the $x(t)$ random signal and it is a function of frequency. It is a real, even function ($S_{xx}(\omega) = S_{xx}(-\omega)$). The cross-spectral density, $\mathbf{S}_{\mathbf{xy}}(\omega)$, is the Fourier transform of the cross-correlation function, and it is a complex quantity. It is given by

$$\mathbf{S}_{\mathbf{xy}}(\omega) = \frac{1}{2\pi} \int_{-\infty}^{\infty} R_{xy}(\tau)\, \mathrm{e}^{-\mathrm{i}\omega\tau}\, \mathrm{d}\tau. \tag{1.121}$$

The auto- and cross-spectral densities are widely used in noise and vibration analysis. It will be shown shortly that the area under an auto-spectrum is the mean-square value of a signal. The cross-spectral density between two signals contains both magnitude and phase information and is very useful for identifying major signals that are common to both the input and output from a linear system. Some of its more important applications will be discussed in chapter 5.

The time histories, auto-correlation functions, and spectral densities for (a) a sine wave, (b) a narrow-frequency-band random noise signal, (c) a wide-frequency-band random noise signal, and (d) a sine wave with a random noise signal superimposed upon it are illustrated in Figure 1.24. The auto-correlation function for the deterministic sine wave is a continuous function – i.e. it is a cosine and therefore does not decay with increasing time delay. The auto-correlation functions of the two random signals decay to zero with increasing time delay – the narrowband decay envelope is more spread out than the wideband (i.e. broadband) decay envelope. The frequency of the sine wave can be identified from the auto-correlation function of the sine wave with a random noise signal superimposed upon it or from the corresponding auto-spectral density. If the amplitude of the sine wave was less than the random noise signal, then its spectral amplitude would be submerged under the random noise spectra and its presence would only be detectable from the auto-correlation function.

It should be pointed out at this stage that the experimental estimation of spectra from measured data does not follow the above mentioned formal mathematical route of obtaining the spectra from the correlation function. With the development of the fast Fourier transform (FFT) technique, digital estimates of spectra can be directly obtained from the time histories with suitable computer algorithms. The procedures are very accurate, rapid and efficient, and will be discussed in chapter 5.

1.6.4 Input–output relationships for linear systems

Consider an arbitrary input signal, $x(t)$, to a linear system such that

$$\int_{-\infty}^{\infty} |x(t)|\,\mathrm{d}t < \infty.$$

Its Fourier transform, $\mathbf{X}(\omega)$, is given by

$$\mathbf{X}(\omega) = \frac{1}{2\pi}\int_{-\infty}^{\infty} x(t)\,\mathrm{e}^{-i\omega t}\,\mathrm{d}t.$$

For a linear system, there is a relationship between the Fourier transforms of the input signal, $\mathbf{X}(\omega)$, and the output signal, $\mathbf{Y}(\omega)$. This relationship was, in fact, derived in sub-section 1.5.5 where the ratio of output to input gave the frequency response characteristics of the single oscillator. In general, this relationship is

$$\mathbf{Y}(\omega) = \mathbf{H}(\omega)\mathbf{X}(\omega), \tag{1.122}$$

whre $\mathbf{H}(\omega)$ is the frequency response function of the linear system. Equation (1.122) is valid for both single oscillator systems (as discussed in sub-section 1.5.5) and systems where there are many natural frequencies. The frequency response function, $\mathbf{H}(\omega)$, can be a receptance, a mobility, an impedance etc., and can be extended to

(a)

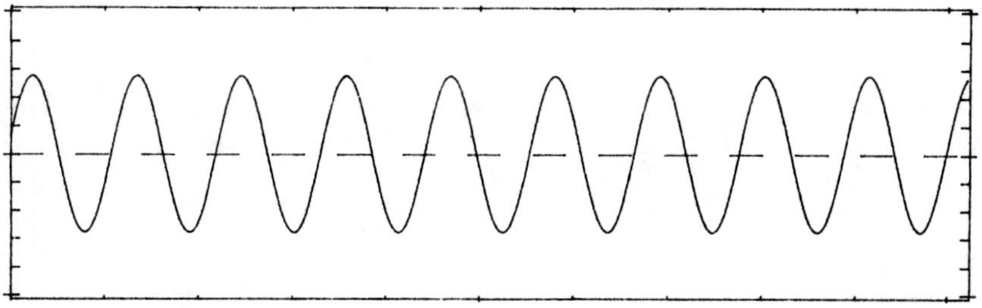

a sine wave

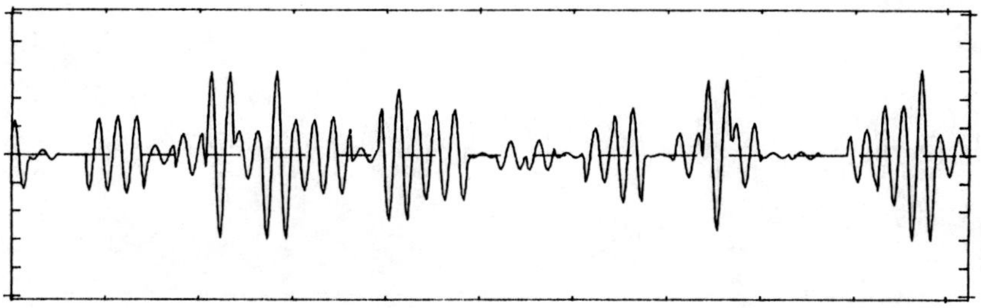

a narrow-frequency-band random noise signal

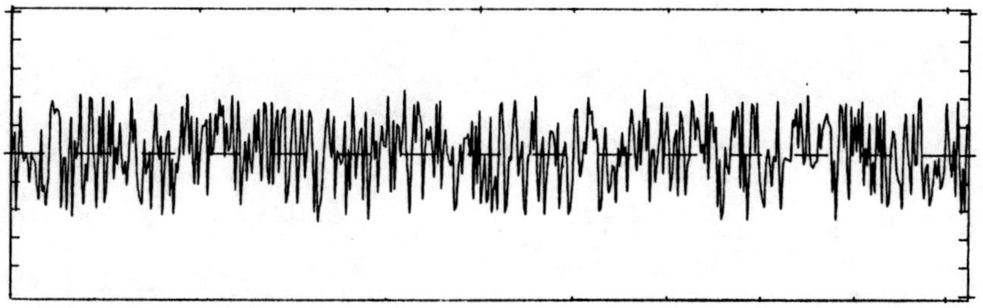

a wide-frequency-band random noise signal

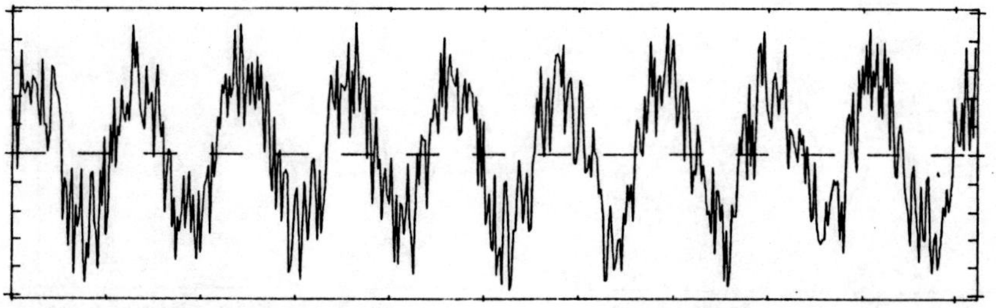

a sine wave with a random noise signal

Fig. 1.24. (a) Time history functions for some typical deterministic and random signals. (b) Auto-correlation functions for some typical deterministic and random signals. (c) Spectral density functions for some typical deterministic and random signals.

(b)

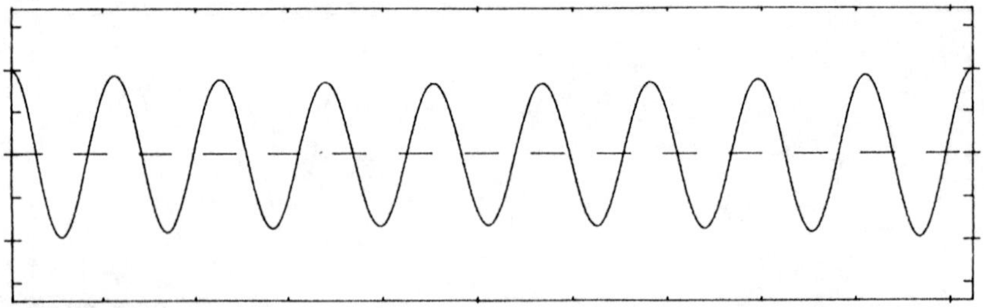

a sine wave

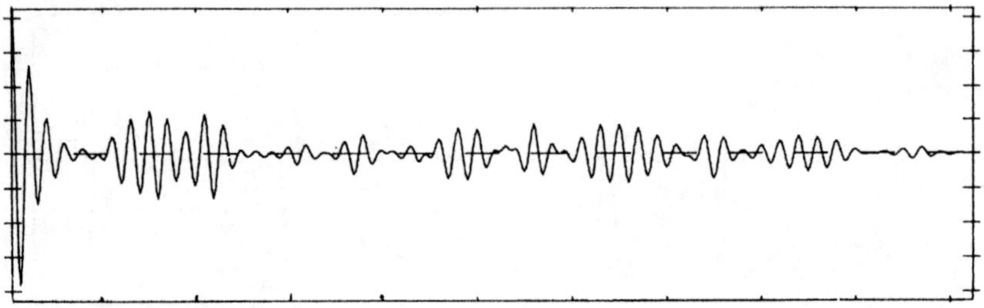

a narrow-frequency-band random noise signal

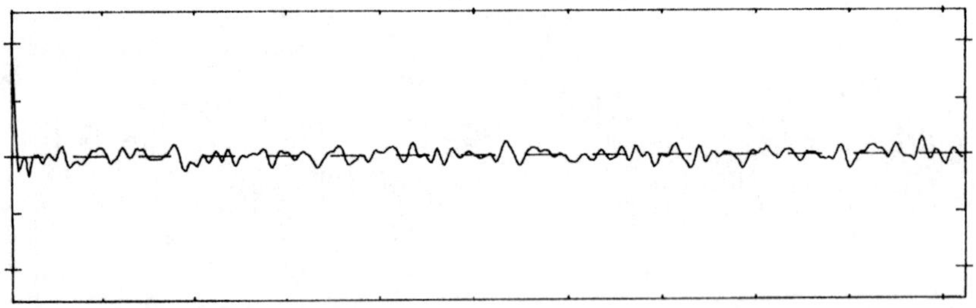

a wide-frequency-band random noise signal

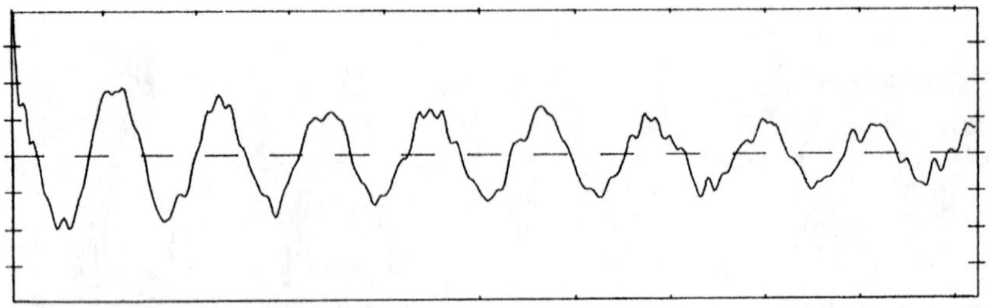

a sine wave with a random noise signal

Fig. 1.24 (*continued*)

(c)

a sine wave

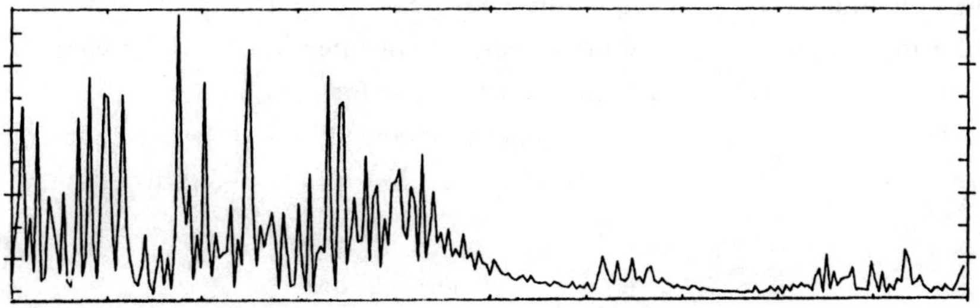

a narrow-frequency-band random noise signal

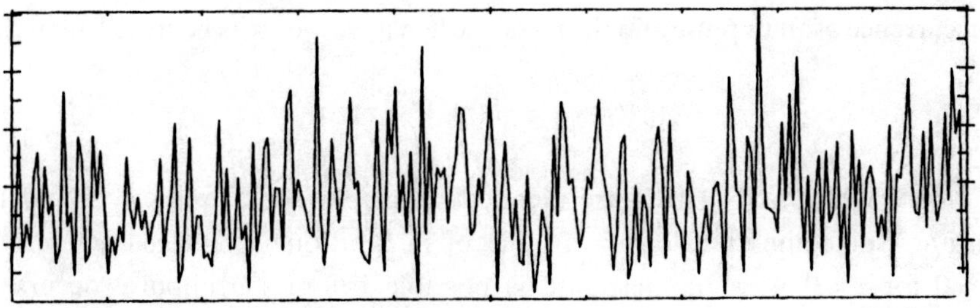

a wide-frequency-band random noise signal

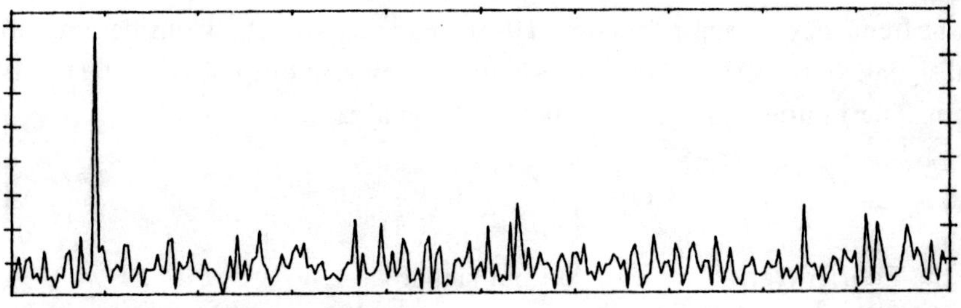

a sine wave with a random noise signal

Fig. 1.24 (*continued*)

systems with a continuum of frequencies. The output signal, $y(t)$, from the linear system can subsequently be obtained by inverse Fourier transforming equation (1.122) – i.e.

$$y(t)=\int_{-\infty}^{\infty}\mathbf{H}(\omega)\left\{\frac{1}{2\pi}\int_{-\infty}^{\infty}x(t)\,\mathrm{e}^{-\mathrm{i}\omega t}\,\mathrm{d}t\right\}\mathrm{e}^{\mathrm{i}\omega t}\mathrm{d}\omega. \tag{1.123}$$

The term inside the curly brackets is the Fourier transform of $x(t)$ – i.e. $\mathbf{X}(\omega)$. Equation (1.23) is a formal solution for the output response. It is not commonly used because the inverse Fourier transform integral with respect to $\mathrm{d}\omega$ is not readily solved. The convolution integral and the impulse response technique discussed in sub-section 1.5.8 are more suitable for obtaining the output, $y(t)$.

The convolution integral (equation 1.110) can be re-arranged for a continuous random input signal, $x(t)$. The upper variable of integration can be changed from $\tau=t$ to $\tau=\infty$. This is because there is no response for $\tau>t$, i.e. for $\tau>t$, $h(t-\tau)=0$. Also, the lower variable of integration can be changed to $-\infty$ because the excitation whose value at time τ is $x(\tau)$ can, in principle, exist from $\tau=-\infty$ to the present, i.e. $\tau=t$. Thus,

$$y(t)=\int_{-\infty}^{\infty}x(t)h(t-\tau)\,\mathrm{d}\tau. \tag{1.124}$$

In this form of the convolution integral, the impulse occurs at time τ, and the output response is evaluated at time t. If τ is defined instead as the time difference between the occurrence of an impulse and the instant when its response is being calculated, then

$$y(t)=\int_{-\infty}^{\infty}h(\tau)x(t-\tau)\,\mathrm{d}\tau. \tag{1.125}$$

Equations (1.124) and (1.125) are identical and both are commonly found in the literature. Sometimes the lower variable of integration is replaced by zero since $h(\tau)=0$ for $\tau<0$ – i.e. no response is possible before the impulse occurs. Both equations are based on the assumption that the random input signal, $x(t)$, is made up of a continuous series of small impulses.

There is an important relationship between the impulse response function, $h(\tau)$, and the frequency response function, $\mathbf{H}(\omega)$, of a linear system. Consider an impulsive input signal, $x(t)=\delta(t)$, and the corresponding transient output, $y(t)=h(t)$, of a linear system. The Fourier transform of the input signal is

$$\mathbf{X}(\omega)=\frac{1}{2\pi}\int_{-\infty}^{\infty}\delta(t)\,\mathrm{e}^{-\mathrm{i}\omega t}\,\mathrm{d}t=\frac{1}{2\pi}, \tag{1.126}$$

and the Fourier transform of the output signal is

$$\mathbf{Y}(\omega)=\frac{1}{2\pi}\int_{-\infty}^{\infty}h(t)\,\mathrm{e}^{-\mathrm{i}\omega t}\,\mathrm{d}t. \tag{1.127}$$

By substituting for $\mathbf{X}(\omega)$ and $\mathbf{Y}(\omega)$ into equation (1.122),

$$\mathbf{H}(\omega) = \int_{-\infty}^{\infty} h(t)\,\mathrm{e}^{-\mathrm{i}\omega t}\,\mathrm{d}t. \tag{1.128}$$

Hence, the frequency response function, $\mathbf{H}(\omega)$ is the Fourier transform of the impulse response function, $h(t)$, less the $1/2\pi$ factor (using the definition of Fourier transform pairs as given by equation 1.119). This inconsistency is easily overcome by accounting for this factor in the inverse Fourier transform such that

$$h(t) = \frac{1}{2\pi}\int_{-\infty}^{\infty} \mathbf{H}(\omega)\,\mathrm{e}^{\mathrm{i}\omega t}\,\mathrm{d}\omega. \tag{1.129}$$

The impulse response function is thus a very powerful tool in noise and vibration analysis. It is the time domain representation of the frequency response of a system and it is related to the frequency response function via the Fourier transform. Amongst other things, it can be used to identify structural modes of vibration and to determine noise transmission paths. Equation (1.125) is the formal input–output relationship for a linear system in terms of the impulse response function (cf. equation 1.123).

Input–output relationships for a single input–output system can now be derived. Consider a random input signal, $x(t)$, and the corresponding output signal, $y(t)$, from an arbitrary linear system. For such a system,

$$y(t)y(t+\tau) = \int_{0}^{\infty}\int_{0}^{\infty} h(\xi)h(\eta)x(t-\xi)x(t+\tau-\eta)\,\mathrm{d}\xi\,\mathrm{d}\eta, \tag{1.130}$$

and the corresponding input–output, auto-correlation relationship is

$$R_{yy}(\tau) = \int_{0}^{\infty}\int_{0}^{\infty} h(\xi)h(\eta)R_{xx}(\tau+\xi-\eta)\,\mathrm{d}\xi\,\mathrm{d}\eta. \tag{1.131}$$

Similarly,

$$x(t)y(t+\tau) = \int_{0}^{\infty} h(\eta)x(t)x(t+\tau-\eta)\,\mathrm{d}\eta, \tag{1.132}$$

and the corresponding input–output cross-correlation relationship is

$$R_{xy}(\tau) = \int_{0}^{\infty} h(\eta)R_{xx}(\tau-\eta)\,\mathrm{d}\eta. \tag{1.133}$$

Equations (1.130) to (1.133) represent the convolution of the input signal with the appropriate impulse response functions. The lower variables of integration have been replaced by zero since $h(\xi)$ and $h(\eta) = 0$ for ξ and $\eta < 0$. Equations (1.131) and (1.133) can now be Fourier transformed to yield

$$S_{yy}(\omega) = |\mathbf{H}(\omega)|^2 S_{xx}(\omega), \tag{1.134}$$

and

$$\mathbf{S_{xy}}(\omega) = \mathbf{H}(\omega) S_{xx}(\omega). \tag{1.135}$$

Equation (1.134) is a real-valued function and it only contains information about the amplitude, $H(\omega)$, of the frequency response function. $S_{xx}(\omega)$ and $S_{yy}(\omega)$ are the auto-spectra of the input and output signals, respectively. Equation (1.135) is a complex-valued function and it contains both magnitude and phase information.

Equation (1.134) represents the output response of a linear system to random vibrations and can be extended for N different inputs to

$$S_{yy}(\omega) = \sum_{p=1}^{N} \sum_{q=1}^{N} \mathbf{H_p^*}(\omega) \mathbf{H_q}(\omega) S_{x_p x_q}(\omega). \tag{1.136}$$

In the above equation, $\mathbf{H_p^*}(\omega)$ is the complex conjugate of $\mathbf{H_p}(\omega)$. The equation is essentially the main result of random vibration theory and it says that the spectral density of the output from a linear system is the summation of the products of the freqency response functions associated with the various inputs and the corresponding spectral densities of the various inputs. For the general case, the cross-spectral densities between the various inputs (i.e. S_{xx} for $p \neq q$) have to be taken into account. If the various inputs are uncorrelated with each other, the cross-terms drop out and equation (1.136) reduces to

$$S_{yy}(\omega) = \sum_{p=1}^{N} |\mathbf{H}(\omega)|^2 S_{x_p x_p}(\omega). \tag{1.137}$$

For the special case of a single input–output system, the results reduce to equations (1.134) and (1.135). For a given input force (with a spectral density $S_{xx}(\omega)$) to a single oscillator, the output displacement spectral density, $S_{xx}(\omega)$, is therefore given by

$$S_{yy}(\omega) = \frac{S_{xx}(\omega)}{(k_s - m\omega^2)^2 + c_v^2 \omega^2}, \tag{1.138}$$

where

$$|\mathbf{H}(\omega)|^2 = \frac{1}{(k_s - m\omega^2)^2 + c_v^2 \omega^2}. \tag{1.139}$$

The frequency response function, $\mathbf{H}(\omega)$, is obtained from equation (1.56) (i.e. $\mathbf{X}/\mathbf{F}$).

For a random signal, $x(t)$, it can be seen from equation (1.116) that, at $\tau = 0$, $R_{xx}(\tau = 0) = R_{xx}(0) = E[x^2]$. Thus, from the Fourier transform relationship between the auto-correlation function and the spectral density function,

$$R_{xx}(0) = E[x^2] = \int_{-\infty}^{\infty} S_{xx}(\omega)\, e^{i\omega 0}\, d\omega = \int_{-\infty}^{\infty} S_{xx}(\omega)\, d\omega. \tag{1.140}$$

Equation (1.140) is a very important relationship – it shows that the area under the auto-spectral density curve is the mean-square value of the signal. Thus, for a single input–output system, the mean-square response of the output signal is

$$E[y]^2 = \int_{-\infty}^{\infty} S_{yy}(\omega)\, d\omega = \int_{-\infty}^{\infty} |\mathbf{H}(\omega)|^2 S_{xx}(\omega)\, d\omega. \tag{1.141}$$

The auto-spectral densities, $S_{xx}(\omega)$, and the cross-spectral densities, $\mathbf{S}_{\mathbf{xy}}(\omega)$, are commonly referred to as the two-sided spectral densities – i.e. they range from $-\infty$ to $+\infty$. Whilst they are convenient for analytical studies, in reality the frequency range is from 0 to $+\infty$. Therefore, a physically measurable one-sided spectral density, $G(\omega)$, has to be defined such that $G(\omega) = 2S(\omega)$. This is illustrated in Figure 1.25. In

Fig. 1.25. One-sided and two-sided spectral density functions.

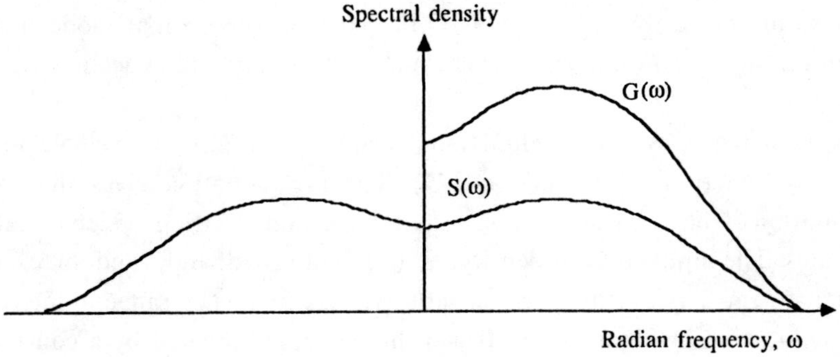

terms of this physically measurable one-sided spectral density, equations (1.134) and (1.135) now become

$$G_{yy}(\omega) = |\mathbf{H}(\omega)|^2 G_{xx}(\omega), \tag{1.142}$$

and

$$\mathbf{G}_{\mathbf{xy}}(\omega) = \mathbf{H}(\omega) G_{xx}(\omega). \tag{1.143}$$

The preceding equations in this section apply to ideal linear systems with no extraneous noise, i.e. there is a perfect correlation at all frequencies between the input and output. This is not the case in practice and a degree of frequency correlation (a coherence function) needs to be defined. The properties of the coherence function and other matters relating to noise and vibration signal analysis techniques will be discussed in chapter 5.

1.6.5 The special case of broadband excitation of a single oscillator

Quite often, the response of a specific resonant mode of a structure to some form of broadband, random, excitation is required, even though the structure would have numerous natural frequencies. Broadband excitation of a resonant mode is defined as an excitation whose spectral density is reasonably constant over the range of frequencies that encompass the resonant response of the mode. At low frequencies (the first few natural frequencies of a structure), the modes of vibration of a structure are generally well separated in frequency, and approximations can be made such as to model each individual mode of vibration as a single-degree-of-freedom system. Estimation procedures can subsequently be developed to determine the modal mean-square response of the particular mode. These procedures are based upon the assumption that equation (1.50) (with the harmonic force term $F \sin \omega t$ replaced by some arbitrary random force $f(t)$) represents the response of a single resonant mode of some continuous system with numerous natural frequencies. This is always the case provided that the modal mass, modal stiffness, modal damping and modal excitations are correctly defined. This is the basis of the normal mode theory of vibrations of linear continuous systems which will be reviewed in section 1.9 in this chapter.

The system frequency response function of displacement/force for a single oscillator is given by equations (1.56) and (1.139). The first equation gives the complex representation of the frequency response, and the second equation, which is real, gives its modulus. The input spectral density, $S_{xx}(\omega)$, of a broadband, random, excitation to such a system is assumed to be constant over the frequency range of interest (i.e. $\sim 0.5 < \omega/\omega_n < 1.5$ in Figure 1.16). It can thus be approximated by a constant, S_0, which is the average value of $S_{xx}(\omega)$ in the region of the resonant mode. The output displacement spectral density, $S_{yy}(\omega)$, from such a system is

$$S_{yy}(\omega) = |\mathbf{H}(\omega)|^2 S_0 = \frac{S_0}{(k_s - m\omega^2)^2 + c_v^2\omega^2}, \tag{1.144}$$

and the mean-square output displacement is

$$E[y^2] = \int_{-\infty}^{\infty} \left| \frac{1}{k_s - m\omega^2 + ic_v\omega} \right|^2 S_0 \, d\omega = \frac{\pi S_0}{k_s c_v}. \tag{1.145}$$

A table of integrals for solving equations such as equation (1.145) above is given by Newland[1.7]. The mean-square output displacement, $E[y^2]$, is also given by

$$E[y^2] = 2S_0 \int_0^{\infty} |\mathbf{H}(\omega)|^2 \, d\omega. \tag{1.146}$$

Note that the lower variable of integration has now been replaced by zero and that a factor of two appears before the integral. This is because the frequency response function associated with the physically measurable one-sided spectral density is required.

Approximate calculations for the response of a single oscillator to broadband excitation can now be made by approximating the frequency response curve for $|\mathbf{H}(\omega)|^2$ by a rectangle with the same area (Newland[1.7]). This is illustrated in Figure 1.26. The exact area under the frequency response curve is obtained by equating equations (1.145) and (1.146) – i.e.

$$\int_0^\infty |\mathbf{H}(\omega)|^2 \, d\omega = \frac{\pi}{2k_s c_v}. \tag{1.147}$$

At resonance, the peak value of $|\mathbf{H}(\omega)|^2$ is $1/c_v^2\omega_n^2 = 1/(4\zeta^2 k_s^2)$, and the bandwidth of the rectangular approximation in Figure 1.26 is therefore $\pi\zeta\omega_n$ since $k_s = \omega_n^2 m$ and $c_v = 2\zeta\omega_n m$. Thus

$$\int_0^\infty |\mathbf{H}(\omega)|^2 \, d\omega = \frac{\pi}{2k_s c_v} \approx (\pi\zeta\omega_n)\left(\frac{1}{c_v^2\omega_n^2}\right). \tag{1.148}$$

This approximation for the area under the frequency response curve can now be substituted into equation (1.146), where now

$$E[y^2] \approx 2S_0(\pi\zeta\omega_n)\left(\frac{1}{c_v^2\omega_n^2}\right)$$

$$\approx 2S_0\{\text{mean-square bandwidth}\}\{\text{peak of } \mathbf{H}(\omega)\}^2. \tag{1.149}$$

Equation (1.149) allows for rapid approximate calculations of $E[y^2]$ whenever the excitation bandwidth includes the natural frequency, ω_n, (i.e. the response is resonant) and is reasonably broadband in regions in proximity to ω_n. It is a very useful approximation.

Fig. 1.26. Mean-square bandwidth for a single oscillator with broadband excitation.

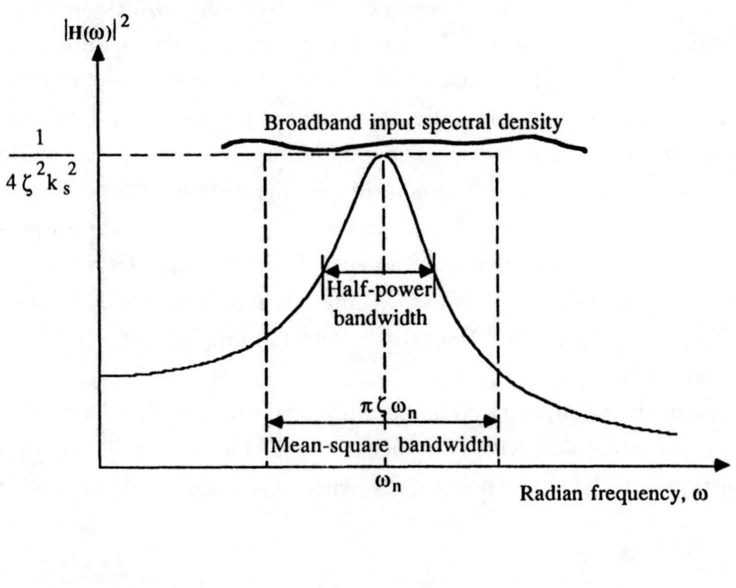

1.6.6 A note on frequency response functions and transfer functions

The term transfer function is commonly used by engineers instead of the term frequency response function when discussing complex ratios such as force/velocity etc. It is worth remembering that this terminology, whilst widely used, is not strictly correct. The transfer function of a system is defined by the Laplace transform and not the Fourier transform. Hence, the transfer function of some process, $x(t)$, is

$$\mathbf{H}'(\mathbf{q}) = \int_{-\infty}^{\infty} x(t)\,\mathrm{e}^{-\mathbf{q}t}\,\mathrm{d}t, \tag{1.150}$$

where $\mathbf{q} = a + ib$. When the variable a is not zero, the transfer function is not equal to the frequency response function. When the variable a is zero, the exponential term is imaginary and the transfer function is equal to the frequency response function. Hence, the transfer function is only equal to the frequency response function along the imaginary axis. It is therefore worth remembering that transfer functions relate to Laplace transforms, and that frequency response functions relate to Fourier transforms.

1.7 Energy and power flow relationships

Having reviewed the dynamics of a single oscillator for various excitation types, including random excitation, it is useful to expand on some of the comments that have been made in relation to energy and power flow. The main reason for this is that a thorough appreciation of these two parameters is very important for a clear understanding of the interactions between mechanical vibrations and noise. Engineers concerned with vibrational displacements on machinery generally utilise frequency response functions of displacement/force – i.e. receptances. Noise and vibration engineers, on the other hand, are concerned with structure-borne sound, and utilise impedances (force/velocity) or mobilities (velocity/force) to obtain information about energy and power flow. Also, because the main concern here is the relationships between structural vibrations and noise, the viscous-damping ratio, ζ, is now replaced by the structural loss factor, η. It is worth remembering that $\eta = 2\zeta$ (see equation 1.90).

The two types of energies in a system are (i) the kinetic energy, T, and (ii) the potential energy, U. Their sum, $T + U$, is the total energy of vibration, and their difference, $T - U$, is called the Lagrangian of the system. Generally, it is the time-averaged energy values, $\langle T \rangle$ and $\langle U \rangle$, that are required. It was shown in section 1.5.2 that $\langle T \rangle = \langle U \rangle$ and that $E = m\langle v^2 \rangle$. The Lagrangian, $\langle L \rangle = \langle T \rangle - \langle U \rangle$, is zero in this instance.

When damping is introduced into the equation of motion (with $\eta/2 < 1$), the solution is given by equation (1.40), with ζ replaced by $\eta/2$. The energy in the system is no longer constant – it decays exponentially with time. The mean-square velocity is obtained by differentiating equation (1.40) (with ζ replaced by $\eta/2$) and subsequently integrating the square value over a time interval, T. It is

$$\langle v^2 \rangle \approx \frac{V^2 \mathrm{e}^{-\eta\omega_n t}}{2}, \tag{1.151}$$

where V is the maximum velocity level. The corresponding mean-square displacement is

$$\langle x^2 \rangle \approx \frac{\langle v^2 \rangle}{\omega_n^2}. \tag{1.152}$$

The above equations are approximations and assume small damping, i.e. $\omega_\mathrm{d} \approx \omega_n$. As for the case of the undamped oscillator, $\langle T \rangle = \langle U \rangle$, $\langle E \rangle = m\langle v^2 \rangle$, and the Lagrangian $\langle L \rangle = 0$. Therefore, the time-averaged power dissipation (see equation 1.48) is

$$\langle -\mathrm{d}E/\mathrm{d}t \rangle = \langle \Pi \rangle = c_\mathrm{v}\langle v^2 \rangle = \eta\omega_n m\langle v^2 \rangle = \eta\omega_n\langle E \rangle. \tag{1.153}$$

Hence, the structural loss factor is

$$\eta = \frac{\langle \Pi \rangle}{\omega_n \langle E \rangle}. \tag{1.154}$$

The structural loss factor is thus related to the time-averaged power dissipation and the time-averaged energy of vibration – it is proportional to the fraction of total energy lost per cycle. Equation (1.154) is a very useful one for the experimental evaluation of structural loss factors and will be used in chapter 6.

The concepts of mechanical impedance ($\mathbf{Z}_\mathbf{m} = \mathbf{F}/\mathbf{V}$) and mobility ($\mathbf{Y}_\mathbf{m} = \mathbf{V}/\mathbf{F}$) were introduced briefly in sub-section 1.5.5. Both these parameters are used frequently, both experimentally and theoretically, to obtain information about energy levels and power flow in complex structures. Generally, **F** is real and **V** is complex – both are represented as complex numbers here for consistency because situations can arise where **F** is complex. As in sub-section 1.5.5, consider a force $F(t) = F \sin \omega t$ producing a displacement $x(t) = X \sin(\omega t - \phi)$ and a velocity $v(t) = V \cos(\omega t - \phi)$. Also, as discussed previously, all three can be represented in complex notation. Only force and velocity are relevant here, thus $F(t) = \mathrm{Im}\,[\mathbf{F}\,\mathrm{e}^{\mathrm{i}\omega t}]$ and $v(t) = \mathrm{Re}\,[\mathbf{V}\,\mathrm{e}^{\mathrm{i}\omega t}]$.

Some general comments are required regarding the usage of complex notation in time-averaging. In practice, the general convention is to use the real part of force and velocity in obtaining the time-averaged power (see equations 1.77 and 1.78). This is not consistent with the above definitions of force and velocity where $F(t)$ is the imaginary part of the complex force and $v(t)$ is the real part of the complex velocity. It can, however, be shown that the imaginary part is only a quarter of a period out of phase with the real part of any complex representation of a harmonic signal.

Consider, for instance, the complex force $\mathbf{F}\,e^{i\omega t}$. The actual force $F(t)$ is

$$F(t) = \text{Im}\,[F\,e^{i\phi}\,e^{i\omega t}] = F \sin(\omega t + \phi), \tag{1.155}$$

whilst the real part of the complex force $\mathbf{F}\,e^{i\omega t}$ is

$$\text{Re}\,[F\,e^{i\phi}\,e^{i\omega t}] = F \cos(\omega t + \phi). \tag{1.156}$$

Now,

$$F \sin(\omega t + \phi) = F \cos\left\{\omega\left(t - \frac{\pi}{2\omega}\right) + \phi\right\}, \tag{1.157}$$

where $\pi/2\omega = T/4$, i.e. a quarter period. Whilst this phase difference is relevant in any calculations involving instantaneous values, it is of no real significance when computing time-averaged values. Hence, the general convention is to always use the real part of the complex quantities when computing time-averaged values.

The mean-square values of force and velocity can now be obtained. Hence

$$\langle F^2(t)\rangle = \langle \text{Re}\,[\mathbf{F}\,e^{i\omega t}]^2\rangle = \tfrac{1}{2}\,\text{Re}\,[\mathbf{F}\mathbf{F}^*] = \tfrac{1}{2}|\mathbf{F}|^2, \tag{1.158}$$

and

$$\langle v^2(t)\rangle = \langle \text{Re}\,[\mathbf{V}\,e^{i\omega t}]^2\rangle = \tfrac{1}{2}\,\text{Re}\,[\mathbf{V}\mathbf{V}^*] = \tfrac{1}{2}|\mathbf{V}|^2. \tag{1.159}$$

It is useful to note that the real time-averaged power, $\langle \Pi \rangle$, is

$$\langle \Pi \rangle = \tfrac{1}{2}|\mathbf{V}|^2\,\text{Re}\,[\mathbf{Z_m}]. \qquad \text{(equation 1.79)}$$

The time-averaged reactive power can now be obtained by considering the product $\mathbf{Z_m}\langle v^2\rangle$ (note that $\langle v^2(t)\rangle$ is simply replaced with $\langle v^2\rangle$). Here,

$$\mathbf{Z_m}\langle v^2\rangle = \{c_v + i(m\omega - k_s/\omega)\}\langle v^2\rangle, \tag{1.160a}$$

and

$$\mathbf{Z_m}\langle v^2\rangle = c_v\langle v^2\rangle + i(m\omega - k_s/\omega)\langle v^2\rangle, \tag{1.160b}$$

thus

$$\mathbf{Z_m}\langle v^2\rangle = \langle \Pi \rangle + im\omega\langle v^2\rangle(1 - \omega_n^2/\omega^2), \tag{1.160c}$$

and the reactive power is given by the imaginary term. It is only zero for a resonant oscillator. The product $\mathbf{Z_m}\langle v^2\rangle$ is termed the complex power. It can be represented in terms of either impedance or mobility and

$$\mathbf{Z_m}\langle v^2\rangle = \tfrac{1}{2}\mathbf{Z_m}|\mathbf{V}|^2 = \tfrac{1}{2}\mathbf{F}\mathbf{V}^* = \tfrac{1}{2}\mathbf{Y_m^*}|\mathbf{F}|^2 = \mathbf{Y_m^*}\langle F^2(t)\rangle. \tag{1.161}$$

The real power $\langle \Pi \rangle$ is the most significant component as it represents the rate at which energy flows out of the system. It too can be represented in terms of either impedance or mobility and

$$\langle \Pi \rangle = \mathrm{Re}\,[\mathbf{Z_m}]\langle v^2 \rangle = \tfrac{1}{2}\,\mathrm{Re}\,[\mathbf{F}\mathbf{V}^*] = \tfrac{1}{2}|\mathbf{F}|^2\,\mathrm{Re}\,[\mathbf{Y}^*_\mathbf{m}], \tag{1.162a}$$

or

$$\langle \Pi \rangle = \tfrac{1}{2}|\mathbf{F}|^2\,\mathrm{Re}\,[\mathbf{Y_m}] = \tfrac{1}{2}\,\mathrm{Re}\,[\mathbf{F}^*\mathbf{V}] = \tfrac{1}{2}|\mathbf{V}|^2\,\mathrm{Re}\,[\mathbf{Z}^*_\mathbf{m}], \tag{1.162b}$$

or

$$\langle \Pi \rangle = \tfrac{1}{2}|\mathbf{V}|^2\,\mathrm{Re}\,[\mathbf{Z_m}] = \langle F^2(t) \rangle\,\mathrm{Re}\,[\mathbf{Y_m}]. \tag{1.162c}$$

Whilst the reactive power is not generally of interest in power flow (which is either to be dissipated as heat or to be transferred to another system), it is relevant for the determination of the amplitude of the system's response (for instance, see equations 1.57 and 1.74).

The relationships discussed in this section have been limited so far to single oscillators with harmonic excitation, and the ratios of the complex amplitudes **F** and **V** have defined the impedance and mobility. From equation (1.161) it can be seen that the impedance and mobility can be defined in terms of the mean-square values of force and velocity, i.e.

$$\frac{\langle v^2 \rangle}{\langle F^2(t) \rangle} = \mathbf{Y}^*_\mathbf{m}\mathbf{Z}^{-1} = \mathbf{Y}^*_\mathbf{m}\mathbf{Y_m} = |\mathbf{Y_m}|^2, \tag{1.163}$$

and

$$\langle v^2 \rangle = |\mathbf{Y_m}|^2 \langle F^2(t) \rangle. \tag{1.164}$$

It was shown in the last section that the mean-square value of a random signal is the area under the spectral density curve. Hence, for random excitation,

$$S_{vv}(\omega) = |\mathbf{Y_m}|^2 S_{FF}(\omega), \tag{1.165}$$

where $S_{vv}(\omega)$ and $S_{FF}(\omega)$ are the auto-spectral densities of the velocity and force, respectively. Equation (1.165) is of the same form as equation (1.134).

For harmonic excitation, the displacement, velocity and acceleration of a system are given by

$$x = \mathbf{X}\,\mathrm{e}^{\mathrm{i}\omega t}, \tag{1.166a}$$

$$v = \dot{x} = \mathrm{i}\omega\mathbf{X}\,\mathrm{e}^{\mathrm{i}\omega t}, \tag{1.166b}$$

and

$$a = \ddot{x} = (\mathrm{i}\omega)^2\mathbf{X}\,\mathrm{e}^{\mathrm{i}\omega t}. \tag{1.166c}$$

Each time derivative is equivalent to multiplication by $i\omega$, i.e. the maximum velocity is ωX and the maximum acceleration is $\omega^2 X$. The auto-spectral densities of displacement, velocity and acceleration are also related in the same way except that mean-square terms are now involved. Hence,

$$S_{aa}(\omega) = \omega^2 S_{vv}(\omega) = \omega^4 S_{xx}(\omega). \tag{1.167}$$

The energy and power flow relationships presented in this section can be readily extended to multiple oscillator systems and continuous systems for either periodic or random excitation.

1.8 Multiple oscillators – a review of some general procedures

The mass–spring–damper model considered so far has been constrained to move in a single axial direction. Most 'real life' systems involve multiple, if not numerous, degrees of freedom and therefore more complex models are required to model their vibrational characteristics. When only the first few natural frequencies are of interest, a system can be modelled in terms of a finite number of oscillators. For instance, mechanical engineers are sometimes concerned with estimating flexural and torsional natural frequencies and the corresponding mode shapes for a range of shaft type configurations, e.g. the drive-shaft of a multi-stage, turbo-alternator set. Alternatively, they might be concerned with isolating the vibrations due to a larger rotating machine which is mounted on a suspended floor, e.g. a centrifuge unit in a wash plant, or estimating the first few flexural (bending) natural frequencies of a large turbine exhaust system on an off-shore oil rig. In each of these examples there is more than one degree of freedom present. The engineers are, however, only concerned with a limited number of natural frequencies. In situations such as these, it is therefore appropriate to use the lumped-parameter, multiple-degree-of-freedom approach. Numerous text books are available on the subject of mechanical vibrations of lumped-parameter systems, some of which are referenced at the end of this chapter, and a range of calculation procedures, including numerical techniques, are presented. Most of these low-order natural frequencies do not themselves generate sound very efficiently (the reasons for this will be discussed in chapter 3) – they might, however, excite other structures which do. Hence, it is instructive to devote some time to multiple oscillator systems.

1.8.1 A simple two-degree-of-freedom system

It is useful to consider a two-degree-of-freedom system as this will furnish information which is easily extrapolated to systems with many degrees of freedom. A system without damping will be initially considered because (i) the mathematics is easier, (ii) in practice the damping is often small, (iii) the prediction of natural frequencies and mode shapes is not too dependent on damping, and (iv) damping can be considered later, either qualitatively or quantitatively.

Consider the two-degree-of-freedom system illustrated in Figure 1.27. The two co-ordinates x_1 and x_2 uniquely define the position of the system if it is constrained to move axially. The equations of motion for the two masses can be obtained from the free-body diagrams by considering the deflected position at some time t. The equations of motion for the two masses are

$$m_1\ddot{x}_1 = -k_{s1}x_1 - k_{s2}(x_1 - x_2), \qquad (1.168a)$$

and

$$m_2\ddot{x}_2 = k_{s2}(x_1 - x_2). \qquad (1.168b)$$

Assuming sinusoidal motion such that $x_1(t) = X_1 \sin \omega t$, and $x_2(t) = X_2 \sin \omega t$ and substituting into the above equations yields

$$X_1(m_1\omega^2 - k_{s1} - k_{s2}) + k_{s2}X_2 = 0, \qquad (1.169a)$$

and

$$k_{s2}X_1 + X_2(m_2\omega^2 - k_{s2}) = 0. \qquad (1.169b)$$

The pair of simultaneous equations can be solved for in terms of X_1. This gives

$$X_1\{-m_1m_2\omega^4 + (m_2k_{s1} + m_2k_{s2} + m_1k_{s2})\omega^2 - k_{s1}k_{s2}\} = 0. \qquad (1.170)$$

The term in the curly brackets is a quadratic equation in ω^2 and thus gives two frequencies at which sinusoidal and non-decaying motion may occur without being forced. That is, there are two natural frequencies ω_1 and ω_2. As a particular example, consider the situation where $m_1 = m_2 = m$ and $k_{s1} = k_{s2} = k$. Equation (1.170) now becomes

$$m^2\omega^4 - 3mk_s\omega^2 + k_s^2 = 0. \qquad (1.171)$$

Fig. 1.27. Two-degree-of-freedom, mass–spring system.

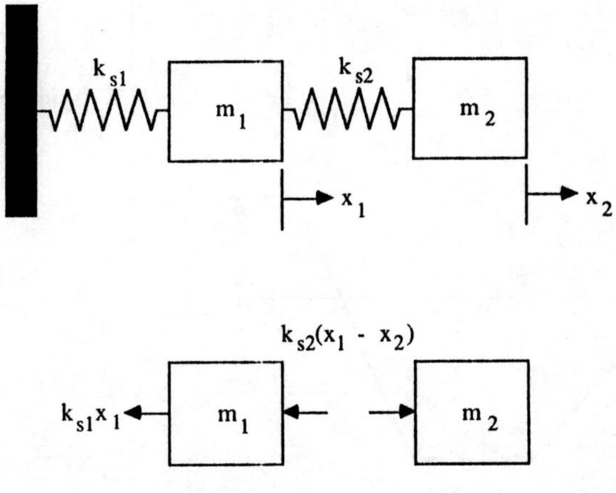

Solving this quadratic equation gives the two natural frequencies as

$$\omega_1 = 0.618(k_s/m)^{1/2}, \qquad \text{and } \omega_2 = 1.618(k_s/m)^{1/2}. \tag{1.172}$$

It should be noted that equation (1.172) is only valid for $m_1 = m_2 = m$, and $k_{s1} = k_{s2} = k$.

For each radian frequency, ω, there is an associated amplitude ratio,X_1/X_2, obtained from the equations of motion with $m_1 = m_2 = m$ and $k_{s1} = k_{s2} = k$. Here,

$$\frac{X_1}{X_2} = \frac{k_s}{2k_s - m\omega^2}, \tag{1.173}$$

where, for $\omega = \omega_1$, $X_1/X_2 = 0.618$ and for $\omega = \omega_2$, $X_1/X_2 = -1.618$. These ratios are called mode shapes or eigenvectors, and can be represented as mode plots. The mode plots for these two modes are illustrated in Figure 1.28. Hence, this simple, two-degree-of-freedom system has two natural frequencies ω_1 and ω_2 with the associated model shapes. When vibrating at the first natural frequency, ω_1, the two masses vibrate in phase, and when vibrating at the second natural frequency, ω_2, they vibrate

Fig. 1.28. Mode shapes for two-degree-of-freedom system illustrated in Figure 1.27.

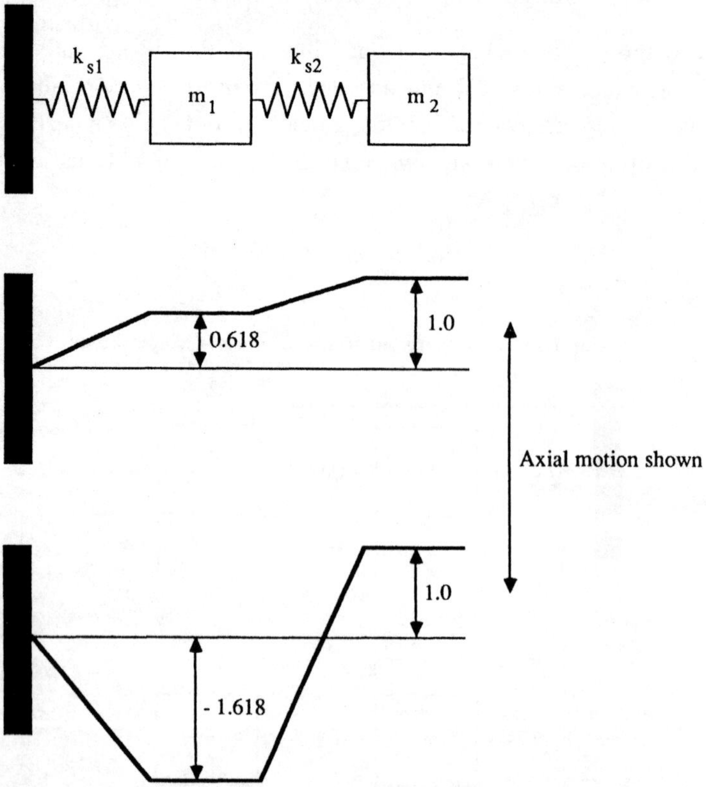

out of phase. It is important to note that the numerical values (0.618 and -1.618) are unique to this particular problem, i.e. $m_1 = m_2 = m$, and $k_{s1} = k_{s2} = k$.

1.8.2 *A simple three-degree-of-freedom system*

Many structures such as beams, plates and shells are often modelled as being free–free – i.e. their boundaries are not clamped or pinned or simply supported etc. It is therefore instructive to analyse a simple, free–free, three-degree-of-freedom system as illustrated in Figure 1.29 to obtain a qualitative understanding of the vibrational

Fig. 1.29. Free–free, three-degree-of-freedom system.

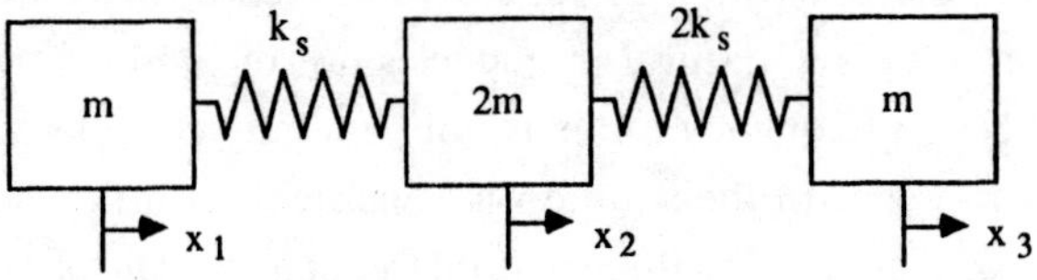

characteristics of such a system. Three co-ordinates, x_1, x_2 and x_3, uniquely define the position of the system if it is constrained to move axially – hence it is a three-degree-of-freedom system. The equations of motion are

$$m\ddot{x}_1 = -k_s(x_1 - x_2), \tag{1.174a}$$

$$2m\ddot{x}_2 = k_s(x_1 - x_2) - 2k_s(x_2 - x_3), \tag{1.174b}$$

and

$$m\ddot{x}_3 = 2k_s(x_2 - x_3). \tag{1.174c}$$

Assuming sinusoidal motion such that $x_1(t) = X_1 \sin \omega t$, $x_2(t) = X_2 \sin \omega t$, $x_3(t) = X_3 \sin \omega t$, and substituting into the above equations yields

$$X_1(k_s - m\omega^2) = k_s X_2, \tag{1.175a}$$

$$X_1(-k_s) + X_2(3k_s - 2m\omega^2) + X_3(-2k_s) = 0, \tag{1.175b}$$

and

$$X_3(2k_s - m\omega^2) = 2k_s X_2. \tag{1.175c}$$

The equations can be solved for in terms of X_1, X_2, or X_3. Solving for X_2 gives

$$X_2\{-2m^3\omega^6 + 9k_s m^2\omega^2 - 8k_s^2 m\omega^2\} = 0. \tag{1.176}$$

This is a cubic equation in ω^2 and thus gives three frequencies at which sinusoidal and non-decaying motion may occur without being forced. Solving for the three natural frequencies gives

$$\omega_1 = 0, \qquad \omega_2 = 1.10(k_s/m)^{1/2}, \qquad \text{and } \omega_3 = 1.81(k_s/m)^{1/2}. \tag{1.177}$$

For each of these frequencies there is an associated mode shape given by

$$\frac{X_1}{X_2} = \frac{k_s}{k_s - m\omega^2}, \tag{1.178a}$$

and

$$\frac{X_3}{X_2} = \frac{2k_s}{2k_s - m\omega^2}. \tag{1.178b}$$

For $\omega = \omega_1$, $X_1/X_2 = 1.0$ and $X_3/X_2 = 1.0$; for $\omega = \omega_2$, $X_1/X_2 = -4.55$ and $X_3/X_2 = 2.27$; for $\omega = \omega_3$, $X_1/X_2 = -0.44$ and $X_3/X_2 = -1.56$. The mode plots for these three modes of vibration are illustrated in Figure 1.30. Hence, this simple, three-degree-of-freedom system has three natural frequencies ω_1, ω_2 and ω_3 with the associated mode shapes. The zero frequency mode is not generally considered to be a mode of vibration, but its presence in the solution is consistent with the fact that the system has three masses and therefore has three natural frequencies. The physical interpretation of the result is that a free–free system without any damping would continue moving along in the absence of any boundary condition. In practice, however, it is the non-zero modes of vibration that are of engineering interest.

Fig. 1.30. Mode shapes for free–free, three-degree-of-freedom system illustrated in Figure 1.29.

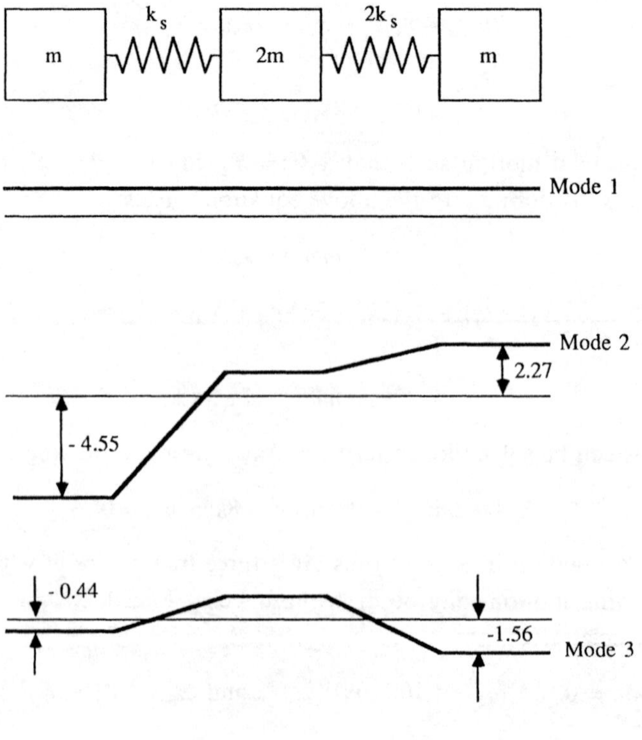

1.8.3 Forced vibrations of multiple oscillators

Consider again the two-degree-of-freedom system illustrated in Figure 1.27, but let the base (abutment) excitation be $x_B(t) = \mathbf{x_B}(t) = \mathbf{X_B}\, e^{i\omega t}$. The equations of motion are

$$m\ddot{x}_1 + 2k_s x_1 - k_s x_2 = k_s \mathbf{X_B}\, e^{i\omega t}, \tag{1.179a}$$

and

$$m\ddot{x}_2 + k_s x_2 - k_s x_1 = 0. \tag{1.179b}$$

In the steady-state, $x_1(t) = \mathbf{x_1}(t) = \mathbf{X_1}\, e^{i\omega t}$ and $x_2(t) = \mathbf{x_2}(t) = \mathbf{X_2}\, e^{i\omega t}$, hence

$$(2k_s - m\omega^2)\mathbf{X_1} - k_s\mathbf{X_2} = k_s\mathbf{X_B}, \tag{1.180a}$$

and

$$-k_s\mathbf{X_1} + (k_s - m\omega^2)\mathbf{X_2} = 0. \tag{1.180b}$$

Thus,

$$\frac{X_1}{X_B} = \frac{\mathbf{X_1}}{\mathbf{X_B}} = \frac{(k_s - m\omega^2)k_s}{(m^2\omega^4 + 3k_s m\omega^2 + k_s^2)}, \tag{1.181a}$$

and

$$\frac{X_2}{X_1} = \frac{\mathbf{X_2}}{\mathbf{X_1}} = \frac{k_s}{(k_s - m\omega^2)}. \tag{1.181b}$$

The amplitude ratios X_1/X_B and X_2/X_B can be expanded in partial fractions and it is quite instructive to interpret the results. It can be shown that

$$\frac{X_1}{X_B} = \frac{0.724}{(1 - \omega^2/\omega_1^2)} + \frac{0.276}{(1 - \omega^2/\omega_2^2)}, \tag{1.182}$$

and that

$$\frac{X_2}{X_B} = \frac{1.17}{(1 - \omega^2/\omega_1^2)} - \frac{0.17}{(1 - \omega^2/\omega_2^2)}. \tag{1.183}$$

The response of a single-degree-of-freedom system to base excitation is

$$\frac{X}{X_B} = \frac{\{1 + (2\zeta\omega/\omega_n)^2\}^{1/2}}{[\{1 - (\omega/\omega_n)^2\}^2 + \{2\zeta\omega/\omega_n\}^2]^{1/2}}, \tag{1.184}$$

and for $\zeta = 0$ it simplifies to

$$\frac{X}{X_B} = \frac{1}{(1 - \omega^2/\omega_n^2)}. \tag{1.185}$$

Thus, equations (1.182) and (1.183) represent the linear superposition of the response of two single-degree-of-freedom systems with different natural frequencies. The response of the components and the superposition is shown in Figure 1.31 for the case of X_1/X_B and in Figure 1.32 for the case of X_2/X_B. The ratio X_1/X_2 of the components at the frequency ω_1 is $0.724/1.17 = 0.618$, i.e. the first mode shape, and at the frequency ω_2 is $0.276/-0.17 = -1.618$, i.e. the second mode shape. The response of the system is thus the superposition of two modes of vibration with their associated mode shapes where each mode responds as a single-degree-of-freedom system.

It now remains to examine the effects of damping. For free vibration, the transients decay and the motion is very complex and it depends upon the initial conditions. The steady-state solution to forced vibration is somewhat easier to obtain, as using an excitation and a solution involving $e^{i\omega t}$ will give terms of the form $ic_v\omega e^{i\omega t}$ for each of the viscous-damping terms. As there is normally a spring and a viscous damper in parallel, these will produce terms of the form $(k_s + ic_v\omega)e^{i\omega t}$. Thus,

Fig. 1.31. Amplitude response ratio X_1/X_B versus ω/ω_n.

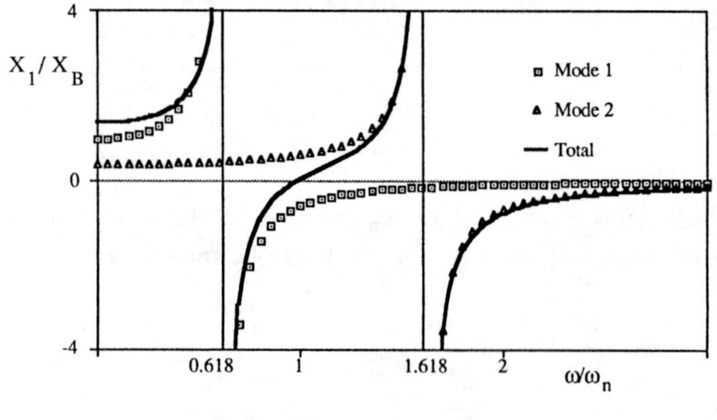

Fig. 1.32. Amplitude response ratio X_2/X_B versus ω/ω_n.

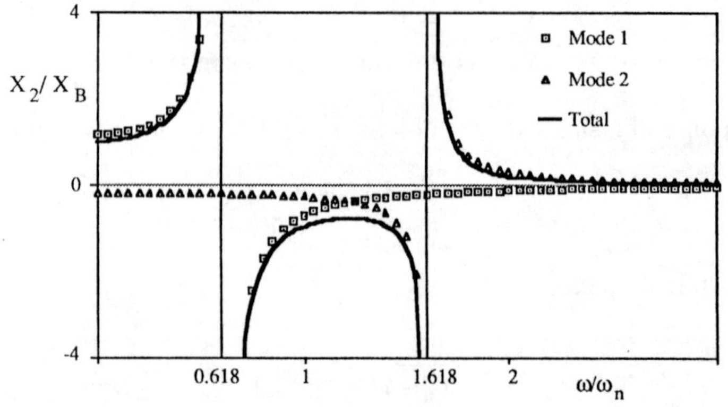

compared to the undamped case, it is only necessary to replace k_s with $(k_s + \mathrm{i}c_v\omega)$ in the final solution. Consider the system in Figure 1.27 with the particular values considered previously ($m_1 = m_2 = m$, and $k_{s1} = k_{s2} = k_s$) and also with viscous dampers c_v in parallel with each of the springs. If base excitation, $\mathbf{X}_\mathbf{B}\,\mathrm{e}^{\mathrm{i}\omega t}$, is again considered the steady-state solution is obtained from equation (1.181) by replacing k_s with $(k_s + \mathrm{i}c_v\omega)$. Thus,

$$\frac{\mathbf{X}_1}{\mathbf{X}_\mathbf{B}} = \frac{(k_s - m\omega^2 + \mathrm{i}c_v\omega)(k_s + \mathrm{i}c_v\omega)}{\{m^2\omega^4 + 3(k_s + \mathrm{i}c_v\omega)m\omega^2 + (k_s + \mathrm{i}c_v\omega)^2\}}. \qquad (1.186)$$

The examples presented so far in this sub-section and in the previous two sub-sections illustrate that the equations of motion are coupled, i.e. the motion $x_1(t)$ is influenced by the motion $x_2(t)$ and vice versa. When there are more than two degrees of freedom present, the equations of motion can be represented in matrix form, i.e.

$$M\{\ddot{x}\} + C_v\{\dot{x}\} + K_s\{x\} = \{F(t)\}. \qquad (1.187)$$

Here, $M = m_{ij}$ is the mass matrix, $C_v = c_{vij}$ is the damping matrix, and $K_s = k_{sij}$ is the stiffness matrix. It is possible to modify the equations of motion and to select a set of independent, orthogonal, co-ordinates called principal co-ordinates such that the mass and stiffness matrices are diagonal – i.e. they are uncoupled and generalised. The various modes of vibration are therefore independent of each other and are referred to as normal modes. The concepts of normal modes and principal co-ordinates are discussed in many texts on mechanical vibrations (e.g. Tse *et al.*[1.5]). When damping is neglected, it is a relatively straightforward job to uncouple the modes of vibration and this was illustrated earlier in this sub-section when the solution for X_1/X_B (equation 1.181) was uncoupled (equation 1.182). The uncoupled equations of motion in principal co-ordinates and generalised masses (m_{nn}) and stiffnesses (k_{snn}) for an undamped, multi-degree-of-freedom system are

$$\begin{bmatrix} m_{11} & 0 \\ 0 & m_{nn} \end{bmatrix}\begin{bmatrix} \ddot{q}_1 \\ \ddot{q}_n \end{bmatrix} + \begin{bmatrix} k_{s11} & 0 \\ 0 & k_{snn} \end{bmatrix}\begin{bmatrix} q_1 \\ q_n \end{bmatrix} = \begin{bmatrix} F_1 \\ F_n \end{bmatrix}. \qquad (1.188)$$

The equations are uncoupled because the off-diagonal terms in the mass and stiffness matrices are zero. The q_n's are the principal co-ordinates and they are obtained by co-ordinate transformation and normalisation. They represent a set of co-ordinates which are orthogonal to each other. Each principal co-ordinate, q_n, thus gives the relative amplitude of displacement, velocity and acceleration of the total system at a given natural frequency, ω_n, and the linear sum of all the principal co-ordinates gives the total response. The concepts of principal co-ordinates are used in the normal mode vibration analysis of continuous structures, and this is discussed in the next section.

When damping is considered, a damping matrix, C_v, has to be included in the equations of motion. In general, the introduction of damping couples the equations of motion because the off-diagonal terms in the damping matrix are not zero – i.e. coupled sets of ordinary, differential equations result. Often, because damping is generally small in mechanical and structural systems, approximate solutions are obtained by considering the coupling due to damping to be of a second order, i.e. $c_{vij} \ll c_{vjj}$ for $i \neq j$. Techniques for the modal analysis of damped, multiple-degree-of-freedom systems are described in many texts on mechanical vibrations (e.g. Tse *et al.*[1.5]).

1.9 Continuous systems – a review of wave-types in strings, bars and plates

At the very beginning of this book it was pointed out that engineers tend to think of vibrations in terms of modes and of noise in terms of waves, and that quite often it is forgotten that the two are simply different ways of looking at the same physical phenomenon! When considering the interactions between noise and vibration, it is important for engineers to have a working knowledge of both physical models.

Any continuous system, such as an aircraft structure, a pipeline, or a ship's hull, has its masses and elastic forces continuously distributed (as opposed to the rigid masses and massless springs discussed in previous sections). The structure generally comprises coupled cables, rods, beams, plates, shells, etc., all of which are neither rigid nor massless. These systems consist of an infinitely large number of particles and hence require an infinitely large number of co-ordinates to describe their motion – i.e. an infinite number of natural frequencies and an infinite number of natural modes of vibration are present. Thus, a continuous system has to be modelled with distributed mass, stiffness and damping such that the motion of each point in the system can be specified as a function of time. The resulting partial differential equations which describe the particle motion are called wave equations and they also describe the propagation of waves in solids (or fluids).

A fundamental understanding of wave propagation in solids and fluids is very important in engineering noise and vibration and it is therefore very instructive to start with a very simple (but not very practical from an engineering viewpoint) example, i.e. a string. The physics of wave propagation in strings yields a basic understanding of wave propagation phenomena.

1.9.1 The vibrating string

Consider a flexible, taut, string of mass ρ_L per unit length, stretched under a tension, T, as illustrated in Figure 1.33. Several simplifying assumptions are now made before attempting to describe the vibrational motion of the string. They are:

(i) the material is homogeneous and isotropic;

(ii) Hooke's law is obeyed;
(iii) energy dissipation (damping) is initially ignored;
(iv) the vibrational amplitudes are small – i.e. the motion is linear;
(v) there are no shear forces in the string, and no bending moments acting upon it;
(vi) the tension applied to the ends is constant and is evenly distributed throughout the string.

The lateral deflection, u, is assumed to be small and the change in tension with the deflections is negligible. The equation of motion in the lateral (transverse) direction is obtained from Newton's second law by considering an element, dx, of the string, and assuming small deflections and slopes. Thus,

$$T\left(\theta + \frac{\partial\theta}{\partial x}\,\mathrm{d}x\right) - T\theta = \rho_{\mathrm{L}}\,\mathrm{d}x\,\frac{\partial^2 u}{\partial t^2}. \tag{1.189}$$

$\theta = \partial u/\partial x$ is the slope of the string and the term $\theta + (\partial\theta/\partial x)\,\mathrm{d}x$ is the Taylor series expansion of the angle θ at the position $x + \mathrm{d}x$. Hence $\partial\theta/\partial x = \partial^2 u/\partial x^2$ and therefore

$$\frac{\partial^2 u}{\partial x^2} = \frac{1}{c_s^2}\frac{\partial^2 u}{\partial t^2}, \qquad \text{where } c_s = \left(\frac{T}{\rho_{\mathrm{L}}}\right)^{1/2}. \tag{1.190}$$

Equation (1.190) is the one-dimensional wave equation. The constant c_s has units of ms^{-1} and is the speed of propagation of the small lateral (transverse) particle displacements during the motion of the string – it is the velocity of wave propagation along the string and is perpendicular to the particle displacement and velocity (see Figure 1.1b) – it is also called the phase velocity of the wave. The wave equation is a second-order, partial differential equation and its most general solution contains two arbitrary independent functions G_1 and G_2 with arguments $(c_s t - x)$ and $(c_s t + x)$, respectively – both equations satisfy the wave equation by themselves. The function G_1 represents a travelling wave of constant shape in the positive x-direction and the

Fig. 1.33. Lateral (transverse) vibrations of a flexible, taut string segment.

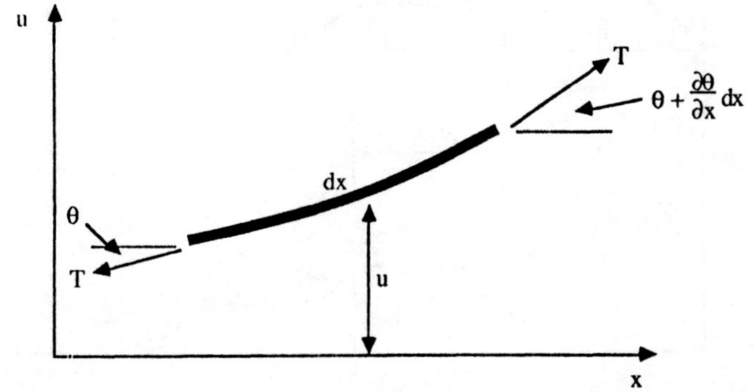

function G_2 represents a travelling wave of constant shape in the negative x-direction. Both waves travel at the same speed c_s. The complete general solution of the wave equation is thus

$$u(x, t) = G_1(c_s t - x) + G_2(c_s t + x). \tag{1.191}$$

The substitution of equation (1.191) into the wave equation for any arbitrary functions G_1 and G_2 (e.g. sine or cosine functions, exponential functions, logarithmic functions or linear functions) readily demonstrates that it is indeed a general solution.

Consider a travelling wave, $G_1(c_s t - x)$, in the positive x-direction at some time t_1. The shape of the wave is illustrated in Figure 1.34(a). At time t_2 the wave has travelled a distance $c_s(t_2 - t_1)$ to the right whilst retaining its shape – i.e. both $G_1(c_s t_1 - x_1)$ and $G_1(c_s t_2 - x_2)$ satisfy the wave equation. This is illustrated in Figure 1.34(b). It must be remembered that damping has been neglected so far and the above description of wave propagation is an idealisation – in practice a small amount of distortion will result as the waves propagate along the string. As for the case of the single oscillator, it turns out that this assumption is quite acceptable for engineering type structures because they are generally lighted damped.

As yet, nothing has been mentioned about the boundaries of the string. In reality, all structures have boundaries and corresponding boundary conditions – i.e. the structures are finite. Before considering finite strings, it is useful to consider a string which starts at $x = 0$ and extends to infinity in the positive x-direction. Whilst not being terribly practical in itself, this serves as a useful, simple introduction to wave propagation in finite structures and to the propagation of sound waves from a source

Fig. 1.34. Illustration of the nature of the solution to the wave equation.

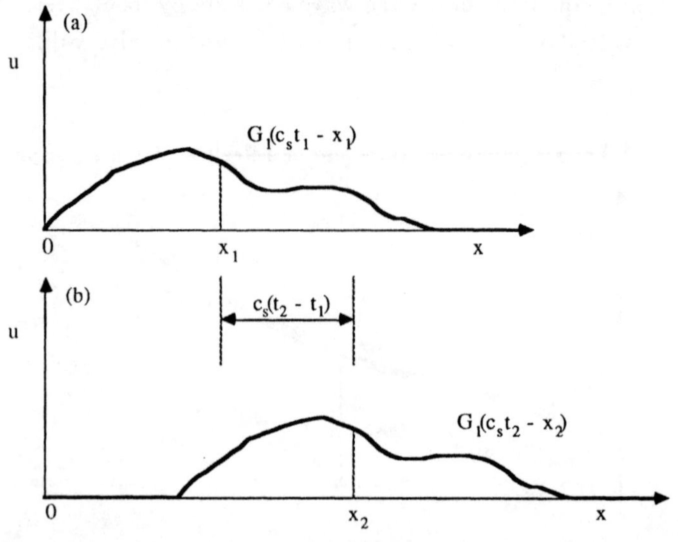

– in open spaces, sound waves do indeed propagate over very large distances and can therefore be modelled as travelling waves propagating to infinity.

Consider such a semi-infinite string which starts at $x = 0$ and extends to infinity in the positive x-direction. A harmonic force $\mathbf{F}\,e^{i\omega t}$ is applied, in the transverse direction, to the string at $x = 0$. Because the string extends to infinity in the positive x-direction and starts at $x = 0$, there is only one wave in the general solution (equation 1.191). Hence, $\mathbf{u}(x, t) = \mathbf{G}_1(c_s t - x)$. Note that $\mathbf{u}$ and $\mathbf{G}$ are now represented as complex quantities. Because the applied force is harmonic, the particle displacement at $x = 0$ also has to be harmonic. Therefore, $\mathbf{u}(0, t) = \mathbf{A}\,e^{i\omega t}$, where $\mathbf{A}$ is a complex constant which is related to the applied force. Thus, $\mathbf{G}_1(c_s t) = \mathbf{A}\,e^{i\omega t}$. The concepts of wavenumbers were introduced in section 1.2 (equation 1.2), thus ω can be replaced by kc_s, where k is the wavenumber. Thus,

$$\mathbf{G}_1(c_s t) = \mathbf{A}\,e^{ikc_s t}, \tag{1.192a}$$

and

$$\mathbf{u}(x, t) = \mathbf{G}_1(c_s t - x) = \mathbf{A}\,e^{ik(c_s t - x)} = \mathbf{A}\,e^{i(\omega t - kx)}. \tag{1.192b}$$

The complex representation $\mathbf{u}(x, t) = \mathbf{A}\,e^{i(\omega t - kx)}$ of the particle displacement is a very important representation of a propagating wave and it is widely used to represent wave propagation both in solids and in fluids.

The complex constant $\mathbf{A}$ can be evaluated by considering a force balance at $x = 0$ – i.e. at the point of application of the force. Summing the forces in the vertical direction yields $\mathbf{F}\,e^{i\omega t} = F\,e^{i0}\,e^{i\omega t} = -T\sin\theta \approx -T\theta \approx -T\,\partial\mathbf{u}/\partial x$ evaluated at $x = 0$. Thus, $\mathbf{A} = F/(iTk)$ and therefore

$$\mathbf{u}(x, t) = \frac{F}{iTk}\,e^{i(\omega t - kx)}, \tag{1.193a}$$

and

$$\mathbf{v}(x, t) = \frac{\partial\mathbf{u}}{\partial t} = \frac{F}{\rho_L c_s}\,e^{i(\omega t - kx)}. \tag{1.193b}$$

$\mathbf{v}(x, t)$ is the particle velocity and it is another important parameter in the analysis of wave propagation in solids and fluids. Recalling the definition of impedance as force/velocity (equation 1.71), the drive-point mechanical impedance of the string can be evaluated at $x = 0$. It is

$$\mathbf{Z}_m = \frac{\mathbf{F}\,e^{i\omega t}}{\dfrac{\mathbf{F}}{\rho_L c_s}\,e^{i\omega t}} = \rho_L c_s. \tag{1.194}$$

This drive-point mechanical impedance is resistive (i.e. it is real) and it is independent of the driving force – i.e. energy continuously propagates away from the driving point. It is commonly referred to as the characteristic mechanical impedance (Z_c) of the string since it is only a function of the physical properties of the string. The average power input into the string can be obtained from equation (1.162), i.e.

$$\langle \Pi \rangle = \tfrac{1}{2}|\mathbf{V}|^2 \operatorname{Re}[\mathbf{Z_m}]. \qquad \text{(equation 1.162)}$$

Thus,

$$\langle \Pi \rangle = \frac{1}{2}\frac{F^2}{\rho_L c_s} = \tfrac{1}{2}Z_m V^2 = \tfrac{1}{2}Z_c V^2, \tag{1.195}$$

where $V = |\mathbf{v}(0, t)| = F/(\rho_L c_s)$. It should be noted that the string has been assumed to possess no damping, hence the energy propagates away (to +ve infinity) from the driving point.

Now consider the same string forced at $x = 0$, except that it is now finite and clamped at $x = L$. The travelling wave in the positive x-direction is now reflected at the boundary and the process of reflection repeats itself at both ends. At any moment in time, the complete motion of the string is described by the linear superposition of a positive and a negative travelling wave – i.e. equation (1.191) with the arbitrary function being replaced by a harmonic function. As a general point, in noise and vibration it is convenient to represent waves as summations of harmonic components. This procedure is similar to the procedures adopted for the macroscopic lumped-parameter models. Hence for the finite, clamped string harmonically excited (in the transverse direction) at $x = 0$, the response, $\mathbf{u}(x, t)$ is given by

$$\mathbf{u}(x, t) = \mathbf{A}_1\, e^{i(\omega t - kx)} + \mathbf{A}_2\, e^{i(\omega t + kx)}. \tag{1.196}$$

The complex constants are evaluated from the two boundary conditions. They are:

(i) At the forced end, $\mathbf{F}\, e^{i\omega t} = F\, e^{i0}\, e^{i\omega t} = -T \sin\theta \approx -T\theta \approx -T\, \partial\mathbf{u}(0, t)/\partial x$.

(ii) At the clamped end, the displacement, $\mathbf{u}(L, t) = 0$.

From the first boundary condition, it is a relatively straightforward exercise to show that

$$F = ikT\mathbf{A}_1 - ikT\mathbf{A}_2. \tag{1.197}$$

From the second boundary condition

$$\mathbf{A}_1\, e^{-ikL} + \mathbf{A}_2\, e^{ikL} = 0. \tag{1.198}$$

These two equations can now be simultaneously solved to obtain solutions for $\mathbf{A}_1$ and $\mathbf{A}_2$. Noting that $2\cos kL = e^{ikL} + e^{-ikL}$,

$$\mathbf{A}_1 = \frac{F\, e^{ikL}}{i2kT \cos kL}, \tag{1.199a}$$

and

$$\mathbf{A}_2 = \frac{-F\, e^{-ikL}}{i2kT \cos kL}. \tag{1.199b}$$

The displacement $\mathbf{u}(x, t)$ is thus given by

$$\mathbf{u}(x, t) = \frac{F}{i2kT \cos kL}\{e^{i(\omega t + k(L-x))} - e^{i(\omega t - k(L-x))}\}. \qquad (1.200)$$

Equation (1.200) describes the displacement of the string in terms of the summation of two travelling waves of equal amplitude but propagating in opposite directions. It can be re-arranged in the following way:

$$\mathbf{u}(x, t) = \frac{F\,e^{i\omega t}}{i2kT \cos kL}\{e^{ik(L-x)} - e^{-ik(L-x)}, \qquad (1.201)$$

and since

$$\sin k(L-x) = \frac{\{e^{ik(L-x)} - e^{-ik(L-x)}\}}{2i}, \qquad (1.202)$$

therefore

$$\mathbf{u}(x, t) = \frac{F \sin k(L-x)\, e^{i\omega t}}{kT \cos kL} = \frac{F \sin k(L-x)\, e^{i\omega t}}{\rho_L c_s \omega \cos kL}. \qquad (1.203)$$

Equation (1.203) is mathematically identical to equation (1.200). It does, however, describe the displacement of the string in terms of a standing wave – i.e. the string oscillates with a spatially varying amplitude within the confines of a specific stationary waveform. A basic, but important, physical phenomenon has been illustrated here – a standing wave is a combination of two waves of equal amplitude travelling in opposite directions.

The drive-point mechanical impedance, $\mathbf{Z_m}$, can now be obtained by first evaluating the particle velocity, $\mathbf{v}(x, t)$, at $x = 0$, and then dividing the applied force, $\mathbf{F}$, by it. It is

$$\mathbf{Z_m} = -i\rho_L c_s \cot kL. \qquad (1.204)$$

The impedance is imaginary and therefore purely reactive. This suggests that there is no nett energy transfer between the driving force and the string – power is not absorbed by the string and the time-averaged power flow is zero. This is to be expected since the string does not possess any damping! It is important to recognise at this point that a resistive impedance implies energy dissipation (see equation 1.71). The form that equation (1.204) takes is presented in Figure 1.35. The minima in $\mathbf{Z_m}$ correspond to when $\cos kL = 0$. This is consistent with equation (1.203), where for $\cos kL = 0$ the displacement $\mathbf{u}(x, t)$ goes to infinity – i.e. there is a maximum displacement. The conditions of minimum impedance are thus the resonance frequencies of the system. At these frequencies the forcing frequency coincides with a natural frequency of the string and $\cos kL = 0$ is commonly referred to as the frequency equation of the string. Thus, for $\cos kL = 0$,

$$\frac{\omega L}{c_s} = n\pi - \frac{\pi}{2}, \tag{1.205}$$

for $n = 1, 2, 3$, etc., and thus

$$\omega_n = \frac{c_s \pi}{L}\{n - \tfrac{1}{2}\}. \tag{1.206}$$

The concepts developed in the preceding paragraph illustrate the significance of the drive-point impedance of a structure in identifying its natural frequencies. This procedure is widely used to experimentally identify natural frequencies on complex, built-up structures. The experimental procedures and their limitations are discussed in chapter 6. It is, however, worth noting at this point that the mechanical impedance of the transducer that is used to measure the drive-point mobility has to be accounted for.

So far in this sub-section, wave-type solutions to the wave equation (equation 1.190) have been sought. A wave–mode duality, as discussed at the beginning of the book, does exist and the string can also be looked upon as a system comprising an infinitely large number of particles. Its displacement response is thus the summation of the response of all the individual particles, each one of which has its own natural frequency and mode of vibration.

Equation (1.190) can now be solved in a different way. By separation of variables, the displacement $u(x, t)$ can now be represented as

$$u(x, t) = \phi(x)q(t). \tag{1.207}$$

Note that the complex displacement used in the earlier analysis has now been replaced by the real transverse displacement. It is convenient when seeking this form of solution

Fig. 1.35. Drive-point mechanical impedance for a string which is harmonically excited at $x = 0$ and clamped at $x = L$.

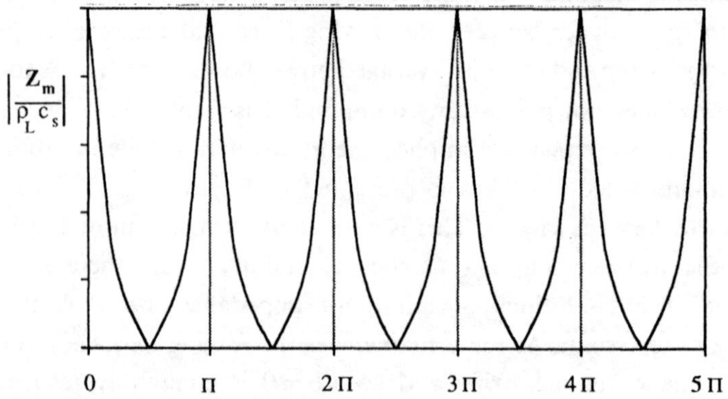

to deal with real numbers only. Some text books (e.g. reference 1.3) prefer to retain the complex notation. Both procedures produce the same final answers. Substituting equation (1.207) into equation (1.190) yields

$$\phi^{-1}\frac{d^2\phi}{dx^2}=q^{-1}c_s^{-2}\frac{d^2q}{dt^2}. \tag{1.208}$$

The left hand side of equation (1.208) is independent of time and the right hand side is independent of spatial position. For the equation to be valid, both sides therefore have to be equal to a constant which relates to the frequency of the vibration. Let this constant be $-k^2$, where k is the wavenumber (i.e. $k=\omega/c_s$). Hence,

$$\frac{d^2\phi}{dx^2}+k^2\phi=0, \tag{1.209}$$

and

$$\frac{d^2q}{dt^2}+\omega^2 q=0. \tag{1.210}$$

The solutions to these linear differential equations are

$$\phi(x)=A\sin kx+B\cos kx, \tag{1.211}$$

and

$$q(t)=C\sin\omega t+D\cos\omega t. \tag{1.212}$$

The arbitrary constants A, B, C and D depend upon the boundary and initial conditions. For a string stretched between two fixed points, the boundary conditions are (i) $u(0,t)=0$, and (ii) $u(L,t)=0$. The first boundary condition suggests that the constant $B=0$, and the second boundary condition suggests that $\sin kL=0$. The frequency equation for the clamped–clamped string is thus

$$\sin kL=0, \qquad \text{or } \frac{\omega L}{c_s}=\frac{\omega_n L}{c_s}=n\pi \text{ for } n=1, 2, 3, \text{etc.} \tag{1.213}$$

The suggestion here is that a continuous system has an infinite number of natural frequencies. This is what one would intuitively expect.

Since the constant $B=0$, the spatial parameter, $\phi(x)$, is now

$$\phi_n(x)=\sin k_n x=\sin\frac{\omega_n x}{c_s}=\sin\frac{n\pi x}{L}. \tag{1.214}$$

Equation (1.214) is conceptually very important. It represents the mode shape for the nth mode of vibration of the string. The displacement $u(x,t)$ is thus

$$u(x,t)=\sum_{n=1}^{\infty}\{C_n\sin\omega_n t+D_n\cos\omega_n t\}\sin\frac{n\pi x}{L}, \tag{1.215}$$

where

$$\omega_n = \frac{n\pi c_s}{L}. \tag{1.216}$$

The constants C_n and D_n are evaluated from the initial conditions. They are generally obtained by Fourier decomposing the initial conditions. Given $u(x, 0) = a(x)$ and $\partial u(x, 0)/\partial t = b(x)$, then

$$D_n = \frac{2}{L}\int_0^L a(x)\sin\frac{n\pi x}{L}\,\mathrm{d}x, \tag{1.217}$$

and

$$C_n\omega_n = \frac{2}{L}\int_0^L b(x)\sin\frac{n\pi x}{L}\,\mathrm{d}x. \tag{1.218}$$

D_n and $\omega_n C_n$ are the Fourier coefficients of the Fourier series expansion of $a(x)$ and $b(x)$, respectively (see equation 1.94).

In the above modal analysis, two important points have emerged. They are: (i) the boundary conditions determine the mode shapes and the natural frequencies of a system, and (ii) the initial conditions determine the contribution of each mode to the total response. The parameters $\phi_n(x)$ and $q_n(t)$ are the basis of the normal mode analysis of more complex continuous systems.

1.9.2 Quasi-longitudinal vibrations of rods and bars

Pure longitudinal waves can only exist in solids where the dimensions of the solids are very large compared with a longitudinal wavelength. The longitudinal type waves that can propagate in bars, plates and shells are generally referred to as being quasi-longitudinal – i.e. the direction of particle displacement is not purely in the direction of wave propagation and Poisson contraction occurs. A detailed discussion of wave-types in solids is given in Fahy[1.2] and Cremer *et al.*[1.12].

Consider a homogeneous, thin, long, bar with a uniform cross-section which is subjected to a longitudinal force. The same assumptions that were made when describing the vibrational motion of the string hold here. The one additional assumption is that the width of the bar is much less than its length. A wave-type equation for the longitudinal displacement, $u(x, t)$, can be obtained by considering a bar element as illustrated in Figure 1.36. The following points should be noted in relation to Figure 1.36.

(i) u is the longitudinal displacement at position x;
(ii) $u + (\partial u/\partial x)\,\mathrm{d}x$ is the longitudinal displacement at position $x + \mathrm{d}x$;
(iii) the element $\mathrm{d}x$ has changed in length by $(\partial u/\partial x)\,\mathrm{d}x$;
(iv) the unit strain is $\varepsilon = \delta/L = \{(\partial u/\partial x)\,\mathrm{d}x\}/\mathrm{d}x = \partial u/\partial x$.

From Hooke's law, the modulus of elasticity, E, is the ratio of unit stress to unit strain, i.e.

$$\frac{F/A}{\partial u/\partial x} = E, \qquad \text{or} \quad \frac{\partial u}{\partial x} = \frac{F}{AE}, \tag{1.219}$$

where A is the cross-sectional area of the bar. Newton's second law can now be applied to the element in Figure 1.36. Hence,

$$\rho A \, \mathrm{d}x \frac{\partial^2 u}{\partial t^2} = \left\{ F + \frac{\partial F}{\partial x} \mathrm{d}x \right\} - F, \tag{1.220}$$

where $F + (\partial F/\partial x)\,\mathrm{d}x$ is the Taylor series expansion of F at the position $x + \mathrm{d}x$, and ρ is the mass per unit volume (i.e. $\rho_L = \rho A$). By substituting equation (1.219) into equation (1.220),

$$\frac{\partial^2 u}{\partial x^2} = \frac{1}{c_L^2} \frac{\partial^2 u}{\partial t^2}, \qquad \text{where } c_L = \left(\frac{E}{\rho} \right)^{1/2}. \tag{1.221}$$

c_L is the velocity of propagation of the quasi-longitudinal displacement (stress wave) in the bar. Equation (1.221) is the one-dimensional wave equation for the propagation of longitudinal waves in solids and it is similar to equation (1.190). Its general solution is therefore also given by equation (1.191) or equation (1.196).

It is useful at this stage to evaluate the characteristic mechanical impedance – i.e. the ratio of force to velocity at any position along the stress wave in the solid bar. It is also known as the wave impedance of the solid material. Consider an arbitrary travelling wave $G(c_L t - x)$. From equation (1.219),

$$F = AE \frac{\partial u}{\partial x}. \tag{1.222}$$

Also,

$$\frac{\partial u}{\partial x} = -G'(c_L t - x), \qquad \text{and} \quad \frac{\partial u}{\partial t} = c_L G'(c_L t - x). \tag{1.223}$$

Fig. 136. Longitudinal displacement of a bar element.

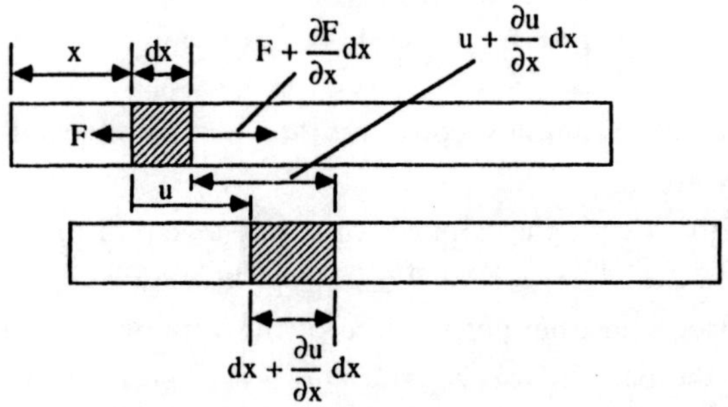

The characteristic mechanical or wave impedance, Z_c, is thus

$$Z_c = |\mathbf{Z}_c| = \frac{AE}{c_L} = \rho A c_L = \rho_L c_L. \tag{1.224}$$

As for the string (equation 1.194), it is real and is only a function of the physical properties of the material. It is worth pointing out here that c_L is the velocity of propagation (phase velocity) of a quasi-longitudinal wave. It is commonly referred to in the literature as the longitudinal wave velocity in a solid. The wave velocity of a 'pure' longitudinal wave is in fact (see Fahy[1.2])

$$c_L' = \left(\frac{B}{\rho}\right)^{1/2}, \qquad \text{where } B = \frac{E(1-\nu)}{(1+\nu)(1-2\nu)}, \tag{1.225}$$

and ν is Poisson's ratio.

Now consider the situation where the bar is harmonically excited at $x = 0$ and rigidly clamped at $x = L$. The problem is analogous to that of the forced, clamped string except that now the string tension, T, is replaced by EA in all the relevant equations (equations 1.197, 1.199, 1.200, 1.201 and 1.203). Hence, using equations (1.196)–(1.203) with the appropriate substitutions, the drive-point mechanical impedance is

$$\mathbf{Z}_m = -i\rho_L c_L \cot kL. \tag{1.226}$$

This equation is similar to equation (1.204) except that now c_s has been replaced by c_L.

In most situations in practice, the boundary conditions are neither free nor rigidly clamped but are somewhere in between. In these instances the supports act like masses – i.e. they possess inertance and subsequently have a finite mechanical impedance themselves. This finite impedance has to be accounted for in any dynamic analysis. Also, as mentioned previously, the mechanical impedance of measurement transducers has to be accounted for in any experimental set-up. Consider the same bar as in the previous paragraph except that now the clamped end has a finite mechanical impedance, $\mathbf{Z}_{mf}$ (a rigidly clamped end would have an infinite mechanical impedance). The longitudinal response of the bar is

$$\mathbf{u}(x, t) = \mathbf{A}_1 \, e^{i(\omega t - kx)} + \mathbf{A}_2 \, e^{i(\omega t + kx)}. \tag{1.227}$$

As for the string, the complex constants are evaluated from the two boundary conditions. They are:

(i) at the forced end, the applied force has to equal the dynamic force in the bar. Hence, $\mathbf{F}\, e^{i\omega t} = F\, e^{i0}\, e^{i\omega t} = -\rho_L c_L^2\, \partial\mathbf{u}(0, t)/\partial x$;

(ii) at the fixed end, the inertia force of the support has to equal the dynamic force in the bar. Hence, $\mathbf{Z}_{mf}\mathbf{v}(L, t) = -\rho_L c_L^2\, \partial\mathbf{u}(L, t)/\partial x$.

Substituting these boundary conditions into equation (1.227), solving for $\mathbf{A}_1$ and $\mathbf{A}_2$, and evaluating the drive-point mechanical impedance, $\mathbf{Z}_m$, yields

$$\mathbf{Z}_m = \frac{(\mathbf{Z}_{mf}/\rho_L c_L) + i \tan kL}{1 + i(\mathbf{Z}_{mf}/\rho_L c_L) \tan kL} \rho_L c_L. \tag{1.228}$$

As the impedance of the fixed end, $\mathbf{Z}_{mf}$, approaches infinity, equation (1.228) approximates to equation (1.226) – i.e. the boundary condition becomes rigid. In practice, the natural frequencies can be identified by the condition of minimum mechanical impedance. Another point worth considering is the power flow. The bar is assumed to possess no internal damping at this stage, hence the nett energy transfer between the driving force and the beam is dependent upon $\mathbf{Z}_{mf}$ – i.e. if $\mathbf{Z}_{mf}$ has a real (resistive) component there will be some energy transfer, and if it is imaginary (reactive) there will be none. Kinsler *et al.*[1.3] discuss the physical significance of this equation in some detail and draw analogies with mass and resistance-loaded strings.

The effects of damping have been neglected so far in this section. Most engineering type structures are lightly damped ($2.5 \times 10^{-4} < \eta < 5.0 \times 10^{-2}$) and damping can therefore be neglected when determining mode shapes and natural frequencies – this point was illustrated for the cases of the single and multiple oscillators, and it is also valid for continuous systems. In practice, the drive-point impedances of real structures have both real and imaginary components – the real components relating to power flow and energy dissipation. Damping can be included in the analysis of continuous systems by replacing the modulus of elasticity, E, by its complex equivalent $\mathbf{E}'$, where

$$\mathbf{E}' = E(1 + i\eta). \tag{1.229}$$

The parameter η is the structural loss factor (equation 1.90). The wave equation (equation 1.221) is now modified – i.e.

$$E(1 + i\eta)\frac{\partial^2 u}{\partial x^2} = \rho_L \frac{\partial^2 u}{\partial t^2}. \tag{1.230}$$

Because the modulus of elasticity is now complex, it follows that the wavenumber, k, is also complex. It takes the form

$$\mathbf{k}' = k(1 - i\chi). \tag{1.231}$$

The solution to the wave equation for a positive travelling wave thus takes the form

$$\mathbf{u}(x, t) = \mathbf{A}\, e^{i(\omega t - \mathbf{k}' x)}. \tag{1.232}$$

Substituting of equations (1.231) and (1.232) into equation (1.230) and separating real and imaginary parts yields

$$\chi = \frac{\eta}{2}, \tag{1.233}$$

hence $\mathbf{k}' = k(1 - \mathrm{i}\eta/2)$ and therefore

$$\mathbf{u}(x, t) = \mathbf{A}\, \mathrm{e}^{\mathrm{i}(\omega t - kx)}\, \mathrm{e}^{-kx\eta/2}. \tag{1.234}$$

The real part of the exponential thus represents the decaying component in the travelling wave. Similar relationships can thus be obtained for the drive-point impedances with E replaced with $\mathbf{E}'$ and k replaced with $\mathbf{k}'$ and the energy of the travelling waves decreases as they propagate through the bar. For lightly damped systems this decrease in energy is small and the waves would be continuously reflected from the boundaries and the bar will exhibit resonant behaviour. If, however, the damping was significant then the reflections would not be efficient and the drive-point impedance would approach the characteristic mechanical impedance – i.e. that of an infinite bar – and the response would be non-resonant.

A modal-type solution similar to equation (1.215) can also be readily obtained for rods and bars for a range of different types of boundary conditions, by separating variables and solving equation (1.221). As for the string, the boundary conditions determine the mode shapes and the natural frequencies, and the initial conditions determine the contributions of each mode to the total response. As an example, consider a uniform bar clamped at one end with a concentrated mass, M, attached at the other. This is illustrated in Figure 1.37.

The general solution for longitudinal vibrations of the bar is

$$u(x, t) = \phi(x)q(t),$$

$$= \sum_{n=1}^{\infty} \{A_n \sin k_n x + B_n \cos k_n x\}\{C_n \sin \omega_n t + D_n \cos \omega_n t\}. \tag{1.235}$$

Fig. 1.37. A uniform clamped bar with a concentrated mass attached to the free end.

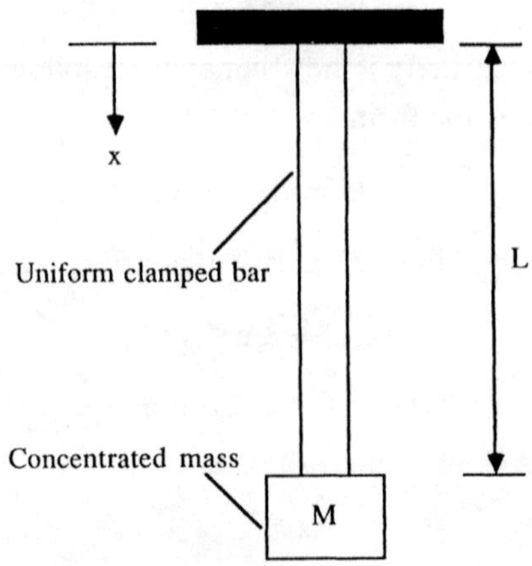

The boundary conditions for this particular problem are:

(i) there is no displacement at the fixed end, thus $u(0, t) = 0$;

(ii) the dynamic force in the bar at the free end is equal to the inertia force of the concentrated mass – i.e. $EA\,\partial u(L, t)/\partial x = -M\,\partial^2 u(L, t)/\partial t^2$.

From the first boundary condition, it can be shown that the coefficient B_n is zero. The second boundary condition yields the frequency equation

$$\frac{\rho AL}{M} = \frac{\omega_n L}{c_L} \tan \frac{\omega_n L}{c_L}, \tag{1.236}$$

where ρ is the mass per unit volume of the bar, M is the concentrated mass at the tip and L is the length of the bar. This equation is a transcendental equation in terms of $k_n L$. Two special cases arise: (i) $\rho AL \ll M$, and (ii) $M \ll \rho AL$. For the first case

$$\tan \frac{\omega_n L}{c_L} \approx \frac{\omega_n L}{c_L}, \qquad \text{and } \omega_n = \left(\frac{EA}{ML}\right)^{1/2}. \tag{1.237}$$

Here, there is only one natural frequency and it is equivalent to that of a single-degree-of-freedom, spring–mass system with stiffness EA/L. For the second case,

$$\frac{\rho AL}{M} \to \infty, \qquad \text{hence } \cos \frac{\omega_n L}{c_L} = 0, \tag{1.238}$$

is the frequency equation. This is equivalent to the vibrations of a bar fixed at one end and free at the other, thus

$$\omega_n = \frac{n\pi c_L}{2L} \qquad \text{for } n = 1, 3, 5, \text{etc.} \tag{1.239}$$

The natural frequencies for the general case are obtained from equation (1.236) by plotting $\tan k_n L$ and $(\rho AL)/(Mk_n L)$ on the same graph – the points of intersection yield the natural frequencies. This is illustrated in Figure 1.38 for various values of $\rho AL/M$.

1.9.3 Transmission and reflection of quasi-longitudinal waves

Low frequency vibration isolation (see chapter 4) is generally achieved by modelling the system in terms of lumped parameters and selecting suitable springs. High frequency vibration isolation in structures is often achieved by wave impedance mismatching. It is therefore useful to analyse the transmission and reflection of quasi-longitudinal waves at a step discontinuity in cross-section and material, as illustrated in Figure 1.39. When a quasi-longitudinal stress wave meets a boundary (i.e. encounters an impedance change) part of the wave will be transmitted and part will be reflected. There has to be continuity of longitudinal particle velocity and continuity of longitudinal force on the particles adjacent to each other on both sides of the boundary.

Consider an incident longitudinal wave

$$\mathbf{u}_i(x, t) = \mathbf{A}_i \, e^{i(\omega t - kx)}. \tag{1.240}$$

A reflected and a transmitted wave are generated at the discontinuity. They are

$$\mathbf{u}_r(x, t) = \mathbf{A}_r \, e^{i(\omega t + kx)}, \qquad \text{and } \mathbf{u}_t(x, t) = \mathbf{A}_t \, e^{i(\omega t - kx)}. \tag{1.241}$$

Now, the corresponding particle velocities are

$$\mathbf{v}_i = i\omega \mathbf{u}_i; \qquad \mathbf{v}_r = i\omega \mathbf{u}_r; \qquad \mathbf{v}_t = i\omega \mathbf{u}_t. \tag{1.242}$$

Continuity of longitudinal particle velocity implies

$$\mathbf{v}_i + \mathbf{v}_r = \mathbf{v}_t. \tag{1.243}$$

Continuity of longitudinal force implies

$$\mathbf{F}_i + \mathbf{F}_r = \mathbf{F}_t, \tag{1.244}$$

Fig. 1.38. Graph of tan k_nL and $(\rho AL)/(Mk_nL)$ versus k_nL – the points of intersection yield the natural frequencies.

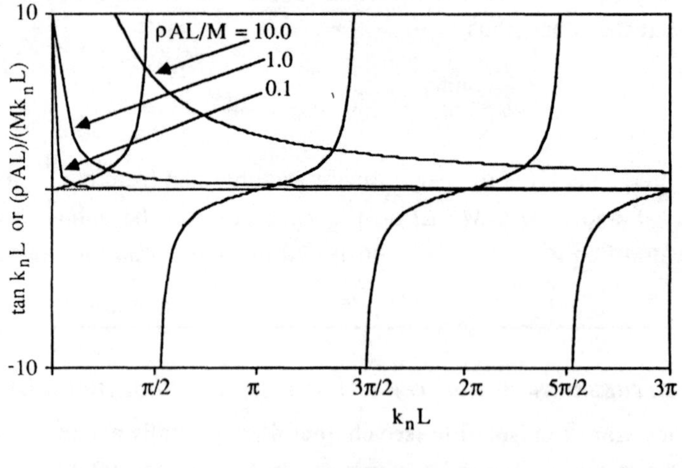

Fig. 1.39. Transmission and reflection of quasi-longitudinal waves at a step-discontinuity.

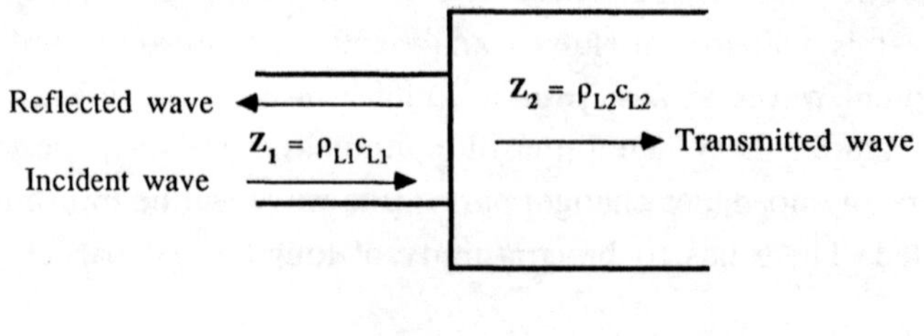

Hence, from equation (1.244)

$$\mathbf{Z}_1\mathbf{v}_i - \mathbf{Z}_1\mathbf{v}_r = \mathbf{Z}_2\mathbf{v}_t. \tag{1.245}$$

where $\mathbf{Z}_1$ and $\mathbf{Z}_2$ are the characteristic mechanical impedances (wave impedances) of mediums 1 and 2, respectively. Solving equations (1.240)–(1.245) yields

$$\frac{\mathbf{A}_t}{\mathbf{A}_i} = \frac{2\mathbf{Z}_1}{\mathbf{Z}_1 + \mathbf{Z}_2}, \tag{1.246}$$

and

$$\frac{\mathbf{A}_r}{\mathbf{A}_i} = \frac{\mathbf{Z}_1 - \mathbf{Z}_2}{\mathbf{Z}_1 + \mathbf{Z}_2}. \tag{1.247}$$

Equation (1.246) represents the ratio of the amplitude of the transmitted wave to that of the incident wave, and equation (1.247) represents the ratio of the amplitude of the reflected wave to that of the incident wave. They can be used to establish a relationship between the reflection and transmission coefficients of the discontinuity. This will be discussed in chapter 6.

The relationships presented in this section (in particular equations 1.246 and 1.247) are approximations and do not necessarily apply to all junctions and step discontinuities. This is especially true when damping is present and energy is absorbed at the discontinuity. They do, however, allow for an order of magnitude estimation of the reflected and transmitted energy. It is also useful to point out that an analogy exists between reflected and transmitted waves in solids, and sound waves in fluids or gases. The main difference is that for sound waves there is continuity of acoustic pressure across the interface.

1.9.4 Transverse bending vibrations of beams

Many types of waves can exist in solids (Cremer *et al.*[1.12]) but the two most important are the quasi-longitudinal waves discussed previously, and bending (flexural) waves. Bending waves play an important part in the radiation of sound from structures and therefore need to be given careful consideration. The equation of motion for beam bending vibrations can be developed in much the same way as the wave equation for strings and bars. Several assumptions, in addition to those made for the vibrating string, have first got to be made. They are:

(i) the effects of rotary inertia and shear deformation are neglected;
(ii) the cross-sectional area of the beam is constant;
(iii) *EI* is constant and the beam is symmetric about its neutral axis;
(iv) no nett longitudinal forces are present.

Consider a beam element of mass ρ_L per unit length as illustrated in Figure 1.40. V is the shear force and M is the bending moment. From Newton's second law,

$$\rho_L \, dx \frac{\partial^2 u}{\partial t^2} = -\left\{V + \frac{\partial V}{\partial x} dx\right\} + V, \tag{1.248}$$

Hence,

$$\rho_L \frac{\partial^2 u}{\partial t^2} = -\frac{\partial V}{\partial x}. \tag{1.249}$$

Summation of moments about the right face of the elemental unit (assuming the clockwise directon to be +ve) yields

$$-M + V\,dx + \left\{M + \frac{\partial M}{\partial x} dx\right\} = 0, \tag{1.250}$$

and thus

$$\frac{\partial M}{\partial x} = -V. \tag{1.251}$$

From beam deflection theory (Euler–Bernoulli or thin beam theory) the curvature and moment for a beam bending in a plane of symmetry are related by

$$EI \frac{\partial^2 u}{\partial x^2} = -M, \tag{1.252}$$

where EI is the flexural stiffness of the beam and I is the second moment of area of the cross-section about the neutral plane axis (the axis into the plane of the diagram in relation to Figure 1.40). The sign of equation (1.252) has to be consistent with the choice of co-ordinate axis and the definition of positive bending moment. It can

Fig. 1.40. Bending moments and shear forces in a beam element subjected to transverse vibrations.

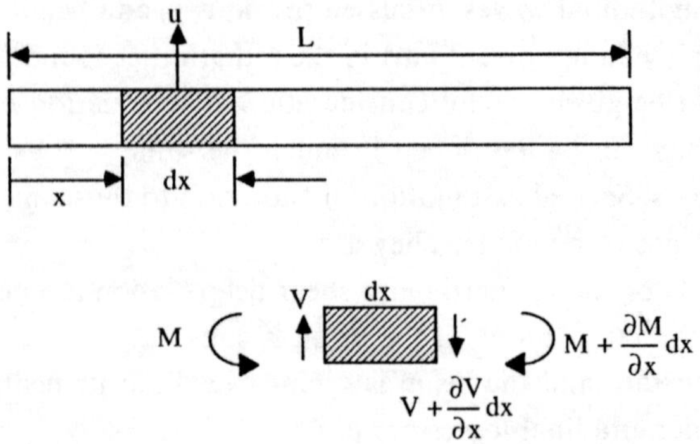

be shown that $\partial^2 u/\partial x^2$ is always opposite in sign to M (Timoshenko[1.13]). Hence,

$$\rho_L \frac{\partial^2 u}{\partial t^2} = -\frac{\partial^2 \left\{ EI \dfrac{\partial^2 u}{\partial x^2} \right\}}{\partial x^2}, \tag{1.253}$$

and therefore

$$\frac{\partial^2 u}{\partial t^2} + a^2 \frac{\partial^4 u}{\partial x^4} = 0, \qquad \text{where } a^2 = \frac{EI}{\rho_L}. \tag{1.254}$$

Equation (1.254) is the Euler beam equation for bending motion in the transverse direction. It is different from the wave equation for transverse string vibrations and quasi-longitudinal waves in bars, in that it is a fourth-order partial differential equation and the constant a^2 is not the bending wave speed. This is because bending waves are a combination of shear and longitudinal waves.

Now consider a solution of the form

$$\mathbf{u}(x, t) = \mathbf{A}\, \mathrm{e}^{\mathrm{i}(\omega t - kx)}. \tag{1.255}$$

Substitution of this equation into the Euler equation yields

$$k^4 = \frac{\rho_L}{EI} \omega^2, \tag{1.256}$$

where k has four roots, two of which are complex. They are

$$k = \pm \left(\frac{\rho_L \omega^2}{EI} \right)^{1/4}, \qquad \text{and } \pm \mathrm{i} \left(\frac{\rho_L \omega^2}{EI} \right)^{1/4}. \tag{1.257}$$

The parameter k has units of m^{-1} and it is the bending wavenumber – it shall therefore be referred to from now on as k_B. The complete solution to equation (1.254) thus has four components, and it is

$$\mathbf{u}(x, t) = \{\mathbf{A}_1\, \mathrm{e}^{-\mathrm{i}k_B x} + \mathbf{A}_2\, \mathrm{e}^{\mathrm{i}k_B x} + \mathbf{A}_3\, \mathrm{e}^{-k_B x} + \mathbf{A}_4\, \mathrm{e}^{k_B x}\}\, \mathrm{e}^{\mathrm{i}\omega t}. \tag{1.258}$$

Equation (1.258) is the solution for transverse bending vibrations of beams. From the form of the solution it can be seen that there are two exponentially decaying, non-propagating wave motions and two propagating wave motions. The non-propagating waves are referred to as evanescent waves and they do not transport nett energy. The two propagating wave components represent wave propagation in the +ve and −ve x-directions.

The bending wave velocity, c_B, can be obtained from the bending wavenumber, k_B. Thus,

$$c_B = \frac{\omega}{k_B} = \omega^{1/2} \{EI/\rho_L\}^{1/4} = \{1.8 c_L t f\}^{1/2}. \tag{1.259}$$

Equation (1.259) is an important one and it illustrates that the bending wave velocity, unlike the longitudinal wave velocity c_L, is not constant for a given material. It is a function of frequency and increases with it – i.e. different frequency components of bending waves travel at different wave speeds and they are therefore dispersive. It will be shown in chapter 3 that the bending wave velocity plays an important role in the radiation of sound from structures. The effects of shear deformation and rotary inertia on the bending wave velocity are discussed in some detail in Fahy[1.2] and Cremer *et al.*[1.12]. Shear deformation restricts the upper limit of the bending wave velocity (i.e. it does not go to infinity at high frequencies).

The concept of group velocity (c_g) was introduced in section 1.2 (equation 1.4). When a wave is non-dispersive the relationship between ω and k is linear and the wave or phase velocity and the group velocity are equal. For bending waves, the relationship between ω and k is non-linear and therefore the group and wave velocities are not equal – i.e. the energy transported by the wave does not travel at the same speed as the phase. For bending waves in solids, $c_g = 2c_B$.

The real transverse displacement of the beam, $u(x, t)$ is obtained from the real part of the complex solution (equation 1.258). It can also be obtained directly from the Euler beam equation by separation of variables and this procedure is the one commonly adopted in books on mechanical vibrations. Separation of variables yields two independent linear differential equations. They are

$$\frac{d^4\phi}{dx^4} - k_B^4\phi = 0, \tag{1.260}$$

and

$$\frac{d^2q}{dt^2} + \omega^2 q = 0. \tag{1.261}$$

Their respective solutions are

$$\phi(x) = A_1 \sin k_B x + A_2 \cos k_B x + A_3 \sinh k_B x + A_4 \cosh k_B x, \tag{1.262}$$

and

$$q(t) = C \sin \omega t + D \cos \omega t. \tag{1.263}$$

The constants A_1–A_4 are obtained from the boundary conditions and the constants C and D are obtained from the initial conditions. As for the case of the vibrating string, the $\phi(x)$'s represent the mode shapes and the $q(t)$'s determine the contribution of each mode to the total response The total response is thus the sum of all the individual modes i.e.

$$u(x, t) = \sum_{n=1}^{\infty} \phi_n(x) q_n(t). \tag{1.264}$$

As an example, consider the free transverse vibrations of a beam that is clamped at $x = 0$ and free at $x = L$ as illustrated in Figure 1.41. The boundary conditions are:

(i) at the fixed end there is no displacement or strain – thus $u(0, t) = 0$ and $\partial u(0, t)/\partial x = 0$;

(ii) at the free end there is no moment or shear force – thus $\partial^2 u(L, t)/\partial x^2 = 0$ and $\partial^3 u(L, t)/\partial x^3 = 0$.

Applying the four boundary conditions to equations (1.262)–(1.264) yields the following transcendental frequency equation in terms of $k_{Bn}L$ (note that k_{Bn} is the wavenumber for the nth mode).

$$\operatorname{sech} k_{Bn}L = -\cos k_{Bn}L. \tag{1.265}$$

The corresponding mode shapes are

$$\phi_n(x) = A_n\left[\cosh k_{Bn}x - \cos k_{Bn}x - \left\{\frac{\cosh k_{Bn}L + \cos k_{Bn}L}{\sinh k_{Bn}L + \sin k_{Bn}L}\right\}(\sinh k_{Bn}x - \sin k_{Bn}x)\right]. \tag{1.266}$$

The first four mode shapes are illustrated in Figure 1.42.

Fig. 1.41. A beam clamped at $x = 0$ and free at $x = L$.

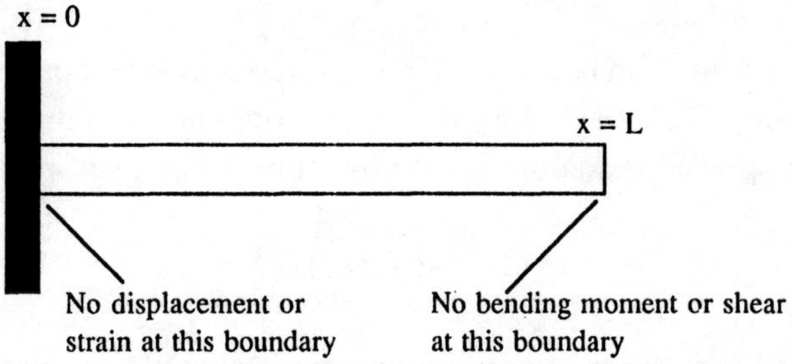

Fig. 1.42. The first four bending mode shapes for a clamped–free beam.

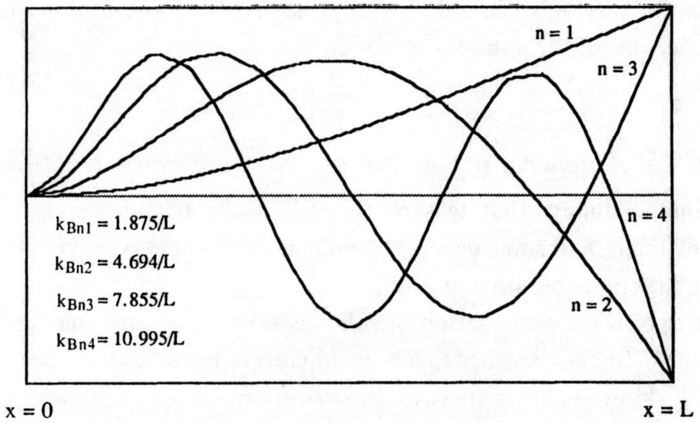

The drive-point mechanical impedance, $\mathbf{Z}_{\mathbf{m}}$, of beam elements can also be evaluated by using the same procedures that were adopted for strings and bars. Two specific results, obtained from Fahy[1.2], are presented here. Reference should be made to Fahy[1.2] or Cramer *et al.*[1.12] for further details. The first result is for point excitation of an infinite beam into its flexural or bending modes of vibration, and the second result is for point excitation of a finite simply supported beam. As was the case previously, damping is neglected at this stage. For point excitation of an infinite beam,

$$\mathbf{Z}_{\mathbf{m}} = \frac{2EIk_{\mathrm{B}}^3}{\omega}(1 + \mathrm{i}), \tag{1.267}$$

and for point excitation of a finite beam

$$\mathbf{Z}_{\mathbf{m}} = \frac{\mathrm{i}4EIk_{\mathrm{B}}^3}{\omega}(\tanh k_{\mathrm{B}}L - \tan k_{\mathrm{B}}L)^{-1}. \tag{1.268}$$

For the case of the infinite beam, the real part of the impedance (i.e. the resistance) is associated with the energy that propagates away from the excitation point and the imaginary part (i.e. the reactance) is associated with mass because it is positive (see equation 1.74). If it were negative, it would be associated with stiffness. Because damping is neglected in the analysis, the impedance for the finite beam is imaginary as was the case previously. Part of this impedance is associated with mass and part is associated with stiffness.

The effects of damping can be included in the analysis by incorporating the complex modulus of elasticity, $\mathbf{E}'$, in the beam equation and obtaining a complex wavenumber, $\mathbf{k}'$. Following the procedures adopted in sub-section 1.9.2 it can be shown that

$$\mathbf{k}' = k\left(1 - \frac{\eta}{4}\right). \tag{1.269}$$

It should be noted that the imaginary component is now $\eta/4$ and not $\eta/2$ as was the case for quasi-longitudinal waves. The distinction between the propagating and non-propagating waves (see equation 1.258) is somewhat more complex now because of the introduction of the complex wavenumber.

1.9.5 A general discussion on wave-types in structures

The wave–mode duality that was mentioned at the beginning of the book has been demonstrated, and some general comments on the different types of waves that can exist in structures are now in order.

The subject of wave propagation in solids is a very complex one and only those waves that are of direct relevance to noise and vibration studies have been considered. Cremer *et al.*[1.12] present an extensive discussion on a survey of different wave-types and their associated characteristics. A brief summary of these different wave-types is presented below.

(1) Pure longitudinal waves: these wave-types have particle displacements only in the direction of wave propagation and they generally occur in large solid volumes – e.g. seismic waves are pure longitudinal waves.

(2) Quasi-longitudinal waves: these wave-types maintain particle displacements which are not purely in the direction of wave propagation – longitudinal waves within the audible frequency range in engineering type structures are quasi-longitudinal.

(3) Transverse plane waves: these wave-types exist in solid bodies because of the presence of shear stresses – the modulus of elasticity, E, is replaced by the shear modulus, G, in the equation for the quasi-longitudinal wave speed.

(4) Torsional waves: these wave-types exist when beams are excited by torsional moments – the wave velocity is identical to that of transverse plane waves.

(5) Pure bending waves: these wave-types exist when the bending wavelength is large compared with the dimensions of the structural cross-sectional area.

(6) Corrected bending waves: the effects of rotary inertia and shear deformation are included in these wave-types.

(7) Rayleigh waves: these wave-types occur at high frequencies and in large, thick structures. They are essentially surface waves with the amplitude decreasing beneath the surface – e.g. ocean waves. Their wave velocities are of the same order as the transverse plane waves.

The two wave-types that are of importance in noise and vibration are the quasi-longitudinal waves and the pure bending waves. The quasi-longitudinal waves have very fast wave velocities ($c_L \sim 5200\ \mathrm{ms}^{-1}$ for steel) and are therefore high impedance waves. The bending wave velocities are a function of frequency (equation 1.259) and are generally significantly lower than the longitudinal wave velocities. For example, a 5 mm thick steel plate has a bending wave speed, c_B, of $\sim 150\ \mathrm{ms}^{-1}$ at 500 Hz and $\sim 485\ \mathrm{ms}^{-1}$ at 5000 Hz. Bending waves are thus low impedance waves, and this low impedance allows for a matching with sound wave impedances in any adjacent fluid. An efficient exchange of energy results with subsequent sound radiation – these concepts will be discussed in chapter 3. Transverse plane waves, torsional waves and Rayleigh waves on the other hand are all high impedance waves. Generally speaking, high impedance waves whilst being very efficient at transmitting vibrational energy do not transmit sound energy.

1.9.6 Mode summation procedures

The modal analysis procedures described in sub-section 1.9.1 (equations 1.207–1.218), in sub-section 1.9.2 (equations 1.235–1.239), and in sub-section 1.9.4 (equations 1.260–1.266) can be generalised in terms of orthogonal, principal co-ordinates such that the modes are uncoupled. The mode shapes of continuous systems are orthogonal if their scalar (vector dot) products are zero. It is always possible to obtain a set of orthogonal independent co-ordinates by linear transformation. The general procedures involved in this process are as follows.

(i) The equations of motion are uncoupled by means of eigenfunctions (the mode shapes, $\phi_n(x)$).

(ii) The uncoupled equations are expressed in terms of generalised mass, stiffness, damping and force,

(iii) The initial conditions are applied to evaluate the time-dependent Fourier coefficients (the generalised co-ordinates, $q_n(t)$).

(iv) The general solution is obtained by the superposition of all the modes.

The generalised co-ordinate of each mode is assumed to satisfy the relationship

$$M_n\ddot{q}_n(t) + C_{vn}\dot{q}_n(t) + K_{sn}q_n(t) = F_n(t), \tag{1.270}$$

where

$M_n = \int_0^L \phi_n^2(x)\rho_L(x)\,dx$ is the generalised mass,

$C_{vn} = \int_0^L \phi_n^2(x)c_v(x)\,dx$ is the generalised damping,

$K_{sn} = \omega_n^2 M_n$ is the generalised stiffness,

and

$F_n(t) = \int_0^L \phi_n(x)p(x, t)\,dx$ is the generalised force.

In the above equations $\rho_L(x)$ is the mass per unit length at location x on the structure, $c_v(x)$ is the viscous-damping coefficient per unit length at location x on the structure, and $p(x, t)$ is the applied load per unit length at location x on the structure at time t.The equations are valid for one-dimensional structures and can be extended to two and three dimensions – only one-dimensional structures (rods and beams) will be considered here. Several points can be made regarding equation (1.270). They are:

(i) the generalised mass, damping, stiffness and force are all functions of the mode shapes and they take on different values for every different mode;

(ii) the form of equation (1.270) is identical to that of a mass–spring–damper system;

(iii) the frequency response at any point x for a given normal mode is the same as for a system with a single degree of freedom;

(iv) the frequency response of the structure at any point x is the weighted sum of the frequency responses of all the normal modes each with its own different natural frequency;

(v) the equation does not contain any generalised cross-terms (e.g. q_{mn}) because the normal modes are orthogonal to each other;

(vi) the damping is assumed to be small – coupling due to damping is assumed to be of a second order.

The application of the method of normal modes is best illustrated by means of an example. Consider the free longitudinal vibrations of an undamped bar. The wave equation (equation 1.221) can be re-written as

$$\rho A\ddot{u} = EAu'', \qquad \text{where } \ddot{u} = \frac{\partial^2 u}{\partial t^2}, \qquad \text{and } u'' = \frac{\partial^2 u}{\partial x^2}. \tag{1.271}$$

For the nth mode, the displacement $u_n(x, t)$ is given by (see equation 1.235)

$$u_n(x, t) = \phi_n(x)\{C_n \sin \omega_n t + D_n \cos \omega_n t\}. \tag{1.272}$$

Substituting equation (1.272) into equation (1.271) yields

$$EA\phi_n''(x) + \rho A\omega_n^2\phi_n(x) = 0, \tag{1.273}$$

where EA is the flexural rigidity, and ρA is the mass per unit length (i.e. $\rho A = \rho_{\mathrm{L}}$).

Equation (1.273) can be re-written as

$$\phi_n''(x) = \lambda_n\phi_n(x), \tag{1.274}$$

where

$$\lambda_n = \frac{-\rho A\omega_n^2}{EA} = \frac{-\omega_n^2}{c_{\mathrm{L}}^2} = -k^2. \tag{1.275}$$

This is a form of the mathematical eigenvalue problem where the λ's are the eigenvalues and the ϕ's are the eigenfunctions. The orthogonality of the eigenfunctions can be investigated by considering the mth and nth modes. From equation (1.274),

$$\phi_m'' = \lambda_m\phi_m, \qquad \text{and } \phi_n'' = \lambda_n\phi_n. \tag{1.276}$$

Multiplying the first equation by ϕ_n and the second by ϕ_m yields

$$\phi_n\phi_m'' = \lambda_m\phi_m\phi_n, \tag{1.277a}$$

and

$$\phi_m\phi_n'' = \lambda_n\phi_m\phi_n. \tag{1.277b}$$

The above products are now integrated over the length of the bar, thus

$$\int_0^L \phi_m''\phi_n \,\mathrm{d}x = \lambda_m \int_0^L \phi_m\phi_n \,\mathrm{d}x, \tag{1.278}$$

and

$$\int_0^L \phi_n''\phi_m \,\mathrm{d}x = \lambda_n \int_0^L \phi_m\phi_n \,\mathrm{d}x. \tag{1.279}$$

The left hand side of equations (1.278) and (1.279) are now integrated by parts and the resultant equations subtracted from each other to yield

$$[\phi_m'\phi_n]_0^L - [\phi_n'\phi_m]_0^L = (\lambda_m - \lambda_n) \int_0^L \phi_m\phi_n \,\mathrm{d}x. \tag{1.280}$$

The integrated terms on the left hand side are zero because of the boundary conditions – i.e. the strains at the free ends are zero and therefore $\phi_m'(0) = \phi_n'(0) = \phi_m'(L) = \phi_n'(L) = 0$. Thus,

$$(\lambda_m - \lambda_n) \int_0^L \phi_m\phi_n \,\mathrm{d}x = 0. \tag{1.281}$$

Equation (1.281) is the general orthogonality relationship for a continuous system without any inertia load. For $\lambda_m \neq \lambda_n$ the integral is zero. When $m = n$ the inegral is a constant and therefore

$$\int_0^L \phi_n^2 \,\mathrm{d}x = \alpha_n. \tag{1.282}$$

Equations (1.281) and (1.282) illustrate that the modes of vibration are orthogonal to each other and it is a straightforward exercise to show that orthogonality relationships also exist amongst their derivatives.

When an inertia load such as a concentrated mass at the end of the beam is included (see Figure 1.37) then the boundary condition becomes an eigenvalue problem itself and this has to be included in the derivation of the orthogonality relationship. Here, the orthogonality relationship is (see Tse *et al.*[1.5])

$$\rho A \int_0^L \phi_m\phi_n \,\mathrm{d}x + M\phi_m(L)\phi_n(L) = 0, \tag{1.283}$$

and

$$\rho A \int_0^L \phi_n^2 \,\mathrm{d}x + M\phi_n^2(L) = \text{constant}. \tag{1.284}$$

Now, remembering that the displacement of the bar can be expressed in terms of a time function and a displacement function, i.e.

$$u(x, t) = \sum_{n=1}^{\infty} \phi_n(x)q_n(t), \tag{1.285}$$

and substituting into the wave equation (equation 1.221) yields

$$\sum_{n=1}^{\infty} \{\rho A \phi_n \ddot{q}_n - EA\phi_n'' q_n\} = 0. \tag{1.286}$$

The above sets of equations can be simplified by (i) multiplying by an orthogonal function, ϕ_m, (ii) integrating over the length of the bar, (iii) using the properties of the orthogonal relationships to eliminate terms, and (iv) using the relationships for generalised mass and stiffness that were derived in equation (1.270). Thus equation (1.286) reduces to

$$M_n \ddot{q}_n + K_{sn} q_n = 0 \qquad \text{for } n = 1, 2, 3, \text{etc.} \tag{1.287}$$

The generalised mass, M_n, is thus $\rho A \alpha_n$ from equation (1.282) and K_{sn} is $\omega_n^2 M_n$.

Equation (1.287) is a typical equation of motion for free vibrations in principal or generalised co-ordinates, and M_n is a principal or generalised mass for the nth mode, whilst K_{sn} is a principal or generalised stiffness for the nth mode. The equations can be normalised – i.e. a set of equations in normal, principal co-ordinates results. For evenly distributed continuous systems, it is convenient to normalise the equations with respect to the mass per unit length (ρA or ρ_L). Hence, the principal mass is unity (i.e. $\alpha_n = 1/\rho_L$) and the principal stiffness is ω_n^2. The equations of motion are now

$$\ddot{q}_n + \omega_n^2 q_n = 0 \qquad \text{for } n = 1, 2, 3, \text{etc.} \tag{1.288}$$

It is worth summarising the procedures for obtaining the normalised equations of motion for a continuous system before proceeding. The equations are transformed into normal co-ordinates by:

(i) expressing the motion of the structure in terms of a spatial displacement function, $\phi_n(x)$, and a time function, $q_n(t)$;
(ii) multiplying by an orthogonal mode, $\phi_m(x)$;
(iii) integrating over the surface;
(iv) normalising the eigenfunctions (mode shapes).

The initial displacement and velocity conditions are required to solve equation (1.288). For a given initial displacement $u(x, 0) = a(x)$ and a given initial velocity $\partial u(x, 0)/\partial x = b(x)$, it is a relatively straightforward matter to show that when $m = n$

$$q_n(0) = \int_0^L a(x)\phi_n(x)\, \mathrm{d}x, \tag{1.289}$$

and

$$\dot{q}_n(0) = \int_0^L b(x)\phi_n(x)\, \mathrm{d}x. \tag{1.290}$$

Thus,

$$q_n(t) = \frac{\dot{q}_n(0)}{\omega_n} \sin \omega_n t + q_n(0) \cos \omega_n t. \tag{1.291}$$

The complete general solution for free longitudinal vibrations of a bar can now be obtained by substituting equation (1.291) into equation (1.272) with the appropriate solution for ϕ_n (obtained from the boundary conditions). It is

$$u(x, t) = \sum_{n=1}^{\infty} \cos \frac{\omega_n x}{c_L} \left\{ \frac{\dot{q}_n(0)}{\omega_n} \sin \omega_n t + q_n(0) \cos \omega_n t \right\}. \tag{1.292}$$

It now remains to apply the method of normal modes to forced vibrations of structures. Once again, consider the longitudinal vibrations of an undamped bar as an example, and assume that the bar is subjected to an applied load $p(x, t)$ per unit length. For an element dx, equation (1.271) now becomes

$$\rho A \ddot{u} - EAu'' = p(x, t), \tag{1.293a}$$

or

$$\ddot{u} - c_L^2 u'' = \frac{p(x, t)}{\rho A}. \tag{1.293b}$$

This equation can now be transformed into normal co-ordinates using the same procedures that were adopted for the free vibration case. Substituting equation (1.293) into equation (1.285) yields

$$\sum_{n=1}^{\infty} \{\phi_n \ddot{q}_n - c_L^2 \phi_n'' q_n\} = \frac{p(x, t)}{\rho A}. \tag{1.294}$$

Multiplying by an orthogonal mode, ϕ_m, integrating over the length of the bar, eliminating terms by using the orthogonality relationships, and normalising yields

$$\ddot{q}_n + \omega_n^2 q_n = \frac{1}{M_n} \int_0^L \phi_n p(x, t) \, \mathrm{d}x. \tag{1.295}$$

This equation is identical to equation (1.270) (without the damping term of course), and it is the equation of motion for forced vibrations in normal co-ordinates. The integral on the right hand side is the nth normal mode load. The solution to this integral gives the forced response for the nth vibrational mode. It is given by the Duhamel convolution integral (Tse *et al.*[1.5])

$$q_n(t) = \frac{1}{M_n \omega_n} \int_0^L \phi_n(x) \int_0^t p(x, t') \sin \omega_n (t - t') \, \mathrm{d}t' \, \mathrm{d}x. \tag{1.296}$$

It should be noted that if the initial conditions are not zero then the complementary solution (equation 1.291) has to be added to obtain the total time response. For forced vibrations of structures it is generally acceptable to assume zero initial conditions and concentrate on the forced response. Finally, substitution of the above function into equation (1.285) gives the total vibrational response $u(x, t)$.

Quite often, the load distribution on a structure can be separated into a time and a space function, i.e.

$$p(x, t) = \frac{P_0}{L} p(x)p(t). \tag{1.297}$$

When this is the case, the Duhamel convolution integral can be separated into a mode participation factor and a dynamic load factor (Thomson[1.6]). The mode participation factor, H_n, is

$$H_n = \frac{1}{L} \int_0^L p(x)\phi_n(x)\,\mathrm{d}x, \tag{1.298}$$

and the dynamic load factor is

$$D_n(t) = \omega_n \int_0^t p(t') \sin \omega_n(t - t')\,\mathrm{d}t'. \tag{1.299}$$

Hence, equation (1.296) becomes

$$q_n(t) = \frac{P_0 H_n \omega_n}{\omega_n^2 M_n} \int_0^t p(t') \sin \omega_n(t - t')\,\mathrm{d}t'. \tag{1.300}$$

The effects of damping have been neglected in this section. Damping, provided that it is light, does not have a significant effect on the natural frequencies and the mode shapes. When damping is significant, the generalised damping terms, C_{vn}'s, (see equation 1.187) are coupled for different values of n – i.e. C_{vn} is not independent of C_{vm} for $m \neq n$ and the modes are no longer orthogonal. Approximate steady-state solutions can be obtained by neglecting the coupling due to damping (i.e. the off-diagonal terms in the damping matrix – see equations 1.187 and 1.188) and simply including damping in the equation of motion (as is the case for a single oscillator). In this case, equation (1.270) adequately describes the independent motion of all the modes in the system and the damped, steady-state response is readily obtained by linear summation. This point is illustrated in the next sub-section in relation to random vibrations of continuous systems. When the total response of the structure (including the time response) is required, and it is felt that the effects of damping need to be included, numerical techniques are usually adopted, especially if the damping is coupled. The procedures discussed in this sub-section, however, provide for a conservative upper estimate of the total response of the structure.

1.9.7 The response of continuous systems to random loads

Consider now the steady-state response of a beam to a single point random loading as illustrated in Figure 1.43. The beam of length L has a transverse point force, $F(t)$, acting at a position x_F, and $u(x, t)$ is the transverse displacement at some arbitrary position x.

The first step in the analysis is to evaluate the frequency response function (receptance) of the displacement at x to a force at $x = x_F$. For the purposes of evaluating the receptance, $\mathbf{H}_{\mathbf{xxF}}$, it is convenient to replace the random point force with a harmonic force (for linear systems the form of the inputs and outputs does not affect the frequency response function). Hence, let

$$\mathbf{F}(x, t) = F_0 \, e^{i\omega t} \, \delta(x - x_F). \tag{1.301}$$

The generalised force, $\mathbf{F}_\mathbf{n}(t)$, is thus

$$\mathbf{F}_\mathbf{n}(t) = \int_0^L F_0 \, e^{i\omega t} \, \delta(x - x_F)\phi_n(x) \, dx = F_0 \, e^{i\omega t} \phi_n(x_F). \tag{1.302}$$

The equation of motion for the beam is thus

$$M_n \ddot{\mathbf{q}}_\mathbf{n} + C_{vn} \dot{\mathbf{q}}_\mathbf{n} + K_{sn} \mathbf{q}_\mathbf{n} = F_0 \, e^{i\omega t} \phi_n(x_F), \tag{1.303}$$

and its form is similar to equation (1.270) – i.e. the mass, damping and stiffness terms are generalised and the time-dependent variables are the normal co-ordinates.

For a single oscillator, the receptance, $\mathbf{X}/\mathbf{F}$, is given by equation (1.56), i.e.

$$\frac{\mathbf{X}}{\mathbf{F}} = \frac{1}{\{k_s - m\omega^2 + ic_v\omega\}}. \qquad \text{(equation 1.56)}$$

For a normal mode of a continuous system, the time-dependent displacement variable, $\mathbf{q}_\mathbf{n}(t)$, is

$$\mathbf{q}_\mathbf{n}(t) = \frac{\phi_n(x_F) F_0 \, e^{i\omega t}}{(K_{sn} - \omega^2 M_n) + iC_{vn}\omega} = \frac{\phi_n(x_F) F_0 \, e^{i\omega t}}{M_n(\omega_n^2 - \omega^2) + iC_{vn}\omega}. \tag{1.304}$$

The total displacement is given by equation (1.285), hence

$$\mathbf{u}(x, t) = F_0 \, e^{i\omega t} \sum_{n=1}^{\infty} \frac{\phi_n(x)\phi_n(x_F)}{M_n(\omega_n^2 - \omega^2) + iC_{vn}\omega}. \tag{1.305}$$

Fig. 1.43. A beam with a single point random load.

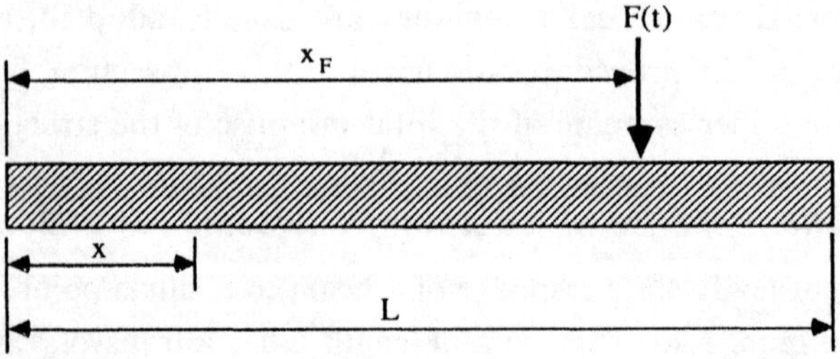

The receptance, $\mathbf{H}_{\mathbf{xxF}}$, is defined as

$$\mathbf{H}_{\mathbf{xxF}}(\omega) = \frac{\mathbf{u}(x, t)}{F_0\, e^{i\omega t}}. \tag{1.306}$$

Hence,

$$\mathbf{H}_{\mathbf{xxF}}(\omega) = \sum_{n=1}^{\infty} \frac{\phi_n(x)\phi_n(x_F)}{M_n} \frac{1}{(\omega_n^2 - \omega^2) + i\omega C_{vn}/M_n}. \tag{1.307}$$

For most structural systems hysteretic damping is more appropriate than viscous damping, hence the generalised viscous damping term has to be replaced. From equations (1.83) and (1.86),

$$\eta_n = \frac{\omega C_{vn}}{K_{sn}}, \tag{1.308}$$

hence

$$\frac{\omega C_{vn}}{M_n} = \eta_n \omega_n^2. \tag{1.309}$$

Thus,

$$\mathbf{H}_{\mathbf{xxF}}(\omega) = \sum_{n=1}^{\infty} \frac{\phi_n(x)\phi_n(x_F)}{M_n} \frac{\omega_n^2 - \omega^2 - i\eta_n\omega_n^2}{(\omega_n^2 - \omega^2)^2 + \eta_n^2\omega^4}. \tag{1.310}$$

Equation (1.310) is the formal solution for the frequency response function (receptance) of the displacement at some position, x, on the beam to a point force at $x = x_F$. It can be conveniently re-expressed as

$$\mathbf{H}_{\mathbf{xxF}}(\omega) = \sum_{n=1}^{\infty} \mu_n(A_n - iB_n), \tag{1.311}$$

where

$$\mu_n = \frac{\phi_n(x)\phi_n(x_F)}{M_n},$$

$$A_n = \frac{\omega_n^2 - \omega^2}{(\omega_n^2 - \omega^2)^2 + \eta_n^2\omega_n^4},$$

and

$$B_n = \frac{\eta_n\omega_n^2}{(\omega_n^2 - \omega^2)^2 + \eta_n^2\omega_n^4}.$$

For a random point force, $F(t)$, with an auto-spectral density, $S_{FF}(\omega)$, the auto-spectral density, $S_{xx}(\omega)$, of the displacement response at position x is given by equation (1.134) – i.e.

$$S_{xx}(\omega) = |\mathbf{H}_{\mathbf{xxF}}(\omega)|^2 S_{FF}(\omega)$$
$$= \left\{\left(\sum_{n=1}^{\infty} \mu_n A_n\right)^2 + \left(\sum_{n=1}^{\infty} \mu_n B_n\right)^2\right\} S_{FF}(\omega). \tag{1.312}$$

Equation (1.312) contains cross-product terms such as $\mu_m\ \mu_n\ A_m\ A_n$ etc. For lightly damped structures, however, the peaks in the receptance function are well defined, and provided that there is no modal overlap (i.e. the natural frequencies are well separated), the response in regions in proximity to a resonance frequency is dominated by a single term in the summation. At regions away from a resonance this is not the case, but the response magnitudes are much smaller here and can therefore be ignored. Hence, for light damping and for well separated natural frequencies, the product terms in equation (1.312) can be neglected and the expression simplifies to

$$S_{xx}(\omega) = \sum_{n=1}^{\infty} \mu_n^2 (A_n^2 + B_n^2) S_{FF}(\omega). \tag{1.313}$$

The preceding section can be extended to the motions of an arbitrarily shaped body with three-dimensional normal modes by using vectors.

For a continuous system with two random point loads, $F(t)$ and $W(t)$, equation (1.136) can be used, with the appropriate receptances, to obtain the output spectral density, $S_{xx}(\omega)$, at position x. Here, the output spectral density is

$$S_{xx}(\omega) = \mathbf{H}^*_{\mathbf{xxF}}\mathbf{H}_{\mathbf{xxF}} S_{FF} + \mathbf{H}^*_{\mathbf{xxF}}\mathbf{H}_{\mathbf{xxW}}\mathbf{S}_{\mathbf{FW}}$$
$$+ \mathbf{H}^*_{\mathbf{xxW}}\mathbf{H}_{\mathbf{xxF}}\mathbf{S}_{\mathbf{WF}} + \mathbf{H}^*_{\mathbf{xxW}}\mathbf{H}_{\mathbf{xxW}} S_{WW}. \tag{1.314}$$

When there is no correlation between $F(t)$ and $W(t)$ the cross-spectral density terms can be neglected and

$$S_{xx}(\omega) = |\mathbf{H}_{\mathbf{xxF}}|^2 S_{FF} + |\mathbf{H}_{\mathbf{xxW}}|^2 S_{WW}, \tag{1.315}$$

i.e the spectral density is the sum of the two response spectral densities obtained with the forces acting separately.

If the two point forces are directly correlated such that $F(t) = \alpha W(t)$, where α is a constant, then from the definitions of the auto- and cross-correlation functions (equations 1.116 and 1.118)

$$R_{FW}(\tau) = E[F(t)\alpha F(t+\tau)] = \alpha R_{FF}(\omega), \tag{1.316a}$$

$$R_{WF}(\tau) = E[\alpha F(t) F(t+\tau)] = \alpha R_{FF}(\tau), \tag{1.316b}$$

and

$$R_{WW}(\tau) = E[\alpha F(t)\alpha F(t+\tau)] = \alpha^2 R_{FF}(\tau). \qquad (1.316c)$$

Thus

$$\mathbf{S}_{\mathbf{FW}}(\omega) = \alpha S_{FF}(\omega), \qquad (1.317a)$$

$$\mathbf{S}_{\mathbf{WF}}(\omega) = \alpha S_{FF}(\omega), \qquad (1.317b)$$

and

$$S_{WW}(\omega) = \alpha^2 S_{FF}(\omega). \qquad (1.317c)$$

Hence,

$$\begin{aligned} S_{xx}(\omega) &= \{\mathbf{H}^*_{\mathbf{xxF}} + \alpha\mathbf{H}^*_{\mathbf{xxW}}\}\{\mathbf{H}_{\mathbf{xxF}} + \alpha\mathbf{H}_{\mathbf{xxW}}\}S_{FF}(\omega) \\ &= |\mathbf{H}_{\mathbf{xxF}} + \alpha\mathbf{H}_{\mathbf{xxW}}|^2 S_{FF}(\omega). \end{aligned} \qquad (1.318)$$

Equation (3.18) illustrates that for this particular case of direct correlation between $F(t)$ and $W(t)$, the output spectral density depends upon the modulus of a vector sum of the two receptances, and on the relative phase between them.

For the special case of $\alpha = 1$, $F(t) = W(t)$ and therefore $S_{FF}(\omega) = S_{WW}(\omega) = S(\omega)$. Thus,

$$S_{xx}(\omega) = \{|\mathbf{H}_{\mathbf{xxF}}|^2 + |\mathbf{H}_{\mathbf{xxW}}|^2 + 2|\mathbf{H}_{\mathbf{xxF}}|\,|\mathbf{H}_{\mathbf{xxW}}|\cos\phi\}S(\omega). \qquad (1.319)$$

Cos ϕ is the phase difference between the two receptances, and when $\phi = \pi/2$, $\cos\phi = 0$ and the output response spectral density $S_{xx}(\omega)$ is the linear sum of the squares of the magnitudes of the two separate inputs. This is an important result in that whilst both inputs have identical auto-spectral densities there is not necessarily any correlation between them – i.e. when $\cos\phi = 0$ the two inputs are uncorrelated.

The basic principles of the steady-state response of continuous systems to random loads have been illustrated in this sub-section, and the effects of damping have been included. The analysis has been limited to point loads, and the subject of distributed loads has not been discussed. Specialist text books (e.g. Newland[1.7]) are available on the subject and the reader is referred to these for a detailed analysis.

1.9.8 Bending waves in plates

This text book is about bringing noise and vibration together. It would therefore be seriously lacking if a section was not included on the vibrations of plate-type structural components, because the bending vibrations of thin plates radiate sound very efficiently – i.e. there is good impedance matching between the bending waves and the fluid. Machine covers, wall partitions, floors etc. are typical two-dimensional thin, plate-type structures and information is required about their modes of vibration, impedances, etc., for any noise and vibration analysis. A detailed analysis of the different wave-types and vibrations that can exist in membranes plates and shells is a specialist subject in its own right and is therefore beyond the scope of this book (Cremer *et al.*[1.12], Leissa[1.14,1.15] and Soedel[1.16] are excellent references on the topic). The wave equation for transverse vibrations of a thin plate can, however, be obtained by extending the one-dimensional beam equation into two dimensions.

Whereas the bending stiffness of a beam is EI, the corresponding bending stiffness of a thin plate is $EI/(1-\nu^2)$. The term $(1-\nu^2)$ is included because of the Poisson contraction effects which are neglected in thin beam analyses. The two-dimensional bending wave equation for transverse vibrations of thin plates is (e.g. see Reynolds[1.4])

$$\rho_S \frac{\partial^2 u}{\partial t^2} + \frac{Et^3}{12(1-\nu^2)} \left\{ \frac{\partial^4 u}{\partial x^4} + 2\frac{\partial^4 u}{\partial x^2\,\partial y^2} + \frac{\partial^4 u}{\partial y^4} \right\} = 0, \tag{1.320}$$

where ρ_S is the mass per unit area, ν is Poisson's ratio, t is the thickness of the plate, and the displacement, u, is a function of x and y. Two important differences arise between this equation and the beam equation (equation 1.254). They are:

(i) the bending wavenumber is now a two-dimensional vector and $\mathbf{k_B} = \mathbf{k_x} + \mathbf{k_y}$, or $k_B^2 = k_x^2 + k_y^2$;

(ii) the longitudinal wave velocity is now

$$c_L = \left\{ \frac{E}{\rho(1-\nu^2)} \right\}^{1/2}. \tag{1.321}$$

Hence, the bending wave velocity is

$$c_B = \omega^{1/2} \left\{ \frac{Et^3}{12(1-\nu^2)\rho_S} \right\}^{1/4} = \{1.8 c_L t f\}^{1/2}. \tag{1.322}$$

Equation (1.322) is similar to equation (1.259) for beams except for the $(1-\nu^2)$ term in the denominator for the plate longitudinal wave velocity.

The normal modes of vibration of a simply supported thin plate can be estimated by assuming a two-dimensional, time-dependent, harmonic solution to the plate equation (e.g. see Beranek[1.17]). They are

$$\omega_{m,n} = 2\pi(1.8 c_L t) \left\{ \left(\frac{m}{2L_x}\right)^2 + \left(\frac{n}{2L_y}\right)^2 \right\}, \tag{1.323}$$

where L_x and L_y are the plate dimensions in the x- and y-directions, respectively, and

$$k_x = \frac{m\pi}{L_x} \text{ for } m = 1, 2, 3, \text{ etc.} \quad \text{and} \quad k_y = \frac{n\pi}{L_y} \text{ for } n = 1, 2, 3, \text{ etc.} \tag{1.324}$$

Thus, there is a mode of vibration corresponding to every particular value of m and n. The integers m and n represent the number of half-waves in the x- and y-directions, respectively, and for clamped end conditions they should be replaced by $(2m + 1)$ and $(2n + 1)$.

Equations (1.321)–(1.324) will be used in chapter 3 when discussing the interactions of sound waves with structures. They are important equations and are used extensively in practice.

References

1.1 Bishop, R. E. D. 1979. *Vibration*, Cambridge University Press.
1.2 Fahy, F. J. 1985. *Sound and structural vibration: radiation, transmission and response*, Academic Press.
1.3 Kinsler, L. E., Frey, A. R., Coppens, A. B. and Sanders, V. J. 1982. *Fundamentals of acoustics*, John Wiley & Sons (3rd edition).
1.4 Reynolds, D. D. 1981. *Engineering principles of acoustics – noise and vibration*, Allyn & Bacon.
1.5 Tse, F. S., Morse, I. E. and Hinkle, R. T. 1979. *Mechanical vibrations – theory and applications*, Allyn & Bacon (2nd edition).
1.6 Thomson, W. T. 1981. *Theory of vibrations with applications*, George Allen & Unwin (2nd edition).
1.7 Newland, D. E. 1984. *An introduction to random vibrations and spectral analysis*, Longman (2nd edition).
1.8 Bendat, J. S. and Piersol, A. G. 1980. *Engineering applications of correlation and spectral analysis*, John Wiley & Sons.
1.9 Smith, P. W. and Lyon, R. H. 1965. *Sound and structural vibration*, NASA Contractor Report CR-160.
1.10 Papoulis, A. 1965. *Probability, random variables and stochastic processes*, McGraw-Hill.
1.11 Stone, B. J. 1985. *A summary of basic vibration theory*, Department of Mechanical Engineering, University of Western Australia, Lecture Note Series.
1.12 Cremer, L., Heckl, M. and Ungar, E. E. 1973. *Structure-borne sound*, Springer-Verlag.
1.13 Timoshenko, S. 1968. *Elements of strength of materials*, Van Nostrand Reinhold.
1.14 Leissa, A. W. 1969. *Vibration of plates*, NASA Special Report, SP-160.
1.15 Leissa, A. W. 1973. *Vibrations of shells*, NASA Special Report, SP-288.
1.16 Soedel, W. 1981. *Vibrations of shells and plates*, Marcel Dekker.
1.17 Beranek, L. L. 1971. *Noise and vibration control*, McGraw-Hill.

Nomenclature

a	acceleration
a_0, a_n	Fourier coefficients
$a(t)$	arbitrary time function
$a(x)$	initial displacement condition at time $t = 0$
A	surface area, cross-sectional area, arbitrary constant

$\mathbf{A}$, $\mathbf{A}_1$ etc.	arbitrary complex constants
$\mathbf{A}_i$	complex constant associated with incident waves
A_n	arbitrary constant, variable associated with beam receptances (see equation 1.311)
$\mathbf{A}_r$	complex constant associated with reflected waves
$\mathbf{A}_t$	complex constant associated with transmitted waves
b_n	Fourier coefficient
$b(x)$	initial velocity condition at time $t = 0$
B	$E(1-\nu)/\{(1+\nu)(1-2\nu)\}$
$\mathbf{B}$	arbitrary complex constant
B_1, B_2, etc.	arbitrary constants
B_n	variable associated with beam receptances (see equation 1.311)
c	speed of sound
c_B	bending wave velocity ($c_B = \omega/k_B$)
c_g	group velocity
c_L	quasi-longitudinal wave velocity ($c_L = \{E/\rho\}^{1/2}$ for beams and $\{E/\rho(1-\nu)^2\}^{1/2}$ for plates)
c'_L	longitudinal wave velocity ($c'_L = \{B/\rho\}^{1/2}$)
c_s	wave velocity in a vibrating string ($c_s = \{T/\rho_L\}^{1/2}$)
c_v	viscous-damping coefficient
c_{vc}	critical viscous-damping coefficient
c_{veq}	equivalent viscous damping
C_D	drag coefficient
C_F	$\rho C_D A/2$
C_n	Fourier coefficient, arbitrary constant
C_v	damping matrix ($C_v = c_{vij}$)
C_{vn}	generalised damping ($C_{vn} = c_{vnn}$)
D_n	Fourier coefficient, arbitrary constant
$D_n(t)$	dynamic load factor
E	energy of vibration, Young's modulus of elasticity
$\mathbf{E}'$	complex modulus of elasticity ($\mathbf{E}' = E(1 + i\eta)$)
$E[x(t)], E[x]$	expected or mean value of a function $x(t)$
$E[x^2(t)], E[x^2]$	mean-square value of a function $x(t)$

f	frequency
f_n	natural frequency of vibration
F	excitation force, impulse magnitude
$\mathbf{F}$, $\mathbf{F}(x, t)$	complex excitation force
F_0	excitation force amplitude
$F_1, F_2, \ldots, F_n$	excitation forces
$\mathbf{F_i}$	complex incident force
F_N	normal force
$\mathbf{F_r}$	complex reflected force
$\mathbf{F_t}$	complex transmitted force
F_v	viscous-damping force
$F(t)$	point force
$F_n(t)$	generalised force
g	gravitational acceleration
G_1, G_2	arbitrary independent functions which satisfy the wave equation
$G_{xx}(\omega)$	one-sided auto-spectral density function of a function $x(t)$
$\mathbf{G_{xy}}(\omega)$	one-sided cross-spectral density function of functions $x(t)$ and $y(t)$ (complex function)
$h(t), h(t-\tau)$	unit impulse response functions
H_n	mode participation factor
$\mathbf{H}(\omega)$	arbitrary frequency response function (Fourier transform of $h(t)$; complex function)
$\mathbf{H_{xxF}}(\omega)$, $\mathbf{H_{xxW}}(\omega)$	complex beam receptances (frequency response functions)
$\mathbf{H^*_{xxF}}(\omega)$, $\mathbf{H^*_{xxW}}(\omega)$	complex conjugates of beam receptances (frequency response functions)
$\mathbf{H^*}(\omega)$	complex conjugate of $\mathbf{H}(\omega)$
$\mathbf{H'(q)}$	transfer function with $\mathbf{q} = a + ib$ (Laplace transform of a function $x(t)$; complex function)
i	integer
I	second moment of area of a cross-section about the neutral plane axis
j	integer
k	wavenumber ($k = \omega/c$)
$\mathbf{k_B}$, k_B	bending wavenumber (bold signifies complex)
k_s, $\mathbf{k_s}$, k_{s1}, k_{s2}, etc.	spring stiffness (bold signifies complex)

k_x, k_y, $\mathbf{k}_x$, $\mathbf{k}_y$	x- and y-components of two-dimensional bending wavenumbers (bold signifies complex)
$\mathbf{k}'$	complex wavenumber ($\mathbf{k}' = k(1 - i\chi)$)
K_s	stiffness matrix ($K_s = k_{sij}$)
K_{sn}	generalised stiffness ($K_{sn} = k_{snn}$)
L	Lagrangian, length
m, m_1, m_2, etc.	masses, integers
m_x, m_y	mean values of functions $x(t)$ and $y(t)$
M	mass matrix ($M = m_{ij}$), mass, bending moment
M_n	generalised mass ($M_n = m_{nn}$)
n	integer
p	integer
$p(t)$	load distribution
$p(x)$	probability density function of a function $x(t)$, load distribution
$p(x_1, x_2)$	second-order probability density function of functions $x_1(t)$ and $x_2(t)$
$p(x, t)$	load distribution
P_0	constant applied load
q	principal or generalised co-ordinate, integer
$q_1, q_2, \ldots, q_n$	principal or generalised co-ordinates
$q(t)$	time-dependent Fourier coefficient (principal or generalised co-ordinate)
Q	quality factor
$R_{FF}(\tau)$, $R_{WW}(\tau)$	auto-correlation functions of forces $F(t)$ and $W(t)$
$R_{FW}(\tau)$, $R_{WF}(\tau)$	cross-correlation functions of forces $F(t)$ and $W(t)$
$R_{xx}(\tau)$	auto-correlation function of a function $x(t)$
$R_{xy}(\tau)$	cross-correlation function of functions $x(t)$ and $y(t)$
Re ($\mathbf{z}$)	real part of a complex number
s	arbitrary constant
s_1, s_2	roots of the characteristic equation
$S_{aa}(\omega)$	two-sided auto-spectral density function of acceleration
$S_{FF}(\omega)$, $S_{WW}(\omega)$	two-sided auto-spectral density functions of forces $F(t)$ and $W(t)$

$\mathbf{S_{FW}}(\omega)$, $\mathbf{S_{WF}}(\omega)$	two-sided cross-spectral density functions of forces $F(t)$ and $W(t)$ (complex function)
$S_{vv}(\omega)$	two-sided auto-spectral density function of velocity
$S_{xx}(\omega)$	two-sided auto-spectral density function of a function $x(t)$, two-sided auto-spectral density function of displacement
$\mathbf{S_{xy}}(\omega)$	two-sided cross-spectral density function of functions $x(t)$ and $y(t)$ (complex function)
t	time, plate thickness
T	temporal period, kinetic energy, string tension
T_{max}	maximum kinetic energy
$u(x, t)$, $\mathbf{u}(x, t)$	lateral displacement of a vibrating string (bold signifies complex)
$\mathbf{u_i}(x, t)$	complex incident displacement wave
$\mathbf{u_r}(x, t)$	complex reflected displacement wave
$\mathbf{u_t}(x, t)$	complex transmitted displacement wave
$\ddot{u}$	acceleration ($\ddot{u} = \partial^2 u/\partial t^2$)
u''	$\partial^2 u/\partial x^2$
U	potential energy
U_d	cyclic energy dissipated by a damping force
U_{max}	maximum potential energy
v	velocity
v_0	initial velocity
$\mathbf{v_i}$	complex particle velocity associated with an incident displacement wave
$\mathbf{v_r}$	complex particle velocity associated with a reflected displacement wave
$\mathbf{v_t}$	complex particle velocity associated with a transmitted displacement wave
$v(x, t)$, $\mathbf{v}(x, t)$	particle velocity ($\mathbf{v} = \partial \mathbf{u}/\partial t$, bold signifies complex)
V, $\mathbf{V}$	velocity amplitude (bold signifies complex), shear force
$W(t)$	point force
x	displacement
x_0	initial displacement
x_{rms}	root-mean-square value of x
$x(t)$	arbitrary time function, input function to a linear system

$x_B(t)$, $\mathbf{x_B}(t)$	base/abutment excitation (bold signifies complex)
$\langle x^2 \rangle$	mean-square value of a signal (time-averaged)
$\langle \overline{x^2} \rangle$	mean-square value of a signal (space-and time-averaged)
$\dot{x}$	velocity
$\ddot{x}$	acceleration
X, $\mathbf{X}$	amplitude of motion (bold signifies complex)
X_B, $\mathbf{X_B}$	amplitude of base/abutment motion (bold signifies complex)
X_0	F/k_s
X_{max}	X_r/X_0
X_r	amplitude resonance
X_T, $\mathbf{X_T}$	amplitude of transient damped motion (bold signifies complex)
$\mathbf{X}(\omega)$	Fourier transform of a function $x(t)$ (complex function)
$y(t)$	arbitrary time function, output function from a linear system
Y_m, $\mathbf{Y_m}$	mobility ($\mathbf{V}/\mathbf{F}$; bold signifies complex)
$\mathbf{Y}(\omega)$	Fourier transform of a function $y(t)$ (complex)
$\mathbf{z}$	complex number
$\mathbf{z}^*$	complex conjugate of $\mathbf{z}$
Z_1, Z_2, $\mathbf{Z_1}$, $\mathbf{Z_2}$	characteristic mechanical impedances or wave impedances ($Z = \rho_L c_L$; bold signifies complex)
Z_m, $\mathbf{Z_m}$	mechanical impedance, drive-point mechanical impedance ($\mathbf{F}/\mathbf{V}$; bold signifies complex)
$\mathbf{Z_{mf}}$	mechanical impedance of fixed end of a bar (complex function)
α	constant of proportionality
α_n	arbitrary constant
β	decay frequency ($\beta = 1/\tau$)
γ	arbitrary constant
$\Delta\tau$	short time duration

δ	incremental increase in distance
δ_{static}	static deflection of a spring
$\delta(t)$, $\delta(t-\tau)$	unit impulse functions
ε	unit strain
ζ	damping ratio ($c_{\text{v}}/c_{\text{vc}}$)
η	structural loss factor ($\eta = 2\zeta = 1/Q$), integration variable
θ	slope of vibrating string ($\theta = \partial u/\partial x$)
λ	wavelength
$\lambda_1, \lambda_2, \ldots, \lambda_n$	eigenvalues
μ	coefficient of friction
μ_n	variable associated with a beam receptance (see equation 1.311)
ν	Poisson's ratio
ξ	integration variable
π	3.14 . . .
Π	power
ρ	density, fluid density
ρ_{L}	mass per unit length
ρ_{S}	mass per unit area (surface mass)
$\rho_{xx}(\tau)$	auto-correlation coefficient
σ	standard deviation
τ	decay time, time delay
ϕ	phase angle between input and output, eigenfunction, mode shape
$\phi(x)$	eigenfunction, mode shape
$\phi_1, \phi_2, \ldots, \phi_n$	normal modes
χ	constant associated with complex wavenumbers ($\chi = \eta/2$ for quasi-longitudinal waves and $\eta/4$ for bending waves)
ψ	initial phase angle for a signal
ω	radian (circular) frequency
ω_{d}, $\boldsymbol{\omega}_{\text{d}}$	damped radian (circular) frequency (bold signifies complex)
ω_n	natural radian (circular) frequency
$\langle\ \rangle$	time-average of a signal
$\bar{\ }$	space-average of a signal (overbar)

2
Sound waves: a review of some fundamentals

2.1 Introduction

Sound is a pressure wave that propagates through an elastic medium at some characteristic speed. It is the molecular transfer of motional energy and cannot therefore pass through a vacuum. For this wave motion to exist, the medium has to possess inertia and elasticity. Whilst vibration relates to such wave motion in structural elements, noise relates to such wave motion in fluids (gases and liquids). Two fundamental mechanisms are responsible for sound generation. They are:

(i) the vibration of solid bodies resulting in the generation and radiation of sound energy – these sound waves are generally referred to as structure-borne sound;

(ii) flow-induced noise resulting from pressure fluctuations induced by turbulence and unsteady flows – these sound waves are generally referred to as aerodynamic sound.

With structure-borne sound, the regions of interest are generally in a fluid (usually air) at some distance from the vibrating structure. Here, the sound waves propagate through the stationary fluid (the fluid has a finite particle velocity due to the sound wave, but a zero mean velocity) from a readily identifiable source to the receiver. The region of interest does not therefore contain any sources of sound energy – i.e. the sources which generated the acoustic disturbance are external to it. A simple example is a vibrating electric motor. Classical acoustical theory (analysis of the homogeneous wave equation) can be used for the analysis of sound waves generated by these types of sources. The solution for the acoustic pressure fluctuation, p, describes the wave field external to the source. This wave field can be modelled in terms of combinations of simple sound sources. If required, the source can be accounted for in the wave field by considering the initial, time-dependent conditions.

With aerodynamic sound, the sources of sound are not so readily identifiable and the regions of interest can be either within the fluid flow itself or external to it. When the regions of interest are within the fluid flow, they contain sources of sound energy because the sources are continuously being generated or convected with the flow (e.g.

turbulence, vortices etc.). These aerodynamic sources therefore have to be included in the wave equation for any subsequent analysis of the sound waves in order that they can be correctly identified. The wave equation is now inhomogeneous (because it includes these source terms) and its solution is somewhat different to that of the homogeneous wave equation in that it now describes both the source and the wave fields.

It is very important to be aware of and to understand the difference between the homogeneous and the inhomogeneous wave equations for the propagation of sound waves in fluids. The vast majority of engineering noise and vibration control relates to sources which can be readily identified, and regions of interest which are outside the source region – in these cases the homogeneous wave equation is sufficient to describe the wave field and the subsequent noise radiation. Most machinery noise, for instance, is associated with the vibration of solid bodies. Engineers should, however, be aware of the existence of the inhomogeneous wave equation and of the instances when it has to be used in place of the more familiar (and easier to solve!) homogeneous wave equation.

A good industrial example that combines both types of noise generation mechanisms is high speed gas flow in a pipeline. Such pipelines are typically found in oil refineries and liquid-natural-gas plants. A typical section of pipeline is illustrated in Figure 2.1. Inside the pipe there are pressure fluctuations which are caused by turbulence,

Fig. 2.1. A schematic model of internal aerodynamic sound and external structure-borne sound in a gas pipeline.

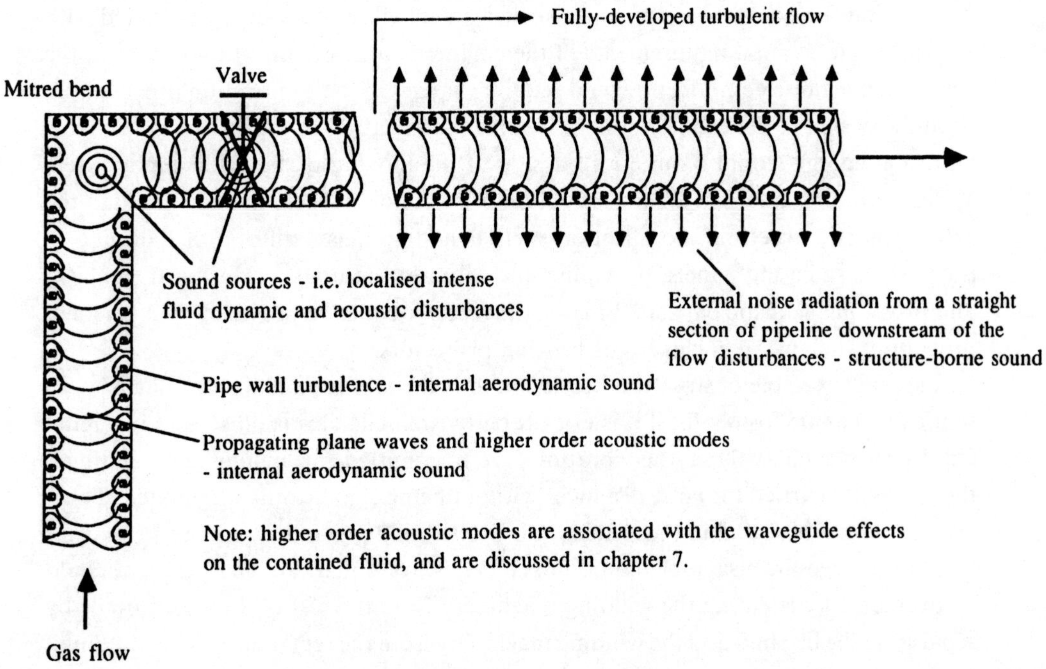

sound waves generated at the flow discontinuities (e.g. bends, valves etc.), and vortices which are convected downstream of some buff body such as a butterfly valve splitter-plate. These pressure fluctuations result in internal pipe flow noise which is aerodynamic in nature – the source of sound is distributed along the whole length of the inside of the pipe. If an analysis of the sound sources within the pipe is required, the inhomogeneous wave equation would have to be used. This internal aerodynamic noise excites the structure internally and the vibrating structure subsequently radiates noise to the surrounding external medium. The source of sound is not in the region of space under analysis for the external noise radiation, and this problem can therefore be handled with the homogeneous wave equation. A knowledge of the internal source field (the wall pressure fluctuations) allows for a prediction of the external sound radiation; the converse is, however, not true. A point which is sometimes overlooked is that a description of the external wave field does not contain sufficient information for the source to be identified, but, once the source has been identified and described, the sound field can be predicted. Pipe flow noise is discussed in chapter 7 as a case study.

Typical machinery noise control problems in industry involve (i) a source, (ii) a path, and (iii) a receiver. There is always interaction and feedback between the three, and there are generally several possible noise and vibration energy transmission paths for a typical machinery noise source. An internal combustion engine, for instance, generates both aerodynamic and mechanical energy, each with several possible transmission paths. This is illustrated schematically in Figure 2.2. The two sources of sound energy are (i) the aerodynamic energy associated with the combustion process and the exhaust system, and (ii) mechanical vibration energy associated with the various functional requirements of the engine. Source modification to reduce the aerodynamic noise component would require changes in the combustion process itself or in the design of the exhaust system. Source modification to reduce the mechanical vibration energy would require a re-design of the moving parts of the engine itself. Various options are open for the reduction of path noise. These include muffling the exhaust noise, structural modification such as adding mass, stiffness or damping to the various radiating panels, providing anti-vibration mounts, enclosing the engine, and providing acoustic barriers. Finally, the receiver could be provided with personal protection such as an enclosure or hearing protectors.

A specific example of structure-borne sound is the vibration and stop–start shocks that can emanate from a lift if it is not properly isolated. This is illustrated in Figure 2.3. In cases such as these, the vibrations are transmitted throughout the building – the waves are carried for large distances without being significantly attenuated. There are no sources of sound in the ambient air in the building, and any acoustic analysis would only require usage of the homogeneous wave equation. The problem could be overcome by isolating the winding machinery from the rest of the structure or by separating the lift shaft and the winding machinery from the remainder of the building.

A specific example of aerodynamic noise is the formation of turbulence in the mixing region at the exhaust of a jet nozzle such as the nozzle of a jet used for cleaning machine components with compressed air. The jet noise increases with flow velocity and the strength of the turbulence is related to the relative speed of the jet in relation to the ambient air. By introducing a secondary, low velocity, air-stream, as illustrated in Figure 2.4, and thus reducing the velocity profile across the jet, significant reductions in radiated noise levels can be achieved. Hence, compound nozzles are sometimes used in industry – here, the velocity of the core jet remains the same but its noise radiating characteristics are reduced by the introduction of a slower outer stream.

Fig. 2.2. Typical noise and vibrations paths for a machinery source. (Adapted from Pickles[2.9].)

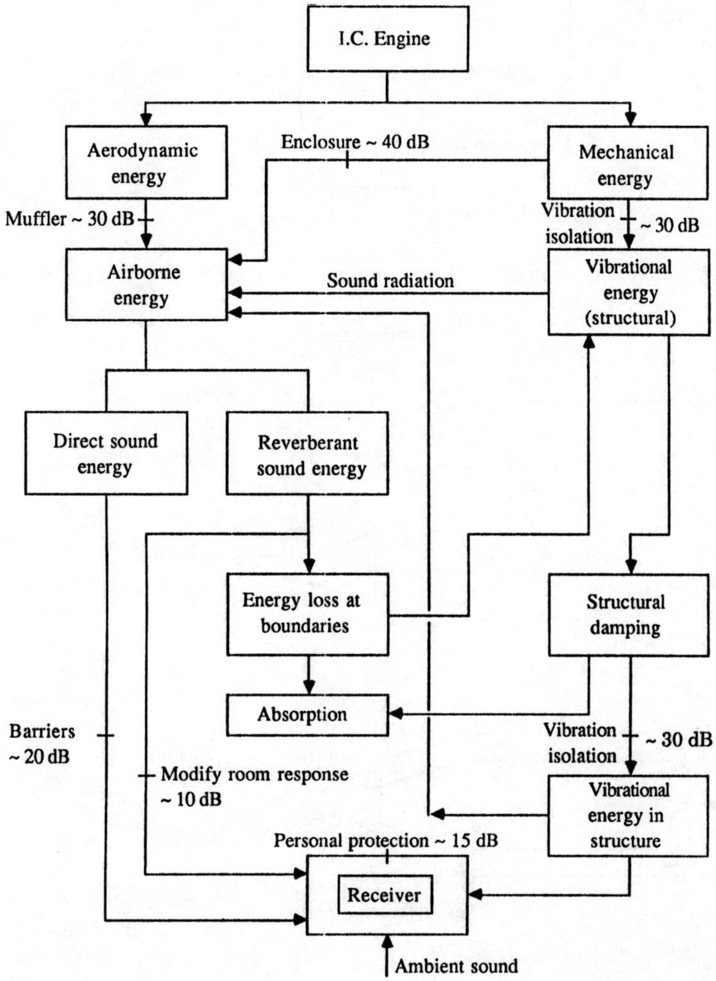

Fig. 2.3. Schematic illustration of structure-borne sound in a building.

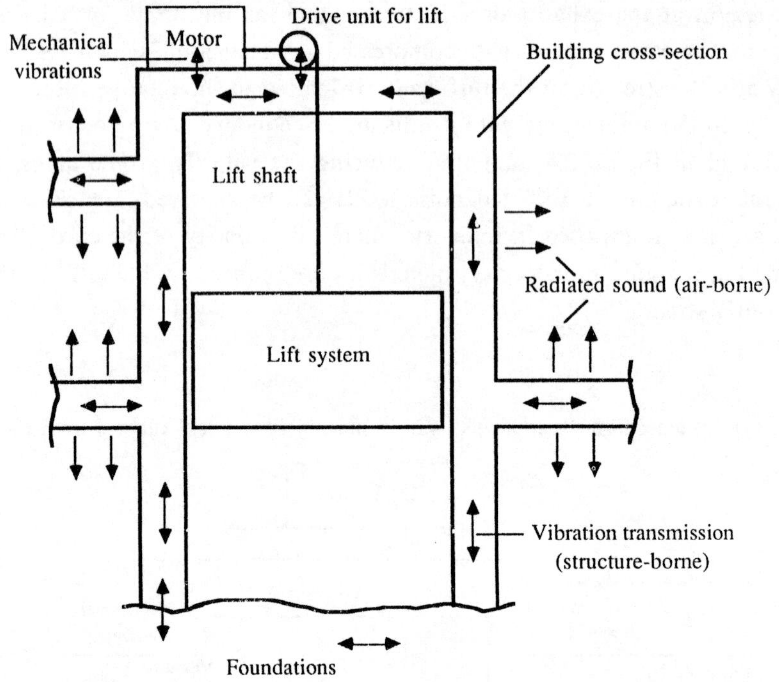

Fig. 2.4. Schematic illustration of aerodynamic sound emanating from a jet nozzle.

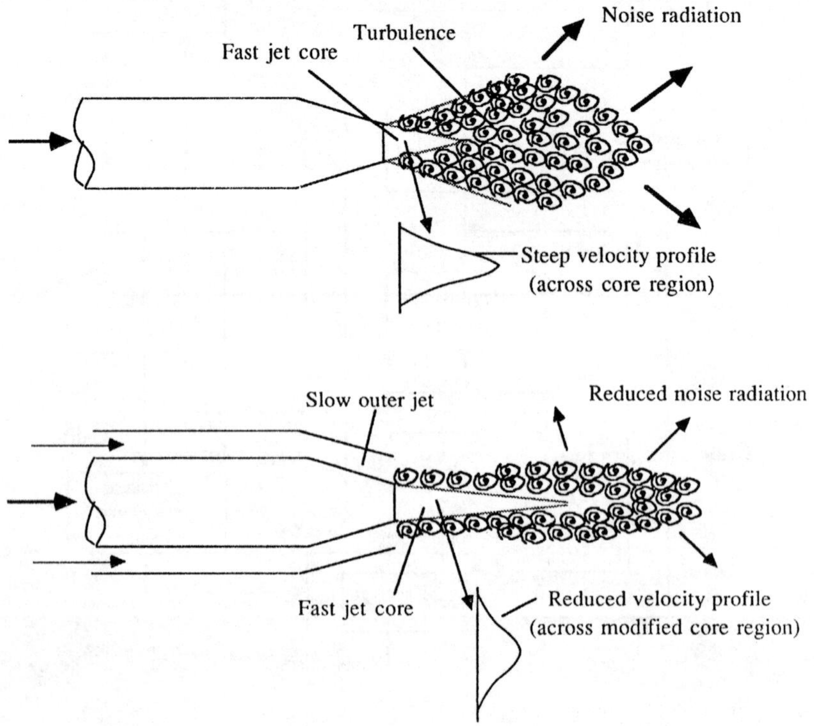

2.2 The homogeneous acoustic wave equation – a classical analysis

Three methods are available for approaching problems in acoustics. They are (i) wave acoustics, (ii) ray acoustics, and (iii) energy acoustics.

Wave acoustics is a description of wave propagation using either molecular or particulate models. The general preference is for the particulate model, a particle being a fluid volume large enough to contain millions of molecules and small enough such that density, pressure and temperature are constant. Ray acoustics is a description of wave propagation over large distances, e.g. the atmosphere. Families of rays are used to describe the propagation of sound waves and inhomogeneities such as temperature gradients or wind have to be accounted for. Over large distances, the ray tracing procedures are preferred because they approximate and simplify the exact wave approach. Finally, energy acoustics describes the propagation of sound waves in terms of the transfer of energy of various statistical parameters where techniques referred to as statistical energy analysis (or S.E.A.) are used.

The wave acoustics approach is probably the most fundamental and important approach to the study of all disciplines of acoustics. The ray acoustics approach generally relates to outdoor or underwater sound propagation over large distances and is therefore not directly relevant to industrial noise and vibration control. The S.E.A. approach is fast becoming popular for quick and effective answers to complex industrial noise and vibration problems. The wave acoustics approach will thus be adopted for the better part of this book, and this chapter is devoted to some of the more important fundamental principles of sound waves. The subject of ray acoustics is not discussed in this book, but the concepts and applications of statistical energy analysis techniques are discussed in chapter 6.

Sound waves in non-viscous (inviscid) fluids are simply longitudinal waves and adjacent regions of compression and rarefaction are set up – i.e. the particles oscillate to and fro in the wave propagation direction, hence the acoustic particle velocity is in the same direction as the phase velocity. The pressure change that is produced as the fluid compresses and expands is the source of the restoring force for the oscillatory motion. There are four variables that are of direct relevance to the study of sound waves. They are pressure, P, velocity, $\vec{U}$, density, ρ, and temperature, T. Pressure, density and temperature are scalar quantities whilst velocity is a vector quantity (i.e. an arrow over a symbol denotes a vector quantity). Each of the four variables has a mean and a fluctuating component. Thus,

$$P(\vec{x}, t) = P_0(\vec{x}) + p(\vec{x}, t), \tag{2.1a}$$

$$\vec{U}(\vec{x}, t) = \vec{U}_0(\vec{x}) + \vec{u}(\vec{x}, t), \tag{2.1b}$$

$$\rho(\vec{x}, t) = \rho_0(\vec{x}) + \rho'(\vec{x}, t), \tag{2.1c}$$

$$T(\vec{x}, t) = T_0(\vec{x}) + T'(\vec{x}, t), \tag{2.1d}$$

The wave equation can thus be set up in terms of any one of these four variables. In acoustics, it is the pressure fluctuations, $p(\vec{x}, t)$, that are of primary concern – i.e. noise radiation is a fluctuating pressure. Thus it is common for acousticians to solve the wave equation in terms of the pressure as a dependent variable. It is, however, quite valid to solve the wave equation in terms of any of the other three variables. Also, generally, $\vec{U}_0(\vec{x})$ is zero (i.e. the ambient fluid is stationary) and therefore $\vec{U}(\vec{x}, t) = \vec{u}(\vec{x}, t)$.

As for wave propagation in solids, several simplifying assumptions need to be made. They are:

(1) the fluid is an ideal gas;
(2) the fluid is perfectly elastic – i.e. Hooke's law holds;
(3) the fluid is homogeneous and isotropic;
(4) the fluid is inviscid – i.e. viscous-damping and heat conduction terms are neglected;
(5) the wave propagation through the fluid media is adiabatic and reversible;
(6) gravitational effects are neglected – i.e. P_0 and ρ_0 are assumed to be constant;
(7) the fluctuations are assumed to be small – i.e. the system behaves linearly.

In order to develop the acoustic wave equation, equations describing the relationships between the various acoustic variables and the interactions between the restoring forces and the deformations of the fluid are required. The first such relationship is referred to as continuity or the conservation of mass; the second relationship is referred to as Euler's force equation or the conservation of momentum; and the third relationship is referred to as the thermodynamic equation of state. In practice, sound waves are generally three-dimensional. It is, however, convenient to commence with the derivation of the above equations in one dimension and to subsequently extend the results to three dimensions.

2.2.1 Conservation of mass

The equation of conservation of mass (continuity) provides a relationship between the density, $\rho(\vec{x}, t)$, and the particle velocity, $\vec{u}(\vec{x}, t)$ – i.e. it relates the fluid motion to its compression.

Consider the mass flow of particles in the x-direction through an elemental, fixed, control volume, dV, as illustrated in Figure 2.5. For mass to be conserved, the time rate of change of the elemental mass has to equal the nett mass flow into the elemental volume. Because the flow is one-dimensional, the vector notation is temporarily dropped. It will be re-introduced later on when the equations are extended to three-dimensional flow. Note that

$$\vec{u} = u_x\vec{i} + u_y\vec{j} + u_z\vec{k},$$

where u_x, u_y and u_z are the particle velocities in the x-, y- and z-directions, respectively.

For flow in the x-direction only:

(i) the elemental mass is $\rho A\,\mathrm{d}x$ (where $A = \mathrm{d}y\,\mathrm{d}z$);

(ii) the mass flow into the elemental volume is $(\rho u A)_x$;

(iii) the mass flow out of the elemental volume is $(\rho u A)_{x+\mathrm{d}x}$.

For the conservation of mass,

$$\frac{\partial(\rho A\,\mathrm{d}x)}{\partial t} = (\rho u A)_x - (\rho u A)_{x+\mathrm{d}x}. \tag{2.2}$$

Using a Taylor series expansion,

$$\frac{\partial(\rho A\,\mathrm{d}x)}{\partial t} = \left\{(\rho u A)_x - (\rho u A)_x - \frac{\partial(\rho u A)_x}{\partial x}\,\mathrm{d}x\right\}. \tag{2.3}$$

Hence,

$$\frac{\partial \rho}{\partial t} + \frac{\partial(\rho u_x)}{\partial x} = 0. \tag{2.4}$$

Equation (2.4) represents the one-dimensional conservation of mass in the x-direction. It can be extended to three dimensions, and the three-dimensional equation of conservation of mass is therefore

$$\frac{\partial \rho}{\partial t} + \vec{\nabla}\cdot\rho\vec{u} = 0, \tag{2.5}$$

where $\vec{\nabla}$ is the divergence operator, i.e.

$$\vec{\nabla} = \left\{\frac{\partial}{\partial x}\vec{i} + \frac{\partial}{\partial y}\vec{j} + \frac{\partial}{\partial z}\vec{k}\right\}. \tag{2.6}$$

Fig. 2.5. Mass flow of particles in the x-direction through an elemental, fixed, control volume.

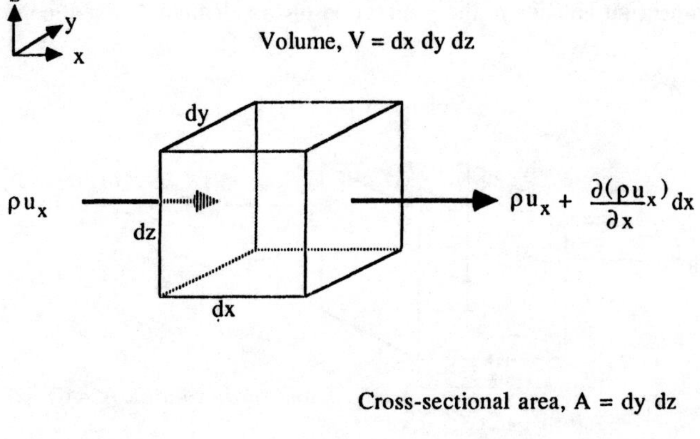

Equation (2.5) is thus a vector representation for

$$\frac{\partial \rho}{\partial t} + \frac{\partial(\rho u_x)}{\partial x} + \frac{\partial(\rho u_y)}{\partial y} + \frac{\partial(\rho u_z)}{\partial z} = 0. \tag{2.7}$$

The equation of conservation of mass (equations 2.5 or 2.7) is a scalar quantity. It is also non-linear because the mass flow terms involve products of two small fluctuating components ($\vec{u}$ and ρ'). These terms have second-order effects as far as the propagation of sound waves is concerned – i.e. the equation can be linearised. Substituting for $\rho(\vec{x}, t) = \rho_0(\vec{x}) + \rho'(\vec{x}, t)$ into equation (2.7) and deleting second- and higher-order terms yields

$$\frac{\partial \rho'}{\partial t} + \rho_0 \frac{\partial u_x}{\partial x} + \rho_0 \frac{\partial u_y}{\partial y} + \rho_0 \frac{\partial u_z}{\partial z} = 0, \tag{2.8}$$

or

$$\frac{\partial \rho'}{\partial t} + \rho_0 \vec{\nabla} \cdot \vec{u} = 0. \tag{2.9}$$

Equation (2.9) is the linearised equation of conservation of mass (continuity).

2.2.2 *Conservation of momentum*

The equation of conservation of momentum provides a relationship between the pressure, $P(\vec{x}, t)$, the density, $\rho(\vec{x}, t)$, and the particle velocity, $\vec{u}(\vec{x}, t)$. It can be obtained either by observing the stated law of conservation of momentum with respect to an elemental, fixed, control volume, dV, in space, or by a direct application of Newton's second law with respect to the fluid particles that move through the elemental, fixed, control volume. To a purist both procedures are identical! It is, however, instructive to consider them both.

Consider the first approach – consider the momentum flow through an elemental, fixed control volume, dV, as illustrated in Figure 2.6. For momentum to be conserved

Fig. 2.6. Momentum balance in the *x*-direction for an elemental, fixed, control volume.

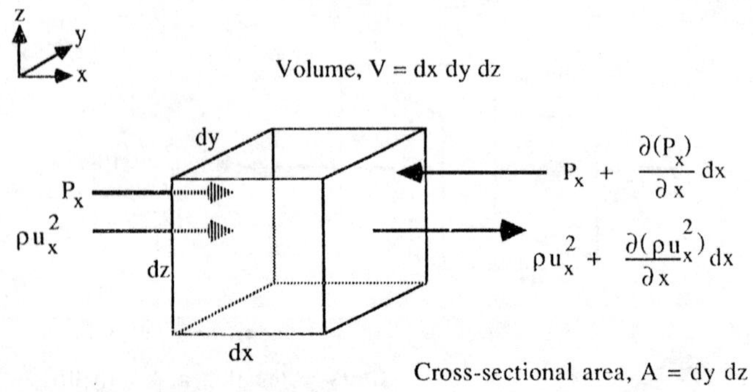

the time rate of change of momentum contained in the fixed volume plus the nett rate of flow of momentum through the surfaces of the volume are equal to the sum of all the forces acting on the volume. Once again, because the flow is one-dimensional, the vector notation is temporarily dropped. Also, body forces are neglected – i.e. only pressure forces act on the body. For flow in the x-direction only:

(i) the momentum of the control volume is $\rho u_x A\,\mathrm{d}x$;

(ii) the momentum flow into the control volume is $(\rho u^2 A)_x$;

(iii) the momentum flow out of the control volume is $(\rho u^2 A)_{x+\mathrm{d}x}$;

(iv) the force at position x is $(PA)_x$;

(v) the force at position $x+\mathrm{d}x$ is $-(PA)_{x+\mathrm{d}x}$.

For the conservation of momentum

$$\frac{\partial(\rho u_x A)\,\mathrm{d}x}{\partial t} = (\rho u^2 A)_x - (\rho u^2 A)_{x+\mathrm{d}x} + (PA)_x - (PA)_{x+\mathrm{d}x}. \tag{2.10}$$

If a Taylor series expansion is used for $(\rho u^2 A)_{x+\mathrm{d}x}$ and $(PA)_{x+\mathrm{d}x}$, equation (2.10) simplifies to

$$\frac{\partial(\rho u_x)}{\partial t} = -\frac{\partial(\rho u_x^2)}{\partial x} - \frac{\partial P}{\partial x}. \tag{2.11}$$

Equation (2.11) can be re-arranged as

$$\rho\frac{\partial u_x}{\partial t} + u_x\left\{\frac{\partial\rho}{\partial t} + u_x\frac{\partial\rho}{\partial x} + \rho\frac{\partial u_x}{\partial x}\right\} + \rho u_x\frac{\partial u_x}{\partial x} + \frac{\partial P}{\partial x} = 0, \tag{2.12}$$

where the term in brackets is the continuity equation. Thus, equation (2.12) simplifies to

$$\rho\frac{\partial u_x}{\partial t} + \rho u_x\frac{\partial u_x}{\partial x} + \frac{\partial P}{\partial x} = 0. \tag{2.13}$$

Equation (2.13) represents the one-dimensional conservation of momentum in the x-direction. Similar expressions can be obtained for the y- and z-directions. The three-dimensional equation of conservation of momentum is therefore obtained by introducing the divergence operator, $\vec{\nabla}$. It is

$$\rho\left\{\frac{\partial\vec{u}}{\partial t} + (\vec{u}\cdot\vec{\nabla})\vec{u}\right\} + \vec{\nabla}P = 0. \tag{2.14}$$

This equation is the non-linear, inviscid momentum equation or Euler's equation.

Equation (2.14) can also be obtained by a direct application of Newton's second law with respect to the fluid particles that move through the elemental, fixed, control volume. The control volume in Figure 2.6 contains a mass, $\mathrm{d}m$, of fluid at any instant in time. The nett force on the volume element is

$$d\vec{f} = \vec{a}\, dm, \tag{2.15}$$

from Newton's second law. It is very important to recognise that the acceleration, $\vec{a}$, is the rate of change of velocity of a given fluid particle as it moves about in space and it is not the rate of change of fluid velocity at a fixed point in space.

The particle velocity, $\vec{u}$, is a function of space and time. Thus, at some time t, a particle is at position (x, y, z) and it has a particle velocity $\vec{u}(x, y, z, t)$. At some further time, $t + dt$ the particle is at position $(x + dx, y + dy, z + dz)$ and it has a particle velocity $\vec{u}(x + dx, y + dy, z + dz, t + dt)$. The particle acceleration is

$$\vec{a} = \lim_{dt \to \infty} \frac{\vec{u}(x + dx, y + dy, z + dz, t + dt) - \vec{u}(x, y, z, t)}{dt}. \tag{2.16}$$

The particle velocity at time $t + dt$ can be re-expressed as $\vec{u}(x + u_x\, dt, y + u_y\, dt, z + u_z\, dt, t + dt)$, where $dx = u_x\, dt$, $dy = u_y\, dt$, and $dz = u_z\, dt$. Thus by re-expressing the particle velocity at time $t + dt$ and using a Taylor series expansion

$$\vec{u}(x + dx, y + dy, z + dz, t + dt) = \vec{u}(x, y, z, t) + \frac{\partial \vec{u}}{\partial x} u_x\, dt + \frac{\partial \vec{u}}{\partial y} u_y\, dt + \frac{\partial \vec{u}}{\partial z} u_z\, dt + \frac{\partial \vec{u}}{\partial t} dt. \tag{2.17}$$

Hence the particle acceleration is

$$\vec{a} = \frac{\partial \vec{u}}{\partial t} + u_x \frac{\partial \vec{u}}{\partial x} + u_y \frac{\partial \vec{u}}{\partial y} + u_z \frac{\partial \vec{u}}{\partial z}. \tag{2.18}$$

Equation (2.18) can be re-expressed in vector notation by using the vector operation $\vec{u} \cdot \vec{\nabla}$. Hence,

$$\vec{a} = \frac{\partial \vec{u}}{\partial t} + (\vec{u} \cdot \vec{\nabla})\vec{u}, \tag{2.19}$$

where

$$\vec{u} \cdot \vec{\nabla} = u_x \frac{\partial}{\partial x} + u_y \frac{\partial}{\partial y} + u_z \frac{\partial}{\partial z}. \tag{2.20}$$

Equation (2.19) is the total acceleration of a fluid particle in an Eulerian frame of reference. It has both a convective part and a local time rate of change when the flow is unsteady.

Neglecting viscosity, the nett force on the elemental fluid volume in the x-direction is

$$d\vec{f} = \left\{ P - \left(P + \frac{\partial P}{\partial x} \right) dx \right\} dy\, dz = -\frac{\partial P}{\partial x} dV\, \vec{i}. \tag{2.21}$$

The complete, three-dimensional, vector force is thus

$$-\vec{\nabla}P\,\mathrm{d}V = -\left\{\frac{\partial P}{\partial x}\vec{i} + \frac{\partial P}{\partial y}\vec{j} + \frac{\partial P}{\partial z}\vec{k}\right\}\mathrm{d}V. \tag{2.22}$$

Hence, from Newton's second law,

$$\rho\,\mathrm{d}V\left\{\frac{\partial \vec{u}}{\partial t} + (\vec{u}\cdot\vec{\nabla})\vec{u}\right\} = -\vec{\nabla}P\,\mathrm{d}V, \tag{2.23}$$

where $\rho\,\mathrm{d}V = \mathrm{d}m$. Thus,

$$\rho\left\{\frac{\partial \vec{u}}{\partial t} + (\vec{u}\cdot\vec{\nabla})\vec{u}\right\} + \vec{\nabla}P = 0. \tag{2.24}$$

Equation (2.24) is identical to equation (2.14) – it is the non-linear, inviscid momentum or Euler's equation. Like the equation of conservation of mass, it can be simplified by linearisation – second- and higher-order terms can be neglected for the propagation of sound waves. Substituting for $P(\vec{x}, t) = P_0(\vec{x}) + p(\vec{x}, t)$, and $\rho(\vec{x}, t) = \rho_0(\vec{x}) + \rho'(\vec{x}, t)$ and deleting the second- and higher-order terms yields

$$\rho_0\left\{\frac{\partial u_x}{\partial t}\vec{i} + \frac{\partial u_y}{\partial t}\vec{j} + \frac{\partial u_z}{\partial t}\vec{k}\right\} + \left\{\frac{\partial p}{\partial x}\vec{i} + \frac{\partial p}{\partial y}\vec{j} + \frac{\partial p}{\partial z}\vec{k}\right\} = 0, \tag{2.25}$$

or

$$\rho_0\frac{\partial \vec{u}}{\partial t} + \vec{\nabla}p = 0. \tag{2.26}$$

Equation (2.26) is the linear inviscid force equation (conservation of momentum). Like the linearised equation of conservation of mass, it is valid for small amplitude sound waves ($\sim < 140$ dB).

2.2.3 The thermodynamic equation of state

The thermodynamic equation of state relates the pressure, density and absolute temperature of a fluid. For a perfect fluid it is

$$P = \rho R T_k. \tag{2.27}$$

P is the absolute pressure, ρ is the density of the gas, R is the gas constant, and T_k is the absolute temperature. It should be noted that the universal gas constant, G, is equal to RM, where M is the molecular weight of the particular gas.

The propagation of sound waves in air does not generally produce any significant changes of thermal energy between particles, and the entropy of the gas is constant. In addition, the thermal conductivity of the gas is very small. The propagation of sound waves can therefore be assumed to be nearly adiabatic. This assumption is valid for linear (small amplitude) sound waves within the audio-frequency range. Any loss of thermal energy would result in an attenuation of the sound waves with time and distance. Thus the adiabatic equation of state for a perfect gas is

$$P/P_0 = (\rho/\rho_0)^\gamma, \tag{2.28}$$

where γ is the ratio of specific heats.

For gases which are not perfect, an adiabatic equation of state can be obtained from a Taylor series expansion of an experimentally determined isentropic relationship between the pressure and density fluctuations. Provided that those fluctuations are assumed to be small, a linear relationship can be establshed between them (Kinsler *et al.*[2.1]). It is

$$p(\vec{x}, t) = B\left\{\frac{\rho'}{\rho_0}\right\}, \tag{2.29}$$

where B is the adiabatic bulk modulus – i.e.

$$B = \rho_0\left\{\frac{\partial P}{\partial \rho}\right\}_{\rho_0}. \tag{2.30}$$

The partial derivative in equation (2.30) is evaluated for an adiabatic process – i.e. it establishes the adiabatic compression and expansion of the gas about its mean density.

2.2.4 The linearised acoustic wave equation

Equation (2.9) (conservation of mass), equation (2.26) (conservation of momentum) and equations (2.29)–(2.30) can now be combined into a single equation with one dependent variable. The dependent variable of interest in acoustics is the fluctuating pressure.

The time derivative of the equation of conservation of mass is

$$\frac{\partial^2 \rho'}{\partial t^2} + \rho_0 \frac{\partial(\vec{\nabla}\cdot\vec{u})}{\partial t} = 0, \tag{2.31a}$$

or

$$\frac{\partial^2 \rho'}{\partial t^2} + \rho_0 \vec{\nabla}\cdot\frac{\partial \vec{u}}{\partial t} = 0. \tag{2.31b}$$

The divergence of the equation of conservation of momentum is

$$\rho_0 \vec{\nabla}\cdot\frac{\partial \vec{u}}{\partial t} + \nabla^2 p = 0. \tag{2.32}$$

Subtracting equation (2.31) from (2.32) yields

$$\nabla^2 p = \frac{\partial^2 \rho'}{\partial t^2}. \tag{2.33}$$

Equation (2.29) can now be substituted into equation (2.33) to eliminate ρ'. Thus,

$$\nabla^2 p = \frac{\rho_0}{B}\frac{\partial^2 p}{\partial t^2} = \frac{1}{c^2}\frac{\partial^2 p}{\partial t^2}. \tag{2.34}$$

Equation (2.34) is the linearised, homogeneous acoustic wave equation with the fluctuating pressure, $p(\vec{x}, t)$ as the dependent variable. The constant, c, is the velocity of propagation of the wave and is therefore the speed of sound. It is

$$c = \left(\frac{B}{\rho_0}\right)^{1/2} = \left\{\left(\frac{\partial P}{\partial \rho}\right)_{\rho_0}\right\}^{1/2}. \tag{2.35}$$

Some useful approximations can now be made in relation to the speed of sound by assuming that the sound propagation medium is a perfect gas. From equation (2.28)

$$\mathrm{P} = \frac{P_0 \rho^\gamma}{\rho_0^\gamma}, \tag{2.36}$$

thus

$$\frac{\partial P}{\partial \rho} = \frac{\gamma P}{\rho}. \tag{2.37}$$

Thus, the speed of sound is

$$c = \left(\frac{\gamma P}{\rho}\right)^{1/2}. \tag{2.38}$$

Also, by substituting equation (2.27) into equation (2.38) yields

$$c = (\gamma R T_k)^{1/2}. \tag{2.39}$$

For small fluctuations,

$$c \approx c_0 \approx \left(\frac{\gamma P_0}{\rho_0}\right)^{1/2}. \tag{2.40}$$

The absolute temperature, T_k, in equation (2.39) is in Kelvin, and c_0 is the speed of sound at atmospheric conditions. The gas constant for air is 0.287 kJ kg^{-1} K^{-1}. The equations demonstrate that the wave speed is constant for a given pressure and medium.

2.2.5 The acoustic velocity potential

Using vector theory, it can be demonstrated that the acoustic particle velocity, $\vec{u}$, is irrotational. From vector theory it can be shown that if a vector function is the gradient of a scalar function, its curl is the zero vector – i.e. for some scalar function ϕ, $\vec{\nabla} \times \vec{\nabla}\phi = 0$.

The curl of the momentum equation (equation 2.26) is

$$\rho_0 \frac{\partial(\vec{\nabla} \times \vec{u})}{\partial t} + (\vec{\nabla} \times \vec{\nabla} p) = 0, \tag{2.41}$$

and

$$\vec{\nabla} \times \vec{u} = 0, \tag{2.42}$$

since (i) p is a scalar quantity, and (ii) the constant associated with the time integral is zero. The constant has to be zero since the acoustic quantities would disappear if an acoustic disturbance was not present. Equation (2.42) therefore confirms that the acoustic particle velocity, $\vec{u}$, is irrotational. Now, because the acoustic particle velocity is irrotational, it can be expressed as the gradient of a scalar function (if a vector function is the gradient of a scalar function, its curl is the zero vector, i.e. it is irrotational). Hence the introduction of the concept of the acoustic velocity potential, ϕ, and

$$\vec{u} = \vec{\nabla}\phi. \tag{2.43}$$

The above result is a very important one. Its physical interpretation is that the acoustical excitation of an inviscid fluid does not produce rotational flow – i.e. there are no boundary layers, shear stresses or turbulence generated. When the effects of viscosity cannot be completely neglected, the particle velocity is not curl free everywhere and there is some rotational flow. For example, when a flat plate is mechanically excited in the presence of a mean fluid flow, in addition to the plate radiating sound, vorticity is generated on the surface of the plate (Soria and Norton[2.2]). These rotational effects are generally confined to the vicinity of boundaries but situations can arise where they exert some influence on sound propagation. These concepts will be discussed later on in this chapter when discussing aerodynamic noise. For the present purposes it is sufficient to assume that, when the mean fluid flow is zero, the particle velocity is irrotational.

Substituting the equation for the velocity potential (equation 2.43) into the momentum equation (equation 2.26) yields

$$\vec{\nabla}\left\{\rho_0 \frac{\partial \phi}{\partial t} + p\right\} = 0, \tag{2.44}$$

The acoustic quantities inside the brackets have to vanish if there is no acoustic disturbance present, thus the integration constant has to be zero – i.e.

$$p = -\rho_0 \frac{\partial \phi}{\partial t}. \tag{2.45}$$

Substituting for p into the wave equation (equation 2.34) yields

$$\nabla^2\left\{\frac{\partial\phi}{\partial t}\right\}=\frac{1}{c^2}\left\{\frac{\partial^3\phi}{\partial t^3}\right\}. \tag{2.46}$$

Thus

$$\nabla^2\phi=\frac{1}{c^2}\frac{\partial^2\phi}{\partial t^2}, \tag{2.47}$$

and ϕ satisfies the wave equation.

It is useful to point out that, when rotational flow components (e.g. near boundaries) have to be accounted for, a vector velocity potential, $\vec{\psi}$, can be introduced such that

$$\vec{U}=\vec{\nabla}\phi+\vec{\nabla}\times\vec{\psi}. \tag{2.48}$$

The first term represents the irrotational fluctuating flow and the second term represents the total (mean plus fluctuating) rotational component (see secton 2.4.4).

The velocity potential concept is commonly used in fluid dynamics where solutions for the particle velocity are usually sought. The particle velocity for the three-dimensional wave equation is a vector quantity, and the introduction of the scalar velocity potential allows for the wave equation to be solved in terms of a scalar. Acousticians generally use the pressure variable, which is also a scalar quantity. The particle velocity can be related to the acoustic pressure fluctuation via the momentum equation (equation 2.26). Hence,

$$\vec{u}=\vec{\nabla}\phi=-\int\frac{1}{\rho_0}\vec{\nabla}p\,\mathrm{d}t. \tag{2.49}$$

2.2.6 The propagation of plane sound waves

Consider a plane, one-dimensional, sound wave propagating in the x-direction. The one-dimensional homogeneous wave equation is

$$\frac{\partial^2 p}{\partial x^2}=\frac{1}{c^2}\frac{\partial^2 p}{\partial t}, \tag{2.50}$$

where $p=p(\vec{x},t)$. As was the case in chapter 1, it is convenient to represent the solution to the wave equation in terms of complex, harmonic functions. The general solution involves waves travelling in both the positive and negative x-directions. It is

$$\mathbf{p}(\vec{x},t)=\mathbf{A}_1\,\mathrm{e}^{\mathrm{i}(\omega t-kx)}+\mathbf{A}_2\,\mathrm{e}^{\mathrm{i}(\omega t+kx)}. \tag{2.51}$$

The complex particle velocity, $\mathbf{\hat{u}}(\vec{x},t)$, and the complex velocity potential, $\boldsymbol{\phi}(\vec{x},t)$, can be obtained by substituting equation (2.51) into equation (2.49). They are

$$\mathbf{\hat{u}}(\vec{x},t)=\left\{\frac{\mathbf{A}_1}{\rho_0 c}\,\mathrm{e}^{\mathrm{i}(\omega t-kx)}-\frac{\mathbf{A}_2}{\rho_0 c}\,\mathrm{e}^{\mathrm{i}(\omega t+kx)}\right\}\vec{i}, \tag{2.52}$$

and

$$\boldsymbol{\phi}(\vec{x}, t) = -\frac{\mathbf{A}_1}{\mathrm{i}\rho_0\omega}\,\mathrm{e}^{\mathrm{i}(\omega t - kx)} - \frac{\mathbf{A}_2}{\mathrm{i}\rho_0\omega}\,\mathrm{e}^{\mathrm{i}(\omega t + kx)}. \tag{2.53}$$

The particle velocity is a vector quantity and the term outside the brackets in equation (2.52) is the unit vector, $\vec{i}$ (i.e. one should not be confused between the complex number i and the unit vector $\vec{i}$). Also, because complex numbers have been introduced, the acoustic variables are now represented as complex numbers. Now, if one lets

$$\mathbf{p}_+ = \mathbf{A}_1\,\mathrm{e}^{\mathrm{i}(\omega t - kx)}, \quad \text{and } \mathbf{p}_- = \mathbf{A}_2\,\mathrm{e}^{\mathrm{i}(\omega t + kx)}, \tag{2.54}$$

then,

$$\hat{\mathbf{u}}(\vec{x}, t) = \left\{\frac{\mathbf{p}_+}{\rho_0 c} - \frac{\mathbf{p}_-}{\rho_0 c}\right\}\vec{i}, \tag{2.55}$$

and

$$\boldsymbol{\phi}(\vec{x}, t) = -\frac{\mathbf{p}_+}{\mathrm{i}\rho_0\omega} - \frac{\mathbf{p}_-}{\mathrm{i}\rho_0\omega}. \tag{2.56}$$

Once again, the term outside the brackets in equation (2.55) is the unit vector, $\vec{i}$.

In acoustics it is common to consider the waves travelling in the positive direction when one is concerned with the propagation of sound waves away from a source into some free field. Both positive and negative travelling waves have to be considered when considering the propagation of sound waves in confined spaces, e.g. ducts, rooms etc. Equation (2.55) illustrates an important point – for a plane, harmonic wave travelling in either the positive or the negative x-directions, the acoustic pressure fluctuations are in phase with the particle velocities. The phase relationship between these two variables is very important for a fundamental understanding of the propagation of different sound wave-types.

The concepts of impedance were introduced in chapter 1. The ratio of the acoustic pressure fluctuations, $\mathbf{p}(\vec{x}, t)$ in a medium to the associated particle velocity, $\hat{\mathbf{u}}(\vec{x}, t)$ is termed the specific acoustic impedance, $\mathbf{Z}_\mathrm{a}$, i.e.

$$\mathbf{Z}_\mathrm{a} = \frac{\mathbf{p}(\vec{x}, t)}{\hat{\mathbf{u}}(\vec{x}, t)}. \tag{2.57}$$

For a plane sound wave,

$$\mathbf{Z}_\mathrm{a} = Z_\mathrm{a} = \pm\rho_0 c. \tag{2.58}$$

The specific acoustic impedance is generally complex – for wave types other than plane sound waves, the acoustic pressure fluctuations and the particle velocities are not always in phase, i.e. the waves diverge. The quantity ρ_0c is often called the characteristic impedance (resistance) of the medium. For air, at 20°C and 1 atm., ρ_0 is $\sim$1.21 kg m^{-3}, c is $\sim$343 m s^{-1} and $\rho_0 c$ is 415 Pa s m^{-1}.

2.2.7 *Sound intensity, energy density and sound power*

Three additional parameters that play an important role in acoustics are the sound intensity, the sound energy density and the radiated sound power.

The sound intensity is defined as the rate of flow of energy through a unit area which is normal to the direction of propagation. From basic dynamics, power = force × velocity. For an acoustic process, the instantaneous power is

$$\Pi = \vec{F} \cdot \vec{u}. \tag{2.59}$$

Here, $\vec{F}$ is the vector force acting on a particle and $\vec{u}$ is the associated particle velocity. The power per unit normal area is the instantaneous sound intensity vector, I', where

$$\vec{I}' = p\vec{u}. \tag{2.60}$$

The time average of the instantaneous power flow through a unit area is the mean intensity vector, $\vec{I}$, where

$$\vec{I} = \frac{1}{T}\int_0^T p\vec{u}\,\mathrm{d}t = \tfrac{1}{2}\,\mathrm{Re}\,[\mathbf{p}\vec{\mathbf{u}}^*]. \tag{2.61}$$

The second representation of equation (2.61) is used when the acoustic pressure fluctuations and the particle velocities are treated as complex, harmonic variables.

For a plane wave travelling in the positive x-direction

$$p(\vec{x}, t) = \mathrm{Re}[\mathbf{A}_1\,\mathrm{e}^{\mathrm{i}(\omega t - kx)}] = \hat{p}\cos(\omega t - kx), \tag{2.62}$$

and

$$u(x, t) = \mathrm{Re}\left[\frac{\mathbf{A}_1}{\rho_0 c}\,\mathrm{e}^{\mathrm{i}(\omega t - kx)}\right] = \frac{\hat{p}}{\rho_0 c}\cos(\omega t - kx). \tag{2.63}$$

The mean sound intensity, I, is obtained by substitution into equation (2.61) and evaluating the integral. Hence,

$$I = \frac{\hat{p}^2}{2\rho_0 c} = \frac{p_{\mathrm{rms}}^2}{\rho_0 c}. \tag{2.64}$$

The vector notation is omitted in the above equations for particle velocity and intensity because of the one-dimensional nature of the travelling wave.

The sound energy density is the sound energy per unit volume of space. The energy transported by a sound wave comprises kinetic energy of the moving particles and potential energy of the compressed fluid. Using the plane wave approximation, a simple relationship can be derived for the sound energy density. It will be seen later on that this approximation is valid because most sound waves approximate to one-dimensional plane waves at large distances from the source.

Consider a fluid element with an undisturbed volume V_0. Its kinetic energy per unit volume is

$$\frac{T}{V_0} = \tfrac{1}{2}\rho_0 u^2 = \frac{p^2}{2c^2\rho_0}, \tag{2.65}$$

since $u = p/(\rho_0 c)$. When the volume changes from V_0 to V_1 there is a change in potential energy. It is

$$U = -\int_{V_0}^{V_1} p\,\mathrm{d}V, \tag{2.66}$$

where the negative sign indicates that a positive acoustic fluctuating pressure produces a decrease in the fluid volume. Now, since $\rho = m/V$,

$$\mathrm{d}\rho = -\frac{m}{V^2}\,\mathrm{d}V, \tag{2.67}$$

and combining with equation (2.37) yields

$$\mathrm{d}V = -\frac{V}{\gamma P}\,\mathrm{d}P. \tag{2.68}$$

For small changes in pressure and volume this approximates to

$$\mathrm{d}V = -\frac{V_0}{\gamma P_0}\,\mathrm{d}p. \tag{2.69}$$

Substituting into equation (2.66) and integrating from 0 to p yields the potential energy per unit volume. It is

$$\frac{U}{V_0} = \frac{p^2}{2\gamma P_0} = \frac{p^2}{2c^2\rho_0}, \tag{2.70}$$

since $c^2 = \gamma P_0/\rho_0$.

The total sound energy per unit volume of space is the sum of the kinetic and potential energies per unit volume. Thus, the instantaneous sound energy density, D', is

$$D' = \frac{T}{V_0} + \frac{U}{V_0}, \tag{2.71a}$$

$$= \frac{p^2}{\rho_0 c^2}. \tag{2.71b}$$

The mean energy density, D, is obtained by integrating equation (2.71*b*) with respect to time. It is

$$D = \frac{\hat{p}^2}{2\rho_0 c^2} = \frac{p_{\mathrm{rms}}^2}{\rho_0 c^2}. \tag{2.72}$$

Thus,

$$D = \frac{I}{c}. \tag{2.73}$$

The fluctuating acoustic pressure and the sound intensity decrease with distance from the source (this will become apparent when spherical sound waves are discussed in the next sub-section). It will also be shown that they are a function of the environment – i.e. the reverberant effects of a room enclosure will alter the intensity and sound pressure due to an acoustic source in the room. The sound power of an acoustic source is independent of distance and essentially independent of location – in some instances, the effects of reflecting surfaces have to be accounted for. The sound power, Π, is the integral of the intensity at some point in space over a surface area which is perpendicular to the flow of sound energy. It is

$$\Pi = \int_S \vec{I} \cdot \mathrm{d}\vec{S}. \tag{2.74}$$

2.3 Fundamental acoustic source models

This section is devoted to the description of a range of fundamental acoustic source models. Most noise sources that are of concern to engineers (e.g. vehicles, construction equipment, industrial machinery, appliances, flow-duct systems etc.) can be modelled in terms of simple sources such as spheres, pistons in an infinite baffle, cylinders or combinations thereof. It is therefore instructive to analyse the characteristics of some of these idealised sound sources.

2.3.1 Monopoles – simple spherical sound waves

A monopole is a single, spherical sound source which radiates sound waves that are only a function of the radial distance, r, from the source. The wave equation has therefore got to be set up in spherical co-ordinates before any analysis can proceed. In sub-secton 2.2.6, the wave equation was solved in terms of the pressure variable, and related to the particle velocity and the velocity potential. For a change, the wave equation will be solved here in terms of the velocity potential, and subsequently related to the pressure fluctuations and the particle velocity.

The one-dimensional spherical wave equation (in terms of the acoustic velocity potential) is

$$\frac{\partial^2(r\phi(r,t))}{\partial r^2} - \frac{1}{c^2}\frac{\partial^2(r\phi(r,t))}{\partial t^2} = 0. \tag{2.75}$$

It represents an omni-directional wave which radiates outwards from or inwards towards a source. The product $r\phi(r, t)$ is treated as a single variable and the equation is of the same form as that for a plane wave. The general solution is

$$r\phi(r,t) = G_1(ct - r) + G_2(ct + r), \tag{2.76}$$

thus

$$\phi(r,t) = \frac{G_1(ct - r)}{r} + \frac{G_2(ct + r)}{r}. \tag{2.77}$$

The first term represents a spherical sound wave travelling radially outwards from the sound source, and the second term represents a spherical sound wave travelling towards the sound source.

In engineering noise and vibration analysis, one is generally only concerned with sound waves that travel away from a source. Hence, one is only concerned with the first part of the solution. In principle, the function G_1 can be any arbitrary function. It is, however, convenient and conventional to assume a complex harmonic solution of the form

$$\phi(r,t) = \frac{\mathbf{A}}{r}\, \mathrm{e}^{\mathrm{i}(\omega t - kr)}. \tag{2.78}$$

The term $\mathbf{A}$ is a constant which is determined by the boundary conditions specified at the surface of the monopole.

Consider an oscillating sphere of radius a, with a normal surface velocity $\mathbf{u_a} = U_a\,\mathrm{e}^{\mathrm{i}\omega t}$. From equation (2.43),

$$\frac{\partial\phi}{\partial r} = u_r. \tag{2.79}$$

Thus, by substituting equation (2.78) into equation (2.79) with $r = a$ and solving for the constant $\mathbf{A}$ yields

$$\mathbf{A} = -U_a\left\{\frac{a^2}{1 + ika}\right\}\mathrm{e}^{ika}. \tag{2.80}$$

Thus,

$$\phi(r,t) = -\frac{U_a}{r}\left\{\frac{a^2}{1 + ika}\right\}\mathrm{e}^{\mathrm{i}(\omega t - k(r - a))}. \tag{2.81}$$

The concept of a source strength, $Q(t)$, can now be introduced. The source strength of a monopole is defined as its surface area multiplied by its surface velocity. Thus,

$$\mathbf{Q}(t) = 4\pi a^2 U_a \,\mathrm{e}^{i\omega t} = Q_p \,\mathrm{e}^{i\omega t}. \tag{2.82}$$

Hence, the complex representation of velocity potential is

$$\phi(r, t) = -\frac{\mathbf{Q}(t)}{4\pi r}\left\{\frac{1}{1 + ika}\right\}\mathrm{e}^{-ik(r-a)}. \tag{2.83}$$

The complex representaton of the fluctuating acoustic pressure, $\mathbf{p}(r, t)$ and the particle velocity, $\mathbf{u}(r, t)$, can now be obtained from equation (2.49) (the vector notation is now dropped because of the omni-directional wave propagation) – i.e.

$$p(r, t) = -\rho_0\{\partial\phi/\partial t\}, \qquad \text{and } u(r, t) = \partial\phi/\partial r. \tag{2.84}$$

Thus,

$$\mathbf{p}(r, t) = -\mathrm{i}k\rho_0 c\phi(r, t) = \frac{\mathbf{Q}(t)}{4\pi r}\left\{\frac{ik\rho_0 c}{1 + ika}\right\}\mathrm{e}^{-ik(r-a)}, \tag{2.85}$$

and

$$\mathbf{u}(r, t) = -\left\{\frac{1 + ikr}{r}\right\}\phi(r, t) = \frac{\mathbf{Q}(t)}{4\pi \mathrm{r}^2}\left\{\frac{1 + ikr}{1 + ika}\right\}\mathrm{e}^{-ik(r-a)}. \tag{2.86}$$

The specific acoustic impedance, $\mathbf{Z_a}$, can now be evaluated using equation (2.57). It is

$$\mathbf{Z_a} = \frac{\mathrm{i}\rho_0 ckr}{1 + ikr} = \rho_0 c\left\{\frac{k^2r^2}{1 + k^2r^2} + \mathrm{i}\frac{kr}{1 + k^2r^2}\right\}. \tag{2.87}$$

Unlike the specific acoustic impedance of a plane wave, the specific acoustic impedance of a spherical wave has both a resistive and a reactive component. When the resistive component dominates, the acoustic pressure fluctuations are in phase with the particle velocities; when the reactive component dominates, they are out of phase with each other. As with energy and power flow in structures (see section 1.7 in chapter 1) one would expect that the in-phase components of the sound waves dominate the radiated sound power and that the out-of-phase components produce some near-field reactive exchange of sound energy. These phase relationships are therefore the basis of the far-field/near-field concepts which are so commonly used in noise control engineering.

The sound intensity of a spherical sound wave can now be estimated. The mean radiated intensity is a real quantity and it can be evaluated from equation (2.61). It is

$$I(r) = \frac{Q_\mathrm{p}^2 k^2 \rho_0 c}{32\pi^2 r^2 (1 + k^2a^2)} = \frac{Q_\mathrm{rms}^2 k^2 \rho_0 c}{16\pi^2 r^2 (1 + k^2a^2)}. \tag{2.88}$$

The sound power radiated by the spherical source can now be evaluated from equation (2.74). It is

$$\Pi = 4\pi r^2 I(r) = \frac{Q_{rms}^2 k^2 \rho_0 c}{4\pi(1 + k^2 a^2)}. \tag{2.89}$$

The radiated mean-square acoustic pressure fluctuations can be readily evaluated from equations (2.61) and (2.85). It is

$$p_{rms}^2 = \tfrac{1}{2}\,\mathrm{Re}\,[\mathbf{p}(r, t)\mathbf{p}^*(r, t)] = \frac{Q_{rms}^2 k^2 (\rho_0 c)^2}{16\pi^2 r^2 (1 + k^2 a^2)} = I(r)\rho_0 c. \tag{2.90}$$

Thus the relationship between the radiated sound pressure and the mean sound intensity is the same as that for a plane wave (see equation 2.64). It is also important to note that it is proportional to the inverse square of the distance from the source.

Returning to equation (2.87), one can see that there are two limiting situations. They are (i) $kr \ll 1$, and (ii) $kr \gg 1$. When $kr \ll 1$,

$$\mathbf{u}(r, t) = \frac{\mathbf{Q}(t)}{4\pi r^2 (1 + ika)} \mathrm{e}^{-ik(r-a)}, \tag{2.91}$$

and

$$\mathbf{Z_a} = ikr\rho_0 c. \tag{2.92}$$

Here, the acoustic pressure fluctuations and the particle velocities are out of phase, the impedance is reactive and no sound power is radiated from the source. When $kr \gg 1$,

$$\mathbf{u}(r, t) = \frac{ik\mathbf{Q}(t)}{4\pi r(1 + ika)} \mathrm{e}^{-ik(r-a)}, \tag{2.93}$$

and

$$\mathbf{Z_a} = \rho_0 c. \tag{2.94}$$

Here, the acoustic pressure fluctuations and the particle velocities are in phase, the impedance is resistive (i.e. it is equal to the characteristic acoustic impedance) and sound power is radiated from the source. Combining equations (2.89) and (2.90), the radiated sound power can also be obtained in terms of the acoustic pressure fluctuations. It is

$$\Pi = \frac{4\pi r^2 p_{rms}^2}{\rho_0 c}. \tag{2.95}$$

Equations (2.89) and (2.95) are identical. The former expresses the radiated sound power of a monopole in terms of its source strength, wavenumber, and dimensions, whilst the latter expresses it in terms of the radiated sound pressure at some distance, r, from the source. The former expression is very useful for obtaining a physical understanding of the effects of the source parameters (source strength, wavenumber, and dimensions) on the radiated noise, whilst the latter allows for an estimate of the sound power from a simple measurement of the mean-square pressure fluctuations. A word of caution should be made regarding equation (2.95). It is only valid in the far-field – i.e. at sufficient distances from the source such that the pressure fluctuations and the particle velocities are in phase. It is also only valid in regions of free space where there are no environmental effects such as reverberation (reflected sound) on the measured levels.

It is now instructive to return once again to equation (2.87) to attempt to define a transition point between the two regions ($kr \ll 1$, and $kr \gg 1$). The resistive and reactive components of the specific acoustic impedance can be plotted as a function of kr. This information is presented in Figure 2.7.

The main characteristic of the far-field of a radiating sound source is that $kr \gg 1$ and that $\mathbf{Z_a}$ is essentially resistive. From Figure 2.7, the resistive component dominates for $kr \approx 10$. Now, λ is the wavelength of the sound source ($\lambda = 2\pi/k$), thus the transition radius is

$$r \approx 1.6\lambda. \tag{2.96}$$

It is very important to note that this transition point is not a constant but is a function of frequency – i.e. at higher frequencies the transition radius is shorter than at lower frequencies.

The concepts presented in this sub-section form an important basis for the modelling of sound sources. Many practical sound sources, including both vibrating bodies and

Fig. 2.7. The resistive and reactive components of the specific acoustic impedance of a spherical sound source.

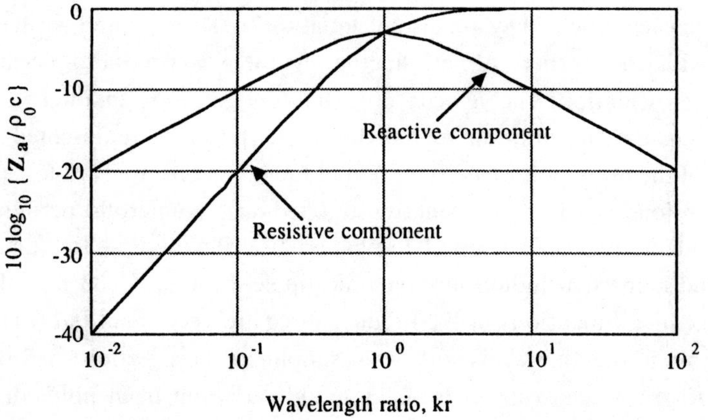

aerodynamic sources, can be approximated as monopoles. If, for instance, the monopole source was of an aerodynamic nature rather than some vibrating body with a surface velocity, the radiated sound power could then be related to the mean flow velocity – i.e. the mean addition or subtraction of mass from a source region. To do this, one has to re-consider equation (2.89). The source strength, Q_{rms}, scales as L^2U, where L is a typical dimension of the region of fluid flow, such as the width of a square duct, and U is the mean flow velocity. The characteristic frequency in the flow scales as U/L, and this has to equal the frequency of aerodynamic sound generation for dimensional consistency. Also, the source dimension, a, is assumed to be very much smaller than the geometrical dimension, L, i.e. $a \ll L$. Thus,

$$\Pi \approx \frac{L^4U^2\rho_0 cU^2}{4\pi L^2c^2} \approx \frac{L^2U^4\rho_0}{4\pi c}. \tag{2.97}$$

Equation (2.97) suggests that the sound power radiated by an aerodynamically generated monopole scales with the fourth power of the flow velocity. This is a very important statement – one which has significant practical consequences. These practical consequences will be discussed later on in this chapter once some other sound source types have been studied, so that comparisons can be made. It will also be seen later on in this chapter that this scaling law only holds for sub-sonic flows – i.e. where the mean flow velocity is less than the speed of sound, c.

Some typical examples of monopole sound sources associated with vibrating structures are small electric motors, pumps and certain types of traffic noise (at medium distances away from a single vehicle – at larger distances the source can sometimes appear to be a dipole). Aerodynamic monopole sound sources include unsteady combustion from a furnace, sirens, pulsed jets and cavitation.

2.3.2 *Dipoles*

A dipole is a sound source model that is composed of two monopoles in close proximity to each other. They are also of equal source strength and oscillate 180° out of phase with each other. A nett fluctuating force is produced because of this out-of-phase oscillation. The velocity potential function for a dipole is a function of the polar angle, θ, in addition to the radius, r. It can be represented as a space derivative of the monopole velocity potential function. The analysis for dipoles with finite dimensions is somewhat lengthy and tedious. Numerous permutations are possible and the reader is referred to Pierce[2.3] or Reynolds[2.4] for further details. Many sound sources, including aerodynamic dipole sources, can be modelled as point dipoles – i.e. the dimensions of the dipole source are very small ($ka \ll 1$), and, with this assumption, the analysis is somewhat simplified. This section shall therefore be restricted to the special case of far-field sound radiation from point dipole sound sources. The analytical techniques involved here can be readily extended to the cases of both near-field and far-field sound radiation from finite dipole sources.

Consider two point monopoles in free space, separated by some small distance $2d$ as illustrated in Figure 2.8. The total velocity potential at some point, X, in space is the sum of the two separate velocity potentials, i.e.

$$\phi(r, t) = \phi_1(r, t) + \phi_2(r, t). \tag{2.98}$$

Fig. 2.8. Point dipole model – two point monopoles in free space separated by some small distance 2*d*.

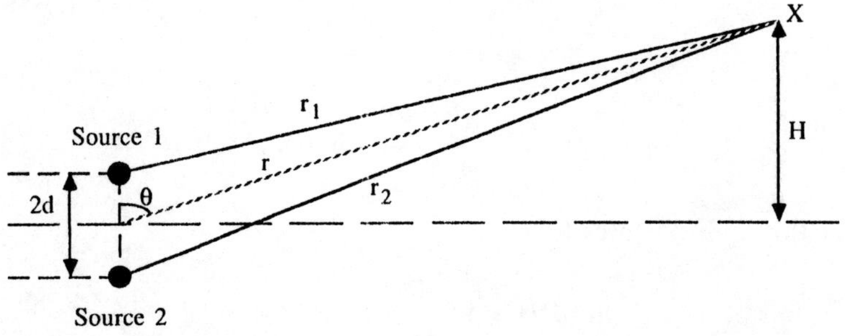

The separate distances from each of the point sources to the point, X, is given by

$$r_1^2 = (r^2 + d^2 - 2rd \cos \theta), \qquad \text{and } r_2^2 = (r^2 + d^2 + 2rd \cos \theta). \tag{2.99}$$

In the far field, $r \gg d$, thus the above geometrical relationships can be simplified to

$$r_1 \approx r - d \cos \theta, \qquad \text{and } r_2 \approx r + d \cos \theta. \tag{2.100}$$

The velocity potential for a point monopole can be obtained from equation (2.83) by letting $ka \to 0$. It is

$$\phi(r, t) = -\frac{\mathbf{Q}(t)}{4\pi r} \mathrm{e}^{-ikr}. \tag{2.101}$$

The combined velocity potential at X due to both point monopoles is thus

$$\phi(r, t) = -\frac{\mathbf{Q}_1(t)}{4\pi r_1} \mathrm{e}^{-ikr_1} - \frac{\mathbf{Q}_2(t)}{4\pi r_2} \mathrm{e}^{-ikr_2}. \tag{2.102}$$

In the far-field, $r \approx r_1 \approx r_2$ and the denominators in the above equation can be replaced by r. The numerators, however, should not be replaced because the phase difference between the two oscillating monopoles has to be accounted for. By sùbstituting equation (2.100) into equation (2.102), the combined velocity potential approximates to

$$\phi(r, \theta, t) \approx -\frac{\mathrm{e}^{-ikr}}{4\pi r} \{\mathbf{Q}_1(t)\, \mathrm{e}^{ikd\cos\theta} + \mathbf{Q}_2(t)\, \mathrm{e}^{-ikd\cos\theta}\}. \tag{2.103}$$

For a point dipole, the two monopole sources are of equal source strength but are out of phase by 180°. Thus

$$\mathbf{Q}_1(t) = Q_p\, e^{i(\omega t + \pi/2)}, \tag{2.104}$$

and

$$\mathbf{Q}_2(t) = Q_p\, e^{i(\omega t - \pi/2)}. \tag{2.105}$$

Hence,

$$\phi(r, \theta, t) = -\frac{Q_p\, e^{i(\omega t - kr)}}{4\pi r}\{e^{i(kd\cos\theta + \pi/2)} + e^{-i(kd\cos\theta + \pi/2)}\}, \tag{2.106}$$

or

$$\phi(r, \theta, t) = \frac{Q_p\, e^{i(\omega t - kr)}}{4\pi r} 2\sin(kd\cos\theta). \tag{2.107}$$

For $kd \ll 1$ this approximates to

$$\phi(r, \theta, t) = \frac{Q_p\, e^{i(\omega t - kr)}}{4\pi r} 2kd\cos\theta. \tag{2.108}$$

The sound intensity of a dipole can now be estimated either by following the same procedures adopted in the previous sub-section on monopoles or by recognising that the far-field dipole velocity potential is equivalent to the monopole velocity potential with an additional term which is a function of the polar angle θ – i.e. a directivity factor $(-2kd\cos\theta)$. It is

$$I(r, \theta) = \frac{Q_p^2 k^2 \rho_0 c}{32\pi^2 r^2}(2kd\cos\theta)^2. \tag{2.109}$$

Thus,

$$I(r, \theta) = \frac{Q_p^2 k^4 d^2 \rho_0 c}{8\pi^2 r^2}\cos^2\theta = \frac{Q_{rms}^2 k^4 d^2 \rho_0 c}{4\pi^2 r^2}\cos^2\theta. \tag{2.110}$$

It is important to point out at this stage that the dipole velocity potential described by equation (2.108) is a far-field approximation. In reality a dipole velocity potential has two terms – one which is associated with the near-field components, and one which is associated with the far-field components. The complete dipole velocity potential can be described in terms of a dipole source strength, $\mathbf{Q}_d$, (see Reynolds[2.4], or Dowling and Ffowcs Williams[2.5]) with units of $m^4\, s^{-1}$, where

$$\phi(r, \theta, t) = \frac{\cos\theta}{4\pi}\left\{\frac{1}{cr}\frac{\partial \mathbf{Q}_d(t - r/c)}{\partial t} + \frac{\mathbf{Q}_d(t - r/c)}{r^2}\right\}. \tag{2.111}$$

The radiated sound field has a cos θ dependence – i.e. at 90° to the dipole axis the sound fields cancel each other. Also, the near-field component varies with the inverse square of distance whilst the far-field component varies with the inverse of distance. This point will be discussed again later on in this chapter in relation to flow noise.

The far-field sound power radiated by the dipole can be evaluated by integrating equation (2.110) over an arbitrary spherical surface at a given radius, r. The polar angular dependence has to be accounted for in this integration. The sound power radiated is

$$\Pi = \frac{Q_{\mathrm{rms}}^2 k^4 d^2 \rho_0 c}{3\pi}. \tag{2.112}$$

A comparison can be made between the sound radiating efficiencies of monopoles and dipoles by comparing equations (2.89) and (2.112). The ratio of the sound power radiated by a dipole to that radiated by a monopole is

$$\frac{\Pi_{\mathrm{D}}}{\Pi_{\mathrm{M}}} = \frac{4k^2 d^2}{3} \sim \left\{\frac{d}{\lambda}\right\}^2. \tag{2.113}$$

Π_{D} is the sound power radiated by the dipole, Π_{M} is the sound power radiated by the monopole and $k = 2\pi/\lambda$, where λ is the wavelength. It can be seen from the equation that at low freqencies (long wavelengths), the dipole is a very much less efficient radiator of noise than at high frequencies (shorter wavelengths) when compared with a monopole of the same source strength.

A dipole is produced by the fluctuating pressure forces acting along the axis between the two sources. These pressure forces are generated by the out-of-phase oscillations of the two monopoles in close proximity to each other. It is therefore useful to describe the sound power radiated by a dipole in terms of the fluctuating force (along the dipole axis) which is a function of the source strength and the source separation. The derivation of the radiated sound power in terms of the fluctuating force is available in Reynolds[2.4]. It is estimated by evaluating the r.m.s. fluctuating force, F_{rms}, on a spherical surface containing the dipole, and is

$$\Pi = \frac{3F_{\mathrm{rms}}^2 k^2}{4\pi\rho_0 c}. \tag{2.114}$$

The sound power radiated by aerodynamic dipole sources can be estimated from either equation (2.112) or (2.114) by using the same procedures developed in the previous sub-section for aerodynamic monopoles. The source strength, Q_{rms}, scales as $(2d)^2 U$, where $2d$ is the separation between the two monopole sources in the region of fluid flow, and U is the mean flow velocity. The fluctuating force, F_{rms}, scales as $\rho_0 U^2 d^2$. The characteristic frequency in the flow scales as $U/(2d)$, and this has to equal the frequency of aerodynamic sound generation for dimensional consistency.

Thus,

$$\Pi = \frac{Q_{\text{rms}}^2 \omega^4 d^2 \rho_0 c}{3\pi c^4} \approx \frac{(2d)^4 U^2 U^4 d^2 \rho_0 c}{(2d)^4 3\pi c^4} \approx \frac{\rho_0 d^2 U^6}{3\pi c^3}. \tag{2.115}$$

Equation (2.115) suggests that the sound power radiated by an aerodynamically generated dipole scales with the sixth power of the flow velocity (a monopole scales with the fourth power of the flow velocity).

The directivity patterns of monopoles and dipoles in free space are schematically illustrated in Figure 2.9. Whilst the monopoles are omni-directional, the dipoles have a typical 'figure 8' sound radiation pattern and there is no sound radiation perpendicular to the dipole axis.

2.3.3 Monopoles near a rigid, reflecting, ground plane

Quite often in practice, most industrial noise sources are mounted on a ground plane or in close proximity to it. In the far-field, they can often be approximated as single point sources. The effects of the ground plane have to be accounted for though. These effects are particularly pronounced when the sound source is less than one acoustic wavelength (λ) from the ground plane. The analysis which follows is not to be confused with the concept of directivity which the reader might be familiar with. The directional effects of floors, intersecting walls, corners etc. on the sound radiation characteristics of an omni-directional noise source are well known and documented – they are reviewed in chapter 4 in this book. This section relates to a point which is often omitted in the literature on noise control engineering – that the sound power of a source can be affected by rigid, reflecting planes. This effect is in addition to the conventional and well known directivity effects.

Fig. 2.9. Directivity patterns of monopoles and dipoles in free space.

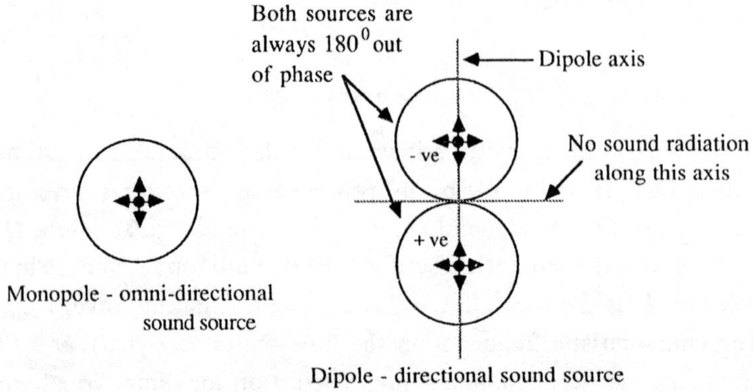

Consider the case of a monopole near a rigid, reflecting, ground plane as illustrated in Figure 2.10. At some point in the far-field, the sound pressure will be the sum of two sound waves – i.e. a direct and a reflected wave. The reflected wave can be modelled by an image monopole below the reflecting surface. The problem thus reduces to that of two interfering monopoles. In practice, the ground plane will have some finite reflection coefficient (not all the sound will be necessarily reflected) and there will be some finite phase difference between the two waves. If one assumes that the ground plane is a hard reflecting surface, as an upper limit, then the reflection coefficient is unity and the phase difference between the two waves is zero. The problem thus reduces to two in-phase monopoles of equal source strength (a dipole was modelled in the previous sub-section as two out-of-phase monopoles of equal source strength).

The combined velocity potential at the observer position (some point, X, in space) can be obtained from equation (2.103) with $\mathbf{Q}_1(t)$ and $\mathbf{Q}_2(t)$ being of equal strength and phase. It is

$$\phi(r,\theta,t) = -\frac{Q_p \, e^{i(\omega t - kr)}}{4\pi r} 2\cos(kd\cos\theta). \tag{2.116}$$

When $d \ll \lambda$, $kd \ll 1$ and the above equation simplifies to

$$\phi(r,t) = -\frac{2Q_p \, e^{i(\omega t - kr)}}{4\pi r}. \tag{2.117}$$

Equation (2.117) is simply double the far-field velocity potential for a monopole sound source (see equation 2.83)! The hard, reflecting ground plane has resulted in a doubling of the velocity potential. This result is very important. It is often overlooked by noise control engineers when estimating noise radiation levels from sound sources.

Fig. 2.10. Monopole near a rigid, reflecting ground plane.

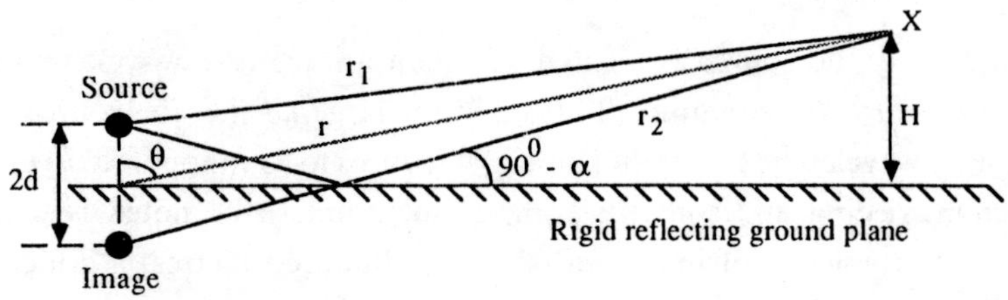

$$\Gamma = \frac{Z_s \cos\alpha - \rho_0 c}{Z_s \cos\alpha + \rho_0 c}$$

Γ - reflection coefficient
Z_s - specific acoustic impedance of reflecting surface
α - oblique angle of incidence
$\rho_0 c$ - characteristic acoustic impedance

This doubling of the velocity potential produces a fourfold increase in the sound intensity (equations 2.61, 2.85 and 2.86 with the appropriate velocity potential). In turn, there is a twofold increase in the radiated sound power because the intensity has only got to be integrated over half space (the other half is baffled by the rigid ground plane). The sound intensity is

$$I(r) = \frac{Q_{\text{rms}}^2 k^2 \rho_0 c}{4\pi^2 r^2}, \tag{2.118}$$

and the radiated sound power is

$$\Pi = 2\pi r^2 I(r) = \frac{Q_{\text{rms}}^2 k^2 \rho_0 c}{2\pi}. \tag{2.119}$$

The interesting result to come out of this approximate (i.e. limit) analysis is that the sound power of the monopole has been doubled. This is essentially because whilst the strength, Q_{rms}, and the surface vibrational velocity of the source have not changed (from when it is radiating into free space), the reflecting plane has produced a velocity potential, or a pressure, doubling. So, instead of having a constant sound power, the source has a constant volume velocity. These concepts of constant volume velocity sources, as opposed to the more commonly referred to constant power sources, will be discussed from an engineering noise control point of view in chapter 4. They can be regarded as an upper limit – in practice the effects on non-perfect reflection from the ground plane will reduce the effect of the image source.

Reynolds[2.4] performs a more rigorous analysis for the effects of a reflecting plane on a monopole and integrates the complete expressions for sound intensity to obtain an expression for the radiated sound power. For the case of perfect reflection and zero phase difference, it is

$$\Pi = \frac{Q_{\text{rms}}^2 k^2 \rho_0 c}{4\pi}\left\{1 + \frac{\sin 2kd}{2kd}\right\}. \tag{2.120}$$

For small kd (i.e. the source is located less than an acoustic wavelength from the surface) this reduces to equation (2.119) and for large kd (i.e. the source is located many acoustic wavelengths from the surface) it reduces to a monopole in free space.

Another interesting and somewhat important point to be noted is that at very large distances (typically of the order of several hundred metres) a critical oblique angle of incidence is reached beyond which the reflection coefficient changes sign (Bies[2.6]). When this happens, the source and its image are out of phase rather than in phase – i.e. the source is now a dipole. This has important consequences for the far-field noise radiation. Referring back to equation (2.110) for a dipole far-field sound intensity, the angle $\cos\theta$ can be approximated by H/r, where H is the distance from the measurement point to the ground plane (see Figure 2.10). Thus the intensity at the point of interest in the far-field is

$$I(r) = \frac{Q_{\text{rms}}^2 H^2 k^4 d^2 \rho_0 c}{4\pi^2 r^4}. \tag{2.121}$$

The important practical point to come out of this is that the intensity of the single sound source now scales as r^{-4} rather than r^{-2}. It will be seen in chapter 4 (where decibels and other noise measurement units are defined) that this produces a 12 dB decay of sound pressure level per doubling of distance rather than the usual 6 dB per doubling of distance for a point source. Bies[2.6] discusses this phenomenon in some detail.

2.3.4 Sound radiation from a vibrating piston mounted in a rigid baffle

The sound radiation from a vibrating piston mounted in an infinite baffle is a classical problem – one which is covered in numerous books on fundamental acoustics (e.g. Kinsler *et al.*[2.1], Reynolds[2.4], Ford[2.7]). The vibrating piston can be either a vibrating surface or a vibrating layer of air. The primary assumption in the analysis (one which is not strictly correct in practice for 'real' surfaces) is that all parts of the piston vibrate in phase and with the same amplitude. Its relevance to engineering noise control is that it serves as an introduction to the sound radiation from different types of surfaces, e.g. loudspeakers, open ends of flanged pipes, plates and shells etc.

Consider a flat, circular piston of radius z which is mounted in an infinite, rigid baffle as illustrated in Figure 2.11. The noise radiated by the vibrating piston can be modelled in terms of numerous point monopoles (monopoles where $ka \ll 1$) radiating together. Each of the monopoles is, however, radiating from a rigid, reflecting, ground plane and not from free space. The sound pressure due to any one of the baffled

Fig. 2.11. Piston mounted in a rigid baffle.

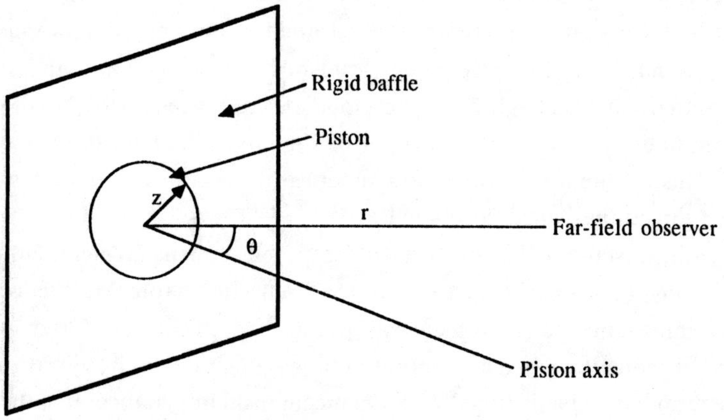

monopoles is therefore twice that of an equivalent monopole in free space. It is

$$\mathbf{p}(r, t) = \frac{ik\rho_0 c}{2\pi r} Q_p \, e^{i(\omega t - kr)}. \tag{2.122}$$

In this equation, Q_p represents the source strength of the elemental monopole on the piston surface and it is equal to $U_p\,\delta S$, where U_p is the peak surface velocity of the monopole and δS is the elemental surface area. The total acoustic pressure fluctuations due to the vibrating piston is simply the resultant pressure due to all the point monopoles vibrating in phase and it is obtained by integration over the whole surface area. It is [2.1,2.4,2.7]

$$\mathbf{p}(r, \theta, t) = \frac{ik\rho_0 c\pi z^2 U_p \, e^{i(\omega t - kr)}}{2\pi r} \left\{ \frac{2J_1(kz \sin\theta)}{kz \sin\theta} \right\}. \tag{2.123}$$

In the above equation, $U_p\, e^{i\omega t}$ is the surface velocity of the piston (i.e. each of the monopoles has the same surface velocity and phase). The radiated sound pressure has a similar form to that of a monopole in a reflecting ground plane with the exception of the term in brackets which is a directivity factor. J_1 is the first order Bessel function and it can be readily evaluated from tables.

The corresponding sound intensity in the far-field is[2.4]

$$I(r, \theta) = \frac{\rho_0 c k^2 U_{rms}^2 \pi^2 z^2}{4\pi^2 r^2} \left\{ \frac{2J_1(kz \sin\theta)}{kz \sin\theta} \right\}^2. \tag{2.124}$$

Once again, the sound intensity has a similar form to that of a monopole in a reflecting ground plane with the exception of the term in brackets which is a directivity factor. The form of the directivity factor is presented in Figure 2.12. It is quite clear that the sound radiation from the piston is quite directional (except along the axis) and that it increases with frequency. There are several pressure nodes, and this results in a beam pattern of sound radiation. This is schematically illustrated in Figure 2.13. At low frequencies ($kz \ll 1$) the intensity distribution is approximately constant, whereas at high frequencies there are several nodal points and corresponding lobes of radiated sound. Hence, low frequency loudspeakers can be large and still remain omni-directional whereas high frequency loudspeakers need to be small and to be relatively omni-directional. The concepts of directivity will be discussed again in chapter 4. This example illustrates how a series of omni-directional sound sources can become directional when combined.

The preceding discussion has been restricted to the acoustic far-field. Now consider an observation point on the piston surface itself. In chapter 1, the concepts of mechanical impedance (force/velocity at point of application of the force) were introduced. In acoustics, when a structure radiates sound due to its vibration, another impedance term has to be included with the mechanical impedance. It is the radiation impedance of the fluid (air) in proximity to the vibrating surface – i.e. the fluid loads the vibrating surface and this alters its vibrational response. The total sound pressure at any arbitrary element on the piston surface is a sum of the pressure due to the vibrating element itself and the radiated pressures from all the other elements on the piston. The piston velocity, $\mathbf{U} = U_p e^{i\omega t}$ is thus given by

$$\mathbf{U} = \frac{\mathbf{F}_m}{\mathbf{Z}_m + \mathbf{Z}_r}, \tag{2.125}$$

where $\mathbf{F}_m$ is the applied mechanical force, $\mathbf{Z}_m$ is the mechanical impedance of the piston, and $\mathbf{Z}_r$ is its radiation impedance. $\mathbf{F}_m$ is not to be confused with the force on the piston due to the acoustic pressure fluctuations, $\mathbf{F}_p$. The mechanical impedance is associated with the mechanical driving force and the radiation impedance is associated with the acoustic driving force. It can be seen from equation (2.125) that the radiation impedance 'fluid-loads' the surface vibrations of the piston. Fluid loading concepts are very relevant when analysing the vibrational characteristics of structures

Fig. 2.12. Functional form of the directivity factor for a circular piston in a rigid baffle.

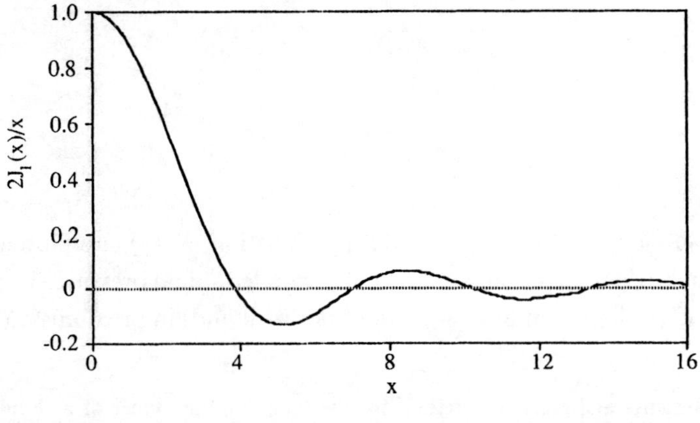

Fig. 2.13. Low and high frequency sound radiation patterns for a circular piston in a rigid baffle.

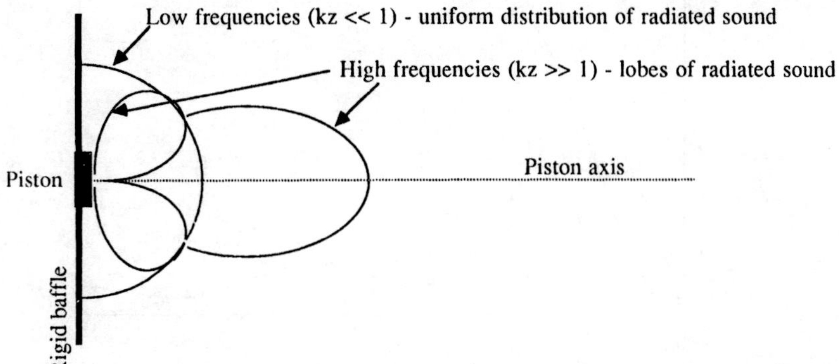

immersed in fluids – e.g. piping systems filled with liquids, submarines etc. The radiation impedance is thus given by

$$\mathbf{Z_r} = \frac{\mathbf{F_p}}{\mathbf{U}}. \tag{2.126}$$

The radiation impedance of a vibrating surface is sometimes defined as the ratio of the sound pressure averaged over the surface to the volume velocity through it (units of $\mathrm{N\,s\,m^{-5}}$ as opposed to $\mathrm{N\,s\,m^{-1}}$ for mechanical impedance) – i.e. some books would define $\mathbf{Z_r}$ as $(F_p/\pi z^2)/\pi z^2 U$ or $F_p/(\pi z^2)^2 U$. In this book, $\mathbf{Z_r}$ is defined, for convenience, in similar units to the mechanical impedance since the radiating surface area, πz^2 is common to both variables (pressure and volume velocity).

The radiation impedance of a piston can be obtained by integrating the elemental pressure distribution over the surface area of the piston to obtain the total sound pressure at a point and subsequently integrating this again over the surface to obtain the force, $\mathbf{F_p}$. The radiation impedance is thus obtained and given by[2.1,2.4]

$$\mathbf{Z_r} = \rho_0 c \pi z^2 \{R_1(2kz) + iX_1(2kz)\}, \tag{2.127}$$

where

$$R_1(x) = \frac{x^2}{2\cdot 4} - \frac{x^4}{2\cdot 4^2\cdot 6} + \frac{x^6}{2\cdot 4^2\cdot 6^2\cdot 8} - \cdots, \tag{2.128}$$

and

$$X_1(x) = \frac{4}{\pi}\left\{\frac{x}{3} - \frac{x^3}{3^2\cdot 5} + \frac{x^5}{3^2\cdot 5^2\cdot 7} - \cdots\right\}, \tag{2.129}$$

The resistive function, $R_1(x)$, and the reactive function, $X_1(x)$, are plotted in Figure 2.14. The resistive part is real and is due to the radiated sound pressure. The imaginary part is a mass loading term due to the fluid (air or liquid) in proximity to the piston.

Fig. 2.14. Resistive and reactive functions for the radiation impedance of a circular piston.

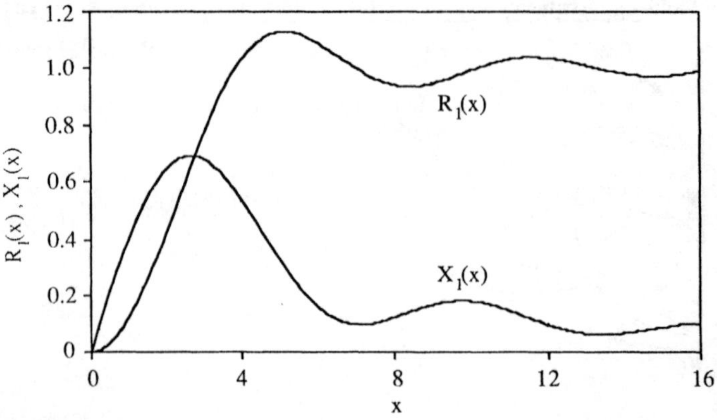

This mass loading term can become significant when the structure is radiating into liquids.

The mechanical properties of the piston also affect its sound radiation properties. The equation of motion of the piston is (from Newton's second law)

$$M\frac{\partial \mathbf{U}}{\partial t}+C_{\mathrm{v}}\mathbf{U}+K_{\mathrm{s}}\int \mathbf{U}\,\mathrm{d}t=\mathbf{F}_{\mathrm{m}}-\mathbf{F}_{\mathrm{p}},\tag{2.130}$$

where M is the piston mass, C_v is its damping, and K_s is its stiffness. Thus equation (2.125) can be re-written as

$$\mathbf{U}=\frac{\mathbf{F}_{\mathrm{m}}}{C_{\mathrm{v}}+\mathrm{i}(M\omega-K_{\mathrm{s}}/\omega)+\rho_0 c\pi z^2\{R_1(2kz)+\mathrm{i}X_1(2kz)\}}.\tag{2.131}$$

Equation (2.131) clearly illustrates how the vibrational velocity of the piston is a function of (i) its structural damping, (ii) its mass, (iii) its stiffness, (iv) the acoustic radiation resistance, and (v) the acoustic radiation reactance. As mentioned earlier, the acoustic radiation resistance is due to the radiated sound pressure and the acoustic radiation reactance is due to mass loading of the piston by the fluid. For low frequencies, $2kz \ll 1$, thus

$$R_1(2kz)\approx\frac{k^2z^2}{2},\tag{2.132}$$

and

$$X_1(2kz)\approx\frac{8kz}{3\pi}.\tag{2.133}$$

For high frequencies, $2kz \gg 1$, thus

$$R_1(2kz)\approx 1,\tag{2.134}$$

and

$$X_1(2kz)\approx\frac{2}{\pi kz}.\tag{2.135}$$

The effects of mass loading can now be estimated. The amplitude of the fluid loaded mass is

$$M_{\mathrm{r}}=\frac{X_{\mathrm{r}}}{\omega}=\frac{\rho_0 c\pi z^2 X_1(2kz)}{\omega}=\frac{\rho_0\pi z^2 X_1(2kz)}{k}.\tag{2.136}$$

Thus at low frequencies ($2kz \ll 1$)

$$M_{\mathrm{r}}=\frac{8\rho_0 z^3}{3},\tag{2.137}$$

and at high frequencies ($2kz \gg 1$)

$$M_r = \frac{2\rho_0 z}{k^2}. \tag{2.138}$$

The sound power radiated by the piston can now be estimated from the real part of the radiation impedance – i.e. the acoustic radiation resistance. It could also be obtained by integrating the far-field sound intensity (equation 2.124). From equation (1.162) the real power (rate of energy flow) of the piston is

$$\Pi = \tfrac{1}{2}U_p^2 \,\mathrm{Re}\,[\mathbf{Z}_m + \mathbf{Z}_r]. \tag{2.139}$$

Thus,

$$\Pi = \tfrac{1}{2}U_p^2\{C_v + \rho_0 c\pi z^2 R_1(2kz)\}, \tag{2.140}$$

where the first term inside the brackets represents the mechanical power that is dissipated and the second term represents the sound power that is radiated into the surrounding medium. Low and high frequency estimates of the radiated sound power can be readily obtained by substituting equations (2.132) and (2.134) into equation (2.140).

2.3.5 Quadrupoles – lateral and longitudinal

Monopoles were modelled in sub-section 2.3.1 as single oscillating spheres, and dipoles were modelled in sub-section 2.3.2 as two equal spheres oscillating out of phase. A natural extension to these acoustic source models is two dipole sources in close proximity to each other and oscillating 180° out of phase with each other. Such a sound source is called a quadrupole. Whereas a dipole has one axis (i.e. the fluctuating pressure forces act along the axis between the two sources), a quadrupole has two. The two dipoles oscillating out of phase with each other results in no nett addition or subtraction of mass away from the source, and no resultant force – i.e there is no physical mechanism available for the mass or the momentum to vary. The quadrupole does, however, apply a stress to the medium and it is this fluctuating stress that generates the sound (monopoles generate sound via fluctuating surface velocities or the addition/subtraction of mass from a source region, and dipoles generate sound via fluctuating forces). In gas flows, for instance, quadrupoles are generated by the viscous stresses within the gas.

The relationships for the intensity and sound power radiated by a dipole were obtained by considering the interference between two point monopoles in close proximity to each other. The relationships for the intensity and sound power radiated by a quadrupole can be obtained in a similar manner by considering the interference between two point dipoles in close proximity to each other or by expressing the quadrupole velocity potential as a space derivative of the monopole velocity potential.

The mathematical manipulations required to obtain the answers are fairly extensive and are not presented here. The interested reader is referred to Pierce[2.3] or Reynolds[2.4]. Two possible combinations of quadrupoles exist. The first is when the two dipole axes do not lie on the same line, and the second is when they both lie on the same line. The former is termed a lateral quadrupole and the latter is termed a longitudinal quadrupole. The two types of quadrupoles and their associated directivity patterns are illustrated in Figure 2.15.

The sound power radiated by a lateral quadrupole is[2.3,2.4]

$$\Pi = \frac{4Q_{\rm rms}^2 \rho_0 c d^4 k^6}{15\pi}, \tag{2.141}$$

and that radiated by a longitudinal quadrupole is

$$\Pi = \frac{4Q_{\rm rms}^2 \rho_0 c d^4 k^6}{5\pi}. \tag{2.142}$$

Fig. 2.15. Lateral and longitudinal quadrupoles and their associated directivity patterns.

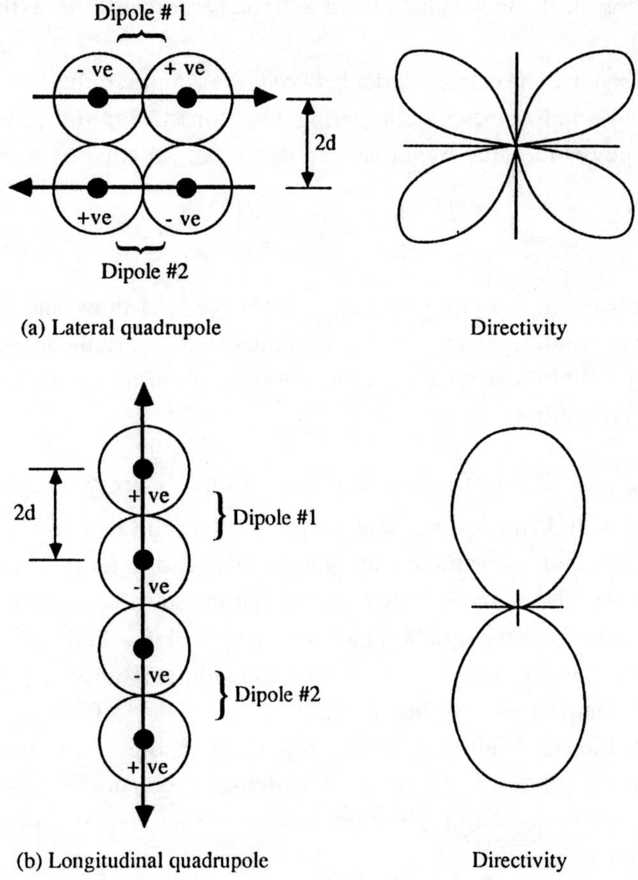

In the above equation, Q_{rms} is the source strength of one of the four monopoles that makes up the quadrupole, and $2d$ is the separation distance between any two monopoles.

The sound power radiated by an aerodynamic quadrupole can be estimated by using the same procedures that were used for monopoles and dipoles. The source strength, Q_{rms}, scales as $(2d)^2U$, where U is the mean flow velocity, and the characteristic frequency in the flow scales as $U/(2d)$. Thus, for a lateral quadrupole,

$$\Pi = \frac{4Q_{rms}^2\rho_0 c d^4\omega^6}{15\pi c^6} \approx \frac{\rho_0 d^2 U^8}{15\pi c^5}, \tag{2.143}$$

and for a longitudinal quadrupole

$$\Pi = \frac{4Q_{rms}^2\rho_0 c d^4\omega^6}{5\pi c^6} \approx \frac{\rho_0 d^2 U^8}{5\pi c^5}. \tag{2.144}$$

The equations suggest that the sound power generated by both lateral and longitudinal quadrupoles scale with the eighth power of the flow velocity (a monopole scales with the fourth power of the flow velocity and a dipole scales with the sixth power of the flow velocity).

A comparison can now be made between the sound radiating efficiencies of monopoles and quadrupoles by comparing equations (2.89) and (2.141). The ratio of the sound power radiated by a quadrupole to that radiated by a monopole is

$$\frac{\Pi_Q}{\Pi_M} \sim d^4k^4 \sim \left\{\frac{d}{\lambda}\right\}^4. \tag{2.145}$$

Comparisons between this equation and equation (2.113) show that monopoles are the most efficient radiators of sound whereas quadrupoles are the least efficient. This important point is also deduced from the scaling relationships with flow velocity for aerodynamic type sources.

2.3.6 Cylindrical line sound sources

These types of sound sources are of some practical importance in that long lengths of pipeline, trains and traffic noise can often be modelled in terms of infinite or finite cylinders. For the case of the infinite cylinder, the sound radiation problem becomes two-dimensional. The wave equation has to be described in cylindrical co-coordinates and the velocity potential turns out to be a function of Hankel functions (Hankel functions are complex Bessel functions). Kinsler *et al.*[2.1], Pierce[2.3], Reynolds[2.4], Dowling and Ffowcs Williams[2.5], and Norton and Bull[2.8] all provide detailed discussions on the solutions to a range of analytical problems relating to cylindrical radiation. Pipe flow noise and vibration is discussed in chapter 7 in this book as a special case study.

An important feature of cylindrical radiation is that the sound intensity varies with r^{-1} instead of r^{-2} for monopole sources or r^{-4} for dipoles. This suggests a 3 dB per doubling of distance. For an infinitely long cylinder of radius a, with a uniform pulsating harmonic surface velocity $\mathbf{U}_r = U_p\,e^{i\omega t}$, the far-field sound intensity is approximated by[2.4]

$$I(r) = U_{\text{rms}}^2 \rho_0 c \,\frac{\pi k a^2}{2r}. \tag{2.146}$$

Infinite and finite cylindrical line sources will also be discussed again in chapter 4 when considering the far-field propagation of sound waves.

2.4 The inhomogeneous acoustic wave equation – aerodynamic sound

So far in this chapter, the solutions to the wave equation have been sought in regions of space that do not contain any sources of sound. As pointed out in the introduction (section 2.1), this is generally the case in most industrial engineering type applications. Quite often, even when the source of sound is aerodynamically generated, one is only concerned with regions of space exterior to the source region, and classical acoustics is adequate to describe the behaviour of the wave field (e.g. external sound radiation from gas flow in a pipeline). When the source of sound is due to some vibrating body (structure-borne sound) it is generally readily identified. If, however, specific information is required about the source region itself, or the source is within the region of space of interest, then the classical, homogeneous, wave equation is inadequate. This is because different source types can in fact produce the same pressure distribution in the wave field exterior to the source region – this is an important point and it will be illustrated later on in this section.

In situations involving flow, and turbulence in particular, the identification of the sound sources is somewhat more difficult – they convect and interact with the fluid flow. Typical examples of instances where the inhomogeneous wave equation is required include unsteady duct flows, fan noise, jet noise, noise from aerofoils, noise from wall cavities, boundary layer noise etc. In these and other examples, in order to understand the mechanisms by which the sound is generated within the flow, and in order to understand the principles of the interactions of the flow with the acoustics one needs to obtain some basic information about the sources of sound. Quite often, this is easier said than done! The field of aerodynamic sound is a relatively new and complex one, and it is the subject of continued research. Its main impetus has been the jet aircraft industry.

Lighthill's[2.10,2.11] work which was published in the early 1950s provided the first general theory of aerodynamic sound. Lighthill reformulated the equations of fluid motion such as to include the source functions which drive the sound wave field. It was postulated that all the non-linearities in the motion of matter act as sources of sound, and they were grouped as non-linear forcing terms on the right hand side of the wave equation. The sound radiated is thus estimated as if it were in a uniform medium with zero mean flow – i.e. convection effects, turbulence etc. are all incorporated within the source function on the right hand side. Lighthill's now famous 'acoustic analogy' states that the sources of sound in a fluid motion are simply the difference between the exact equations of fluid motion and the acoustical approximations.

The main problem which has limited researchers for many years since is that the source function (i.e. the difference between the exact equations of fluid motion and the acoustical approximations) is generally very difficult to evaluate. This is primarily because sufficient information about the fluid flow is not always readily available. Lighthill, in his acoustic analogy, reduced the problem of aerodynamically generated sound to an analogous classical acoustical problem. In that form, whilst the wave equation is exact, its main limitation is that it does not shed any light on any subsequent interaction between the aerodynamically generated sound and the fluid flow which caused it in the first place. It has been shown that Lighthill's analogy allows for a precise analysis of the sound field for low Mach number flows (i.e. compact source regions where the typical source dimension is significantly smaller than the corresponding acoustic wavelength). However, at higher flow speeds (including supersonic flows), it has been shown that 'excess noise' is often present due to flow–acoustic interactions. There have been several attempts to reformulate Lighthill's source terms firstly to include the effects of solid boundaries (i.e. obstacles in the flow), secondly to minimise the requirements for a detailed description of the fluid flow in the source region (as far as noise radiation is concerned), and thirdly to shed some light on the effects that the interaction between the flow field and the sound field has on the radiated sound.

Curle[2.12], Powell[2.13] and Ffowcs Williams and Hall[2.14] (also see Dowling and Ffowcs Williams[2.5]) extended Lighthill's theory to incorporate the effects of solid bodies. Most of this work is now generally available in the research literature. However, only a few textbooks are currently available on the subject of aerodynamic sound. The reader is referred to three recent books – the first by Goldstein[2.15], the second by Dowling and Ffowcs Williams[2.5], and the third by Blake[2.16]. Blake's book (two volumes) in particular is the most recent comprehensive publication on aerodynamic sound; it is an up-to-date research monograph on the mechanics of flow-induced sound and vibration and it deals with general concepts, elementary sources and complex flow–structure interactions. Ffowcs Williams[2.17] also provides a comprehensive review article on aerodynamic sound in the *Annual Review of Fluid Mechanics*.

A major advance in the field of aerodynamic sound occurred in 1975 when in terms of entropy and vorticity variations. He subsequently described the interactions between flow and sound fields in terms of fluid vorticity and acoustical particle velocities associated with the generated sound field. The concepts relating to the generation of aerodynamic sound by the motion of vortices in an unsteady fluid flow relate back to Powell[2.20]. In his review paper, Ffowcs Williams[2.17] points out that Howe's[2.18] identification of acoustic sources within vortical regions of isentropic flow is in fact a formalisation of the steps originally taken by Powell[2.20]. It is now widely recognised amongst researchers in the field that Lighthill's theory of aerodynamic sound (1952) and the Powell–Howe theory of vortex sound (1964, 1975) are the two singularly most important advances in the field of aerodynamic sound to date. Howe's contribution, in particular, is a successful attempt to separately describe both the flow–acoustic interactions and the mean flow effects.

Whilst it is felt that it is necessary for noise control engineers to be aware of the existence of the inhomogeneous wave equation, and its relevance to the control of aerodynamically generated sound, a detailed discussion of the topic is well beyond the scope of this textbook. In the remaining sections of this chapter, the inhomogeneous acoustic wave equation is derived, and the basic solutions for some simple acoustic source processes are discussed, such as to provide the reader with sufficient background information to pursue the matter further if required. Lighthill's acoustic analogy is derived and discussed in relation to fluid dynamically generated monopoles, dipoles and quadrupoles, but some of the more recent advanced work in the area is only qualitatively discussed.

2.4.1 Solutions to the inhomogeneous wave equation for simple sources

In section 2.3, solutions to the homogeneous wave equation were obtained for a range of simple sources. These solutions were obtained for far-field approximations and the instantaneous values were limited to constant frequencies – i.e. the solutions for instantaneous pressure, particle velocity etc. were obtained in terms of single frequency components using harmonic waves and complex algebra in very much the same way that instantaneous values of displacement, velocity, and acceleration were obtained in terms of single frequency components using harmonic waves and complex algebra in chapter 1. The resulting equations for mean-square pressure, intensity etc. were steady-state, time-averaged values. The procedures adopted in section 2.3 cannot be used if one requires instantaneous solutions for the acoustic variables (for all frequencies) and that the source be accounted for in the wave field. A somewhat different approach has to be taken – i.e. the initial time-dependent conditions have to be considered to describe the transient behaviour of the source.

Reconsider the simple spherical sound source (monopole) of section 2.3.1. The solution for the acoustic pressure (equation 2.85) is valid everywhere except at the origin where a singularity occurs. It was stated at the outset that the homogeneous wave equation is only valid in the wave field and that it is not valid in the source field. The singularity at the origin of a simple spherical source is thus consistent with this argument!

A spherical sound source is a fluctuating source of mass at the origin. Its instantaneous source strength ($m^3\,s^{-1}$), $Q(t)$ is given by

$$Q(t) = 4\pi a^2 u_a(t), \tag{2.147}$$

where $u_a(t)$ is the instantaneous normal surface velocity, and a is the radius. The mass flux per unit time ($kg\,s^{-1}$) through the volume is

$$\mathrm{Q}(t) = 4\pi a^2 \rho u_a(t), \tag{2.148}$$

where ρ is the density of the fluid. The rate of change of mass flux ($kg\,s^{-2}$) is thus

$$\mathrm{Q}'(t) = 4\pi a^2 \frac{\partial(\rho u_a)}{\partial t}. \tag{2.149}$$

From the momentum equation (equation 2.11) neglecting the viscous stress terms

$$\frac{\partial(\rho u_a)}{\partial t} = -\frac{\partial P}{\partial x}, \tag{2.150}$$

and since $p = c^2\rho'$ (equations 2.29, 2.30 and 2.35),

$$\frac{\partial P}{\partial x} = \frac{\partial p}{\partial x} = c^2 \frac{\partial \rho'}{\partial x}. \tag{2.151}$$

Thus, in spherical (radial) co-ordinates

$$\mathrm{Q}'(t) = -4\pi a^2 c^2 \frac{\partial \rho'}{\partial r}. \tag{2.152}$$

It was shown in section 2.3 (and in chapter 1) that $\{G(ct - r)\}/r$ satisfies the wave radiation condition and that it is a solution to the spherical, homogeneous wave equation. It is convenient in this section to represent the function as $\{G(t - r/c)\}/r$ instead. It is straightforward to show that it is also a general solution to the homogeneous wave equation. The term r/c is the retardation time – i.e. the time it takes the sound wave to travel from the source to the observer. The concepts of retardation times are important in aerodynamic sound – this will become evident later on in this section. Thus,

$$\rho'(r, t) = \frac{1}{r} G\left(t - \frac{r}{c}\right) \tag{2.153}$$

is a solution to the homogeneous wave equation (in terms of the density fluctuations) for outward travelling waves. Hence, the rate of change of mass flux, $\mathrm{Q}'(t)$, can be estimated by evaluating $\partial\rho'/\partial r$ at $r = a$ from equation (2.153) for $a \to 0$. Thus,

$$\mathrm{Q}'(t) = 4\pi c^2 G(t), \tag{2.154}$$

and hence

$$G\left(t-\frac{r}{c}\right)=\frac{1}{4\pi c^2}\mathrm{Q}'\left(t-\frac{r}{c}\right). \tag{2.155}$$

Thus, the solution to the homogeneous wave equation (in terms of the density fluctuations) for outward travelling waves is (from equation 2.153)

$$\rho'(r,t)=\frac{1}{4\pi c^2 r}\mathrm{Q}'\left(t-\frac{r}{c}\right). \tag{2.156}$$

The acoustic pressure fluctuations are thus given by

$$p(r,t)=\frac{1}{4\pi r}\mathrm{Q}'\left(t-\frac{r}{c}\right). \tag{2.157}$$

Equation (2.157) is consistent with equation (2.85) and

$$\mathrm{Q}'\left(t-\frac{r}{c}\right)=\mathrm{Re}\,[ik\rho_0 c\mathbf{Q}(t)\,\mathrm{e}^{-ikr}]. \tag{2.158}$$

The exponential factor e^{-ikr} in the above equation is analogous to the time shift r/c, since $\mathbf{Q}(t)=Q_\mathrm{p}\,\mathrm{e}^{i\omega t}$.

Equation (2.157) is: (i) only valid in a region exterior to the source because of the singuarity at the origin, (ii) a function of the rate of change of mass flux, and (iii) a function of the retardation time. It can be extended to a volume region in space where there is some rate of flux of mass per unit volume, $q(\vec{y},t)$, with units of kg $\mathrm{m}^{-3}\,\mathrm{s}^{-1}$. Note that the vector $\vec{y}$ represents spatial positions which are within the source region. The equation of conservation of mass (equation 2.5) is now

$$\frac{\partial\rho}{\partial t}+\vec{\nabla}\cdot\rho\vec{u}=q(\vec{y},t). \tag{2.159}$$

It can be combined in the usual manner with the momentum equation to yield the wave equation, which is now inhomogeneous with the time derivative of q (i.e. $q'(\vec{y},t)$) as the forcing term (note that q' has units of kg $\mathrm{m}^{-3}\,\mathrm{s}^{-2}$). It is

$$\frac{1}{c^2}\frac{\partial^2 p}{\partial t^2}-\nabla^2 p=q'(\vec{y},t). \tag{2.160}$$

Equation (2.160) is the inhomogeneous wave equation. The function q' is only non-zero within the source region. In regions exterior to the source, it vanishes, and the wave equation reverts to being homogeneous.

One can readily see from equation (2.160) that in the wave field the pressure distributions can be the same for different forcing functions since $q'(\vec{y},t)$ is zero outside the source region – i.e. different source distributions of equal strength will produce the same wave field. Information obtained in the wave field (from the homogeneous wave equation) will therefore not provide any information about the source distribution.

A general solution to equation (2.160) for the pressure in the sound field (exterior to the source region) can be obtained by an integration of terms similar to equation (2.157) over the whole source region. It is

$$p(\vec{x}, t) = \int_V \frac{q'\left(\vec{y}, t - \frac{|\vec{x} - \vec{y}|}{c}\right)}{4\pi|\vec{x} - \vec{y}|} \, d^3\vec{y}. \tag{2.161}$$

The term $\vec{y}$ represents position vectors within the source region; the term $\vec{x}$ represents position vectors in the sound field which is exterior to the source region; and $|\vec{x} - \vec{y}|$ represents the distance between the source and the observer. This point is illustrated in Figure 2.16. Equation (2.161) is an important general equation in the field of aerodynamic noise. It can be re-written as

$$p(\vec{x}, t) = \frac{1}{4\pi} \frac{\partial}{\partial t} \int_V \frac{q\left(\vec{y}, t - \frac{|\vec{x} - \vec{y}|}{c}\right)}{|\vec{x} - \vec{y}|} \, d^3\vec{y}, \tag{2.162}$$

where it can be clearly seen that the acoustic field, in this instance, is due to the rate of change of mass flux per unit volume – i.e. steady mass flux will not produce any sound; it is the unsteady mass flux that generates the sound!

Equation (2.161) is the formal solution to the inhomogeneous wave equation for a given arbitrary source function and it can take on many forms (e.g. equation 2.162 for mass flux). Equation (2.162) is thus the solution to the inhomogeneous wave equation for a monopole type sound source. It will be seen later on in this chapter that equation (2.161) can also be used to describe dipoles and quadrupoles. It has been derived heuristically (by comparison with equation 2.157) and it is instructive to also outline its derivation in a more formal sense.

Fig. 2.16. Schematic illustration of source and observer positions.

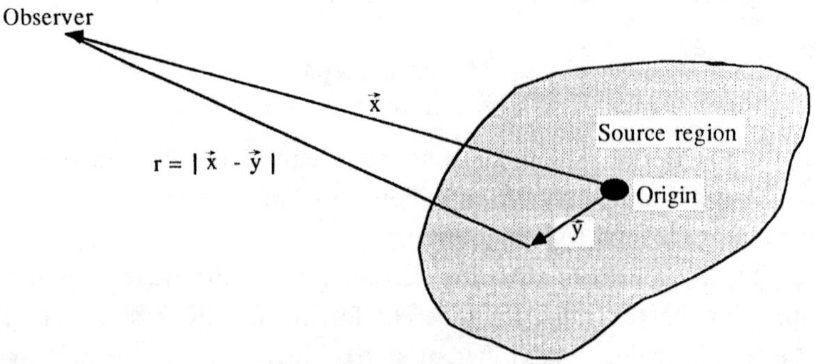

The general solution to the inhomogeneous wave equation can be obtained by superposition of the acoustic fields due to simple point sources. In order to do this, a function called a Green's function needs to be introduced. Green's functions satisfy the inhomogeneous wave equation, and if the medium surrounding the source is stationary (zero mean flow) and is not restricted by any boundaries, they are commonly referred to as free space Green's functions. Green's functions are also used in classical acoustics, and in fact were used earlier on in this chapter (this will become obvious shortly!) The reason why free space Green's functions are so relevant to aerodynamic sound is that Lighthill simply reduced the problem of predicting sound in a moving medium to that of the equivalent classical acoustical problem in a stationary fluid! – this too will be demonstrated in a short while. The free space Green's function is described by the transform pair[2.3,2.15].

$$\mathbf{G}_{\omega}(\vec{y}, \omega|\vec{x}, \omega) = \frac{\mathrm{e}^{ikr}}{4\pi r}, \tag{2.163}$$

and

$$G_t(\vec{y}, \tau|\vec{x}, t) = \frac{1}{4\pi r}\,\delta(\tau - t + r/c), \tag{2.164}$$

where

$$G_t = \frac{1}{2\pi}\int_{-\infty}^{\infty} \mathbf{G}_{\omega}\,\mathrm{e}^{\mathrm{i}(\omega t - \tau)}\,\mathrm{d}\omega. \tag{2.165}$$

From equation (2.163) one can readily see how the Green's function was used (inadvertently) earlier on in this chapter. The Green's function in equation (2.163) can be interpreted as the solution to the wave equation in free space for an applied unit, time-harmonic, point source, and the Green's function in equation (2.164) can be interpreted as the solution to the wave equation in free space for a unit, impulsive, point source. An important property of the Green's function is that it is symmetrical – i.e. it remains the same when the source and the receiver position are interchanged. This property, commonly known as reciprocity, has important applications in noise and vibration control, and it will be discussed in some detail in chapters 3 and 6.

The Green's function for an impulsive, unit, point source located at $\vec{y}$ and set off at time τ satisfies the inhomogeneous wave equation – i.e.

$$\left\{\frac{1}{c^2}\frac{\partial^2}{\partial t^2} - \nabla^2\right\} G_t(\vec{y}, \tau|\vec{x}, t) = \delta(\tau - t)\,\delta(\vec{x} - \vec{y}). \tag{2.166}$$

The function G_t must vanish for $t < \tau$ (this is referred to as the causality condition), and it must exhibit outgoing wave behaviour (this is referred to as the radiation condition). Thus, for some source function $f(\vec{y}, \tau)$,

$$\left\{\frac{1}{c^2}\frac{\partial^2}{\partial t^2} - \nabla^2\right\} f(\vec{y}, \tau) G_t(\vec{y}, \tau|\vec{x}, t) = f(\vec{y}, \tau)\,\delta(\tau - t)\,\delta(\vec{x} - \vec{y}). \tag{2.167}$$

Hence,

$$\left\{\frac{1}{c^2}\frac{\partial^2}{\partial t^2} - \nabla^2\right\} \int_T \int_V f(\vec{y}, \tau) \frac{\delta(\tau - t + r/c)}{4\pi r} \, d\tau \, d^3\vec{y}$$

$$= \int_T \int_V f(\vec{y}, \tau) \, \delta(\tau - t) \, \delta(\vec{x} - \vec{y}) \, d\tau \, d^3\vec{y}. \quad (2.168)$$

Thus,

$$\left\{\frac{1}{c^2}\frac{\partial^2}{\partial t^2} - \nabla^2\right\} \int_V \frac{f(\vec{y}, t - r/c)}{4\pi r} \, d^3\vec{y} = \int_V f(\vec{y}, t) \, \delta(\vec{x} - \vec{y}) \, d^3\vec{y}. \quad (2.169)$$

Now, in terms of the pressure fluctuations, the inhomogeneous wave equation is

$$\left\{\frac{1}{c^2}\frac{\partial^2}{\partial t^2} - \nabla^2\right\} p = f(\vec{x}, t) = \int_V f(\vec{y}, t) \, \delta(\vec{x} - \vec{y}) \, d^3\vec{y}, \quad (2.170)$$

where $f(\vec{x}, t)$ is an arbitrary source function. A general solution can be obtained for $p(\vec{x}, t)$ by equating equations (2.169) and (2.170). Hence, for $r = |\vec{x} - \vec{y}|$,

$$p(\vec{x}, t) = \int_V \frac{f\left(\vec{y}, t - \frac{|\vec{x} - \vec{y}|}{c}\right)}{4\pi|\vec{x} - \vec{y}|} \, d^3\vec{y}. \quad (2.171)$$

Equation (2.171) is the formal solution for the acoustic pressure fluctuations (in a region where there is zero mean flow and no solid boundaries) for an arbitrary source function, f. It is identical in form to equation (2.161) and is a superposition of all the waves generated by all the elements within the source region.

For a sound source that only represents mass flux into a volume, the function f in equation (2.171) is replaced by the rate of change of mass flux per unit volume, q' (kg m^{-3} s^{-2}). The rate of change of mass flux through the whole volume region is

$$Q'(t) = \int_V q'(\vec{y}, t) \, d^3\vec{y}. \quad (2.172)$$

If it is assumed that the source region is compact, small phase differences between the various source elements can be neglected – i.e. retardation times can be neglected as far as a far-field observer is concerned, and

$$p(r, t) = \frac{1}{4\pi r} \int_V q'(\vec{y}, t - r/c) \, d^3\vec{y}. \quad (2.173)$$

Hence,

$$p(r, t) = \frac{1}{4\pi r} Q'\left(t - \frac{r}{c}\right), \quad (2.174)$$

where $r = |\vec{x} - \vec{y}| \approx |\vec{x}|$. Equation (2.174) is identical to equation (2.157) and it represents a point monopole radiation field with uniform directivity – the amplitude of the pressure field is proportional to the nett rate of change of mass flux over the whole source region.

For a sound source that only represents some externally applied force $f_i(\vec{x}, t)$, the function q' is replaced by the divergence of the vector force – i.e.

$$q'(\vec{x}, t) = -\text{div}\, f_i(\vec{x}, t) = -\frac{\partial f_i(\vec{x}, t)}{\partial x_i}. \tag{2.175}$$

The function $f_i(\vec{x}, t)$ represents an external applied body force per unit volume – e.g. a solid body in the flow – f_i is a vector in the ith direction, thus $\partial f_i/\partial x_i$ is a scalar. The momentum equation (neglecting viscous stresses) becomes

$$\frac{\partial(\rho u_i)}{\partial t} + \frac{\partial P}{\partial x_i} = f_i(\vec{x}, t), \tag{2.176}$$

the equation of conservation of mass remains as per equation (2.5), and the inhomogeneous wave equation therefore becomes

$$\frac{1}{c^2}\frac{\partial^2 p}{\partial t^2} - \nabla^2 p = -\frac{\partial f_i(\vec{x}, t)}{\partial x_i} = -\text{div}\, f_i(\vec{x}, t). \tag{2.177}$$

The source term on the right hand side of equation (2.177) has to be such that it vanishes outside some finite region of space in proximity to the source. Hence, by analogy with equations (2.170) and (2.171) the solution $p(\vec{x}, t)$ is

$$p(\vec{x}, t) = -\int_V \frac{\text{div}\, f_i\left(\vec{y}, t - \dfrac{|\vec{x} - \vec{y}|}{c}\right)}{4\pi|\vec{x} - \vec{y}|}\, \mathrm{d}^3\vec{y}, \tag{2.178}$$

or

$$p(\vec{x}, t) = -\frac{\partial}{\partial x_i}\int_V \frac{f_i\left(\vec{y}, t - \dfrac{|\vec{x} - \vec{y}|}{c}\right)}{4\pi|\vec{x} - \vec{y}|}\, \mathrm{d}^3\vec{y}. \tag{2.179}$$

There is a subtle difference between equations (2.178) and (2.179). In the former equation, the source is represented as a series of monopoles of strength $\partial f_i/\partial y_i$. In the latter, it is represented as a dipole of strength f_i – the force exerted on the fluid volume is equal to the rate at which the momentum changes, since the mass flux in and out of the fluid volume is the same. By retaining the solution in the form of equation (2.178), all the various retardation times between the various source elements and the observer position would have to be known. If the retardation times were neglected, the volume integral would reduce to zero and there would be no sound produced! This can be proved mathematically[2.5] but it is to be intuitively expected because there is, by definition, a zero rate of mass flux (i.e. $Q' = 0$). The various monopole elements within the source region effectively cancel each other, and if it were not for the retardation time differences between all the various source elements, the nett sound radiation would be zero. Thus, by transforming equation (2.178) into the dipole distribution of equation (2.179), the radiated sound pressure can be estimated from a knowledge of the force distribution and the requirement for a detailed knowledge of the retardation times between source and observer positions is thus avoided.

Equation (2.179) can be further simplified for engineering approximations. Assuming $r = |\vec{x} - \vec{y}|$, it can be re-written as

$$p(\vec{x}, t) = -\int_V \frac{\partial r}{\partial x_i} \frac{\partial}{\partial r} \frac{f_i\left(\vec{y}, t - \frac{|\vec{x} - \vec{y}|}{c}\right)}{4\pi r} \, d^3\vec{y}. \tag{2.180}$$

Hence,

$$p(\vec{x}, t) = \frac{1}{4\pi} \int_V \frac{\partial r}{\partial x_i} \left\{ \frac{f_i}{r^2} + \frac{1}{cr} \frac{\partial f_i}{\partial t} \right\} d^3\vec{y}, \tag{2.181}$$

since

$$\frac{\partial f_i}{\partial r} = -\frac{1}{c} \frac{\partial f_i}{\partial t}. \tag{2.182}$$

Now, $\partial r/\partial x_i$ can be interpreted as the dipole directivity factor – i.e. it is the cosine of the angle between the dipole axis (the line along which the fluctuating force is acting) and the distance between the source and the observer. Thus,

$$p(\vec{x}, t) = \frac{1}{4\pi} \int_V \left\{ \frac{f_i}{r^2} + \frac{1}{cr} \frac{\partial f_i}{\partial t} \right\} \cos\theta_i \, d^3\vec{y}. \tag{2.183}$$

The radiated pressure has both a near- and a far-field component in addition to the $\cos\theta$ angular dependence.

2.4.2 Lighthill's acoustic analogy

In the previous sub-section it has been illustrated that (i) for a monopole type aerodynamic sound source, the amplitude of the radiated sound pressure field is a function of the rate of change of mass flux over the whole source region, and (ii) for a dipole type aerodynamic sound source, the amplitude of the radiated sound pressure field is a function of an externally applied body force. Acoustic sources can also be generated in flow situations where the flow velocities are large enough not to comply with the linear acoustic equations – i.e. high Reynold's number flows.

Lighthill[2.10] considered an unbounded region of space, a part of which included a fluctuating fluid flow, and grouped all the non-linear terms in the fluid dynamic equations of motion together as acoustic source terms on the right hand side of the inhomogeneous wave equation (i.e. the function $f(\vec{x}, t)$ in equation 2.170). In so doing, the radiated sound pressure due to the high Reynold's number flows could be calculated by using the procedures described previously (equation 2.171) provided that information is available about the source terms. Lighthill's acoustic analogy is thus based on the fact that the sources of sound are the difference between the exact laws of fluid motion and the linearised acoustical approximations – it is the non-linearities that generate the sound!

Consider an unbounded region of space in which there is a fluctuating fluid flow. Conservation of mass states that the time rate of change of mass within a region has to equal the nett mass flow into it. Thus, from sub-section 2.2.1 (equation 2.5) for mass conservation

$$\frac{\partial \rho}{\partial t} + \vec{\nabla} \cdot \rho \vec{u} = 0. \tag{2.184}$$

It is conventional and convenient to use tensor rather than vector notation in aerodynamic sound and the above equation can be re-written as

$$\frac{\partial \rho}{\partial t} + \frac{\partial (\rho u_i)}{\partial x_i} = 0. \tag{2.185}$$

Here, the suffix i can take on any value from 1 to 3. Hence a product term such as $u_i u_j$ has nine possible permuations since $i = 1, 2, 3$ and $j = 1, 2, 3$.

Conservation of momentum states that the time rate of change of momentum in a region equals the sum of all the forces acting on the region plus the nett rate of flow of momentum through the region. For non-linear flow, where viscous effects associated with the shear forces due to the interaction of fluid particles are taken into account, conservation of momentum implies that

$$\frac{\partial (\rho \vec{u})}{\partial t} + \rho (\vec{u} \cdot \vec{\nabla}) \vec{u} + \vec{\nabla} P = v \nabla^2 \vec{u}, \tag{2.186}$$

where v is the coefficient of shear viscosity. In tensor notation, this is

$$\frac{\partial (\rho u_i)}{\partial t} + \frac{\partial (\rho u_i u_j)}{\partial x_j} + \frac{\partial (p \delta_{ij})}{\partial x_j} - \frac{\partial \tau_{ij}}{\partial x_j} = 0, \tag{2.187}$$

where

$$v \nabla^2 \vec{u} = \frac{v \partial^2 u_i}{\partial x_j \partial x_j}, \qquad \tau_{ij} = v \frac{\partial u_i}{\partial x_j}, \tag{2.188}$$

$\delta_{ij}=1$ if $i=j$ and 0 if $i\neq j$, and τ_{ij} is the viscous shear stress. Thus,

$$\frac{\partial(\rho u_i)}{\partial t}+\frac{\partial(p\delta_{ij}-\tau_{ij}+\rho u_i u_j)}{\partial x_j}=0. \tag{2.189}$$

Equation (2.189) is the exact momentum equation in tensor notation and $\rho u_i u_j$ is the unsteady Reynold's shear stress.

The wave equation can now be derived in the usual manner by taking the time derivative ($\partial/\partial t$) of the equation of conservation of mass, and the divergence ($\partial/\partial x_i$) of the equation of conservation of momentum. Thus,

$$\frac{\partial^2\rho}{\partial t^2}+\frac{\partial^2(\rho u_i)}{\partial x_i\,\partial t}=0, \tag{2.190}$$

and

$$\frac{\partial^2(\rho u_i)}{\partial x_i\,\partial t}+\frac{\partial^2(p\delta_{ij}-\tau_{ij}+\rho u_i u_j)}{\partial x_i\,\partial x_j}=0. \tag{2.191}$$

Subtracting equation (2.191) from (2.190) yields

$$\frac{\partial^2\rho'}{\partial t^2}=\frac{\partial^2(p\delta_{ij}-\tau_{ij}+\rho u_i u_j)}{\partial x_i\,\partial x_j}, \tag{2.192}$$

since $\rho=\rho_0+\rho'$, and ρ_0 is time invariant. Equation (2.192) can now be forced to take the form of the inhomogeneous wave equation by introducing the factor $c^2\nabla^2\rho'$ and subtracting it from both sides. Thus,

$$\frac{\partial^2\rho'}{\partial t^2}-c^2\nabla^2\rho'=\frac{\partial^2(p\delta_{ij}-\tau_{ij}+\rho u_i u_j-c^2\rho'\delta_{ij})}{\partial x_i\,\partial x_j}, \tag{2.193}$$

since

$$c^2\nabla^2\rho'=c^2\frac{\partial^2\rho'\delta_{ij}}{\partial x_i\,\partial x_j}. \tag{2.194}$$

The Lighthill stress tensor is defined as

$$T_{ij}=p\delta_{ij}-\tau_{ij}+\rho u_i u_j-c^2\rho'\delta_{ij}; \tag{2.195}$$

hence, Lighthill's non-linear, viscous, inhomogeneous wave equation is

$$\frac{1}{c^2}\frac{\partial^2 p}{\partial t^2}-\nabla^2 p=\frac{\partial^2 T_{ij}}{\partial x_i\,\partial x_j}, \tag{2.196}$$

with p replacing $c^2\rho'$ (i.e. see equations 2.29, 2.30 and 2.35).

The source term on the right hand side of Lighthill's equation represents a double divergence – i.e. there is a double tendency for the various source elements within the fluid to cancel. It was seen in the previous sub-section that a dipole is produced as a consequence of cancelling monopoles – the instantaneous dipole strength is zero and the sound radiation is due to the retardation time effects, i.e. the cancellation is not complete because of the time delays between the various sound waves reaching the observer position. The same arguments apply here and the sound radiation is once again due to the retardation time effects between the various source elements. This double divergence of a monopole field (or divergence of a dipole field) is a quadrupole. Because of the double tendency for cancellation one would qualitatively expect quadrupole sound radiation to be less efficient than dipole sound radiation, and dipole sound radiation to be less efficient than monopole sound radiation.

The solution to Lighthill's equation for the radiated sound pressure can be readily obtained from equation (2.179). It is

$$p(\vec{x}, t) = \frac{\partial^2}{\partial x_i \, \partial x_j} \int_V \frac{T_{ij}\left(\vec{y}, t - \dfrac{|\vec{x} - \vec{y}|}{c}\right)}{4\pi|\vec{x} - \vec{y}|} \, \mathrm{d}^3\vec{y}. \tag{2.197}$$

The qualitative arguments that were used in the previous section to justify proceeding from equation (2.178) to equation (2.179) also apply here – i.e. the formal solution has the second derivative inside the volume integral (i.e. $\partial^2/\partial y_i \, \partial y_j$) and it is transformed into equation (2.197) with the derivative outside the integral. Goldstein[2.15] provides the formal mathematical procedures that have to be followed in proceeding with the transformation. This transformation procedure is the key to the correct estimation of aerodynamic sound because it is the small retardation time differences between the different source elements that in fact produce the radiated sound. The dipole and quadrupole representations of the source elements account for the cancellations that occur, and thus allow for a more accurate estimate of the radiated sound.

In aerodynamic sound predictions, the main problem is therefore to suitably identify the components within the Lighthill stress tensor, T_{ij}, that are dominant. The $p\delta_{ij} - c^2\rho'\delta_{ij}$ component represents the effects of heat conduction which affect the speed of sound within the fluid. For low Mach number flows this is a second-order effect. The τ_{ij} component is the viscous shear stress, and this is generally very small when compared with the Reynold's shear stress component, $\rho u_i u_j$. Hence, for low Mach number flows, it is the Reynold's shear stresses, $\rho u_i u_j$, which are dominant. For linear, inviscid flows, the wave equation is homogeneous – this is consistent with Lighthill's wave equation since $p = c^2\rho'$, $\tau_{ij} = 0$, $\rho u_i u_j$ is negligible, and therefore $T_{ij} \sim 0$.

Equation (2.197) can now be re-arranged such as to identify the quadrupole nature of the sound source. Like the dipole, one would intuitively expect a near and a far-field radiation term. Equation (2.197) has to be differentiated in a similar manner to which equation (2.179) was differentiated to yield equation (2.183) for the dipole.

The procedures can be found in Goldstein[2.15] or Richards and Mead[2.21]. Thus, equation (2.197) can be differentiated to yield

$$p(\vec{x}, t) = \frac{1}{4\pi} \int_V \frac{\partial^2 r}{\partial x_i \, \partial x_j} \left\{ \frac{1}{c^2 r} \frac{\partial^2 T_{ij}}{\partial t^2} + \frac{2}{cr^2} \frac{\partial T_{ij}}{\partial t} + \frac{2}{r^3} T_{ij} \right\} \mathrm{d}^3 \vec{y}. \qquad (2.198)$$

The term in brackets represents the near- and far-field radiation terms – i.e. there are two near-field terms (r^{-2} and r^{-3}) and one far-field term. As for the dipole, the $\partial^2 r/\partial x_i \, \partial x_j$ term is a directivity factor. For a longitudinal quadrupole, $i = j$ and $\partial^2 r/\partial x_i \, \partial x_j \approx \cos^2 \theta$; for a lateral quadrupole $i \neq j$ and $\partial^2 r/\partial x_i \, \partial x_j \approx \cos \theta \sin \theta$.

2.4.3 *The effects of the presence of solid bodies in the flow*

Solid body interactions with flows are an important part of aerodynamic sound generation. Turbulent flows interacting with solid bodies produce increased sound over free space turbulence – e.g. a butterfly valve arrangement in an industrial gas pipeline, aircraft engine turbine blades etc. The main reason for this is that the presence of solid bodies allows for the existence of monopoles (due to the fluctuating motion or dilatations of the solid body resulting from the unsteady flow) and dipoles (due to the fluctuating forces on the solid body resulting from the unsteady flow), both of which are more efficient sound radiators than quadrupoles.

Lighthill's inhomogeneous wave equation is exact and it is therefore possible to use it to estimate the effects of the presence of solid bodies in the flow. Dowling and Ffowcs Williams[2.5], and Goldstein[2.15] derive the necessary fundamental equations. In summary, the procedure is as follows:

(i) it is assumed that the observer and the solid body are clearly separated;

(ii) the quadrupoles contained within the interior of the solid body are transformed into boundary sources over the surface of the solid body – this procedure converts the interior quadrupole source field into an equivalent monopole (fluctuating motion) and dipole (fluctuating force) field on the surface of the solid body;

(iii) the equations are solved for the density or pressure fluctuations and added to the solution for free turbulence to obtain the complete solution in some region exterior to the solid body.

For sound volume, V, bounded by a solid body of surface, S, the far-field radiated sound pressure is given by

$$p(\vec{x}, t) = \frac{\partial^2}{\partial x_i \, \partial x_j} \int_V \frac{T_{ij}\left(\vec{y}, t - \frac{|\vec{x} - \vec{y}|}{c}\right)}{4\pi|\vec{x} - \vec{y}|} \mathrm{d}^3 \vec{y} - \frac{\partial}{\partial x_i} \int_S \frac{f_i\left(\vec{y}, t - \frac{|\vec{x} - \vec{y}|}{c}\right)}{4\pi|\vec{x} - \vec{y}|} \mathrm{d}S(\vec{y})$$

$$+ \frac{\partial}{\partial t} \int_S \frac{\rho\vec{u}\left(\vec{y}, t - \frac{|\vec{x} - \vec{y}|}{c}\right) \cdot \vec{n}}{4\pi|\vec{x} - \vec{y}|} \mathrm{d}S(\vec{y}). \qquad (2.199)$$

The first term in the equation represents the free turbulence component; the second term represents the component due to fluctuating body forces; and the third term represents fluctuating motions of the solid body resulting from the unsteady flow. In the third term, $\hat{u}$ is the surface velocity and $\vec{n}$ is the unit normal vector. The equation can be used to develop useful dimensional parametric relationships for aerodynamic sound. Only low subsonic Mach number flows will be analysed here – the reader is referred to Dowling and Ffowcs Williams[2.5], and Goldstein[2.15] for the analyses relating to supersonic flows.

If the free turbulence component is the dominant term (e.g. flow exhausting from a jet nozzle), then the turbulent eddies scale with the mean flow, U, and a boundary layer thickness or a jet nozzle diameter, D. The time scale of the sound field is thus D/U. Thus, the sound wavelength scale is $\lambda = cD/U = D/M$, where c is the speed of sound and M is the Mach number of the flow. Thus,

$$M = \frac{D}{\lambda} = \frac{\text{typical source dimension}}{\text{acoustical wavelength}}. \tag{2.200}$$

An important point is now made. For low Mach number flows, $\lambda \gg D$ and the source region is acoustically compact. This means that the retardation times between the different points within the source region can be neglected. This allows for a considerable simplification in the analysis. Alternatively, for high subsonic and supersonic flows, $D \sim \lambda$ or $D \gg \lambda$ and the retardation times are critical to the analysis.

The various terms associated with the free turbulence component can now be suitably scaled. $\partial/\partial x_i$ scales with $1/\lambda$ or M/D; the volume integral scales with D^3; and the Lighthill stress tensor T_{ij} scales with ρU^2 since it is the Reynold's shear stresses which are dominant. Thus, from the free turbulence component of equation (2.199)

$$p(\vec{x}, t) \sim \left\{\frac{M}{D}\right\}^2 \frac{D^3 \rho U^2}{4\pi r} = \frac{\rho D}{4\pi r c^2} U^4, \tag{2.201}$$

and

$$I(r) \sim \frac{p^2(\vec{x}, t)}{(\rho c)_0} = \frac{\rho^2 D^2}{16\pi^2 r^2 c^4 (\rho c)_0} U^8. \tag{2.202}$$

Equation (2.202) suggests that the radiated sound power scales with the eighth power of the flow velocity – this is Lighthill's famous U^8 law for turbulence generated sound. It is similar to the U^8 relationship that was derived in section 2.3.5 for quadrupoles. It should be noted that the term $(\rho c)_0$ is the density and speed of sound at the observer position. In principle, it could be different from the density, ρ, and speed of sound, c, in the source region, particularly if hot gases etc. were involved.

The total fluctuating force scales as $\rho U^2 D^2$, thus from the fluctuating force component of equation (2.199)

$$p(\vec{x}, t) \sim \begin{Bmatrix} M \\ D \end{Bmatrix} \frac{D^2 \rho U^2}{4\pi r} = \frac{\rho D}{4\pi c r} U^3, \tag{2.203}$$

and

$$I(r) \sim \frac{p^2(\vec{x}, t)}{(\rho c)_0} = \frac{\rho^2 D^2}{16\pi^2 c^2 r^2 (\rho c)_0} U^6. \tag{2.204}$$

Equation (2.203) suggests that the radiated sound power scales with the sixth power of the flow velocity. It is similar to the U^6 relationship that was derived in section 2.3.2 for dipoles.

Finally the sound power radiation associated with the third term (fluctuating motions of the solid bodies which produce volume dilatations) in equation (2.199) can be evaluated. Here, $\partial/\partial t$ scales as the acoustic frequency, i.e. U/D and the surface integral scales as D^2. Thus

$$p(\vec{x}, t) \sim \frac{U}{D} \frac{\rho D^2 U}{4\pi r} = \frac{\rho D}{4\pi r} U^2, \tag{2.205}$$

and

$$I(r) \sim \frac{p^2(\vec{x}, t)}{(\rho c)_0} = \frac{\rho^2 D^2}{16\pi^2 r^2 (\rho c)_0} U^4. \tag{2.206}$$

Equation (2.206) suggests that the radiated sound power scales with the fourth power of the flow velocity. It is similar to the U^4 relationship that was derived in section 2.3.1 for monopoles.

Several important observations can be made in relation to equations (2.201)–(2.206). Firstly, the radiated sound intensity (or the mean-square sound pressure) is proportional to the inverse square of distance; secondly, the intensity is proportional to the square of the typical source dimension, e.g. boundary layer thickness or jet nozzle diameter; thirdly, the velocity dependence ranges from U^8 for quadrupoles to U^4 for monopoles.

2.4.4 The Powell–Howe theory of vortex sound

Lighthill's theory of aerodynamic sound has been extensively used, both in the subsonic and supersonic flow regimes, for jet noise prediction studies, boundary layer noise studies, unsteady flows over stationary and moving solid bodies, etc. When the complete details of the flow are unavailable and the only available parameters are the intensities and length scales, Lighthill's stress tensor, T_{ij}, is a very powerful tool.

If, however, detailed knowledge of the flow–acoustic interactions is required, then the source terms in Lighthill's equation have to be re-defined.

In 1964, Powell[2.20] postulated that the origin of aerodynamic sound might be attributed to the process of forming eddies or vortices – the very action that causes the formation of vortices, simultaneously gives rise to the sound radiation. Howe[2.18] reformulated Lighthill's theory for low Mach number flows in terms of Powell's concept of vortex sound and associated the aerodynamic sound sources with certain regions in the flow where the total (i.e. mean plus unsteady) vorticity vector, $\vec{\omega}$, is non-vanishing. It should be noted here that, whilst vorticity is a necessary condition for an aerodynamic sound source, it is not a sufficient condition.

From fluid dynamics, when the flow contains both rotational and irrotational components, the total velocity, $\vec{U}$, is given by equation (2.48) – i.e.

$$\vec{U} = \vec{\nabla}\phi + \vec{\nabla} \times \vec{\psi}, \tag{2.207}$$

where ϕ is the irrotational velocity potential, and $\vec{\psi}$ is the rotational, vector velocity potential. In turbulent flow regimes, $\vec{v} = \vec{\nabla} \times \vec{\psi}$ is the total (mean plus unsteady) rotational, incompressible component of the velocity field, and $\vec{u} = \vec{\nabla}\phi$ is the irrotational unsteady component. The irrotational, unsteady component is only non-zero when the fluid is compressible and it is therefore associated with the acoustic particle velocity. The vorticity is thus only related to the vector velocity potential and it is

$$\vec{\omega} = \vec{\nabla} \times \vec{v} = \vec{\nabla} \times (\vec{\nabla} \times \vec{\psi}). \tag{2.208}$$

For vortical regions located in free space, Howe[2.18] re-formulated Lighthill's wave equation to yield

$$\frac{1}{c^2}\frac{\partial^2 p}{\partial t^2} - \nabla^2 p = \rho_0 \vec{\nabla} \cdot (\vec{\omega} \times \vec{v}). \tag{2.209}$$

The term $\rho_0(\vec{\omega} \times \vec{v})$ is referred to in the literature as the Lamb vector and it is the unsteady vortical lifting force. Howe went on to generalise the equation, with the stagnation enthalpy rather than the density as the acoustic variable, to account for interactions with solid bodies etc. His classical analysis resulted in the following analogy with the Lighthill stress tensor –

$$\frac{\partial^2 T_{ij}}{\partial x_i \, \partial x_j} = \rho_0 \vec{\nabla} \cdot \{(\omega \times \vec{v}) - T\vec{\nabla}S\}, \tag{2.210}$$

where T is the temperature and S is the entropy. The equation illustrates that the sources of aerodynamic sound are contained in the regions of flow where the vorticity vector and the entropy-gradient vector are non-zero. Howe's analysis has been successfully applied to a range of complex problems of sound generation in inhomogeneous flows[2.18,2.19,2.23]. It is useful to note that it is not necessarily restricted to potential flows; in fact distributed vorticity, as implied by equation (2.211) below, negates potential flow. Also, viscosity does not disable the theory. Crighton[2.22] points out, however, that the Powell–Howe vortex theory, whilst associating the sound sources with regions of vorticity, is still non-locally and non-linearly related to the vorticity. No universal procedures are yet available to linearly relate the sound sources to vorticity. Whilst the Lighthill theory allows for accurate general results when the stress tensor, T_{ij}, can be readily evaluated, the Powell–Howe theory allows for specific problems relating to flow–acoustic interactions to be solved.

Howe[2.19,2.23] subsequently established a general momentum balance relationship describing the rate of 'dissipation' of sound energy by the generation of vorticity in the presence of low sub-sonic mean flows The term 'dissipation' is used loosely as it can refer to both negative dissipation (generation of sound) and positive dissipation (absorption of sound) – i.e. the generation of sound by vorticity produces negative dissipation, and the generation of vorticity by sound produces positive dissipation. Howe's dissipation formula is

$$\Pi_{\mathrm{D}} = \rho_0 \int_V \{(\vec{\omega} \times \vec{v}) \cdot \vec{u}\} \, \mathrm{d}V, \tag{2.211}$$

where $\vec{\omega}$ is the total (i.e. mean plus unsteady) vorticity vector, $\vec{v}$ is the total (mean plus unsteady) rotational, incompressible component of the velocity field – it is the vorticity convection velocity – and $\vec{u}$ is the acoustic particle velocity. This relationship is especially useful in that it allows for both a qualitative and a quantitative description of different types of flow–acoustic interactions; in particular, regions of energy transfer between the sound and the flow field can be explicitly identified. Howe[2.19] illustrates the positive dissipation (sound absorption) aspect in relation to the influence of mean flows on the diffraction of sound from semi-infinite plates, and the attenuation of sound by grazing flow perforated screens. He also discusses how negative dissipation (sound generation) can occur at certain flow speeds and directions of propagation. The mean flow energy can also be converted into sound if an aerodynamic acoustic resonance is set up, i.e. sound waves in a duct synchronising with some internal, flow-induced, vortex shedding phenomena. Welsh *et al.*[2.24–2.26] have recently studied several such types of flow–acoustic interactions and successfully applied Howe's dissipation formula to them.

Several important observations[2.24] can be made in relation to equation (2.211). Firstly, $\vec{\omega}$, $\vec{v}$ and $\vec{u}$ must have large angles between them – i.e., if the acoustic particle velocity is parallel to the vortex path, there will be no interactions between the sound and flow fields. Secondly, $\vec{\omega}$, $\vec{v}$ and $\vec{u}$ must all have large magnitudes – in regions where the acoustic particle velocity is small there will be very little absorption or generation of sound. Thirdly, because $\vec{u}$ is oscillatory, the time-averaged nett sound energy will approximate to zero (because of cancellations) unless either the magnitude or direction of the vectors changes during an oscillatory cycle. Welsh *et al.*[2.24–2.26] refer to this condition as 'imbalance'. In a recent paper[2.25], they discuss how the energy is transferred from the mean flow to sustain an aerodynamic acoustic resonance condition, for four specific examples. One of those examples, that of a Helmholtz cavity resonator, is discussed here.

When a flowing fluid approaches a cavity, as illustrated in Figure 2.17, a vortex motion is set up at the point where the shear layer separates at the edge of the cavity.

Fig. 2.17. Schematic illustration of vortex generation in a Helmholtz resonator. (Adapted from Stokes *et al.*[2.25].)

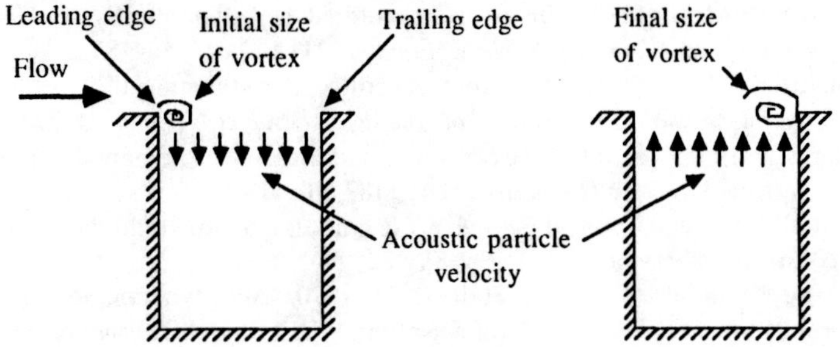

The vortex propagates across the cavity to the opposite edge before being convected away. The vortex grows across the cavity and reaches its full strength as it approaches the trailing edge of the cavity; it takes one cycle for the vortex to be generated and convected from the leading to the trailing edge. The acoustic particle velocities within the resonating cavity change sign during one cycle (because they are oscillatory), hence there is a sign change in the triple-vector product over any one cycle. Now, since the vortex has grown during this cycle, there is a nett generation of sound energy – i.e. the time-average of the Howe integral (equation 2.211) is not zero over any one cycle, and there is a significant angle between the acoustic particle velocity and the vortex convection velocity.

The above example qualitatively illustrates the application of the vortex theory of sound to the identification of both regions and mechanisms of flow-induced sound generation.

References

2.1 Kinsler, L. E., Frey, A. R., Coppens, A. B. and Sanders, V. J. 1982. *Fundamentals of acoustics*, John Wiley & Sons (3rd edition).

2.2 Soria, J. and Norton, M. P. 1986. *The response of a laminar shear layer on a flat plate to transverse surface vibrations*, Proceedings 9th Australasian Fluid Mechanics Conference, Auckland, New Zealand, pp. 610–13.

2.3 Pierce, A. D. 1981. *Acoustics: an introduction to its physical principles and applications*, McGraw-Hill.
2.4 Reynolds, D. D. 1981. *Engineering principles of acoustics – noise and vibration*, Allyn & Bacon.
2.5 Dowling, A. P. and Ffowcs Williams, J. E. 1983. *Sound and sources of sound*, Ellis Horwood.
2.6 Bies, D. A. 1982. *Noise control for engineers*, University of Adelaide, Mechanical Engineering Department Lecture Note Series.
2.7 Ford, F. D. 1970. *Introduction to acoustics*, Elsevier.
2.8 Norton, M. P. and Bull, M. K. 1984. 'Mechanisms of the generation of external acoustic radiation from pipes due to internal flow disturbances', *Journal of Sound and Vibration* **94**(1), 105–46.
2.9 Pickles, J. M. Personal Communication, University of Adelaide, Mechanical Engineering Department.
2.10 Lighthill, M. J. 1952. 'On sound generated aerodynamically. I. General theory', *Proceedings of the Royal Society (London)* **211A**, 1107, 564–87.
2.11 Lighthill, M. J. 1954. 'On sound generated aerodynamically. II. Turbulence as a source of sound', *Proceedings of the Royal Society (London)* **222A**, 1148, 1–32.
2.12 Curle, N. 1955. 'The influence of solid boundaries on aerodynamic sound', *Proceedings of the Royal Society (London)* **231A**, 1187, 505–14.
2.13 Powell, A. 1960. 'Aerodynamic noise and the plane boundary', *Journal of the Acoustical Society of America* **32**(8), 982–90.
2.14 Ffowcs Williams, J. E. and Hall, L. H. 1970. 'Aerodynamic sound generation by turbulent flow in the vicinity of a scattering half plane', *Journal of Fluid Mechanics* **40**(4), 657–70.
2.15 Goldstein, M. E. 1976. *Aeroacoustics*, McGraw-Hill.
2.16 Blake, W. K. 1986. *Mechanics of flow-induced sound and vibration*, Academic Press.
2.17 Ffowcs Williams, J. E. 1977. 'Aeroacoustics', *Annual Review of Fluid Mechanics* **9**, 447–68.
2.18 Howe, M. S. 1975. 'Contributions to the theory of aerodynamic sound, with applications to excess jet noise and the theory of the flute', *Journal of Fluid Mechanics* **71**(4), 625–73.
2.19 Howe, M. S. 1984. 'On the absorption of sound by turbulence and other hydrodynamic flows', *I.M.A. Journal of Applied Mathematics* **32**, 187–209.
2.20 Powell, A. 1964. 'Theory of vortex sound', *Journal of the Acoustical Society of America* **36**(1), 177–95.
2.21 Richards, E. J. and Mead, D. J. 1968. *Noise and acoustic fatigue in aeronautics*, John Wiley.
2.22 Crighton, D. G. 1981. 'Acoustics as a branch of fluid mechanics', *Journal of Fluid Mechanics* **106**, 261–98.
2.23 Howe, M. S. 1980. 'The dissipation of sound at an edge', *Journal of Sound and Vibration* **70**(4), 407–11.
2.24 Welsh, M. C. and Stokes, A. N. 1985. *Transient vortex modelling of flow-induced acoustic resonances near cavities or obstructions in ducts*, Aero and hydro-acoustics, IUTAM Symposium, Lyon, Springer-Verlag, pp. 499–506.
2.25 Stokes, A. N., Welsh, M. C. and Hourigan, K. 1986. *Sound generated by separated flows around bluff bodies*, Proceedings 9th Australasian Fluid Mechanics Conference, Auckland, New Zealand, pp. 164–7.
2.26 Welsh, M. C., Stokes, A. N. and Parker, R. 1984. 'Flow-resonant sound interaction in a duct containing a plate, part I: semi-circular leading edge', *Journal of Sound and Vibration* **95**(3), 305–23.

Nomenclature

a	radius of an oscillating sphere, source dimension
$\vec{a}$	particle acceleration (vector quantity)
A	surface area
$\mathbf{A}, \mathbf{A}_1, \mathbf{A}_2$	arbitrary complex constants
B	adiabatic bulk modulus
c	speed of sound
c_0	speed of sound for constant pressure and density
C_v	piston mechanical damping
d	distance, separation between point sources, distance between source and reflecting plane etc.
$\mathrm{d}\vec{f}$	incremental increase in force (vector quantity)
$\mathrm{d}m$	incremental increase in mass
$\mathrm{d}V$	incremental increase in volume
D	mean sound energy density, diameter
D'	instantaneous sound energy density
$\vec{f}$	force (vector quantity)
$f(\vec{y}, t), f(\vec{y}, t - r/c)$	arbitrary source functions
$f_i(\vec{x}, t), f_i(\vec{y}, t - r/c)$	externally applied force ($r = \lvert\vec{x} - \vec{y}\rvert$)
$\mathbf{F}$	complex force
$\vec{F}$	force (vector quantity)
$\mathbf{F}_\mathbf{m}$	complex applied mechanical force
$\mathbf{F}_\mathbf{p}$	complex force on piston due to acoustic pressure fluctuations)
F_{rms}	root-mean-square fluctuating force
G	universal gas constant
G_1, G_2	arbitrary independent functions which satisfy the wave equation
$\mathbf{G}_\omega(\vec{y}, \omega \vert \vec{x}, \omega)$	free space Green's function for a unit, time-harmonic, point source – i.e. frequency domain Green's function (complex function)
$G_t(\vec{y}, \tau \vert \vec{x}, t)$	free space Green's function for a unit, impulsive, point source – i.e. time domain Green's function

H	height
$I, \vec{I}$	mean sound intensity (arrow denotes vector quantity)
$I(r)$	mean sound intensity as a function of radial distance
$I(r, \theta)$	mean sound intensity as a function of radial and angular distance
$\vec{I}'$	instantaneous sound intensity vector
J_1	first-order Bessel function
k, k_1, k_2, etc.	wavenumbers
K_s	piston stiffness
L	length dimension
m	mass
M	molecular weight, piston mass, Mach number
M_r	amplitude of fluid-loaded mass
$\vec{n}$	unit normal vector
p	pressure fluctuation
p_{rms}	root-mean-square pressure
$p(r, t), \mathbf{p}(r, t)$	pressure fluctuation as a function of radial distance (bold signifies complex)
$p(r, \theta, t), \mathbf{p}(r, \theta, t)$	pressure fluctuation as a function of radial and angular distance (bold signifies complex)
$p(\vec{x}, t), \mathbf{p}(\vec{x}, t)$	pressure fluctuation (bold signifies complex)
$\mathbf{p}^*(r, t), \mathbf{p}^*(\vec{x}, t)$	complex conjugates of $\mathbf{p}(r, t), \mathbf{p}(\vec{x}, t)$
$\mathbf{p}_+, \mathbf{p}_-$	complex pressure amplitudes
$\hat{p}$	pressure amplitude
$P(\vec{x}, t)$	total pressure
$P_0(\vec{x})$	static (mean) pressure
$q(\vec{y}, t)$	rate of mass flux per unit volume ($kg\ m^{-3}\ s^{-1}$)
$q'(\vec{y}, t), q'(\vec{y}, t - r/c)$	rate of change of mass flux per unit volume ($kg\ m^{-3}\ s^{-2}$)
$\mathbf{Q_d}$	complex dipole source strength
Q_p	peak source strength
Q_{rms}	root-mean-square source strength
$Q(t), \mathbf{Q}(t), \mathbf{Q}_1(t)$, etc.	source strength ($m^3\ s^{-1}$; bold signifies complex)
$\mathbf{Q_d}(t - r/c)$	complex dipole source strength as a function of retarded time

$Q(t)$	mass flux per unit time ($kg\,s^{-1}$)
$Q'(t)$	rate of change of mass flux ($kg\,s^{-2}$)
r	radius, radial distance
r_1, r_2, etc.	radial distances
R	gas constant
$R_1(2kz), R_1(x)$	resistive function associated with the radiation impedance of a piston
S	surface, entropy
t	time
T	time, kinetic energy, temperature
$T_{ij}, T_{ij}(\vec{y}, t - r/c)$	Lighthill stress tensor ($r = \lvert\vec{x} - \vec{y}\rvert$)
$T(\vec{x}, t)$	total temperature
$T_0(\vec{x})$	mean temperature
T_k	absolute temperature
$T'(\vec{x}, t)$	temperature fluctuation
u	fluid velocity fluctuation, acoustic particle velocity (scalar quantity)
$\vec{u}, \hat{\mathbf{u}}$	fluid velocity fluctuation, acoustic particle velocity, irrotational unsteady velocity component (vector quantity, bold signifies complex)
$\mathbf{u_a}$	complex normal surface velocity of an oscillating sphere
u_i, u_j	fluid velocity fluctuations and/or acoustic particle velocities in tensor notation
u_r	fluid velocity fluctuation and/or acoustic particle velocity in the *r*-direction
u_x	fluid velocity fluctuation and/or acoustic particle velocity in the *x*-direction
u_y	fluid velocity fluctuation and/or acoustic particle velocity in the *y*-direction
u_z	fluid velocity fluctuation and/or acoustic particle velocity in the *z*-direction
$u(r, t), \mathbf{u}(r, t)$	fluid velocity fluctuation and/or acoustic particle velocity as a function of radial distance (scalar quantity, bold signifies complex)
$u_a(t)$	instantaneous normal surface velocity of an oscillating sphere
$u(x, t)$	fluid velocity fluctuation, acoustic particle velocity (scalar quantity)

$\vec{u}(\vec{x}, t)$, $\hat{\mathbf{u}}(\vec{x}, t)$	fluid velocity fluctuation, acoustic particle velocity (vector quantity, bold signifies complex)
$\hat{\mathbf{u}}^*$	complex conjugate of $\hat{\mathbf{u}}$ (vector quantity)
$\vec{\mathrm{u}}$	surface velocity (vector quantity)
U	potential energy, mean fluid velocity
$\mathbf{U}$	complex piston surface velocity
U_a	peak normal surface velocity of an oscillating sphere
U_{p}	peak piston surface velocity, peak cylindrical line source surface velocity
$\mathbf{U}_{\mathbf{r}}$	complex surface velocity of a cylindrical line source
$\vec{U}(\vec{x}, t)$	total fluid velocity (vector quantity)
$\vec{U}_0(\vec{x})$	mean fluid velocity (vector quantity)
$\vec{v}$	total (mean plus unsteady) rotational incompressible component of the velocity field in a turbulent flow regime, vorticity convection velocity (vector quantity)
V	volume
$\vec{x}$	position vector in the x-direction
$X_1(2kz)$, $X_1(x)$	reactive function associated with the radiation impedance of a piston
X_{r}	reactive component of the radiation impedance of a piston
$\vec{y}$	position vector in the y-direction
z	piston radius
$\vec{z}$	position vector in the z-direction
Z_{a}, $\mathbf{Z}_{\mathbf{a}}$	specific acoustic impedance (bold signifies complex)
$\mathbf{Z}_{\mathbf{s}}$	specific acoustic impedance of a reflecting surface (complex function)
$\mathbf{Z}_{\mathbf{m}}$	mechanical impedance (complex function)
$\mathbf{Z}_{\mathbf{r}}$	radiation impedance (complex function)
α	oblique angle of incidence
γ	ratio of specific heats
$\delta(\tau - t)$, $\delta(\tau - t + r/c)$, $\delta(\vec{x} - \vec{y})$	unit impulse functions (delta functions)
δ_{ij}	unit impulse function in tensor notation
θ, θ_i	angle (spherical co-ordinates)
λ	wavelength

ν	coefficient of shear viscosity
π	3.14 . . .
Π	power, sound power
Π_D	sound power radiated by a dipole
Π_M	sound power radiated by a monopole
Π_Q	sound power radiated by a quadrupole
$\rho, \rho(\vec{x}, t)$	total fluid density
$\rho_0, \rho_0(\vec{x})$	mean fluid density
$\rho', \rho'(\vec{x}, t)$	fluid density fluctuation
τ_{ij}	viscous shear stress (in tensor notation)
Γ	reflection coefficient
ϕ	acoustic velocity potential, scalar velocity potential
$\phi(r, t), \boldsymbol{\phi}(r, t)$	acoustic velocity potential as a function of radial distance (bold signifies complex)
$\phi(r, \theta, t), \boldsymbol{\phi}(r, \theta, t)$	acoustic velocity potential as a function of radial and angular distance (bold signifies complex)
$\phi(\vec{x}, t), \boldsymbol{\phi}(\vec{x}, t)$	acoustic velocity potential (bold signifies complex)
$\vec{\psi}$	vector velocity potential
ω	radian (circular) frequency
$\vec{\omega}$	total vorticity (mean plus unsteady; vector quantity)
$\vec{\nabla}$	divergence operator (vector quantity)

3
Interactions between sound waves and solid structures

3.1 Introduction

Wave–mode duality concepts were introduced and discussed in some detail in chapter 1. It was pointed out that, whilst the lumped-parameter approach to mechanical vibrations is adequate to describe mode shapes and natural frequencies, it is not suitable for relating vibrations to radiated noise. One therefore has to use the fundamental wave approach to obtain an understanding of the essential features of mechanical vibrations as they relate to sound radiation and sound transmission. These interactions between sound waves and the mechanical vibrations of solid structures form a very important part of engineering noise and vibration control.

Because solids can store energy in shear and compression, all types of waves can be sustained in structures – i.e. compressional (longitudinal) waves, flexural (transverse or bending) waves, shear waves and torsional waves. On the other hand, since fluids can only store energy in compression, they can only sustain compressional (longitudinal) waves. For reasons which will become evident later on on this chapter, flexural (bending) waves are the only type of structural wave that plays a direct part in sound radiation and transmission. At this stage it is sufficient to note that the primary reason for this is that the bending wave particle velocities are perpendicular to the direction of wave propagation (see Figure 1.1b) resulting in an effective exchange of energy between the structure and the fluid.

Fluctuating pressures in close proximity to any arbitrary surface will generate an acoustic radiation load on that surface. This is in addition to any mechanical excitation of the surface which could be the primary source of vibration in the first instance (e.g. the baffled piston which was considered in the preceding chapter). If the fluid medium is air (as is usually the case in engineering noise control), then this acoustic radiation load is generally very small and the sound pressure field at regions away from the source can be estimated from the bending wave particle velocities on the surface of the structure. If, however, the fluid medium is a liquid, then the acoustic radiation load can become very significant and has to be accounted for – the radiation load modifies the forces acting on the structure, a feedback coupling between the fluid and the structure is set up, and the structure subsequently becomes 'fluid-loaded'. This chapter is, however, mainly concerned with structure-borne sound in the audio frequency range with air as the fluid medium. These conditions are often representative of typical engineering noise control problems such as sound radiation from plates, shells and cylinders in industrial type environments and sound transmission through building partitions. As such, for the larger part of this chapter, fluid-loading effects can be neglected.

3.2 Fundamentals of fluid–structure interactions

At its most fundamental level, the radiation of sound from an arbitrary vibrating body can be formulated in terms of an integral equation involving Green's functions with an imposed radiation condition – i.e. the radiation condition ensures that the integral equation for the radiated sound pressure represents outward travelling sound waves. Green's functions were introduced in chapter 2 (see sub-section 2.4.1) and they represent solutions to the wave equation – they can also be considered to be either frequency response functions or impulse response functions between the source and receiver. In its most general form, the integral equation is attributable to Kirchhoff, although Helmholtz modified it for single frequency (harmonic) applications. The derivation of the integral, and a discussion about the radiation condition is provided in the advanced literature (Junger and Feit[3.1], Pierce[3.2]). The integral is sometimes referred to as the Kirchhoff–Helmholtz integral equation, and Fahy[3.3,3.4] provides a useful discussion on its physical significance together with some examples. The Kirchhoff–Helmholtz integral equation relates harmonic surface vibrational motion on any arbitrary body to the radiated sound pressure field in the surrounding fluid. It is

$$\mathbf{p}(\vec{r})=\int_S\left\{\mathbf{p}(\vec{r}_0)\frac{\partial\mathbf{G}_\omega(\vec{r},\omega|\vec{r}_0,\omega)}{\partial\vec{n}}+\mathrm{i}\omega\rho_0\dot{\mathbf{u}}_\mathbf{n}(\vec{r}_0)\mathbf{G}_\omega(\vec{r},\omega|\vec{r}_0,\omega)\right\}\mathrm{d}\vec{S}, \tag{3.1}$$

where $\vec{r}$ is a position vector at some receiver position in the sound field, $\vec{r}_0$ is a position vector on the vibrating body, $\vec{n}$ is the unit normal vector, $\mathbf{p}(\vec{r}_0)$ is the surface pressure on the body, and $\mathrm{i}\omega\dot{\mathbf{u}}_\mathbf{n}(\vec{r}_0)$ is the normal surface acceleration. $\mathbf{G}_\omega$ is the frequency domain Green's function – it is a solution to the wave equation for a harmonic source. For a point source, it is given by equation (2.163) – i.e. $\mathrm{e}^{\mathrm{i}kr}/4\pi r$, where $r=|\vec{r}-\vec{r}_0|$ is the modulus of the distance between the source and receiver positions. It should be noted that the acoustic pressure fluctuations are a function of both space and time thus, $\mathbf{p}(\vec{r},t)=\mathbf{p}(\vec{r})\,\mathrm{e}^{\mathrm{i}\omega t}=p_{\max}\,\mathrm{e}^{-\mathrm{i}kr}\,\mathrm{e}^{\mathrm{i}\omega t}$ etc. Equation (3.1) can thus be interpreted as representing the radiating sound pressure field of a vibrating body by a distribution of point sources and forces on the surface of the body. The point sources and the forces are functions of surface pressure and surface acceleration, respectively. It is important to note that the surface pressure and the normal surface vibrational velocity are inter-related and not independent of each other. In practice, analytical Green's functions for bodies other than those that are suitably represented by combinations of point sources can only be constructed for geometries such as plates, cylinders etc. Examples relating to source configurations of practical interest may be found in the advanced literature[3.1–3.3]. By an appropriate selection of co-ordinates, the normal derivative of the Green's function can be forced to be zero, thus eliminating the requirement for a knowledge of the surface pressure distributions – i.e. only a knowledge of the surface vibrational velocity is required. On arbitrary, complicated, three-dimensional bodies, such as large industrial machinery etc., analytical solutions are generally not possible, and the usual procedure is either to use numerical techniques to solve the integral equation[3.3], or to use experimental techniques to establish the Green's function.

Rayleigh modified equation (3.1) for the specific case of a planar source located in an infinite baffle and illustrated that it is equivalent to a distribution of point sources. Rayleigh's equation for the radiated sound pressure from a planar source located in an infinite baffle is[3.1–3.3]

$$\mathbf{p}(\vec{r}) = \int_S \mathrm{i}\omega\rho_0\hat{\mathbf{u}}_\mathbf{n}(\vec{r}_0)\frac{\mathrm{e}^{ikr}}{2\pi r}\,\mathrm{d}\vec{S}, \tag{3.2}$$

where $r = |\vec{r} - \vec{r}_0|$. Once again, $\mathbf{p}(\vec{r}, t) = \mathbf{p}(\vec{r})\,\mathrm{e}^{\mathrm{i}\omega t} = p_{\max}\,\mathrm{e}^{-ikr}\,\mathrm{e}^{\mathrm{i}\omega t}$ etc. There is a factor of two on the denominator instead of four because of the pressure doubling that occurs due to the presence of the planar surface (see chapter 2, sub-section 2.3.3). Equation (3.2) was used in sub-section 2.3.4 in the previous chapter to estimate the far-field radiated sound pressure from a piston in an infinite baffle – in that case it was derived from the solution for a point source.

It was mentioned in the previous chapter that the Green's function is symmetrical – source and receiver positions can be interchanged. This property, commonly referred to as the principle of reciprocity, has very important applications in engineering noise and vibration analysis – for instance, reciprocal experiments can be set up to experimentally determine the Green's function. Lyamshev[3.5] is largely responsible for extending the principle of reciprocity to linear, acoustically coupled, structures and fluids. The concept is best illustrated by means of a simple example – the circular piston vibrating in an infinite baffle.

The total impedance of a circular piston of radius z vibrating in an infinite baffle is given by (see chapter 2, sub-section 2.3.4)

$$\mathbf{Z} = \mathbf{Z}_\mathbf{m} + (\mathbf{F}_\mathbf{p}/\mathbf{U}) = \mathbf{Z}_\mathbf{m} + \mathbf{Z}_\mathbf{r} = \frac{\mathbf{F}_\mathbf{m}}{\mathbf{U}}, \tag{3.3}$$

where $\mathbf{Z}_\mathbf{m}$ is the mechanical impedance of the piston, $\mathbf{Z}_\mathbf{r}$ is the aoustic radiation impedance, $\mathbf{U}$ is the piston velocity, $\mathbf{F}_\mathbf{m}$ is the applied mechanical force, and $\mathbf{F}_\mathbf{p}$ is the force on the piston due to the surface acoustic pressure. The far-field acoustic pressure at some distance r from the vibrating piston can be obtained from equation (2.123). Only situations where $kz \ll 1$ – i.e. low frequencies and compact piston radii will be considered in this example in order to simplify the mathematics so that attention can be focused on the concepts involved. In this instance, the radiated sound pressure is

$$\mathbf{p}(\vec{r}, t) = \frac{ik\rho_0 c\pi z^2 \mathbf{U}\, e^{-ikr}}{2\pi r}, \tag{3.4}$$

and by replacing $\mathbf{U}$ by equation (3.3) it becomes

$$\mathbf{p}(\vec{r}, t) = \frac{i\omega\rho_0 z^2 \mathbf{F_m}\, e^{-ikr}}{2r(\mathbf{Z_m} + \mathbf{Z_r})}, \tag{3.5}$$

where $\mathbf{F_m} = F_m\, e^{i\omega t}$. Thus, for a mechanically excited piston in an infinite baffle, the ratio of the radiated sound pressure to the mechanical excitation force is

$$\frac{\mathbf{p}(\vec{r}, t)}{\mathbf{F_m}} = \frac{i\omega\rho_0 z^2\, e^{-ikr}}{2r(\mathbf{Z_m} + \mathbf{Z_r})}. \tag{3.6}$$

Now consider the reciprocal experiment where the piston is excited acoustically by an omni-directional point source, $\mathbf{Q}(t) = Q_p\, e^{i\omega t}$, which is placed at the receiver position. This reciprocal experiment is illustrated in Figure 3.1. If the piston was held stationary such that it could not vibrate and radiate sound, the 'blocked' pressure on the piston surface, due to the omni-directional point source at the receiver position, would be double the pressure at that position on the piston surface if the piston were removed. This pressure doubling is simply a result of the effects of the rigid boundary. The blocked pressure is thus (see equation 2.85)

$$\mathbf{p_b}(\vec{r}, t) = \frac{i\omega\rho_0 \mathbf{Q}(t)\, e^{-ikr}}{2\pi r}. \tag{3.7}$$

Fig. 3.1. Schematic illustration of the reciprocal experiment.

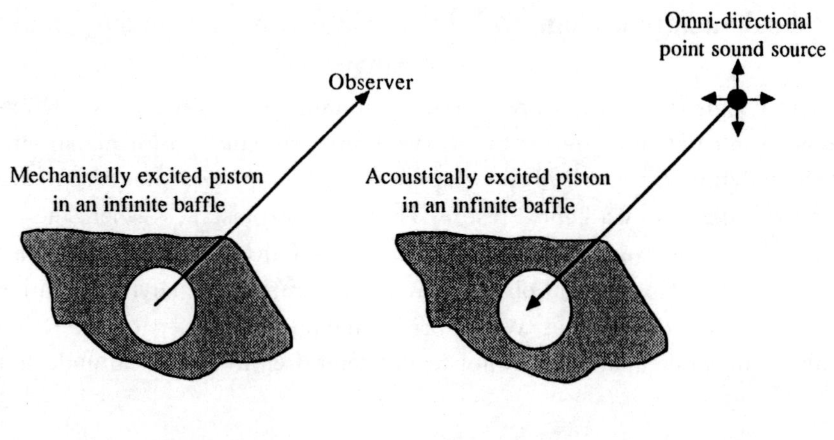

In this instance, the only forces acting upon the piston are produced by the blocked pressure. Thus, the piston velocity that would be expected to be produced by this blocked pressure is

$$\mathbf{U} = \frac{\pi z^2 \mathbf{p}_b(\vec{r}, t)}{\mathbf{Z}_m + \mathbf{Z}_r} = \frac{i\omega\rho_0 z^2 \mathbf{Q}(t)\, e^{-ikr}}{2r(\mathbf{Z}_m + \mathbf{Z}_r)}. \tag{3.8}$$

Hence, for an acoustically excited piston in an infinite baffle, the ratio of the piston surface velocity to the volume velocity of the point source is

$$\frac{\mathbf{U}}{\mathbf{Q}(t)} = \frac{i\omega\rho_0 z^2\, e^{-ikr}}{2r(\mathbf{Z}_m + \mathbf{Z}_r)}. \tag{3.9}$$

Equation (3.9), which is related to the acoustic excitation of the piston, is identical to equation (3.6), which is related to the mechanical excitation. Thus

$$\frac{\mathbf{p}(\vec{r}, t)}{\mathbf{F}_m} = \frac{\mathbf{U}}{\mathbf{Q}(t)}. \tag{3.10}$$

Equation (3.10) illustrates that the two reciprocal experiments are identical! Several important comments can now be made. Firstly, it should be recognised that the ratio of the blocked pressure to the volume velocity of the point source yielded the free space Green's function (see equation 3.7). This suggests that the free space Green's function for some complicated structure or machine could be evaluated experimentally by setting up a controlled experiment. Secondly, the sound pressure produced at some point in space by mechanical excitation at some point on a structure can be estimated by simply measuring the vibrational level at that point (on the structure) when the structure is excited by an omni-directional acoustic source which is located at the point of interest in space. Thus, by conducting a series of simple experiments in this fashion, an optimum position for the application of the mechanical force can be found such as to minimise the sound radiation. This very powerful tool is widely used in engineering noise control and will be discussed again later on in this chapter.

3.3 Sound radiation from an infinite plate – wave/boundary matching concepts

Many problems in acoustics are solved by suitable application of the Kirchhoff–Helmholtz integral equation (or the Rayleigh integral equation for planar surfaces) – either analytically, numerically or experimentally. For instance, it is used to evaluate the sound radiation by a baffled piston. There are also instances where the integral is intractable. As an example, Junger and Feit[3.1] and Fahy[3.3] use the Rayleigh integral equation for a rectangular flat plate – they demonstrate that analytical solutions are only available for the far-field. Whilst being fundamental and exact, the Kirchhoff–Helmholtz integral equation does not readily lend itself to a physical understanding of how structures radiate sound. It is more appropriate, and quite consistent with the fundamental Kirchhoff–Helmholz integral equation, to analyse the waves that travel along a structure–fluid interface and to match the acoustic particle velocities in the fluid to the normal surface vibrational velocities in the structure. For a planar surface, the Green's function is such that the normal derivative on the surface of the structure is zero; furthermore when the fluid medium is air (as is usually the case in engineering noise and vibration control), the radiation load due to the fluctuating surface pressures is very small and can therefore be neglected – hence wave/boundary matching can also be applied to non-planar surfaces.

The fundamental basis of wave/boundary matching is (i) the Kirchhoff–Helmholtz integral equation, and (ii) the fact that it is the bending waves in a structure that radiate sound. Both these points are consistent with each other – the bending wave particle velocities are perpendicular to the direction of wave propagation, and this is consistent with the normal surface acceleration terms dominating the Kirchhoff–Helmholtz integral equation.

Before proceeding with a discussion on how sound waves and structural waves interact with each other it is worth recapitulating some of the fundamentals of wave propagation in structures. An elementary analysis of wave motion in a bar or a plate (using the wave equation) readily shows that

$$c_B = (1.8 c_L t f)^{1/2}, \tag{3.11}$$

and

$$\lambda_B = \frac{c_B}{f} = \left(\frac{1.8 c_L t}{f}\right)^{1/2}, \tag{3.12}$$

where c_B is the bending wave velocity, λ_B is the corresponding bending wavelength, t is the thickness of the plate or bar, f is the frequency in hertz, and c_L is the longitudinal wave velocity. The equation for plate motion yields similar results to that of a bar, and the relationships between the frequency and velocity of a propagating free wave are the same for both cases. Equation (3.11) was derived from first principles in chapter 1 (sub-section 1.9.4 for a beam/bar, and sub-section 1.9.8 for a plate). For a bar, c_L is given by $(E/\rho)^{1/2}$, and for a plate c_L is given by $\{E/\rho(1-\nu^2)\}^{1/2}$, where ν is the Poisson's ratio – Poisson contraction effects are neglected in the thin beam/bar analysis. Now, the very important result to be noted here is that the bending wave velocity, c_B, is dispersive – it varies with frequency for a given material and thickness. An appreciation of this dependence of wave velocity on frequency is critical to an understanding of how structures radiate sound.

The study of the interactions between simple plate-type structures and sound is important for an appreciation of structure-borne sound in more complex geometries. From the point of view of sound radiation, the most important parameter is the ratio of the surface displacement (bending) wavelength, λ_B, to the corresponding acoustic wavelength, λ, at the same frequency. Consider an undamped, infinite, thin plate which can sustain such a structural wavelength, λ_B, at some frequency, f. The corresponding acoustic wavelength, $\lambda = c/f$, at the same frequency (c is the speed of sound) can be sustained in the fluid medium surrounding the plate. The wavelength ratio between a structural wave and a sound wave at the same frequency is thus

$$\frac{\lambda_B}{\lambda} = \left(\frac{1.8c_L tf}{c^2}\right)^{1/2}. \tag{3.13}$$

One would thus expect different efficiencies of sound radiation from the structure depending on whether $\lambda_B/\lambda > 1$ or < 1. This concept of an 'efficiency' of sound radiation from a structure is an important one in structure-borne sound. It leads to the development of a 'radiation ratio' for a given structural element such as a plate, a cylinder, a shell, etc. Radiation ratios will be defined and discussed later on in this chapter, and they will be used throughout the remainder of this book. Returning for the moment to equation (3.13), it can be seen that $\lambda_B = \lambda$ when $c_B = c$, i.e. when the bending wave velocity in the structure equals the speed of sound in the fluid. The critical frequency at which this occurs can be obtained from equation (3.11) by equating c_B to c. Thus,

$$f_C = \frac{c^2}{1.8c_L t}, \tag{3.14}$$

where f_C is the plate critical frequency. Intuitively, one would expect very efficient sound radiation from the plate at frequencies greater than or equal to f_C. This will be demonstrated shortly. If equation (3.14) is substituted into equation (3.13), another important fundamental relationship is obtained, i.e.

$$\frac{\lambda_B}{\lambda} = \left(\frac{f}{f_C}\right)^{1/2}. \tag{3.15}$$

It is important to remember that λ_B is the structural wavelength at a frequency f, λ is the corresponding acoustic wavelength at the same frequency, f_C is the critical frequency of the structure, and f is the frequency of interest corresponding to both λ_B and λ.

Now consider an undamped, infinite plate mechanically driven to carry a plane bending wave of constant amplitude and propagation speed c_B. The plate is illustrated in Figure 3.2. The wave fronts of the plane sound wave that is radiated outwards into free space are such that

$$\lambda = \lambda_B \sin\theta, \tag{3.16}$$

i.e. the plane sound wave propagates in a direction which is perpendicular to the wave fronts. It is useful to re-introduce the concept of the wavenumber ($k = \omega/c$) in the analysis of structure-borne sound. The wavenumber can be looked upon as a spatial frequency parameter which is inversely proportional to wavelength. The bending wavenumber of the plate is

$$k_B = \frac{2\pi}{\lambda_B} = \frac{\omega}{c_B} = 2\pi\left(\frac{f}{1.8c_L t}\right)^{1/2}. \tag{3.17}$$

The corresponding acoustic wavenumber for the plane sound wave that is radiated into free space is

$$k = \frac{2\pi}{\lambda} = \frac{\omega}{c}. \tag{3.18}$$

The directions of the bending and acoustic wavenumbers correspond to their respective directions of propagation. This is illustrated in Figure 3.2.

It can be seen from Figure 3.2 that for sound radiation into the fluid medium (gas or liquid – usually air for typical engineering noise control problems), the angle θ must be such that $\lambda_B > \lambda$ or $f > f_C$. The bending wave travels faster than the speed of sound in the fluid, and it is only above the critical frequency that the free waves in the mechanically driven infinite plate radiate sound efficiently. It is important to note that the direction and magnitude of the sound wave in the ambient fluid are governed by the bending wavenumber, k_B, in the plate. This is evident from the wavenumber vector triangle in Figure 3.2. This observation is consistent with the Kirchhoff–Helmholtz integral equation – i.e. the sound pressure is a function of the

Fig. 3.2. Sound radiation from an infinite plate.

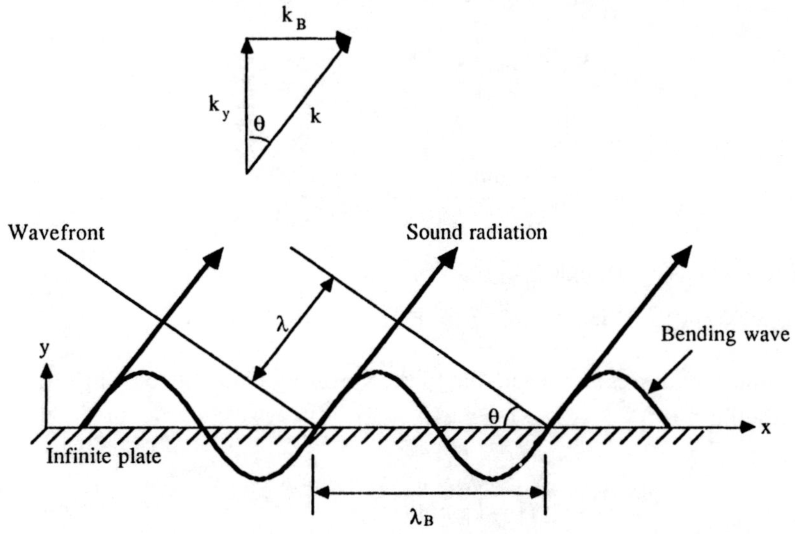

normal surface vibrational velocities and the surface pressure distribution due to any acoustic radiation load. The acoustic radiation load component is neglected if the fluid medium is air. The infinite plate model is a good approximation for finite plates provided that $\lambda_B \ll l$, where l is the plate length. Also, in relation to the above discussion, the converse is also true – i.e. a sound wave incident upon a plate at an angle θ can excite bending waves in it (see sections 3.5 and 3.9).

From fundamental acoustics (see chapter 2, sub-section 2.2.6), the sound pressure level, due to the vibrating plate, at some arbitrary point (x, y) in the fluid is

$$\mathbf{p}(x, y, t) = \mathbf{p}(x, y)\,e^{i\omega t} = p_{max}\,e^{-ik_B x}\,e^{-ik_y y}\,e^{i\omega t}. \tag{3.19}$$

The above equation satisfies the two-dimensional wave equation – i.e. it is a function of both x and y and represents an undamped, plane sound wave. Now, the acoustic particle velocity, $\vec{u}$, in a fluid is related to the sound pressure, p, by (see equation 2.49)

$$\vec{u} = -\int \frac{1}{\rho_0} \vec{\nabla} p \, dt, \tag{3.20}$$

or for a harmonic wave it is

$$\vec{u} = -\frac{1}{i\omega\rho_0} \vec{\nabla} p. \tag{3.21}$$

The wave/boundary matching condition is that the component of the acoustic particle velocity which is perpendicular to the plate has to equal the normal plate vibrational velocity at the surface. Thus,

$$(\mathbf{u}_{y\,\mathrm{fluid}})_{y=0} = \mathbf{u}_{y\,\mathrm{plate}} = \mathbf{u}_{yp} = -\frac{1}{i\omega\rho_0}\left\{\frac{\partial p}{\partial y}\right\}_{y=0}. \tag{3.22}$$

Thus,

$$\mathbf{u}_{yp} = u_{yp\max}\,e^{-ik_B x} = \frac{k_y p_{max}\,e^{-ik_B x}}{\omega\rho_0}. \tag{3.23}$$

Thus,

$$p_{max} = \frac{\omega\rho_0 u_{yp\max}}{k_y} = \frac{kc\rho_0 u_{yp\max}}{(k^2 - k_B^2)^{1/2}}, \tag{3.24}$$

since from the vector triangle

$$k_y^2 = k^2 - k_B^2. \tag{3.25}$$

Hence, the sound pressure level, due to the vibrating plate, at some arbitrary point (x, y) in the fluid is

$$\mathbf{p}(x, y, t) = \frac{c\rho_0 u_{yp\max}}{(1 - k_B^2/k^2)^{1/2}}\,e^{i\omega t}\,e^{-ik_B x}\,e^{-iy(k^2 - k_B^2)^{1/2}}. \tag{3.26}$$

Equation (3.26) illustrates that the sound wave generated by a bending wave in a mechanically driven, infinite plate is a plane wave. The wave fronts do not spread with increasing distance from the source, and therefore any decay in sound pressure with distance from the source is only a function of any resistance or damping in the fluid (in air this is very small). It therefore takes a very long distance and time for true plane waves to decay.

The preceding analysis demonstrates that a harmonically excited, infinite plate can only generate plane sound waves in the adjacent fluid. When $k_B < k$ (i.e. $\lambda_B > \lambda$) the radiated sound pressure is positive and real – i.e. plane sound waves are radiated from the plate. However, when $k_B > k$ (i.e. $\lambda_B < \lambda$) the third exponential term in the above equation is real and decays exponentially as the distance, y, from the plate increases – i.e. no sound waves are radiated and only a near-field exists. When $k_B = k$, the theory suggests that the radiated sound pressure level goes to infinity. This is of course not possible in practice as all real surfaces are finite and not infinite as conveniently assumed here! It is sufficient to note that in practice on real, finite structures the sound radiation at $k_B = k$ is very high. Also, for finite structures the radiated sound decays with distance from the source; in the above example, there is no decay of sound with distance from the source! The boundary conditions that are associated with real, finite structures produce standing, structural waves, and the associated natural frequencies and mode shapes. These mode shapes produce pockets of oscillations which can be interpreted as being oppositely phased point sources. As was seen in chapter 2, point sound sources produce spherical sound waves – i.e. the wave fronts spread with increasing distance from the source. This spherical extension of the wave front produces a drop in the level of the pressure fluctuations associated with the wave. For a simple point source, it varies with r^{-1} (where r is the distance from the source). Thus, finite structures are in fact arrays of point sources with surface velocity distributions which are generally rather complicated, and not harmonic as tacitly assumed here. Generally, there are also complicated phase relationships between them. Radiated sound pressure distributions associated with the vibration of finite structures can be modelled (more realistically) in terms of arrays of point sources via the Kirchhoff–Helmholtz or the Rayleigh integral equation. The concepts relating to sound radiation from finite structures will be discussed in section 3.5. For the moment, it is sufficient to note that, for finite structures, the radiated sound decreases with distance from the structure because the sound waves are no longer plane. Figure 3.3 illustrates the difference between sound radiation from planar and spherical sources – at large distances from the source, spherical waves approximate to plane waves.

The analysis in this section, whilst being restricted to infinite plates, illustrates that if $k_B < k$ (i.e. $\lambda_B > \lambda$), the plate radiates a sound wave into the ambient fluid at some angle θ which is defined by the relevant wavenumber vectors. On the other hand, if

$k_B > k$ (i.e. $\lambda_B < \lambda$), then no nett sound is radiated away from the plate. This is a very important conclusion, and the concept of a critical frequency is very relevant to sound radiation from finite structures. As a general rule, there is very efficient sound radiation from finite structures when $k_B < k$ (i.e. $\lambda_B > \lambda$). Unlike infinite plates, however, there can also be significant sound radiation below the critical frequency when $k_B > k$ (i.e. $\lambda_B < \lambda$). For mechanical excitation of the structure, this is primarily because of the existence of end or boundary conditions; for acoustic excitation it is due to both the boundary conditions, and the forced response of the structure at the frequency of excitation. These mechanisms of sound radiation from finite structures at frequencies below the critical frequency will be discussed in section 3.5.

3.4 Introductory radiation ratio concepts

Consider a large, rigid piston (i.e. all parts of the piston vibrate in phase) vibrating in an infinite baffle. If the piston's dimensions are such that its circumference is very much larger than the corresponding acoustic wavelength in the fluid, then the particle velocity of the fluid has to equal the normal surface vibrational velocity – the air cannot be displaced. In this instance, the sound that is radiated from the vibrating piston is normal to its surface. The sound power that is radiated by the piston into the surrounding medium is simply the force times velocity – i.e.

$$\Pi = \pi z^2 p_{rms} u_{rms}, \tag{3.27}$$

Fig. 3.3. The difference between sound radiation from planar and spherical sound sources.

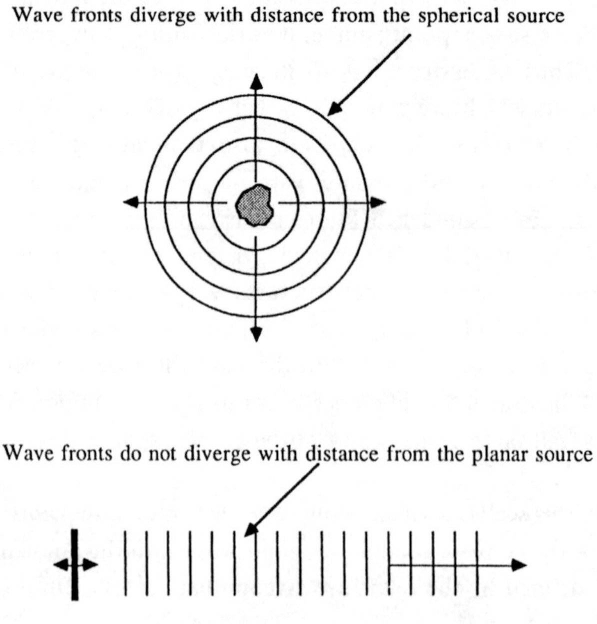

Wave fronts diverge with distance from the spherical source
Wave fronts do not diverge with distance from the planar source

where p_{rms} is the root-mean-square radiated pressure at some point in space, u_{rms} is the corresponding root-mean-square acoustic particle velocity at the same point, and z is the radius of the vibrating piston. From fundamental acoustics (see equation 2.63 or 3.24 and noting that $k_y = k \cos\theta$ and $u_{yp} = u \cos\theta$),

$$u = \frac{p}{\rho_0 c}. \tag{3.28}$$

Thus, for the large, rigid piston

$$\Pi = \rho_0 c S \langle \overline{u^2} \rangle, \tag{3.29}$$

where $S = \pi z^2$, $\langle\ \rangle$ represents a time average and $\overline{}$ represents a space average (also see chapter 2, sub-section 2.3.4, equation 2.140).

The radiation ratio, σ, of an arbitrary structure is defined as the sound power radiated by the structure into half space (i.e. one side of the structure) divided by the sound power radiated by a large piston with the same surface area and vibrating with the same r.m.s. velocity as the structure. The radiation ratio thus describes the efficiency with which the structure radiates sound as compared with a piston of the same surface area, i.e. the piston has a radiation ratio of unity. Hence, for an arbitrary structure, with some time- ($\langle\ \rangle$) and space-averaged ($\overline{}$) mean-square vibrational velocity, v^2, the radiated sound power is

$$\Pi = \sigma \rho_0 c S \langle \overline{v^2} \rangle, \tag{3.30}$$

where S is the radiating surface area of the structure, ρ_0 is the density of the fluid medium into which the structure radiates, and c is the speed of sound in the fluid medium. It should be noted that the mean-square space- and time-averaged vibrational velocity is in fact the averaged normal surface velocity. The radiation ratio, σ, thus provides a powerful relationship between the structural vibrations and the associated radiated sound power. The radiation ratio can be either greater or less than unity, hence it is more appropriate to use the term ratio rather than the term efficiency which is sometimes used in the literature. If values or relationships for radiation ratios of different types of structural elements can be established, then the estimation of the noise radiation and any subsequent noise control is a relatively easier process – i.e. radiated sound power can be estimated directly from surface vibration levels which can be obtained either theoretically or experimentally.

Consider the infinite, flat plate of the last section. In that example, the radiation ratio can be obtained from an analysis of the velocity of the plate and the velocity of the associated sound wave. Sound radiation is only defined for those waves where $\lambda_B > \lambda$ or $f > f_C$. If $\lambda_B < \lambda$, a near-field which attenuates very rapidly is present and the sound pressure is out of phase with the plate velocity. Subsequently no sound is radiated and the radiation ratio is zero. For the first case, ($\lambda_B > \lambda$ or $f > f_C$), the normal plate velocity has to be equal to the component of the acoustic particle velocity which is perpendicular to the plate surface. From fundamental acoustics, the acoustic particle velocity, u, which is given by $p/\rho_0 c$ (see equation 2.63), is perpendicular to the wave front. This is illustrated in Figure 3.4. From the figure, it can be seen that

$$u_{\mathrm{yp}} = \frac{p}{\rho_0 c} \cos\theta. \tag{3.31}$$

The sound power radiated by the infinite plate is

$$\Pi = S p_{\mathrm{rms}} u_{\mathrm{yp_{rms}}} = S u_{\mathrm{rms}} \rho_0 c u_{\mathrm{yp_{rms}}}. \tag{3.32}$$

Equating this to equation (3.30) and solving for the radiation ratio, σ, yields

$$\sigma = \frac{S u_{\mathrm{rms}} \rho_0 c u_{\mathrm{yp_{rms}}}}{\rho_0 c S u^2_{\mathrm{yp_{rms}}}} = \frac{u_{\mathrm{rms}}}{u_{\mathrm{yp_{rms}}}}. \tag{3.33}$$

Thus, since $y_{\mathrm{yp}} = u\cos\theta$,

$$\sigma = \frac{1}{\cos\theta} = \frac{k}{k_y}. \tag{3.34}$$

By substituting equation (3.25) into equation (3.34) one gets

$$\sigma = \frac{1}{(1 - k_{\mathrm{B}}^2/k^2)^{1/2}} = \frac{1}{\left(1 - \dfrac{f_{\mathrm{c}}}{f}\right)^{1/2}}. \tag{3.35}$$

Equation (3.35) represents the radiation ratio for an infinite, undamped flat plate. At the critical frequency, a singularity arises and the physical interpretation of this is that if the plate velocity were constant the radiation ratio would approach infinity. At this critical frequency, θ is 90° and the radiated sound wave is parallel to the

Fig. 3.4. Relationship between normal plate velocity and radiated sound wave.

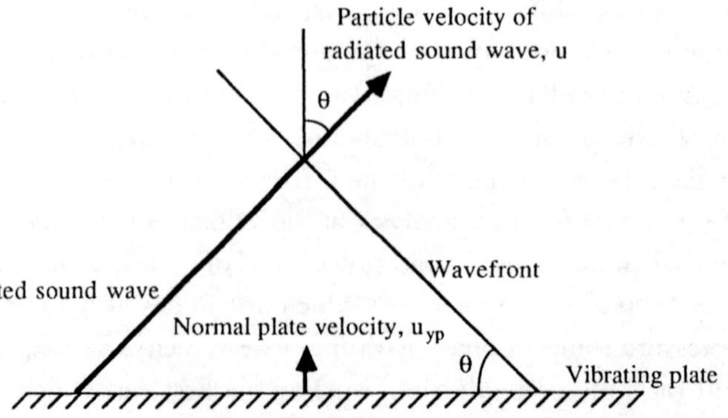

surface of the plate. The wavelengths (λ_B and λ) are equal and the sound is radiated very efficiently. At frequencies above the critical frequency, the radiation ratio approaches unity. This is illustrated in Figure 3.5.

Now consider another fundamental sound radiator – a spherical sound source. Many practical sound sources can be modelled as combinations of spherical oscillators provided that (i) the dimensions of the source are small compared to the wavelength of sound being generated, or (ii) the source is sufficiently far away from the receiver such that it is perceived to be a spherical source. In chapter 2 (sub-section 2.3.1) the sound power radiated by a simple spherical sound source was shown to be

$$\Pi = \frac{Q_{\rm rms}^2 k^2 \rho_0 c}{4\pi(1 + k^2a^2)} = \frac{Q_{\rm rms}^2}{4\pi a^2}\rho_0 c \frac{k^2a^2}{(1 + k^2a^2)}, \tag{3.36}$$

where $Q_{\rm rms} = Q_{\rm p}/\sqrt{2}$ and $Q_{\rm p} = 4\pi a^2 U_a$ (see equation 2.82). The radiation ratio of the spherical sound source is defined in the usual manner by equation (3.30). The mean-square, normal velocity of the oscillating sphere is $U_a^2/2$, thus using equation (3.30) and equation (3.36) it can be readily shown that, for a spherical sound source,

$$\sigma = \frac{k^2a^2}{(1 + k^2a^2)}. \tag{3.37}$$

The radiation ratio for a spherical sound source is illustrated in Figure 3.6. An important observation is that the radiation ratio is not a function of frequency but of the wavenumber multiplied by a typical structural dimension (the radius of the sphere in this instance). It can be shown that the radiation ratios of most bodies resolve into functions of ka. For plate-type structural or machine elements they are also a function of the ratio of the bending wave frequency to the critical frequency,

Fig. 3.5. Radiation ratios for bending waves on an infinite flat plate.

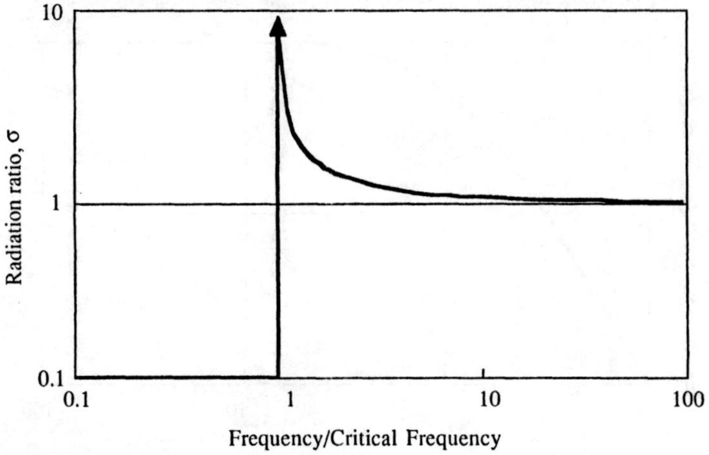

as illustrated for the infinite plate. For small sound sources and low frequencies ($ka \ll 1$), the radiation ratio, and thus the efficiency of sound radiation, increases with the square of frequency. This is equivalent to a 6 dB increase per octave (see chapter 4 for a definition of decibels, octaves etc.). Another important practical observation is that the radiation ratio approaches unity when half the circumference of the source approximates to an acoustic wavelength ($\pi a = \lambda$). Quite often in practice, one finds a situation where efficient sound radiators (in the far-field) have dimensions that match the offending acoustic wavelengths.

3.5 Sound radiation from free bending waves in finite plate-type structures

Sound radiation from free bending waves (bending waves which are not restricted by some structural discontinuity) in a structure can be categorised as (i) modal sound radiation at any given arbitrary frequency including non-resonant frequencies, and (ii) frequency-band-averaged sound radiation. Finite structural elements always allow for the existence of natural frequencies and their associated mode shapes. Thus, when a structure is excited by some broadband force, this generally results in the resonant excitation of numerous structural modes. Therefore, frequency-band-averaged sound radiation is necessarily dominated by resonant structural modes whereas modal sound radiation is not. It is worth reminding the reader that a resonance occurs when an excitation frequency coincides with a structural natural frequency.

Specialist texts such as Junger and Feit[3.1], Fahy[3.3] and Cremer *et al.*[3.6] all provide analytical expressions for the sound radiation from finite, planar surfaces for arbitrary, single frequency excitation. The solutions are generally restricted to the far-field. Cremer *et al.*[3.6] also provide analytical expressions for the near-field sound power radiated at regions in proximity to the excitation point. Rayleigh's equation (equation 3.2) is the starting point for all the above mentioned analyses, and analytical

Fig. 3.6. Radiation ratios for a finite spherical oscillator.

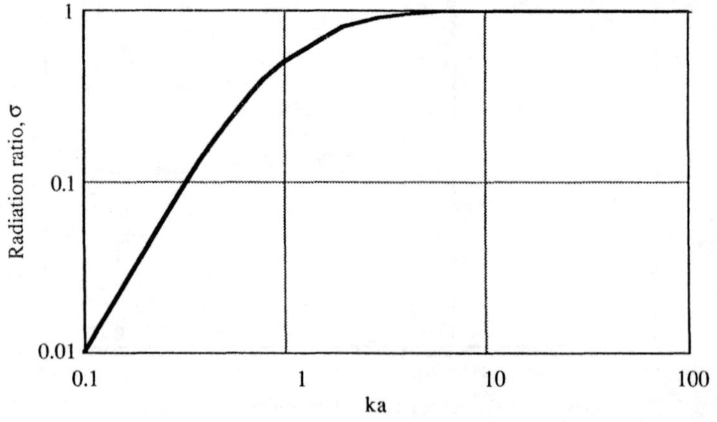

expressions are derived for modal radiation ratios. Two formal routes can be used in the analysis.

The first formal route is a direct approach (using Rayleigh's equation) to obtain expressions for the sound pressure at some point in the far-field. The intensity and radiated sound power are subsequently derived using the formal definitions (see chapter 2), and approximate expressions for radiation ratios are finally obtained for given modal distributions of surface vibrational velocity. It is important to note that the resulting expressions are only valid for modal excitation at any arbitrary frequency (i.e. not necessarily at a resonance).

The second formal route involves the analysis of travelling bending waves along a structure–fluid interface. Once again, the analysis in the literature is restricted to modal sound radiation at any given arbitrary frequency including non-resonant frequencies. The procedures involve wavenumber transforms[3.3,3.6]. The plate velocity distribution is transformed from the space–time domain into the wavenumber domain via the Fourier transform (the procedure is analogous to the more commonly used transformation into the frequency domain). The corresponding surface pressure field (in the wavenumber domain) is obtained by an application of wave/boundary matching at the structure–fluid interface, and the radiated power and radiation ratios are subsequently evaluated. Fahy[3.3] provides a very useful qualitative and quantitative discussion of sound radiation by flexural waves in plates in terms of wavenumber spectra and clearly identifies the various radiating wavenumber components for a range of different practical situations. Again, it is important to emphasise that the resulting expressions are only valid for modal excitation at any arbitrary frequency (i.e. not necessarily at a resonance).

In practice, when structures are mechanically excited by some broadband force they respond in a multi-mode, resonant form; many natural frequencies are excited and they resonate with the applied force. In this instance it is often, but not always, the case that these resonant modes are responsible for most of the sound radiation. It is not always the case because radiation ratios of finite structures generally increase with frequency – a situation could arise where the higher frequency, but non-resonant, modes (i.e. modes above the excitation frequency band) with their associated higher radiation ratios generate more sound than the lower frequency, but resonant, modes. Generally, however, whilst these higher frequency modes have higher radiation ratios, their vibrational levels are significantly reduced because they are non-resonant; thus the nett effect is that they radiate less sound than the lower frequency, lower radiation ratio, resonant modes which are within the excitation band. Hence, as a general rule, resonant structural modes tend to dominate the sound radiation from mechanically excited structures.

The situation is somewhat different for acoustically excited structures. This form of structural excitation will be discussed in some quantitative detail later on in this chapter in relation to sound transmission through structures. At this stage it is worth noting that the vibrational response of finite structures to acoustic excitation (i.e. incident sound waves) comprises (i) a forced vibrational response at the excitation frequency, and (ii) a vibrational response due to the excitation of the various structural natural frequencies. The former is associated with a wave that propagates through the structure at the trace wavelength, $\lambda/\sin\theta$, of the incident sound wave (see Figure 3.2). The latter is associated with the structural waves that are generated when the trace wave interacts with the boundaries; these structural waves are, in effect, free bending waves with corresponding natural frequencies. The important point to be noted at this stage is that the structural response is now both resonant and forced, and the transmission of sound through the structure (e.g. an aircraft fuselage, a partition between two rooms, or a machine cover) can be due to either one of the mechanisms or both.

Returning to the sound radiation from free bending waves in finite plate-type structures, it is the frequency-band-averaged, multi-mode, resonant, sound radiation that is of general practical significance to engineers. The remainder of this section shall therefore be limited to this form of sound radiation from finite plates with the exception of some qualitative comments, where appropriate, relating to sound radiation associated with a forced response due to either mechanical or acoustic excitation. The reader is also referred to references 3.1, 3.3 and 3.6 for detailed qualitative and quantitative discussions on modal mechanical excitation at arbitrary frequencies.

Consider a finite, rectangular, simply-supported plate with sides L_x and L_y, respectively. The natural frequencies, $f_{m,n}$, associated with the modes of vibration of the plate can be obtained from the plate equation (chapter 1, sub-section 1.9.8) by assuming a two-dimensional, time-dependent, harmonic solution. They are given by

$$f_{m,n} = 1.8c_{\mathrm{L}}t\left\{\left(\frac{m}{2L_x}\right)^2 + \left(\frac{n}{2L_y}\right)^2\right\}, \tag{3.38}$$

where m and n represent the number of half-waves in the x- and y-directions, respectively (i.e. $m = 1, 2, 3$ etc., $n = 1, 2, 3$ etc.). For clamped end conditions, m and n should be replaced by $(2m + 1)$ and $(2n + 1)$, respectively. Equation (3.38) can be rearranged in terms of wavelengths such that

$$f_{m,n} = \frac{1.8c_{\mathrm{L}}t}{\lambda_{m,n}^2}, \tag{3.39}$$

where $\lambda_{m,n}$ is the characteristic wavelength of the mode, and

$$\frac{1}{\lambda_{m,n}^2} = \frac{1}{\lambda_x^2} + \frac{1}{\lambda_y^2}, \tag{3.40}$$

with $L_x = m\lambda_x/2$ and $L_y = n\lambda_y/2$.

Each vibrational mode can be represented as a two-dimensional grid with nodal (zero displacement) lines in the x- and y-directions, respectively. The nodal lines sub-divide the plate into smaller rectangular vibrating surfaces each of which displaces the fluid in proximity to it. The resulting fluid motions between adjacent rectangular vibrating surfaces interact with each other and the resulting compressions and rarefactions of the fluid medium generate sound. Because of these interactions, the sound power radiated from the plate is not simply a function of the average plate velocity, as was the case for the infinite plate. The boundary conditions ensure that standing waves (vibrational modes) are now present and the radiated sound power has to be related to the number of these modes that are present. In the case of some forced excitation of the plate, the vibrational modes within the excitation frequency bandwidth would be resonant. The radiation ratios of each of these modes would vary and this would also have to be taken into account in any estimation of the radiated sound power.

The two-dimensional wavenumber of a mode of vibration is now

$$k_{m,n} = \frac{2\pi}{\lambda_{m,n}} = \left\{\left(\frac{m\pi}{L_x}\right)^2 + \left(\frac{n\pi}{L_y}\right)^2\right\}^{1/2}, \tag{3.41}$$

where

$$k_x = \frac{m\pi}{L_x}, \qquad \text{and } k_y = \frac{n\pi}{L_y}. \tag{3.42}$$

Each vibrational mode can thus be represented by a single point in wavenumber space, and this concept is illustrated in Figure 3.7. It should be noted that, because of the two-dimensional nature of the problem, it is possible to have several vibrational

Fig. 3.7. Illustration of the concepts of wavenumber space for a flat plate.

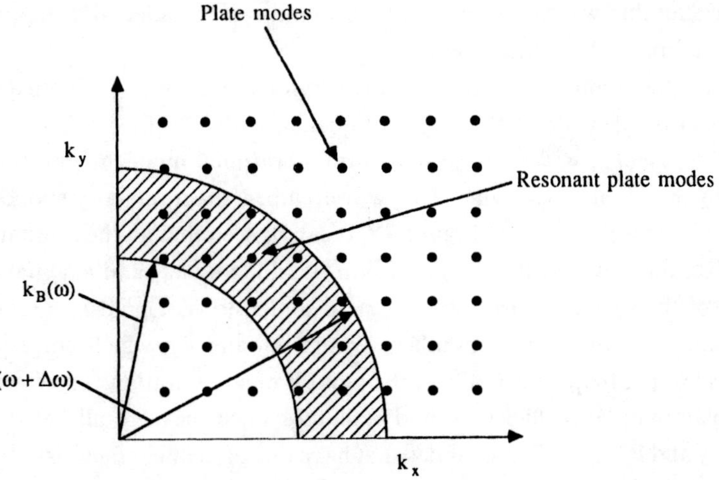

modes at any one frequency. Wavenumber diagrams are a convenient and informative way of representing the vibrational characteristics of a structure, particularly in relation to the interaction of structural and sound waves. The resonances at any given frequency are those points on the wavenumber diagram where the modal wavenumber, $k_{m,n}$, equals the bending wavenumber, k_B, associated with the applied force. Thus, for some pre-defined excitation band, $\Delta\omega$, the resonant vibrational modes are those modes that fall within the two wavenumber vectors defining the frequencies ω and $\omega + \Delta\omega$, respectively – i.e. the resonant modes in Figure 3.7 are shaded. The radius of an arc defining a wavenumber vector is given by

$$r_B = k_B = k_{m,n} = \frac{2\pi}{\lambda_{m,n}} = 2\pi\left(\frac{f_{m,n}}{1.8c_L t}\right)^{1/2}. \quad (3.43)$$

Similar equations can be obtained for the radii of wavenumber vectors at the critical frequency, and radiated (or incident) sound waves at some frequency, f, which coincides with a frequency, $f_{m,n}$ (corresponding to a particular resonance frequency, $f_{m,n}$, there is a sound wave which has the same frequency but a different wavelength and wavenumber because of the different propagation speeds). The equations are

$$r_C = k_C = 2\pi\left(\frac{f_C}{1.8c_L t}\right)^{1/2} = \frac{2\pi c}{1.8c_L t}, \quad (3.44)$$

and

$$r_A = k = \frac{2\pi}{\lambda} = \frac{2\pi}{\lambda_{m,n}}\left(\frac{f_{m,n}}{f_C}\right)^{1/2}. \quad (3.45)$$

For the infinite plate, it is clear that sound is only radiated for those structural waves where $f_{m,n} > f_C$. Unfortunately, the situation is not so simple for finite plates. Firstly, the problem is now two-dimensional, and secondly the plate boundaries generate standing waves. Because of this, sound can be radiated at frequencies both below and above the critical frequency.

Consider the resonant response of a plate excited by some band-limited force (either mechanical or acoustical) the upper frequency limit of which is below the critical frequency – i.e. $f_{m,n} < f_C$. All the structural vibrational modes within this excitation frequency band are resonant. The wavenumber diagram corresponding to this particular case is presented in Figure 3.8. It can be seen that all the resonant structural modes (i.e. those within the shaded region bounded by k_{B1} and k_{B2}) have either one or both of their characteristic wavenumber dimensions, k_x and k_y, greater than the corresponding acoustic wavenumber vectors, k_1 and k_2, which correspond to the lower and upper frequency limits of the band-limited excitation. The resonant modes of the plate will be inefficient sound radiators since they are all below the critical frequency and $k_{m,n} > k$. Those modes that have one of their characteristic wavenumber dimensions (k_x or k_y) greater than the corresponding acoustic wavenumber at the same frequency are referred to as edge modes; those that have both of their characteristic wavenumber dimensions greater than the corresponding acoustic wavenumber at the same frequency are referred to as corner modes. Edge modes are more efficient sound radiators than corner modes. Corner modes also generate some sound even though k_x and k_y are both greater than k. The reasons for this will become evident later on in this section. Edge and corner modes are commonly referred to as being acoustically slow or subsonic (i.e. the bending wave speed is less than the speed of sound).

For the case where the band-limited excitation extends to frequencies above the critical frequency (i.e. $f_{m,n} > f_C$), as illustrated in Figure 3.9, all the resonant structural

Fig. 3.8. Wavenumber diagram for the resonant excitation of acoustically slow plate modes ($k_{m,n} < k_C$).

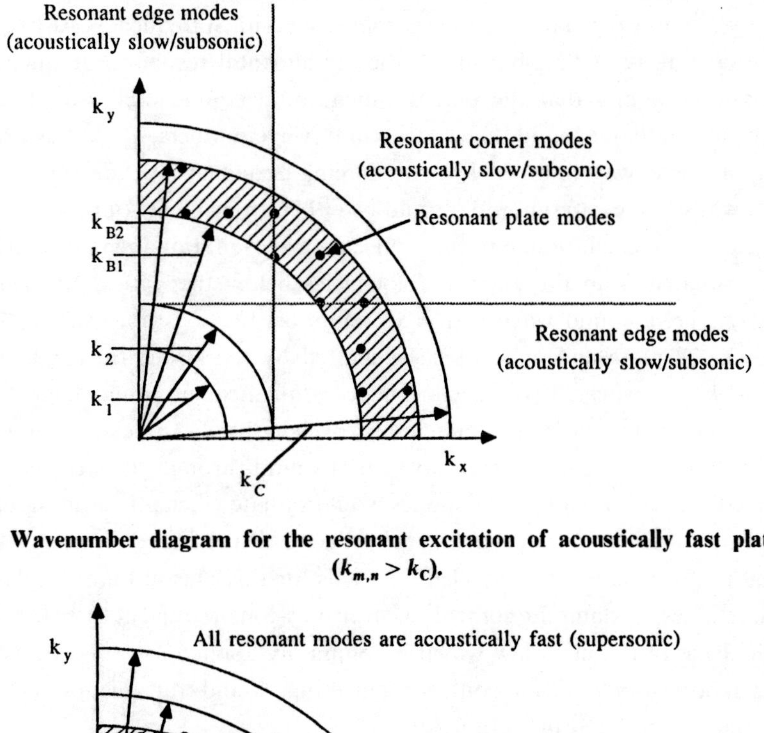

Fig. 3.9. Wavenumber diagram for the resonant excitation of acoustically fast plate modes ($k_{m,n} > k_C$).

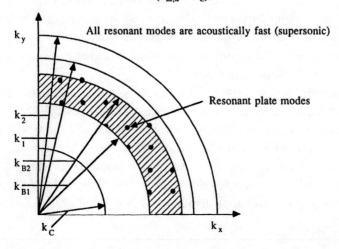

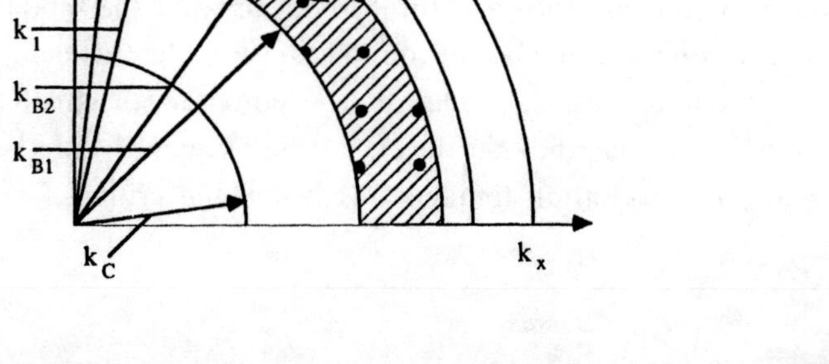

modes (i.e. those within the shaded region bounded by k_{B1} and k_{B2}) become acoustically fast and supersonic (the bending wave speed is greater than the speed of sound). Here, both of their characteristic dimensions, k_x and k_y, are less than the corresponding acoustic wavenumber vectors, k_1 and k_2 which correspond to the lower and upper frequency limits of the band-limited excitation. Under these conditions, the plate radiates sound very efficiently.

When a plate is forced at a particular frequency, its response is the superposition of all its modes driven at the forcing frequency. Here, the vibrational response of the plate is forced rather than resonant. In chapter 1 it was illustrated, for the elementary case of a single-degree-of-freedom system, that the response is mass controlled when $\omega > \omega_n$, damping controlled when $\omega = \omega_n$ and, stiffness controlled when $\omega < \omega_n$. An essentially analogous result can be obtained for plate-type structures, and this is a very important practical observation when considering the transmission of sound through a plate or a panel at frequencies below the critical frequency. At frequencies below the critical frequency, but above the fundamental resonance frequency, it is important to recognise that the modes which could couple well with the forcing frequency and radiate sound have structural wavenumbers, $k_{m,n}$, less than the equivalent acoustic wavenumber, k, at the forcing frequency. It is also important to note that, whilst the excitation force could be either an incident sound field or some form of mechanical excitation, if it were a mechanical excitation then no sound waves would be radiated from the plate at frequencies below the critical frequency (see Figure 3.2) – i.e. a sound wave with a wavenumber, k, does not exist in the fluid surrounding the plate and will not be generated by the structural bending waves since $k_{m,n} < k_C$. However, if the excitation were an incident sound field, then the non-resonant forced modes which match the wavelengths of the incident sound waves would allow for a very efficient transmission of sound through the structure. Hence these 'forced' structural vibrational modes would radiate efficiently even though they are below the critical frequency. Under such an acoustically forced response situation, the modal response is mass controlled, and it is for this reason that the plate mass and not its stiffness or damping controls the transmission of sound at these frequencies. This is the basis of the mass law which is commonly applied to sound transmission problems in noise and vibration control engineering. Sound transmission phenomena will be discussed in detail in section 3.9.

In relation to forced response situations it should be appreciated that the excitation frequency is different from the response frequencies. The 'forced' bending wave in the plate does not have to coincide with a natural frequency and the modal responses can therefore be non-resonant. For efficient sound radiation, the wavelengths of these non-resonant vibrational modes, $\lambda_{m,n}$, have to be equal to or greater than the corresponding acoustic wavelength, λ, at the excitation frequency. This is illustrated in Figure 3.10 where the excitation frequency is below the critical frequency and corresponds to a structural wavenumber, k_B, and an acoustic wavenumber, k. Those vibrational modes within the shaded region have structural wavenumbers, $k_{m,n}$, less than the equivalent acoustic wavenumber, k, at the forcing frequency. They therefore satisfy the criteria that $k_{m,n} < k$, and therefore radiate sound efficiently. Any resonant modes due to the excitation of free bending waves below the critical frequency will not radiate sound as efficiently.

Thus, in practice, the mechanical excitation of plates or panels results in most of the radiated sound being produced by resonant plate modes – the sound radiated by non-resonant forced modes tends not to be very significant. With acoustic excitation, however, it is the non-resonant forced modes, driven by the incident sound field, which match the wavelengths of the sound waves thus transmitting sound very efficiently through the structure at frequencies below the critical frequency (but above the fundamental resonance). At frequencies above the critical frequency, both forced and resonant modes contribute to the radiated sound.

The above discussion illustrates that the sound radiation characteristics of finite plates are somewhat complex, especially at frequencies below the critical frequency. It is, however, obvious that the sound radiation (below and above the critical frequency) depends upon the number of possible vibrational modes that can exist within a given frequency bandwidth. Hence the concept of 'modal density' is relevant to the radiation of sound from vibrating structures. Modal density is defined as the number of vibrational modes per unit frequency. For any plate of arbitrary shape, surface area, S, and thickness, t, it can be approximated by[3.7]

$$n(\omega) = \frac{S}{3.6 c_L t}. \tag{3.46}$$

Fig. 3.10. Forced response of a plate for $k_B < k_C$.

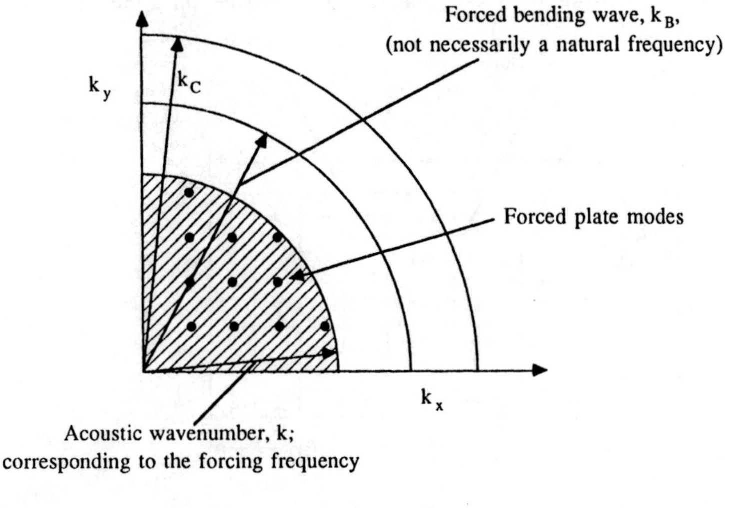

Equation (3.46) is simple and fairly useful as it allows for a rapid estimation of the number of resonant modes to be expected – the vibrational response and the sound radiation from a structure can be directly related to this. Modal density concepts play an important role in the analysis of noise and vibration from complicated structures and will be discussed in further detail in chapter 6.

It is fairly instructive at this stage to qualitatively analyse the radiation of sound from plate modes in some detail and to try to understand how the sound is radiated. The mode shape of a typical mode on a rectangular plate is illustrated in Figure 3.11. For this particular example, the bending wavenumbers, k_x and k_y are greater than the corresponding acoustic wavenumber, k, at the same frequency. Hence, λ_x and λ_y are both smaller than λ. This situation is representative of the corner modes in Figure 3.8. The structural wavelengths in both the x- and y-directions are less than a corresponding acoustic wavelength at the same frequency and, as such, the fluid which is displaced outwards by a positive sub-section is transferred to an adjacent negative sub-section without being compressed. The consequence of this is that very little sound is radiated. The radiated sound can be modelled in terms of monopole, dipole, and quadrupole sound sources. The central regions of the plate are quadrupole sound sources (groups of four sub-sections that essentially cancel each other as they oscillate), the edges of the plate comprise a line of dipole sources (groups of two sub-sections oscillating out of phase and cancelling each other), and the uncancelled oscillating volumes of fluid in the corners are monopole sources. From fundamental acoustics (see chapter 2), the quadrupole sound sources are the least efficient and the monopoles the most efficient. Thus only the corners of the plate radiate sound efficiently.

In the above example, if the lengths of the plate are much less than an acoustic wavelength (L_x and $L_y < \lambda$) then the four corner monopoles will interact with each

Fig. 3.11. Schematic illustration of corner radiation for a finite plate.

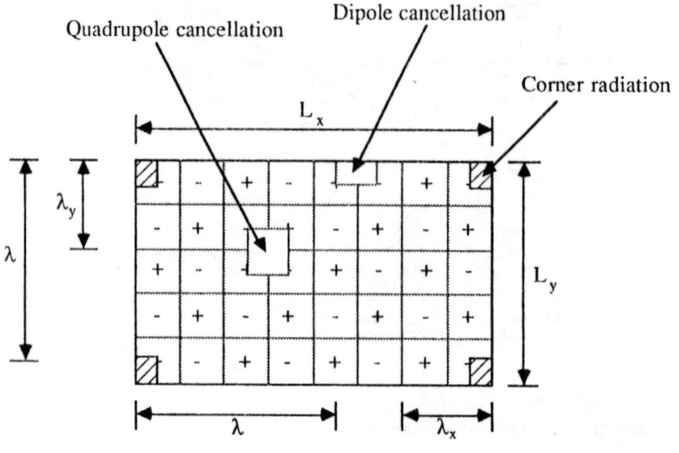

other. This interaction will be dependent upon their respective phases. For instance, for odd values of *m* and *n*, the four corners will radiate in phase with each other and behave like a monopole. For *m* even and *n* odd, adjacent pairs will be in phase but out of phase with the opposite pair, and behave like a dipole. For both *m* and *n* even, all four corners are out of phase with each other and the behaviour is quadrupole like. When L_x and $L_y > \lambda$ the four corners radiate like individual, uncoupled monopoles.

Figure 3.12 illustrates the case where one of the bending wavenumbers, k_y, is less than the corresponding acoustic wavenumber, k, at the same frequency. Hence, λ_x is smaller than λ but λ_y is greater than λ. This situation is representative of the edge modes in Figure 3.8. In this case, the central regions of the plate form long narrow dipoles which cancel each other, but the edges along the *y*-direction do not cancel. The structural wavelength in the *y*-direction is greater than a corresponding acoustic wavelength at the same frequency and, as such, the fluid which is displaced outwards by the positive sub-section (in the *y*-direction) is compressed when it is transferred to the adjacent negative sub-section. Sound is radiated as a consequence of this. These edge modes are more efficient radiators than corner modes.

As the exciting frequency approaches the critical frequency, the cancellation in the central regions starts to diminish. This is because the separation between sub-sections approaches $\lambda/2$. The cancellation breaks down totally at and above the critical frequency and the whole plate radiates sound. These modes are called surface modes, and both λ_x and λ_y are greater than the corresponding acoustic wavelength, λ, at the same frequency (or k_x and $k_y > k$). The fluid which is displaced outwards by positive sub-sections (in both the *x*- and *y*-directions) is compressed as it is transferred to adjacent negative sub-sections since all the sub-sections are greater than a fluid wavelength. Surface modes are very efficient radiators of sound.

Fig. 3.12. Schematic illustration of edge radiation for a finite plate.

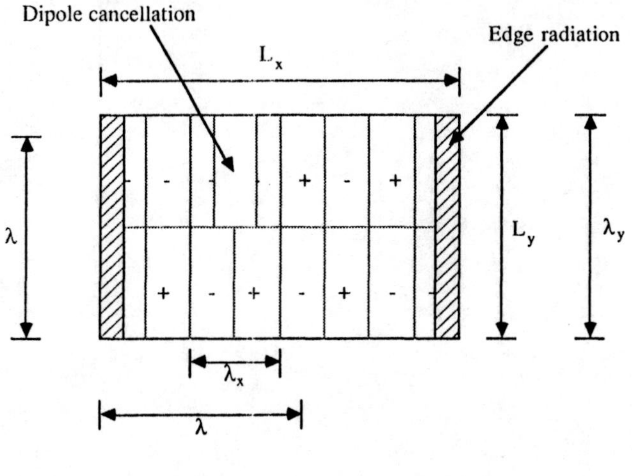

Fig. 3.13. Sound radiation from a plate with a structural discontinuity.

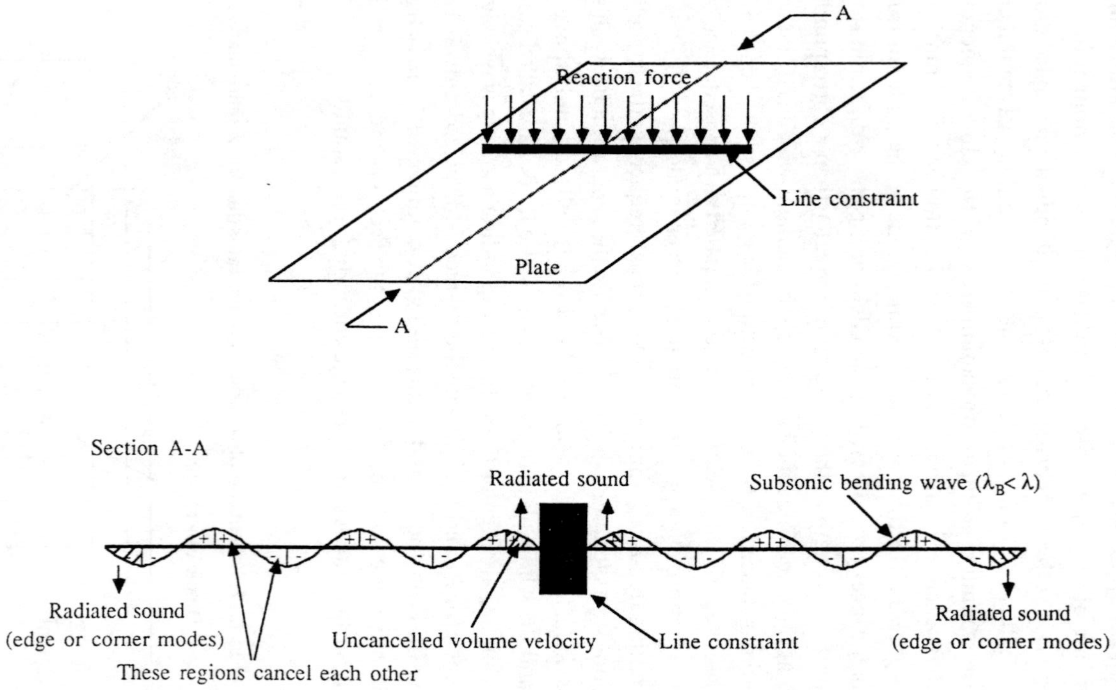

As a result of the preceding discussions relating to flat plates, it is clear that whilst at frequencies above the critical frequency finite plates behave in a similar manner to infinite plates, this is not the case at lower frequencies. Above the critical frequency, the radiation ratio, σ, is the same in both cases, but it is clear that the radiation ratio for finite plates is not zero below the critical frequency; there is some sound radiation which in some instances is very efficient. It should by now be very clear that radiation ratios have a very important role to play in engineering noise and vibration control. The radiation ratios of finite structural elements will be discussed in section 3.7.

3.6 Sound radiation from regions in proximity to discontinuities – point and line force excitations

In the previous section, the sound radiation characteristics of finite plates were qualitatively discussed. It was argued that, for acoustically excited plates, any sound that is radiated or transmitted (at frequencies below the critical frequency) is due to a forced response. On the other hand, for mechanically excited plates, any sound that is radiated at frequencies below the critical frequency is due to a resonant response. It was also illustrated that all the plate modes above the critical frequency are capable of radiating sound. The discussion was, however, limited to regions where the bending waves are free and not restricted by structural discontinuities – i.e. regions far away from any mechanical excitation points or structural constraints such as stiffeners, joints etc.

All real structures have regions where there are structural constraints and discontinuities – e.g. a large machine cover or an aircraft fuselage would have ribs and stiffeners. When subsonic bending waves interact with such a discontinuity, reaction forces are generated on the structure. Also, there might be regions where some external mechanical excitation is transmitted to the structure via either a point or a line. Sound is radiated from regions in proximity to these various types of discontinuities; this sound is in addition to the sound that is radiated from the free bending waves discussed in the previous section. It is due to the near-field bending waves that are generated by the point and line reaction forces associated with some form of external mechanical excitation (a driving force) and/or any structural constraints. This form of sound radiation from plates is schematically illustrated in Figure 3.13 – the sound is produced by the uncancelled volume velocities in regions in proximity to the structural constraint. Quite often, this sound radiation dominates over the sound radiated by the resonant corner and edge modes – as will be seen shortly, the point and line forces produce sound radiation at all frequencies and not only at resonant frequencies.

Junger and Feit[3.1], Fahy[3.3], and Cremer *et al.*[3.6] derive an expression for the sound power radiated, at frequencies below the critical frequency, from a point-excited infinite plate using the wavenumber-transform approach. An infinite plate is used for the analysis because the infinite bending travelling waves do not radiate sound below the critical frequency (see Figure 3.2) and the only radiated sound is due to the point excitation. The wavenumber-transform approach requires firstly that the velocity distribution on the surface of the plate (obtained from the assumed mode shapes) be transformed from the space–time domain into the wavenumber domain via the Fourier transform. The surface pressure transform is then obtained by recognising that there is a surface pressure wavenumber associated with every surface velocity wavenumber (i.e. wave/boundary matching). The sound power radiated from the plate is subsequently obtained from the real part of the product of surface pressure and surface velocity. Fahy[3.3] also derives an expression for the sound power radiated from a point-excited infinite plate using the same wavenumber-transform approach, and extends the analysis to a line excitation of an infinite plate. The wavenumber-transform technique is often used in advanced analyses of structure-borne sound and will not be discussed in this book. For the present purposes, it is sufficient to be informed about the availability of the technique and to utilise some of the more relevant results relating to flat plates.

For an infinite flat plate with point mechanical excitation, the sound power radiated (at frequencies below the critical frequency) from a single side of the plate is[3.3,3.6,3.7]

$$\Pi_{\mathrm{dp}} = \frac{\rho_0 F_{\mathrm{rms}}^2}{2\pi c \rho_{\mathrm{S}}^2}, \tag{3.47}$$

where ρ_0 is the density of the ambient fluid medium, F_{rms} is the root-mean-square value of the applied force, c is the speed of sound in the ambient fluid medium, and ρ_{S} is the surface mass (kg m^{-2}) of the plate. The radiated sound power is not a function of frequency and is only a function of the surface mass (mass per unit area). This is a very important practical result. It can be conveniently re-expressed in terms of the r.m.s. drive-point velocity by replacing F_{rms} by the product of the drive-point mechanical impedance of the infinite plate and the r.m.s. drive-point velocity, $v_{0\mathrm{rms}}$.

The drive-point mechanical impedance of an infinite plate of thickness t and mass per unit area ρ_{S} can be obtained in a similar manner to the string and beam impedances that were derived in chapter 1. It is[3.3,3.6]

$$Z_{\mathrm{m}} = 8\left\{\frac{Et^3\rho_{\mathrm{S}}}{12(1-\nu^2)}\right\}^{1/2} = \frac{8c^2\rho_{\mathrm{S}}}{\omega_{\mathrm{C}}}, \tag{3.48}$$

since the critical frequency is (see chapter 1, equation 1.322)

$$\omega_{\mathrm{C}} = c^2 \rho_{\mathrm{S}}^{1/2}\left\{\frac{12(1-\nu^2)}{Et^3}\right\}^{1/2}, \tag{3.49}$$

where ν is Poisson's ratio, and E is Young's modulus of elasticity. As would be expected, the drive-point mechanical impedance of an infinite plate is real (resistive) – energy flows away from the drive-point and there is no local reactive component. If the plate were finite, then a reactive component would exist. This point was discussed in chapter 1.

The drive-point radiated sound power (equation 3.47) can now be re-written by replacing F_{rms} by $Z_m v_{0\text{rms}}$. Hence,

$$\Pi_{\text{dp}} = \frac{8\rho_0 c^3 \langle v_0^2 \rangle}{\pi^3 f_c^2}, \tag{3.50}$$

where $\langle v_0^2 \rangle$ is the mean-square vibrational velocity at the drive-point. In this book, $\langle\ \rangle$ represents a time average and $\overline{}$ represents a spatial average. Equation (3.50) is a very useful practical result as it relates the radiated sound power (at frequencies below the critical frequency) due to point excitation of the plate to the drive-point vibrational velocity.

It is also useful to compare this drive-point radiated sound power with the sound power that would be radiated from the free-bending waves of the plate associated with all the resonant modes below the critical frequency. The sound power radiated by all the resonant modes is given by equation (3.30). The mean-square vibrational velocity averaged over the surface of the plate can be obtained by equating the input power to the dissipated power during steady-state. The input power is

$$\Pi_{\text{in}} = F_{\text{rms}}^2 \,\text{Re}\,[\mathbf{Z}_{\mathbf{m}}^{-1}], \tag{3.51}$$

where $\mathbf{Z}_{\mathbf{m}}$ is represented in general terms as a complex number and the dissipated power is a function of the loss factor, η and the vibrational energy, E. It was derived in chapter 1 (see section 1.7) and it is

$$\Pi_{\text{dis}} = \omega\eta E = \omega\eta\rho_s S\langle \overline{v^2} \rangle, \tag{3.52}$$

where S is the plate surface area. Equating equations (3.51) and (3.52) and solving for the space- and time-averaged mean-square velocity of the plate with the appropriate substitution for Z_m yields

$$\langle \overline{v^2} \rangle = \frac{f_c F_{\text{rms}}^2}{8c^2 f \rho_s^2 \eta S}. \tag{3.53}$$

Hence, from equation (3.30), the radiated sound power is

$$\Pi_{\text{rad}} = \frac{\rho_0 f_c F_{\text{rms}}^2 \sigma}{8cf\rho_s^2 \eta}. \tag{3.54}$$

This equation represents the sound power radiated by all the resonant modes both below and above the critical frequency. It can be now compared with the sound power radiated at the drive-point (equation 3.47). Thus

$$\frac{\Pi_{\mathrm{dp}}}{\Pi_{\mathrm{rad}}}=\frac{4f\eta}{\pi f_{\mathrm{C}}\sigma}. \tag{3.55}$$

Also, the total sound power radiated by the plate is

$$\Pi=\frac{\rho_0 F_{\mathrm{rms}}^2}{2\pi c\rho_{\mathrm{S}}^2}+\frac{\rho_0 f_{\mathrm{C}} F_{\mathrm{rms}}^2\sigma}{8cf\rho_{\mathrm{S}}^2\eta}. \tag{3.56}$$

It is worth reiterating that Π_{dp} only relates to frequencies below the critical frequency, whereas Π_{rad} is valid at all frequencies. Artificially damping the plate will only reduce the sound radiation associated with the second term in equation (3.56). The drive-point radiation thus represents a lower limit to the radiated sound power and any amount of damping will not reduce this portion of the radiated sound!

Now consider an infinite flat plate with a line source mechanical excitation (e.g. a clamped boundary or a stiffener). The sound power radiated (at frequencies below the critical frequency) from a single side of the plate can be obtained via the wavenumber-transform procedure and it is[3.3,3.7]

$$\Pi_{\mathrm{dl}}=\frac{\rho_0 F_{\mathrm{rms}}^2 l}{2\omega\rho_{\mathrm{S}}^2}, \tag{3.57}$$

where l is the length of the line source, and F_{rms} is the force per unit length (i.e. it is equivalent to a point force). It is assumed that the line force is uniformly distributed over its whole length and that all points are in phase with each other.

The drive-point mechanical impedance of an infinite beam of thickness t, width b, and mass per unit length, ρ_{L}, can be obtained in a similar manner to the string and beam impedances that were derived in chapter 1. Because the beam is not infinite in all directions (i.e. it has a finite thickness and width), its impedance is complex and it has both a resistive and a reactive part. It is[3.6]

$$\mathbf{Z}_{\mathrm{m}}=2c_{\mathrm{B}}\rho_{\mathrm{L}}(1+\mathrm{i}), \tag{3.58}$$

and the square of its modulus is

$$|\mathbf{Z}_{\mathrm{m}}|^2=2\sqrt{2}c_{\mathrm{B}}\rho_{\mathrm{L}}. \tag{3.59}$$

Now, the bending wave velocity for a bar is (see chapter 1, equation 1.259)

$$c_{\mathrm{B}}=\omega^{1/2}\left\{\frac{EI}{\rho_{\mathrm{L}}}\right\}^{1/4}, \tag{3.60}$$

thus

$$|\mathbf{Z}_{\mathrm{m}}|^2=8\omega(EI)^{1/2}\rho_{\mathrm{L}}^{3/2}. \tag{3.61}$$

E is Young's modulus of elasticity and I is the second moment of area ($I = bt^3/12$). At the critical frequency, $c_B = c$, hence equation (3.60) can be re-arranged such that

$$(EI)^{1/2} = \frac{c^2 \rho_L^{1/2}}{\omega_C}. \qquad (3.62)$$

Thus,

$$|\mathbf{Z}_m|^2 = \frac{8\omega c^2 \rho_L^2}{\omega_C} \qquad (3.63)$$

The sound power radiated by the line source excitation (equation 3.57) can now be conveniently re-expressed in terms of the r.m.s. velocity along the line by replacing F_{rms} by the product of the drive-point mechanical impedance of the infinite plate and the r.m.s. line source velocity, v_{rms}. It is assumed that the line force is uniformly distributed over its whole length and that all points are in phase with each other – i.e. it is valid to use the drive-point impedance and velocity to obtain the sound power radiated per unit length and subsequently multiply it by the length of the line source to obtain the total sound power radiated. Thus,

$$\Pi_{dl} = \frac{4\langle v_1^2 \rangle c^2 \rho_0 l}{\omega_C} \frac{\rho_L^2}{\rho_S^2}. \qquad (3.64)$$

Now, assuming (i) similar materials for the plate and the stiffener etc. that generates the line force, and (ii) a unit width, then $\rho_L = \rho_S$ and thus

$$\Pi_{dl} = \frac{2\langle v_1^2 \rangle c^2 \rho_0 l}{\pi f_C}. \qquad (3.65)$$

As for the point excitation case, it is useful to compare the sound power radiated from this line source with the sound power that would be radiated from the free bending waves that are associated with all the resonant modes below the critical frequency. The sound power radiated by all the resonant modes can be obtained from equation (3.30) where the mean-square vibrational velocity averaged over the surface of the plate can be obtained by equating the input power (equation 3.51) to the dissipated power (equation 3.52) during steady-state.

The real part of the reciprocal of the drive-point mechanical impedance of an infinite beam is given by (see equations 3.58 and 3.60).

$$\mathrm{Re}\,[\mathbf{Z}_m^{-1}] = \frac{1}{2\{2(EI)^{1/4}\omega^{1/2}\rho_L^{3/4}\}}. \qquad (3.66)$$

By equating equations (3.51) and (3.52), solving for the space- and time-averaged mean-square vibrational velocity of the plate and substituting into equation (3.30) yields

$$\Pi_{rad} = \frac{\sigma \rho_0 c l F_{rms}^2}{4\eta (EI)^{1/4} \rho_L^{7/4} \omega^{3/2}}. \qquad (3.67)$$

Equation (3.67) represents the sound power radiated by all the resonant modes both below and above the critical frequency and it can be compared with the sound power that is radiated by the line source (equation 3.57). Thus

$$\frac{\Pi_{\text{dl}}}{\Pi_{\text{rad}}} = \frac{2\eta}{\sigma}\left(\frac{f}{f_{\text{C}}}\right)^{1/2}, \tag{3.68}$$

and the total radiated sound power from the plate is

$$\Pi = \frac{\rho_0 F_{\text{rms}}^2 l}{2\omega\rho_{\text{S}}^2} + \frac{\sigma\rho_0 c l F_{\text{rms}}^2}{4\eta(EI)^{1/4}\rho_{\text{L}}^{7/4}\omega^{3/2}}. \tag{3.69}$$

As for the case of point excitation of the plate, the line source excitation represents a lower limit of radiated sound power which is independent of damping, is a function of the surface mass ($\rho_{\text{L}} = \rho_{\text{S}}$ for unit dimensions), and is also inversely proportional to frequency.

3.7 Radiation ratios of finite structural elements

The radiation ratio, σ, was defined in section 3.4 and it was shown that for an arbitrary structure with some time- ($\langle\ \rangle$) and space-averaged ($\overline{\ \ }$) mean-square vibrational velocity, v, the radiated sound power, Π, is

$$\Pi = \sigma\rho_0 c S\langle\overline{v^2}\rangle. \qquad \text{(equation 3.30)}$$

The concept of radiation ratios is an important one, particularly for obtaining engineering estimates of the radiated sound power from vibrating machines or structures. Equation (3.30) clearly illustrates the relationship between sound power radiated from a structure or a machine element, the vibrational level on the structure and, the radiation ratio. It suggests that an estimate of the radiated sound power can be obtained directly from surface vibrational measurements if the radiation ratio, σ, is known. Hence, a knowledge of σ for a given structural component (e.g. a plate, a cylinder, an I-beam, a small compact point source etc.) is indeed very valuable. If values or relationships for radiation ratios of structures can be found, then the estimation of the noise radiation and any subsequent noise control is a relatively easier process.

Equation (3.30) is sometimes expressed in logarithmic form. This is done by taking logarithms on both sides to yield

$$10\log_{10}\Pi = 10\log_{10}(\rho_0 c) + 10\log_{10}S + 10\log_{10}\langle\overline{v^2}\rangle + 10\log_{10}\sigma. \tag{3.70}$$

In this equation, each of the variables is expressed in terms of decibels. Decibels are most commonly associated with sound pressure levels, but are also frequently used for a wide range of other variables where a logarithmically compressed scale is required. Two variables differ by one bel if one is 10^1 times greater than the other, or by two bels if one is 10^2 times greater than the other. The bel is an inconveniently large unit so it is divided into ten parts, hence the decibel. Two variables differ by one decibel if they are in the ratio $10^{0.1}$. Three decibels (3 dB) represents a doubling of the variable, i.e. $10^{0.3} \approx 2.00$. A detailed discussion on decibels, including addition and subtraction, is provided in chapter 4.

The radiation ratios for an infinite flat plate and a spherical pulsating body were derived in section 3.4. For compact bodies (e.g. spherical type sources) the radiation ratios are a function of the parameter ka which corresponds to the number of sound waves that can be sustained within a distance corresponding to a characteristic parameter of the body such as a circumference – i.e. $2\pi a/\lambda = 2\pi af/c = \omega a/c = ka$. For finite, flat, plate-type structures (i.e. structures where bending waves can be set up) the radiation ratios are a function of the parameter ka and of the ratio of the bending wave frequency to the critical frequency (f/f_C). If the sound waves cannot flow around the edges of the plate but can only flow along it (e.g. a wall partition), then the radiation ratios are only a function of the ratio of the bending wave frequency to the critical frequency. The main conclusion that can thus be reached is that the radiation ratios of finite elements are not a direct function of frequency. It is, however, more convenient to have access to radiation ratio curves which are a direct function of frequency for engineering design applications, particularly for monopole- and dipole-type sound sources, and typical structural elements such as flat plates, rods, I-beams, and cylindrical shells – a large number of noise sources encountered in practice by engineers can be classified in this way.

Monopole-type sound sources include emissions from exhaust systems, combustion processes, cavitation and any other forms of 'whole body' pulsation where the pulsations are normal to the body. Also, at large distances from a source ($r \gg \lambda$), the radiation approximates to that of a uniform spherical radiator – typical examples include domestic vacuum cleaners, overhead projectors, hand drills, small electric motors, etc. In these instances, the monopole-type radiation ratios, derived in section 3.4 can be utilised. The radiation ratio for a spherical sound source is given by equation (3.37) – i.e.

$$\sigma = \frac{\left(\dfrac{2\pi fa}{c}\right)^2}{1 + \left(\dfrac{2\pi fa}{c}\right)^2}, \qquad \text{(equation 3.37)}$$

where $2\pi f/c = k$. Thus, design curves can be generated over a range of frequencies for different source dimensions by varying the spherical radius, a. A typical family of such curves is presented in Figure 3.14. A general observation is that smaller bodies have lower radiation ratios at lower frequencies.

Dipole-type sound sources (non-aerodynamic) involve the 'rigid' oscillation of solid bodies – i.e. the bodies do not pulsate and are not in flexure, but oscillate about some mean position without any volume change. The motion of the body thus approximates to a rigid sphere oscillating rectilinearly in an unbounded fluid. Typical industrial examples include diesel engine vibrations and the vibration of large industrial hammer and anvils. The radiation ratios of these types of sound sources can thus be obtained by modelling the source as a rigid oscillating sphere and proceeding to evaluate its radiation resistance in a similar manner to which the radiation resistance of a piston was evaluated in chapter 2. In fact, as should be obvious by now, there is a direct analogy between equation (2.140) and equation (3.30) – i.e. the radiation resistance of a solid vibrating body is in fact its radiation ratio! The radiation ratio of a rigid oscillating sphere is[3.7]

$$\sigma = \frac{(ka)^4}{12 + 4(ka)^4} = \frac{\left(\frac{2\pi fa}{c}\right)^4}{12 + 4\left(\frac{2\pi fa}{c}\right)^4}, \tag{3.71}$$

where a is the radius. Once again, design curves can be generated over a range of frequencies for different source dimensions by varying the radius. A typical family of such curves is presented in Figure 3.15.

Sometimes, only certain portions of a body vibrate, whilst the remainder of the body remains stationary, e.g. loudspeakers or radiation through ducts or orifices in an otherwise solid body. In these instances the vibrations approximate to that of a piston in an infinite baffle. The radiation ratio can thus be given by equation (2.128) (chapter 2, sub-section 2.3.4), and Figure 2.14.

Fig. 3.14. Radiation ratios for monopole-type sound sources.

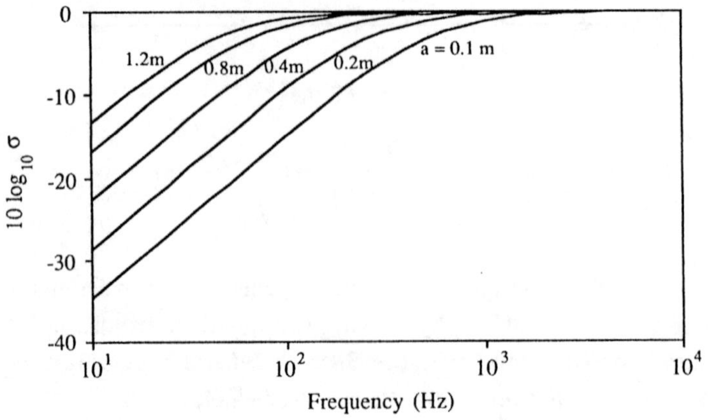

Unfortunately, not all sound sources behave like monopoles, dipoles or pistons, and sometimes their radiation characteristics are a function of both *ka* and f/f_C. Richards[3.8] provides a comprehensive list of theoretically and experimentally determined radiation ratios for a range of typical industrial structural elements. These include steel plates of varying thickness, aluminium plates of varying thickness, long circular beams, steel bars, and I-beams. The data is ideal for engineering design applications and reduces the problem of sound power estimation to one of the estimation of structural vibration levels.

The radiation ratio of finite, flat plates vibrating in their resonant, flexural modes in response to broadband mechanical excitation is a very useful quantity to have readily available. Quite often, machine or engine covers and other types of radiating panels which are so often found within an industrial environment can be modelled as flat plates. Ver and Holmer[3.9] present a very useful empirical relationship for the modal-averaged radiation ratios of simply supported and clamped plates. A design curve based upon their relationships is presented in Figure 3.16. The design curve allows for an estimation of the radiation ratio once the radiating surface area, S, the perimeter of the plate P, and the critical wavelength, λ_C, are established. It is important to remember that the curve is only valid for resonant, broadband, mechanical excitation. The radiation ratios for acoustically excited structures (particularly below the critical frequency) tend to be somewhat larger. Fahy[3.3] provides several examples which are obtained from the research literature.

Another useful geometry is a cylinder. The noise and vibration generated by cylindrical shells is a specialised topic, and several aspects relating to flow-induced noise and vibration will be discussed in chapter 7. Quite often, long runs of pipeline are encountered in industry and radiation ratios are convenient for estimating the radiated noise levels. For a long, uniformly radiating cylinder pulsating at the same

Fig. 3.15. Radiation ratios for dipole-type sound sources.

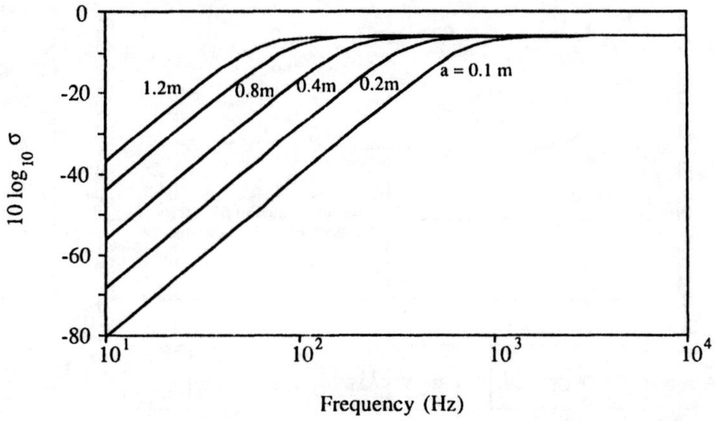

wavenumber and frequency as the excitation (some internal pressure fluctuations), the radiation ratio is given by[3.10]

$$\sigma = \frac{2}{\pi(ka)|H_1^{(1)}(ka)|^2}, \tag{3.72}$$

where $H_1^{(1)}$ is the first-order Hankel function of the first kind, and a is the cylinder radius. The Hankel function is a form of complex Bessel function, details of which can be obtained in any advanced mathematical handbook. The radiation ratios can thus be presented in a generalised form as a function of ka or as a function of frequency for specific pipe radii. The generalised results are presented in Figure 3.17, and specific values at a given frequency and radius can be obtained by replacing the wavenumber, k, by $2\pi f/c$.

When it is the resonant structural modes of a cylinder that are the dominant sources of sound, rather than some forced motion, the radiation ratios of the different shell modes can vary significantly, particularly in regions where the bending waves are acoustically slow (subsonic). Standing waves will be set up both in the axial and in the circumferential directions, and certain modes will radiate more efficiently than others. Radiation ratios for resonant pipe modes resulting from wave motion for which the wave speed is subsonic or supersonic can be obtained from the book by Junger and Feit[3.1]. Norton and Bull[3.10] have computed these radiation ratios for typical industrial-type pipes. Some typical results for a length-to-diameter ratio of

Fig. 3.16. Design curve (adapted from Ver and Holmer[3.9]) for estimating the radiation ratios of broadband mechanical excitation of flat plates. (*P* is the perimeter; *S* is the radiating surface area; λ_c is the critical wavelength.

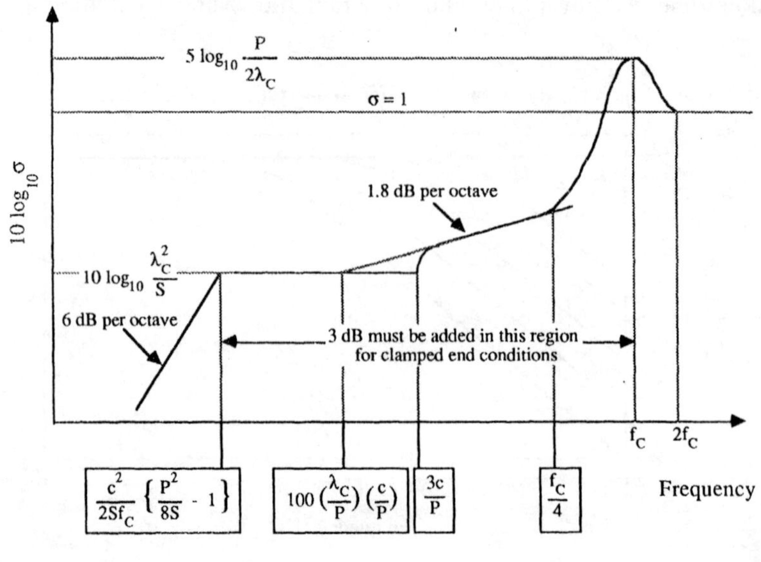

40 are presented in Figure 3.18(a) and (b) as a function of the ratio of bending wave velocity in the shell, c_s, to the speed of sound in the external fluid, c_e, for different values of m and n (m is the number of half-waves along the pipe's axis, and n is the number of full waves along the pipe's circumference).

The radiation ratios associated with the corresponding forced peristaltic motion (a slightly modified form of equation 3.72) are also presented in both figures. When the bending wave velocity equals the speed of sound the radiation ratios approach unity in all cases. The resonant modes are now acoustically fast (supersonic) and they all radiate very efficiently. At the lower frequencies, the lower-order circumferential modes are more efficient sound radiators than the higher-order circumferential modes. The bending wave speed in the pipe wall is given by[3.10]

$$c_s = \frac{2\pi f}{\left\{\left(\frac{m\pi}{l}\right)^2 + \left(\frac{n}{a_m}\right)^2\right\}^{1/2}}, \tag{3.73}$$

where a_m is the mean pipe radius and l is its length. Equation (3.73) allows for the abscissa on Figure 3.18 to be converted into a frequency scale. The behaviour of

Fig. 3.17. Radiation ratios for a long, uniformly radiating, pulsating cylinder.

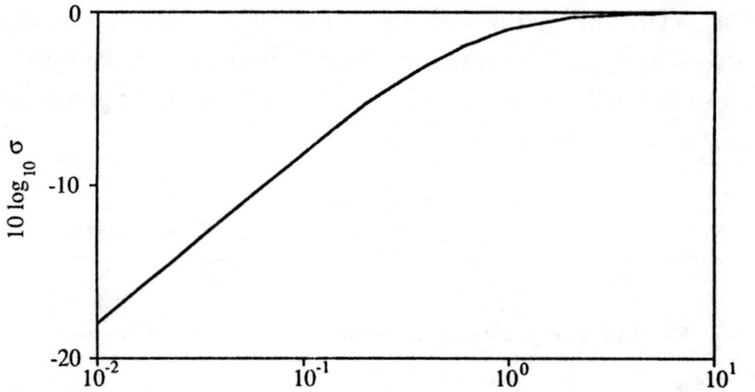

Fig. 3.18. Typical radiation ratios associated with resonant structural modes of a cylinder. Values for forced peristaltic motion.

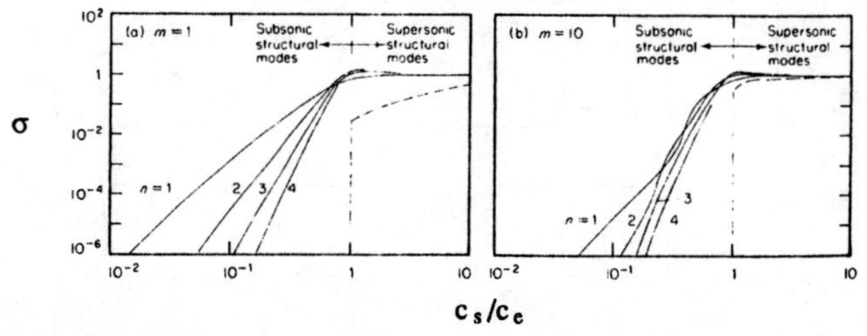

cylinders is somewhat more complex than flat structures, and the data provided in Figure 3.18 relates to specific values of m and n, and is only provided as an illustrative example at this stage. The radiation ratios of cylindrical shells will be discussed again in chapter 7.

3.8 Some specific engineering-type applications of the reciprocity principle

The basic concepts of the principle of reciprocity, as they relate to acoustics, were presented at the beginning of this chapter. Rayleigh, in his classic book on the theory of sound, demonstrated that this principle applies to all systems whose energy can be described in a quadratic form (kinetic and potential energy). In noise and vibration control applications, reciprocity can be used to utilise theoretical and experimental data to estimate some parameter that cannot otherwise be directly measured. The principle of reciprocity is, for example, commonly used in statistical energy analysis applications; some of these procedures will be discussed in chapter 6.

Reciprocity is only valid for linear processes, thus it is valid for the study of noise and vibration. It can be defined as follows: if the force excitation and velocity measurement positions are interchanged in some experiment, the ratio of the excitation force to the measured velocity remains constant. An important condition for reciprocity is that the direction of the applied force in the first experiment and the direction of the measured velocity in the second experiment have to be the same. This point is illustrated schematically in Figure 3.19. A force F_{x-} acting at some position X generates a velocity v_{y+} at some other position Y. If the same force were now applied at position y such that $F_{x-} = F_{y+}$, a velocity $v_{x-} = v_{y+}$ would be produced at position X. Hence, as per equation (3.10),

$$\frac{F_{x-}}{v_{y+}} = \frac{F_{y+}}{v_{x-}}. \tag{3.74}$$

Fig. 3.19. Reciprocity relationship between input and response.

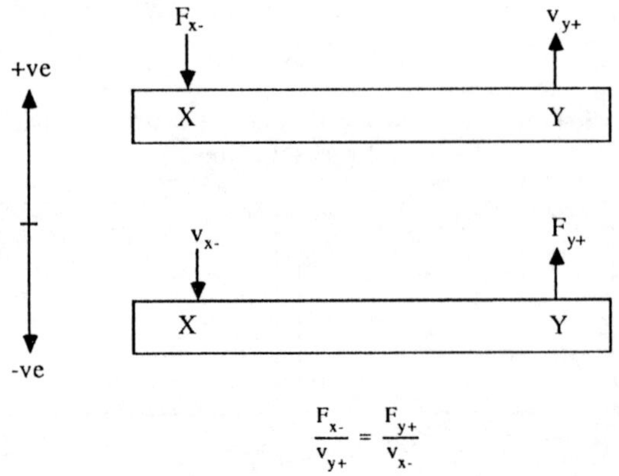

Now, consider a situation in which there are two machines in a room in a factory. Assume that the room is reverberant – i.e. the sound waves reflect off the hard walls and the sound associated with the reflected waves dominates over any direct sound that emanates from either of the sources. Reverberation concepts will be dealt with in detail in the next chapter. If one of the machines was significantly louder than the other, the principle of reciprocity would allow for the noise radiated by the quieter machine to be estimated without having to turn the louder machine off.

To simplify the mathematics in this example, assume that both machine sources are compact ($d \ll \lambda$ or $ka \ll 1$). The analysis can be readily extended to non-compact sources. The sound power radiated by a compact source is given by equations (3.30) and (3.37). Thus,

$$\Pi = (ka)^2 \rho_0 c 4\pi a^2 \langle v^2 \rangle = \frac{Q_{\text{rms}}^2 k^2 \rho_0 c}{4\pi}, \tag{3.75}$$

where $Q_{\text{rms}}^2 = (4\pi a^2)^2 \langle v^2 \rangle$.

Define the louder machine as #1 and the quiet one as #2. Now, firstly switch off the quieter machine, excite it mechanically with a point force, and measure the mean-square vibrational velocity at the drive-point on the structure. The vibrational velocity is proportional to the applied point force – i.e.

$$\langle v_2^2 \rangle = \kappa \langle F_2^2 \rangle. \tag{3.76}$$

Because the louder machine is radiating noise, the sound pressure generated by the quiet one cannot be measured. The sound power radiated by the quiet machine is, however, proportional to the vibrational velocity at the drive-point and also to the mean-square sound pressure in the room – i.e.

$$\Pi_2 = \beta \langle v_2^2 \rangle = \chi \langle \overline{p^2} \rangle. \tag{3.77}$$

Thus,

$$\frac{\langle \overline{p^2} \rangle}{\langle F_2^2 \rangle} = \frac{\kappa\beta}{\chi}. \tag{3.78}$$

In the above equations, κ and β are location-dependent whereas χ is not.

Now measure the vibrational response of the quiet machine (at the same point as the point mechanical excitation) to the sound produced by the louder machine. This vibrational response is also proportional to the mean-square sound pressure in the room (which is different from the mean-square sound pressure due to the first experiment). Thus,

$$\langle v_2'^2 \rangle = \psi \langle \overline{p'^2} \rangle. \tag{3.79}$$

Once again, the constant of proportionality ψ is location-dependent.

If the sound power produced by the louder machine is modelled as a pulsating source with a source strength Q_{rms} as per equation (3.75), then

$$\Pi' = \frac{Q_{1\,\mathrm{rms}}^2 k^2 \rho_0 c}{4\pi} = \chi\langle \overline{p'^2} \rangle, \tag{3.80}$$

since it is also proportional to the mean-square pressure in the room. Hence, dividing equation (3.79) by equation (3.80) yields

$$\frac{\langle v_2'^2 \rangle}{Q_{1\,\mathrm{rms}}^2} = \frac{\psi k^2 \rho_0 c}{4\pi\chi}. \tag{3.81}$$

Equations (3.81) and (3.78) are dimensionally similar and have the same units of m^{-4}. As per equation (3.10) in section 3.2, they describe the ratios of the inputs and the outputs for the reciprocal experiment. Hence, via reciprocity, they can be equated and

$$\frac{\langle \overline{p^2} \rangle}{\langle F_2^2 \rangle} = \frac{\kappa\beta}{\chi} = \frac{\langle v_2'^2 \rangle}{Q_{1\,\mathrm{rms}}^2} = \frac{\psi k^2 \rho_0 c}{4\pi\chi}. \tag{3.82}$$

The parameter β relates the vibration of the quiet machine to its radiated sound power, thus

$$\beta = \frac{\psi k^2 \rho_0 \chi}{4\pi\kappa}. \tag{3.83}$$

The parameters ψ and κ can be readily obtained by experimental measurements of the mean-square vibrational velocities of the quieter machine (at some specific point) firstly for point mechanical excitation at that point, and secondly for acoustic excitation by the louder machine. The applied point force would also have to be measured together with the mean-square sound pressure in the room. Thus, it is important that F_2 and v_2' are measured at the same point. By repeating the experiment at several points on the machine an averaged value of β can thus be obtained.

Having estimated β_{avg}, the sound power radiated by the quiet machine can be estimated simply by measuring its mean-square vibrational velocity (space- and time-averaged) whilst it is running – i.e.

$$\Pi_2 = \beta_{\mathrm{avg}} \langle \overline{v_2^2} \rangle. \tag{3.84}$$

Several assumptions are made in the preceding analysis. Firstly, it is assumed that the vibrational response of the quiet machine to mechanical point excitation is unaffected by the radiated sound field from the louder machine, i.e. $v_2 > v_2'$. Secondly, the mechanical excitation is supplied at a specific point on the structure, i.e. the parameter κ is dependent upon the location of the excitation. Thirdly, the sound field generated by the louder machine produces a diffuse (reverberant) sound field in the room. Finally, the measured response of the structure to the diffuse sound field is also dependent upon locaton – i.e. the parameter ψ is location-dependent since different parts of the machine respond in a different manner.

Reciprocity relationships similar to equations (3.82) and (3.83) can also be obtained for a range of other examples. The principle behind the reciprocity relationship has significant practical applications. For instance, it can be applied to determine locations in a reverberant factory environment which would produce minimum response to point force excitation. By exciting the room with an acoustic source and measuring the point in the room with the smallest vibrational response, one can easily deduce the location at which the sound power radiated due to a point force would be smallest. This location would thus be a suitable one for locating a vibrating machine such as to minimise structure-borne sound!

Further examples of the application of the principle of reciprocity will be presented in chapter 6 on statistical energy analysis.

3.9 Sound transmission through panels and partitions

A fundamental understanding of how sound waves are transmitted through panels and partitions is very important in practical engineering noise and vibration control. Most types of engineering applications of noise and vibration control involve the usage of panels or partitions of one form or the other. Machine covers, wall partitions, aircraft fuselages, windows, etc. all transmit noise and vibration, and, in practice, panels and partitions come in all shapes and sizes. Typical examples include homogeneous panels, double-leaf panels with or without sound absorbent material within the enclosed cavity, stiffened panels, mechanically coupled panels etc. Because of the vast variety of panels and partitions that are available, no one single theory adequately describes their sound transmission characteristics. Basic theories are available for single, uniform panels and for uniform, double-leaf panels. A range of empirical formulae is also available in the literature. The basic theories, whilst limited since they are not always able to provide precise answers to real practical problems, serve to illustrate the important physical characteristics that are involved; they will be reviewed and summarised in this section.

Several important general comments can be made regarding sound transmission through panels and partitions. It is useful to summarise them prior to any detailed discussion. They are as follows.

(1) When considering sound transmission through (and/or sound radiation from) panels, it is necessary to consider the complete frequency range of interest. The sound transmission characteristics would be very different depending on whether the stiffness, the mass, or the damping dominates the panel's response. Any finite structure can sustain natural frequencies and mode shapes, and a simple one-degree-of-freedom model readily illustrates that

when $\omega \ll \omega_n$ the stiffness dominates, when $\omega \approx \omega_n$ the damping dominates, and when $\omega \gg \omega_n$ the mass dominates. Thus, it would not be very sensible to add damping to a panel if the frequency range in which attenuation is required is in the mass-controlled region!

(2) The response of a panel is quite different depending on whether it is mechanically or acoustically excited. When it is mechanically excited, most of the radiated sound is produced by resonant panel modes irrespective of whether the frequency range of interest is below or above the critical frequency.

(3) When a panel is acoustically excited by incident, diffuse sound waves, its vibrational response comprises both a forced vibrational response at the excitation frequencies, and a resonant response of all the relevant structural natural frequencies which are excited due to the interactions of the forced bending waves with the panel boundaries. The non-resonant, forced modes, driven by the incident sound field, tend to transmit most of the sound at frequencies below the critical frequency – this was illustrated in section 3.5. The resonant frequencies below the critical frequency have very low radiation ratios and also have bending wavelengths that are smaller than the incident sound waves – hence they are very poor sound transmitters or radiators. Thus, at frequencies below the critical frequency, it is generally the mass of the panel that controls the reduction in sound transmission since the low frequency resonant structural modes do not radiate or transmit sound. Above the critical frequency, it is the resonant modes that transmit most of the sound. The one qualification to the phenomena discussed here is that the incident sound field has to be diffuse – i.e. no acoustic standing waves are present in the fluid medium adjacent to the panel.

(4) When considering the transmission of sound through a panel separating two rooms in which a diffuse field does not exist (either in one or in both rooms), the acoustic standing waves that are sustained within the enclosed fluid volumes can couple to the structural modes in the panel if their natural frequencies are in close proximity to each other or if there is good spatial matching between the fluid and structural mode shapes. These coupled modes (below or above the critical frequency) will reduce the effectiveness of the reduction in sound transmission through the panel. In situations such as these, both added mass and damping are appropriate. This phenomenon is especially important when considering the transmission of sound into or out of small confined spaces such as motor cars, aircraft fuselages, or cylindrical pipelines. In these instances, it is quite incorrect to use the diffuse-field model for the prediction of the reduction of sound transmission at frequencies below the critical frequency. The coupled structural–acoustic modes dominate the

sound transmission. Some of these concepts will be discussed in chapter 7 in relation to the transmission of sound through cylindrical shells with high speed internal gas flows.

(5) The mechanical properties of a panel (i.e. stiffness, mass, and damping) are only important if the characteristic acoustic impedances of the fluids on either side of the panel are approximately equal (i.e. $\rho_1 c_1 \approx \rho_2 c_2$). If $\rho_1 c_1 \gg \rho_2 c_2$ or vice versa, then the mechanical properties of the panel are relatively unimportant and it is the impedance mismatch between the two fluid media which governs the sound transmission characteristics. Equations similar to those that were derived in chapter 1 (sub-section 1.9.3) for the transmission and reflection of quasi-longitudinal structural waves at a step discontinuity can be readily derived. The main difference is that for sound waves there is continuity of acoustic pressure across the interface.

3.9.1 Sound transmission through single panels

The term 'transmission loss' (*TL*) or 'sound reduction index' (*R*) is commonly used to describe the reduction in sound that is being transmitted through a panel or a partition. The first term (i.e. *TL*) will be used in this book. The transmission loss through a panel is defined as

$$TL = 10 \log_{10}\left(\frac{1}{\tau}\right), \tag{3.85}$$

where τ is the ratio of the transmitted to the incident sound intensities. τ is commonly referred to as the 'transmission coefficient'. The characteristic transmission loss of a bounded homogeneous, single panel is schematically illustrated in Figure 3.20. There are four general regions of interest and they are stiffness controlled, resonance controlled, mass controlled, and coincidence controlled.

Fig. 3.20. Characteristic transmission loss of a bounded, homogeneous, single panel.

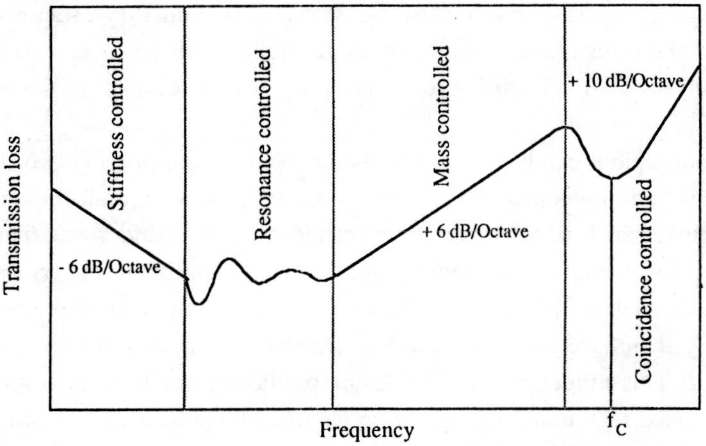

Firstly, because the panel is finite and bounded, it has a series of natural frequencies. It is important to note that these natural frequencies are not always relevant to sound transmission. If the panel is mechanically excited, or if the incident sound field is not diffuse (i.e. coupling occurs between the panel modes and the acoustic modes in the fluid volume), then the resonant structural modes control the sound transmission through the panel. Under these conditions, the addition of suitable damping material would increase the *TL*. If the panel is acoustically excited below the critical frequency and the incident sound field is diffuse, then the forced bending waves at the excitation frequencies dominate the sound transmission through the panel and the resonant structural modes are relatively unimportant.

Secondly, at frequencies well below the first fundamental natural frequency, it is the stiffness of the panel which dominates its sound transmission characteristics. In this region there is a 6 dB decrease in *TL* per octave increase in frequency – this will be quantitatively demonstrated shortly (also, octaves and one-third-octaves are defined in chapter 4). Also, in this region, the addition of mass or damping will not affect the transmission loss characteristics. Doubling the stiffness would increase the transmission loss by 6 dB.

Thirdly, at frequencies above the first few natural frequencies but below the critical frequency, the response is mass controlled. In this region there is a 6 dB increase in transmission loss per octave increase in frequency – this will be quantitatively demonstrated shortly. There is also a 6 dB increase in transmission loss if the mass is doubled. Damping and stiffness do no control the sound transmission characteristics in this region. It is important to note that, although doubling the mass increases the transmission loss, it also reduces the critical frequency! – see equation (3.14).

Finally, at regions in proximity to and above the critical frequency, there is a sharp drop in the transmission loss. In these regions, all the structural modes are coincident ($\lambda_B = \lambda/\sin\theta$) and their resonant responses are damping controlled. At frequencies above the critical frequency all the resonant structural modes have wavelengths greater than the corresponding acoustic wavelengths and they radiate sound very efficiently. The transmission loss increases at about 10 dB per octave in this region; the resonant response is damping controlled and the non-resonant response is stiffness controlled.

These four regions can be quantitatively discussed by considering two simple panel models. The first model involves a finite, bounded panel with uniform mass, stiffness and damping, which is subjected to an incident plane sound wave. This model is appropriate for predicting the transmission loss in regions below the critical freqency. The second model involves the transmission of sound through an unbounded, flexible partition – there are no panel natural frequencies in this model because it is unbounded. This model is appropriate for predicting the transmission loss in the mass controlled region and in regions above the critical frequency. Both models are consistent with each other, meeting in the mass controlled region and providing similar results there. The same boundary conditions apply for both models, the primary difference between them being in the modelling of the respective mechanical impedances.

Two boundary conditions have to be satisfied in both instances. They are as follows.

(i) The total pressure that acts on the panel comprises contributions from the incident, reflected and transmitted sound waves – i.e.

$$\mathbf{p} = \mathbf{p}_I + \mathbf{p}_R - \mathbf{p}_T. \tag{3.86}$$

(ii) The components of the acoustic particle velocities normal to the surface on both sides of the panel have to equal the plate velocity (see equation 3.31). Since the angle θ is common to the incident, reflected and transmission waves if the fluid medium is the same on both sides of the plate, as illustrated in Figure 3.21, this simplifies to

$$\mathbf{u}_I - \mathbf{u}_R = \mathbf{u}_T. \tag{3.87}$$

Fig. 3.21. An unbounded, flexible partition subjected to obliquely incident sound waves.

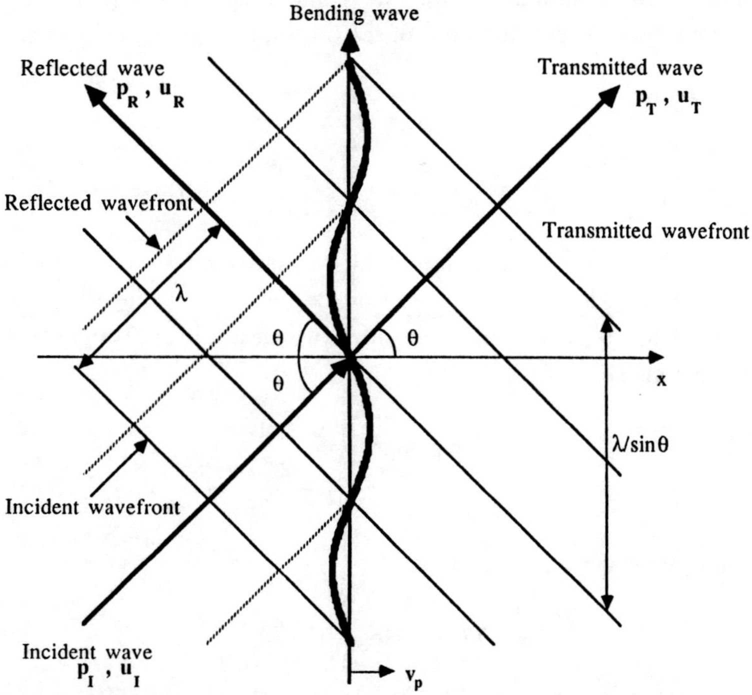

Note: (i) The angle θ is common to the incident, reflected and transmitted waves since the fluid medium is assumed to be the same on both sides of the plate; (ii) the incident and reflected wavefronts are not necessarily perpendicular to each other.

$$p = p_I + p_R - p_T$$

$$u_I \cos\theta - u_R \cos\theta = u_T \cos\theta = v_p$$

The ratio of the total pressure acting on the panel to the panel velocity ($\mathbf{p}/\mathbf{v}_\mathbf{p}$) is the impedance per unit area, $\mathbf{Z}'_\mathbf{m}$, since the pressure is simply the force per unit area. In general, it is complex. Thus,

$$\mathbf{Z}'_\mathbf{m} = \frac{\mathbf{p}}{\mathbf{v}_\mathbf{p}}, \tag{3.88}$$

where from equation (3.31)

$$\mathbf{v}_\mathbf{p} = \frac{\mathbf{p}_\mathbf{T}}{\rho_0 c}\cos\theta = \mathbf{u}_\mathbf{T}\cos\theta. \tag{3.89}$$

Also,

$$\mathbf{u}_\mathbf{I} = \frac{\mathbf{p}_\mathbf{I}}{\rho_0 c}, \quad \text{and } \mathbf{u}_\mathbf{R} = \frac{\mathbf{p}_\mathbf{R}}{\rho_0 c}. \tag{3.90}$$

Thus by substituting equations (3.89) and (3.90) into equation (3.87) one gets

$$\mathbf{p}_\mathbf{I} - \mathbf{p}_\mathbf{T} = \mathbf{p}_\mathbf{R}. \tag{3.91}$$

By substituting equations (3.86), (3.89) and (3.91) into equation (3.88) and re-arranging terms one gets the ratio of the transmitted to the incident sound pressure. It is

$$\frac{\mathbf{p}_\mathbf{T}}{\mathbf{p}_\mathbf{I}} = \frac{1}{1 + \dfrac{\mathbf{Z}'_\mathbf{m}\cos\theta}{2\rho_0 c}}. \tag{3.92}$$

The transmission coefficient, τ, and the transmission loss, TL, can now be obtained from equation (3.92). The transmission coefficient is defined as the ratio of the transmitted to incident sound intensities (which are proportional to the square of the pressures). Thus

$$\tau = |\mathbf{p}_\mathbf{T}/\mathbf{p}_\mathbf{I}|^2 = \frac{1}{\left|1 + \dfrac{\mathbf{Z}'_\mathbf{m}\cos\theta}{2\rho_0 c}\right|^2}. \tag{3.93}$$

Thus, the transmission loss, TL, is

$$TL = 10\log_{10}\left|1 + \frac{\mathbf{Z}'_\mathbf{m}\cos\theta}{2\rho_0 c}\right|^2 \tag{3.94}$$

Equation (3.94) is valid for both the bounded and the unbounded panel models. The only variable is the impedance, $\mathbf{Z}'_\mathbf{m}$.

First, consider the bounded panel model with a uniform distribution of mass, stiffness and damping. Its impedance can be given by equations (1.71) or (2.131) (with the acoustic radiation damping term neglected since air is the common fluid medium on both sides of the panel in most industrial type applications). Thus, the transmission loss becomes

$$TL = 10\log_{10}\left\{\left(1 + \frac{C_v \cos\theta}{2\rho_0 c}\right)^2 + \left(\frac{\rho_S\omega - K_s/\omega}{2\rho_0 c}\cos\theta\right)^2\right\}. \tag{3.95}$$

In the above equation, C_v, ρ_S and K_s are the viscous damping, mass and stiffness per unit area, respectively. Three regions of interest can be readily identified. If $\omega \ll \omega_n$ (where $\omega_n = (K_s/\rho_S)^{1/2}$) then

$$TL \approx 10\log_{10}\left\{1 + \left(\frac{K_s/\omega}{2\rho_0 c}\cos\theta\right)^2\right\}. \tag{3.96}$$

Here, the stiffness of the panel dominates the transmission loss. Doubling the frequency (an octave increase) produces a fourfold decrease in transmission loss – i.e. a 6 dB decrease. Doubling the stiffness produces a fourfold increase in transmission loss – i.e. a 6 dB increase.

If $\omega = \omega_n$, then

$$TL \approx 10\log_{10}\left(1 + \frac{C_v \cos\theta}{2\rho_0 c}\right)^2. \tag{3.97}$$

At these resonance frequencies the transmission loss is damping controlled.

If $\omega \gg \omega_n$, then

$$TL \approx 10\log_{10}\left\{1 + \left(\frac{\rho_S\omega}{2\rho_0 c}\cos\theta\right)^2\right\}. \tag{3.98}$$

Now, the panel mass dominates the transmission loss. Doubling the frequency (an octave increase) produces a fourfold increase in transmission loss – i.e. a 6 dB increase. Doubling the mass also produces a fourfold increase in transmission loss – i.e. a 6 dB increase. Equation (3.98) is commonly referred to as the mass law for oblique incidence. For normal incidence, $\cos\theta = 1$.

For mechanical excitation of panels, all three regions (stiffness controlled, damping controlled and mass controlled) are relevant. However, when a panel is acoustically excited by a diffuse sound field it has been shown that forced bending waves govern its sound transmission characteristics. Under these conditions, the mass law equation (equation 3.98) is the governing equation for the prediction of transmission loss characteristics at frequencies below the critical frequency.

The analysis for obtaining the impedance, $\mathbf{Z}'_m$ is not so straightforward for the unbounded flexible partition, particularly if damping is to be taken into account, and it will not be derived here. Ver and Holmer[3.9] present an expression for the transmission coefficient, τ, for an unbounded, damped, flexible partition subjected to an incident sound wave at some angle θ, as illustrated in Figure 3.21. The corresponding transmission loss is

$$TL = 10\log_{10}\left\{1 + \eta\left(\frac{\rho_S\omega}{2\rho_0 c}\cos\theta\right)\left(\frac{B\omega^2}{\rho_S c^4}\sin^4\theta\right)\right\}^2$$
$$+ 10\log_{10}\left\{\left(\frac{\rho_S\omega}{2\rho_0 c}\cos\theta\right)\left(1 - \frac{B\omega^2}{\rho_S c^4}\sin^4\theta\right)\right\}^2. \qquad (3.99)$$

In the above equation, ρ_S is the surface mass of the panel, and B is the bending stiffness per unit width (N m). It should be noted that the damping is now represented in terms of the structural loss factor, η, which is related to the complex bending stiffness and the corresponding complex modulus of elasticity (see equation 1.229) – i.e. the viscous-damping coefficient C_v is replaced by $\eta\omega\rho_S$. Fahy[3.3] also derives a similar expression for the transmission loss and the associated impedance. The bending stiffness (real or complex) is related to the corresponding modulus of elasticity by

$$B = \frac{Et^3}{12(1-\nu^2)}, \qquad (3.100)$$

where t is the plate thickness (note that whilst B and E are real in the above equation, they can be replaced by their complex equivalents). The complex quantities subsequently disappear in the transmission loss expression since it is related to the modulus of the impedance (see equation 3.94).

At frequencies below the critical frequency

$$\frac{B\omega^2}{\rho_S c^4} \ll 1, \qquad (3.101)$$

since from equation (1.322)

$$\omega_C^2 = \frac{\rho_S c^4}{B}. \qquad (3.102)$$

Thus, in this frequency range equation (3.99) simplifies to

$$TL \approx 10\log_{10}\left\{1 + \left(\frac{\rho_S\omega}{2\rho_0 c}\cos\theta\right)^2\right\}. \qquad (3.103)$$

This equation is identical to equation (3.98) and it confirms that the panel mass controls the sound transmission through it at frequencies below the critical frequency.

Equation (3.103) is only valid for a specific angle of incidence ranging from 0° to 90°. When the incident sound field is diffuse, as is generally the case in practice with the exception of certain confined spaces, an empirical field-incidence mass law is commonly used in place of the oblique-incidence mass law. It is

$$TL = 10\log_{10}\left\{1 + \left(\frac{\rho_S\omega}{2\rho_0 c}\right)^2\right\} - 5\text{ dB}. \qquad (3.104)$$

This equation is valid for normal-incidence transmission losses greater than 15 dB and it represents an incident diffuse field with a limiting angle of 78° (see reference 3.9). A random-incidence mass law can also be obtained by averaging equation (3.103) over all angles from 0° to 90°. If the normal-incidence transmission loss ($\theta = 0$ in equation 3.103) is defined as TL_0 then the random-incidence transmission loss is

$$TL_R = TL_0 - 10\log_{10}(0.23TL_0). \tag{3.105}$$

Likewise, the field-incidence transmission loss (equation 3.104) can be re-expressed as

$$TL_F = TL_0 - 5\text{ dB}. \tag{3.106}$$

A comparison of the three transmission loss equations (normal incidence, random incidence, and field incidence) for the mass controlled region is presented in Figure 3.22. Experimental results, collated over the years by researchers and product manufacturers etc., suggest that the field-incidence mass law equation is the most appropriate equation for estimating sound transmission characteristics through single panels subjected to diffuse sound fields at frequencies below the critical frequency.

Equation (3.99) can also be used to obtain a qualitative understanding of the behaviour of panels at frequencies above the critical frequency. It cannot, however, be readily used in practice because incident sound waves generally involve a broad range of frequencies and angles of incidence; the latter are generally indeterminate. A close examination of equation (3.99) indicates that the transmission loss is a minimum when

$$\frac{B\omega^2}{\rho_S c^4}\sin^4\theta = 1. \tag{3.107}$$

Fig. 3.22. Transmission loss for panels in the mass-controlled region.

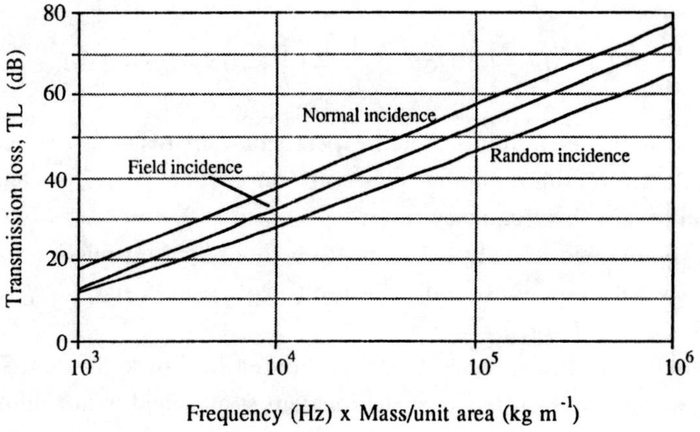

This condition is referred to as the coincidence condition and it corresponds to a situation where the trace wavelength ($\lambda/\sin\theta$) of the incident sound wave equals a free bending wavelength, λ_B, at the same frequency. For finite panels, free bending waves only occur at natural frequencies; for infinite panels they can occur at any frequency. Thus, for finite panels there will be certain coincidence angles, θ_C's, and corresponding coincidence frequencies, ω_{CO}, at frequencies above the critical frequency for which there is very efficient transmission of sound. For finite flat panels, the coincidence frequencies are in fact natural frequencies. From equations (3.102) and (3.107),

$$\sin\theta_{CO} = \left(\frac{\omega_C}{\omega}\right)^{1/2}, \qquad \text{or } \omega_{CO} = \frac{\omega_C}{\sin^2\theta}. \tag{3.108}$$

At these coincidence angles, the panel transmission loss is obtained by substituting equation (3.107) into equation (3.99). It is

$$TL = 10\log_{10}\left\{1 + \eta\left(\frac{\rho_S\omega}{2\rho_0 c}\cos\theta_{CO}\right)\right\}^2. \tag{3.109}$$

At the critical frequency, $\theta = 90°$, and the panel offers no resistance to incident sound waves. At other coincidence angles, the transmission loss is limited by the amount of damping that is present. At angles of incidence that do not correspond to a coincidence angle, the transmission loss is obtained from equation (3.99). Here, both stiffness and damping limit the transmission of sound through the panel.

The above discussion qualitatively illustrates the complex manner in which the transmission of sound can be controlled at frequencies above the critical frequency. In practice, because of the random nature of the frequency composition of the incident sound waves and the associated angles of incidence, equation (3.99) must be solved by numerical integration procedures to obtain a field-incidence transmission loss for frequencies above the critical frequency. Alternatively, an empirical relationship developed by Cremer (see Fahy[3.3]) can be used. It is

$$TL_R = TL_0 + 10\log_{10}\left(\frac{f}{f_C} - 1\right) + 10\log_{10}\eta - 2\text{ dB}. \tag{3.110}$$

The equation indicates a 10 dB increase per octave increase in frequency. It also suggests that structural damping plays an important part in maximising the transmission loss in this frequency range.

In summary, the two relevant equations to be used for transmission loss estimates for panels exposed to diffuse sound fields are (i) equation (3.104) for $f < f_C$, and (ii) equation (3.110) for $f \geqslant f_C$.

The discussions in this section have been limited to diffuse incident sound fields. Sometimes, as already mentioned, if the incident sound field is not diffuse, acoustic standing waves that are sustained within the enclosed fluid volume can couple to the structural panel modes The coupling can be either resonant or non-resonant. When it is resonant, there is both spatial and frequency matching between the fluid and the structural modes; when it is non-resonant there is only spatial matching but no frequency matching. This phenomenon can occur at frequencies below and above the critical frequency. When it occurs, there is a significant reduction in the transmission loss as compared with that predicted by the diffuse sound field relationships. Fahy[3.3] derives the following relationship for the transmission loss below the critical frequency ($f < f_C$):

$$TL = TL_0 - 10\log_{10}\left\{\left[1.5 + \ln\left(\frac{2\omega}{\Delta\omega}\right)\right] + \frac{16c^2}{\eta\omega_C(\omega\omega_C)^{1/2}}\left[\frac{L_x^2 + L_y^2}{L_x^2 L_y^2}\right]\left[1 + \frac{2\omega}{\omega_C} + 3\left(\frac{\omega}{\omega_C}\right)^2\right]\right\}. \quad (3.111)$$

L_x and L_y are the panel dimensions and $\Delta\omega$ is the frequency bandwidth. It turns out that non-resonant coupling produces transmission loss values that are similar to the diffuse field values, and that resonant coupling produces transmission loss values that are about 3–6 dB lower. At frequencies above the critical frequency the transmission loss values, obtained by accounting for the coupling modes, are very similar to Cremer's equation (equation 3.110)[3.3].

The major aspects of the various points raised so far in this sub-section are summarised in Figure 3.23. The main observations are: (i) a 6 dB increase in transmission loss per doubling of stiffness at low frequencies (for mechanical excitation); (ii) a reduction in low frequency resonant responses (increase in transmission loss) with damping treatment; (iii) a 6 dB increase in transmission loss

Fig. 3.23. Sound transmission characteristics of a single, homogeneous panel.

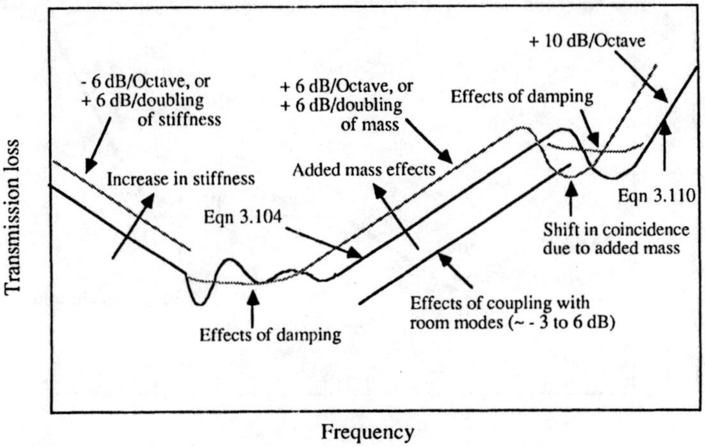

per doubling of mass (below the critical frequency); (iv) a 3–6 dB decrease in transmission loss when structure–acoustic couplings are present (below the critical frequency); (v) a lowering of the coincidence frequency with an increase of mass; (vi) an increase in transmission loss at the critical frequency with the addition of damping; and (vii) an increase in transmission loss with added damping and stiffness at high frequencies.

In practice, many types of complex 'single' panels are available including two- and three-ply laminates, orthotropic panels, ribbed panels, and various other forms of composite barriers. The common denominator in each of these cases is that the panel is solid and therefore behaves essentially as a single panel. Ver and Holmer[3.9] and Reynolds[3.11] provide detailed empirical information on a range of complex solid panels and partitions. A more subtle means of improving transmission loss characteristics without significantly increasing mass is to utilise double-leaf panels with an enclosed air gap. The performance characteristics of these types of panels are discussed in the next sub-section.

A commonly used empirical procedure for estimating the field-incidence transmission loss characteristics of some common building materials is the 'plateau method'. The method is applicable to frequencies below and above the critical frequency. It approximates the transmission loss of single panels and assumes that a diffuse field exists on both sides of the panel. The length and width of the panel have to be at least twenty times the panel thickness. A typical plateau method design chart is presented in Figure 3.24. Firstly, the mass law region is determined using equation (3.104). Then, the coincidence region is approximated by a horizontal line whose height is obtained from Table 3.1. Point A lies at the intersection of the horizontal coincidence line and the mass law line, and point B is determined relative to point A from the frequency ratio in the table. The transmission loss in regions above B is

Fig. 3.24. The plateau method for the estimation of single panel transmission loss characteristics.

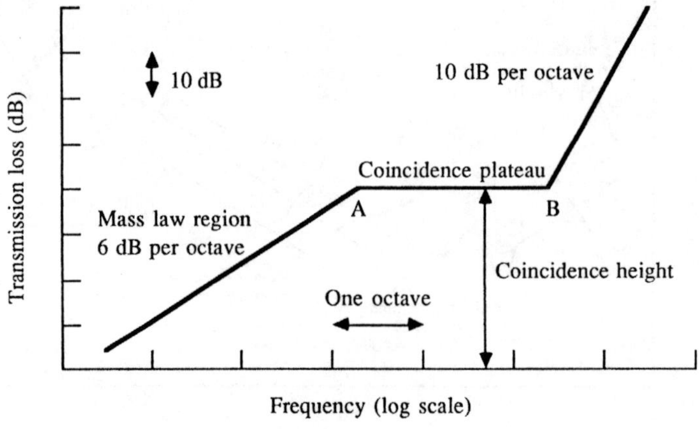

subsequently estimated by projecting a line upwards from point B with a slope of 10 dB per octave.

3.9.2 Sound transmission through double-leaf panels

When weight restrictions are critical and substantial transmission losses are required (e.g. aircraft bodies, multi-storey buildings etc.) single panels are generally not adequate. Doubling the surface mass of a single panel only produces a 6 dB increase in transmission loss. Double-leaf panels can produce significantly larger transmission losses, and are generally used these days to overcome some of the limitations of single panels. Double-leaf panels comprise two separate single panels separated by an air gap. Generally, the two panels are also mechanically connected, and some form of absorption material is contained within the cavity. A typical cross-section of a double-leaf panel is schematically illustrated in Figure 3.25. The two main sound transmission paths through the double-leaf panel are (i) direct transmission via the panel–fluid–panel path, and (ii) structure-borne transmission through the mechanical couplings.

Table 3.1. *Data for use with the plateau method for the estimation of the transmission loss of some common materials.*

Material	Surface density ($kg\ m^{-2}$ per mm thickness)	Coincidence height (dB)	Frequency ratio, B/A
Aluminium	2.66	29	11.0
Brick	2.10	37	4.5
Concrete	2.28	38	4.5
Glass	2.47	27	10.0
Lead	11.20	56	4.0
Plaster	1.71	30	8.0
Plywood	0.57	19	6.5
Steel	7.60	40	11.0

Fig. 3.25. Schematic illustration of a cross-section of a typical double-leaf panel.

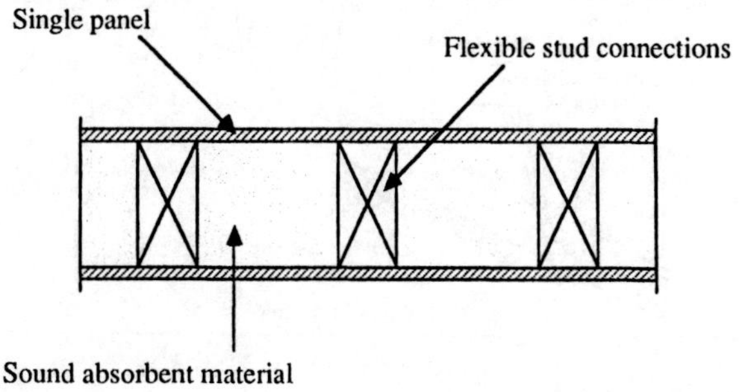

Fahy[3.3] provides a detailed theoretical analysis for the transmission of normally and obliquely incident plane waves through an unbounded double-leaf partition. Ver and Holmer[3.9] also provide some empirical relationships and qualitative discussions. Most of the detailed information that is currently available is, however, only published in the research literature – Fahy provides numerous recent references in his book; also, a large range of products are commercially available, each with their own specific transmission loss characteristics. A detailed analysis of the performance of double-leaf panels is beyond the scope of this book and only the more important fundamental principles will be discussed.

The behaviour of a typical double-leaf partition is schematically illustrated in Figure 3.26. Two important features of the transmission loss performance of double-leaf panels are (i) a double-leaf panel resonance (also known as a mass–air–mass resonance), and (ii) air-gap resonances (also known as cavity resonances). The double-leaf panel resonance is a low frequency resonance which is due to the panels behaving like two masses coupled by an air-spring. It is a function of the panel masses and the air gap. Fahy[3.3] derives the double-leaf panel resonance frequency (also see Ver and Holmer[3.9]). It is

$$f_0 = \frac{1}{2\pi}\left\{\left(\frac{\rho_0 c^2}{d}\right)\left(\frac{\rho_{S1}+\rho_{S2}}{\rho_{S1}\rho_{S2}}\right)\right\}^{1/2}, \tag{3.112}$$

where d is the air-gap separation between the two panels, ρ_{S1} is the surface mass ($\mathrm{kg\,m^{-2}}$) of the first panel, and ρ_{S2} is the surface mass of the second panel. The

Fig. 3.26. Schematic illustration of the behaviour of a typical double-leaf panel.

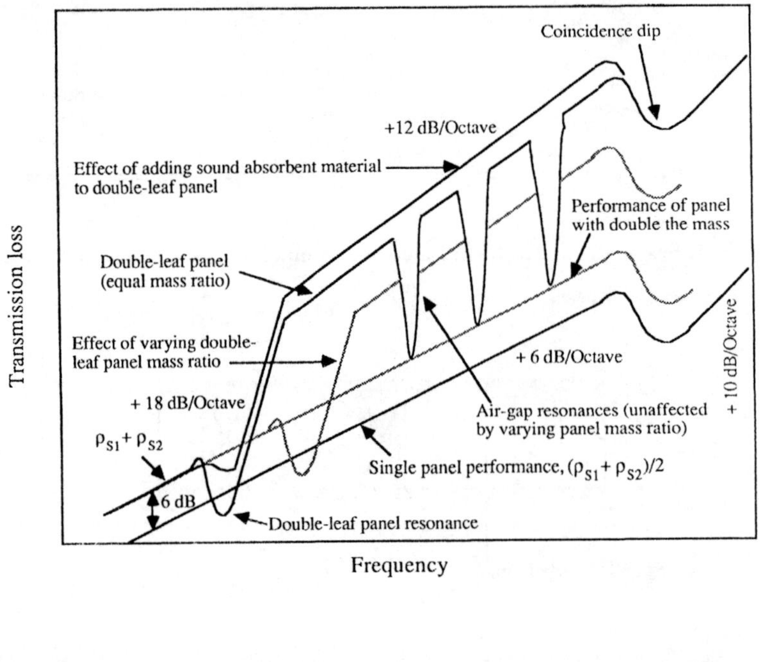

air-gap resonances, on the other hand, are high frequency resonances ($kd > 1$) and they are associated with the cavity dimensions. The identification of the air-gap resonance frequencies is somewhat complicated – for the present purposes it is sufficient to note that the transmission loss of a double-leaf panel is reduced to that of a single panel of surface mass $\rho_{S1} + \rho_{S2}$ at the air-gap resonances.

Some general comments and observations relating to the performance of a typical double-leaf panel are listed below.

(1) At frequencies below the double-leaf panel resonance the transmission loss is equivalent to that of a single panel of surface mass $\rho_{S1} + \rho_{S2}$ (i.e. there is a 6 dB increase in transmission loss over a single panel of average surface mass $\rho_{S1}/2 + \rho_{S2}/2$).

(2) There is a significant reduction in transmission loss at the double-leaf panel resonance. The addition of damping improves this.

(3) There is a sharp increase in transmission loss ($\sim$ 18 dB per octave) after the double-leaf panel resonance. This increase is maintained until the first air-gap resonance is encountered.

(4) At the air-gap resonances the transmission loss of a double-leaf panel is reduced to that of a single panel of surface mass $\rho_{S1} + \rho_{S2}$. These air-gap resonances can be minimised and significant improvements can be achieved by the inclusion of suitable sound absorbent material within the cavity. The absorbent material has the added effect of damping the double-leaf panel resonance and sometimes completely decoupling the individual partitions.

(5) In the general air-gap resonance region the transmission loss increases at $\sim$12 dB per octave up to the critical frequency at which point the usual coincidence dip occurs.

(6) The transmission loss performance at the double-leaf panel resonance can be improved by increasing the surface mass ratio (i.e. $\rho_{S1} \neq \rho_{S2}$). This has, however, a twofold negative effect. Firstly, the double-leaf panel resonance is shifted upwards in the frequency domain, and secondly the high frequency transmission loss is reduced. The minima at the air-gap resonances remain the same.

(7) Optimum high frequency performance is achieved when $\rho_{S1} = \rho_{S2}$.

(8) The sound absorbent material that is used in the cavity should have as high a flow resistance as possible without producing any unnecessary mechanical coupling between the two panels.

(9) Mechanical coupling should be minimised wherever possible by using flexible stud connections (or flexible studs) between partitions. Rigid connections always substantially compromise the transmission loss performance of double-leaf panels, particularly at frequencies in proximity to the double-leaf panel resonance.

3.10 The effects of fluid loading on vibrating structures

Fluid loading of vibrating structures as discussed here and in other texts on noise and vibration control only relates to small amplitude motions that do not affect the excitation forces. Fluid loading problems relating to various forms of dynamic instabilities are a separate issue.

Fluid loading has two main effects on vibrating structures. Firstly, the fluid mass-loads the structure, and this alters the structural natural frequencies. Secondly, the fluid medium provides acoustic radiation damping, and this affects the sound radiation characteristics of the structure. When the fluid medium is air, which is generally the case for most engineering noise and vibration control applications, the mass loading effects of the fluid are generally of a second order since fluid forces are proportional to density. An exception occurs in small confined spaces where even air can fluid-load a surface; a typical example being the double-leaf panel in the previous section where the air-gap acts like a spring and produces a double-leaf resonance. Dense fluids (e.g. water) have significant effects on the vibrational and sound radiation characteristics of structures. When the fluid volume is unbounded (e.g. a vibrating plate submerged in a large volume of fluid) it cannot sustain standing waves and it simply mass-loads the structure, and provides acoustic radiation damping; when the fluid volume is bounded (e.g. dense liquids contained within cylindrical shells), the problem is more complex because now both the structure and the fluid can sustain standing waves and natural frequencies, and there is feedback between the structure and the fluid. When this occurs the system is referred to as being strongly coupled. Fortunately, there are many instances when the fluid natural frequencies can be neglected and the subsequent feedback ignored. Whilst only dense fluids mass-load structures, all fluids (including air) possess acoustic radiation damping characteristics – energy is dissipated from the vibrating structure in the form of radiated sound.

The subject of fluid loading of vibrating structures is a complex one – one which is addressed in the specialist literature. Junger and Feit[3.1] and Fahy[3.3] provide a fairly extensive coverage of the subject. Fahy[3.12] also provides a general review on structure–fluid interactions which includes numerous references to the recent research literature. This section will only cover some of the fundamental principles involved. It is important that engineers dealing with noise and vibration control problems are aware of the different effects that fluid loading can have on the results.

Fluid loading concepts were introduced briefly in chapter 2 (sub-section 2.3.4) in relation to sound radiation from a vibrating piston mounted in a rigid baffle. The vibrating piston serves as a useful example to illustrate the effects of fluid loading on vibrating structures. As already mentioned, fluctuating pressures which are in close proximity to a vibrating surface will generate an acoustic radiation load on that surface. This acoustic radiation load is in addition to any mechanical excitation of the surface which could be the primary source of vibration in the first instance. Any mechanical load on a vibrating surface manifests itself as a mechanical impedance; likewise, any acoustic radiation load manifests itself as a radiation impedance. The total impedance to any surface motion would thus be the linear sum of the mechanical and acoustic radiation impedances. If, however, the fluid volume was confined (e.g. the inside of a duct or a small container) then it could sustain natural frequencies and mode shapes and these would couple to the structural modes – the resultant coupled natural frequencies would not necessarily be the same as the uncoupled natural frequencies of the fluid and structural systems, and the coupled impedance would not necessarily be the linear sum of the mechanical and acoustic radiation impedances. Such strongly coupled systems which involve feedback between the structure and the fluid are not discussed here.

In chapter 2 (sub-section 2.3.4) it was shown that the total impedance of a piston vibrating in a rigid baffle is

$$\mathbf{Z} = \mathbf{Z}_{\mathbf{m}} + \mathbf{Z}_{\mathbf{r}} = \frac{\mathbf{F}_{\mathbf{m}}}{\mathbf{U}} = C_{\mathrm{v}} + \mathrm{i}(M\omega - K_{\mathrm{s}}/\omega) + \rho_0 c\pi z^2\{R_1(2kz) + \mathrm{i}X_1(2kz)\}. \quad (3.113)$$

It is worth reminding the reader that the acoustic radiation impedance, $\mathbf{Z}_{\mathbf{r}}$, is defined in this book in similar units to the mechanical impedance since the radiating surface area is common to both pressure and volume velocity (see the paragraph preceding equation 2.127 in chapter 2 for a detailed explanation). Also, the resistive and reactive functions R_1 and X_1 are defined in chapter 2 (equations 2.127 and 2.128).

Several important points can be made regarding equation (3.113). Firstly, the mechanical impedance, $C_{\mathrm{v}} + \mathrm{i}(M\omega - K_{\mathrm{s}}/\omega)$, has both real and imaginary components. The structural damping is resistive and real; the mass is reactive and positive imaginary; and the stiffness is reactive and negative imaginary. Secondly, the acoustic radiation impedance has two terms: the first, $\rho_0 c\pi z^2 R_1(2kz)$ is resistive and real, and is therefore associated with acoustic radiation damping; the second, $\rho_0 c\pi z^2 \mathrm{i}X_1(2kz)$, is reactive and positive imaginary, and is therefore associated with mass. The fluid thus (i) provides additional damping, and (ii) mass-loads the structure.

Acoustic radiation damping plays a very important part in the sound radiation of structures, even in light fluid media such as air. This is particularly so for lightweight structures with high radiation ratios. Recent work by Rennison and Bull[3.13] and Clarkson and Brown[3.14] on the estimation of damping in lightweight structures demonstrates this. Quite often the acoustic radiation damping dominates over the *in vacuo* structural damping. Fahy[3.3] also shows that acoustic radiation damping and the associated sound radiation depends upon the average distribution of vibration over the whole structure, the exception being at very high frequencies.

In addition to providing acoustic radiation damping, dense fluids also mass-load structures – this is apparent from the reactive component of the radiation impedance in equation (3.113). Unlike the resistive component which depends upon the average distribution of vibration over the whole surface, this reactive component is highly dependent upon local motions[3.3]. The inertial mass associated with these reactive components also has the effect of reducing the natural frequencies of the fluid-loaded structure. Fahy[3.3] provides a useful relationship, derived from some previous work by Davies[3.15], for estimating the natural frequencies of fluid-loaded structures. The relationship is restricted to frequencies below the critical frequency. It is

$$f'_m \approx f_m\left(1 + \frac{\rho_0}{\rho_S k_m}\right)^{-1/2}, \tag{3.114}$$

where ρ_0 is the fluid density (kg m^{-3}), ρ_S is the surface mass per unit area of the structure (kg m^{-2}), and k_m is the primary (*in vacuo*) structural wavenumber component. As the wavenumber increases, the fluid loading has a smaller effect on the structural natural frequencies.

Fluid loading also affects the sound radiated from structures. The topic is too complex for inclusion in this book. However, one important observation can be made in relation to sound radiation at frequencies below the critical frequency – the directivity and source characteristics of point and line sources on structures are modified by the presence of significant fluid loadings such that point and line monopoles become point and line dipoles, respectively, with the dipole axis coincident with the applied force.

In summary, fluid loading has the following general effects on vibrating structures:

(1) The natural frequencies of the structure are altered – this is associated with the fluid mass loading effects. The greatest effects occur at low wavenumbers.

(2) The acoustic radiation damping associated with sound waves radiating from the structure varies with the fluid density – radiation damping is also important in light fluid media.

(3) When the fluid volume is confined, the possibility of strong coupling between fluid and structural modes exists.

(4) The impedance of the structure is altered – numerous relationships are available for point and line forces and moments for plates and shells[3.3].

(5) The directivity and source characteristics of fluid-loaded radiators are modified[3.3].

3.11 Impact noise

Impact noise is a very common occurrence in the industrial environment, and typical examples include punch presses, drop forges, impacting gears etc. Until recently, very little has been known about the various mechanisms involved. Some pioneering research by Richards[3.8] has led to a better understanding of impact noise mechanisms (a comprehensive list of Richards's earlier work, and the work of other researchers, is provided in reference 3.8).

When two bodies are impacted together (e.g. a hammer and a sheet of metal), sound is created by two processes. The first process, known as acceleration (or deceleration) noise, is due to the rapid change in velocity of the moving body (e.g. the hammer) during the impact process – i.e. the sound emanates from the impacting elements. The second process, sometimes known as ringing noise, is more conventional and is simply due to sound radiation from resonant structural modes of the workpiece or any other attached structures. The sound radiation associated with ringing noise is dependent upon radiation ratios, mean-square surface vibrational velocities, damping, etc., and it can be predicted by utilising the various procedures described earlier on in this chapter. It is the first process (acceleration noise) which requires special attention and which is therefore the subject of this section.

When a single body of mass M moves through a fluid (e.g. air) with a velocity v_0, the virtual mass of the fluid displaced by the body possesses kinetic energy; the virtual mass being the mass of fluid equal to half that displaced by the body. When the body is brought to rest instantaneously, the kinetic energy of the mass is lost immediately. The energy contained in the fluid is subsequently lost in the generation of sound (assuming that the fluid is non-viscous); if the body were brought to rest slowly, most of the fluid energy would be returned to the body. Richards[3.8] shows that the energy associated with the virtual mass of the fluid displaced by a single body is

$$E_v = \frac{\rho_0 V v_0^2}{4}, \tag{3.115}$$

where ρ_0 is the fluid density, V is the volume of the single body, and v_0 is its velocity prior to impact. Equation (3.115) thus represents the energy content of the radiated sound.

In practice, a body will take some finite time to stop and some of the virtual energy will be radiated as sound and some will be returned to the body. An acceleration noise efficiency, μ_{accn}, can be defined such as to relate the actual noise energy radiated during an impact process (involving a moving body of volume V and mass M) to the energy that would be radiated if two equal bodies (each of volume V and mass M) were brought to rest immediately upon impact. Thus,

$$\mu_{accn} = \frac{E_{accn}}{2E_v} = \frac{2E_{accn}}{\rho_0 V v_0^2}, \tag{3.116}$$

where E_{accn} is the radiated noise energy during the actual impact process. Unlike the radiation ratio, σ, which can be greater than unity, the acceleration noise efficiency is always less than unity. It is a function of the contact time between the moving mass and the workpiece – the shorter the contact time the greater the radiated noise energy. A non-dimensional contact time, δ, can be defined such that it is the reciprocal of ka, where k is the wavenumber and a is a typical body dimension (e.g. radius of a sphere). It is essentially the number of typical body dimensions travelled by the sound wave during the deceleration process, and it is given by[3.8]

$$\delta = \frac{ct_0}{V^{1/3}}, \tag{3.117}$$

where t_0 is the duration of the impact time (i.e. the mass M decelerates from a velocity v_0 to zero in a short time interval t_0).

Equations (3.116) and (3.117) can be combined to provide a relationship between radiated noise energy and contact time. A useful empirical relationship based on sizeable quantities of experimental data is (Richards[3.8]),

$$\mu_{\text{accn}} = 0.7 \quad \text{for } \delta < 1, \tag{3.118a}$$

and

$$\mu_{\text{accn}} = 0.7\delta^{-3.2} \quad \text{for } \delta > 1. \tag{3.118b}$$

Equations (3.118*a* and *b*) are very useful ready-reckoners for predicting the radiated noise energy associated with industrial impact processes. They are presented in graphical form in Figure 3.27. For most metal–metal impact processes $\delta < 1$, and the acceleration noise is significant. Thus, it is highly desirable to increase impact times or to increase contact times during any impact process in order to reduce acceleration noise. Preloading workpieces, dense fluid lubrication, etc., are some of the practical ways of increasing contact times.

The information provided by equations (3.116)–(3.118*a*, *b*) relates to the energy associated with the radiated noise. Often it is the sound pressure level that is of more

Fig. 3.27. Acceleration efficiencies for arbitrary bodies subject to rapid deceleration and subsequent impact excitation.

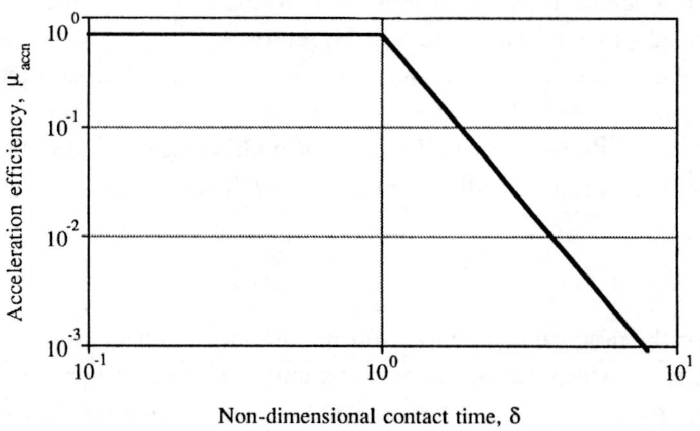

direct relevance in industrial noise control (decibels and sound pressure levels are defined in the next chapter). Richards provides two very useful empirical formulae for predicting the peak sound pressure levels at some distance, r, from an impact process. They are

$$L_p = 143 + 20 \log_{10} v_0 - 20 \log_{10} r + 6.67 \log_{10} V \quad \text{for } \delta < 1, \quad (3.119)$$

and

$$L_p = 143 + 20 \log_{10} v_0 - 20 \log_{10} r + 6.67 \log_{10} V - 40 \log_{10} \delta \quad \text{for } \delta > 1. \quad (3.120)$$

Hence, in industrial situations where both acceleration and ringing noises are present, engineers should be able to ascertain as to which of the two is dominant by separately estimating both components. Ringing noise levels (resonant structural modes) associated with most machine components can be evaluated by utilising the radiation ratio approach (equation 3.30) or by using reciprocity; acceleration noise levels associated with impact processes can be estimated by using the procedures described in this section.

References

3.1 Junger, M. C. and Feit, D. 1972. *Sound, structures, and their interaction*, M.I.T. Press.

3.2 Pierce, A. D. 1981. *Acoustics: an introduction to its physical principles and applications*, McGraw-Hill.

3.3 Fahy, F. J. 1985. *Sound and structural vibration: radiation, transmission and response*, Academic Press.

3.4 Fahy, F. J. 1986. *Sound and structural vibration – a review*, Proceedings Inter-Noise '86, Cambridge, USA, pp. 17–38.

3.5 Lyamshev, L. M. 1960. 'Theory of sound radiation by thin elastic shells and plates', *Soviet Physics Acoustics* **5**(4), 431–8.

3.6 Cremer, L., Heckl, M. and Ungar, E. E. 1973. *Structure-borne sound*, Springer-Verlag.

3.7 Temkin, S. 1981. *Elements of acoustics*, John Wiley & Sons.

3.8 Richards, E. J. 1982. 'Noise from industrial machines', chapter 22 in *Noise and vibration*, edited by R. G. White and J. G. Walker, Ellis Horwood.

3.9 Ver, I. L. and Holmer, C. I. 1971. 'Interaction of sound waves with solid structures', chapter 11 in *Noise and vibration control*, edited by L. L. Beranek, McGraw-Hill.

3.10 Norton, M. P. and Bull, M. K. 1984. 'Mechanisms of the generation of external acoustic radiation from pipes due to internal flow disturbances, *Journal of Sound and Vibration* **94**(1), 105–46.

3.11 Reynolds, D. D. 1981. *Engineering principles of acoustics – noise and vibration*, Allyn & Bacon.

3.12 Fahy, F. J. 1982. 'Structure–fluid interactions', chapter 11 in *Noise and vibration*, edited by R. G. White and J. G. Walker, Ellis Horwood.

3.13 Rennison, D. C. and Bull, M. K. 1977. 'On the modal density and damping of cylindrical pipes', *Journal of Sound and Vibration* **54**(1), 39–53.

3.14 Clarkson, B. L. and Brown, K. T. 1985. 'Acoustic radiation damping', *Journal of Vibration, Acoustics, Stress, and Reliability in Design* **107**, 357–60.
3.15 Davies, H. G. 1971. 'Low frequency random excitation of water loaded rectangular plates', *Journal of Sound and Vibration* **15**(1), 107–20.

Nomenclature

a	radius of an oscillating sphere
a_m	mean pipe radius
b	width of a beam
B	bending stiffness per unit width
c, c_1, c_2, etc.	speeds of sound
c_B	bending wave velocity
c_e	speed of sound in the fluid external to a cylindrical shell
c_L	quasi-longitudinal wave velocity
c_s	bending wave velocity in a cylindrical shell
C_v	viscous damping per unit area, piston mechanical damping
d	air-gap separation between two panels
E	Young's modulus of elasticity
E_{accn}	radiated noise energy during impact
E_v	energy associated with the virtual mass of fluid displaced by a body
f	frequency
f_0	double-leaf panel resonance frequency
f_C	critical frequency
f_m	*in vacuo* natural frequency associated with fluid loading
$f_{m,n}$	natural frequencies of a rectangular plate
f'_m	fluid-loaded natural frequency
F	excitation force
$\mathbf{F}$	complex excitation force
$\mathbf{F_m}$	complex applied mechanical force
$\mathbf{F_p}$	complex force on piston due to acoustic pressure fluctuations
F_{rms}	root-mean-square applied force
F_{x-}	$-$ve force at position x
F_{y+}	$+$ve force at position y
$\mathbf{G}_{\omega}(\vec{r}, \omega \vert \vec{r}_0, \omega)$	free space Green's function for a unit, time-harmonic point source – i.e. frequency

	domain Green's function (complex function)
$H_1^{(1)}$	first-order Hankel function of the first kind
I	second moment of area of a cross-section about the neutral plane axis
k, k_1, k_2, etc.	wavenumbers, acoustic wavenumbers
k_B, k_{B1}, k_{B2}, etc.	bending wavenumbers
k_C	critical wavenumber
k_m	primary structural wavenumber component associated with fluid loading
$k_{m,n}$	characteristic wavenumber of the (m, n)th plate mode
k_x	wavenumber in the x-direction on a rectangular plate
k_y	acoustic wavenumber in the y-direction, wavenumber in the y-direction on a rectangular plate
K_s	stiffness per unit area, piston stiffness
l	length of a line source, length of a cylindrical pipe
L_p	sound pressure level
L_x	length of a rectangular plate in the x-direction
L_y	length of a rectangular plate in the y-direction
m	integer number of half-waves in the x-direction on a rectangular plate, number of half-waves along a pipe axis
M	piston mass, mass of an arbitrary body
n	integer number of half-waves in the y-direction on a rectangular plate, number of full waves along a pipe circumference
$n(\omega)$	modal density
$\vec{n}$	unit normal vector
p, $\mathbf{p}$	sound pressure (bold signifies complex)
$\mathbf{p}_I$	complex incident sound pressure
p_{max}	maximum amplitude of radiated sound pressure
$\mathbf{p}_R$	complex reflected sound pressure
$\mathbf{p}_T$	complex transmitted sound pressure

$\mathbf{p}(\vec{r})$	complex radiated sound pressure in the sound field
$\mathbf{p}(\vec{r}, t)$	complex radiated sound pressure in the sound field
$\mathbf{p}(x, y, t), \mathbf{p}(x, y)$	complex radiated sound pressure in the sound field
$\mathbf{p}(\vec{r}_0)$	complex surface pressure on a vibrating body
$\mathbf{p}_b(\vec{r}, t)$	complex blocked pressure on a piston surface
P	perimeter
Q_p	peak source strength
Q_{rms}	root-mean-square source strength
$\mathbf{Q}(t)$	complex source strength ($m^3 s^{-1}$)
r	radius, radial distance
$\vec{r}$	position vector at a receiver position in the sound field
$\vec{r}_0$	position vector on a vibrating body
r_A	radius of an arc defining an acoustic wavenumber
r_B	radius of an arc defining a bending wave-number vector
r_C	radius of an arc defining the critical wave-number
$R_1(2kz)$	resistive function associated with the radiation impedance of a piston
S	surface area
t	thickness of a plate or bar
t_0	duration of impact time
TL	transmission loss
TL_0	normal-incidence transmission loss
TL_F	field-incidence transmission loss
TL_R	random-incidence transmission loss
$u, \vec{u}$	acoustic particle velocity (arrow denotes vector quantity)
$\mathbf{u}_I$	complex incident acoustic particle velocity
$\mathbf{u}_R$	complex reflected acoustic particle velocity
u_{rms}	root-mean-square acoustic particle velocity
$\mathbf{u}_T$	complex transmitted acoustic particle velocity
$\mathbf{u}_{y\,fluid}$	complex acoustic particle velocity perpen-

	dicular to a plate
$\mathbf{u}_{y\,plate}$, $\mathbf{u}_{yp}$	complex normal plate surface vibrational velocity
$u_{yp_{max}}$	maximum normal plate surface vibrational velocity
$u_{yp_{rms}}$	root-mean-square normal plate surface vibrational velocity
$\vec{\mathbf{u}}_n(\vec{r}_0)$	complex normal surface velocity (vector quantity)
$\langle \overline{u^2} \rangle$	mean-square acoustic particle velocity (space- and time-averaged)
$\mathbf{U}$	complex piston surface velocity
U_a	peak normal surface velocity of an oscillating sphere
v	vibrational velocity of an arbitrary structure
v_0	velocity of a body prior to impact
v_{0rms}	root-mean-square drive-point vibrational velocity
v_{lrms}	root-mean-square line source vibrational velocity
$\mathbf{v}_p$	complex plate velocity
v_{y+}	+ve velocity at a position y
v_{x-}	−ve velocity at a position x
$\langle v^2 \rangle$	time-averaged mean-square vibrational velocity
$\langle \overline{v^2} \rangle$	mean-square vibratonal velocity (space- and time-averaged)
$\langle v_0^2 \rangle$	mean-square drive-point vibrational velocity
$\langle v_l^2 \rangle$	mean-square drive-line vibrational velocity
V	volume
X	arbitrary position
$X_1(2kz)$	reactive function associated with the radiation impedance of a piston
Y	arbitrary position
z	piston radius
$\mathbf{Z}$	impedance (complex function)
Z_m, $\mathbf{Z}_m$	mechanical impedance (bold signifies complex)
$\mathbf{Z}'_m$	mechanical impedance per unit area (complex function)
$\mathbf{Z}_r$	radiation impedance (complex function)

β, β_{avg}	constants of proportionality
δ	non-dimensional contact time
$\Delta\omega$	incremental increase in radian frequency
η	structural loss factor
θ	angle
θ_C	coincidence angle
κ	constant of proportionality
λ	acoustic wavelength
λ_B	bending wavelength
λ_C	critical wavelength
$\lambda_{m,n}$	characteristic wavelength of the (m n)th plate mode
λ_x	wavelength in the x-direction on a rectangular plate
λ_y	wavelength in the y-direction on a rectangular plate
μ_{accn}	acceleration noise efficiency
ν	Poisson's ratio
π	3.14 . . .
Π	sound power
Π_{dis}	dissipated power from a vibrating structure
Π_{dl}	drive-line radiated sound power
Π_{dp}	drive-point radiated sound power
Π_{in}	input power to a vibrating structure
Π_{rad}	radiated sound power
ρ, ρ_1, ρ_2, *etc.*	densities
ρ_0	mean fluid density
ρ_L	mass per unit length
ρ_S, ρ_{S1}, ρ_{S2}, etc.	masses per unit area (surface masses)
σ	radiation ratio
τ	sound transmission coefficient (wave transmission coefficient)
χ	constant of proportionality
ψ	constant of proportionality
ω	radian (circular) frequency
ω_C	radian (circular) critical frequency
ω_{CO}	coincidence frequency
ω_n	natural radian (circular) frequency
$\vec{\nabla}$	divergence operator (vector quantity)
$\langle\ \rangle$	time-average of a signal
$\overline{}$	space-average of a signal (overbar)

4
Noise and vibration measurement and control procedures

4.1 Introduction

A vast amount of applied technology relating to noise and vibration control has emerged over the last twenty years or so. It would be an impossible task to attempt to cover all this material in a text book aimed at providing the reader with a fundamental basis for noise and vibration analysis, let alone in a single chapter! This chapter is therefore only concerned with some of the more important fundamental considerations required for a systematic approach to engineering noise and vibration control, the main emphasis being the industrial environment. The reader is referred to Harris[4.1] for a detailed engineering-handbook-type coverage of existing noise control procedures, and to Harris and Crede[4.2] for a detailed engineering-handbook-type coverage of existing shock and vibration control procedures. Beranek[4.3] also covers a wide range of practical noise and vibration control procedures. Some of the more recent advances relating to specific areas of noise and vibration control are obviously not available in the handbook-type literature, and one has to refer to specialist research journals. A list of major international journals that publish research and development articles in noise and vibration control is presented in Appendix 1.

This chapter commences with a discussion on noise and vibration measurement units. The emphasis is on the fundamental principles involved with the selection of objective and subjective sound measurement scales, vibration measurement scales, frequency analysis bandwidths, and the addition and subtraction of decibels. A brief section is included on the appropriate selection of noise and vibration measurement instrumentation; a wide range of detailed application notes is readily available from the various product manufacturers.

Useful relationships for the measurement of omni-directional spherical and cylindrical free-field sound propagation are developed; the relationships are based upon the sound source models developed in chapter 2 and are useful for predicting noise levels from individual sources, from strings of sources such as a row of cars on a highway, or from line sources such as trains. The relationships are necessarily limited to regions in free space where there are no reverberation effects. The directional characteristics of noise sources are subsequently accounted for.

The concepts of different types of sound power models are introduced. In chapter 2, it was illustrated that hard reflecting surfaces result in a pressure doubling, a fourfold increase in sound intensity and a subsequent doubling in radiated sound power of monopoles. This is a very important point; one which is often overlooked by noise control engineers – instead of having a constant sound power (as is commonly assumed), the source has a constant volume velocity. A section is also included on the measurement of sound power. Knowledge of the sound power characteristics of a source allows for subsequent engineering noise control analysis for different environments. Free-field, reverberant-field, semi-reverberant, and sound intensity techniques are described. The sound intensity technique, in particular, is one which is still the subject of much research. It has significant advantages over the others and is expected to become the recommended international standard in the foreseeable future.

The 'control' section of the chapter commences with some general comments on the basic sources of industrial noise and vibration, existing industrial noise and vibration control methods, and the economics of industrial noise and vibration control. Sound transmission between rooms, acoustic enclosures, acoustic barriers, and sound-absorbing materials have been selected as appropriate topics for discussion since they are all widely used in engineering practice.

A section is devoted to vibration control procedures. Low frequency vibration isolation for both single- and multi-degree-of-freedom systems is discussed together with vibration isolation in the audio-frequency range where the flexibility of the supporting structure has to be accounted for. Different types of vibration isolation materials currently used are also discussed. Dynamic absorption by the attachment of a secondary mass to a vibrating structure is reviewed, and the chapter ends with a brief discussion on different types of damping materials.

A range of noise and vibration control topics including mufflers, acoustic transmission lines and filters, outdoor sound propagation over large distances, architectural acoustics, noise and vibration control criteria and regulations, hearing loss and the psychological effects of noise and community noise are not covered in this book. A suitable list of references is provided at the end of the chapter. Chapter 8 also follows on from where this chapter ends, and deals with the usage of noise and vibration signals as a diagnostic tool for a range of industrial machinery. The subject of 'machine condition monitoring' is increasingly becoming more and more relevant to industry. It has been amply demonstrated that considerable economic advantages are to be had.

4.2 Noise and vibration measurement units – levels, decibels and spectra

4.2.1 Objective noise measurement scales

Pressure fluctuation amplitudes are by far the most easily measured parameters at a single point in a sound field. Hence, noise levels are often quantified in terms of sound pressure levels (usually r.m.s.). The human perception of sound ranges from a lower limit of 20 micropascals (μPa) to an upper limit of about 200 Pa. This represents a considerable linear dynamic range – i.e. about 10^7. Because of this, it is more convenient to firstly work with relative measurement scales rather than with absolute measurement scales, and secondly to logarithmically compress them.

Two variables differ by one bel if one is ten (10^1) times greater than the other, and by two bels if one is one hundred (10^2) times greater than the other. The bel is still a very large unit and it is more convenient to divide it into ten parts – hence the decibel. Two variables differ by one decibel (1 dB) if they are in the ratio $10^{1/10}$ (≈ 1.26) or by three decibels if they are in the ratio $10^{3/10}$ (≈ 2.00). Three decibels (3 dB) thus represent a doubling of the relative quantity (e.g. sound power, sound intensity, sound pressure, etc.).

Decibel scales are commonly used to quantify both noise and vibration levels, and, since they represent relative values, they have to be constructed with reference values which are universally accepted. Consider the sound power radiated by a sound source. Let Π_0 be the reference sound power, and Π the radiated sound power such that

$$\frac{\Pi}{\Pi_0} = 10^n = (10^{0.1})^{10n}, \tag{4.1}$$

where n is a number. The sound source has a sound power level of n bels re Π_0 or $10n$ decibels re Π_0. Taking logarithms on both sides yields

$$\log_{10} \frac{\Pi}{\Pi_0} = n, \tag{4.2}$$

or

$$10 \log_{10} \frac{\Pi}{\Pi_0} = 10n \text{ dB} = L_\Pi. \tag{4.3}$$

Thus,

$$L_\Pi = 10 \log_{10} \frac{\Pi}{\Pi_0} \text{ dB re } \Pi_0. \tag{4.4}$$

L_Π is the sound power level of a sound source relative to the reference sound power, Π_0. It is important to note that the sound power of a sound source refers to the absolute value of power in watts etc., whereas the sound power level refers to the magnitude of the power (in dB) relative to a reference sound power. The same argument applies when describing sound intensity, sound pressure or even vibrations in terms of decibels. Because of the relative nature of the decibel scale it is critical that each variable has a unique reference value. The internationally accepted reference sound power is

$$\Pi_0 = 10^{-12} \text{ W} = 1 \text{ pW}. \tag{4.5}$$

A 0.5 W sound source would thus have a sound power level, L_Π, of 117 dB.

Like sound power, sound intensity can also be expressed in terms of a sound intensity level by dividing it by a reference value and taking logarithms. The sound intensity level, L_I, is defined as

$$L_I = 10 \log_{10} \frac{I}{I_0} \text{ dB re } I_0, \tag{4.6}$$

where I_0 is an internationally accepted value. It is

$$I_0 = 10^{-12} \text{ W m}^{-2} = 1 \text{ pW m}^{-2}. \tag{4.7}$$

In chapter 2 it was demonstrated that spherical waves approximate to plane waves in the far-field. When this is the case, the sound intensity, I, of a sound field is proportional to the mean-square pressure fluctuation, p^2. Since $I \propto p^2$, the sound intensity level, L_I, can be converted into a sound pressure level, L_p – as mentioned earlier, pressure is a quantity that is readily measurable. Thus,

$$L_p = 10 \log_{10} \frac{p^2}{p_{ref}^2} \text{ dB} = 20 \log_{10} \frac{p}{p_{ref}} \text{ dB re } p_{ref}, \tag{4.8}$$

where p_{ref} is an internationally accepted value. It is

$$p_{ref} = 2 \times 10^{-5} \text{ N m}^{-2} = 20\ \mu\text{Pa}. \tag{4.9}$$

Since $I = p^2/\rho_0 c$ for a plane wave (see equation 2.64), by taking logarithms and substituting the appropriate reference values on both sides of equation (2.64) yields

$$L_I = L_p + 10 \log_{10} \left\{ \frac{(2 \times 10^{-5})^2}{\rho_0 c \times 10^{-12}} \right\}. \tag{4.10}$$

The last term in equation (4.10) is pressure and temperature dependent. At 20°C and 1 atm. it is ~0.16 dB. Hence, for all intents and purposes, $L_I \approx L_p$. The decibel scale thus reduces the audible pressure range from 10^7:1 to 0:140 dB.

4.2.2 *Subjective noise measurement scales*

The objective noise measurement scales described in the previous sub-section are suitable for a physical description of noise and are commonly used by engineers, for instance, to quantify sound transmission through partitions, etc. However, the linear logarithmic scales are not suitable for evaluating the subjective reaction of humans. This is essentially because the human ear does not have a linear frequency response – it filters certain frequencies and amplifies others. The mechanical and physiological processes of the hearing mechanism produce a mental reaction which is non-linear; a doubling of the intensity of a noise is not interpreted by the human brain as a doubling of intensity. The human response to a given change in sound pressure level is therefore very subjective. This subjective response is tabulated in Table 4.1.

A need has arisen over the years for a range of subjective assessment procedures for noise. Various factors have to be included in these subjective measurement scales, including: (i) loudness levels; (ii) the degree of annoyance; (iii) the frequency spectrum; (iv) the degree of interference with speech communication; and (v) the degree of intermittency (e.g. continuous or impulsive noise, etc.). Hence, different subjective assessment procedures are required for different situations, and subjective measurement scales are based upon a statistical average of the response of a large sample population.

The unit of loudness level is the phon (P), and the scale of loudness is the sone (S). The relationship between the two is

$$S = 2^{(P-40)/10}. \qquad (4.11)$$

A value of n phons indicates that the loudness level of a sound is equal in loudness to a pure tone at 1000 Hz, with a sound pressure level, L_p, of n dB. The sone scale is chosen such that the ratio of loudness of two sounds is equal to the ratio of the sone value of the sounds. Thus, $2n$ sones is twice as loud as n sones.

Variations of the loudness of a sound with frequency and with sound pressure level can be accounted for by using weighting or filter networks. The shape of loudness level contours varies with loudness and single weighting networks cannot therefore account for the characteristic of the ear at all values of sound intensity. Hence there are several weighting networks (i.e. A, B, C and D) available. The A-weighted network is the most common network and dB(A) sound levels are commonly referred to in industrial noise control. It is important to note that the weighted readings (e.g. dB(A), dB(C), etc.) are sound levels and not sound pressure (or power or intensity) levels. The numerical values associated with the commonly used weighting levels are presented in sub-section 4.2.5.

The A-weighting network approximates the human ear's response at a 40 phon loudness level, whilst the B and C weighting networks approximate the human ear's response at higher levels (70 and 90, respectively). The D-weighting network amplifies high frequencies and produces a better measure of human subjective evaluations of high frequency noise. In practice, the dB(A) level correlates well with the human

Table 4.1. *Subjective response of humans to changes in sound pressure levels.*

Change in L_p (dB)	Pressure fluctuation ratio	Subjective response
3	1.4	Just perceptible
5	1.8	Clearly noticeable
6	2.0	
10	3.2	Twice as loud
20	10	Much louder

response in a wide range of situations, and all general industrial type noise measurements utilise the A-weighting network.

As the reader might well appreciate by now, subjective acoustics is a subject in its own right! Besides weighting networks, a wide variety of subjective measurement scales are available for a range of different situations including industrial noise, traffic noise, aircraft noise, and railway noise. Some of these include:

(1) preferred speech interference levels;
(2) preferred noise criteria (PNC) curves;
(3) noise criteria (NC) curves;
(4) noise rating (NR) curves;
(5) noise pollution level (NPL) curves;
(6) equivalent continuous sound levels (L_{eq} or L_{Aeq}).

The interested reader is referred to Rice and Walker [4.4] for a detailed discussion on the subject of subjective acoustics.

4.2.3 Vibration measurement scales

The three vibration measurement units are (i) displacement, (ii) velocity, and (iii) acceleration. The form and frequency content of a vibration signal is the same whether it is the displacement, velocity or acceleration of the vibrating body that is being measured. There is, however, a time shift (or a phase difference) between the three. The velocity signal is obtained by multiplying the displacement signal by $i\omega$, and the accleration signal is obtained by multiplying the velocity signal by $i\omega$. This multiplication is generally performed electronically.

Because of the nature of the relationship between displacement, velocity and acceleration, the choice of parameter is very important when making a vibration measurement, particularly when it includes a wide frequency band. The nature of most mechanical systems is such that large displacements only occur at low frequencies. Thus, measurement of displacement will give the low frequency components most weight. Likewise, high accelerations generally occur at high frequencies; hence acceleration measurements are weighted towards high frequency vibration components. As it turns out, the severity of mid frequency vibrations is best described with velocity measurements.

It is always best to select a vibration measurement parameter which allows for an accurate measure of both the smallest and the largest values that need to be measured. The difference between the smallest and largest parameters that can be measured is the dynamic range. Because the dynamic range of electronic instrumentation is limited, one should always aim to minimise the difference between the smallest and the largest parameters that one wishes to measure. Thus, if a large frequency range is required, velocity is the appropriate vibration measurement parameter to select. The dynamic range characteristics of displacement, velocity and acceleration are schematically illustrated in Figure 4.1. For low frequencies (<100 Hz) displacement measurements are appropriate; for mid frequencies (50–2000 Hz) velocity measurements are appropriate; and for high frequencies (>2000 Hz) acceleration measurements are appropriate.

Vibration levels can be expressed in terms of decibels in a similar manner to noise levels. The vibration displacement level, L_d, is

$$L_d = 20 \log_{10} \frac{d}{d_0} \text{ dB re } d_0, \tag{4.12}$$

where d_0 is an internationally accepted value. It is

$$d_0 = 10^{-11} \text{ m} = 10 \text{ pm}. \tag{4.13}$$

Fig. 4.1. Dynamic range characteristics of displacement, velocity and acceleration.

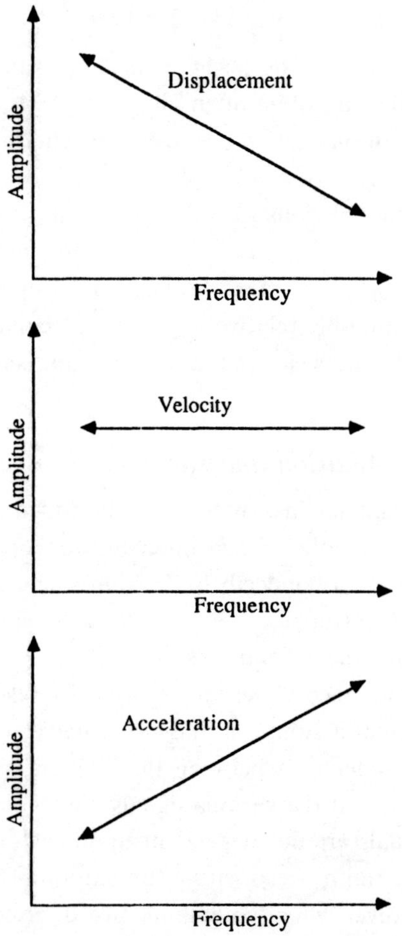

The vibration velocity level, L_v, is

$$L_v = 20 \log_{10}\frac{v}{v_0} \text{ dB re } v_0, \tag{4.14}$$

where v_0 is an internationally accepted value. It is

$$v_0 = 10^{-9} \text{ m s}^{-1} = 1 \text{ nm s}^{-1}. \tag{4.15}$$

It is useful to note that 10^{-9} m s^{-1} corresponds to 10^{-6} mm s^{-1} – vibration velocity levels are often quoted in dB re 10^{-6} mm s^{-1}.

The vibration acceleration level, L_a, is

$$L_a = 20 \log_{10}\frac{a}{a_0} \text{ dB re } a_0, \tag{4.16}$$

where a_0 is an internationally accepted value. It is

$$a_0 = 10^{-6} \text{ m s}^{-2} = 1 \text{ }\mu\text{m s}^{-2}. \tag{4.17}$$

Unlike noise measurement scales, where the internationally accepted reference values are strictly adhered to, quite often vibration levels are expressed in decibels relative to a range of alternatives. Some common alternatives are $d_0 = 1$ m; $v_0 = 1$ m s^{-1}; $v_0 = 10^{-8}$ m s^{-1}; $a_0 = 1$ m s^{-2}; $a_0 = 10^{-5}$ m s^{-2}; and $a_0 = 9.81$ m s^{-2}. These alternative reference values obviously produce different dB values. Until recently, the $v_0 = 10^{-8}$ m s^{-1} and the $a_0 = 10^{-5}$ m s^{-2} were more widely used than the current recommended values of $v_0 = 10^{-9}$ m s^{-1} and $a_0 = 10^{-6}$ m s^{-2}. This clearly illustrates the point that decibels are only relative values; when comparing different vibration levels in dB, one should always ensure that they are all relative to the same reference value.

4.2.4 Addition and subtraction of decibels

Decibel levels cannot be added linearly but must be added on a ratio basis. Provided that the various signals are incoherent, the procedure for combining decibel levels is:

(i) convert the values of the decibel levels into the corresponding linear values by taking anti-logarithms;

(ii) add the resulting linear quantities;

(iii) re-convert the summed value into a decibel level by taking the logarithm.

An important fundamental assumption has been made in the above procedure for adding decibels. Phase differences between the different signals have been ignored and it has been assumed that the various signals are incoherent – i.e. the frequency distributions of the signals are not dependent upon each other. This is usually the case in practice and one can proceed with a summation of the linear quantities – i.e. $p^2 = p_1^2 + p_2^2 +$ etc. However, when combining two discrete pure tones of the same frequency, the phase difference beween the two signals has to be taken into account. Now, $p^2 = p_1^2 + p_2^2 + 2p_1p_2 \cos\theta$, where θ is the phase angle between the signals.

Sound energy is proportional to p^2, hence addition of sound pressure levels requires a linear addition of p^2 for different sound sources. Thus,

$$\frac{p_T^2}{p_{ref}^2} = \frac{1}{p_{ref}^2}\{p_1^2 + p_2^2 + \cdots + p_n^2\}. \tag{4.18}$$

Hence,

$$L_{pT} = 10 \log_{10}\left\{\frac{p_T^2}{p_{ref}^2}\right\}. \tag{4.19}$$

Equations (4.18) and (4.19) can be re-written as

$$L_{pT} = 10 \log_{10}\{10^{L_{p1}/10} + 10^{L_{p2}/10} + \cdots\}, \tag{4.20}$$

where L_{p1}, etc. are the sound pressure levels of the individual sources, and L_{pT} is the total. Equation (4.20) is a universal equation for the addition of decibels; the L_p's can be replaced by L_Π's, L_a's, etc.

The addition of decibels is simplified by the usage of a chart giving the difference between two dB levels. This is illustrated in Figure 4.2. The addition of two equal decibel levels provides a total which is 3 dB above the original signal levels. Also, if two dB levels are separated by 10 dB or more, the sum is less than 0.5 dB – i.e. the lower level can be neglected if the difference is $\gg 10$ dB.

Sometimes it is required to subtract a background or ambient sound pressure level (or another variable) from some total value. Having said this, it should be noted that it is not possible to make any meaningful measurement of a variable associated with a specific source unless the background level, L_{pB}, is at least 3 dB below that

Fig. 4.2. Chart for the addition of decibels.

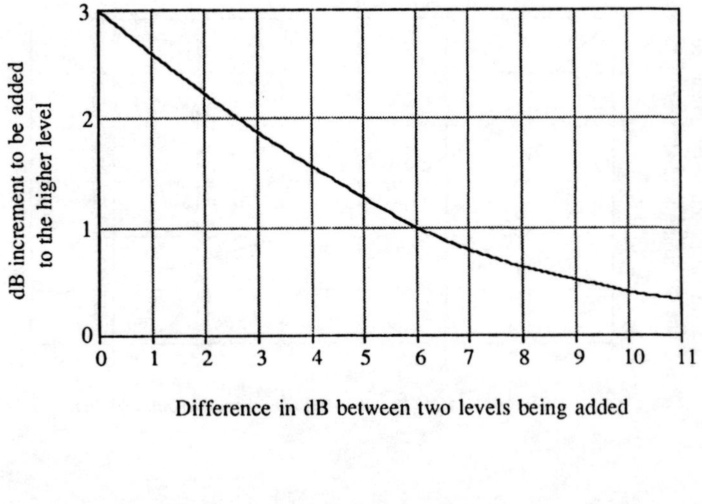

of the source acting alone. Consider the subtraction of sound pressure levels. Just as with the addition of levels, the intensities must be considered. The source intensity is obtained by subtracting the background intensity from the total (source + background). The decibel level due to the source, L_{pS}, is subsequently obtained by taking logarithms. Thus,

$$L_{pS} = 10 \log_{10}\{10^{L_{pT}/10} - 10^{L_{pB}/10}\}. \tag{4.21}$$

Once again, the subtraction of decibels is simplified by the usage of a chart giving the difference between two dB levels. This is illustrated in Figure 4.3. The subtraction of two equal decibel levels provides a value which is 3 dB below the total level. Also, if two dB levels are separated by 10 dB or more, the correction to be subtracted from the total is less than 0.5 dB – i.e. the lower level can be neglected if the difference is $\gg 10$ dB.

In addition to addition and subtraction, quite often one needs to establish some average noise level from a series of measurements. The procedure to obtain an averaged sound pressure level, etc. is similar to the decibel addition procedure. The average is obtained by dividing the linear sum by the number of measurements and subsequently taking logarithms, i.e.

$$\bar{L}_p = 10 \log_{10}\left\{\frac{1}{N}\sum_{i=1}^{N} 10^{L_{pi}/10}\right\} \text{dB}. \tag{4.22}$$

4.2.5 Frequency analysis bandwidths

The frequency range for audio acoustics extends from about 20 Hz to 18 kHz. The ultrasonic region starts at about 18 kHz, and the average human ear is totally insensitive to higher frequencies. Vibration signals of interest to engineers can extend right down to frequencies very close to 0 Hz (e.g. ~0.1 Hz). Noise and vibration

Fig. 4.3. Chart for the subtraction of decibels.

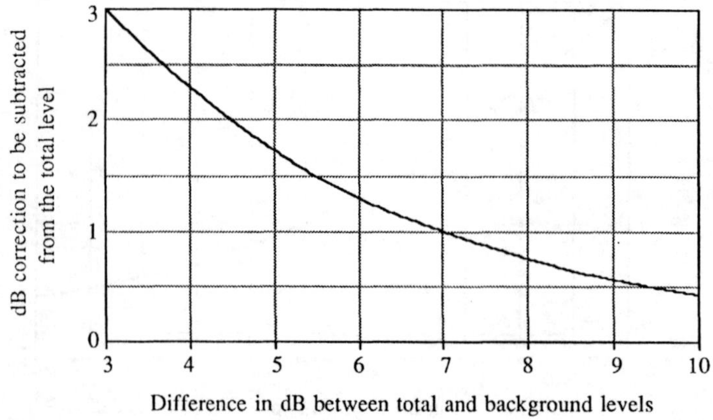

signals are always analysed in terms of their frequency components. A pure tone of sound has a simple harmonic pressure fluctuation of constant frequency and amplitude; a complex harmonic wave has several frequency components which could be either harmonically or non-harmonically related; and a random noise signal has either a broadband or a narrowband frequency spectrum. Most industrial type noise and vibration signals are either complex, deterministic signals or random signals, and therefore have to be analysed in frequency bands.

Octave bands are the widest bands that are used for frequency analysis. The word octave implies halving or doubling a frequency. 1000 Hz is the internationally accepted reference frequency and is the centre frequency of an octave band. Centre frequencies of other octave bands are obtained by multiplying or dividing previous centre frequencies by $10^{3/10}$ (a factor of two), starting at 1000 Hz. The frequency limits of each band are obtained by multiplying or dividing the centre frequencies by $10^{3/20}$ (a factor of $\sqrt{2}$). Thus, the upper frequency limit is equal to twice the lower frequency limit for an octave band.

Frequency bandwidths can be generalised. This is fairly useful because it is sometimes more convenient to use narrower frequency bands. If the centre frequency is defined as f_0, the upper frequency limit is defined as f_u, and the lower frequency limit is defined as f_l, then

$$f_u = 2^n f_l, \tag{4.23}$$

where n is any number. For an octave band $n = 1$. One-third-octave bands are commonly used in noise control studies, and in this instance $n = 1/3$. The centre frequency, f_0, is thus the geometric mean of the upper and lower frequency limits, hence

$$f_0 = (f_l f_u)^{1/2}. \tag{4.24}$$

A table of octave and one-third-octave band centre frequencies and lower and upper frequency limits is presented in Table 4.2.

Frequency bandwidths such as octave and one-third-octave bands are constant percentage bandwidths since the bandwidth is always a constant percentage of the centre frequency. Thus, as seen from Table 4.2, the frequency bandwidths increase with frequency. Octave and one-third-octave band analyses are adequate when the amplitudes of the frequency components within the various bands are relatively constant. When this is not the case and certain frequencies dominate over others, a narrowband spectral analysis is required. Here, it is more appropriate to use a constant, narrow bandwidth analysis – i.e. the frequency analysis bandwidth is constant throughout the frequency spectrum. All modern digital signal analysers are constant bandwidth analysers with a variable range of constant bandwidths.

Constant bandwidth frequency analysis (spectral analysis) techniques will be discussed in some detail in the next chapter. In spectral analysis, the mean-square pressure (or vibration) is determined in each band of a set of contiguous frequency bands, and it is plotted as a function of the band centre frequency. Each frequency band (constant percentage or constant bandwidth) is divided up into a number of smaller increments each with its own mean-square pressure, and the total band mean-square pressure is obtained by summation. Generally, each of these small sub-band increments has a width of 1 Hz, and if the mean-square pressure on the average is p_1^2 then the band mean-square pressure is

$$p_{\text{band}}^2 = p_1^2 \, \Delta f, \tag{4.25}$$

Table 4.2. *Preferred frequency bands.*

Octave band centre frequency (Hz)	One-third-octave band centre frequency (Hz)	Band frequency limits (Hz)	
		Lower	Upper
	25	22	28
31.5	31.5	28	35
	40	35	44
	50	44	57
63	63	57	71
	80	71	88
	100	88	113
125	125	113	141
	160	141	176
	200	176	225
250	250	225	283
	315	283	353
	400	353	440
500	500	440	565
	630	565	707
	800	707	880
1000	1000	880	1130
	1250	1130	1414
	1600	1414	1760
2000	2000	1760	2250
	2500	2250	2825
	3150	2825	3530
4000	4000	3530	4400
	5000	4400	5650
	6300	5650	7070
8000	8000	7070	8800
	10 000	8800	11 300
	12 500	11 300	14 140
16 000	16 000	14 140	17 600
	20 000	17 600	22 500

where Δf is the width of the parent band. Thus, by taking logarithms on both sides,

$$L_{\text{p band}} = 10 \log_{10} \frac{p_1^2}{p_{\text{ref}}^2} + 10 \log_{10} \frac{\Delta f}{\Delta f_0}, \tag{4.26}$$

or

$$L_{\text{p band}} = L_{\text{p1}}(f) + 10 \log_{10} \Delta f, \tag{4.27}$$

where $\Delta f_0 = 1$ Hz, and $L_{\text{p1}}(f)$ is the spectrum level at frequency f. The spectrum level is thus the average value in the sub-band which when added to $10 \log_{10} \Delta f$ gives the band level.

In sub-section 4.2.2 it was mentioned that sound pressure levels are often weighted for subjective acoustics. The three most common weighting networks are the A, B and C networks. The D network is sometimes used to assess aircraft noise and other high frequency noises. The attenuations (positive and negative) associated with each of the weighting networks are presented in Table 4.3. and in Figure 4.4.

Table 4.3. *Attenuation levels associated with the A, B, C and D weighting networks.*

One-third-octave centre frequency (Hz)	A-network (dB)	B-network (dB)	C-network (dB)	D-network (dB)
31.5	−39.4	−17.1	−3.0	−16.0
40	−34.6	−14.2	−2.0	−14.0
50	−30.2	−11.6	−1.3	−12.8
63	−26.2	−9.3	−0.8	−10.9
80	−22.5	−7.4	−0.5	−9.0
100	−19.1	−5.6	−0.3	−7.2
125	−16.1	−4.2	−0.2	−5.5
160	−13.4	−3.0	−0.1	−4.0
200	−10.9	−2.0	0	−2.6
250	−8.9	−1.3	0	−1.6
315	−6.6	−0.8	0	−0.8
400	−4.8	−0.5	0	−0.4
500	−3.2	−0.3	0	−0.3
630	−1.9	−0.1	0	−0.5
800	−0.8	0	0	−0.6
1000	0	0	0	0
1250	0.6	0	0	2.0
1600	1.0	0	−0.1	4.9
2000	1.2	−0.1	−0.2	7.9
2500	1.3	−0.2	−0.3	10.6
3150	1.2	−0.4	−0.5	11.5
4000	1.0	−0.7	−0.8	11.1
5000	0.5	−1.2	−1.3	9.6
6300	−0.1	−1.9	−2.0	7.6
8000	−1.1	−2.9	−3.0	5.5
10 000	−2.5	−4.3	−4.4	3.4
12 500	−4.3	−6.1	−6.2	1.4
16 000	−6.6	−8.4	−8.5	−0.5
20 000	−9.3	−11.1	−11.2	−2.5

4.3 Noise and vibration measurement instrumentation

The measurement and analysis of noise and vibration requires the utilisation of transducers to convert the mechanical signal (pressure fluctuation or vibration) into an electrical form. A basic noise or vibration measurement signal includes (i) a transducer, (ii) a preamplifier and, (iii) a means of analysing, displaying, measuring and recording the electrical output from the transducer.

4.3.1 Noise measurement instrumentation

Microphones are the measurement transducers that are used for the measurement of noise. Three types of microphone are readily available. They are (i) condenser microphones, (ii) dynamic microphones, and (iii) ceramic microphones.

Condenser microphones are the most commonly used type of microphone because they have a very wide frequency range. The sensing element is a capacitor with a diaphragm which deflects with variations in the pressure difference across it. The change in capacitance is subsequently converted into an electrical signal for recording or analysis. As a general rule, the smaller the diameter of the diaphragm, the higher is the frequency response of the microphone. There is a trade-off in that the smaller microphones have a lower sensitivity. Condenser microphones are very stable, have a wide frequency range, can be used in extreme temperatures and are very insensitive to vibrations. They are, however, very expensive and very sensitive to humidity and moisture.

Dynamic microphones involve the generation of an electrical signal via a moving coil in a magnetic field. The moving coil is connected to a diaphragm which deflects with variations in the pressure difference across it. Dynamic microphones have excellent sensitivity characteristics and are relatively insensitive to extreme variations

Fig. 4.4. ***A*, *B*, *C* and *D* frequency weighting networks.**

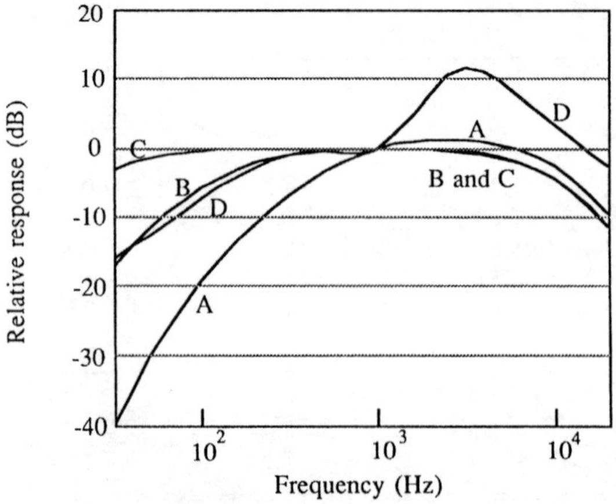

in humidity. They are also generally cheaper than condenser microphones. Dynamic microphones should not be used in environments where magnetic fields are present. They also have a lower frequency response than condenser microphones.

Ceramic microphones are often referred to as piezoelectric microphones because the sensing element is a piezoelectric crystal. Ceramic transducers have a high frequency response, a very high dynamic range, are very cheap and can often be custom built in-house. They are ideal for research applications where very small microphones are required. For instance, they are used to measure aerodynamically generated wall-pressure fluctuations on vibrating surfaces. The piezoelectric crystal becomes electrically polarised as the crystal is strained due to the pressure differential across it. When mounted on a vibrating structure so as to measure wall-pressure fluctuations (e.g. the internal wall-pressure fluctuations in a piping system), care has to be taken to isolate the transducer from the mechanical vibrations of the piping system because the piezoelectric element is equally sensitive to vibrations.

The condenser microphone is the most suitable transducer that is available for the measurement of sound pressures. Unlike the ceramic microphone, it is very insensitive to vibrations and this is a distinct advantage in an industrial environment. Hence, most commercially available noise measurement transducers are of the condenser microphone type. A variety of condenser microphones are commercially available, and sound pressures can be measured at frequencies as low as 0.01 Hz and as high as 140 kHz. Dynamic ranges of up to 140 dB can also be attained. The microphones are generally directly connected to a high input impedance, low output impedance preamplifier with a cable leading to the analysing/recording instrumentation. The preamplifier has two important functions: it amplifies the transducer signal, and it acts as an impedance mismatch (isolation device) between the transducer and the processing equipment. A typical condenser microphone is schematically illustrated in Figure 4.5.

Condenser microphones are available with three different types of response characteristics: free-field, pressure, and random incidence. Free-field condenser

Fig. 4.5. Schematic illustration of a typical condenser microphone.

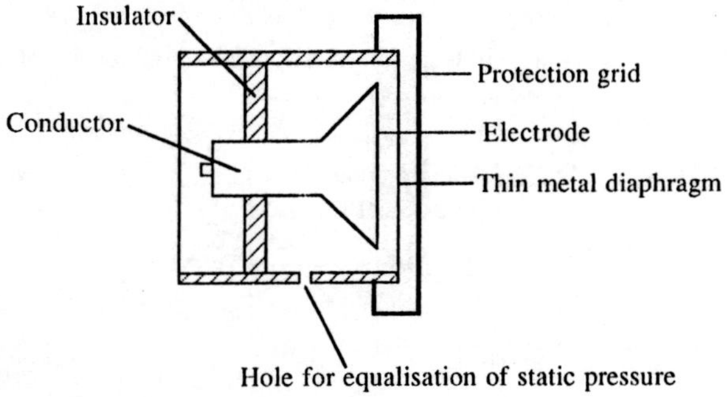

microphones are designed to compensate for the disturbance that they create due to their presence in the sound field and they produce a uniform frequency response for the sound pressure that existed prior to their insertion in the sound field. Free-field microphones can thus be pointed directly at the sound source. Pressure microphones are specifically designed to have a uniform frequency response to the actual sound pressure. Their diaphragms should thus be perpendicular to the sound source such as to achieve grazing incidence. Pressure microphones are often flush-mounted on surfaces for the measurement of flow noise. Random incidence microphones are omnidirectional microphones which are designed to respond uniformly to sound pressures in diffuse fields. Free-field microphones can be adapted for usage as random incidence microphones by fitting them with suitable correctors (manufacturers usually provide such correctors with their free-field microphones). The three different types of condenser microphones are schematically illustrated in Figure 4.6.

The most common instrument for the measurement of noise is the sound level meter. It combines the transducer, preamplifier, amplifier/attenuator and analysis electronics within the one instrument. The sound pressure level can thus be directly obtained from a readout meter (analogue or digital). A typical sound level meter is illustrated schematically in Figure 4.7. Generally, sound level meters include a selection of weighting networks, a wide amplification/attenuation range, an octave or a one-third-octave filter set, a variable r.m.s. averaging facility, a direct A.C. output prior to r.m.s. averaging for tape recording the signal, and an internal voltage calibration facility. Some sound level meters also allow for the measurement of the

Fig. 4.6. Different types of condenser microphones.

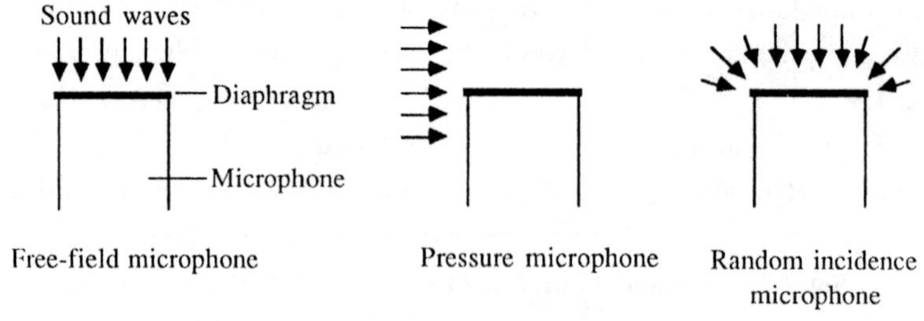

Fig. 4.7. Schematic illustration of a typical sound level meter.

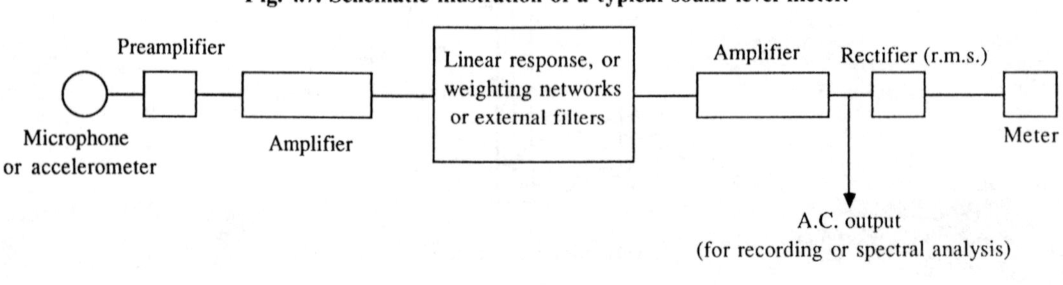

peak response of signals. This is especially useful for the measurement of impulsive sounds, e.g. punch presses, gun shots, etc.

The environment limits the capacity of condenser microphones – wind, temperature, humidity, dust and reflections from adjacent surfaces all have some effect on their response characteristics. Manufacturers usually provide data sheets with each microphone to provide the necessary information about its response characteristics. Generally, a windshield is provided to reduce the effects of air movements over the microphone. Also, an extension rod is sometimes provided to isolate the microphone from the measurement instrumentation (and the operator) to minimise any reflections. Temperature and humidity limitations are generally provided on the data sheet. Microphones (including sound level meters) can be calibrated either by using an acoustic calibrator (generally called a pistonphone), comprising a small loudspeaker which generates a precise sound pressure level in a cavity into which the microphone is placed, or by providing an electrical signal with a known frequency and amplitude.

4.3.2 Vibration measurement instrumentation

Several types of vibration transducers are available, including eddy current probes, moving element velocity pick-ups, and accelerometers. The accelerometer is the most commonly used vibration transducer; it has the best all-round characteristics and it measures acceleration and converts the signal into velocity or displacement as required. The electrical signal from the accelerometer (or any other vibration transducer) is passed through a preamplifier and subsequently sent to processing and display equipment. The instrumentation which is used for the processing, etc. of vibration signals varies considerably in range from a simple analogue device which yields a root-mean-square value of the signal, to one that yields an instantaneous analysis of the entire vibration frequency spectrum. Frequency analysis of noise and vibration signals is discussed in chapter 5.

Eddy current probes measure displacement, are non-contacting, have no moving parts (i.e. no wear) and work right down to D.C. The upper frequency range is limited to about 400 Hz because displacement decreases with frequency. Thus, the dynamic range of eddy current probes is small (about 100:1). Their main advantage is that they are non-contacting and go down to zero frequency. They are generally used with rotating machinery where it is impossible to mount a conventional accelerometer. Their main disadvantage is that geometric irregularities or variations in the magnetic properties of the rotating shaft result in erroneous readings.

Moving element velocity pick-ups have a typical dynamic range of 100:1, and measure velocity. They operate above their mounted resonance frequencies and this limits their lower frequencies to about 10 Hz. They are generally large and this is sometimes a problem in that the mass of the transducer modifies the response of the vibrating structure. These transducers are also very sensitive to orientation and magnetic fields, and the moving parts are prone to wear.

Accelerometers are the most widely used vibration transducer. They measure acceleration and have a very large dynamic range (30×10^6:1). They come in all shapes and sizes, are very rugged and have a wide frequency range. The main limitation of accelerometers is that they do not have a D.C. response. The most common type of accelerometer available is the piezoelectric accelerometer, where the sensing element is a piezoelectric crystal which functions in a manner similar to that of the ceramic microphone. A piezoelectric accelerometer (or microphone) generates an electric charge across a polarised, ferroelectric ceramic element when it is mechanically stressed either in tension, compression or shear, as illustrated in Figure 4.8.

The basic construction of an accelerometer is outlined in Figure 4.9. It essentially comprises a spring-mounted mass in contact with a piezoelectric element. The components are encased in a metal housing attached to a base. The mass applies a dynamic force to the piezoelectric element, and the force is proportional to the acceleration level of the vibration. Two types of accelerometers are commercially available. They are: (i) the compression type where a compressive force is exerted

Fig. 4.8. Generation of an electrical charge across a polarised piezoelectric crystal.

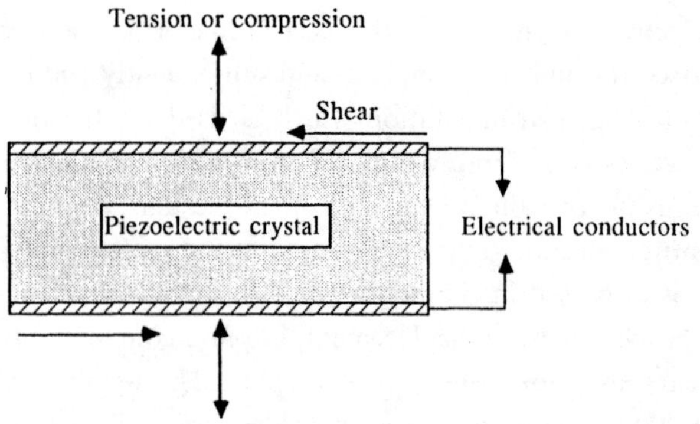

Fig. 4.9. Schematic illustration of the basic construction of an accelerometer.

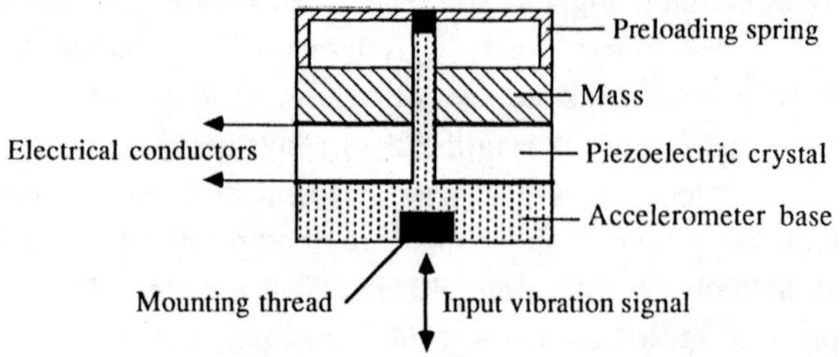

on the piezoelectric element, and (ii) the shear type where a shear force is exerted instead. Compression type accelerometers are generally used for measuring high shock levels and the shear accelerometer is used for general purpose applications. Most manufacturers produce a wide range of accelerometers of different sizes and specifications, some being more sensitive to small vibrations than others. Tri-axial (three directions) accelerometers are also available.

The mass of an accelerometer can significantly distort the true vibration level on a structure. This 'mass loading' is generally a problem on lightweight structures and at higher frequencies. One also has to ensure that the frequency range of an accelerometer can cover the range of interest. There is a trade-off between sensitivity and frequency range. This is illustrated in Figure 4.10. Larger accelerometers have lower resonant frequencies and smaller useful frequency ranges. Manufacturers generally provide a frequency range chart with every accelerometer.

The mounting of an accelerometer on a vibrating structure is very important to obtaining reliable results – large errors can result if it is not solidly mounted to the vibrating surface. Accelerometers should also always be mounted such that the designed measuring direction coincides with the main sensitivity axis. Five common ways of mounting accelerometers are: (i) via a connecting threaded stud; (ii) via a cementing stud; (iii) via a thin layer of wax; (iv) via a magnet; and (v) via a hand held probe. The type of mounting affects the frequency response – methods (i)–(iii) produce very good frequency responses; method (iv) limits the frequency response to about 6000 Hz but it provides good electromagnetic isolation – a closed magnetic path is used and there is no magnetic field at the accelerometer position; method (v) limits the frequency response to about 1000 Hz but is very convenient for quick measurements.

Fig. 4.10. Frequency response characteristics of accelerometers.

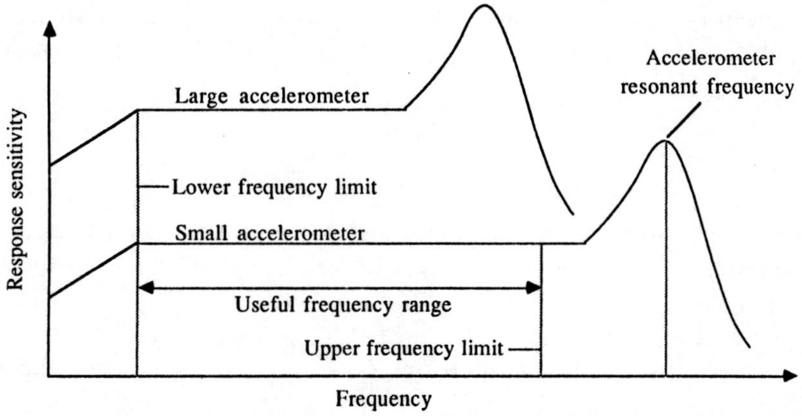

The environmental influences that can affect the accuracy of an accelerometer include humidity, temperature, ground loops, base strains, electromagnetic interferences, and cable noise. Moisture can only enter an accelerometer through the connector since it is a sealed unit. Silicone rubber sealants are commonly used to overcome this problem. Generally, temperatures of up to about 250°C can be sustained. Ground loops can be overcome by suitable earthing and isolating via a mica washer between the accelerometer and the connecting stud. If the measurement surface is undergoing large strain variations, this will contaminate the output of the accelerometer – shear accelerometers are usually recommended to minimise this problem. Care has also got to be taken to avoid 'cable whip'. Severe cable whip will ultimately produce fatigue failure at the connecting terminal and also generate cable noise.

All commercially available accelerometers are supplied with individual calibration charts – the information provided includes frequency response, resonant frequency, accelerometer mass, maximum allowable operating temperature, etc. Provided that the accelerometer is not subjected to excessive shock or temperature its calibration should not change over a very long period (several years). Accelerometer calibrators which provide a reference vibration level of 1 g (9.81 m s^{-2}) are available.

A range of different types of vibration measuring instrumentation is available. These include simple analogue r.m.s. meters and frequency analysers. Most sound level meters can also be adapted to measure vibration levels. Some typical set-ups for the measurement of vibration levels are illustrated schematically in Figure 4.11.

4.4 Relationships for the measurement of free-field sound propagation

This section is concerned with the propagation of sound waves in open spaces where there are no reflecting surfaces. Three commonly encountered sound sources, namely (i) point sources, (ii) line sources, and (iii) plane sources, are considered.

A sound source can generally be modelled as a point spherical sound source if its diameter is small compared with the wavelength that is generated, or if the measurement (receiver) position is at a large distance away from the source. It was shown in chapter 2, from the solution to the wave equation, that the far-field sound intensity, I, is

$$I(r)=\frac{\langle p^2(r)\rangle}{\rho_0 c}=\frac{\Pi}{4\pi r^2}, \tag{4.28}$$

where r is the distance from the source and $\langle\ \rangle$ represents a time-average. It is convenient to represent these relationships in decibels by taking logarithms on both sides. Thus,

$$L_I=L_p+10\log_{10}\left\{\frac{(2\times 10^{-5})^2}{\rho_0 c\times 10^{-12}}\right\}, \qquad \text{(equation 4.10)}$$

and

$$L_{\Pi} = L_{\mathrm{I}} + 10 \log_{10} \frac{4\pi r^2}{S_0}. \tag{4.29}$$

The last term in equation (4.10) is ≈ 0.16 dB at normal temperatures and at 1 atmosphere and can therefore be neglected – i.e. $L_{\mathrm{I}} \approx L_{\mathrm{p}}$. Also, the reference radiating surface area S_0 is 1 m^2. Thus,

$$L_{\Pi} = L_{\mathrm{p}} + 10 \log_{10} 4\pi r^2, \tag{4.30}$$

and

$$L_{\mathrm{p}} = L_{\Pi} - 20 \log_{10} r - 11 \text{ dB}. \tag{4.31}$$

Equation (4.31) provides a relationship between the sound pressure level at some distance r from a point source in a free-field and its sound power. The sound pressure level at some other distance can be computed since the sound power is constant (the effects of reflecting surfaces on sound power are discussed in section 4.6) – i.e.

$$L_{\mathrm{p2}} = L_{\mathrm{p1}} - 20 \log_{10} \left\{\frac{r_2}{r_1}\right\}. \tag{4.32}$$

Fig. 4.11. Some typical set-ups for the measurement of vibration levels.

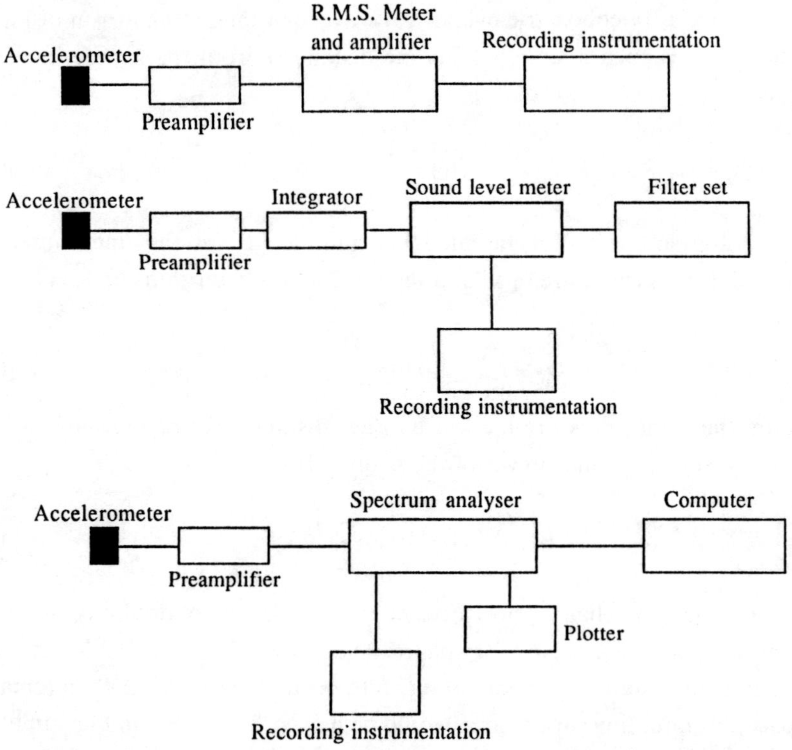

Equation (4.32) illustrates that the variations in sound pressure level between different distances from a source can be estimated without any knowledge of the sound power of the source. It is the inverse square law relationship which states that a doubling of the distance from a source produces a 6 dB drop in sound pressure level.

If the spherical point source was located in the ground plane, it would radiate sound energy through a hemispherical surface centred on the source. This would reduce the radiating surface area by half (i.e. $2\pi r^2$ instead of $4\pi r^2$) and equation (4.31) becomes

$$L_p = L_\Pi - 20 \log_{10} r - 8 \text{ dB}. \tag{4.33}$$

There is a corresponding increase of 3 dB in L_p at the radius r, assuming that L_Π remains the same. As already mentioned, in certain instances hard reflecting surfaces affect the sound power characteristics of sound sources, and this has also got to be taken into account. This important point was demonstrated in chapter 2 (sub-section 2.3.3) and will be discussed again in section 4.6. As was the case for spherical propagation, a doubling of the distance from the hemispherical, ground plane source produces a drop in L_p of 6 dB. It should also be noted that at very large distances from a ground plane source, a 12 dB drop per doubling of distance can occur instead (see Chapter 2, sub-section 2.3.3).

Now, consider a uniform infinite line source in free space with sound waves radiating as a series of concentric cylindrical waves – a long straight run of pipeline can be modelled as such a source. At some distance r from the source, the sound intensity is

$$I(r) = \frac{\Pi_l}{2\pi r}, \tag{4.34}$$

where Π_l is the sound power radiated per unit length of the line source, and $2\pi r$ is the radiating surface area per unit length. Taking logarithms and replacing L_I with L_p,

$$L_p = L_{\Pi_l} - 10 \log_{10} r - 8 \text{ dB}. \tag{4.35}$$

Once again, the sound pressure level at another distance can be evaluated without any knowledge of the sound power of the source. It is

$$L_{p2} = L_{p1} - 10 \log_{10} \left\{\frac{r_2}{r_1}\right\}. \tag{4.36}$$

It is important to note that L_p now decays by 3 dB for every doubling of distance rather than 6 dB as was the case for spherical sources.

If the infinite line source was brought down to ground level (rather than remaining in free space), its radiating surface area would be halved (i.e. πr per unit length instead of $2\pi r$ per unit length). Thus

$$L_p = L_{\Pi_l} - 10\log_{10}r - 5\text{ dB}. \tag{4.37}$$

The decay rate is still 3 dB per doubling of distance.

The semi-cylindrical infinite line source model described above can be used for uniform traffic flow on a straight road. An improved model can be obtained by representing the stream of traffic by an infnite row of point sources each separated by some distance x, as illustrated in Figure 4.12. Rathe[4.5] and Pickles[4.6] provide expressions for the mean-square pressure at some arbitrary point y, due to a line of equally spaced incoherent point sources, each of power Π. The mean-square pressure at the observer position is

$$\langle p^2\rangle = \frac{\Pi\rho_0 c}{4\pi x^2}\left\{\frac{\pi x}{y}\coth\frac{\pi y}{x}\right\}. \tag{4.38}$$

When y is small, $\coth \pi y/x$ approaches $x/\pi y$, and

$$\langle p^2\rangle = \frac{\Pi\rho_0 c}{4\pi y^2}. \tag{4.39}$$

When y is large, $\coth \pi y/x$ approaches unity, and

$$\langle p^2\rangle = \frac{\Pi\rho_0 c}{4xy}. \tag{4.40}$$

Thus, for small y, the measured sound pressure level is dominated by a single point source, the attenuation is spherical, and the decay rate is 6 dB per doubling of distance; for large y the infinite row of point sources behaves like an infinite line source, the attenuation is cylindrical, and the decay rate is reduced to 3 dB per doubling of distance. The demarcation between the two decay rates is given by $y = x/\pi$.

Fig. 4.12. An infinite row of point sources.

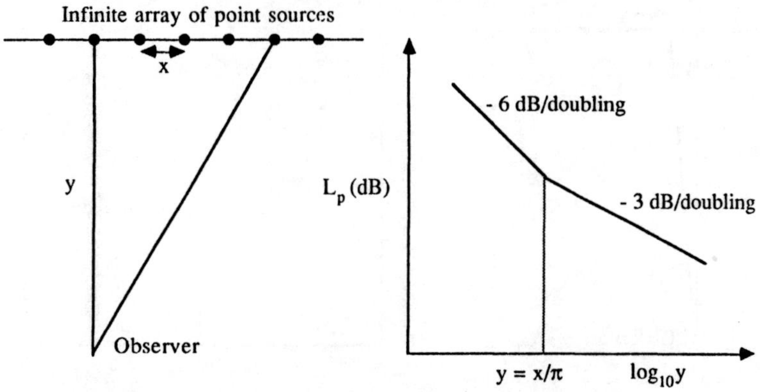

Now consider a finite, uniformly radiating straight line source of length x and total power $\Pi = \Pi_l x$, as illustrated in Figure 4.13. The angles θ_1 and θ_2 are in radians. The mean-square pressure at the observer position is[4.5,4.6]

$$\langle p^2 \rangle = \frac{\Pi \rho_0 c}{4\pi x y}(\theta_2 - \theta_1). \tag{4.41}$$

When y is small, $\theta_2 - \theta_1$ approaches π, and

$$\langle p^2 \rangle = \frac{\Pi \rho_0 c}{4 x y}. \tag{4.42}$$

When y is large, $\theta_2 - \theta_1$ approaches x/y, and

$$\langle p^2 \rangle = \frac{\Pi \rho_0 c}{4\pi y^2}. \tag{4.43}$$

Thus, initially the attenuation is cylindrical and the finite line source decays at 3 dB per doubling of distance. When $y > x/\pi$ a transition occurs; the source behaves like a point source, the attenuation reverts to being spherical and the sound pressure level decays at 6 dB per doubling of distance.

Now consider the sound radiation from a large plane surface of dimensions x and y in a free-field (e.g. the wall of an enclosure), as illustrated in Figure 4.14. The mean-square pressure at some observer position z is[4.5,4.6]

$$\langle p^2 \rangle = \frac{\Pi \rho_0 c}{\pi x y}\left\{\tan^{-1}\frac{x}{2z}\tan^{-1}\frac{y}{2z}\right\}. \tag{4.44}$$

When $z \ll x$ and $z \ll y$,

$$\langle p^2 \rangle = \frac{\Pi \rho_0 c \pi}{4 x y}, \tag{4.45}$$

Fig. 4.13. A finite, uniformly radiating straight line source.

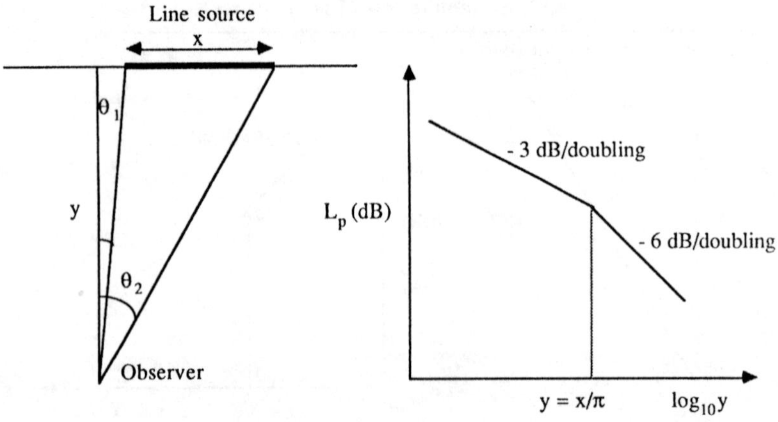

and there is no variation of sound pressure level with distance from the source. This is only strictly correct if the surface vibrates like a piston – all points vibrate with the same amplitude and phase. In practice, the sound field near to a large plane vibrating surface is not uniform. Different sections will vibrate with different amplitudes and phases – i.e. acoustic pressures in the near-field vary in both time and space. When $z \gg y$ but $z \ll x$,

$$\langle p^2 \rangle = \frac{\Pi \rho_0 c}{4zx}. \tag{4.46}$$

Here, the source behaves like a line source and there is a 3 dB decay per doubling of distance. When $z \gg x$ and $z \gg y$,

$$\langle p^2 \rangle = \frac{\Pi \rho_0 c}{4\pi z^2}, \tag{4.47}$$

and the source behaviour is analogous to that of a point source – there is a 6 dB decay per doubling of distance.

4.5 The directional characteristics of sound sources

Sound sources whose dimensions are small compared with the wavelengths of sound that they are radiating are generally omni-directional; sound sources whose dimensions are large compared with the wavelengths of sound that they are radiating are directional. Thus, in practice, most sound sources are directional and this has to be taken into account in any analysis.

A directivity factor, Q_θ, defines the ratio of the sound intensity, I_θ, at some distance r from the source and at an angle θ to a specified axis, of a directional noise source of sound power Π, to the sound intensity, I_S, produced at some distance r from a uniformly radiating sound source of equal sound power. Thus,

$$Q_\theta = \frac{I_\theta}{I_S} = \frac{\langle p_\theta^2 \rangle}{\langle p_S^2 \rangle}. \tag{4.48}$$

Fig. 4.14. A finite, large plane radiating source.

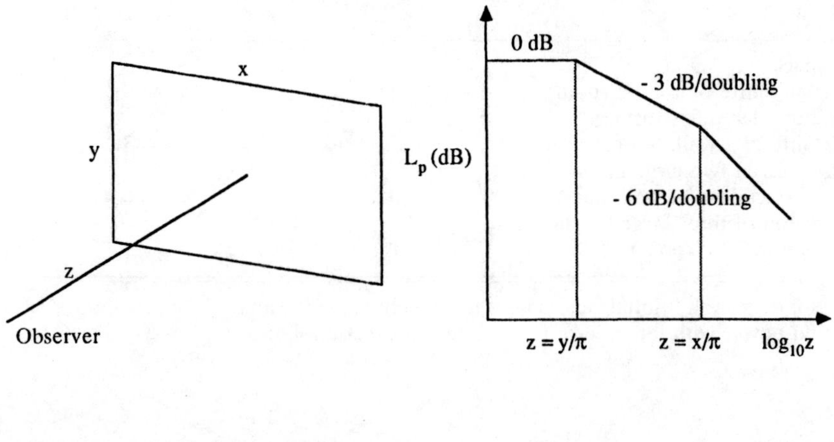

Q_θ is a function of both angular position and frequency. A directivity index, DI_θ, is defined as

$$DI_\theta = 10 \log_{10} Q_\theta, \tag{4.49}$$

thus

$$DI_\theta = L_{p\theta} - L_{pS}. \tag{4.50}$$

A relationship can be obtained between the sound power, L_Π, the sound pressure level, $L_{p\theta}$, at some given angle, θ, and the directivity factor, Q_θ, of a spherical sound source in free space by substituting equations (4.49) and (4.50) into equation (4.30). Some re-arrangement yields

$$L_\Pi = L_{p\theta} + 10 \log_{10} \frac{4\pi r^2}{Q_\theta}. \tag{4.51}$$

From equations (4.10), (4.28), (4.29) and (4.51) it can be shown that

$$\Pi = \frac{4\pi r^2 I_\theta}{Q_\theta}. \tag{4.52}$$

Equation (4.52) illustrates the relationship between sound power, sound intensity and directivity. If a sound source was omni-directional but was placed at some position other than in free space (e.g. on a hard reflecting floor or in a corner, etc.), the rigid boundaries would force it to radiate in some preferential direction – it would become directional. The mean-square pressure and the intensity would fold back upon itself. The directivity factors, Q's, and indices, DI's, for a simple omni-directional source placed near to one or more bounding planes are summarised in Table 4.4. It

Table 4.4. *Values of directivity factors and directivity indices for an omni-directional sound source.*

Position	Directivity factor, Q	Directivity index, DI (dB)
Free space (e.g. near centre of a large room)	1	0
Centre of a large flat surface (e.g. centre of a wall, floor, or ceiling)	2	3
Intersection of two large flat surfaces (e.g. intersection of a wall and a floor)	4	6
Intersection of three large flat surfaces (e.g. a corner of a room)	8	9

Note: sometimes additional factors have to be included in the analysis to account for variations in sound power with location of the source – see section 4.6.

should be noted that, in addition to directivity, sometimes the hard reflecting surfaces also affect the amount of sound power that is produced – i.e. the sound power, Π, is not constant! This was demonstrated in chapter 2 (sub-section 2.3.3) for the case of a monopole near a rigid reflecting ground plane, and is quantified in the next section.

4.6 Sound power models – constant power and constant volume sources

Sound pressure levels depend upon distance from the source and the environment (i.e. free or reverberant fields), hence the sound power level of a source provides a better description. Given the sound power level, L_Π, and the directivity index, DI, the sound pressure level, L_p, can be evaluated at any position relative to the source. The specification of sound power levels is thus generally preferred for noise control problems, but it should always be remembered that the sound pressure level is the quantity which is related to human response and is therefore the quantity which has to be eventually controlled.

Most noise control books make the tacit assumption that the sound power of a sound source is constant. This assumption is based upon the approximation that the acoustic radiation impedance of a source in a free-field remains the same when the source is relocated in some environment other than a free-field. This is not always the case, especially for machine surfaces in close proximity to rigid boundaries such as floors, walls, corners, etc. Often for vibrating and radiating structures, a better approximation is to assume that the sources are constant volume sources – i.e. the motion of the vibrating surface is unaffected by the acoustic radiation load, implying an infinite internal impedance. Bies[4.7] discusses the effects of variations in acoustic radiation impedance on the sound power of various types of sound sources.

For a simple omni-directional sound source,

$$\Pi = \frac{4\pi r^2 I}{Q}, \tag{4.53}$$

where I is the sound intensity, r is the distance from the source, and Q is the directivity factor. Now, for a constant power source, $\Pi = \Pi_0 =$ a constant; hence, as Q increases, p^2 and I increase. If, for argument, the source were a constant pressure source, $p^2 =$ a constant, and, as Q increases, Π would decrease. The concept of a constant pressure source is a theoretical one (Bies[4.7]) and, as will become evident shortly, it represents a lower limit of variations in sound power. If the source were a constant volume source, Π would increase as Q increases; thus an increase in p^2 (and I) is a function of both Q and Π.

It was shown in chapter 2 (sub-section 2.3.3) that when a monopole is placed close ($d \ll \lambda$, where d is the distance from the monopole to the surface) to a rigid, reflecting surface, the far-field velocity potential doubles. This doubling of the velocity potential produces a fourfold increase in sound intensity and a twofold increase in the radiated sound power. There is only a twofold increase in sound power (rather than a fourfold increase) because the intensity has only got to be integrated over half space, the other half being baffled by the rigid ground plane (see equations 2.88, 2.89, 2.118 and 2.119). Thus by considering velocity potentials and analysing the problem from fundamentals it is evident that, instead of a twofold increase in intensity (as would be expected if a directivity factor of two was allocated to the baffled source), there is an additional factor to be accounted for – the radiated sound power of the source has increased! The velocity potential (and hence the acoustic pressure) everywhere has now doubled. For a constant power source, the effect of the ground reflector is to fold the sound field back onto itself; for a constant volume source, in addition to this the pressure is doubled. Thus, for a constant volume source,

$$\Pi = \Pi_0 Q = \frac{4\pi r^2 I}{Q}, \tag{4.54}$$

and

$$I = \frac{\Pi_0 Q^2}{4\pi r^2}. \tag{4.55}$$

By taking logarithms on both sides

$$L_p = L_{\Pi_0} + 10 \log_{10} Q^2 - 10 \log_{10} 4\pi r^2. \tag{4.56}$$

Based on the preceding discussions, in principle, three sound power models can be postulated: constant power; constant volume; and constant pressure. The effects of source position on these sound power models are summarised in Table 4.5. From the table it can be seen that, if a sound source is modelled as a constant power source, the source position does not affect its radiated sound power; if a sound source is modelled as a constant volume source, reflecting surfaces increase the radiated sound

Table 4.5. *Variations in sound power for different sound power models.*

		Sound power model		
Source position	Directivity, Q	Const. power, $\Pi = \Pi_0$	Const. volume, $\Pi = \Pi_0 Q$	Const. pressure, $\Pi = \Pi_0/Q$
Free space	1 (+0 dB)	Π_0	Π_0	Π_0
Centre of a large flat surface	2 (+3 dB)	Π_0	$2\,\Pi_0$ (+3 dB)	$\Pi_0/2$ (−3 dB)
Intersection of two large flat surfaces	4 (+6 dB)	Π_0	$4\,\Pi_0$ (+6 dB)	$\Pi_0/4$ (−6 dB)
Intersection of three large flat surfaces	8 (+9 dB)	Π_0	$8\,\Pi_0$ (+9 dB)	$\Pi_0/8$ (−9 dB)

power of the source; if a sound source is modelled as a constant pressure source, reflecting surfaces decrease the radiated sound power of the source. As already mentioned, the constant pressure model is only a theoretical concept and it represents a lower limit to the sound power radiated by sound sources. The constant volume model, on the other hand, is a conservative model and it represents an upper limit. In reality, most practical sources fall somewhere in between the constant power model and the constant volume model – i.e. hard reflecting surfaces do have an effect on the sound power radiated by the source at frequencies where the distance, d, from the acoustic centre of the source to the reflecting surface is smaller than the acoustic wavelength ($d \ll \lambda$).

Some recent experiments by Norton and Drew[4.8] using sound intensity measurement techniques have illustrated that the sound power of common domestic appliances such as vacuum cleaners and power tools is dependent upon the environment. When the distance, d, from the acoustic centre of the source to the reflecting plane is less than the acoustic wavelength, λ, the radiated sound power is not constant. The general trend is for the increases to be somewhat less than that predicted by the constant volume model. Typical increases in radiated sound power for small compact domestic appliances, positioned in a corner, over the corresponding free-field values are of the order of 6–8 dB.

4.7 The measurement of sound power

Sound power levels allow for a comparison of the noise producing properties of different machines and allow for the prediction of expected noise levels in free-fields and in reverberant spaces, when the directivity is known. Sound power can only be accurately computed in two limiting cases: (i) in a free-field region away from the near-field of the source (e.g. an anechoic chamber); (ii) in a diffuse sound field (a reverberation room). Under field conditions such as semi-reverberant conditions, approximations have to be made. There is an exception to this rule, however, and if the sound intensity at some distance from a radiating source can be accurately measured, then the sound power of the source can be deduced *in situ*. In recent times, sound intensity meters have become commercially available – the principles involved in the sound intensity technique for the measurement of sound power are discussed in this section.

There are a variety of national and international standards available for the determination of sound power levels of noise sources for a range of different test environments ranging from precision environments such as anechoic chambers or reverberation rooms to engineering and survey environments. As yet, standards are not available for the sound intensity technique; it is anticipated that they will become available in the near future.

4.7.1 Free-field techniques

Free-field techniques are required for estimating the sound power of any machine producing sound which contains prominent discrete frequency components or narrowband spectra or if the directional characteristics of the sound field are required. Normally a large anechoic chamber would be used, but when this is impossible measurements can be made in a free-field above a reflecting plane. The test procedure involves making a number of sound pressure level measurements on the surface of an imaginary sphere or hemisphere surrounding and centred on the machine which is being tested. Also, depending upon the degree of directionality of the sound field, the number of microphone positions required for the measurements has to be varied. Once the average sound pressure level at some specified distance from the source is established, the sound power of the source is computed using equations (4.31) or (4.33) depending on whether the test surface is a sphere or a hemisphere.

4.7.2 Reverberant-field techniques

In a completely reverberant (diffuse) sound field, the sound waves are continuously being reflected from the bounding surfaces and the sound pressure field is essentially independent of distance from the source – the flow of sound energy is uniform in all directions and the sound energy density is uniform. The sound power of a source in a reverberant sound field can be calculated from (i) the acoustic characteristics of the room, and (ii) the sound pressure level in the room. As for the case of free-field testing, national and international standards provide detailed specifications for reverberant-field testing.

Sound power measurements can be readily made in a reverberation room provided that the source does not produce any prominent discrete frequency components or narrowband spectra. If it does, a rotating diffuser should be used and the lowest discrete frequency which can be reliably measured is about 200 Hz. The free-field technique is recommended for discrete noise sources below 200 Hz.

Consider a directional sound source of total sound power Π, placed in the centre of a reverberation room. The contribution of the direct (unreflected) field to the sound intensity in the room is

$$\frac{\langle p_\theta^2 \rangle}{\rho_0 c} = \frac{\Pi Q_\theta}{4\pi r^2}, \tag{4.57}$$

where

$$Q_\theta = \frac{I_\theta}{I_\mathrm{S}}; \qquad I_\theta = \frac{\langle p_\theta^2 \rangle}{\rho_0 c}; \qquad \text{and } I_\mathrm{S} = \frac{\Pi}{4\pi r^2}.$$

The sound field produced by the reflected waves has now got to be determined. Before proceeding, the concept of sound absorption must be introduced. The sound transmission coefficient, τ, was introduced in chapter 3 (sub-section 3.9.1) – it is the ratio of transmitted to incident sound intensities (or energies) on a surface. The sound absorption coefficient, α, is the ratio of absorbed to incident sound intensities (or energies) on a surface. In principle, when a sound wave is incident upon a surface, part of the sound energy is reflected (Π_R), part of it is transmitted through the surface (Π_T), and part of it is dissipated within the surface (Π_D). Thus,

$$\Pi_I = \Pi_R + \Pi_T + \Pi_D. \tag{4.58}$$

Now, by definition, all the energy which is not reflected is 'absorbed' – i.e. it is either transmitted through the material or dissipated in the material as heat via flow constrictions and vibrational motions of the fibres in the material. Hence, the absorbed sound energy (Π_A) is given by

$$\Pi_A = \Pi_T + \Pi_D. \tag{4.59}$$

Thus, an open window, for instance, has a sound absorption coefficient, α, of unity because it 'absorbs' all the sound impinging on it. Sound absorbing materials are discussed in section 4.12. The difference between the transmission and absorption coefficients of materials should be appreciated. When a material has a small transmission coefficient, it implies that the incident sound is either reflected or dissipated. When a material has a large transmission coefficient, it implies that most of the incident sound is neither reflected nor dissipated, but transmitted through the material. When a material has a small sound absorption coefficient, the incident sound is neither transmitted nor dissipated but is reflected back instead. Finally, when a material has a large sound absorption coefficient, most of the incident sound is either dissipated as heat within the material, or transmitted through it. Porous sound absorbing materials have large sound absorption coefficients and most of the sound energy is dissipated within the material – however, they do not possess mass and therefore do not make good barriers. Building materials such as brick or concrete walls have small transmission coefficients (large transmission losses) and small absorption coefficients since there is negligible dissipation within the material – they are massive and therefore make good sound barriers. Thus, when one is concerned with the transmission of sound through a partition, it is the transmission coefficient, τ, which is relevant; when one is concerned with the reflection and absorption of sound within an enclosed volume of space, it is the absorption coefficient, α, which is relevant. The absorption coefficient for any given material is always greater than its transmission coefficients since $\Pi_A = \Pi_T + \Pi_D$.

Now, returning to the sound field produced by the reflected waves in the reverberation room, assuming each surface, S_n, of the room has a different sound absorption coefficient, α_n, the space-average absorption coefficient in the room is given by

$$\alpha_{avg} = \frac{S_1\alpha_1 + S_2\alpha_2 + \cdots + S_n\alpha_n}{S_1 + S_2 + \cdots + S_n}. \tag{4.60}$$

Equation (4.60) represents the average sound absorption coefficient of all the various materials within the room. At high frequencies (> 1500 Hz), and in rooms with large volumes, absorption of sound in the air space has to be accounted for. The average absorption coefficient, α_T, (including air absorption) is given by

$$S\alpha_T = S\alpha_{avg} + 4mV, \tag{4.61}$$

where S is the total absorbing surface area in the room, V is the room volume and m is an energy attenuation constant with units of m^{-1}. Values of $4m$ for different relative humidities and frequencies are given in Table 4.6[4.3].

The proportion of incident energy which is reflected back into the room is $(1 - \alpha_{avg})$, thus

$$\Pi_{rev} = \Pi(1 - \alpha_{avg}). \tag{4.62}$$

This is the rate at which energy is supplied (power input) to the reverberant field. In the steady-state, i.e. a constant sound pressure level in the room, this has to equal the rate at which energy is absorbed by the walls in subsequent reflections.

The sound energy per unit volume (energy density) of a reverberant field is (see chapter 2, sub-section 2.2.7)

$$D = \frac{\langle p^2 \rangle}{\rho_0 c^2}, \tag{4.63}$$

Table 4.6. *Values of the air absorption energy attenuation constant*, 4m, *for varying relative humidity and frequency (units of* m^{-1}).

Relative humidity	Temperature (°C)	2000 Hz	4000 Hz	6300 Hz	8000 Hz
30%	15	0.0143	0.0486	0.1056	0.1360
	20	0.0119	0.0379	0.0840	0.1360
	25	0.0114	0.0313	0.0685	0.1360
	30	0.0111	0.0281	0.0564	0.1360
50%	15	0.0099	0.0286	0.0626	0.0860
	20	0.0096	0.0244	0.0503	0.0860
	25	0.0095	0.0235	0.0444	0.0860
	30	0.0092	0.0233	0.0426	0.0860
70%	15	0.0088	0.0223	0.0454	0.0600
	20	0.0085	0.0213	0.0399	0.0600
	25	0.0084	0.0211	0.0388	0.0600
	30	0.0082	0.0207	0.0383	0.0600

where $\langle p^2 \rangle$ is the time-averaged, mean-square, sound pressure. In principle, no space-averaging is required in a reverberant field because all the different standing wave patterns for the different volume modes tend to average out – the acoustic pressure fluctuations are uniformly distributed throughout the field. In reality, some space-averaging is required. Also, close to the room boundaries, all the standing waves have anti-nodes or pressure maxima; in these regions the r.m.s. pressure is double the r.m.s. pressure elsewhere within the room. The total energy in a room of volume V is DV.

Every time a wave strikes a wall, a quantity of sound energy, αDV, is lost from the reverberant field. Statistically, this reflection occurs $cS/4V$ times per second. Hence, the rate at which energy is lost from the reverberant field is

$$\frac{cS}{4V}\alpha DV = \frac{cS}{4V}\alpha_{\text{avg}}\frac{\langle p^2 \rangle}{\rho_0 c^2}V. \tag{4.64}$$

For a steady-state, this has to equal the input sound power to the reverberant field. Thus,

$$\frac{cS}{4V}\alpha_{\text{avg}}\frac{\langle p^2 \rangle}{\rho_0 c^2}V = \Pi(1-\alpha_{\text{avg}}), \tag{4.65}$$

and

$$\frac{\langle p^2 \rangle}{\rho_0 c} = \frac{4\Pi(1-\alpha_{\text{avg}})}{S\alpha_{\text{avg}}} = \frac{4\Pi}{R}, \tag{4.66}$$

where $R = S\alpha_{\text{avg}}/(1-\alpha_{\text{avg}})$ is the room constant. It is a parameter which is often used in architectural acoustics to describe the acoustical characteristics of a room.

The total sound intensity at any point in the reverberant room is the sum of (i) the direct and (ii) the reverberant contributions. Thus,

$$I_{\text{total}} = \Pi\left\{\frac{Q_\theta}{4\pi r^2} + \frac{4}{R}\right\}, \tag{4.67}$$

and

$$L_{\text{p}} = L_\Pi + 10\log_{10}\left\{\frac{Q_\theta}{4\pi r^2} + \frac{4}{R}\right\}. \tag{4.68}$$

Equation (4.68) is an important equation, one which is extensively used in engineering noise control. It will be used later on in this chapter for sound transmission between rooms, acoustic enclosures, and acoustic barriers. Now, if all the sound pressure level measurements are made far enough from the source such that $Q_\theta/4\pi r^2 \ll 4/R$, then

$$L_{\text{p}} = L_\Pi + 10\log_{10}(4/R). \tag{4.69}$$

The absorption coefficient of the room, $\langle\alpha\rangle$, can be experimentally obtained by measuring the time taken for an abruptly terminated noise in the room to decay to a specified level. This specified level is known as the reverberation time and it corresponds to a decrease of 60 dB in the sound pressure level or sound energy. Sabine derived an empirical relationship relating the reverberation time of a room to its volume and its total sound absorption. The total sound absorption coefficient of a room, α_T, is commonly referred to in the literature as the Sabine absorption coefficient. For a 60 dB decay, the reverberation time as given by the Sabine equation is

$$T_{60} = \frac{60V}{1.086cS\alpha_T} = \frac{0.161V}{S\alpha_T}, \tag{4.70}$$

when $c = 343 \text{ m s}^{-1}$ (1 atm. and 20°C). It should be noted that α_T includes air absorption and the absorption associated with any type of object within the room, including human beings.

For reverberant-field testing of the sound power of a sound source, the absorption coefficient within the reverberant room is very small. Hence, $\alpha \approx \alpha/(1-\alpha)$ and equation (4.69) becomes

$$L_\Pi = L_p + 10\log_{10}V - 10\log_{10}T_{60} - 14 \text{ dB}. \tag{4.71}$$

Equation (4.71) demonstrates how the sound power of a sound source can be obtained in a reverberation room by (i) measuring the sound pressure level in the room, and (ii) measuring the reverberation time in the room. The accuracy of the measurement is dependent upon the diffuseness of the reverberation field. It is generally recommended that, in each frequency band of interest, sound pressure levels are measured in the reverberant field at three positions over a length of one wavelength. These values are then averaged to obtain L_p.

A modified empirical relationship which accounts for the effects of the room bounding surfaces on the diffuseness of the room is

$$L_\Pi = L_p + 10\log_{10}V - 10\log_{10}T_{60} + 10\log_{10}\left(1 + \frac{S\lambda}{8V}\right) - 10\log_{10}\frac{p_{amb}}{1000} - 14 \text{ dB}, \tag{4.72}$$

where S is the total area of all reflecting surfaces in the room, λ is the wavelength of sound at the band centre frequency, and p_{amb} is the barometric pressure in millibars. Under normal atmospheric conditions, the last two terms in equation (4.72) can be replaced by -13.5 dB.

The description of sound fields enclosed within reverberant volumes as presented in this sub-section is discussed in numerous text books. A major assumption, one which is often forgotten, is that the analysis assumes that the walls of the enclosure are locally reactive – i.e. there is no coupling between the structural modes of the walls and the fluid modes in the enclosed volume since the sound field is diffuse. Sometimes this assumption is not valid and there is coupling between the structural and fluid modes. This is particularly so in small volumes such as small rooms, aircraft fuselages, motor vehicles, etc. In these instances, the coupled structural–fluid modes dominate the noise radiation. Structure–fluid coupling in cylindrical shells is discussed in chapter 7.

4.7.3. *Semi-reverberant-field techniques*

When sound power measurements have to be made in ordinary rooms, e.g. factories or laboratory areas, the resulting sound field is neither free nor diffuse. The preferred test method is to substitute the noise source with a calibrated reference source with a known sound power spectrum. The method assumes that the reverberation time in the room will be the same for both the reference and the noise source. The average sound pressure level around the noise source is determined from an array of microphone positions which are uniformly distributed on a spherical (or hemispherical, etc.) surface which is centred on it. When $Q = 1$, twenty measurement positions are recommended; when $Q = 2$, twelve measurement positions are recommended; when $Q = 4$, six measurement positions are recommended; and, when $Q = 8$, three measurement positions are recommended. The sound power level of the noise source is thus obtained by using equation (4.71) or equation (4.72) for both the noise source and the calibrated reference source. Thus,

$$L_\Pi = L_{\Pi r} - L_{pr} + L_p. \tag{4.73}$$

$L_{\Pi r}$ and L_{pr} are the sound power and sound pressure levels of the calibrated reference source.

Sometimes, due to nearby reflecting surfaces or high background levels, additional approximations have to be made for sound power measurement procedures. A method that is commonly used involves making sound pressure level measurements at a number of points suitably spaced around the noise source. The measurement points have to be sufficiently close to the source such that the measurements are not significantly affected by nearby reflecting surfaces or background noise.

The mean sound pressure measured over the prescribed artificial surface (usually a hemisphere) is normalised with respect to an equivalent sound pressure level at some specified reference radius. Thus,

$$L_{pd} = L_p - 10\log_{10}(d/r)^2. \tag{4.74}$$

L_{pd} is the equivalent sound pressure level at the reference radius, d, and L_p is the mean sound pressure level measured over the surface of area S, and radius $r = (S/2\pi)^{1/2}$. An approximate estimate of the sound power of the noise source is

$$L_\Pi \approx L_{pd} + 10\log_{10}(2\pi d^2). \tag{4.75}$$

A technique which is essentially a refinement of the preceding equation is now described. In a semi-reverberant environment, the walls and ceilings generally have very small absorption coefficients, and the noise source, which is typically a machine, is placed on a hard floor. No restrictions are made on the type and shape of the room except that it should be large enough such that the sound pressure levels can be measured in the far-field, and at the same time not be too close to the room boundaries. Standards specify that the microphone should be at least $\lambda/4$ away from any reflecting surface not associated with the machine or any room boundary. The test surface itself can be (i) hemispherical, (ii) a quarter sphere, or (iii) a one-eight sphere depending on where it is located in the room. The test surface radii should always be in the far-field of the source.

Let L_{p1} be the average sound pressure level measured over the smaller test surface of radius r_1, and L_{p2} be the average sound pressure level measured over the larger test surface of radius r_2. L_{p1} and L_{p2} are calculated from equation (4.76) below. When the test surface is a half sphere, $N=12$; when the test surface is a quarter sphere, $N=6$; and, when the test surface is a one-eight sphere, $N=3$.

$$L_p = 10 \log_{10}\left\{\frac{1}{N}\sum_{i=1}^{N} 10^{L_{pi}/10}\right\} \text{ dB}, \tag{4.76}$$

where L_{pi} is the sound pressure level measured at the ith point on the measurement surface. Now, let $D = L_{p1} - L_{p2}$; $x =$ the reciprocal of the area of the smaller test surface; and $y =$ the reciprocal of the area of the larger test surface. Thus, for a hemispherical surface, $x = 1/2\pi r_1^2$ and $y = 1/2\pi r_2^2$; for a quarter sphere, $x = 1/\pi r_1^2$ and $y = 1/\pi r_2^2$; and, for a one-eighth sphere, $x = 2/\pi r_1^2$ and $y = 2/\pi r_2^2$. Using the above relationships, the sound power level of the source can be determined from

$$L_\Pi = L_{p1} - 10 \log_{10}(x - y) + 10 \log_{10}(10^{D/10} - 1) - D. \tag{4.77}$$

Once L_Π has been determined, the room constant, R, can be evaluated from

$$L_\Pi = L_{p1} - 10 \log_{10}\left(\frac{1}{S} + \frac{4}{R}\right), \tag{4.78}$$

where $S = 2\pi r_1^2$ for a hemisphere, πr_1^2 for a quarter sphere, and $\pi r_1^2/2$ for a one-eighth sphere.

It should be noted that the semi-reverberant field technique assumes that the background noise level is at least 10 dB below the source noise level. If this is not the case then the measurements have to be corrected to take account of the background noise. If the noise levels from the source are less than 4 dB above the background noise level, then valid measurements cannot be made.

Sometimes, measurements cannot be made in the far-field for one of several reasons. For instance, the room is too small, or the background noise levels are very high such that reliable far-field measurements cannot be made. Under these conditions the measurements have to be made in the near-field. As a rule of thumb, the test surface is about 1 m from the machine surface but it may need to be closer at times. As before, the average sound pressure level over the test surface is found by measuring the sound pressure level at a discrete number of equally spaced points over the surface. The number of measurement positions is a variable – it is dependent upon the irregularity of the sound field; hence, sufficient measurements should be obtained to account for this. The sound power level is obtained from

$$L_\Pi = L_p + 10\log_{10}S, \tag{4.79}$$

where S is the surface area of the measuring surface. A correction factor, Δ, is recommended[4.7] to account for the absorption characteristics of the room and any nearby reflecting surfaces. Thus,

$$L_\Pi = L_p + 10\log_{10}S - \Delta, \tag{4.80}$$

and Δ is given in Table 4.7 in dB for various ratios of test room volume, V, to the surface area, S, of the measuring surface.

4.7.4 *Sound intensity techniques*

The sound intensity technique for the measurement of sound power of machines is one that is yet to be accepted in terms of international standards. It is, however, a very powerful tool, one which is rapidly gaining acceptance. Several commercial sound intensity meters are now available, and the accurate measurement of sound intensity has significant applications in machinery diagnostics. The measurement of sound intensity allows for the measurement of the sound power produced by a machine in the presence of very high background noise. In fact, the correct utilisation of the technique suggests that anechoic chambers and reverberation rooms are redundant as far as the measurement of sound power is concerned. The technique is also very useful for source identification on machines (e.g. diesel engines). The physical principles associated with the technique have been known since the 1930s, but the electronic instrumentation required to reliably measure sound intensity has only been available since the 1970s.

Table 4.7. *Correction factors for near-field sound power measurements.*

Room type	V/S (m)			
Rooms without highly reflective surfaces	20–50	50–90	90–3000	>3000
Rooms with highly reflective surfaces	50–100	100–200	200–600	>600
Δ (dB)	3	2	1	0

Sound intensity is the flux of sound energy in a given direction – it is a vector quantity and therefore has both magnitude and direction. Sound pressure (which is the most common acoustic quantity, and the easiest to measure) is, on the other hand, a scalar quantity.

In a stationary fluid medium, the sound intensity is the time-average of the product of the sound pressure $p(\vec{x}, t)$ and the particle velocity $\vec{u}(\vec{x}, t)$ at the same position. The instantaneous sound pressure is the same in all directions at any given position in space because it is a scalar. The particle velocity is a vector quantity and it is therefore not the same in all directions. Hence, at some position $\vec{x}$, the sound intensity vector, $\vec{I}$, in a given direction is the time-average of the instantaneous pressure and the corresponding instantaneous particle velocity in that direction. Thus,

$$\vec{I} = \frac{1}{T} \int_0^T p(\vec{x}, t)\vec{u}(\vec{x}, t)\, \mathrm{d}t = \tfrac{1}{2}\, \mathrm{Re}\, [\mathbf{p}\vec{\mathbf{u}}^*]. \tag{4.81}$$

The second representation of equation (4.81) is used when the sound pressure fluctuations and the particle velocities are treated as complex, harmonic variables. Quite often in the literature, the vector notation for sound intensity is omitted when dealing with one-dimensional plane waves, or the far-field of simple sources such as point monopoles, dipoles, etc., where the sound waves radiate away from the source in a radial direction (with or without some superimposed directivity pattern). When using complex representations, the product of pressure and particle velocity has both real and imaginary,parts. Intensity and sound power is associated with the real (or in-phase) part; the imaginary part is reactive (out of phase) and does not produce any nett flow of energy away from the source. Reactive intensity implies equal but opposite energy flow during positive and negative parts of a cycle, the average value being zero.

In noise and vibration control one is generally interested in the in-phase components of the product of sound pressure fluctuations and particle velocity. It can readily be shown from fundamental acoustics (see chapter 2) that the sound pressure and the particle velocity are always in phase for plane waves, hence sound pressure level measurements can be made anywhere in space. For all other types of sound waves, the two acoustic variables are only in phase in the far-field; hence the requirement for far-field testing when attempting to measure sound power with only sound pressure level measurements. Any near-field measurement will inevitably involve out-of-phase components. The sound intensity technique overcomes this limitation by measuring both the sound pressure and the in-phase component of the particle velocity – the out-of-phase component is ignored.

The measurement of sound intensity requires the measurement of (i) the instantaneous sound pressure, and (ii) the instantaneous particle velocity. Whilst the measurement of the sound pressure is a relatively straightforward procedure, the measurement of the particle velocity is not. Hot wire anemometers or lasers would be required, and this is not practical for field conditions. An indirect method utilising the momentum equation (Euler's equation) has proved to be very successful and is the basis for most current techniques.

In the far-field, there is a very simple relationship between the mean-square sound pressure and the intensity. This relationship is only exact for plane waves, but, since at large distances from a source all sources approximate to plane waves, it is generally accepted as being valid. The relationship was derived in chapter 2 (equation 2.64, sub-section 2.2.6), and it is

$$I = \frac{p_{\text{rms}}^2}{\rho_0 c}. \tag{4.82}$$

where p_{rms}^2 is the mean-square sound pressure at some point in the far-field. The sound power, Π, is subsequently obtained by integrating the intensity over an arbitrary surface corresponding to the radius at which the sound-pressures were measured. Equation (4.82) is not valid in the near-field for the reasons discussed earlier, namely that the two variables are not always in phase. Thus, the fundamental relationship (equation 4.81) is the correct starting point for the development of a procedure to utilise sound intensity for sound power measurements.

The linear, inviscid momentum (force) equation which is valid for sound waves of small amplitude (>140 dB) was derived in chapter 2 (equation 2.26). It is

$$\rho_0 \frac{\partial \vec{u}}{\partial t} = -\vec{\nabla} p. \qquad \text{(equation 2.26)}$$

The pressure gradient is proportional to the particle acceleration in any given direction and the particle velocity can thus be obtained by integration – i.e.

$$\vec{u} = -\int_0^t \frac{1}{\rho_0} \vec{\nabla} p \, \mathrm{d}\tau. \qquad \text{(equation 2.49)}$$

Thus, for a one-dimensional flow,

$$u_x = -\frac{1}{\rho_0} \int_0^t \frac{\partial p}{\partial x} \, \mathrm{d}\tau. \tag{4.83}$$

Practically, the pressure gradient along the x-direction can be approximated by the finite difference gradient by using the measured instantaneous fluctuating pressures at two closely spaced microphones, denoted by subscripts 1 and 2, respectively. The microphones are separated by a distance Δx. Thus, the instantaneous particle velocity, u_x, is

$$u_x \approx -\frac{1}{\rho_0 \Delta x} \int_0^t (p_2 - p_1) \, \mathrm{d}\tau. \tag{4.84}$$

This approximation is only valid if the separation, Δx, between the two measurement positions is small compared with the wavelength of the frequencies of interest ($\Delta x \ll \lambda$). The instantaneous fluctuating acoustic pressure is approximated by

$$p \approx \frac{(p_1 + p_2)}{2}, \tag{4.85}$$

where p_1 and p_2 are the instantaneous fluctuating acoustic pressures at positions x and $x + \Delta x$, respectively.

The sound intensity vector component in the x-direction is thus

$$I_x = -\frac{1}{\Delta x \rho_0 T} \int_0^T \left\{ \frac{(p_1 + p_2)}{2} \int_0^t (p_2 - p_1)\, \mathrm{d}\tau \right\} \mathrm{d}t. \tag{4.86}$$

Equation (4.86) is obtained from equation (4.81) with the sound pressure taken to be the mean pressure between the two measurement positions (equation 4.85) and the particle velocity as per equation (4.84).

A practical sound intensity measuring system thus comprises two closely spaced sound pressure microphones, and this provides the pressure and the component of the pressure gradient along a line joining the microphone centre lines. It is a critical requirement that the two microphones are very closely matched in phase. Any phase difference between the two microphones will result in errors. A typical sound intensity microphone arrangement is illustrated in Figure 4.15. The microphone configurations can take any one of three main forms: face to face, side to side, and back to back. The face to face configuration is generally recommended by product manufacturers.

Fig. 4.15. Schematic illustration of the sound intensity measurement technique.

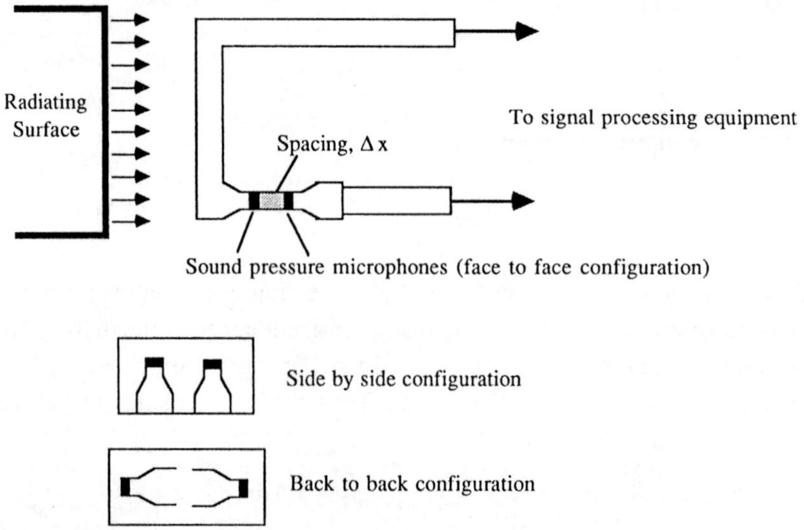

The sound power, Π, of a source can thus be obtained by integrating the components of sound intensity normal to an arbitrary control surface enclosing the noise source. To achieve this, it is essential that the line joining the two microphones is normal to the control surface. Hence,

$$\Pi = \int_S I_x \,\mathrm{d}S. \tag{4.87}$$

Because the sound intensity is averaged over positions normal to a control surface, any noise associated with other machines in the vicinity is eliminated. This is a major advantage of the technique.

Sound intensity can also be measured by using a dual channel signal analyser and F.F.T. procedures. Here,

$$\mathbf{P}(f) = \frac{\{\mathbf{P}_1(f) + \mathbf{P}_2(f)\}}{2}, \tag{4.88}$$

and

$$\mathbf{U}_x(f) = -\frac{1}{\mathrm{i}\omega\rho_0\Delta x}\{\mathbf{P}_2(f) - \mathbf{P}_1(f)\}, \tag{4.89}$$

where the **P**'s and $\mathbf{U}_x$ are the Fourier transforms of the p's and u_x respectively. By substitution, it can be shown that the sound intensity vector component in the x-direction can subsequently be obtained from the imaginary part of the cross-spectrum between the two microphone signals. Thus,

$$I_x(f) = -\frac{1}{2\pi f\rho_0\Delta x}\,\mathrm{Im}\,[\mathbf{G}_{12}(f)], \tag{4.90}$$

where $\mathbf{G}_{12}(f)$ is the cross-spectral density between the pressures $\mathbf{P}_1(f)$ and $\mathbf{P}_2(f)$. The total sound intensity between two frequencies f_1 and f_2 is

$$I_x = -\frac{1}{2\pi\rho_0\Delta x}\int_{f_1}^{f_2}\frac{\mathrm{Im}\,[\mathbf{G}_{12}(f)]}{f}\,\mathrm{d}f. \tag{4.91}$$

The sound power can thus be obtained in the usual manner by integrating the intensity over a surface area as per equation (4.87).

The sound intensity technique does have limitations associated with it. There are high and low frequency limitations together with bias errors in the near-field. These limitations are, however, no different from those associated with the more conventional techniques. Some of the practical problems associated with the measurement of sound intensity are discussed in a collection of papers by Brüel and Kjaer[4.9].

A major application of the sound intensity technique, other than the measurement of sound power, is for source identification on engines and machines. The usual industrial procedure for noise source identification involves the lead wrapping technique, where the whole machine is wrapped in layers of lead sheets and other acoustical absorbing materials. Parts of the machine are then selectively unwrapped and sound pressure level measurements made – noise source identification proceeds in this manner. The technique has many limitations and is very time consuming and expensive. Measurements of sound intensity in the near field allow for rapid identification of 'hot spots' of sound intensity and of directions of sound power flow.

4.8 Some general comments on industrial noise and vibration control

The main emphasis so far in this chapter has been on the measurement of noise and vibration. The remaining sections are now devoted to the control of noise and vibration. In this section, the basic sources of industrial noise and vibration are summarised together with some suitable control methods, taking into account the economic factor.

4.8.1 Basic sources of industrial noise and vibration

Most machinery and manufacturing processes generate noise as an unwanted by-product of their output. Offensive industrial noises can generally be classified into one of four groups. They are: continuous machinery noise; high-speed repetitive actions that create intense tonal sounds; flow-induced noise; and the impact of a working tool on a workpiece. Some typical specific examples of noise and vibration sources in the industrial environs include combustion processes associated with furnaces, impact noise associated with punch presses, motors, generators and other electro-mechanical devices, unbalanced rotating shafts, gear meshing, gas flows in piping systems, pumps, fans, compressors, etc.

It is not physically possible to list each and every source of industrial noise and vibration. There are, however, only a few basic noise producing mechanisms, and recognising this allows for a systematic approach to be adopted. As an example, a punch press is a very noisy machine. The press noise originates from several basic sources such as metal to metal impact, gear meshing, and high velocity air. The noise originating from plastic moulding equipment comes from cooling fans, hydraulic pumps and high velocity air. Empirical sound power estimation procedures are available for all these common industrial machinery components, e.g. fan noise, air compressors, pumps, electric motors, and various other typical machine shop items. It is not the intention of this book to discuss these empirical procedures but only to draw the reader's attention to them. Irwin and Graf[4.10], Bell[4.11] and Hemond[4.12] provide an extensive list of empirical procedures for the estimation of the sound power of typical industrial noise sources. Typical A-weighted overall sound power levels for a range of 'untreated' industrial equipment are provided in Table 4.8. The variation associated with each particular item is due to varying power ratings or sizes. Table 4.9 (adapted from Gibson and Norton[4.13]) provides a list of typical A-weighted overall noise levels, at the operator position, for a range of 'untreated' noise sources.

4.8.2 Basic industrial noise and vibration control methods

A basic understanding of the physics of sound, and an introduction to the techniques available for measuring sound pressure levels and sound power levels, are the essential requirements for the identification and characterisation of major noise sources, and for the determination of the treatment required to meet design and/or legislative requirements. To reduce noise at a receiver, one must (i) lower the noise at the source through redesign or replacement, (ii) modify the propagation path through enclosures, barriers or vibration isolators, and (iii) protect or isolate the receiver. In principle, the reduction of noise at source should always be the primary goal of a noise and vibration control engineer. Quite often, however, this goal is not achievable because of the economic factor – it is generally cheaper for the client to have the offending noise sources boxed in. Too often it is assumed that the source noise level cannot be reduced and the path is modified via an enclosure. The 'boxing in' syndrome persists more often than not, and a significant industry has been developed around this philosophy. Whilst acoustic enclosures have an important part to play in industrial noise control, there are certain situations where a bit of innovative engineering would reduce the cost of the 'fix'. The wider economic consequences of 'boxing in' are discussed by Gibson and Norton[4.13] and are summarised in sub-section 4.8.3.

Table 4.8. *Typical A-weighted sound power levels for a range of industrial noise sources.*

Equipment	A-weighted sound power levels
Compressors (3.5–17 $m^3\ min^{-1}$)	85–120
Pneumatic hand tools	105–123
Axial flow fans (0.05 $m^3\ min^{-1}$–50 $m^3\ min^{-1}$); 10 mm H_2O	61–88
Axial flow fans (0.05 $m^3\ min^{-1}$–50 $m^3\ min^{-1}$); 300 mm H_2O	88–120
Centrifugal fans (0.05 $m^3\ min^{-1}$–50 $m^3\ min^{-1}$); 10 mm H_2O	45–77
Centrifugal fans (0.05 $m^3\ min^{-1}$–50 $m^3\ min^{-1}$); 300 mm H_2O	75–108
Propeller fans (0.05 $m^3\ min^{-1}$–50 $m^3\ min^{-1}$); 10 mm H_2O	62–94
Propeller fans (0.05 $m^3\ min^{-1}$–50 $m^3\ min^{-1}$); 300 mm H_2O	94–125
Centrifugal pumps (> 1600 rpm)	105–132
Screw pumps (> 1600 rpm)	110–137
Reciprocating pumps (> 1600 rpm)	115–138
Pile driving equipment (up to 6 ton drop hammer)	103–131
Electric saws	96–126
Generators (1.25–250 kV A)	99–119
Industrial vibrating screens	100–107
Cooling towers	95–120
Room air-conditioners (up to 2 hp)	55–85
Tractors and trucks	110–130

Table 4.9. *A-weighted noise levels at the operator position for a range of industries.*

General description of industry	Machine or process	Sound level (dB(A))
Boiler shop, machine shop	Punch presses	95–118
	Fabrication (hammering)	110–114
	Power billet saw	98–114
	Tube cutting	87–112
	Air grinder	104–108
	Pedestal grinder	95–106
	Chipping welds	92–106
	Metal cutting jigsaw	102–104
	Circular saw	96–104
Foundry	Moulders	90–102
	Furnaces	95–100
	Knocking out area	92–100
Timber mills, wood working shop, timber joinery	Waste wood	115–118
	Turners	110–116
	Shapers	110–112
	Chipper	94–110
	Docking saw	104–108
	Band saw	100–104
	Automatic contours	100–104
	Line bar resaw	95–104
	Pulp mill	86–104
	Hand planer	95–100
	Circular saw	94–98
Textile mills	Shuttle looms	95–106
	Dye houses	95–102
	Weaving looms	90–100
Can manufacturers	Feed in	106–109
	Body making	98–104
	Bottling line	93–100
	Canning line	90–95
Building and construction	Pneumatic hammers	100–116
	Pavement breakers	90–107
	Jack hammers	100–104
	Motor graders	95–99
Garbage compactors		82–101
Combustion noise	Furnaces, flares	87–120
Engine rooms	Pilot vessel engine room	104–110
	Compressors	94–96
	Boiler house	88–95
Bottle manufacturer	Palletisers	95–110
	Washer units	92–103
	Bottle inspection	86–98
	Decappers	90–94
	Packers	88–92

A systematic approach to a noise and vibration control problem should always involve three stages. They are (i) analysis of the problem and identification of the sources; (ii) an investigation as to whether source modification is possible (technically and economically); (iii) recommendations for appropriate modifications. The first stage involves defining the problem, identifying the noise sources, establishing acceptable limits and restraints. The second stage involves an economic analysis to establish the most cost-effective solution – technical, legal, social and economic factors have to be considered depending upon the severity of the problem. Finally, the technical recommendations should be made. These could include modification of the source, structural damping, vibration isolation, enclosures, barriers, etc. A typical flow-chart for the various stages in industrial noise control is illustrated in Figure 4.16.

Brüel and Kjaer[4.14] provides some useful guidelines for general industrial noise and vibration control. Some of these guidelines are summarised here.

Noise and vibration control measures for machines

(1) Reduce impact and rattle between machine components.
(2) Provide machines with adequate cooling fins to reduce the requirement for cooling fans.
(3) Isolate vibration sources within a machine.

Fig. 4.16. Flow-chart for various stages in industrial noise control.

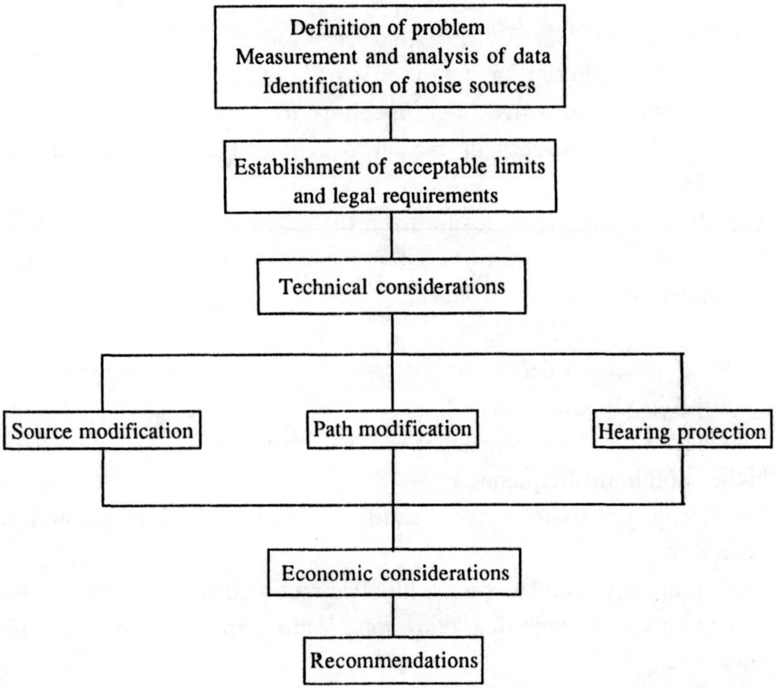

(4) Replace metal components with plastic, nylon or compound components where possible.
(5) Provide correctly designed enclosures for excessively noisy components.
(6) Brake reciprocating movements gently.
(7) Select power sources and transmissions which provide quiet speed regulation.

Noise and vibration control measures for general equipment

(1) Provide sound attenuators for ventilation duct work.
(2) Install dampers in hydraulic lines.
(3) Ensure that the oil reservoirs of hydraulic systems are adequately stiffened.
(4) Provide silencers for all air exhaust systems.
(5) Establish a plan of examining the noise specifications of all new equipment prior to purchase.

Noise and vibration control measures for material handling equipment

(1) Minimise the fall height for items collected in boxes and containers.
(2) Stiffen and dampen panels.
(3) Absorb hard shocks by utilising wear resistant rubber.
(4) Select conveyor belts in preference to rollers for material transport.
(5) Select trolleys with nylon or plastic wheels.

Noise and vibration control measures for enclosures

(1) Use a sealed material for the outer surface of the enclosure.
(2) Install mufflers on any duct openings for the passage of cooling air.
(3) Line the inner surfaces of the enclosure with suitable sound absorbing materials.
(4) Vibration isolate the enclosure from the machine.
(5) Ensure that the inspection hatches have easy access for maintenance personnel.

Some additional general rules to be observed in relation to industrial noise and vibration control are listed below.

(1) Changes in force, pressure or speed lead to noise – rapid changes generate higher dominant frequencies.
(2) Low frequency sound waves readily bend around obstacles and through openings.
(3) High frequency sound waves are highly directional and are very easy to reflect.
(4) Close to a source, high frequency noise is more annoying than low frequency noise.

(5) High frequency noise attenuates quicker with distance than low frequency noise.
(6) Sound sources should be positioned away from reflecting surfaces.
(7) Structure-borne vibrations require large surface areas to be converted into air-borne sound – thus small vibrating objects radiate less noise than large vibrating objects.
(8) Structure-borne sound propagates over very large distances.
(9) Vibrating machinery should be mounted on a heavy foundation wherever possible.
(10) Damped mechanically excited structures produce less noise radiation.
(11) Resonances transferred to a higher frequency (via stiffening a structure) are easier to damp.
(12) Correctly chosen flexible mountings isolate machine vibrations.
(13) Free edges on panels allow pressure equalisation around them and reduce radiated noise levels – thus when covers are only used for protection, perforated mesh panels are more desirable than solid covers.

Some typical noise reductions that are achievable are as follows.

(1) Mufflers – 30 dB.
(2) Vibration isolation – 30 dB.
(3) Screens and barriers – 15 dB.
(4) Enclosures – 40 dB.
(5) Absorbent ceilings – 5 dB.
(6) Damping – 10 dB.
(7) Hearing protectors – 15 dB.

4.8.3 The economic factor

Despite an increasing public awareness of the environmental, psychological and physiological hazards of excessively high noise and vibration levels, noise control is generally regarded by industry as being uneconomic and a nuisance! Vibration control, particularly for low frequency vibrations which often lead to structural damage to expensive equipment, on the other hand, is always accepted by industry as an economic necessity. Industrial noise control, as it is known today, is not only expensive but also no general solution exists. Often, new technology, whilst it is available for many specific instances, cannot be introduced because of economic restrictions due to the limitations of both the overall market and competition. There is a marked difference between the industrial noise control market and other types of noise control markets (e.g. the consumer product market), which are essentially much larger. Besides being generally cost-effective, these larger markets provide competition, product variation, and result in technological innovations. Some typical examples of cost-effective noise control are the damping treatment that is now commonly applied to automobile bodies and circular saws, noise reduction in commercial aircraft, and of course automobile mufflers.

Gibson and Norton[4.13] looked at hazardous industrial noise and estimated the 'worth' of industrial noise control (with specific reference to Australia). The paper focused on (i) an examination of the important sources of industrial noise and the corresponding worker exposure levels; (ii) techniques for predicting the degree of hearing impairment to be expected from exposure to various noise levels; (iii) comparisons beween cost and claim statistics and the noise exposure studies; and (iv) an assessment of the social and economic consequences of the noise problem and the present incentive for change.

The risk of hearing loss increases rapidly as noise levels rise. Table 4.10 illustrates how the risk increases for a sample working population (Australian age distribution in 1980) which is exposed to noise for forty hours a week in a fifty week working year. The table is a simple illustration of the total percentage of a working population with some form of hearing impairment and does not provide any information on the hearing level or percentage loss of hearing associated with the hearing damage. The reader is referred to Gibson and Norton[4.13] for further details. The effects of years of exposure on the hearing impairment of people exposed to 90 dB(A) is summarised in Table 4.11. A qualitative scale of the severity of various noise levels is provided in Table 4.12.

The outcome of the economic assessment of the study was both unexpected and unpleasant. It was found that there is very little financial incentive for most industry to reduce noise levels to 90 dB(A) essentially because noise was not considered in the design of most equipment presently installed. As a consequence, industrial noise will remain for at least the economic life of present machinery. Until these machines are replaced, industry generally has to resort to remedial 'band-aid' measures such as 'boxing in' a noisy machine. It has been estimated that, in the long run, with close collaboration between researchers and machinery manufacturers, the cost of noise

Table 4.10. *Compensable hearing damage as a function of noise level.*

Noise level	Percentage of working population with some form of compensable hearing damage
80 dB(A)	18
85 dB(A)	28
90 dB(A)	39
95 dB(A)	54
100 dB(A)	70
105 dB(A)	86

reduction at source will be about one-tenth of the present 'boxing in' costs. Whilst supporting this goal, it is important to recognise that 'boxing in' is the most common noise control treatment currently used. Putting a machine in an enclosure is very costly, hence the conclusion in the study by Gibson and Norton[4.13] that there is very little financial incentive for industry to act. It was also found, in the study, that the incentive to provide a comprehensive hearing protection programme is often marginal – the total payments for industrial deafness are generally a very small fraction of the total workers' compensation payments, and the existing levels of compensation are of the same order of magnitude as hearing conservation programmes incorporating hearing protectors, audiometric testing, etc. Without a change in monetary incentive, industrial noise will be reduced only as far and fast as community or industrial legislation requires. Legislation varies from country to country and often also from state to state, some being more effective than others. The technological challenge to engineers is therefore to develop more cost-effective methods of noise

Table 4.11. *The effects of years of exposure on the hearing impairment of people exposed to 90 dB(A).*

Years of exposure to 90 dB(A)	Percentage of working population with some form of compensable hearing damage
2.5	7
12.5	24
22.5	40
32.5	67
42.5	90

Table 4.12. *Qualitative scale of the severity of various noise levels.*

dB(A)	Qualitative scale
140	Jet take-off at 25 m, threshold of pain
130	Painfully loud
120	Jet take-off at 60 m
110	Car horn at 1 m
100	Shouting into an ear
90	Heavy truck at 15 m
80	Pneumatic drill at 15 m
70	Road traffic at 15 m
60	Room air-conditioner at 6 m
50	Normal conversation at 3 m
40	Background wind noise
30	Soft whisper at 4 m
0	Threshold of hearing

control – with technological innovation, particularly at the design stage, the economic argument could very well change.

4.9 Sound transmission from one room to another

Quite often, a situation arises where one has to reduce sound transmission from a noise source in a large reverberant or semi-reverberant room by partitioning off the section of the room that contains the source. To do this, one has to consider the steady-state sound power relations between a sound source room and a receiving reverberant room as illustrated in Figure 4.17. Flanking transmission via mechanical connections or air gaps is neglected in the following analysis.

During steady-state conditions, the sound power Π_{12} flowing from the source room to the receiving room must equal the sound power Π_{21} flowing back into the source room from the receiving room plus the sound power Π_a, that is absorbed within the receiving room. Thus

$$\Pi_{12} = \Pi_{21} + \Pi_a. \tag{4.92}$$

The sound power, Π_1, incident upon the source side of the partition is

$$\Pi_1 = I_{1w} S_w, \tag{4.93}$$

where S_w is the surface area of the partition between the two rooms, and I_{1w} is the sound intensity at the wall. Now,

$$I_{1w} = D_{1w} c = \frac{D_1 c}{4}, \tag{4.94}$$

where D_{1w} is the energy density at the wall and D_1 is the energy density in the source room. The energy density at the wall is not the same as the energy density in the room because the total sound intensity (in a room with a diffuse field) from all angles of incidence impinging upon any unit surface element is a quarter of the total intensity

Fig. 4.17. Sound transmission from one room to another.

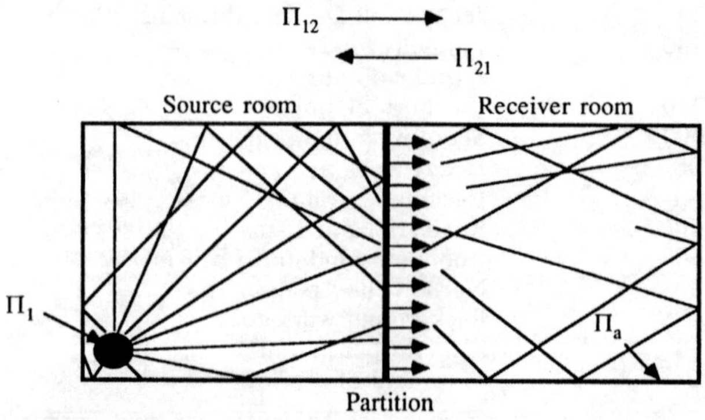

in the volume – i.e. only a quarter of the energy flow is outwards. Kinsler *et al.*[4.15] derive this relationship between the energy density and the power flow across the boundaries of a room using ray acoustics. It is assumed that, at any point within the room, energy is transported along individual ray paths with random phase, the energy density at a point in the room being the linear sum of all the energy densities of the individual rays. Likewise, the sound power incident upon the receiving room side of the partition is

$$\Pi_2 = I_{2\mathrm{w}} S_{\mathrm{w}}, \tag{4.95}$$

where $I_{2\mathrm{w}} = D_{2\mathrm{w}} c = D_2 c/4$.

The sound power which is transmitted from the source room to the receiving room is

$$\Pi_{12} = \Pi_1 \tau = I_{1\mathrm{w}} S_{\mathrm{w}} \tau, \tag{4.96}$$

where τ is the transmission coefficient of the partition. The sound power being transmitted from the receiving room back to the source room is

$$\Pi_{21} = \Pi_2 \tau = I_{2\mathrm{w}} S_{\mathrm{w}} \tau. \tag{4.97}$$

The sound power absorbed by the receiving room is

$$\Pi_{\mathrm{a}} = \frac{D_2 c}{4} S_2 \alpha_{2\,\mathrm{avg}} = I_{2\mathrm{w}} S_2 \alpha_{2\,\mathrm{avg}}, \tag{4.98}$$

where S_2 is the total surface area of the receiving room and $\alpha_{2\,\mathrm{avg}}$ is the average absorption coefficient in the receiving room.

By substituting equations (4.96), (4.97) and (4.98) into equation (4.92) one gets

$$I_{1\mathrm{w}} S_{\mathrm{w}} \tau = I_{2\mathrm{w}} S_{\mathrm{w}} \tau + I_{2\mathrm{w}} S_2 \alpha_{2\,\mathrm{avg}}. \tag{4.99}$$

Replacing the intensities with the corresponding mean-square pressures and taking logarithms yields

$$10 \log_{10}(1/\tau) = L_{\mathrm{p}1} - L_{\mathrm{p}2} + 10 \log_{10}\left\{\frac{S_{\mathrm{w}}}{S_2 \alpha_{2\,\mathrm{avg}} + \tau S_{\mathrm{w}}}\right\}, \tag{4.100}$$

or

$$NR = TL - 10 \log_{10}\left\{\frac{S_{\mathrm{w}}}{S_2 \alpha_{2\,\mathrm{avg}} + \tau S_{\mathrm{w}}}\right\}, \tag{4.101}$$

where $NR = L_{\mathrm{p}1} - L_{\mathrm{p}2}$ is the noise reduction and $TL = 10 \log_{10}(1/\tau)$ is the transmission loss of the partition.

The term $S_2 \alpha_{2\,\mathrm{avg}} + \tau S_{\mathrm{w}}$ represents the total absorption of the receiving room, and equation (4.101) clearly illustrates that the noise reduction that results from placing a partition between the two rooms is not only a function of the transmission loss across the wall but also a function of both the total absorption of the receiving room and the surface area of the partition. For rooms with small absorption coefficients and for partitions with small transmission losses the noise reduction is generally less than the transmission loss of the partition material. This is a very important consideration in noise control procedures. Also, in practice, the noise reduction is generally lower (by a few dB) than the value predicted by equation (4.101) because of flanking transmission via mechanical connections and air leaks.

4.10 Acoustic enclosures

Acoustic enclosures are commonly used in industry to box noise sources in. It is therefore useful to analyse some of the main principles involved in the design of enclosures for the purposes of controlling machinery noise. It should always be remembered, however, that enclosures do not eliminate or reduce the source of the noise – they just constrain it. Hence, it is good engineering practice to only consider enclosures as a last resort.

When an enclosure is mounted around a machine, is performance is restricted by (i) the transmission loss of the panels which are used to construct it, (ii) the extent of the vibration isolation between the noise source and it, and (iii) the presence of air gaps and leaks. With careful design and construction, enclosures can attenuate machinery noise by ~40–50 dB.

Provided that an enclosure is not close-fitting (i.e. it is not fitted directly on to a machine and it is at least 0.5 m from any major machine surface), then the mathematical relationships governing the performance of the enclosures are relatively simple. Close-fitting enclosures produce complicated physical effects such as cavity or air-gap resonances which can significantly impair their performance characteristics. Close-fitting enclosures will be qualitatively discussed at the end of this section. In the main, this section will be concerned with large enclosures where cavity resonances do not arise.

Any enclosure increases the noise levels within itself by establishing an internal reverberant field. The sound pressure level inside an enclosure at any arbitrary point away from the walls thus comprises both a direct-field component and a reverberant-field component, and is given by equation (4.68) – i.e.

$$L_{\rm p} = L_{\Pi} + 10\log_{10}\left\{\frac{Q_\theta}{4\pi r_{\rm E}^2} + \frac{4}{R_{\rm E}}\right\}, \tag{4.102}$$

where $R_{\rm E}$ is the room constant of the inside of the enclosure, as defined earlier, $r_{\rm E}$ is the distance from the source to the measurement point inside the enclosure, and Q_θ is the directivity of the noise source inside the enclosure. If the noise source is omni-directional, Q_θ is replaced by Q as per Table 4.4. When considering the design of enclosures, it is the reverberant term, $4/R_{\rm E}$, which is generally the dominant one.

The sound energy density inside a reverberant enclosure is related to the mean-square sound pressure, $\langle p^2 \rangle$ inside the enclosure by

$$D_R = \frac{\langle p^2 \rangle}{\rho_0 c^2}, \tag{4.103}$$

and the energy density at an enclosure wall is $D_w = D_R/4$. This is because at the inside surface of the enclosure only one quarter of the energy flow is outwards, whereas at the outside surface of the enclosure wall all the sound power flow is outwards. Thus,

$$I_{OE} S_E = I_w \tau_w S_E, \tag{4.104}$$

where the subscript OE refers to the outside surface of the enclosure, and S_E is its external radiating surface area. Hence,

$$\langle p_{OE}^2 \rangle = \frac{\langle p^2 \rangle \tau_w}{4}. \tag{4.105}$$

At this stage of the analysis, it is assumed that the enclosed sound source is in free space and that the sound contribution due to any reverberant field in the surrounding room is negligible. It should also be noted that τ_w is the transmission coefficient of the enclosure walls. Taking logarithms,

$$L_{pOE} = L_p - TL - 6 \text{ dB}. \tag{4.106}$$

The total sound power radiated by the enclosure is therefore

$$\Pi_E = \frac{\langle p_{OE}^2 \rangle}{\rho_0 c} S_E. \tag{4.107}$$

Thus,

$$L_{\Pi E} = L_{pOE} + 10 \log_{10} S_E. \tag{4.108}$$

$L_{\Pi E}$ thus approximates the sound power of the radiating enclosure. If the enclosure was located outdoors (or in a free-field environment), the sound pressure level at some point p2 at a distance r from the enclosure is

$$L_{p2} = L_{pOE} + 10 \log_{10} S_E + 10 \log_{10} \frac{Q_\theta}{4\pi r^2}. \tag{4.109}$$

If the enclosure is located indoors, then the reverberant sound field due to the enclosing room must be considered and the sound pressure level at some point p2 in the room is given by

$$L_{p2} = L_{pOE} + 10 \log_{10} S_E + 10 \log_{10} \left\{ \frac{Q_\theta}{4\pi r^2} + \frac{4}{R} \right\}, \tag{4.110}$$

where R is the room constant of the reverberant environment, and $L_{pOE} + 10 \log_{10} S_E = L_{\Pi E}$.

The value of the sound pressure level at point p2 in the room without any enclosure over the noise source is

$$L'_{p2} = L_{\Pi} + 10\log_{10}\left\{\frac{Q_\theta}{4\pi r^2} + \frac{4}{R}\right\}, \tag{4.111}$$

where L_{Π} is the sound power of the noise source itself as per equation (4.102).

The reduction in noise that the enclosure would provide is simply the difference between equations (4.111) and (4.110). It is defined as the insertion loss, IL – i.e. it is the difference between the sound pressure levels at a given point with and without the enclosure (the noise reduction NR, defined in the previous section, is the difference in sound pressure level at two specific points, one inside and one outside the enclosure). The difference between the insertion loss, IL, the noise reduction, NR, and the transmission loss, TL, is illustrated schematically in Figure 4.18. Thus, from equations (4.110) and (4.111),

$$IL = L'_{p2} - L_{p2} = L_{\Pi} - L_{\Pi E} = L_{\Pi} - L_{pOE} - 10\log_{10}S_E. \tag{4.112}$$

Fig. 4.18. Schematic illustration of the difference between *IL*, *NR* and *TL*.

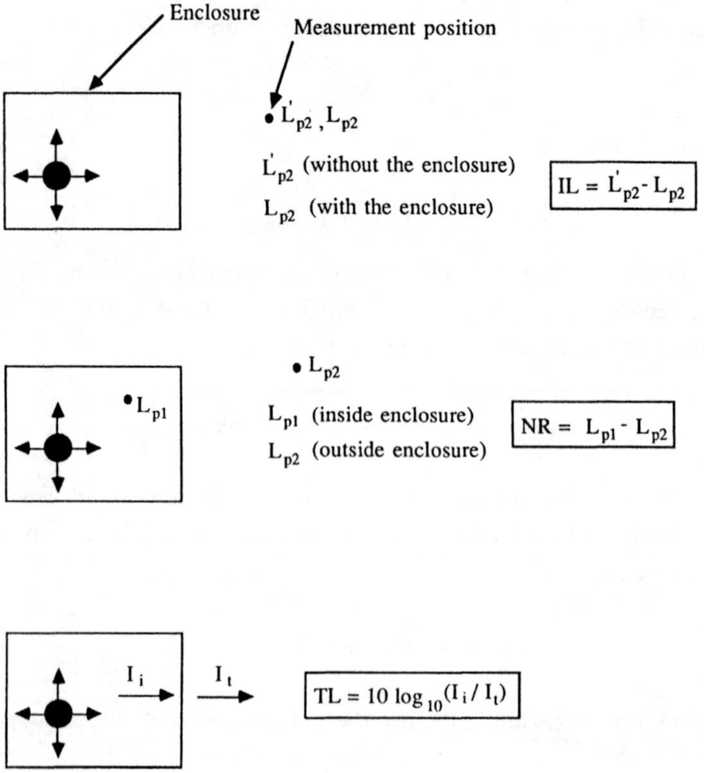

L_{pOE} and L_{Π} can now be eliminated by using equations (4.102) and (4.106). Hence,

$$IL = TL - 10\log_{10}S_E + 6 - 10\log_{10}\left\{\frac{Q_\theta}{4\pi r_E^2} + \frac{4}{R_E}\right\}. \quad (4.113)$$

Assuming that the inside of the enclosure is a reverberant space, the direct field term can be neglected and

$$R_E = \frac{A_E\alpha_{E\,avg}}{1-\alpha_{E\,avg}}, \quad (4.114)$$

where A_E is the total internal surface area inside the enclosure – it includes both the surface area of the inside surfaces of the enclosure and the surface area of the machine, and $\alpha_{E\,avg}$ is the average absorption coefficient inside the enclosure. Thus,

$$IL = TL - 10\log_{10}S_E + 10\log_{10}\frac{A_E\alpha_{E\,avg}}{1-\alpha_{E\,avg}}, \quad (4.115)$$

Equation (4.115) is the primary design equation for large fitting enclosures. Information is required about the transmission loss characteristics and the absorption coefficients of the panels that are used to construct the enclosure and the absorption coefficients of any absorbing materials that are inserted on the inside walls. The transmission loss characteristics can be evaluated from the empirical procedures outlined in chapter 3 (the plateau method, etc.) or by referring to tables. A list of typical transmission loss coefficients and absorption coefficients for some common building materials is presented in Appendix 2.

Various other factors have to be taken into account in the design of enclosures. They include enclosure resonances, structure-borne sound due to flanking transmission, air-gap leakages, vibrations, and ventilation. Crocker and Kessler[4.16] discuss many of the practical aspects associated with enclosure design.

There are three types of enclosure resonances. The first is due to the structural resonances in the panels that make up the enclosure, the second is due to standing wave resonances in the air gap between the machine and the enclosure, and the third is the double-leaf panel resonance described in chapter 3. At each of these resonant frequencies, the insertion loss due to the enclosure is significantly reduced. To avoid problems associated with panel resonances, the enclosure panels should be designed such that their resonant frequencies are higher than or lower than the frequency range in which the maximum sound attenuation is desired. Hence, a low frequency sound source would require the enclosure panels to have high resonant frequencies – i.e. the enclosure should be stiff but not massive. Alternatively, a high frequency sound source would require the enclosure panels to have low resonant frequencies – i.e. the enclosure should have a significant mass. The air-gap resonant frequencies occur at frequencies where the average air-gap size is an integral multiple of a half-wavelength of sound. These resonances can be eliminated by using absorptive treatment on the enclosure walls and by ensuring that the enclosure is not close-fitting. The double-leaf panel resonance is controlled by the mass of the walls and the stiffness of the air gap.

Mechanical paths between the machine to be isolated and the enclosure must be avoided as far as possible to minimise structure-borne sound due to flanking transmission. Flexible vibration breaks and correctly designed vibration isolators should be used whenever necessary.

The presence of air gaps around an enclosure reduces its effectiveness. Leaks occur frequently in practice and they can pose a serious problem. The air gaps usually occur around removable panels or where services (electricity, ventilation, etc.) enter the enclosure. Air paths are significantly more efficient than mechanical paths. Empirical design charts are available for estimating the reduction in transmission loss due to air gaps. They are essentially based on an equation similar to (4.60) with the absorption coefficients replaced by transmission coefficients, and the transmission coefficient of the leak assumed to be unity. Thus,

$$\tau_{avg} = \frac{S_1\tau_1 + S_2\tau_2 + \cdots + S_n\tau_n}{S_1 + S_2 + \cdots + S_n}. \tag{4.116}$$

The equation is used in practice to estimate the average trasmission coefficient, τ_{avg}, of an enclosure that is constructed with different panels; if the surface area of the leak is known, its effect on the overall transmission loss can be readily evaluated.

Most enclosures generally require some form of forced ventilation to cool the machinery inside them. The openings, such as air flow ducts or ventilation fans, need to be silenced. Various techniques are available in the handbook literature[4.1,4.3].

Finally, if sufficient space is left inside an enclosure for normal maintenance on all sides of the machine, the enclosure need not be regarded as being close-fitting. When this space is not available, the transmission loss of the enclosure has to be increased by up to 10 dB, particularly at low frequencies, to overcome the reduction in effectiveness due to the enclosure resonances. When enclosures are close-fitting, the internal sound field is neither diffuse nor reverberant, and the sound waves generally impinge on the enclosure walls at normal rather than random incidence. At each of the three resonant frequencies described earlier, the noise reduction can be significantly attenuated. In fact, at the double-leaf panel resonance, the sound can even be amplified! Advanced theories for close-fitting enclosures based on work by Ver[4.17] are reviewed by Crocker and Kessler[4.16].

4.11 Acoustic barriers

Acoustic barriers are placed between a noise source and a receiver such as to reduce the direct-field component of the sound pressure levels at the receiver position. As will be seen shortly, barriers do not reduce reverberant-field noise. Well designed barriers simply diffract the sound waves around their boundaries, hence they alter the effective directivity of the source.

Consider a barrier which is inserted into a room. Before the barrier is placed in position, the mean-square pressure at the receiver is p_0^2, and

$$L_{p0} = L_{\Pi} + 10\log_{10}\left\{\frac{Q_\theta}{4\pi r^2} + \frac{4}{R}\right\}, \tag{4.117}$$

where L_{Π}, Q_θ, r and R are defined in the usual manner.

Assuming that the mean-square sound pressure, p_2^2, at the receiver (with the barrier in place) is the sum of the square of the pressures due to the diffracted field around the barrier, p_{b2}^2, and the average reverberant field of the room, p_{r2}^2, then

$$p_2^2 = p_{r2}^2 + p_{b2}^2, \tag{4.118}$$

and

$$L_{p2} = L_{r2} + L_{b2}. \tag{4.119}$$

The barrier insertion loss, IL, is defined as

$$IL = 10\log_{10}\left(\frac{p_0^2}{p_2^2}\right) = L_{p0} - L_{p2}. \tag{4.120}$$

It now remains to establish a relationship between the barrier and the room parameters. Assume that the barrier is placed in a rectangular room, and that the surface area of the barrier is small compared to the planar cross-section of the room (this restriction will be lifted later on in this section). Hence, as illustrated in Figure 4.19,

Fig. 4.19. Schematic illustration of a room with a barrier between the source and receiver.

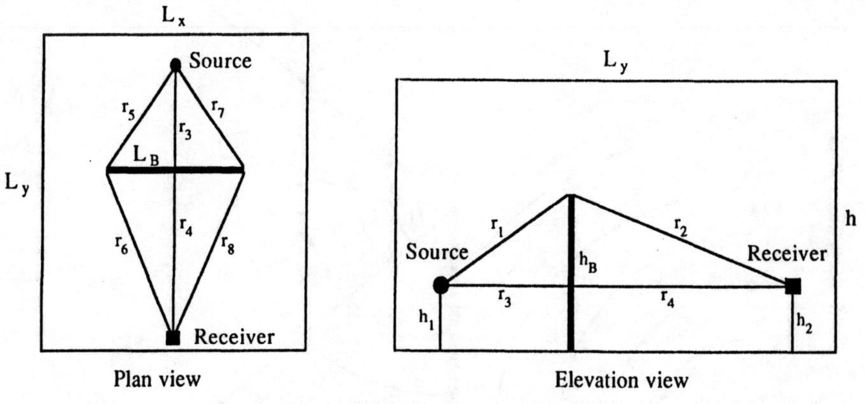

Note: $h_1 = h_2$ in this illustration for convenience - in reality, $h_1 \neq h_2$ and therefore r_3 - r_8 have to be evaluated by using the relevant angles.

$L_x h \gg L_B h_B$. Under this condition, it can be assumed that the reverberant field in the shadow zone of the barrier is the same with and without the barrier. Thus, the sound pressure level in the shadow zone of the barrier is never less than that due to just the reverberant field by itself. The reverberant mean-square sound pressure in the room is

$$p_{r2}^2 = \frac{4\Pi \rho_0 c}{R}. \tag{4.121}$$

The barrier performance depends upon Fresnel diffraction of the sound waves from the source. These diffractions are incident along the edges of the walls. This is illustrated in Figure 4.20. To observers in the shadow region, the diffracted sound field is being radiated by a line source along the edges of the barrier.

According to Fresnel diffraction theory, only that portion of a wave field due to a sound source that is incident upon the edges of a barrier contribute to the wave field that is diffracted over the barrier. The mean-square pressure in the diffracted field is given by

$$p_{b2}^2 = p_{d0}^2 \sum_{i=1}^{n} \left(\frac{1}{3 + 10N_i} \right), \tag{4.122}$$

where p_{d0}^2 is the mean-square pressure due to the direct field prior to the insertion of the barrier, and the Fresnel number, N_i, for diffraction around the ith edge is given by

$$N_i = \frac{2\delta_i}{\lambda}, \tag{4.123}$$

Fig. 4.20. Schematic illustration of a barrier diffracted field.

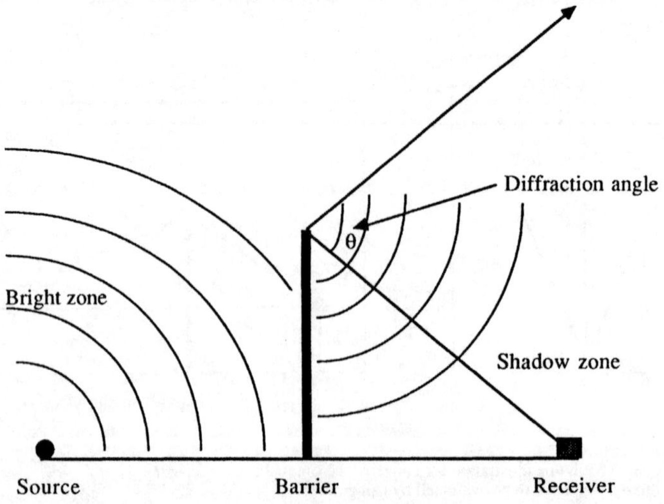

with λ being the wavelength of the sound frequency being considered, and δ_i being the difference between the ith diffracted path and the direct path between the source and the receiver when the barrier is absent. For the example in Figure 4.19,

$$\delta_1 = (r_1 + r_2) - (r_3 + r_4), \tag{4.124a}$$

$$\delta_2 = (r_5 + r_6) - (r_3 + r_4), \tag{4.124b}$$

and

$$\delta_3 = (r_7 + r_8) - (r_3 + r_4). \tag{4.124c}$$

The mean-square pressure due to the direct field (prior to the insertion of the barrier) is

$$p_{\mathrm{d}0}^2 = \frac{\Pi \rho_0 c Q_\theta}{4\pi r^2}, \tag{4.125}$$

thus the diffracted mean-square pressure is

$$p_{\mathrm{b}2}^2 = \frac{\Pi \rho_0 c Q_\theta}{4\pi r^2} \sum_{i=1}^{n} \left(\frac{1}{3 + 10N_i} \right). \tag{4.126}$$

Now, if the effective directivity, Q_{B}, of the source in the direction of the shadow zone of the barrier is defined such that

$$Q_{\mathrm{B}} = Q_\theta \sum_{i=1}^{n} \left(\frac{1}{3 + 10N_i} \right), \tag{4.127}$$

then the total mean-square sound pressure at the receiver position in the presence of the barrier is

$$p_2^2 = \Pi \rho_0 c \left(\frac{Q_{\mathrm{B}}}{4\pi r^2} + \frac{4}{R} \right), \tag{4.128}$$

or

$$L_{\mathrm{p}2} = L_\Pi + 10 \log_{10} \left(\frac{Q_{\mathrm{B}}}{4\pi r^2} + \frac{4}{R} \right). \tag{4.129}$$

The barrier insertion loss, IL, is $L_{\mathrm{p}0} - L_{\mathrm{p}2}$ as per equation (4.120), thus

$$IL = 10 \log_{10} \left(\frac{\left(\dfrac{Q_\theta}{4\pi r^2} + \dfrac{4}{R} \right)}{\left(\dfrac{Q_{\mathrm{B}}}{4\pi r^2} + \dfrac{4}{R} \right)} \right). \tag{4.130}$$

This equation represents the general equation for the insertion loss of a barrier with a receiver in the shadow zone. The fundamental assumption in this derivation is that the reverberant field in the shadow zone of the barrier is the same with and without the barrier. A more rigorous theory for estimating the insertion loss due to a barrier in an enclosed room has been developed by Moreland and Musa[4.18] and Moreland and Minto[4.19], and the modified design equation is

$$IL = 10 \log_{10}\left(\frac{\left(\dfrac{Q_\theta}{4\pi r^2} + \dfrac{4}{S_0\alpha_0}\right)}{\left(\dfrac{Q_B}{4\pi r^2} + \dfrac{4k_1k_2}{S(1-k_1k_2)}\right)}\right), \tag{4.131}$$

where $S_0\alpha_0$ is the room absorption for the original room before inserting the barrier, S_0 is the total room surface area, α_0 is the mean room absorption coefficient, S is the open area between the barrier perimeter and the room walls and ceiling, and k_1 and k_2 are dimensionless numbers related to the room absorption on the source side $(S_1\alpha_1)$ and the receiver side $(S_2\alpha_2)$ of the barrier, respectively, as well as the open area. These numbers are given by

$$k_1 = \frac{S}{S + S_1\alpha_1}, \quad \text{and } k_2 = \frac{S}{S + S_2\alpha_2}. \tag{4.132}$$

Two special cases of barrier insertion loss arise. They are (i) when the barrier is located in a free field, and (ii) when the barrier is located in a highly reverberant field. For the first case, the sound absorption coefficient is unity and thus the room constant, R, approaches infinity. Hence, from equation (4.130),

$$IL = 10 \log_{10}\frac{Q_\theta}{Q_B} = -10 \log_{10}\sum_{i=1}^{n}\frac{\lambda}{3\lambda + 20\delta_i}. \tag{4.133}$$

For the second case, $4/R \gg Q_\theta/4\pi r^2$ or $Q_B/4\pi r^2$ and therefore

$$IL = 10 \log_{10}1 = 0 \text{ dB}. \tag{4.134}$$

This is a very important point and it illustrates that barriers are ineffective in highly reverberant environments. The exception to the rule is when the barrier is treated with sound absorbing material and the overall sound absorption of the room is increased.

A simplified expression can be derived for a semi-infinite barrier (e.g. a wall) in a free-field environment. Consider the barrier in Figure 4.21. Neglecting Fresnel diffraction along the edges of the barrier and assuming that $D \gg R \geqslant H$, it can be shown that $\delta \approx H^2/2R$. Thus,

$$IL = -10 \log_{10}\frac{\lambda}{3\lambda + 10H^2/R}. \tag{4.135}$$

It has also been shown[4.16] that when the noise source approximates to an incoherent line source (e.g. a string of traffic) then the insertion loss is about 5 dB lower than the theory which assumes a point source.

It is useful to note that sometimes when barriers are placed outdoors ground absorption effects can reduce the effectiveness of the barrier, particularly at low frequencies. The barrier can reduce destructive interactions between the ground plane and the direct sound waves. The exact frequencies at which this phenomenon might occur is dependent upon the particular geometry being considered. As a rule of thumb it generally occurs between 300 and 600 Hz. The reader is referred to Beranek[4.3] and Crocker and Kessler[4.16] for a list of references dealing with a range of issues relating to the performance of barriers including the effects of buildings, streets, depressed and elevated highways, ground effects, wind and temperature gradients, etc.

4.12 Sound-absorbing materials

The concept of a sound absorption coefficient, α, was introduced in sub-section 4.7.2. It is the ratio of the absorbed to incident sound intensities or energies and it has a value somewhere between zero and unity. As already mentioned, an open window absorbs all the sound energy impinging on it and therefore has an absorption coefficient of unity. Materials with high absorption coefficients are often used for the control of reverberant noise – they absorb the sound waves and significantly reduce the reflected energy. Porous or fibrous materials generally have high absorption coefficients, two good examples being open-cell foam rubber and fibreglass. The absorption coefficient of a given material varies with frequency and with the angle of incidence of the impinging sound waves; it is a function of the fibre or pore size, the thickness of the material, and the bulk density. The two mechanisms responsible for sound absorption are viscous dissipation in the air cavities, and friction due to the vibrating fibres – both mechanisms convert the sound energy into heat energy. Thus, it is important that the material is porous or fibrous, i.e. the sound waves have to be able to move about in the material.

The two most common methods of measuring the sound absorption coefficients of a given material are (i) the measurement of the normal incidence sound absorption

Fig. 4.21. One-dimensional (infinite) barrier.

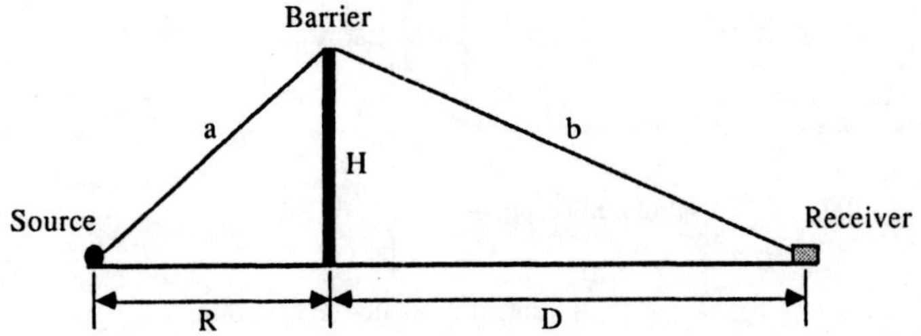

coefficient, α_n, using a device called an impedance tube, and (ii) the measurement of the random incidence sound absorption coefficient, α, using a reverberation room. The random incidence sound absorption coefficient is the one that is most commonly used in engineering noise control.

The first technique, whilst restricted to normal incidence sound waves, is simple and inexpensive and it can provide an order of magnitude approximation. A small loudspeaker, which generates a sinusoidal sound wave which travels down the tube, is placed at one end. The test material is placed at the other end of the tube. The sound field in the tube is a standing wave which is a resultant of the incident and reflected waves. The standing wave ratio (ratio of r.m.s. pressure maxima to pressure minima) can be readily obtained by traversing a probe microphone connected to a carriage as illustrated in Figure 4.22. The reflection coefficient can be obtained directly from the standing wave ratio, and the absorption coefficient is subsequently obtained from the reflection coefficient. If the incident sound wave has a complex pressure amplitude $\mathbf{P_I}$, and the reflected sound wave has a complex pressure amplitude $\mathbf{P_R}$, then the reflection coefficient, $\mathbf{r}$, is given by

$$|\mathbf{r}| = \frac{|\mathbf{P_R}|}{|\mathbf{P_I}|}. \tag{4.136}$$

Thus, the standing wave ratio, s, is given by

$$s = \frac{|\mathbf{P_I}| + |\mathbf{P_R}|}{|\mathbf{P_I}| - |\mathbf{P_R}|} = \frac{1 + |\mathbf{r}|}{1 - |\mathbf{r}|}, \tag{4.137}$$

and the reflection coefficient is

$$|\mathbf{r}| = \frac{s - 1}{s + 1}. \tag{4.138}$$

Fig. 4.22. Schematic illustration of an impedance tube for measuring normal incidence sound absorption coefficients.

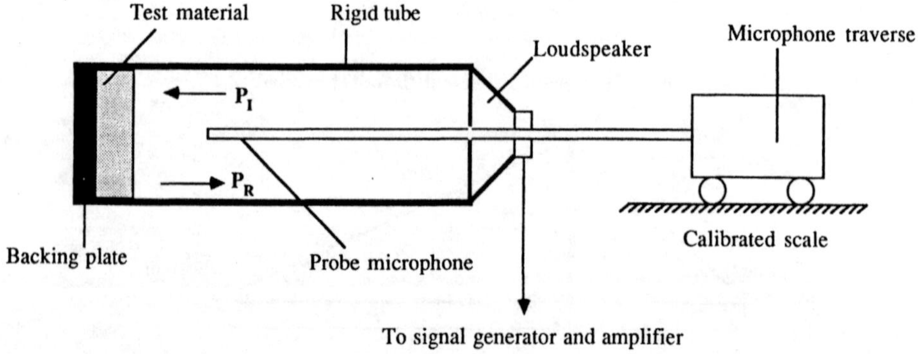

The normal incidence absorption coefficient is subsequently obtained from the relationship

$$\alpha_n = 1 - |\mathbf{r}|^2. \tag{4.139}$$

The normal impedance of the test surface is given by

$$\mathbf{Z}_s = \frac{1+\mathbf{r}}{1-\mathbf{r}} \rho_0 c. \tag{4.140}$$

The impedance tube method is restricted because it only allows for the normal incidence absorption coefficient to be evaluated. The normal incidence absorption coefficient is always slightly less than the random incidence absorption coefficient, thus it allows for a conservative estimate. The method is attractive because of its simplicity and relatively low cost. It is ideal for comparative measurements between different types of material. Also, the effects of varying the material thickness, air gaps, sealing surfaces, perforated plates, etc., can be readily investigated by using this method. Because it is essential that the sound waves travelling in the tube are only plane waves, it is important that the walls of the impedance tube must be rigid and massive and that its cross-sectional area must be uniform. Also, the diameter of the tube limits the upper frequency that can be tested. Above a given diameter, higher-order acoustic cross-modes are generated (higher-order acoustic modes in a circular pipe are discussed in chapter 7). The relationship between maximum frequency and impedance tube diameter, D, is

$$f_{\max} = \frac{c}{1.7D}. \tag{4.141}$$

A more general form of equation (4.141) is derived in chapter 7. Equation (4.141) itself is limited to the case where there is no mean air flow in the tube.

The random incidence sound absorption coefficient, α, is obtained by conducting sound absorption tests in a reverberation room. The sound field is generated by a loudspeaker which is placed in the corner of the room to excite as many room acoustic modes as possible. The first measurement is made with the room empty and the random incidence absorption coefficient, α_0, of the reverberation room is obtained from equation (4.70). Hence,

$$\alpha_0 = \frac{0.161V}{S_0 T_0}, \qquad \text{(equation 4.70)}$$

where the subscript 0 refers to the empty room. A sample of the test material is then placed in the room. It should be about 10 to 12 m^2 with a length to breadth ratio of about 0.7 to 1.0. The reverberation time of the room is now

$$T_M = \frac{0.161V}{(S_0 - S_M)\alpha_0 + S_M\alpha_M}, \tag{4.142}$$

and the random incidence sound absorption coefficient, α_M, of the test material is obtained by solving equations (4.142) and (4.70). It is

$$\alpha_M = \frac{1}{S_M}\left\{\frac{0.161V}{T_M} - \frac{(S_0 - S_M)0.161V}{S_0 T_0}\right\}. \tag{4.143}$$

Details of the recommended experimental procedures for reverberation room testing are provided in the handbook literature and in various national and international standards. It is worth remembering that the parameter $0.161V$ relates to a speed of sound of 343 m s^{-1} (1 atm. and 20°C) – for significantly higher or lower ambient conditions, the variations in the speed of sound have to be accounted for.

A parameter that is a useful guide for optimising the absorption coefficient of a given porous material is its flow resistance. The specific (unit area) flow resistance of a given porous material is the ratio of the applied air pressure differential across the test specimen to the particle velocity through and perpendicular to the two faces of the test specimen. The particle velocity is obtained experimentally by dividing the volume velocity of the airflow by the surface area of the sample. It is important to note that the tests have to be conducted under conditions of a very slow steady airflow. Bies[4.20] discusses the relationships between flow resistivity and the acoustical properties of porous materials in some detail. It is generally accepted that optimum acoustic absorption will be achieved if the flow resistance of a given material is between $2\rho_0 c$ and $5\rho_0 c$. If the flow resistance is too low, the sound waves will pass through the material and reflect off the rigid backing which is generally used to support the sound absorbing material. If the flow resistance is too high, the sound waves will reflect off the absorbing material itself.

The sound intensity technique, described in sub-section 4.7.4, can also be used for the measurement of sound absorbing characteristics of different materials. The technique has been successfully applied by numerous researchers for both impedance tube and reverberation room techniques. Maling[4.21] reviews the various applications of sound intensity measurements to noise control engineering. With the impedance tube method, two closely spaced microphones are flush mounted on the impedance tube wall, and the cross-spectral method is used to determine the maximum and minimum intensity of the sound wave in the tube. This method has proved so reliable that it has recently been standardised by the American Society of Testing and Materials. With reverberation room testing, the sound power absorbed by a test surface is simply measured by moving an intensity probe over the surface and measuring the flow of intensity into the surface. It appears that the sound intensity technique overcomes some of the problems associated with reverberation room techniques which sometimes yield sound absorption coefficients greater than one due to edge effects, diffraction or non-diffuse sound fields.

Porous or fibrous materials generally have good sound absorbing characteristics at high frequencies (> 1000 Hz), with a rapid deterioration at low frequencies. At very low frequencies (< 250 Hz) the sound absorption coefficient decreases rapidly. Increasing the thickness of the material generally improves the low frequency sound absorption characteristics. The particle velocity of a sound wave is zero at a rigid interface such as a backing wall for a sound absorbing material. For effective absorption of sound energy to occur, the sound wave has to pass through the absorbing material during a particle velocity maximum. Hence, the thickness of the sound absorbing material should be a quarter-wavelength ($\lambda/4$) of the lowest frequency of interest. This important point is illustrated in Figure 4.23. A second alternative is to use a thinner material with an air gap between the sound absorbing material and the rigid backing wall. The combination of the air-gap thickness and the absorbing material thickness should always be a quarter-wavelength. When curtains are used to reduce low frequency sound they should always be hung slightly away from the wall and not touching the wall! The frequency range of absorption can also be increased by staggering the material.

Fig. 4.23. Schematic illustration of the quarter-wavelength effect.

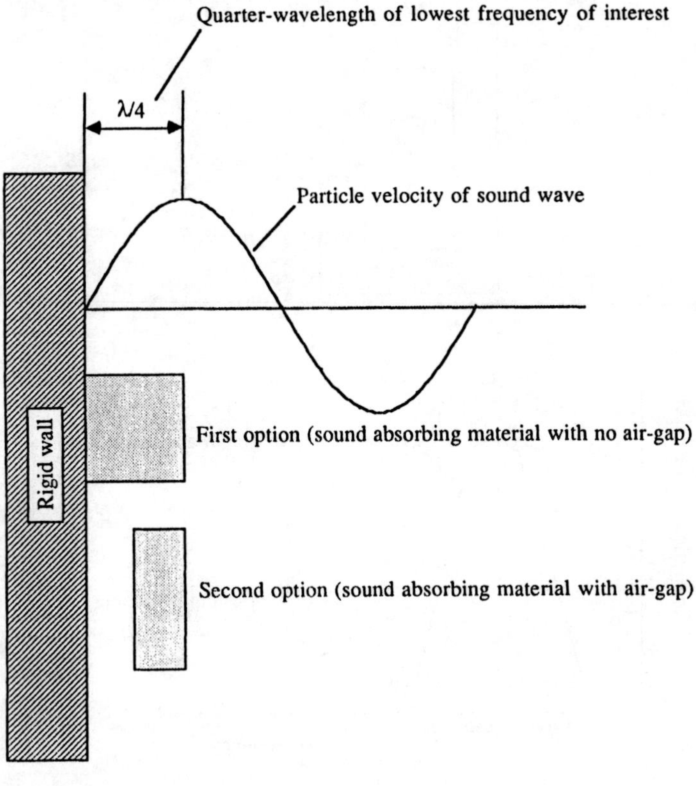

Good low frequency absorption can be achieved with resonant absorbers which involve volumes of air. Two types of resonant absorber principles are commonly used for low frequency sound absorption. They are Helmholtz (or cavity) resonators and panel absorbers.

A Helmholtz resonator is a cavity of air which acts like a spring – it is forced in and out of the cavity by a periodic air flow which behaves like a mass. The cavity volume and the neck can be tuned to a specific low frequency. Its sound absorption characteristics can be broadened over the frequency range by lining the cavity with sound absorbing material. This has the effect of reducing the sound absorption at the resonant frequency of the cavity but significantly increasing it at other frequencies. This is illustrated schematicaly in Figure 4.24. The resonant frequency of a Helm'· ɔltz resonator is given by

$$f_{res} = \frac{c}{2\pi}\left(\frac{S}{lV}\right)^{1/2}, \tag{4.144}$$

where S is the cross-sectional area of the neck, l is the effective length of the neck, and V is the enclosed air volume. The effective neck length is given by

$$l = L + 0.8S^{1/2}, \tag{4.145}$$

Fig. 4.24. Characteristics of a Helmholtz resonator.

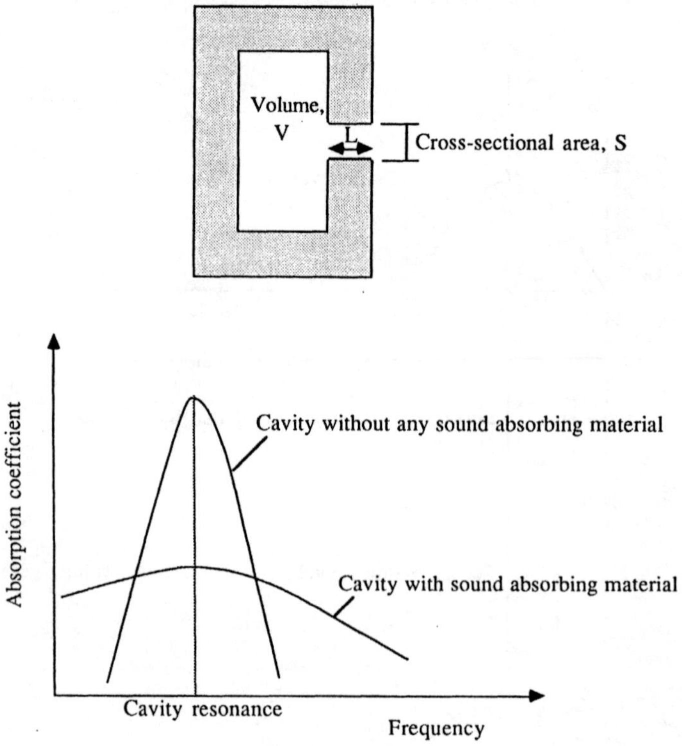

where L is the actual length of the neck. Helmholtz resonators are commonly used in muffling devices; the sound waves that enter the volume cavity are returned out of phase to the duct, resulting in cancellation of sound.

Panel absorbers mounted on walls, etc., with an air gap between the panel and the wall, also provide low frequency sound absorption. The mechanism of sound absorption is the resonant mass–spring behaviour of the system which dissipates sound energy. The sound absorption characteristics are increased by filling the air gap behind the absorber with sound absorbing material. Maximum sound absorption is achieved at the resonant frequency of the panel absorber which is

$$f_{\text{res}} = \frac{60}{(\rho_S l)^{1/2}}, \tag{4.146a}$$

where ρ_S is the surface mass of the panel (kg m^{-2}) and l is the depth of the air gap behind the panel (m).

The Helmholtz resonator principle can be applied to panel absorbers by covering the sound absorbing material with a perforated panel – this provides a large number of small Helmholtz resonators. Provided that the percentage open area is less than ~30%, low frequency Helmholtz resonant behaviour is attained; if the percentage open area is significantly larger, the panel behaves like porous absorbent material acting alone. An empirical relationship for estimating the frequency of maximum resonant absorption is

$$f_{\text{res}} = 5000\left\{\frac{P}{l(t + 0.8d)}\right\}^{1/2}, \tag{4.146b}$$

where P is the percentage of open area of the panel, t is the panel thickness in mm, d is the perforation diameter in mm, and l is the depth of air gap occupied by the sound absorbent material behind the panel.

Panel absorbers by themselves tend to provide excellent very low frequency sound absorption (<250 Hz) – their mid-frequency (~500–1000 Hz) performance is very poor. Perforated panel absorbers (multiple Helmholtz resonators) tend to reduce the effectiveness of the very low frequency absorption, but they significantly increase the mid frequency performance – they also reduce the effectiveness of the high frequency performance of the sound absorbing material. The high frequency performance can be improved by perforating with many small holes rather than a lesser number of large holes (with the same percentage open area); this, however, reduces the resonant effect of the panel. The performance characteristics of sound absorbing material, panel absorbers and perforated panel absorbers are schematically illustrated in Figure 4.25. The selection of a suitable practical sound absorbing system thus depends very much upon the specific attenuation requirements (i.e. low, mid or high frequencies).

Sound absorbing materials are commonly used to reduce reverberant sound in rooms; as a rule of thumb, noise level reductions of up to 10 dB are readily achievable by their correct utilisation. They have very little effect on direct sound such as the direct noise reaching the operator of a machine. The room acoustic equation (equation 4.68) derived in sub-section 4.7.2 can be used to establish a critical distance which provides a demarcation between the effects of the direct (free) and reverberant fields. Beyond the critical distance, the strength of the reverberant field is larger than that of the direct field, hence sound absorption treatment is appropriate. The critical distance for an omni-directional sound source can be obtained by equating the

Fig. 4.25. Performance characteristics of sound absorbing materials, panel absorbers and perforated panel absorbers.

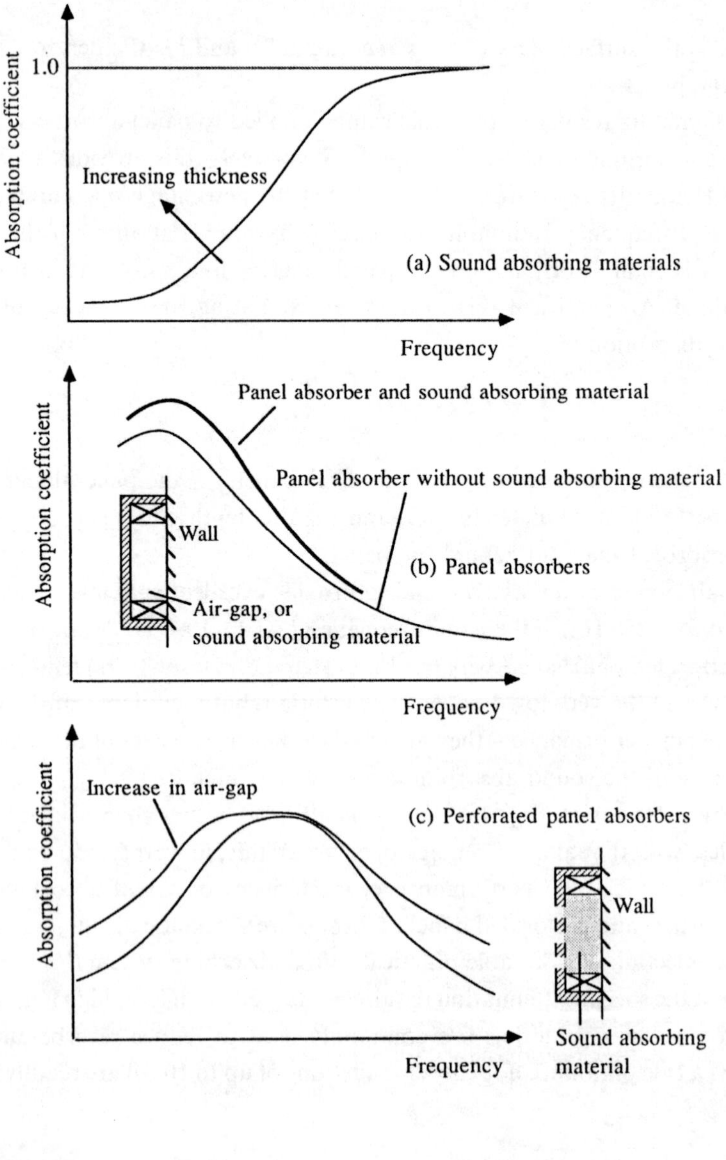

free-field and reverberant field terms in equation (4.68) – i.e. $Q/4\pi r^2 = 4/R$. Thus,

$$r_C = \left(\frac{QR}{16\pi}\right)^{1/2}, \tag{4.147}$$

where R is the room constant of the room, Q is the directivity associated with the omni-directional source position, and r_C is the critical distance of the 'reverberation radius'. When $r < r_C$, sound absorption treatment will not reduce the noise levels in the room; when $r > r_C$, sound absorption treatment will reduce the noise levels in the room.

In addition to wall, ceiling or floor mountings, space absorbers are sometimes used. Space absorbers take on a variety of shapes and sizes; they typically take the form of free hanging cylinders, cones, cubes, hexagons, etc. They are readily positioned at regions in proximity to noise sources. The total sound absorption of a space absorber is equal to its total surface area at frequencies ranging from ~500 Hz to 4000 Hz, since the absorption coefficient of porous materials at these frequencies is very high. When rooms are irregularly shaped (e.g. long rooms where one of the floor dimensions is significantly larger than the ceiling height), ceiling–floor reflections dominate the reverberant sound field and sound absorption treatment, such as space absorbers, should be confined to the ceilings (and the floors where possible). Factory spaces are common examples of such rooms and it is useful to note that under these circumstances, wall absorption is often not required.

Typical values of random incidence sound absorption coefficients for a range of materials are provided in Table 4.13. Further information is provided in Appendix 2.

Table 4.13. *Absorption coefficients for some typical construction and acoustical materials.*

	Octave band centre frequency (Hz)					
Surface	125	250	500	1000	2000	4000
Exposed brick	0.05	0.04	0.02	0.04	0.05	0.05
Normal carpet	0.02	0.06	0.14	0.37	0.60	0.66
Thick pile carpet	0.15	0.25	0.50	0.60	0.70	0.70
Concrete	0.01	0.01	0.02	0.02	0.02	0.02
Fibrous glass (25 mm)	0.07	0.23	0.48	0.83	0.88	0.80
Fibrous glass (100 mm)	0.39	0.91	0.99	0.97	0.94	0.89
Plate glass	0.25	0.25	0.18	0.12	0.07	0.05
Hardboard	0.10	0.10	0.15	0.15	0.10	0.10
Plasterboard	0.30	0.20	0.15	0.05	0.05	0.05
Plasterboard ceiling	0.20	0.20	0.15	0.10	0.05	0.05
Open cell polyurethane (25 mm)	0.14	0.30	0.63	0.91	0.98	0.91
Open cell polyurethane (50 mm)	0.35	0.51	0.82	0.98	0.97	0.95
Wood	0.15	0.11	0.10	0.07	0.09	0.03

4.13 Vibration control procedures

The two fundamental mechanisms responsible for sound generation are (i) structure-borne sound associated with vibrating structural components, and (ii) aerodynamic sound. Vibration control is thus not only important in terms of minimising structural vibrations and any associated fatigue, but it is also important in terms of noise control. It is, however, very important to recognise that there does not have to be a one to one relationship between vibration reduction and noise reduction. As was demonstrated in chapter 3, some modes of vibration are more efficient radiators of sound than others; generally the modes which radiate sound are at the higher frequencies, and they do not have the high vibration levels that the low frequency vibration modes have. As a general rule of thumb, low frequency modes produce structural fatigue and failure, whilst high frequency modes produce noise. Sometimes however, a complex interaction between the two can exist; flow-induced noise and vibration in a pipeline is such an example where high frequency modes can produce both noise and structural fatigue – it will be discussed in chapter 7.

This section is essentially concerned with the control of structural vibration levels generated by vibrational forces associated with machines or engines. These vibrational forces are often unavoidable, but they can be minimised by the application of correct vibration control procedures. Vibration control procedures generally involve either isolation of the vibrational forces, or the application of damping to the structure. Vibration isolation is the reduction of vibration transmission from one structure to another via some elastic device; it is a very important and common part of vibration control. It can be conveniently sub-divided into three regions – low freqency, single--degree-of-freedom vibration isolation; low frequency, multiple-degree-of-freedom vibration isolation; and audio-frequency vibration isolation. The first region involves uni-directional low frequency vibrational forces on a machine or structure; the second region involves multi-directional low frequency vibrational forces; and the third region involves frequencies where the vibrational wavelengths are significantly smaller than the thickness of the isolator, hence reducing the problem to one of wave transmission through the material. Resonant structural vibrations can also be reduced by the application of damping. This can take the form of a dynamic absorber (a secondary mass attached to the vibrating component via a spring) or layers of damping material applied to the surfaces of the structure. Vibration dampers dissipate the vibrational energy at regions in proximity to resonance by converting the vibrational motion into heat.

4.13.1 Low frequency vibration isolation – single-degree-of-freedom systems

Quite often in practice, machines or pieces of equipment are mounted on four isolators (springs, rubber pads, or air bags) and the primary vibrational force is both uni-directional and harmonic (single frequency). Under these conditions it is quite common for the system to be modelled as a single-degree-of-freedom system. Whilst the model is strictly only valid if the supporting base is rigid (i.e. the base does not vibrate in flexure, but moves as a rigid body), it serves as a useful introduction to vibration isolation procedures.

Consider the one-degree-of-freedom, mass–spring–damper system in Figure 4.26 which is subjected to a harmonic excitation force. The equation of motion is

$$m\ddot{x} + c_v\dot{x} + k_s x = F \sin \omega t. \qquad \text{(equation 1.50)}$$

The exciting force is transmitted to the foundations via the spring and the damper. It is

$$f_T(t) = k_s x + c_v \dot{x}. \qquad (4.148)$$

The ratio of the amplitude of the transmitted force, F_T, to that of the driving force, F, is commonly known as the transmissibility, TR; it is obtained by simultaneously solving the equation of motion of the system and equation (4.148) for steady-state conditions (it is convenient to use the complex algebra procedures outlined in sub-section 1.5.5, chapter 1). It is

$$TR = \frac{F_T}{F} = \frac{\{1 + (2\zeta\omega/\omega_n)^2\}^{1/2}}{[\{1 - (\omega/\omega_n)^2\}^2 + \{2\zeta\omega/\omega_n\}^2]^{1/2}}. \qquad (4.149)$$

The transmissibility expresses the fraction of the exciting force or displacement which is transmitted through the isolating system and equation (4.149) is a generalised equation for a vibrational system which can be modelled as a single-degree-of-freedom system. It is presented in Figure 4.27 for various damping ratios, ζ. All the curves

Fig. 4.26. Single-degree-of-freedom forced excitation model for vibration isolation.

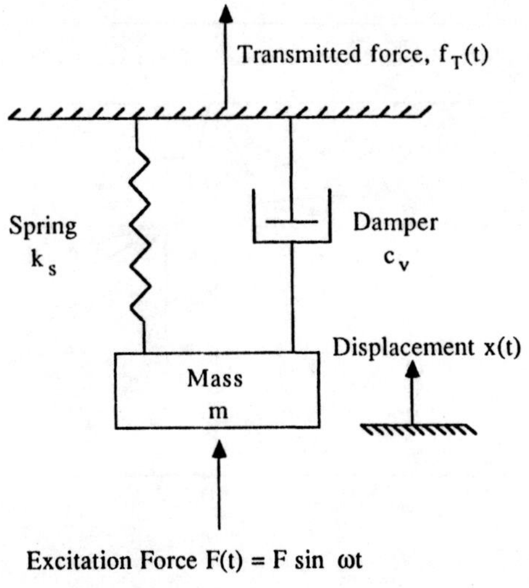

cross the $TR = 1.0$ line at $\omega/\omega_n = \sqrt{2}$. Hence the transmitted force is greater than the driving force at frequencies below this frequency ratio and less than the driving force at frequencies above it. This is a very important point, one which is fundamental to vibration isolation. Thus, vibration isolation by mounting a machine on a spring–damper system is only possible for $\omega/\omega_n > \sqrt{2}$. The results also illustrate that for $\omega/\omega_n > \sqrt{2}$, damping actually reduces the efficiency of the vibration isolator! Some damping is, however, required to allow machines to pass through their mounted resonance region during start up. If no damping was present, severe damage could result due to excessive vibrations when $\omega \approx \omega_n$. It is also useful to note that the problem of isolating a mass (e.g. a piece of electronic equipment) from a base motion is identical to that of isolating the disturbing force of a vibrating machine from being transmitted to other structural components.

When the speed of rotation of a machine is not constant but is a variable, the excitation force, F, varies as a function of ω^2 (i.e. $F = me\omega^2$). It is thus desirable to look at the amplitude of the transmitted force relative to some constant force because even though the transmissibility, TR, may be small, the amplitude of the transmitted force ($F_T = me\omega^2 TR$) may be large at the higher frequencies. It is convenient to replace F_T/F by

$$\frac{F_T}{F_n} = \frac{F_T}{F}\frac{F}{F_n} = TR\frac{me\omega^2}{me\omega_n^2} = \frac{TR\omega^2}{\omega_n^2}. \tag{4.150}$$

The amplitude of the transmitted force is now non-dimensionalised relative to some constant force, $F_n = me\omega_n^2$ (note: e is the eccentricity of any rotating mass which is producing the vibration). Equation (4.150) is presented in Figure 4.28 for various values of ζ. Depending on the frequency of operation, the magnitude of the transmitted force can be high in spite of the low transmissibility. Also, once again for $\omega/\omega_n > \sqrt{2}$, increasing the damping decreases the isolation achieved.

Fig. 4.27. Transmissibility versus frequency ratio for a single-degree-of-freedom system.

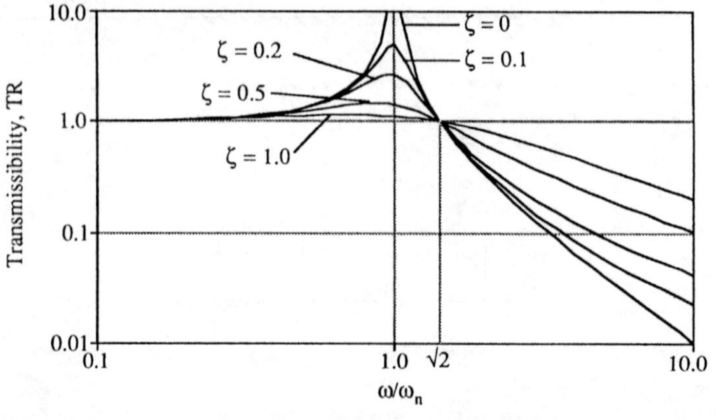

Because very little damping is required for vibration isolation of machines, design equations are often presented in the literature without the damping term, ζ. In addition, ω_n^2 can be replaced by g/δ_{static} (equation 1.15). Thus, equation (4.149) reduces to

$$TR = \frac{1}{\dfrac{(2\pi f)^2\delta_{\text{static}}}{g} - 1}, \tag{4.151}$$

and the disturbing frequency, f, can be obtained by re-arranging terms such that

$$f = \frac{1}{2\pi}\left\{\frac{g}{\delta_{\text{static}}}\left(\frac{1}{TR} + 1\right)\right\}^{1/2}. \tag{4.152}$$

The isolation efficiency of flexibly mounted systems is generally obtained by using nomograms based upon equation (4.152), where the disturbing frequency is plotted against static deflection for a range of different transmissibilities. It is very important to recognise that the equation (and the preceding theory presented in this sub-section) is restricted to bodies with translation along a single co-ordinate. In general, a rigid body has six degrees of freedom; translation along the three perpendicular co-ordinate axes, and rotation about them. If the body vibrates in more than one direction, then the natural frequencies associated with each of the six degrees of freedom should be examined. For instance, in rotation the transmissibility is the ratio of the transmitted torque to the disturbing torque.

4.13.2 Low frequency vibration isolation – multiple-degree-of-freedom systems

When a rigid body is free to move in more than one direction it immediately becomes a multi-degree-of-freedom system. The three translational and three rotational degrees

Fig. 4.28. Force ratio versus frequency ratio for a single-degree-of-freedom system.

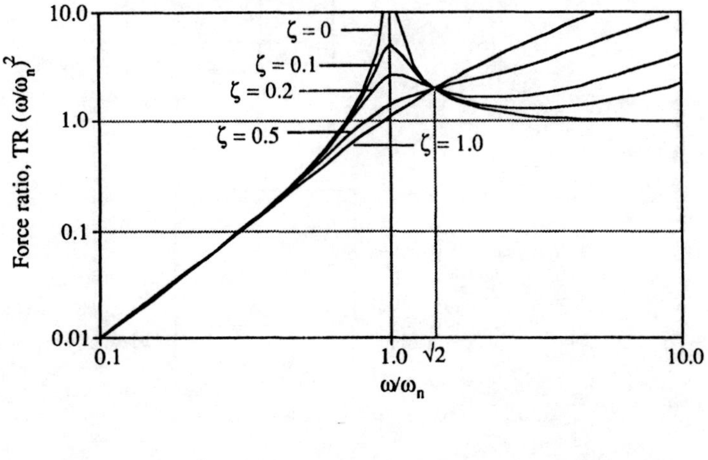

of freedom result in six possible natural frequencies (instead of the single natural frequency for a single-degree-of-freedom system). For a mass supported on four springs of equal stiffness, the six natural frequencies are: (i) a vertical translational mode, (ii) a rotational mode about the vertical axis, and (iii) four rocking modes – i.e. two in each plane. The six possible natural frequencies are illustrated schematically in Figure 4.29.

The vertical translational natural frequency, f_z, can be readily obtained from equation (1.15) in chapter 1, which is for a single-degree-of-freedom system, by replacing the mass m by the total seismic mass of the rigid body, and k_s by the total dynamic stiffness in the vertical direction. The natural frequency of the rotational mode about the vertical axis can be obtained from a knowledge of the angular stiffness

Fig. 4.29. Schematic illustration of the six natural frequencies of a rigid body multi-degree-of-freedom system.

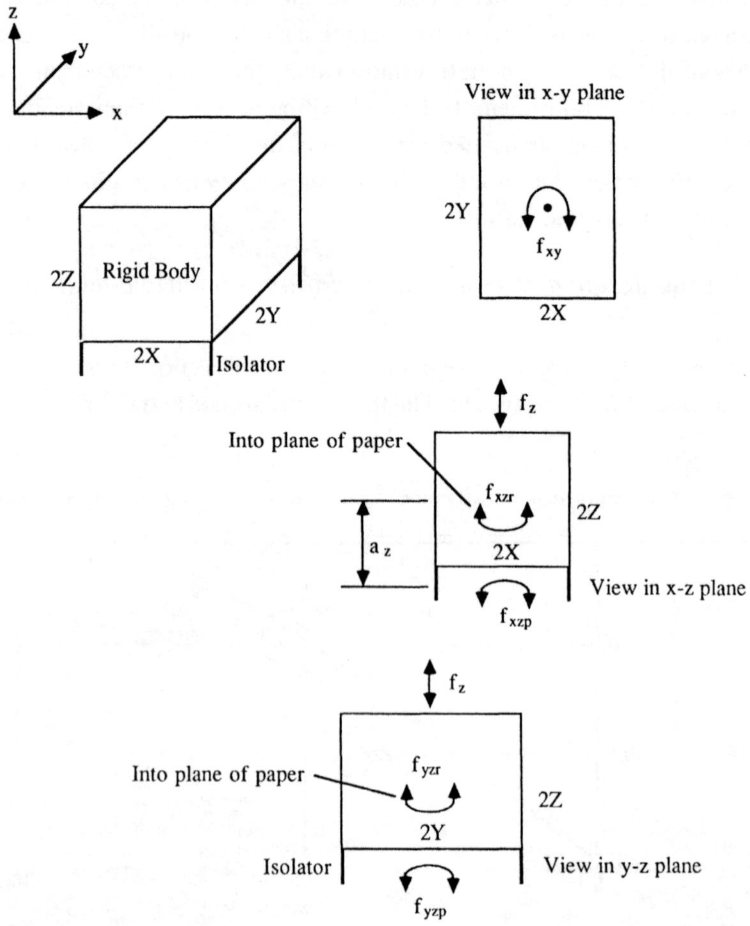

about the vertical axis, and the mass moment of inertia of the seismic mass about the vertical axis. For a rectangular shaped body as illustrated in Figure 4.29, the natural frequency of the rotational mode in the x-y plane (about the vertical z axis), f_{xy}, is

$$f_{xy} = \frac{1}{2\pi}\left\{\frac{nk_{xy}(X^2 + Y^2)}{I_z}\right\}^{1/2}, \tag{4.153}$$

where n is the number of isolators, k_{xy} is the horizontal stiffness of an individual isolator in the x-y plane, I_z is the mass moment of inertia (kg m^2) of the seismic mass about the vertical axis ($I_z = Mr_z^2$, where r_z is the radius of gyration of the mass about the vertical axis, and M is the seismic mass), and the dimensions of the body are $2X \times 2Y \times 2Z$. For a rectangular section, $2X \times 2Y$, in the x-y plane, the radius of gyration, r_z, about an axis which is perpendicular to the plane of the section and located at the centre of the plane is

$$r_z = \left\{\frac{(X^2 + Y^2)}{3}\right\}^{1/2}. \tag{4.154}$$

The horizontal stiffness characteristics of different isolator materials are generally provided by the manufacturers as some percentage of the vertical stiffness characteristics of the material.

The four rocking modes of a rigid body, as illustrated in Figure 4.29, can be obtained from a nomogram originally developed by Harris and Crede [4.2], and discussed in some detail by Macinante[4.22]. The procedure involves the usage of various non-dimensional ratios. They are (i) the ratios of the various rocking mode natural frequencies, f_{xzp}, f_{xzr}, f_{yzp}, f_{yzr}, to the decoupled vertical translational natural frequency, f_z, of the body, (ii) the ratio of the radius of gyration, r_z, of the mass about the vertical axis to the half-distance, X, between isolators in the x-direction, (iii) the ratio of the height of the centre of gravity of the seismic mass above the horizontal elastic plane of the isolators, a_z, to the radius of gyration, r_z, and (iv) the ratio of the horizontal to vertical stiffness of an individual isolator, k_{xy}/k_z. The two rocking modes in each vertical plane (pitch and roll) can be obtained from the nomogram, presented in Figure 4.30, by first evaluating, r_z, X, Y, f_z, a_z, k_{xy}, and k_z. Macinante[4.22] has recently extended Crede's earlier work to allow for the estimation of the six natural frequencies of an unsymmetrically mounted rigid body by replacing the half-distances X and Y by the root-mean-square values of the x and y co-ordinates of the isolator positions.

It is good engineering practice to use identical vibration isolators on each of the four corners of a machine as far as possible and to locate them symmetrically in relation to the centre of gravity of the machine. This procedure minimises the possibility of the coupled rocking modes being excited. It is also good practice to attempt to ensure that the six natural frequencies are no more than about 40% of the excitation frequency. Macinante[4.22] provides an extensive coverage of the practical requirements of good vibration isolation systems.

4.13.3 Vibration isolation in the audio-frequency range

The single-degree-of-freedom and the multiple-degree-of-freedom rigid body models described in the previous two sub-sections are adequate for vibration isolation calculations for predicting transmissibility at low (infrasonic) frequencies. At higher frequencies (generally in the audio-frequency range), practical experience has conclusively demonstrated that the models are inadequate and that they can considerably underestimate transmissibility. There are three possible reasons for this. Firstly, the foundations upon which the seismic mass is mounted are not always perfectly rigid (as assumed in the model); if the mass is mounted on isolators on a suspended floor, the deflection of the floor plays a significant role in the dynamic characteristics of the overall system. Sometimes, provided that the excitation frequency is relatively low, the system can be modelled as a double mass system, but generally many natural frequencies of the foundation are present and there is increased transmissibility over a wide frequency range. Secondly, in practice, the vibration isolators have a finite mass and this allows for natural frequencies to be sustained within the isolator itself. Generally, as a rule of thumb, if the thickness of the isolator is greater than $\lambda/2$ the isolator can sustain standing waves, and the problem becomes one of transmission loss through the isolator material. Finally, machines are generally distributed systems rather than rigid masses and they also possess many natural frequencies. Once again,

Fig. 4.30. Nomogram for evaluating the two rocking modes in each vertical plane (for modes in *y–z* plane replace *X* by *Y*).

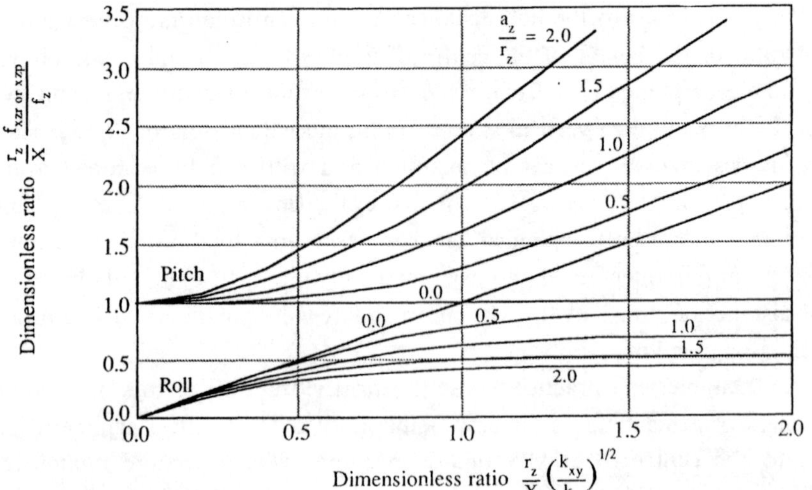

this results in increased transmissibility at high frequencies. The characteristic behaviour of a flexibly mounted system in the audio frequency range is illustrated schematically in Figure 4.31. As a general rule of thumb, the single-degree-of-freedom and the multiple-degree-of-freedom rigid body models considerably underestimate transmissibility for excitation frequencies $\omega > 10\omega_n$.

The effectiveness of an isolator system can be quantified by a simple theoretical model based on an analysis of either the impedances (force/velocity) or the mobilities (velocity/force) of the seismic mass itself, the isolators, and the foundation. Consider the free-body diagram in Figure 4.32(a). The velocity of the seismic mass at the attachment point, $\mathbf{v}_m$, is the sum of the velocity of the mass by itself due to its own internal forces, $\mathbf{v}$, and the additional velocity due to the reaction force at the foundation. Thus,

$$\mathbf{v}_m = \mathbf{v} + \mathbf{Y}_m\mathbf{F}_m, \tag{4.155}$$

where $\mathbf{Y}_m$ is the mobility of the mass, and $\mathbf{F}_m$ is the reaction force on the mass due to the foundation. Now, at the attachment point, $\mathbf{v}_m = \mathbf{v}_f$ where $\mathbf{v}_f$ is the velocity of the foundation, and $\mathbf{F}_m = -\mathbf{F}_f$, $\mathbf{F}_f$ being the reaction force on the foundation. Hence,

$$\mathbf{v}_m = \mathbf{v}_f = \mathbf{Y}_f\mathbf{F}_f = \mathbf{v} - \mathbf{Y}_m\mathbf{F}_f, \tag{4.156}$$

and the force on the foundation (without any isolator between the mass and the foundation) is

$$\mathbf{F}_f = \frac{\mathbf{v}}{\mathbf{Y}_m + \mathbf{Y}_f}. \tag{4.157}$$

Fig. 4.31. Characteristic behaviour of a flexibly mounted system in the audio-frequency range.

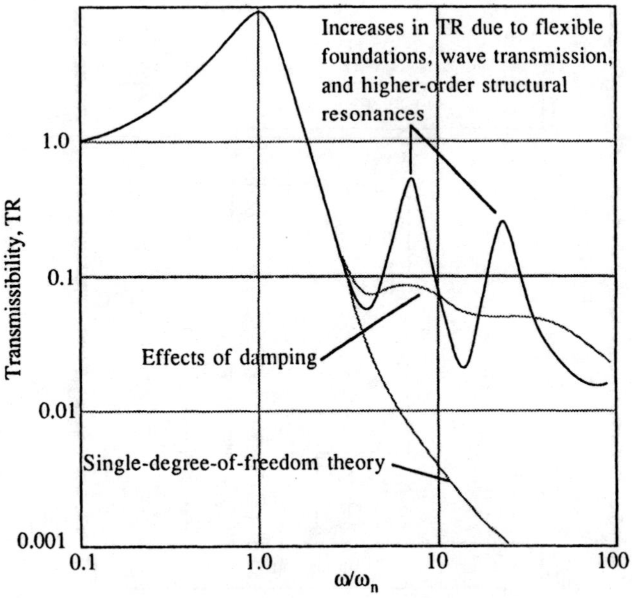

Now consider the free-body diagram in Figure 4.32(b) where an isolator is mounted between the mass and the foundation. As a first approximation, it is convenient to neglect the mass of the isolator so that all the force is transmitted through the isolator. The reaction forces between the mass and the isolator, and between the foundation and the isolator, have to balance. Also, the velocities at the contact point between the mass and the isolator and the foundation and the isolator have to be equal. Thus, $\mathbf{v}_m = \mathbf{v}_{im}$, and $\mathbf{v}_f = \mathbf{v}_{if}$, and $\mathbf{F}_m = -\mathbf{F}_{im} = +\mathbf{F}_{if} = -\mathbf{F}_f$. If the force into the isolator, $\mathbf{F}_i$ is defined as $\mathbf{F}_i = \mathbf{F}_{im} = -\mathbf{F}_{if} = \mathbf{F}_f$, and the relative velocity across the isolator $\mathbf{v}_i$ is defined as $\mathbf{v}_i = \mathbf{v}_{im} - \mathbf{v}_{if}$, then $\mathbf{v}_i = \mathbf{v}_m - \mathbf{v}_f$, and

$$\mathbf{v}_i = \mathbf{Y}_i\mathbf{F}_i = \mathbf{Y}_i\mathbf{F}_f = \mathbf{v} - \mathbf{Y}_m\mathbf{F}_f - \mathbf{Y}_f\mathbf{F}_f. \tag{4.158}$$

Thus, the force on the foundation (with an isolator between the mass and the foundation) is

$$\mathbf{F}_f = \frac{\mathbf{v}}{\mathbf{Y}_i + \mathbf{Y}_m + \mathbf{Y}_f}. \tag{4.159}$$

The transmissibility, TR, is thus obtained by dividing equation (4.157) by equation (4.159) and taking the modulus of the complex quantity. Hence,

$$TR = \left|\frac{\mathbf{Y}_m + \mathbf{Y}_f}{\mathbf{Y}_i + \mathbf{Y}_m + \mathbf{Y}_f}\right|. \tag{4.160}$$

Fig. 4.32. Free-body diagram for interactions between a mass, an isolator and the foundation.

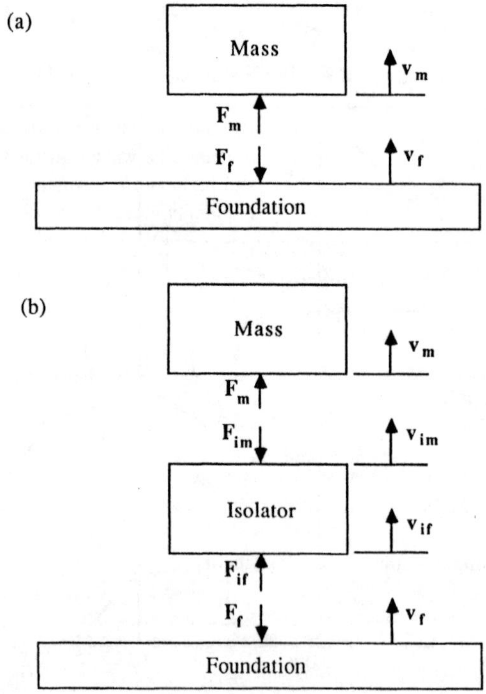

Equation (4.160) illustrates that, for effective vibration isolation, the mobility of the isolator must be larger than the combined mobilities of the mass and the foundation – i.e. softer or flexible isolation is more efficient. If the foundation is not rigid, then its mobility will be high and the transmissibility will be increased. Techniques for measuring the mobilities of structures are discussed in chapter 6.

For a simple single-degree-of-freedom model it is relatively straightforward to show that (i) the mobility of a mass element by itself is $1/i\omega m$; (ii) that the mobility of a damper by itself is $1/c_v$; and (iii) that the mobility of a spring element by itself is $i\omega/k$. Thus, increasing the stiffness, damping or mass of an isolator reduces its mobility and its effectiveness; increasing the stiffness, damping or mass of a structure and the foundation reduces their mobility and subsequently increases the isolation potential of the isolator.

4.13.4 Vibration isolation materials

The most commonly used vibration isolators include felt compression pads, cork compression pads, fibrous glass compression pads, rubber compression or shear pads, metal springs, elastomeric-type mounts, air springs, and inertia blocks. Each type has is own advantage depending upon the degree of isolation required, the weight of the mass to be isolated, the temperature range which it has to function in, and most importantly the dominant excitation frequencies. Macinante[4.22] provides an excellent discussion on the different types of seismic mounts that are used in practice.

Felt pads are generally used for frequencies above 40 Hz, and provide good isolation in the low audio-frequency range. They are only effective in compression and are not generally used where torsional modes (shear) are present. As a rule of thumb, the deflections should not exceed $\sim$25% of the thickness of the felt because the stiffness characteristics increase rapidly if the material is compressed any further. Because of their organic content, they tend to deteriorate when exposed to oils and solvents and should therefore be used with care in industrial situations.

Cork pads can be used both in compression and in shear, and like felt are used for frequencies above 40 Hz. Cork is resistive to corrosion, solvents and moderately high temperatures. It does, however, compress with age. Its stiffness decreases with increasing loads (i.e there is a maximum allowable safe load beyond which it is overstressed), and its dynamic properties are frequency dependent. Most manufacturers provide recommended loads for given deflections.

Fibrous glass pads have vibration isolation characteristics that are similar to felt pads, their main advantage being that the fibrous glass material is inert and very resistive to oils, solvents, etc. The deflection versus static load curve is linear up to about 25% compression and good isolation is not achievable below about 40 Hz.

Rubber is a common material in vibration isolation applications, rubber compression or shear pads and composite elastomeric-type mounts being commonly used. Rubber is useful in both shear and compression, and different types of rubber are used for a variety of applications. They include butyl, silicone, neoprene, and natural rubber. Numerous factors such as thickness, hardness and shape affect the stiffness associated with rubber. Also, the dynamic stiffness of rubber is about 75% of the static stiffness. The damping characteristics of rubber are temperature and frequency dependent. Manufacturers usually provide information about the stiffness and damping characteristics of their rubber products together with recommended loads per unit area. Rubber pads and elastomeric mounts are generally used in the 5 Hz to 50 Hz frequency range.

Metal springs are widely used and are ideal for low frequency vibration isolation (>1.5 Hz) since they can sustain large static deflections (i.e. large loads and low forcing frequencies). They are highly resistant to environmental factors such as solvents, oils, temperature, etc. Their main disadvantage is that they readily transmit high frequency vibrations and possess very little damping. In practice, this problem is overcome by inserting rubber or felt pads between the ends of the springs. A variety of spring mounts and spring types are available; torsional springs, beam springs, leaf springs, etc. When specifying a coil spring mounting system, one has to be careful to ensure that the system is laterally stable. Manufacturers tend to supply information about lateral stability requirements and provide suitable mounting arrangements.

Air springs (air bags) are very useful for vibration isolation at very low frequencies (0.07 Hz to ~5 Hz). Isolation against very low excitation frequencies requires large static deflections (equations 1.15 and 4.152). For instance, a static deflection of 1.5 m with a corresponding mounted natural frequency of 0.4 Hz is required to provide 80% vibration isolation at an excitation frequency of 1 Hz! Obviously, such static deflections are quite unrealistic and unachievable with conventional springs or pads. Air springs enable a mounted system to have a very low natural frequency with very small static deflections. Air springs are generally manufactured out of high-strength rubber air containers, sealed by retainers at each end. They have been successfully used to solve a wide range of low frequency vibration isolation problems, including vibrating shaker screens, presses, textile looms, seat suspensions, jet engine test platforms on aircraft carriers, and rockets in storage, ground handling and transit. Care has to be exercised in relation to the lateral stability of air springs; manufacturers usually provide advice on suitable types of lateral restraining methods (e.g. snubbers, rubber bumper pads, bearing mounts, strap stabilisers, sway cables, etc.).

Inertia blocks involve adding substantial mass to a system in the form of a solid inertia base. They reduce the mounted natural frequency of the system, bring down the centre of gravity, reduce any unwanted rocking motions, and minimise alignment errors because of their inherent stiffness. They are generally 1.5–2 times the mass of the system which is to be isolated (for lightweight machinery it is not uncommon for the inertia base to be up to ten times the original mass). Commercially available inertia blocks come in different shapes and sizes, depending on the specific requirements.

Often, they simply comprise large concrete or steel blocks attached to the vibrating mass. In this instance, they are either mounted on some isolator material, or independently mounted (i.e. directly on the base foundation of a structure). Heavy I-beam type, spring-mounted, rectangular steel frames are also commonly used. If it is desirable to retain the original mounted natural frequency of a system, then the stiffness of the isolators has to be increased in proportion to the mass of the inertia block; this is desirable if the originally required stiffness was unrealistically low. Inertia blocks are ideal for machines with large unbalanced moving parts (e.g. a centrifuge in a salt wash plant).

4.13.5 Dynamic absorption

A dynamic absorber is an alternative form of vibration control. It involves attaching a secondary mass to the primary vibrating component via a spring which can be either damped or undamped. This secondary mass oscillates out of phase with the main mass and applies an inertia force (via the spring) which opposes the main mass – i.e. the natural frequency of the vibration absorber is tuned to the frequency of the excitation force. If it is damped, the secondary system also absorbs the vibrational energy associated with the resonance of the primary mass, and it is therefore essentially a damper which can be used over a narrow frequency range. Den Hartog[4.23] is the accepted classical reference for a dynamic absorption.

It is convenient to analyse the behaviour of a dynamic absorber by considering the dynamics of an undamped two-degree-of-freedom system and qualitatively adding the damping at a later stage; the analysis becomes more complicated with the presence of damping. Now, consider the two-degree-of-freedom system in Figure 4.33. The

Fig. 4.33. Simplified model of a dynamic absorber.

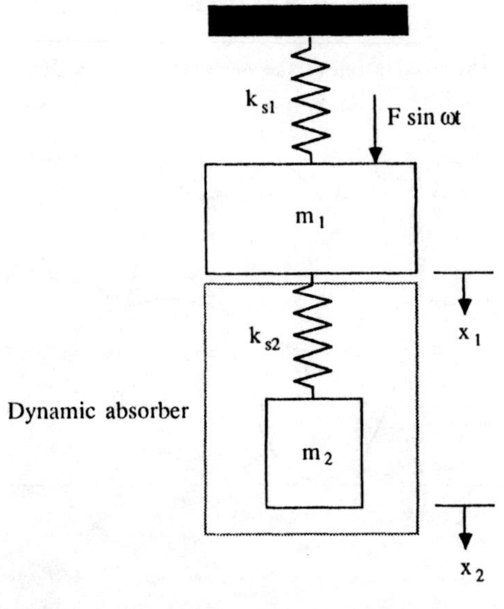

equations of motion for the two masses are

$$m_1\ddot{x}_1 + k_{s1}x_1 + k_{s2}(x_1 - x_2) = F \sin \omega t, \tag{4.161}$$

and

$$m_2\ddot{x}_2 + k_{s2}(x_2 - x_1) = 0. \tag{4.162}$$

By assuming a harmonic solution (see chapter 1), the amplitude X_1 of the primary mass can be obtained. It is

$$X_1 = \frac{(k_{s2} - m_2\omega^2)F}{(k_{s1} + k_{s2} - m_1\omega^2)(k_{s2} - m_2\omega^2) - k_{s2}^2}. \tag{4.163}$$

The main concern is to reduce the vibration of the primary mass, thus in practice it is desirable that X_1 be as small as possible. From equation (4.163), X_1 is zero if $k_{s2} = m_2\omega^2$. This suggests that, if the natural frequency of the absorber is tuned to the excitation frequency, the primary mass will not vibrate! In fact, if the primary mass were being excited near or at its own natural frequency prior to the addition of the dynamic absorber, then if the absorber was chosen such that $\omega^2 = k_{2s}/m_2 = k_{1s}/m_1$, the primary mass would not vibrate at its own resonance! In practice, the primary mass will have some finite vibration level because of the presence of damping. The performance of a dynamic absorber is illustrated schematically in Figure 4.34. The two peaks correspond to the two natural frequencies of the composite system. The addition of damping reduces the resonant peaks and increases the trough. Whilst some damping is desirable, it should be recognised that it limits the effectiveness of the absorber – i.e. only the minimal amount of damping that is required should be used. Dynamic absorbers are generally only used with constant speed machinery because they are limited to narrowband or single frequency excitation forces. It is important to ensure that the operating frequency is sufficiently far away from the

Fig. 4.34. Schematic illustration of the performance of a dynamic absorber.

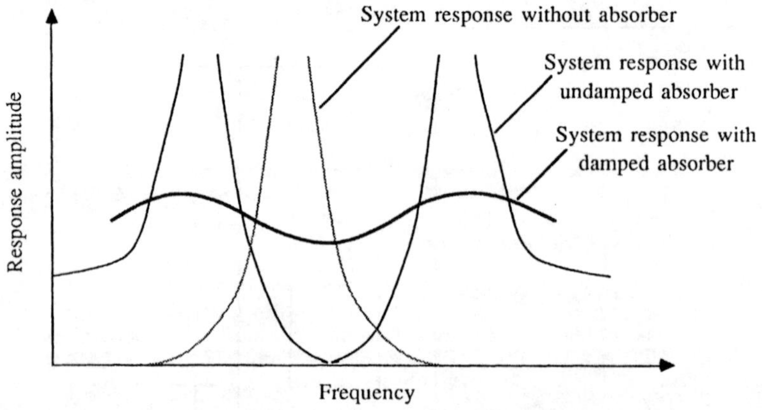

double mass resonances otherwise response amplification might occur; if this is a problem (e.g. due to drift in the operating frequency) then damping is very important.

4.13.6 Damping materials

The trend in most modern constructions is to use welded joints as far as possible rather than bolts and rivets. Such joints tend to have significantly less damping than bolts and rivets which often generate additional damping due to gas pumping at the interfaces. Hence modern composite structures made up of metal plates, panels, shells and cylinders are generally very lightly damped. Because of this, they are prone to the efficient radiation of mechanically induced sound. It is also a fact of life that high strength materials such as steel, aluminium, etc. possess very little damping, whereas low strength materials such as soft plastics, rubber, etc. possess high damping.

A range of commercially available damping products of a viscoelastic polymer nature are now readily available, and in recent times it has become common to apply such damping materials to built-up structures in a variety of ways in order to increase their damping characteristics. The two most common ways of applying damping materials to structures are via (i) free or non-constrained damping layers, and (ii) constrained damping layers. Both types of layers are schematically illustrated in Figure 4.35.

Non-constrained viscoelastic damping layers are applied to the surfaces of structures via an adhesive or spray. As a rule of thumb they are generally about three times the thickness of the structure and the mass of the damping layer has to be greater than ~20% of the structural mass for it to be effective. The viscoelastic material absorbs energy by longitudinal contractions and expansions as the structure vibrates. Hence it is best applied near vibrational antinodes (e.g. the centre of a panel) rather than near stiffeners, etc. Non-constrained layer damping increases with the square of the mass of the damping layer.

Fig. 4.35. Schematic illustration of non-constrained and constrained damping layers.

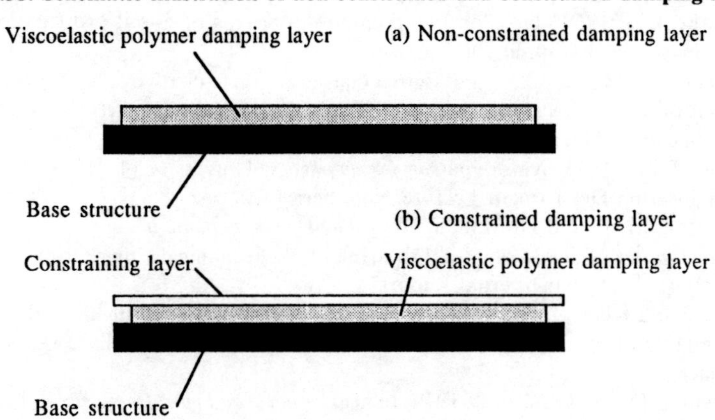

Constrained damping layers are often used when significant weight increases are unacceptable – e.g. the motor car and aerospace industries. They are also used on ships, saw blades, railroad wheels, professional cameras, drill rods, and valve and rocker arm covers. The technique utilises the energy dissipating properties of viscoelastic polymers which are constrained between the vibrating structure and an extensionally stiff constraining layer such as a thin metallic foil. As the constrained layer vibrates, shearing forces generated by the differential strain cause the energy to be dissipated. This shear–strain energy dissipation is in addition to the longitudinal contractions and expansions associated with the non-constrained layers. Thus, the loss factors, η, associated with constrained damping layers are generally larger than the associated non-constrained damping layers. As a rule of thumb, the mass of a constained damping layer required to provide the same amount of damping as a non-constrained layer is $\sim 10\%$ or less of the structural mass. Constrained layer damping increases linearly with the mass of the damping layer. The interest reader is referred to chapter 20 in Harris[4.1], chapter 14 in Beranek[4.3], and to Nashif *et al.*[4.24] for more detailed information on damping materials. Nashif *et al.*[4.24] in particular provide a comprehensive coverage of the characterisation of damping in structures and materials, the behaviour and typical properties of damping materials, discrete damping devices, and surface damping treatments together with design data sheets and numerous references.

References

4.1 Harris, C. M. 1979. *Handbook of noise control*, McGraw-Hill (2nd edition).

4.2 Harris, C. M. and Crede, C. E. 1976. *Shock and vibration handbook*, McGraw-Hill (2nd edition).

4.3 Beranek, L. L. 1971. *Noise and vibration control*, McGraw-Hill.

4.4 Rice, C. J. and Walker, J. G. 1982. 'Subjective acoustics', chapter 28 in *Noise and vibration*, edited by R. G. White and J. G. Walker, Ellis Horwood.

4.5 Rathe, E. J. 1969. 'Note on two common problems of sound propagation', *Journal of Sound and Vibration* **10**(3), 472–9.

4.6 Pickles, J. M. 1973. 'Sound source characteristics', chapter 2 in *Noise control and acoustic design specificiations*, edited by M. K. Bull, Department of Mechanical Engineering, University of Adelaide.

4.7 Bies, D. A. 1982. *Noise control for engineers*, University of Adelaide, Mechanical Engineering Department Lecture Note Series.

4.8 Norton, M. P. and Drew, S. J. 1987. *The effects of bounding surfaces on the radiated sound power of sound sources*, Department of Mechanical Engineering, University of Western Australia, Internal Report.

4.9 Brüel and Kjaer. 1985. *Acoustic intensity*, Papers presented at the 2nd International Congress on Acoustic Intensity (sponsored by CETIM), Senlis, France, Brüel and Kjaer.

4.10 Irwin, J. D. and Graf, E. R. 1979. *Industrial noise and vibration control*, Prentice-Hall.

4.11 Bell, L. H. 1982. *Industrial noise control*, Marcel Dekker.
4.12 Hemond, C. J. 1983. *Engineering acoustics and noise control*, Prentice-Hall.
4.13 Gibson, D. C. and Norton, M. P. 1981. 'The economics of industrial noise control in Australia', *Noise Control Engineering* **16**(3), 126–35.
4.14 Brüel and Kjaer. 1982. *Noise control, principles and practice*, Brüel and Kjaer.
4.15 Kinsler, L. E., Frey, A. R., Coppens, A. B. and Sanders V. J. 1982. *Fundamentals of acoustics*, John Wiley & Sons (3rd edition).
4.16 Crocker, M. J. and Kessler, F. M. 1982. *Noise and noise control, Vol II*, CRC Press.
4.17 Ver, I. L. 1973. *Reduction of noise by acoustic enclosures*, Proceedings ASME Design Engineering Conference on Isolation of Mechanical Vibration, Impact and Noise, Cincinnati, Ohio, pp. 192–220.
4.18 Moreland, J. and Musa, R. 1972. *Performance of acoustic barriers*, Proceedings Inter-Noise '72, Washington D.C., U.S.A., pp. 95–104.
4.19 Moreland, J. and Minto, R. 1976. 'An example of in-plant noise reduction with an acoustical barrier', *Applied Acoustics* **9**, 205–14.
4.20 Bies, D. A. 1971. 'Acoustical properties of porous materials', chapter 10 in *Noise and vibration control*, edited by L. L. Beranek, McGraw-Hill.
4.21 Maling, G. C. 1986. *Progress in the application of sound intensity techniques to noise control engineering*, Proceedings Inter-Noise '86, Cambridge, U.S.A., pp. 41–74.
4.22 Macinante, J. A. 1984. *Seismic mountings for vibration isolation*, John Wiley & Sons.
4.23 Den Hartog, J. D. 1956. *Mechanical vibrations*, McGraw-Hill (4th edition).
4.24 Nashif, A. D., Jones, D. I. G. and Henderson, J. P. 1985. *Vibration damping*, John Wiley & Sons.

Nomenclature

a	vibration acceleration, distance
a_0	reference vibration acceleration
a_Z	distance of the centre of gravity of a seismic mass from the horizontal elastic plane of the isolators
b	distance
c	speed of sound
c_v	viscous-damping coefficient
d	vibration displacement, distance, reference radius, perforation diameter
d_0	reference vibration displacement
D	mean sound energy density, difference in sound pressure level between two positions, distance from a barrier to a receiver, impedance tube diameter
D_1	mean sound energy density in a source room (room 1)
D_{1w}	sound energy density at wall in a source room (room 1)

D_2	mean sound energy density in a receiving room (room 2)
D_{2w}	sound energy density at wall in a receiving room (room 2)
D_R	sound energy density inside a reverberant enclosure
D_w	sound energy density at the inside of an enclosure wall
DI	directivity index
DI_θ	directivity index at an angle θ
e	eccentricity of a rotating mass
f, f_1, f_2, etc.	frequencies
f_0	centre frequency of a band
f_1	lower frequency limit of a band
f_{max}	maximum frequency for impedance tube testing
f_{res}	resonant frequency of a Helmholtz resonator
f_u	upper frequency limit of a band
f_{xy}	natural frequency of the rotational mode in the x-y plane
$f_{xzp}, f_{xzr}, f_{yzp}, f_{yzr}$	rocking mode natural frequencies
f_z	decoupled vertical translational natural frequency
$f_T(t)$	transmitted force
F	excitation force
$\mathbf{F}_f$	complex reaction force on a foundation
$\mathbf{F}_i$	complex force into an isolator
$\mathbf{F}_{if}$	complex reaction force into an isolator at the contact point with a foundation
$\mathbf{F}_{im}$	complex reaction force into an isolator at the contact point with a seismic mass
$\mathbf{F}_m$	complex reaction force on a seismic mass due to a foundation
F_n	constant force (see equation 4.150)
F_T	transmitted force amplitude
g	gravitational acceleration
$\mathbf{G}_{12}(f)$	one-sided cross-spectral density function of functions $\mathbf{P}_1(f)$ and $\mathbf{P}_2(f)$ (complex function)

H	barrier height
i	integer
$I, I_1, I_2, \vec{I}$	mean sound intensities (arrow denotes vector quantity)
IL	insertion loss
I_i	incident sound intensity
I_t	transmitted sound intensity
I_0	reference sound intensity
I_{1w}, I_{2w}	sound intensity at walls in a source room (1) and a receiver room (2)
I_{OE}	sound intensity immediately outside an enclosure surface
I_S	sound intensity of a uniformly radiating sound source
I_{total}	total sound intensity
I_w	sound intensity on the inside of an enclosure wall
I_x	sound intensity in the x-direction
I_z	mass moment of inertia of a seismic mass about the vertical axis
I_θ	sound intensity at an angle θ
$I(r)$	mean sound intensity as a function of radial distance
k_1, k_2	dimensionless numbers relating to the room absorption on the source side and the receiver side of a barrier
k_s, k_{s1}, k_{s2}, etc.	spring stiffnesses
k_{xy}	horizontal stiffness of an individual isolator in the x-y plane
k_z	vertical stiffness of an individual isolator
l	effective length of the neck of a Helmholtz resonator, depth of air gap behind a panel absorber
L	actual length of the neck of a Helmholtz resonator
L_a	vibration acceleration level
L_{Aeq}	equivalent continuous A-weighted sound level
L_{b2}	diffracted sound pressure level at a receiver with a barrier in place

L_d	vibration displacement level
L_{eq}	equivalent continuous sound pressure level
L_I	sound intensity level
L_p, L_{p1}, L_{p2}, etc.	sound pressure levels
L_{p0}	sound pressure level at a receiver location prior to the insertion of a barrier
$L_{p1}(f)$	sound pressure spectrum level at a frequency f
L_{p2}	sound pressure level at a receiver location with a barrier in place
L_{pB}	background sound pressure level
$L_{p\,band}$	sound pressure level in a frequency band
L_{pd}	equivalent sound pressure level at a reference radius d
L_{pi}	sound pressure level of ith component
L_{pOE}	sound pressure level immediately outside an enclosure surface
L_{pr}	sound pressure level of a calibrated reference source
L_{pS}	sound pressure level due to an arbitrary source, sound pressure level due to a uniformly radiating source
L_{pT}	total sound pressure level
$L_{p\theta}$	sound pressure level at an angle θ
L_{r2}	reverberant sound pressure level in a room with a barrier in place
L_v	vibration velocity level
L'_{p2}	sound pressure level at some far-field position without any enclosure over the source (see equation 4.111)
$\bar{L}_p$	average sound pressure level
L_Π	sound power level
$L_{\Pi o}$	sound power level of a sound source in free space
$L_{\Pi E}$	sound power radiated by an enclosure
$L_{\Pi l}$	sound power level per unit length
$L_{\Pi r}$	sound power level of a calibrated reference source
m	energy attenuation constant, mass
m_1, m_2, etc.	masses

n	integer, number of vibration isolators
N	integer
N_i	Fresnel number for diffraction around the ith edge
NR	noise reduction
p, p_1, p_2, etc.	sound pressures
$\mathbf{p}$	complex sound pressure
p_0^2	mean-square sound pressure at a receiver location prior to the insertion of a barrier
p_2^2	mean-square sound pressure at a receiver location with a barrier in place
p_{amb}	barometric pressure
p_{band}	sound pressure in a frequency band
p_{b2}^2	mean-square sound pressure at a receiver location due to the diffracted field around a barrier
p_{d0}^2	mean-square sound pressure at a receiver location due to the direct field, prior to the insertion of a barrier
$\mathbf{p}_{\mathrm{I}}$	complex incident sound pressure
$\mathbf{p}_{\mathrm{R}}$	complex reflected sound pressure
p_{ref}	reference sound pressure
p_{r2}^2	mean-square sound pressure due to the average reverberant field in a room with a barrier in place
p_{T}	total sound pressure (see equation 4.18)
$\mathbf{p}_{\mathrm{T}}$	complex transmitted sound pressure
$p(r)$	sound pressure as a function of radial distance
$p(\vec{x}, t)$	sound pressure as a function of position $\vec{x}$ and time t
$\langle p^2 \rangle$	time-averaged, mean-square sound pressure
$\langle p_{\mathrm{OE}}^2 \rangle$	time-averaged mean-square sound pressure immediately outside an enclosure surface
$\langle p_{\mathrm{S}}^2 \rangle$	time-averaged, mean-square sound pressure for a uniformly radiating sound source
$\langle p_\theta^2 \rangle$	time-averaged mean-square sound pressure at an angle θ
P	phon
$\mathbf{P}(f)$, $\mathbf{P}_1(f)$, $\mathbf{P}_2(f)$, etc.	Fourier transforms of p, p_1, p_2, etc. (complex functions)

Q	directivity factor
Q_B	effective directivity of a source in the direction of the shadow zone of a barrier
Q_θ	directivity factor at an angle θ
r, r_1, r_2, etc.	radial distances, distances between sources and receivers
$\mathbf{r}$	complex reflection coefficient
r_C	critical distance of the reverberation radius
r_E	distance from source to the measurement point inside an enclosure
r_z	radius of gyration of a body about the vertical axis
R	room constant, distance from a barrier to a source
R_E	room constant of enclosure
s	standing wave ratio
S	sone
S	surface area, open area between a barrier perimeter and the room walls and ceiling, cross-sectional area of the neck of a Helmholtz resonator
S_0	reference radiating surface area, total room surface area, empty room surface area (prior to the insertion of test absorption material)
S_1	room surface area on the source side of a barrier
S_2	room surface area on the receiver side of a barrier
S_E	external radiating surface area of an enclosure
S_M	surface area of a room including test absorption material
S_n	surface area of the nth component
S_w	surface area of a partition between two rooms
t	time, panel thickness
T	time
T_0	reverberation time of an empty room with no test absorption material

T_{60}	reverberation time for a 60 dB decay
T_{M}	reverberation time of a room with test absorption material
TL	transmission loss
TR	transmissibility
$\hat{\mathbf{u}}$	complex acoustic particle velocity (vector quantity)
$\hat{\mathbf{u}}^*$	complex conjugate of $\hat{\mathbf{u}}$
u_x	acoustic particle velocity in the x-direction
$\vec{u}(\vec{x}, t)$	acoustic particle velocity (vector quantity)
$\mathbf{U_x}(f)$	Fourier transform of u_x (complex function)
v	vibration velocity
$\mathbf{v}$	complex velocity of a seismic mass due to its own internal forces
v_0	reference vibration velocity
$\mathbf{v_f}$	complex velocity of a foundation
$\mathbf{v_i}$	complex relative velocity across an isolator
$\mathbf{v_{if}}$	complex velocity of an isolator at the contact point with a foundation
$\mathbf{v_{im}}$	complex velocity of an isolator at the contact point with a seismic mass
$\mathbf{v_m}$	complex velocity of a seismic mass at the attachment point
V	room volume, enclosed air volume in a Helmholtz resonator
x	distance
$\dot{x}$	velocity
$\ddot{x}$	acceleration
X	half-distance between isolators in the x-direction
X_1	displacement amplitude of primary mass (see equation 4.163)
y	distance
Y	half-distance between isolators in the y-direction
$\mathbf{Y_f}$	complex mobility of a foundation
$\mathbf{Y_i}$	complex mobility of an isolator
$\mathbf{Y_m}$	complex mobility of a seismic mass
z	distance
$\mathbf{Z_s}$	normal impedance of a test material (complex function)

α	sound absorption coefficient
α_0	mean room absorption coefficient, random incidence absorption coefficient of a room prior to the insertion of test absorption material
α_{avg}	space-average sound absorption coefficient
$\alpha_{E\,avg}$	average sound absorption coefficient inside an enclosure
α_M	random incidence absorption coefficient of a room with test absorption material
α_n	sound absorption coefficient of the nth component, normal incidence absorption coefficient
α_T	average sound absorption coefficient including air absorption
δ_i	difference between the ith diffracted path and the direct path between a source and a receiver
δ_{static}	static deflection of a spring
Δ	correction factor for near-field sound power measurements
Δf	frequency increment
Δf_0	reference frequency increment (usually 1 Hz)
Δx	microphone separation distance
ζ	damping ratio
θ, θ_1, θ_2, etc.	angles
λ	wavelength
π	3.14 . . .
Π	radiated sound power
Π_0	reference sound power, sound power of a sound source in free space
Π_1	sound power incident upon source side of a partition between rooms 1 and 2
Π_2	sound power incident upon the receiving room side of a partition between rooms 1 and 2
Π_{12}, Π_{21}, etc.	sound power flowing from room 1 to room 2, etc.
Π_a	sound power absorbed within a receiving room

Π_A	absorbed sound power
Π_D	dissipated sound power
Π_E	sound power radiated by an enclosure
Π_I	incident sound power
Π_l	radiated sound power per unit length
Π_R	reflected sound power
Π_{rev}	reverberant sound power
Π_T	transmitted sound power
ρ_0	mean fluid density
ρ_S	mass per unit area (surface mass)
τ	time variable, sound transmission coefficient (wave transmission coefficient)
τ_{avg}	average sound transmission coefficient
τ_n	sound transmission coefficient of the *n*th component
ω	radian (circular) frequency
ω_n	natural radian (circular) frequency
$\langle\ \rangle$	time-average of a signal
$\overline{\quad}$	space-average of a signal (overbar)

5
The analysis of noise and vibration signals

5.1 Introduction

A time history of a noise or vibration signal is just a direct recording of an acoustic pressure fluctuation, a displacement, a velocity, or an acceleration waveform with time – it allows a view of the signal in the time domain. A basic noise or vibration meter would thus provide a single root-mean-square level of the time history measured over a wide frequency band which is defined by the limits of the meter itself. These single root-mean-square levels of the noise or vibration signals generally represent the cumulative total of many single frequency waves since the time histories can be synthesised by adding single frequency (sine) waves together using Fourier analysis procedures. Quite often, it is desirable for the measurement signal to be converted from the time to the frequency domain, so that the various frequency components can be identified, and this involves frequency or spectral analysis. It is therefore important for engineers to have a basic understanding of spectral analysis techniques. The appropriate measurement instrumentation for monitoring noise and vibration signals were discussed in section 4.3 in chapter 4. The subsequent analysis of the output signals, in both the time and frequency domains, forms the basis of this chapter.

Just as any noise or vibration signal that exists in the real world can be generated by adding up sine waves, the converse is also true in that the real world signal can be broken up into sine waves such as to describe its frequency content. Figure 5.1 is an elementary three-dimensional schematic illustration of a signal that comprises two sine waves; the frequency domain allows for an identification of the frequency components of the overall signal and their individual amplitudes, and the time domain allows for an identification of the overall waveform and its peak amplitude.

It is pertinent at this stage to stop and ask the question 'why make a frequency or spectral analysis?' Firstly, individual contributions from components in a machine to the overall machine vibration and noise radiation are generally very difficult to identify in the time domain, especially if there are many frequency components involved. This becomes much easier in the frequency domain, since the frequencies of the major peaks can be readily associated with parameters such as shaft rotational frequencies, gear toothmeshing frequencies, etc. This simple, but important, point is illustrated in Figure 5.2. Secondly, a developing fault in a machine will always show up as an increasing vibration at a frequency associated with the fault. However, the fault might be well developed before it affects either the overall r.m.s. vibration level or the peak level in the time domain. A frequency analysis of the vibration will give a much earlier warning of the fault, since it is selective, and will allow the increasing vibration at the frequency associated with the fault to be identified. This is illustrated in Figure 5.3. The usage of noise and vibration as a diagnostic tool for a range of different applications will be discussed in some detail in chapter 8.

Fig. 5.1. Schematic illustration of time and frequency components.

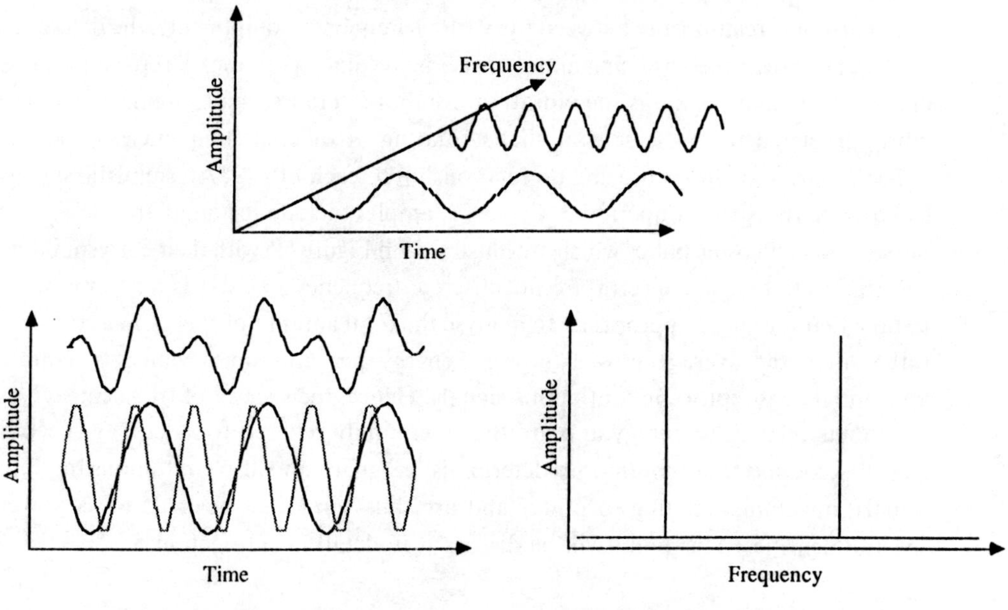

Fig. 5.2. Identification of frequency components associated with meshing gears.

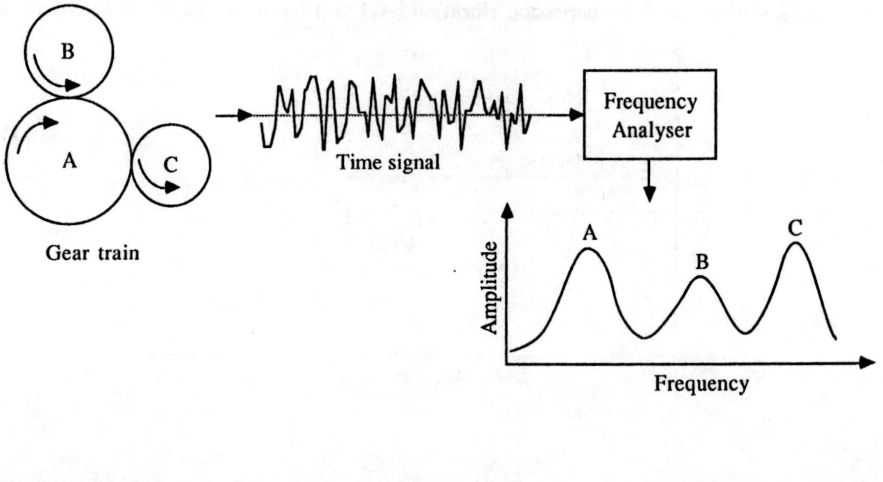

5.2 Deterministic and random signals

Observed time histories of noise and vibration signals can be classified as being either deterministic or random. Deterministic signals can be expressed by explicit mathematical relationships, and random signals must be expressed in terms of probability statements and statistical averages (the concepts of deterministic and random signals were introduced in chapter 1 – see Figure 1.14).

From a practical engineering viewpoint, deterministic signals (with the exception of transients) produce discrete line frequency spectra. This is illustrated in Figure 5.4. When the spectral lines show a harmonic relationship (i.e. they are multiples of some fundamental frequency), the deterministic signal is described as being periodic. A typical example of a periodic signal is the vibration from a rotating shaft. When there is no harmonic relationship between the various frequency components, the deterministic signal is described as being almost periodic or quasi-periodic. A typical example of a quasi-periodic signal is the vibration from an aircraft turbine engine, where the vibration signal from the several shafts rotating at different frequencies produces different harmonic series bearing no relationship to each other. Deterministic signals can also be transient or aperiodic. Typical examples include rectangular pulses, tone bursts, and half-cosine pulses which are illustrated in Figure 5.5 with their corresponding spectra. Note that the spectra are not discrete frequency lines. It is also important to note that it is more appropriate to analyse the total amount of energy in a transient rather than the average power (power = energy per unit time) which is a more appropriate descriptor for continuous signals. Hence, the spectra of transient signals have units relating to energy and are thus commonly referred to as energy spectral densities; the spectra of continuous deterministic signals (and also continuous random signals) have units relating to power and are thus commonly referred to as power spectral densities. This point will be discussed in detail in sub-section 5.3.3.

Fig. 5.3. Identification of an increasing vibration level at a frequency associated with a fault.

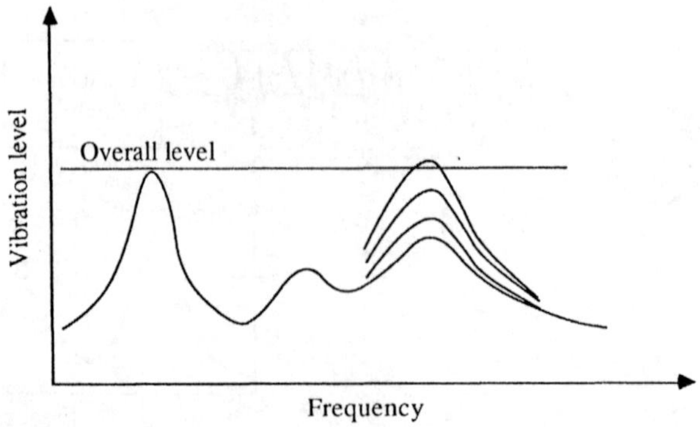

Random vibration signals are continuous signals and they therefore produce continuous spectra as illustrated in Figure 5.6. Because of their random nature, they cannot be described by explicit mathematical relationships and have to be analysed in terms of statistical parameters. The relevant statistical parameters were introduced and defined in section 1.6, chapter 1. They are mean-square values, variances, probability distributions, correlation functions, and power spectral density functions. The reader is referred to Figure 1.24(a)–(c) for the time history functions, auto-correlation functions, and spectral density functions of some typical deterministic and random signals. Because they are continuous functions, the spectra associated with random signals are power spectral densities rather than energy spectral densities. As already mentioned in the first chapter, most random signals of concern to engineers can be approximated as being stationary – i.e. the probability distributions are constant. This implies that the mechanisms producing the stationary signals are time-invariant. Even if the random signals are non-stationary (i.e. the probability

Fig. 5.4. Discrete line frequency spectra associated with periodic and quasi-periodic signals.

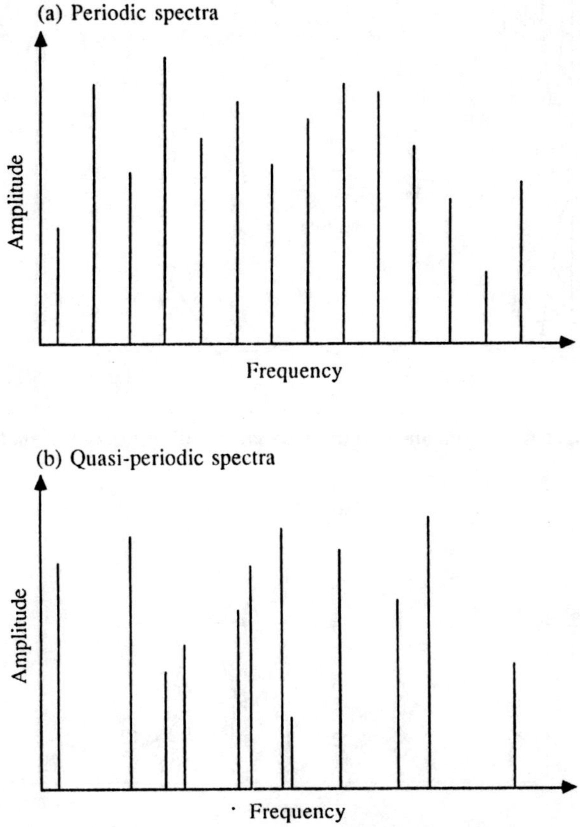

distributions and the mechanisms producing the signals vary with time), they can generally be broken up into smaller quasi-stationary segments or into smaller transient segments. Such procedures are used in speech analysis to separate consonants, vowels, etc., a continuous section of speech being a classical example of a non-stationary process. Typical engineering examples of non-stationary random processes include

Fig. 5.5. Some transient signals and their associated spectra.

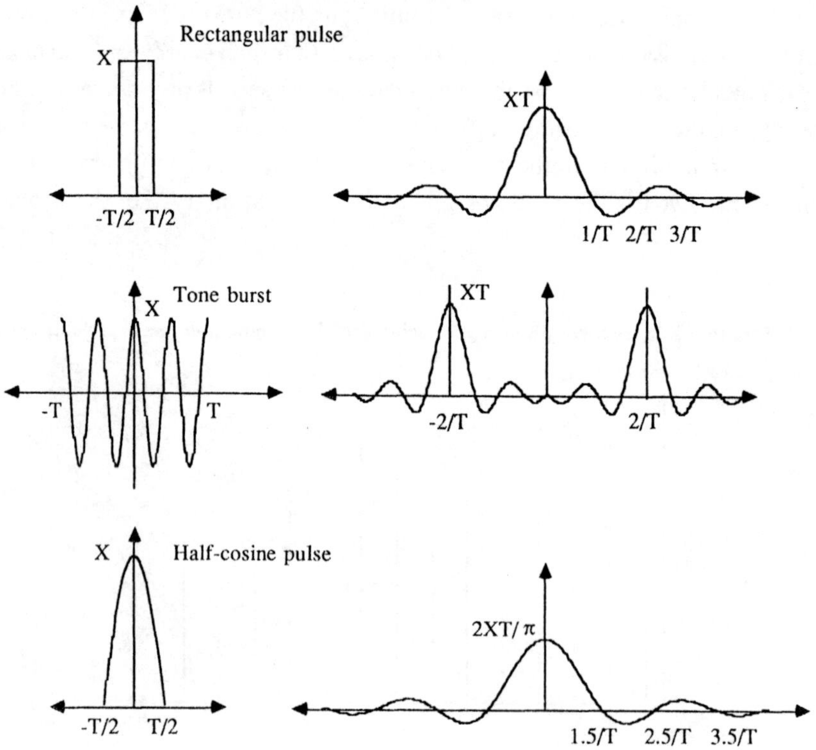

Fig. 5.6. Continuous spectra associated with a random signal.

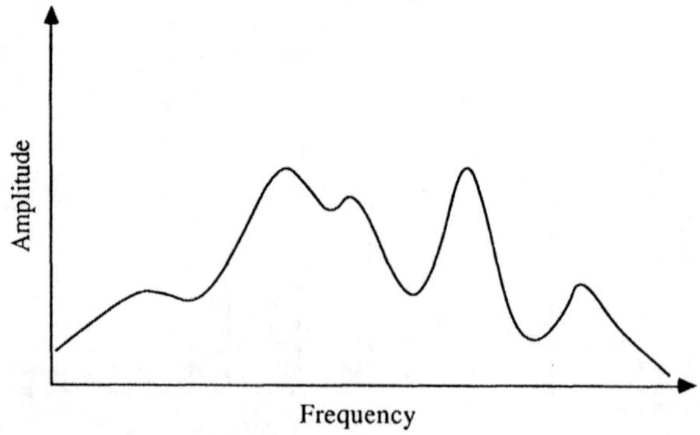

the vibrations associated with a spacecraft during the various stages of the launching process, and atmospheric gust velocities. An analysis of non-stationary random processes is beyond the scope of this book and the reader is referred to Newland[5.1] for a quantitative discussion.

Most industrial noise and vibration signals are either stationary deterministic (i.e. sinusoidal, periodic or quasi-periodic), stationary random, or transient. The discussions in this chapter will therefore be restricted to these three signal types.

5.3 Fundamental signal analysis techniques

Signal analysis techniques can be categorised into four fundamental sub-sections. They are (i) signal magnitude analysis; (ii) time domain analysis of individual signals; (iii) frequency domain analysis of individual signals; and (iv) dual signal analysis in either the time or the frequency domain. Each of the four techniques has its advantages and disadvantages. As a rule of thumb, signal magnitude analysis and time domain analysis provide basic information about the signal and therefore only require inexpensive and unsophisticated analysis instrumentation, whereas frequency domain and dual signal analysis provide very detailed information about the signal and therefore require specialist expertise and reasonably sophisticated analysis instrumentation. Thus, it is very important that the engineer makes an appropriate value judgement as to which technique best meets the necessary requirements for the job. A recent trend has developed for the principles governing dual signal analysis techniques to be extended to situations involving the simultaneous analysis of multiple signals. These specialist techniques are especially useful in noise source identification and will be briefly discussed in chapter 8. Bendat and Piersol[5.2] provide a comprehensive discussion on engineering applications of correlation and spectral analysis of multiple signals.

The signal analysis techniques which are commonly used to quantify an experimentally measured signal are summarised in Figure 5.7.

5.3.1 Signal magnitude analysis

Sometimes, only the overall magnitude (r.m.s. or peak) of a signal is of any real concern to a maintenance engineer. Prior research and/or experience with the performance of the particular piece of machinery often provide sufficient guidelines to allow for the establishment of 'go' and 'no go' confidence levels. Some simple examples include the allowable overall dynamic stress level and the associated vibrational velocity at some critical point on a piece of machinery, the allowable peak sound pressure level due to some impact process, or the allowable r.m.s. overall dB(A) sound level due to some continuous noise source; it is also quite common for r.m.s. and peak vibration levels at various locations on an aircraft to be continuously monitored – when the allowable levels are exceeded, the respective components are inspected and serviced, etc. Under these circumstances, relatively simple analysis equipment for evaluating the overall magnitude of the signals is all that is required. It is common practice for the overall magnitude of a noise or vibration signal to be monitored continuously, and for a spectral analysis to be only periodically obtained.

Signal magnitude analysis thus involves the monitoring and analysis of parameters such as mean signal levels, mean-square and r.m.s. signal levels, peak signal levels, and variances. These four parameters were defined in sub-section 1.6.1, chapter 1 (equations 1.111–1.115, respectively); they all provide information about the signal amplitude.

On occasions, information is also required about additional statistical properties of the signal amplitude in order to establish the relative frequency of occurrences. This requires a knowledge of the probability density functions $p(x)$ and the probability distribution functions $P(x)$ of the signals. The probability density function, $p(x)$, specifies the probability $p(x)\,dx$ that a signal $x(t)$ lies in the range x to $x+dx$. The probability distribution function, $P(x)$, is a cumulative probability function with a maximum value of unity. The two functions are related by

$$P(x) = \int_{-\infty}^{x} p(\alpha)\, d\alpha \leqslant 1, \tag{5.1}$$

where α is an integration variable, and $P(x) = 1$ when the upper limit of integration, x, represents the maximum amplitude of the signal; the total area under the probability

Fig. 5.7. Commonly used signal analysis techniques.

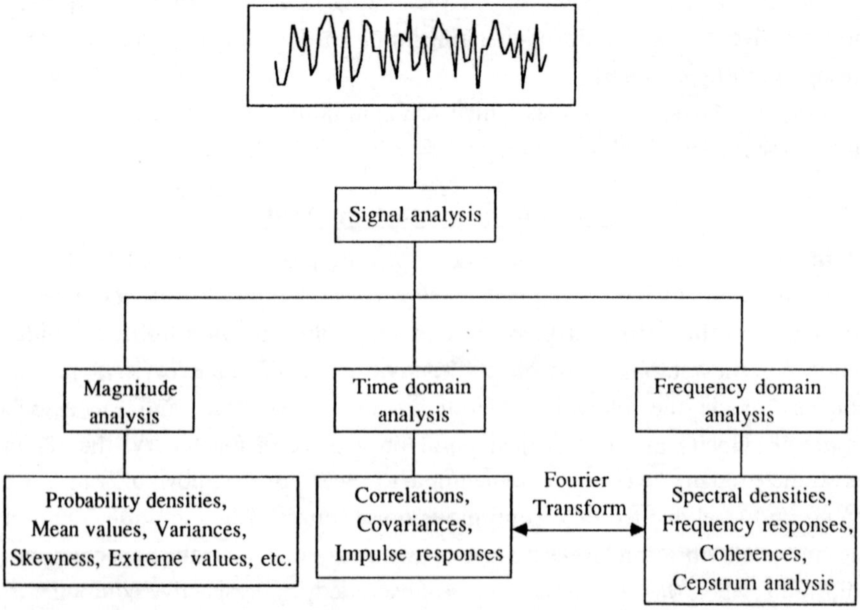

density function must always be unity. This relationship is illustrated in Figure 5.8. Differentiation of equation (5.1) illustrates that the probability density function is the slope of the probability distribution function – i.e.

$$\frac{dP(x)}{dx} = p(x). \tag{5.2}$$

In principle, each physical phenomenon has its own probability density function. Fortunately, however, stationary random processes are generally Gaussian in nature and thus have the well known Gaussian probability density distribution given by

$$p(x) = \frac{1}{\sigma(2\pi)^{1/2}} e^{-(x-m_x)^2/2\sigma^2}, \tag{5.3}$$

where m_x is the mean value of the signal, and σ is its standard deviation. The other type of probability density function that is generally of interest to engineers is that of a sine wave. Its probability density distribution is given by

$$p(x) = \frac{1}{\pi\{(X^2 - x^2)\}^{1/2}}, \tag{5.4}$$

for $-X \leqslant x \leqslant X$. Both probability density distributions are presented in Figure 5.9. It is useful to note that only the mean value and the mean-square value of a stationary random signal are required to compute its probability density distribution.

Another very important application of signal magnitude analysis is a study of the distribution of peaks or extreme values of discrete events. A typical example is the prediction of nuisance damage potential resulting from air blast overpressures associated with some surface mining operation. A large amount of discrete data (e.g. peak sound pressure levels) could be acquired over a long period of time. Typical further examples include wind loading on structures and the fatigue life of various materials. Quite often, under these circumstances, the distributions are not Gaussian and a marked skew can be observed. Statistical information is thus required about

Fig. 5.8. Relationship between probability density and probability distribution.

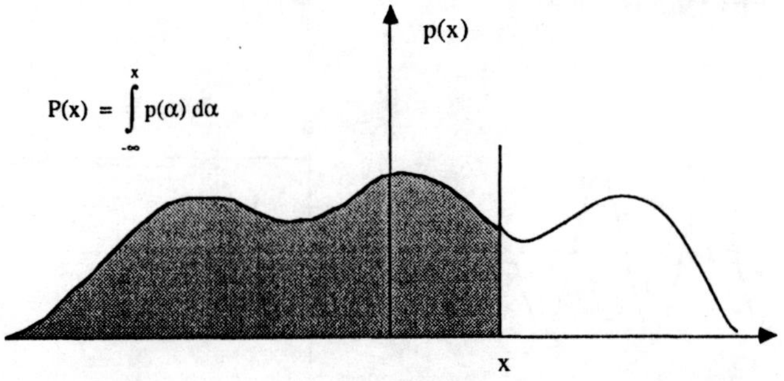

the skewness of the distribution. The mean value of a distribution is its first statistical moment (equation 1.111) and the mean-square value is its second statistical moment (equation 1.112). The skewness of a distribution is its third statistical moment. It is conventionally given in non-dimensional form by

$$\text{skewness} = \frac{E[x^3]}{\sigma^3} = \frac{1}{\sigma^3}\int_{-\infty}^{\infty} x^3 p(x)\,\mathrm{d}x = \frac{1}{\sigma^3 T}\int_0^T x^3\,\mathrm{d}t, \tag{5.5}$$

or

$$\frac{E[x^3]}{\sigma^3} = \lim_{n\to\infty}\frac{1}{\sigma^3 N}\sum_{i=1}^{N} x_i^3(t). \tag{5.6}$$

The skewness is a measure of the symmetry of the probability density function; a function which is symmetric about the mean has a skewness of zero, positive skewness being to the left and negative skewness being to the right, respectively. Various types of probability distribution functions are available for the analysis of skewed distributions. These include log-normal distributions, chi-square distributions, student-t distributions, Maxwell distributions, Weibull distributions, and Gumble distributions amongst others.

Weibull distributions of peaks and Gumble logarithmic relationships are two convenient procedures for estimating the probability of exceedance (or non-exceedance) of a particular level of a non-Gaussian defined event whose probability distribution is significantly skewed. They are particularly useful for the statistical analysis of many separate experimental results and for correlating past results with future outcomes.

Fig. 5.9. Probability density distributions for Gaussian random noise and sine waves.

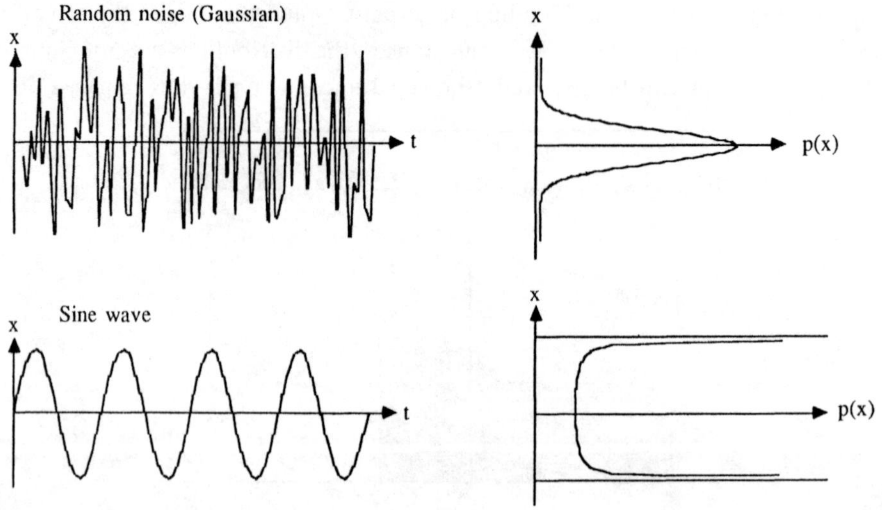

The procedures are also known as extreme value analysis. Newland[5.1], Kennedy and Neville[5.3] and Lawson[5.4] all provide extensive general discussions on various aspects of the topic. Norton and Fahy[5.5] have recently utilised Gumble logarithmic relationships for estimating the probability of non-exceedance of specific ratios of velocity to strain on constrained and unconstrained cylindrical shells with a view to correlating the stress/strain levels with pipe wall vibrations for statistical energy analysis applications. Gumble logarithmic relationships have also been used to predict peak sound pressure levels (at a given location) associated with blast noise from surface mining operations.

5.3.2 Time domain analysis

Individual signals can be analysed in the time domain either by studying the time records by themselves or by generating their auto-correlation functions. Auto-correlation functions were introduced in sub-section 1.6.2, chapter 1, and they provide a measure of the degree of correlation of signals with themselves as a function of time displacement.

Signals can be readily observed in the time domain on an oscilloscope, and this is a useful way of analysing the form of the time histories and of identifying signal peaks, etc. It is also good engineering practice to monitor the time histories of recorded signals prior to performing a frequency analysis so as to get an overall feel for the quality of the signals (i.e. to ensure that clipping, etc. has not occurred), to observe the signal levels, and to detect any pecularities if they exist. If the signal is acquired digitally, time record averaging is a useful means of extracting signals from random noise of about the same frequency content – averaging involves acquiring several independent time records to obtaining an average; this will be discussed in section 5.6. Over a sufficiently long time period, the random noise averages to a mean value of zero, and if an impulse is present it will be detected; time record averaging is used to extract sonar pulses hidden in random ocean noises. The signal to noise ratio for time record averaging is given by

$$\mathrm{S/n\ (dB)} = 10\log_{10} n, \tag{5.7}$$

where n is the number of time records that are averaged – as n increases, the signal to noise ratio improves.

Auto-correlation functions were defined earlier on in this book. In summary, their properties are as follows: (i) for periodic functions, $R_{xx}(\tau)$ is periodic; (ii) for random functions, $R_{xx}(\tau)$ decays to zero for large τ; (iii) $R_{xx}(\tau)$ always peaks at zero time delay; (iv) the value of $R_{xx}(\tau)$ at $\tau = 0$ is the mean-square value. Auto-correlation functions for some typical deterministic and random signals were illustrated in Figure 1.24(b) – they can be used to identify pulses in signals and their associated time delays, and to detect any sinusoidal components that might be submerged in a random noise signal. It is also important to remember that auto-correlation functions do not provide any phase information about a time signal.

Sometimes, auto-correlation functions are defined in terms of their covariances. From chapter 1,

$$R_{xx}(\tau) = E[x(t)x(t+\tau)]. \tag{5.8}$$

Now, the covariance $C_{xx}(\tau)$ is defined as

$$C_{xx}(\tau) = E[x(t)x(t+\tau)] - m_x^2, \tag{5.9}$$

where m_x is the mean value of the signal. Thus,

$$R_{xx}(\tau) = C_{xx}(\tau) + m_x^2. \tag{5.10}$$

When $\tau = 0$, $C_{xx}(0) = E[x^2(t)] - m_x^2 = \sigma_x^2$ and thus the correlation coefficient,

$$\rho_{xx}(\tau) = \frac{R_{xx}(\tau) - m_x^2}{\sigma_x^2}, \qquad \text{(equation 1.117)}$$

is simply a normalised covariance. A value of 1 implies maximum correlation, and a value of 0 implies no correlation (a value of -1 implies that the signal is 180° out of phase with itself).

The time domain analysis of dual signals includes cross-correlation functions and impulse response functions; both functions will be discussed in sub-section 5.3.4.

5.3.3 Frequency domain analysis

In principle, the frequency domain analysis of continuous signals requires a conversion of the time history of a signal into an auto-spectral density function via a Fourier transformation of the auto-correlation function. In practice, digital fast Fourier transform (FFT) techniques are utilised. Prior to the availability of digital signal processing equipment, spectral density functions were obtained experimentally via analogue filtering procedures utilising electronic filters with specified roll-off characteristics. Analogue and digital signal analysis techniques will be discussed in sections 5.4 and 5.5, respectively. This section is concerned with reviewing the fundamental principles of frequency domain analysis.

Auto-spectral density functions ($S_{xx}(\omega)$–double sided, or $G_{xx}(\omega)$–single sided), were introduced in sub-section 1.6.3, chapter 1, and they provide a representation of the frequency content of signals. They are real-valued functions, and it is important to note that the area under an auto-spectrum represents the mean-square value of a signal (i.e. acceleration, velocity, displacement, pressure fluctuation, etc.). Also, because it is a real-valued function, an auto-spectrum does not contain any information about the phase of the signal. Auto-spectra are commonly used by engineers in noise and vibration analysis, and typical examples for deterministic and random signals were illustrated in Figure 1.24(c).

It was pointed out in section 5.2 that the spectra of continuous signals are referred to as power spectral densities because they have units relating to power and that the spectra of transient signals are referred to as energy spectral densities because they have units relating to energy. This is an important point – one which warrants further discussion. A power spectral density has units of (volts)^2 per hertz or V^2s. Thus, the area under a power spectral density curve has units of (volts)^2 which is proportional to power (i.e. electrical power is $\propto V^2$). Now, since energy is equal to power × time, an energy spectral density would have units of V^2s Hz^{-1}, and the area under an energy spectral density curve would have units of V^2s $Hz^{-1} \times Hz = V^2$s. It is more relevant to analyse the total energy in a transient signal rather than the power or average energy per unit time. Thus, for a transient signal of duration T, the energy spectral density, $\mathscr{G}_{xx}(\omega)$, is given by

$$\mathscr{G}_{xx}(\omega) = TG_{xx}(\omega), \tag{5.11}$$

where $G_{xx}(\omega)$ is the power spectral density. The only difference between power spectral densities and energy spectral densities is the factor, T, on the ordinate scale. As is the case for power spectral density functions, both single-sided (i.e. $\mathscr{G}_{xx}(\omega)$) and double-sided (i.e. $\mathscr{S}_{xx}(\omega)$) energy spectral densities can be used.

In recent years, a powerful new spectral analysis technique has emerged. It is referred to as cepstrum analysis. The power cepstrum, $C_{pxx}(\tau)$, is a real-valued function and it is the inverse Fourier transform of the logarithm of the power spectrum of a signal – i.e.

$$C_{pxx}(\tau) = \mathscr{F}^{-1}\{\log_{10} G_{xx}(\omega)\}, \tag{5.12}$$

where $\mathscr{F}^{-1}\{\ \}$ represents the inverse Fourier transform of the term in brackets (likewise, $\mathscr{F}\{\ \}$ would represent a forward Fourier transform). The independent variable, τ, has the dimensions of time (it is similar to the time delay variable of the auto-correlation function) and it is referred to in the literature as 'quefrency'. The advantage that the power cepstrum has over the auto-correlation function is that multiplication effects in the power spectrum become additive in a logarithmic power spectrum – thus, the power cepstrum allows for the separation (deconvolution) of source effects from transmission path or transfer function effects. Deconvolution effects are illustrated at the end of this sub-section.

Sometimes, the power cepstrum is defined as the square of the modulus of the forward Fourier transform of the logarithm of the power spectrum of a signal instead of the inverse Fourier transform – i.e.

$$C_{pxx}(\tau) = |\mathscr{F}\{\log_{10} G_{xx}(\omega)\}|^2. \tag{5.13}$$

It can be shown that both definitions are consistent with each other as the frequency spectral distribution remains the same, the only difference being a scaling factor.

Randall[5.6] and Randall and Hee[5.7] argue that the latter definition is more convenient as it is more efficient to use two forward Fourier transforms.

The power cepstrum has several applications in noise and vibration. It can be used for the identification of any periodic structure in a power spectrum. It is ideally suited to the detection of periodic effects such as detecting harmonic patterns in machine vibration spectra – e.g. the detection of turbine blade failures, and for detecting and separating different sideband families in a spectrum – e.g. gearbox faults. The power cepstrum is also used for echo detection and removal, for speech analysis, and for the measurement of the properties of reflecting surfaces – here, its application is related to its ability to clearly separate source and transmission path effects into readily identifiable quefrency peaks and to provide a deconvolution.

Power cepstrum analysis is generally used as a complementary tool to spectral analysis. It helps identify items which are not readily identified by spectral analysis. Its main limitation is that it tends to suppress information about the overall spectral content of a signal, spectral content which might contain useful information in its own right. It is thus recommended that cepstrum analysis always be used in conjunction with spectral analysis.

Figure 5.10 (from Randall and Hee[5.7]) illustrates the diagnostic potential of power cepstrum analysis for a gearbox vibration signal. The power spectral density of the gearbox vibration signal does not allow for a detection of any periodic structure in the vibration, whereas the power cepstrum clearly identifies the presence of two harmonic sideband families with spacings of 85 Hz and 50 Hz, respectively, corresponding to the rotational speeds of the two gears (note that the harmonics in the cepstrum are referred to as rahmonics). Some further practical examples relating to power cepstrum analysis of bearings with roller defects will be presented in chapter 8.

Another type of cepstrum which is sometimes used in signal analysis is the complex cepstrum. It is defined as the inverse Fourier transform of the logarithm of the forward Fourier transform of a time signal $x(t)$ – i.e.

$$C_{cxx}(\tau) = \mathscr{F}^{-1}\{\log_{10}\mathscr{F}\{x(t)\}\}, \tag{5.14}$$

where

$$G_{xx}(\omega) = \frac{2|\mathscr{F}\{x(t)\}|^2}{T}, \tag{5.15}$$

and T is the finite record length. Equation (5.15) is the digital signal analysis equivalent of the integral transform relationship for continuous signals (equation 1.120) – it is discussed in section 5.5. Despite its name, the complex cepstrum is a real-valued function because $\mathscr{F}\{x(t)\}$ is conjugate even. It does, however, contain information about the phase of the signal. Because the phase information is retained, one can always obtain a complex cepstrum, discard any unwanted quefrency components by

Fig. 5.10. Power cepstrum analysis of a gearbox vibration signal (from Randall and Hee[5.7]).

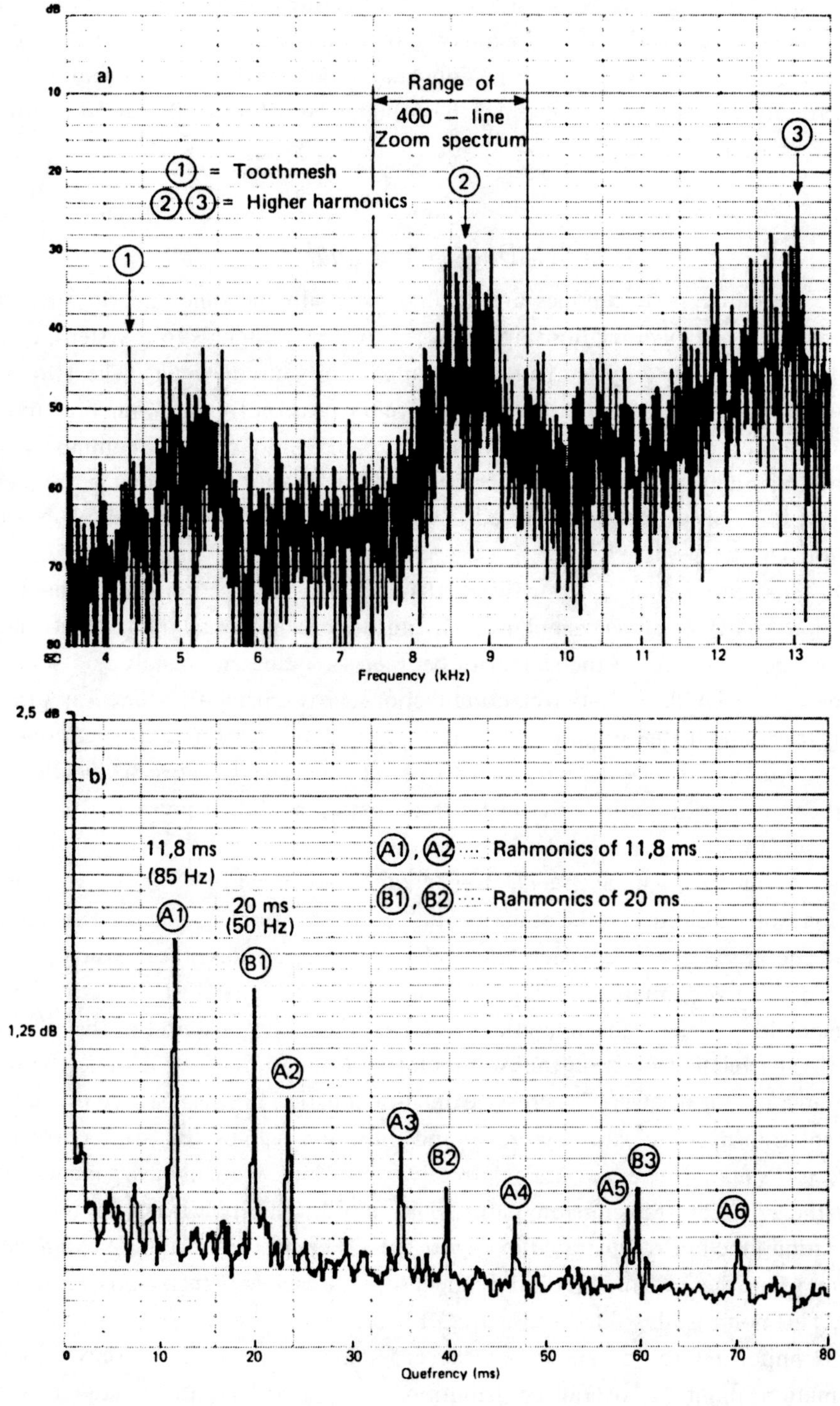

editing the spectrum, and then return from the quefrency domain to the time domain, thus producing the original time signal without the unwanted effects. This procedure is used in echo removal and in the analysis of seismic signals by deconvoluting the seismic wave pulse from the impulse response of the earth at the measurement position. The procedure of echo removal using the complex cepstrum is illustrated in Figure 5.11 (from Randall and Hee[5.7]).

5.3.4 Dual signal analysis

Dual signal analysis techniques are available in both the time and the frequency domains. They involve relationships between two input signals to a system, or two output signals from a system, or an input and an output signal. The two most commonly used time domain relationships are the cross-correlation function and the impulse response function. Three frequency domain relationships are commonly used in spectral analysis. They are the cross-spectral density function, the frequency response function (sometimes referred to as a transfer function), and the coherence function.

The cross-correlation function was introduced in sub-section 1.6.2, chapter 1 (equation 1.118). It is very similar to the auto-correlation function except that it provides an indication of the similarity between two different signals as a function of time shift, τ. Unlike auto-correlation functions, cross-correlation functions are not symmetrical about the origin (see Figures 1.22 and 1.23). As already mentioned in chapter 1, cross-correlations are used to detect time delays between two different signals, transmission path delays in room acoustics, air-borne noise analysis, noise source identification, radar and sonar applications.

If a transmitted signal such as a swept frequency sine wave (from a given source location) was received at some other location after a time delay, τ, and if the received signal comprised the swept sine wave plus extraneous noise, a cross-correlation between the two signals would provide a signal which peaked at a time delay corresponding to the transmission delay. Given the speed of sound in the medium, the cross-correlation function thus allows for an estimation of the distance between the source and the receiver. The cross-correlation function can also be used to establish different transmission paths for noise and vibration signals. By cross-correlation between a single source and a single receiver position, one can easily identify and rank the different transmission paths. This important application of the cross-correlation function is illustrated in Figure 5.12. The procedure can be extended to systems where there are multiple independent sources, each with their own transmission path. This point is illustrated in Figure 5.13.

It is important to recognise that the cross-correlation function only provides information about the overall contribution of a particular path or source to the

Fig. 5.11. Echo removal using the complex cepstrum (from Randall and Hee[5.7]).

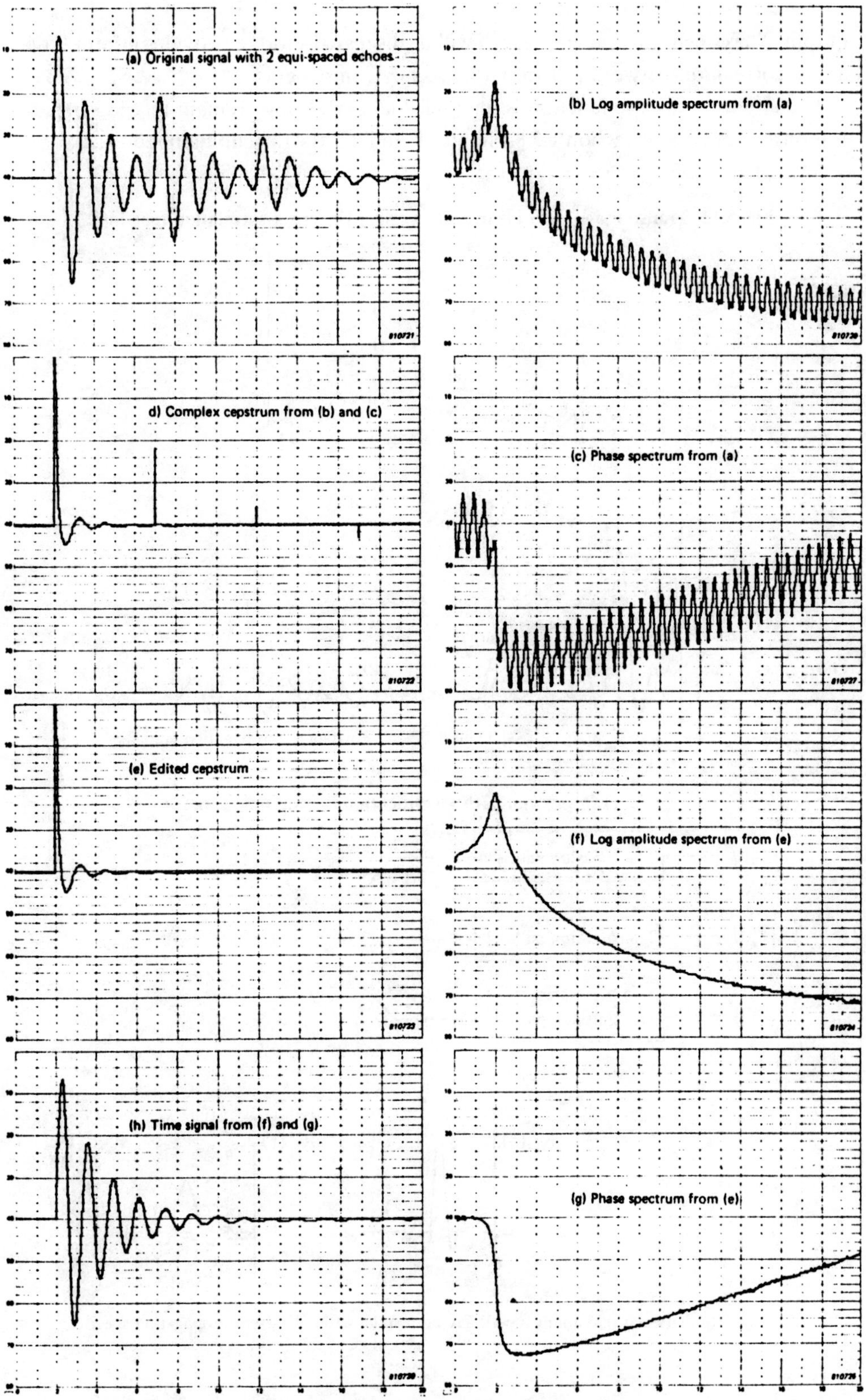

output. The coherence function, which will be discussed shortly, provides information about correlation between individual frequency components.

The impulse response function is another dual signal time domain relationship. It was introduced in sub-section 1.5.8, chapter 1, and it is the time domain representation

Fig. 5.12. Transmission path identification using the cross-correlation function.

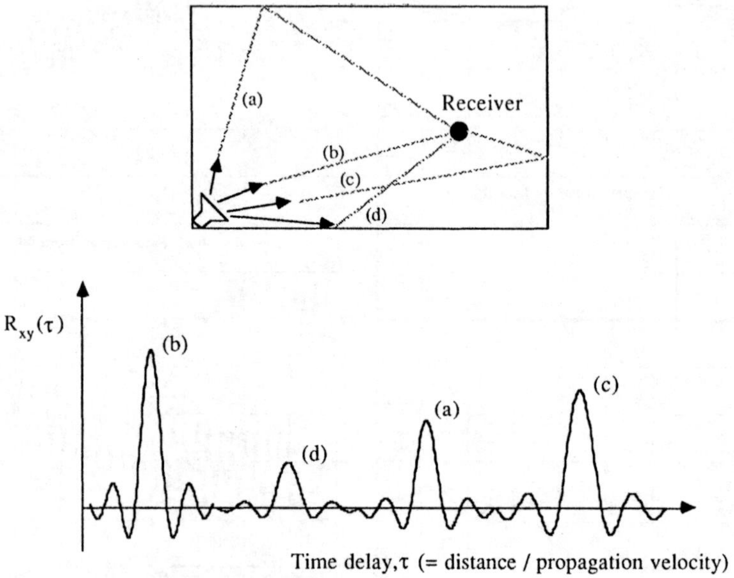

Fig. 5.13. Transmission path identification for multiple sources.

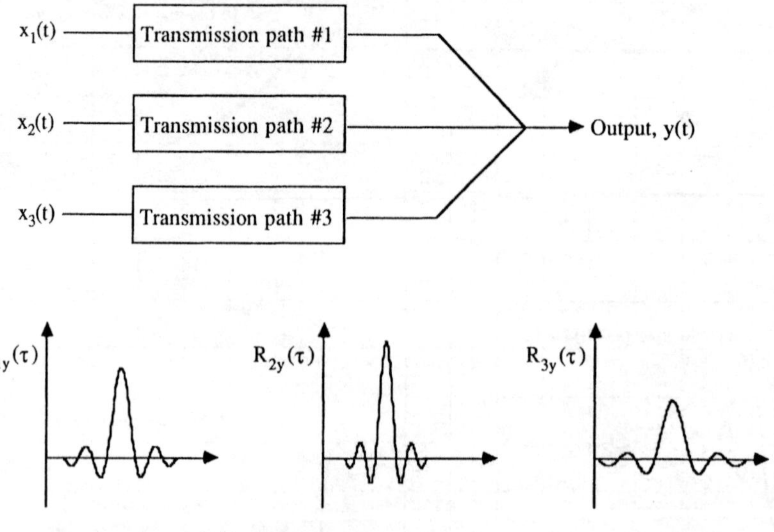

Note that each maximum time delay corresponds to a different propagation speed

of the frequency response function of a system – i.e. it is related to the system frequency response function via the Fourier transform. Like the cross-correlation, it can be used to identify peaks associated with various propagation paths. Sometimes when a signal is dispersive (i.e. its propagation velocity varies with frequency) the cross-correlation function tends to lack the definition of the impulse response function. In these instances, the impulse response function is more useful. Impulse response functions are also used to measure acoustic absorption coefficients, to characterise electronic filters, and to determine noise transmission paths.

Another particularly useful application of the impulse response function is the identification of structural modes of vibration via a Fourier transformation process. The structural modes of vibration of a complex structure or machine can be readily established *in situ* by providing the system with an impulsive input (with a hammer containing a calibrated force transducer), and monitoring the transient output response. The frequency response function of the system is subsequently obtained by Fourier transforming the impulse response function of the system (see equations 1.122, 1.126, 1.127, 1.128, and 1.129). This impact testing procedure is illustrated schematically in Figure 5.14. Its main advantages are that elaborate fixtures are not required for the test structure, the work can be carried out *in situ*, the equipment is relatively easy to use, and the tests can be carried out rapidly. The main disadvantage is that, since there is little energy input into the system, the frequency response of the input signal is limited to about 6000 Hz – i.e. impact testing is not suitable for identifying high frequency structural modes.

Fig. 5.14. Identification of structural modes via impact testing.

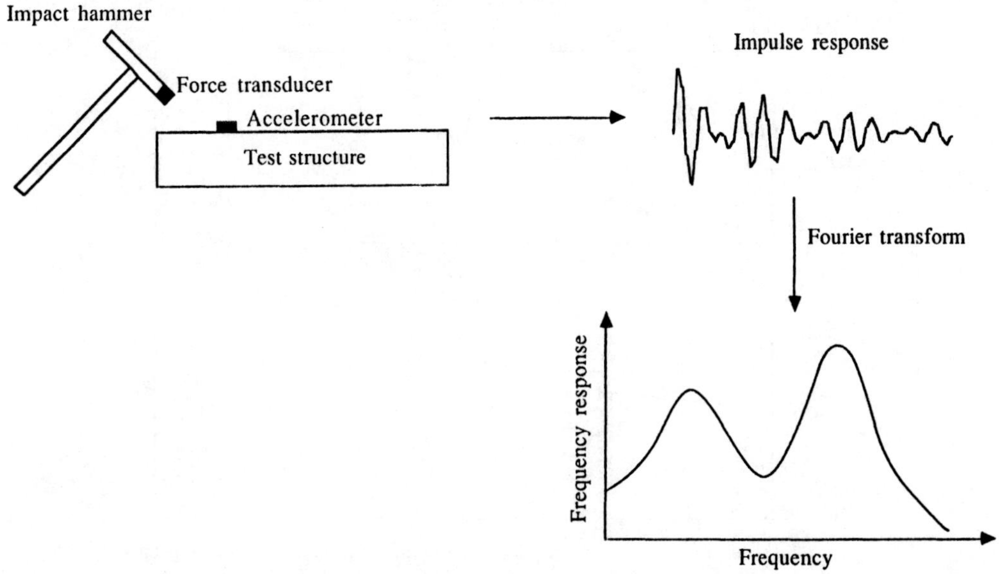

The cross-spectral density is a complex function and it is the Fourier transform of the cross-correlation function. It is a measure of the mutual power between two signals and it contains both magnitude and phase information. It is very useful for identifying major signals that are common to both the input and the output of a system. It is also commonly used to analyse the phase differences between two signals. The phase shifts also help to identify structural modes that are very close together in the frequency domain – it is not always easy to identify closely spaced structural modes from frequency spectra. This point is illustrated in Figure 5.15. The cross-spectral density suggests the presence of two or three structural modes; the information is not very clear, however, because of extraneous noise in the measurement system. The corresponding phase information provides a much clearer picture; the presence of two modes is readily identified by the phase shift at ~1000 Hz.

Cross-spectral densities can also be used to measure power (energy and power flow relationships were discussed in section 1.7, chapter 1). Power is the product of force and velocity – i.e.

$$\langle \Pi \rangle = E[F(t)v(t)]. \tag{5.16}$$

Now,

$$E[F(t)v(t+\tau)] = R_{Fv}(\tau) = \int_0^\infty \mathbf{G}_{\mathbf{Fv}}(\omega)\, \mathrm{e}^{\mathrm{i}\omega\tau}\, \mathrm{d}\omega, \tag{5.17}$$

and

$$\int_0^\infty \mathbf{G}_{\mathbf{Fv}}(\omega)\, \mathrm{e}^{\mathrm{i}\omega\tau}\, \mathrm{d}\omega = \int_0^\infty \mathbf{G}_{\mathbf{Fv}}(\omega) \cos \omega\tau\, \mathrm{d}\omega + \mathrm{i} \int_0^\infty \mathbf{G}_{\mathbf{Fv}}(\omega) \sin \omega\tau\, \mathrm{d}\omega. \tag{5.18}$$

Fig. 5.15. Cross-spectral density and phase for a linear system (fifty averages).

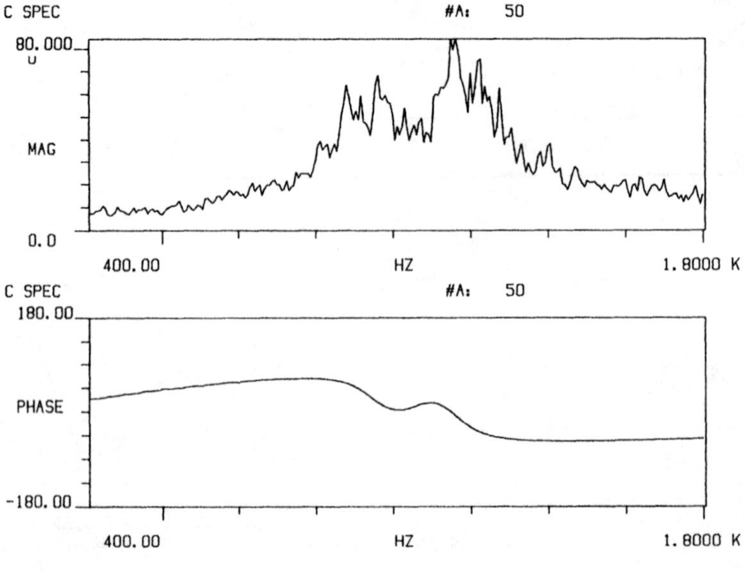

Thus,

$$\langle \Pi \rangle = R_{Fv}(\tau = 0) = \int_0^\infty \mathbf{G}_{\mathbf{Fv}}(\omega)\, \mathrm{d}\omega. \tag{5.19}$$

The total power input (resistive plus reactive) to a structure, or the power output from a system can thus be obtained by integrating the cross-spectral density of force and velocity. The integral of the real part of the cross-spectral density thus represents the power flow away from the source; the imaginary component represents the reactive power in the vicinity of the source. Power flow techniques are used to measure structural loss factors and other parameters required for statistical energy analysis – they will be discussed in more detail in the next chapter.

Frequency response functions (sometimes referred to as transfer functions) play a very important role in the analysis of noise and vibration signals – they describe relationships between inputs and outputs of linear systems. A variety of frequency response functions are available. They include ratios of (i) displacement to force – receptances; (ii) force to displacement – dynamic stiffness; (iii) velocity to force – mobility; (iv) force to velocity – impedance; (v) acceleration to force – inertance; and force to acceleration – apparent mass.

For a single input, single output system as illustrated in Figure 5.16, the frequency response function is defined as the ratio of the forward Fourier transform of the output, $\mathscr{F}\{y(t)\}$ to the forward Fourier transform of the input, $\mathscr{F}\{x(t)\}$ – i.e.

$$\mathbf{H}(\omega) = \frac{\mathscr{F}\{y(t)\}}{\mathscr{F}\{x(t)\}}. \tag{5.20}$$

Thus,

$$|\mathbf{H}(\omega)|^2 = \frac{\mathscr{F}\{y(t)\}\mathscr{F}^*\{y(t)\}}{\mathscr{F}\{x(t)\}\mathscr{F}^*\{x(t)\}} = \frac{G_{yy}(\omega)}{G_{xx}(\omega)}, \tag{5.21}$$

were $\mathscr{F}^*\{y(t)\}$ is the complex conjugate of $\mathscr{F}\{y(t)\}$, etc., $G_{yy}(\omega)$ is the auto-spectral density of the output signal, and $G_{xx}(\omega)$ is the auto-spectral density of the input signal. The factor $2/T$ is omitted because it is common to both the numerator and the denominator.

Fig. 5.16. Single input, single output frequency response function.

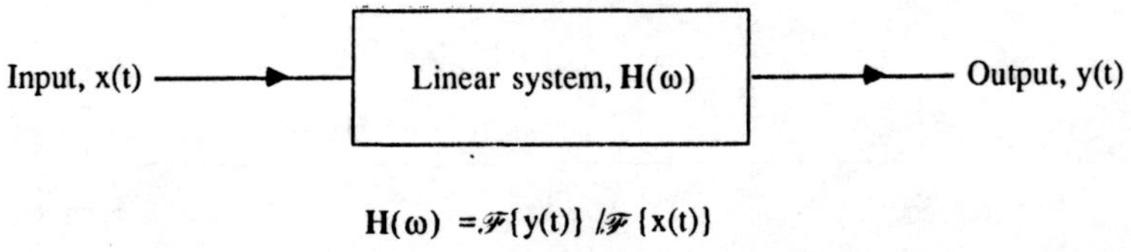

The effects of measurement noise can be reduced by manipulating the frequency response function relationships such that $\mathbf{H}(\omega)$ is obtained from the cross-spectral density. The effects of measurement noise are discussed in section 5.7. At this stage, it is sufficient to note that the forward Fourier transforms can be rearranged such that

$$\mathbf{H}(\omega) = \frac{\mathscr{F}\{y(t)\}\mathscr{F}^*\{x(t)\}}{\mathscr{F}\{x(t)\}\mathscr{F}^*\{x(t)\}} = \frac{\mathbf{G}_{\mathbf{xy}}(\omega)}{G_{xx}(\omega)}. \quad (5.22)$$

Frequency response functions are used for a variety of applications. These include the modal analysis of structures, the estimation of structural damping, the vibrational response of a structure due to an input excitation, and wave transmission analysis (i.e. reflection, transmission, absorption, etc.). Because they are complex functions, they contain information about both magnitude and phase. A typical example of a frequency response function of a linear system with two natural frequencies is illustrated in Figure 5.17. The system is identical to that used in Figure 5.15 for the cross-spectral density. The first point to note is that the frequency response function provides a much clearer picture of the modal response of the system – the two resonant modes are clearly identified both from the magnitude and the phase information.

The impulse response function of a system is the inverse Fourier transform of the frequency response function (see sub-section 1.6.4, chapter 1). The impulse response function for a single resonant mode is illustrated in Figure 5.18 – it was obtained by inverse Fourier transforming the frequency response function.

Fig. 5.17. Frequency response function and phase for a linear system (fifty averages).

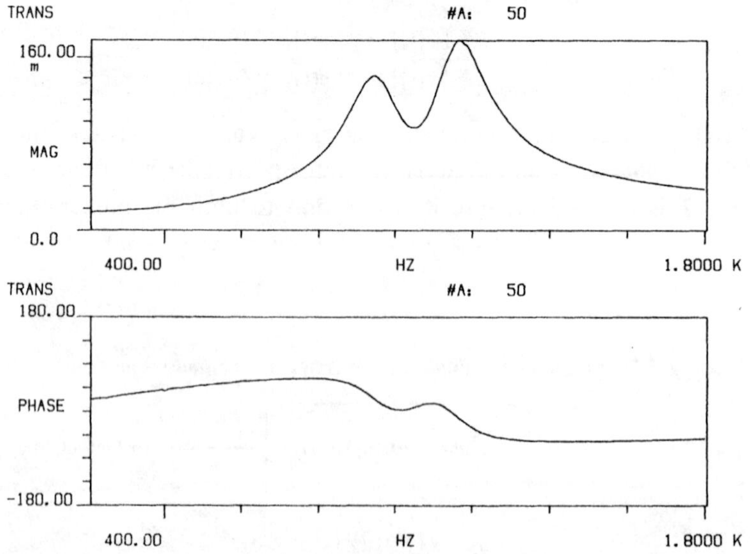

The coherence function, $\gamma^2_{xy}(\omega)$, measures the degree of correlation beween signals in the frequency domain. It is defined as

$$\gamma^2_{xy}(\omega) = \frac{|\mathbf{G}_{\mathbf{xy}}(\omega)|^2}{G_{xx}(\omega)G_{yy}(\omega)}. \tag{5.23}$$

The coherence function is such that $0 < \gamma^2_{xy}(\omega) < 1$, and it provides an estimate of the proportion of the output that is due to the input. For an ideal single input, single output system with no extraneous noise at the input or output stages

$$\gamma^2_{xy}(\omega) = \frac{|\mathbf{H}(\omega)G_{xx}(\omega)|^2}{G_{xx}(\omega)|\mathbf{H}(\omega)|^2 G_{xx}(\omega)} = 1. \tag{5.24}$$

Generally, $\gamma^2_{xy}(\omega) < 1$ because (i) extraneous noise is present in the measurements; (ii) resolution bias errors are present in the spectral estimates; (iii) the system relating $x(t)$ to $y(t)$ is non-linear; or (iv) the output $y(t)$ is due to additional inputs besides $x(t)$.

As an example, consider a system with extraneous noise, $n(t)$, at the output as illustrated in Figure 5.19. Here, $y(t) = v(t) + n(t)$ and

$$G_{yy}(\omega) = G_{vv}(\omega) + G_{nn}(\omega). \tag{5.25}$$

Also, $\mathbf{G}_{\mathbf{xy}}(\omega) = \mathbf{G}_{\mathbf{xv}}(\omega)$ since the extraneous noise can be assumed to be uncorrelated with the input signal (i.e. $R_{xn}(\tau) = 0$ and $\mathbf{G}_{\mathbf{xn}}(\omega) = 0$). Now,

$$G_{vv}(\omega) = |\mathbf{H}(\omega)|^2 G_{xx}(\omega), \tag{5.26}$$

Fig. 5.18. Frequency response and impulse response for a linear system (fifty averages).

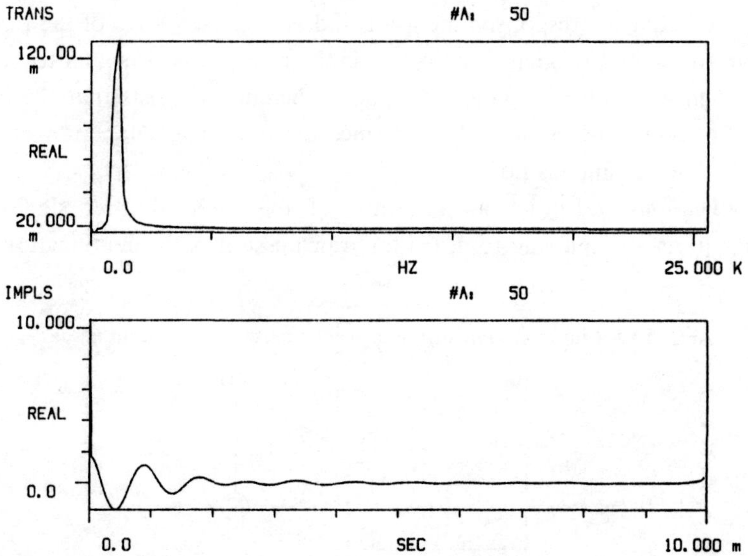

but

$$|\mathbf{H}(\omega)|^2 = |\mathbf{G}_{\mathbf{xv}}(\omega)/G_{xx}(\omega)|^2 = |\mathbf{G}_{\mathbf{xy}}(\omega)/G_{xx}(\omega)|^2, \tag{5.27}$$

and thus

$$G_{vv}(\omega) = |\mathbf{G}_{\mathbf{xy}}(\omega)/G_{xx}(\omega)|^2 G_{xx}(\omega) = \gamma_{xy}^2(\omega) G_{yy}(\omega). \tag{5.28}$$

Hence,

$$\gamma_{xy}^2(\omega) = \frac{1}{1 + \{G_{nn}(\omega)/G_{vv}(\omega)\}}. \tag{5.29}$$

Equation (5.28) represents the coherent output power spectrum – i.e. the output spectral density which is associated with the input. Thus, $G_{yy}(\omega)\{1 - \gamma_{xy}^2(\omega)\}$ is that fraction of the output spectral density which is due to extraneous noise. Equation (5.29) illustrates that the coherence is the fractional portion of the output spectral density which is linearly due to the input.

The signal to noise ratio can be readily evaluated from the coherence function. It is

$$\mathrm{S/n} = \frac{G_{vv}(\omega)}{G_{nn}(\omega)} = \frac{\gamma_{xy}^2(\omega)}{1 - \gamma_{xy}^2(\omega)}. \tag{5.30}$$

Examples of good and bad coherence and the associated frequency response functions are presented in Figures 5.20 and 5.21, respectively. Both figures relate to the same linear system which was used earlier on as an illustration for the cross-spectral density function, the frequency response function, and the impulse response function. The good coherence (~ 1) in Figure 5.20 suggests that the extraneous noise has been eliminated, that the output is completely due to the input, and that the frequency response function is indeed representative of the response of the system to the input signal. Figure 5.21 is the result of poor signal to noise ratio ($n = 2$) and insufficient averaging – the poor coherence suggests that the frequency response function is not clearly defined since the output is a function of both the input and some extraneous noise.

The applications and practical limitations of noise and vibration signal analysis techniques (i) as a diagnostic tool, (ii) for transmission path identification, (iii) for

Fig. 5.19. Linear system with extraneous noise at the output stage.

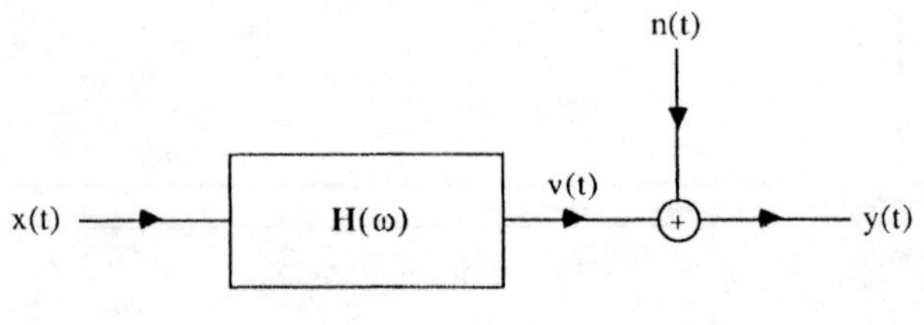

the study of system response characteristics, and (iv) for noise source identification are discussed in chapter 8.

5.4 Analogue signal analysis

Prior to the availability of digital signal analysis equipment, frequency analysis was performed using sets of narrowband analogue filters with unit frequency response

Fig. 5.20. Example of good coherence (linear system, fifty averages).

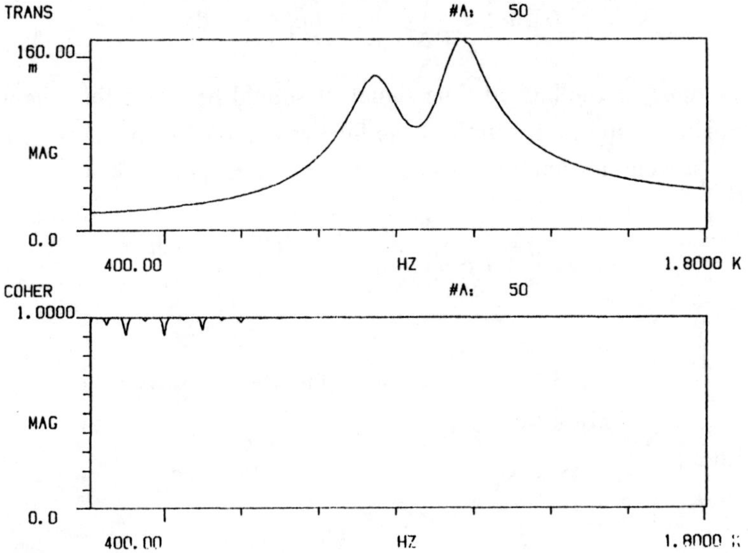

Fig. 5.21. Example of bad coherence (linear system, two averages).

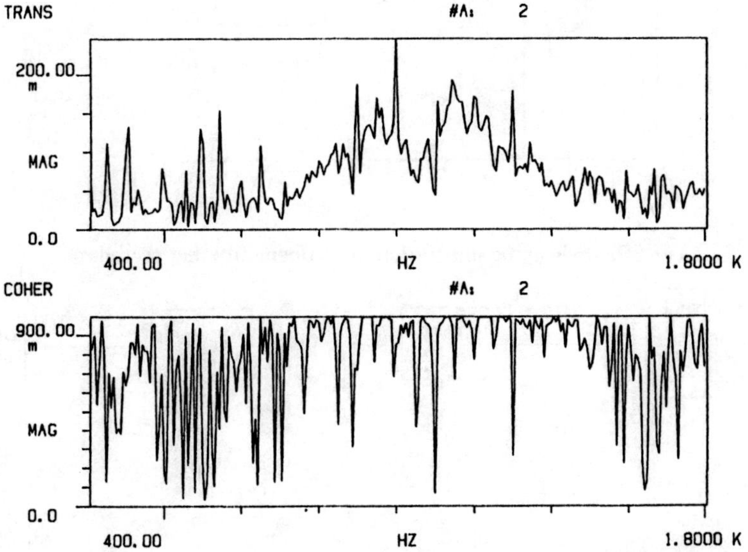

functions. A typical narrowband analogue filter characteristic is illustrated in Figure 5.22.

Analogue signal analysers are still commonly used in practice. A time signal, $x(t)$ is fed into a variable frequency narrowband filter (centre frequency ω and bandwidth $\Delta\omega$). The output from the filter is then fed into a squaring device, an averaging device, and finally divided by the filter bandwidth. This procedure, which is illustrated in Figure 5.23, provides an estimate of the auto-spectral density function. Thus

$$G_{xx}(\omega) \approx \frac{1}{T\Delta\omega}\int_0^T x^2(\omega, \Delta\omega, t)\,dt, \tag{5.31}$$

where $x(\omega, \Delta\omega, t)$ is the filtered time signal. It should be noted that the amplitude of the frequency response function of the filter is assumed to be unity in the above equation. If it were not unity but some arbitrary value $|\mathbf{H}(\omega)|$, then

$$G_{xx}(\omega) = \frac{1}{T\Delta\omega|\mathbf{H}(\omega)|^2}\int_0^T x^2(\omega, \Delta\omega, t)\,dt, \tag{5.32}$$

Fig. 5.22. Typical analogue filter characteristics.

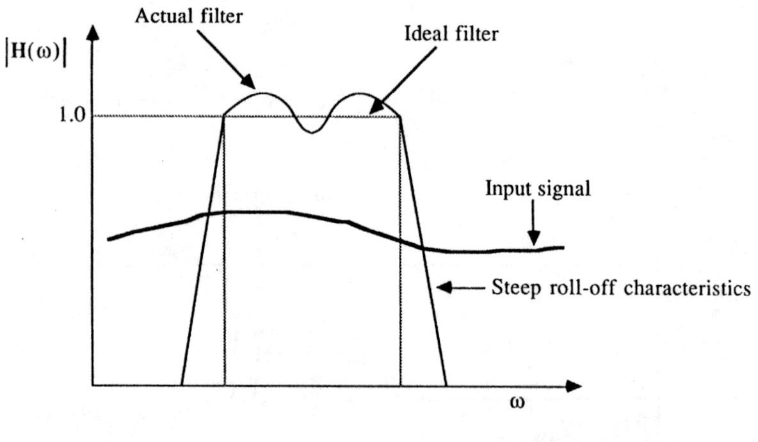

Fig. 5.23. Schematic illustration of analogue filtering procedure.

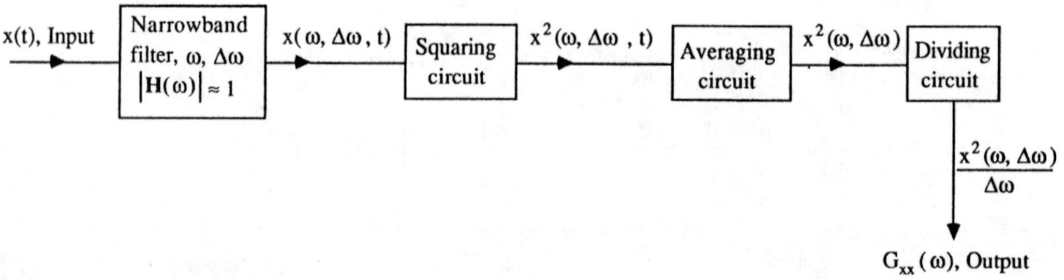

since

$$E[y^2] = \frac{1}{T}\int_0^T x^2(\omega, \Delta\omega, t)\,dt = \int_0^\infty |\mathbf{H}(\omega)|^2 G_{xx}(\omega)\,d\omega$$

$$\approx |\mathbf{H}(\omega)|^2\,\Delta\omega\,G_{xx}(\omega). \qquad (5.33)$$

An accurate estimate of $G_{xx}(\omega)$ is dependent upon (i) the flatness of the filter, (ii) its roll-off characteristics, (iii) the averaging time T, and (iv) the magnitude of any phase shifts between the input and output. Intuitively, better accuracy is to be expected with longer averaging times – in practice, analogue averaging is achieved by using a low-pass RC smoothing filter with a particular time constant. Also, the narrower the bandwidth, $\Delta\omega$, the more accurate is the frequency resolution. Analogue filters are available for variable narrow frequency bands, octave bands and one-third-octave bands.

The statistical errors associated with analogue and digital signal analysis are discussed in section 5.6. The reader is also referred to Randall[5.6] and to Bendat and Piersol[5.8] for a comprehensive discussion on the practical details relating to analogue signal analysis.

5.5 Digital signal analysis

With the ready availability of analogue to digital converters (A/D converters) spectral density functions can be obtained via a Fourier transformation of a discrete time series representation of the original time signal either directly or via the auto-correlation function. This important point is illustrated schematically in Figure 5.24. Averaging for statistical reliability is performed in the frequency domain for the direct transformation procedure and in the time domain when using the auto-correlation/Fourier transformation procedure.

The direct transformation procedure is commonly referred to in the literature as a direct Fourier transform (DFT) and it is performed over a finite, discrete series of sampled values. The discrete time series is generated by a rapid sampling of a finite length of the analogue time signal over a series of regularly spaced time intervals. This procedure is illustrated in Figure 5.25. The subsequent direct Fourier transformation of the signal into the frequency domain has been significantly enhanced by the introduction of the fast Fourier transform (FFT) algorithm. It is the FFT algorithm that is widely used by both commercially available spectrum analysers and computed based signal analysis systems.

A general Fourier transform pair, $\mathbf{X}(\omega)$ and $x(t)$ was defined in sub-section 1.6.3, chapter 1. It is

$$\mathbf{X}(\omega) = \frac{1}{2\pi}\int_{-\infty}^{\infty} x(t)\,e^{-i\omega t}\,dt,$$

and

$$x(t) = \int_{-\infty}^{\infty} \mathbf{X}(\omega)\, e^{i\omega t}\, d\omega. \qquad \text{(equation 1.119)}$$

Because classical Fourier theory is only valid for functions which are absolutely integrable and decay to zero, the transform $X(\omega)$ will only exist for a random signal which is restricted by a finite time interval. Thus the concept of a finite Fourier transform, $\mathbf{X}(\omega, T)$ is introduced. The finite Fourier transform of a time signal $x(t)$ is given by

$$\mathscr{F}\{x(t)\} = \mathbf{X}(\omega, T) = \frac{1}{2\pi}\int_0^T x(t)\, e^{-i\omega t}\, dt, \qquad (5.34)$$

and it is restricted to the time interval $(0, T)$. As noted earlier, $\mathscr{F}\{\ \}$ represents a forward Fourier transform, and $\mathscr{F}^{-1}\{\ \}$ represents an inverse Fourier transform.

For a stationary random signal, the one-sided spectral density $G_{xx}(\omega)$ is given by

$$G_{xx}(\omega) = \lim_{T\to\infty} \frac{2}{T} E[\mathbf{X}^*(\omega, T)\mathbf{X}(\omega, T)]. \qquad (5.35)$$

It can be shown that this equation is identical to the spectral density function defined

Fig. 5.24. Digital signal analysis of a random signal.

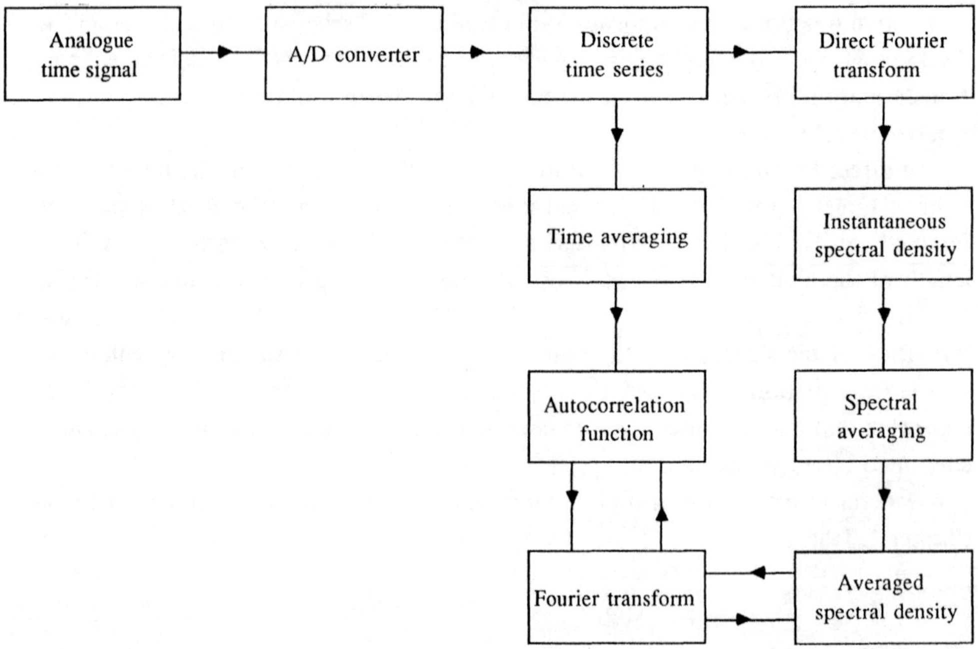

in terms of the auto-correlation function[5.2]. $G_{xx}(\omega)$ can thus be estimated by

$$G_{xx}(\omega) = \frac{2|\mathscr{F}\{x(t)\}|^2}{T} = \frac{2|\mathbf{X}(\omega, T)|^2}{T}. \tag{5.36}$$

Likewise, cross-spectral terms such as $\mathbf{G}_{\mathbf{xy}}(\omega)$, etc. can also be defined in terms of finite Fourier transforms. It is important to note that the spectral density function $\mathbf{G}_{\mathbf{xy}}(\omega)$ is defined by $\mathbf{X}^*\mathbf{Y}$ and not by $\mathbf{X}\mathbf{Y}^*$.

The concept of a Fourier series expansion for a harmonically related periodic signal was introduced in chapter 1. This concept can be extended to a discrete time series of a random time signal (Figure 5.25), and the frequency spectrum is thus approximated by a series of equally spaced (harmonic) frequency lines – i.e. in digital signal analysis procedures the Fourier transform, $\mathbf{X}(\omega, T)$, is obtained from the discrete time series of the time signal.

Fig. 5.25. Schematic illustration of the analogue to digital conversion of a continuous time signal.

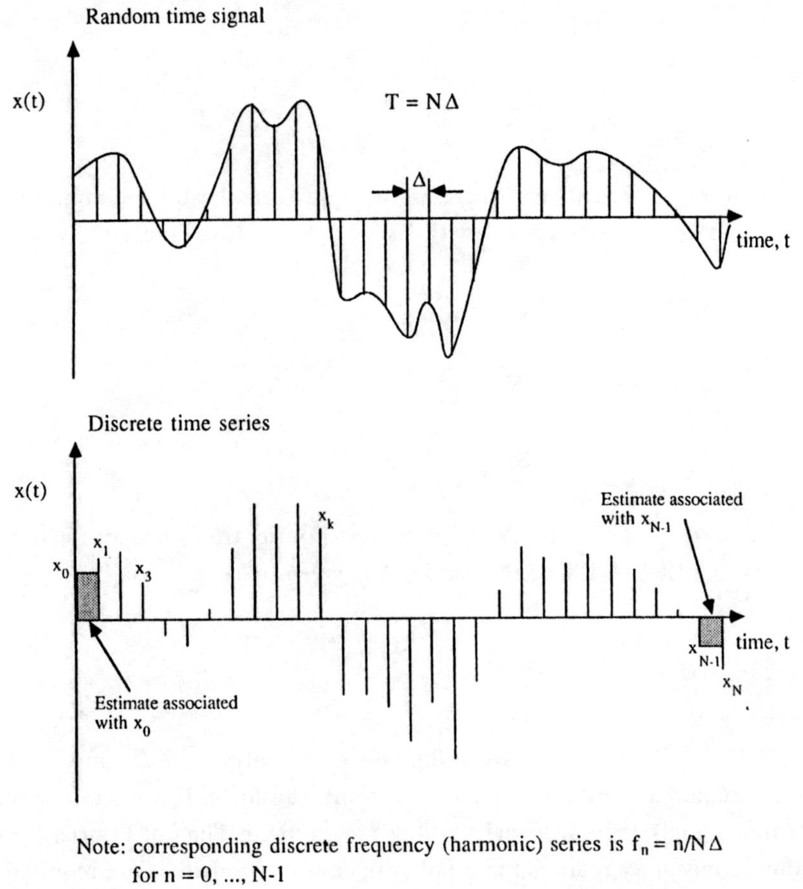

The Fourier series expansion for a harmonically related periodic signal (equations 1.93 and 1.94) can be re-expressed in exponential form as

$$x(t) = \sum_{n=-\infty}^{\infty} \mathbf{X_n}\, e^{i\omega_n t}, \tag{5.37}$$

where $X_0 = a_0/2$, and

$$\begin{aligned} \mathbf{X_n} &= \tfrac{1}{2}(a_n - ib_n) \\ &= \frac{1}{T}\int_0^T x(t)\, e^{-i\omega_n t}\, dt \qquad \text{for } n = \pm 1, 2, \text{etc.} \end{aligned} \tag{5.38}$$

The $\mathbf{X_n}$'s are now the complex Fourier coefficients of the time signal.

Equations (5.34) and (5.37) demonstrate that at the discrete frequencies $f_n = \omega_n/2\pi = n/T$,

$$\mathbf{X}(\omega_n, T) = \frac{T\mathbf{X_n}}{2\pi}. \tag{5.39}$$

Thus,

$$\mathbf{X_n} = \frac{2\pi}{T}\mathbf{X}(\omega_n, T) = \frac{1}{T}\int_0^T x(t)\, e^{-i\omega_n t}\, dt, \tag{5.40}$$

and the Fourier coefficients can therefore be approximated by a summation based upon the discrete time series $x_k(t)$ (with $k = 0, 1, 2, \ldots, N-1$) of $x(t)$. Hence,

$$\mathbf{X_n} = \frac{1}{T}\sum_{k=0}^{N-1} x_k\, e^{-i2\pi f_n k\Delta}\, \Delta, \tag{5.41}$$

where $t = k\Delta$. Now, since $T = N\Delta$ and $f_n = n/T = n/N\Delta$,

$$\mathbf{X_n} = \frac{1}{N}\sum_{k=0}^{N-1} x_k\, e^{-i2\pi nk/N}, \tag{5.42}$$

for $n = 0, \ldots, N-1$. This is the N-point discrete Fourier transform for the time series $x_k(t)$ for $k = 0, \ldots, N-1$. The inverse DFT is given by

$$x_k = \sum_{n=0}^{N-1} \mathbf{X_n}\, e^{i2\pi nk/N}, \tag{5.43}$$

for $k = 0, \ldots, N-1$.

The DFT algorithm is the basis of digital signal analysis – N^2 complex multiplications are required to establish a single N-point transform. If averaging is required over M time signals, then MN^2 calculations are required. The fast Fourier transform algorithm significantly reduces the number of computations that are required – it is essentially a more efficient procedure for evaluating a DFT. Here only $N \log_2 N$ computations are required. For instance, when $N = 1000$, the FFT is 100 times faster, and, when $N = 10^6$, the FFT is ~50 000 times faster. Newland[5.1], Randall[5.6] and Bendat and Piersol[5.8] all provide specific details about the FFT algorithm. The algorithm is also readily available as a commercial package for a range of mainframe, mini and micro computers, and contained within all digital spectrum analysers.

Statistical errors associated with digital signal analysis include random errors due to insufficient averaging, bias errors, aliasing errors, and errors due to inadequate windowing of the signal. These parameters are discussed in the next section.

5.6 Statistical errors associated with signal analysis

It is impossible to analyse an infinite ensemble or a single data record of infinite length. Errors do exist, and they can result from statistical sampling considerations and data acquisition errors. The former are commonly known as random errors, and the latter as bias errors. Random errors are due to the fact that any averaging operation must involve a finite number of sample records, and any analysis will always have a degree of random error associated with it. Bias errors, on the other hand, are systematic errors and they always occur in the same direction.

In addition to random and bias errors, which are common to both analogue and digital signal analysis, there are two additional error types that are peculiar to digital signal analysis. They are aliasing and inadequate windowing. Aliasing is related to the digitising or sampling interval, Δ. Too small a sampling interval produces a large quantity of unnecessary data; too large a sampling interval results in a distortion of the frequency spectra because of high frequency components which fold back onto the lower part of the spectrum. Aliasing can be avoided by selecting an appropriate sampling interval, Δ. All finite time records are windowed functions since their ends are truncated. When this truncation process is abrupt (e.g. a rectangular window), the windowing is inadequate because it produces leakage – i.e. unwanted spectral components are generated and the spectrum is distorted. Suitable windowing functions which avoid the abrupt truncation of the signal are utilised in digital signal analysis to minimise the effects of finite time records.

5.6.1 Random and bias errors

In all practical signal analysis problems, there is a compromise between the analysis frequency bandwidth and the analysis time. A filter with a bandwidth of B Hz takes approximately $1/B$ seconds to respond to a signal that is applied to its input. The relationship between frequency bandwidth and time is considered to be the most important rule in signal analysis. It is[5.1,5.8]

$$BT \geqslant 1, \tag{5.44}$$

where B is the filter bandwidth of the measurement for analogue signal analysis and the resolution bandwidth for digital signal analysis (for digital signal analysis, the resolution bandwidth is commonly defined as B_e), and T is the duration of the measurement. This important relationship says: (i) if a signal lasts for T seconds, the best measurement bandwidth that can be achieved is $1/T$ Hz, or (ii) if the analysing filter bandwidth is B Hz, one would have to wait $1/B$ seconds for a measurement.

Another important aspect of signal analysis is the requirement to average the data over several measurements. Averaging is particularly critical for broadband random signals where sufficient data has to be obtained such that the values are representative of the signal. During averaging, it is also necessary to ensure that the relationship $BT \geqslant 1$ is satisfied and that numerous periods of the lowest frequency of interest are included. For digital signal analysis, the total duration of the signal to be analysed is defined by $T_t = nT$, where n is the number of time records that are sampled; for analogue signal analysis, the total duration of the signal is simply defined by the duration of the recording process.

The interpretation of frequency bandwidths and averaging times depends upon whether one is using analogue or digital equipment. The normalised random error of a measurement obtained via an analogue spectrum analyser can be expressed as[5.1]

$$\varepsilon_r = \frac{\sigma}{m} \approx \frac{1}{(BT)^{1/2}}, \tag{5.45}$$

where σ is the standard deviation, and m is the mean value. Hence, for small standard deviations, $BT \geqslant 1$; this is consistent with equation (5.44). Equation (5.45) highlights the conflicting requirements between the filter bandwidth B and the duration of the measurement; for good resolution B has to be small, and for good statistical reliability B has to be large compared with $1/T$.

If the time record, T, is digitised into a sequence of N equally spaced sampled values, as illustrated in Figure 5.25, the minimum available frequency resolution bandwidth is

$$\Delta f = B_e = \frac{1}{T} = \frac{1}{N\Delta}. \tag{5.46}$$

The resolution bandwidth is thus determined by the individual record length, T, and not by the total amount of data ($T_t = nT$, where n is the number of time records) that is analysed. The normalised random error, ε_r, is, however, a function of the total amount of digitised data, T_t. The relationship is similar to that for an analogue signal and is given by

$$\varepsilon_r = \frac{\sigma}{m} \approx \frac{1}{(B_e T_t)^{1/2}}. \tag{5.47}$$

The normalised random error formulae provided here only relate to auto-spectral measurements (these measurements are most commonly used in engineering applications), and do not relate to correlations, cross-spectral densities or coherence functions. The results thus only constitute a representation of the general form of the error, and should not be used as a quantitative measure for anything other than auto-spectra. However, the normalised random error, ε_r for any signal can always be made smaller by increasing the total record length – i.e. increasing the number of averages, n, for a given frequency resolution, B_e (for example, see Figures 5.20 and 5.21).

The normalised bias error, ε_b, is a function of the resolution bandwidth, B_e, and the half-power bandwidth, $B_r \approx 2\zeta f_d$, of the system frequency response function, where ζ is the damping ratio and f_d is the damped natural frequency. The normalised bias error is approximated by[5.2]

$$\varepsilon_b \approx -\frac{1}{3}\left(\frac{B_e}{B_r}\right)^2. \tag{5.48}$$

Bias errors thus occur at resonance frequencies in spectral estimates; this has specific relevance when using spectral analysis techniques to estimate damping ratios of lightly damped systems. Procedures for estimating damping are discussed in chapter 6. Bias errors have the effect of limiting the dynamic range of an analysis; the spectral peaks are underestimated and the spectral troughs are overestimated. The normalised bias error formula is appropriate for both analogue and digital signals, and for auto- and cross-spectral density measurements. Correlation measurements do not have any bias errors.

The normalised r.m.s error for both analogue and digital signal analysis can be obtained from the random and bias errors. It is given by

$$\varepsilon = (\varepsilon_r^2 + \varepsilon_b^2)^{1/2}. \tag{5.49}$$

The reader is referred to Bendat and Piersol[5.2,5.8] for a detailed discussion on random and bias errors associated with functions other than auto-spectral densities.

5.6.2 *Aliasing*

Aliasing is a problem that is unique to digital signal analysis. Consider a sine wave which is digitised. At least two samples per cycle are required to define the frequency of the sine wave. Hence, for a given sampling interval Δ, the highest frequency which can be reliably defined is $1/2\Delta$. If higher frequency components are present in the signal, they will not be detected and will instead be confused with the lower frequency signal – i.e. the higher frequency components will fold back onto the lower frequency components. This point is illustrated in Figure 5.26 where there are six periods of the high frequency sine wave and three periods of the low frequency sine wave. If nine digitisation points were used for argument sake, the low frequency wave would be adequately defined. However, the high frequency wave would not be adequately defined and instead it would be aliased with the low frequency wave. The effects of aliasing for a broadband spectrum are illustrated schematically in Figure 5.27.

Aliasing can be avoided by (i) digitising the signal at a rate which is at least twice the highest frequency of interest, and/or (ii) removing all high frequency components (i.e. $f > 1/2\Delta$) by suitable analogue filtering. The procedure of applying an analogue low-pass filter prior to digitisation is referred to as anti-aliasing. The cut-off frequency

$$f_c = \frac{1}{2\Delta}, \tag{5.50}$$

Fig. 5.26. Illustration of aliasing.

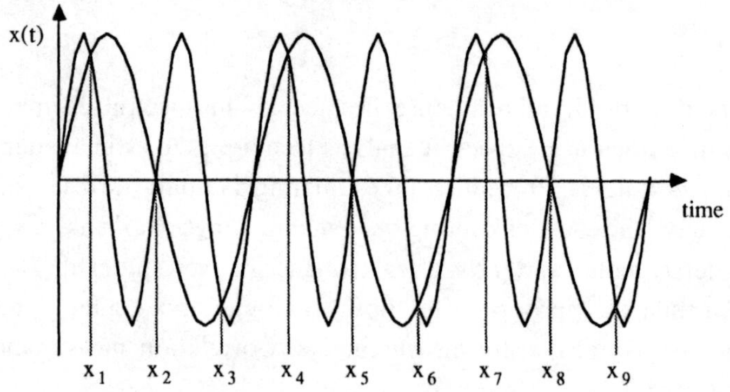

Fig. 5.27. The effects of aliasing for a broadband spectrum.

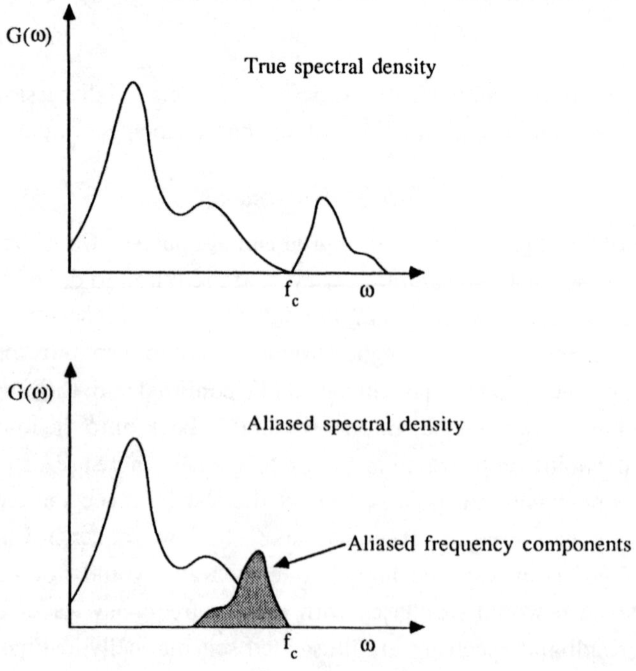

is referred to as the Nyquist cut-off frequency. It represents an upper frequency limit for digital signal analysis – i.e. it is the maximum frequency that can be reliably detected with a sampling interval of Δ.

5.6.3 *Windowing*

Experimental measurements have to be based upon finite time records. Hence, any correlation or spectral density function is by nature only an approximation of the ideal function. When a signal is acquired digitally, the effects of the finite length of the time record can be minimised by applying a suitable window function to the signal. A window function can be thought of as a weighting function which forces the data to zero at its ends, and it can be applied to a time record, a correlation function, or a spectral density function. Time domain windows are commonly referred to as lag windows, and frequency domain windows are commonly referred to as spectral windows.

The true spectral density of a random process $x(t)$ is given by equation (1.120), chapter 1. Since $R_x(\tau)$ is a symmetrical function and $G_{xx}(\omega) = 2S_{xx}(\omega)$, this theoretical relationship can be re-written as

$$G_{xx}(\omega) = \frac{1}{\pi}\int_{-\infty}^{\infty} R_{xx}(\tau)\cos\omega\tau\,\mathrm{d}\tau, \tag{5.51}$$

where $R_x(\tau)$ is the true auto-correlation function of the signal $x(t)$. Now, since any actual digitised time record is finite, an experimentally obtained auto-correlation function is really only an estimate of the theoretically correct function. Hence an experimentally obtained spectral density is also an estimate, and it is given by

$$G_{xx}(\omega) = \frac{1}{\pi}\int_{-\infty}^{\infty} w(\tau)R_{xx}(\tau)\cos\omega\tau\,\mathrm{d}\tau, \tag{5.52}$$

where $w(\tau)$ is an even (symmetrical) weighting function. The experimentally obtained auto-correlation function is thus equivalent to a weighted true auto-correlation function. This weighting function can be modified to suit the experimental data and the application of such an appropiate weighting function (i.e. a lag window) to the auto-correlation function produces a weighted spectral density which compensates for errors due to the finite nature of the signal. The weighting function/lag window can be regarded as a window through which the signal is viewed – it forces the signal to be zero outside the window.

The Fourier transform of the lag window, $w(\tau)$, is the spectral window, $W(\omega)$ which is a real function since $w(\tau)$ is defined here as being even. Hence,

$$W(\omega) = \frac{1}{2\pi}\int_{-\infty}^{\infty} w(\tau)\,\mathrm{e}^{-i\omega\tau}\,\mathrm{d}\tau. \tag{5.53}$$

The weighted estimate of the spectral density can also be obtained by convoluting the spectral window with the true spectral density – i.e.

$$\hat{G}_{xx}(\omega) = \int_{0}^{\infty} G_{xx}(\alpha)W(\omega - \alpha)\,\mathrm{d}\alpha. \tag{5.54}$$

Spectral window functions can be normalised, i.e.

$$\int_{-\infty}^{\infty} W(\omega)\,\mathrm{d}\omega = 1, \tag{5.55}$$

if the window lag function is defined such that $w(\tau = 0) = 1$.

The above discussion illustrates how any digital signal analysis procedure automatically generates a window function simply by the fact that the time signal is truncated. The simplest window function is thus a rectangular lag window (sometimes called a box-car window) in either the time or the correlation (time delay) domain. Consider a rectangular lag window with $w(\tau) = 1$, and with $-T \leqslant \tau \leqslant T$. The spectral window is the Fourier transform of this rectangular lag window and it is obtained from equation (5.53). Thus for $-T \leqslant \tau \leqslant T$

$$W(\omega) = \frac{T}{\pi}\left(\frac{\sin \omega T}{\omega T}\right), \tag{5.56}$$

and both the lag and the spectral window functions are illustrated in Figure 5.28.

Fig. 5.28. Rectangular lag window and corresponding spectral window.

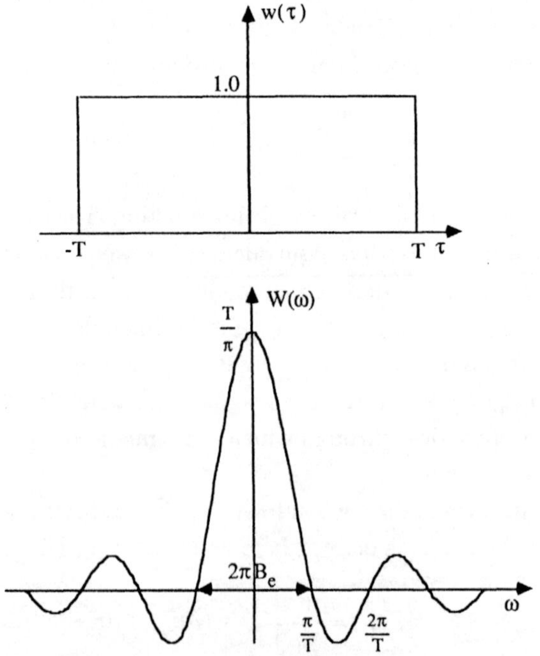

The lobes to the side of the main peak distort the spectrum. This phenomenon is called leakage and it is due to the fact that the time signal is abruptly truncated by the rectangular window. The rectangular lag window is a classical example of how inadequate windowing distorts the true spectrum and produces unwanted spectral components. Leakage is minimised by the clever usage of appropriate window functions. Triangular (tapered) window functions. Hanning (cosine tapering) window functions, Hamming (modified Hanning) window functions, and Gaussian window functions are but some of the variety of window functions that are available. All these window functions smooth the time domain data such that it eventually decays to zero thus minimising leakage from the spectral windows. The ideal lag window would produce a rectangular spectral window (i.e. one with a flat spectrum and no leakage) which would provide a true representation of all the frequency components in the time signal.

As an example, the triangular lag window is given by

$$w(\tau) = 1 - \frac{|\tau|}{T} \qquad \text{for } 0 \leqslant |\tau| \leqslant T$$

$$= 0 \text{ otherwise,} \tag{5.57}$$

where T is the width of the triangle. The spectral window corresponding to this lag window function can be obtained by Fourier transforming equation (5.57). It is

$$W(\omega) = \frac{T}{2\pi}\left\{\frac{\sin(\omega T/2)}{\omega T/2}\right\}^2. \tag{5.58}$$

Both the lag and spectral window functions are illustrated in Figure 5.29. The lobes to the side of the spectrum are now reduced as compared to Figure 5.28 and leakage is minimised. The spectral window is still not ideal (i.e. it is not rectangular) and it applies a weighting to the spectral density estimates. This necessitates the introduction of an effective bandwidth for the spectral window. This effective bandwidth is defined as

$$B_e = \left\{\int_{-\infty}^{\infty} W^2(\omega)\,d\omega\right\}^{-1/2}, \tag{5.59}$$

and it can be approximated by[5.1,5.8]

$$B_e \approx \frac{1}{T}. \tag{5.60}$$

Equation (5.60) illustrates the necessity for time record averaging in digital signal analysis. Since $B_eT \sim 1$, the normalised random error for a single time record would be unity (see equation 5.47)! This is obviously quite unacceptable, and it is overcome by averaging the spectra/time records numerous times.

More sophisticated window functions and advanced analysis techniques, such as zoom analysis and overlap averaging, are available to minimise the weighting effects of spectral windows. Most commercially available digital signal analysers incorporate these features, and the reader is referred to Newland[5.1], Bendat and Piersol[5.2,5.8], and Randal[5.6,5.7] for further details.

5.7 Measurement noise errors associated with signal analysis

Besides the statistical errors associated with the data analysis procedures, there are also errors due to the effects of measurement noise. For instance, the signal to noise ratio in equation (5.30) relates to the coherence function and is therefore associated with measurement noise whereas the signal to noise ratio in equation (5.7) is associated with repeated averaging of a finite time record. In practice, measurement noise is generally due to signal to noise ratio problems in the measurement transducer. This problem was discussed briefly in sub-section 5.3.4 in relation to extraneous noise at the output stage. In reality, there is noise at both input and output stages, and in addition feedback sometimes occurs. The effects of feedback noise on structural measurements will be illustrated in chapter 6. An example of the effects of uncorrelated input and output noise based on some work by Bendat and Piersol[5.2] is presented here. The reader is referred to Bendat and Piersol's text for a wide range of possible sources of measurement error in digital signal analysis.

Fig. 5.29. Triangular lag window and corresponding spectral window.

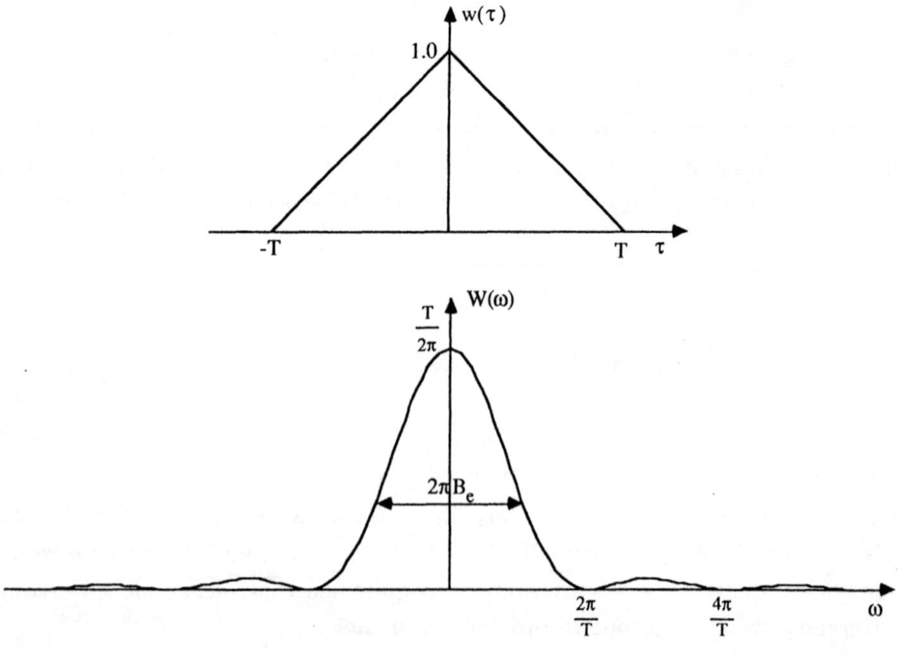

Consider a system where the actual input and output signals are $u(t)$ and $v(t)$, respectively, and the measured input and output signals are $x(t)$ and $y(t)$. There is noise present at both input and output stages, with $m(t)$ and $n(t)$ being the input and output noise, respectively. With this model, the extraneous noise does not pass through the system and is only a function of the measurement instrumentation. The system is illustrated in Figure 5.30.

Because of the presence of the input and output noise, the measured input and output time records are

$$x(t) = u(t) + m(t),$$

and

$$y(t) = v(t) + n(t). \tag{5.61}$$

Because the noise is extraneous and random, it is assumed to be uncorrelated and therefore the cross-spectral terms $\mathbf{G_{mn}}$, $\mathbf{G_{um}}$ and $\mathbf{G_{vn}}$ are all zero. Hence the measured input and output spectral density functions are

$$G_{xx}(\omega) = G_{uu}(\omega) + G_{mm}(\omega),$$

and

$$G_{yy}(\omega) = G_{vv}(\omega) + G_{nn}(\omega). \tag{5.62}$$

It can be seen from equation (5.62) that $G_{xx}(\omega) \geqslant G_{uu}(\omega)$ and $G_{yy}(\omega) \geqslant G_{vv}(\omega)$. Now, from equations (5.22) and (5.26),

$$G_{vv}(\omega) = |\mathbf{H}(\omega)|^2 G_{uu}(\omega),$$

and

$$\mathbf{G_{uv}}(\omega) = \mathbf{H}(\omega) G_{uu}(\omega). \tag{5.63}$$

Also, since the input and output noises are extraneous and uncorrelated,

$$\mathbf{G_{xy}}(\omega) = \mathbf{G_{uv}}(\omega). \tag{5.64}$$

Fig. 5.30. Linear system with measurement noise at the input and output stages.

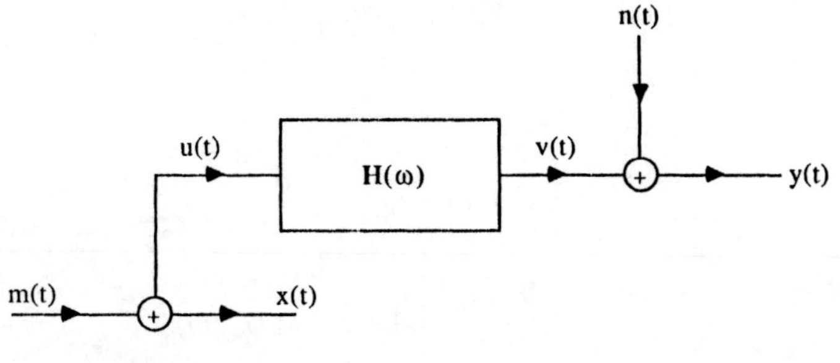

The above relationships can now be used in conjunction with the coherence function (equation 5.23) to study the effects of the measurement noise. The measured coherence is

$$\gamma_{xy}^2(\omega) = \frac{|\mathbf{G_{xy}}(\omega)|^2}{G_{xx}(\omega)G_{yy}(\omega)}, \tag{5.65}$$

and it is obtained directly from the input and output signals ($x(t)$ and $y(t)$). The coherence between the actual input and output signals to the system is given by

$$\gamma_{uv}^2(\omega) = \frac{|\mathbf{G_{uv}}(\omega)|^2}{G_{uu}(\omega)G_{vv}(\omega)}, \tag{5.66}$$

and $\gamma_{xy}^2 < \gamma_{uv}^2$ because of the extraneous noise. Now,

$$|\mathbf{G_{xy}}(\omega)|^2 = |\mathbf{G_{uv}}(\omega)|^2 = |\mathbf{H}(\omega)|^2 G_{uu}^2(\omega) = G_{uu}(\omega)G_{vv}(\omega),$$

hence

$$\gamma_{xy}^2(\omega) = \frac{G_{uu}(\omega)G_{vv}(\omega)}{\{G_{uu}(\omega) + G_{mm}(\omega)\}\{G_{vv}(\omega) + G_{nn}(\omega)\}}, \tag{5.67}$$

and $\gamma_{xy}^2 \leqslant 1$.

From equation (5.28), the coherent output power is

$$\gamma_{xy}^2(\omega)G_{yy}(\omega) = \frac{G_{vv}(\omega)G_{uu}(\omega)}{G_{uu}(\omega) + G_{mm}(\omega)}. \tag{5.68}$$

The coherent output power is only dependent upon the input noise and not upon the output noise. Thus, when attempting to measure the true output power spectrum, $G_{vv}(\omega)$, of the system, one only has to minimise the input noise.

By the auto-spectral density method, the system's frequency response function is given by

$$|\mathbf{H}(\omega)|_{\text{Auto}}^2 = \frac{G_{yy}(\omega)}{G_{xx}(\omega)}, \tag{5.69}$$

and by the cross-spectral density method it is

$$|\mathbf{H}(\omega)|_{\text{Cross}}^2 = \frac{|\mathbf{G_{xy}}(\omega)|^2}{G_{xx}^2(\omega)}. \tag{5.70}$$

Thus,

$$|\mathbf{H}(\omega)|_{\text{Auto}}^2 = \frac{G_{vv}(\omega) + G_{nn}(\omega)}{G_{uu}(\omega) + G_{mm}(\omega)}$$

$$= |\mathbf{H}(\omega)|^2 \frac{1 + G_{nn}(\omega)/G_{vv}(\omega)}{1 + G_{mm}(\omega)/G_{uu}(\omega)}, \tag{5.71}$$

where $\mathbf{H}(\omega)$ is the true frequency response function of the system. Similarly

$$|\mathbf{H}(\omega)|_{\text{Cross}} = \frac{|\mathbf{G_{uv}}(\omega)|}{G_{uu}(\omega) + G_{mm}(\omega)}$$

$$= |\mathbf{H}(\omega)| \frac{1}{1 + G_{mm}(\omega)/G_{uu}(\omega)}. \qquad (5.72)$$

From equations (5.71) and (5.72) it can be seen that the cross-spectral density method provides an estimate of the frequency response function which is independent of the output noise. It is therefore more reliable than the auto-spectral density method which is dependent upon both input and output noise.

References

5.1 Newland, D. E. 1984. *An introduction to random vibrations and spectral analysis*, Longman (2nd edition).
5.2 Bendat, J. S. and Piersol, A. G. 1980. *Engineering applications of correlation and spectral analsis*, John Wiley & Sons.
5.3 Kennedy, J. B. and Neville, A. M. 1976. *Basic statistical methods for engineers and scientists*, Harper & Row.
5.4 Lawson, T. V. 1980. *Wind effects on buildings, Volume 2, Statistics and meteorology*, Applied Science Publishers.
5.5 Norton, M. P. and Fahy, F. J. 1988. 'Experiments on the correlation of dynamic stress and strain with pipe wall vibrations for statistical energy analysis applications', *Noise Control Engineering* **30**(3), 107–11.
5.6 Randall, R. B. 1977. *Application of B&K equipment to frequency analysis*, Brüel & Kjaer.
5.7 Randall, R. B. and Hee, J. 1985. 'Cepstrum analysis', chapter 11 in *Digital Signal Analysis*, Brüel & Kjaer.
5.8 Bendat, J. S. and Piersol, A. G. 1971. *Random data: analysis and measurement procedures*, John Wiley & Sons.

Nomenclature

a_0, a_n	Fourier coefficients
b_n	Fourier coefficient
B	filter bandwidth
B_e	frequency resolution bandwidth
B_r	half-power bandwidth
$C_{cxx}(\tau)$	complex cepstrum
$C_{pxx}(\tau)$	power cepstrum
$C_{xx}(\tau)$	covariance of a function $x(t)$
$E[y^2]$	mean-square value of a function $y(t)$

$E[x^3]$	third statistical moment (skewness) of a function $x(t)$
f_c	Nyquist cut-off frequency
f_d	damped natural frequency
f_n	discrete frequency (n/T)
$F(t)$	force signal
$\mathscr{F}\{\ \}$	forward Fourier transform (complex function)
$\mathscr{F}^{-1}\{\ \}$	inverse Fourier transform (real function)
$\mathscr{F}^*\{\ \}$	complex conjugate of $\mathscr{F}\{\ \}$
$G_{mm}(\omega)$, $G_{nn}(\omega)$	one-sided auto-spectral density functions of noise signals
$G_{uu}(\omega)$	one sided auto-spectral density function of a true input signal to a linear system
$G_{vv}(\omega)$	one-sided auto-spectral density function of a true output signal from a linear system
$G_{xx}(\omega)$, $G_{yy}(\omega)$	one-sided auto-spectral density functions of functions $x(t)$ and $y(t)$
$\hat{G}_{xx}(\omega)$	weighted estimate of the auto-spectral density of a function $x(t)$
$\mathscr{G}_{xx}(\omega)$	one-sided energy spectral density function of a function $x(t)$ ($\mathscr{G}_{xx} = TG_{xx}$)
$\mathbf{G}_{\mathbf{Fv}}(\omega)$	one-sided cross-spectral density function of force and velocity (complex function)
$\mathbf{G}_{\mathbf{mn}}(\omega)$	one-sided cross-spectral density function of functions $m(t)$ and $n(t)$ (complex function)
$\mathbf{G}_{\mathbf{um}}(\omega)$	one-sided cross-spectral density function of functions $u(t)$ and $m(t)$ (complex function)
$\mathbf{G}_{\mathbf{vn}}(\omega)$	one-sided cross-spectral density function of functions $v(t)$ and $n(t)$ (complex function)
$\mathbf{G}_{\mathbf{xn}}(\omega)$	one-sided cross-spectral density function of functions $x(t)$ and $n(t)$ (complex function)
$\mathbf{G}_{\mathbf{xv}}(\omega)$	one-sided cross-spectral density function of functions $x(t)$ and $v(t)$ (complex function)
$\mathbf{G}_{\mathbf{xy}}(\omega)$	one-sided cross-spectral density function of functions $x(t)$ and $y(t)$ (complex function)
$\mathbf{H}(\omega)$	arbitrary frequency response function (complex function)
$\lvert\mathbf{H}(\omega)\rvert_{\mathbf{Auto}}$	estimate of $\mathbf{H}(\omega)$ using the auto-spectral density function

$\lvert \mathbf{H}(\omega) \rvert_{\text{Cross}}$	estimate of $\mathbf{H}(\omega)$ using the cross-spectral density function
i	integer
k	integer
m, m_x	mean value of a function $x(t)$
$m(t)$	noise signal (at the input stage)
n	number of time records, integers
$n(t)$	noise signal (at the output stage)
N	integer number of equally spaced sample values
$p(x)$	probability density function
$P(x)$	probability distributed function
$R_{Fv}(\tau)$	cross-correlation function of force and velocity
$R_{xn}(\tau)$	cross-correlation function of functions $x(t)$ and $n(t)$
$R_{xx}(\tau)$	auto-correlation function of a function $x(t)$
$S_{xx}(\omega)$	two-sided auto-spectral density function of a function $x(t)$
$\mathcal{S}_{xx}(\omega)$	two-sided energy spectral density function of a function $x(t)$ ($\mathcal{S}_{xx} = TS_{xx}$)
S/n	signal to noise ratio
t	time
T	time, duration of a transient signal, duration of a sample of a random time signal
T_t	total duration of a digitised signal ($T_t = nT$)
$u(t)$	true input signal to a linear system
$v(t)$	velocity signal, true output signal from a linear system
$w(\tau)$	weighting function, lag window
$W(\omega)$	spectral window
x, x_i, $x(t)$, $x_i(t)$	input signals, random variables
$x(\omega, \Delta\omega, t)$	filtered signal
X	amplitude
$\mathbf{X}$, $\mathbf{X}(\omega)$	Fourier transform of a function $x(t)$ (complex function)
$\mathbf{X}_n$	complex Fourier coefficient of a time signal
$\mathbf{X}^*$	complex conjugate of $\mathbf{X}$
$\mathbf{X}(\omega, T)$	finite Fourier transform (complex function)

$\mathbf{X}^*(\omega, T)$	complex conjugate of $\mathbf{X}(\omega, T)$
$y(t)$	output signal
$\mathbf{Y}, \mathbf{Y}(\omega)$	Fourier transform of a function $y(t)$ (complex function)
$\mathbf{Y}^*$	complex conjugate of $\mathbf{Y}$
α	integration variable
$\gamma_{uv}^2(\omega)$	true coherence function for a linear system
$\gamma_{xy}^2(\omega)$	measured coherence function for a linear system
Δ	incremental time step (sampling interval)
$\Delta\omega$	incremental increase in radian frequency
ε	normalised r.m.s. error
ε_b	normalised bias error
ε_r	normalised random error
ζ	damping ratio
π	3.14 . . .
$\langle\Pi\rangle$	time-averaged power
$\rho_{xx}(\tau)$	auto-correlation coefficient (normalised covariance)
σ	standard deviation
τ	time delay
ω	radian (circular) frequency
ω_n	discrete radian (circular) frequency $(2\pi n/T)$

6
Statistical energy analysis of noise and vibration

6.1 Introduction

Statistical energy analysis (S.E.A.) is a modelling procedure for the theoretical estimation of the dynamic characteristics of, the vibrational response levels of, and the noise radiation from complex, resonant, built-up structures using energy flow relationships. These energy flow relationships between the various coupled subsystems (e.g. plates, shells, etc.) that comprise the built-up structure have a simple thermal analogy, as will be seen shortly. S.E.A. is also used to predict interactions between resonant structures and reverberant sound fields in acoustic volumes. Many random noise and vibration problems cannot be practically solved by classical methods and S.E.A. therefore provides a basis for the prediction of average noise and vibration levels particularly in high frequency regions where modal densities are high. S.E.A. has evolved over the past two decades and it has its origins in the aero-space industry. It has also been successfully applied to the ship building industry, and it is now being used (i) as a prediction model for a wide range of industrial noise and vibration problems, and (ii) for the subsequent optimisation of industrial noise and vibration control.

Lyon's[6.1] book on the general applicability of S.E.A. to dynamical systems was the first serious attempt to bring the various aspects of S.E.A. into a single volume. It is a useful starting point for anyone with a special interest in the topic. There have been numerous advances in the subject since the publication of Lyon's book, and some of these advances are discussed in review papers by Fahy[6.2] and Hodges and Woodhouse[6.3].

This chapter is specifically concerned with the application of S.E.A. to the prediction of noise and vibration associated with machine structures and industrial type acoustic volumes, such as enclosures, semi-reverberant rooms, etc. To this end, firstly the underlying principles of S.E.A. are developed. The successful prediction of noise and vibration levels of coupled structural elements and acoustic volumes using S.E.A. techniques depends to a large extent on an accurate estimate of three parameters. They are (i) the modal densities of the individual subsystems, (ii) the internal loss factors (damping) of the individual subsystems, and (iii) the coupling loss factors (degree of coupling) between the subsystems. The significance of each of these three parameters and the associated measurement and/or theoretical estimation procedures for their evaluation are discussed in this chapter. Secondly, some of the more recent advances in S.E.A. are critically reviewed. These include (i) the effects of non-conservative coupling (i.e. the introduction of damping at a coupling joint), and (ii) the concepts of steady-state and transient total loss factors of coupled subsystems. Thirdly, relationships between mean-square velocity and mean-square stress in structures subject to broadband excitation are developed. S.E.A. facilitates the rapid evaluation of mean-square vibrational response levels of coupled structures. For any useful prediction of service life as a result of possible fatigue or failure, these vibrational response levels must be converted into stress levels. The ability to predict stress levels in a structure directly from vibrational response levels makes S.E.A. a very powerful prediction/monitoring tool.

S.E.A. is particularly attractive in high frequency regions where a deterministic analysis of all the resonant modes of vibration is not practical. This is because at these frequencies there are numerous resonant modes, and numerical computational techniques such as the finite element method have very little applicability.

6.2 The basic concepts of statistical energy analysis

For most S.E.A. applications, it is assumed that the majority of the energy flow between subsystems is due to resonant structural or acoustic modes – i.e. S.E.A. is generally about energy or power flows between different groups of resonant oscillators, although some work has been done on extending it to non-resonant systems[6.1,6.2].

An excellent conceptual introduction to S.E.A. can be found in a paper by Woodhouse[6.4] who discusses a very simple thermal analogy – i.e. vibrational energy is analogous to heat energy. Heat energy flows from a hotter to a cooler place at a rate proportional to the difference of temperature. The constant of proportionality in this instance is a measure of thermal conductivity. As a simple example, Woodhouse[6.4] considers two identical elements, one of which is supplied by heat from some external source. The model is illustrated in Figure 6.1. The two parameters of primary importance are the radiation losses and the degree of coupling via the thermal conductivity link. In practice, situations of high or low radiation losses and high or low thermal conductivity can arise. High thermal conductivity implies a strong coupling link between the two elements, and low thermal conductivity suggests a weak coupling link. Four possible situations can arise. These situations are illustrated schematically in Figure 6.2.

There is an analogy between the thermal model and certain parameters associated with noise and vibration because the flow of vibrational energy in a structure (or noise in an acoustic volume) behaves in the same way as the flow of heat. Provided that there are sufficient resonant structural or acoustic modes within a frequency band of interest, the mean modal energy can be regarded as being equivalent to a measure of temperature. The modal density (number of modes per hertz) is analogous to the thermal capacity of the thermal model, the internal loss factors (damping) are analogous to the radiative losses of the thermal model, and the coupling loss factors (a measure of the strength of the mechanical coupling between the subsystems) are analogous to the thermal conductivity links between the various elements in the thermal model. For two coupled subsystems, Figure 6.2 shows how the mean-square vibrational levels depend on damping and coupling loss factors, and how mean-square temperature levels depend upon radiation and thermal conduction.

Fig. 6.1. Thermal–vibration/acoustic analogy.

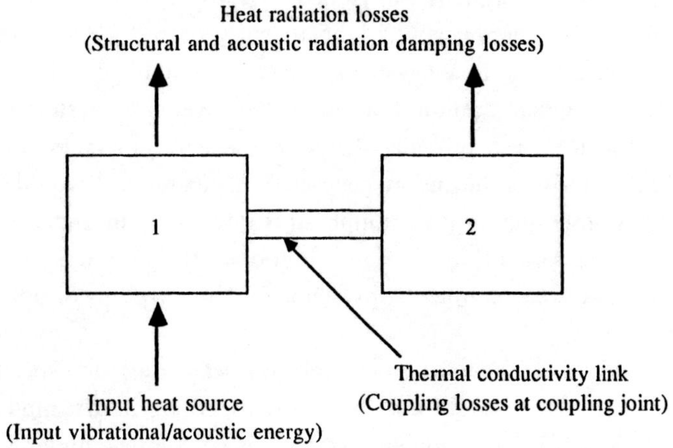

Fig. 6.2. Mean-square temperatures or vibrational energies for various energy loss combinations. (Adapted from Woodhouse[6.4].)

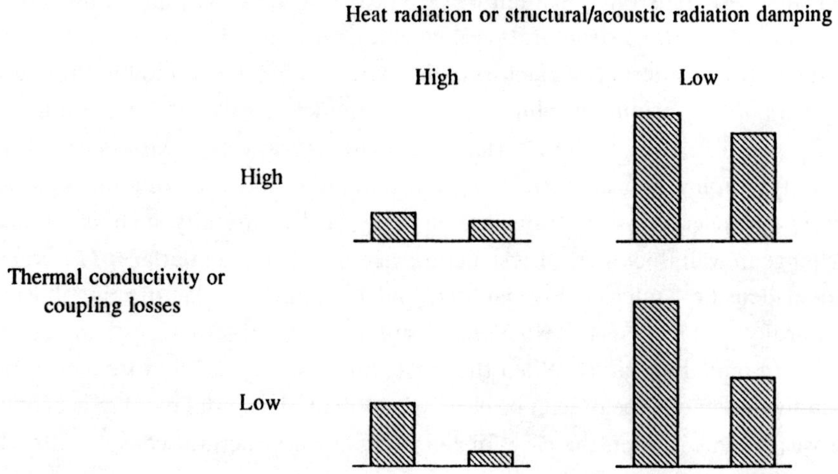

Consider again the two subsystem example in Figure 6.1. If this were a structural system then the input would be some form of vibrational energy, the radiation losses would correspond to internal losses due to structural and acoustic radiation damping, and the conductivity link would be associated with coupling losses at the coupling joint between the two subsystems. Now, assume that (i) the subsystems are strongly coupled; (ii) only subsystem 1 is directly driven; (iii) subsystem 1 is lightly damped; (iv) subsystem 2 is heavily damped, and that one wishes to minimise the vibrational levels transmitted to subsystem 2. Vibration isolation between the two subsystems would not be effective by itself because the vibrational levels in subsystem 1 would rise to a possibly unacceptable level since it is lightly damped (vibration isolation would prevent the vibrational energy from flowing to the more heavily damped subsystem where it could be dissipated). Alternatively, if the vibration isolator was removed and damping treatment was added to subsystem 1 instead, a significant amount of vibrational energy would flow to subsystem 2 because of the strong coupling, and because both subsystems are heavily damped they would both approximately have the same amount of energy. However, if subsystem 1 was damped and vibration isolation was provided between the two subsystems to reduce the coupling link, then most of the energy generated in subsystem 1 would be dissipated at source. This simple qualitative example illustrates how an analysis based upon S.E.A. procedures can provide a very powerful tool for the parametric study of energy flow distributions between coupled subsystems for the purposes of optimising noise and vibration control.

Before proceeding any further, it is desirable to briefly consider a specific structural vibrational problem which could possibly be analysed via S.E.A. modelling. Flow-induced noise and vibration in pipeline systems is such an example. S.E.A. would, however, require the breaking up of a particular piping arrangement into appropriate subsystems. A typical piping arrangement and the associated 'split-up' subsystems are schematically illustrated in Figure 6.3. The S.E.A. modelling procedures require information about three structural parameters: (i) the modal densities of the various subsystems, (ii) the internal loss factors of the various subsystems, and (iii) the coupling loss factors of the various coupling joints. The model density defines the number of modes per unit frequency, the internal loss factor is associated with energy lost by structural damping and acoustic radiation damping, and the coupling loss factor represents the energy lost by transmission across a discontinuity such as a flange, a step change in wall thickness, or a structure–acoustic volume interface. The concepts of modal densities, internal loss factors, and coupling loss factors are illustrated schematically in Figure 6.4. Two specific situations arise with regard to the interpretation of modal densities. When there are numerous modes in a frequency band, if the individual modal peaks can be clearly identified, the modal overlap is defined as being weak – this is often the case for lightly damped structural components. If the

Fig. 6.3. Schematic illustration of S.E.A. subsystems.

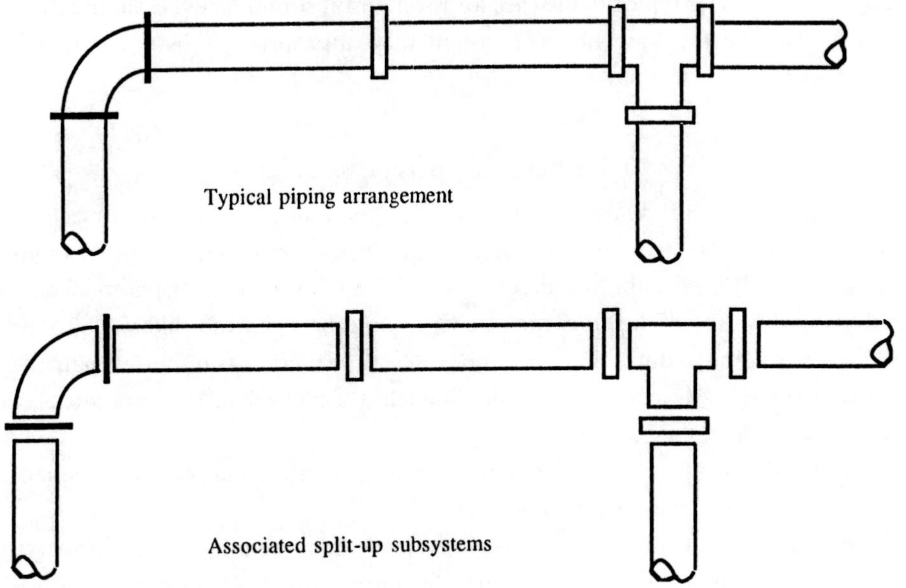

Fig. 6.4. Schematic illustration of modal density, loss factors and coupling loss factors.

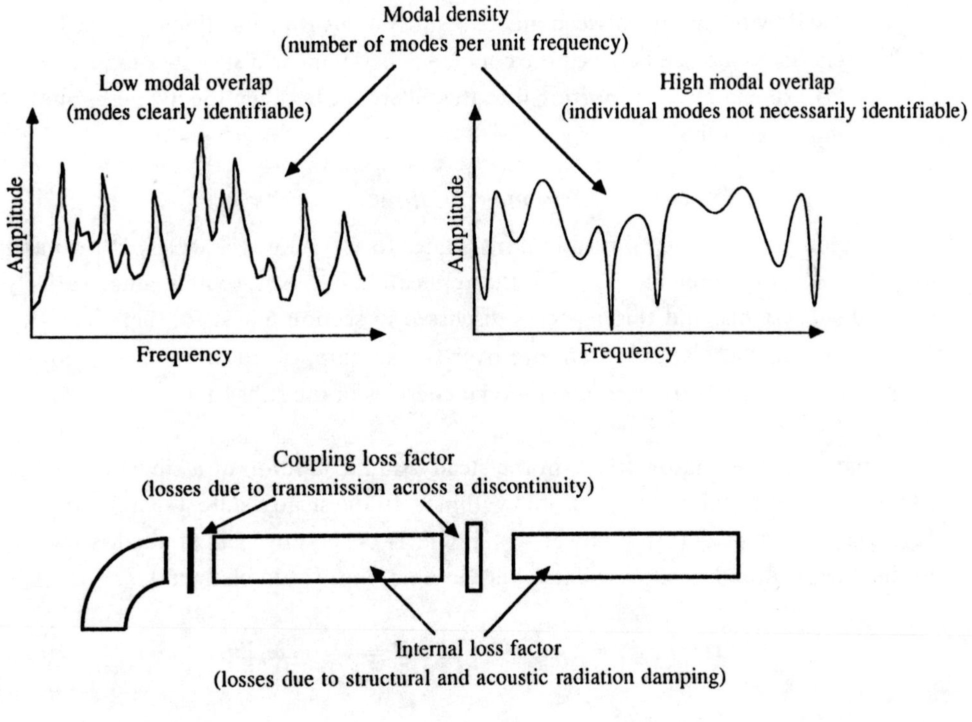

individual modal peaks cannot be clearly identified, the modal overlap is defined as being strong – this is typically the case for reverberant sound fields. It should be clear by now that the breaking up of a system into appropriate subsystems is a very important first step in S.E.A.

6.3 Energy flow relationships

The procedures of S.E.A. can be thought of as the modelling of elastic mechanical systems and fluid systems by subsystems, each one comprising groups of multiple oscillators, with a probabilistic description of the relevant system parameters. The analysis is thus about the subsequent energy flow between the different groups of oscillators. The procedures are based upon several general assumptions, namely:

(i) there is linear, conservative coupling (elastic, inertial and gyrostatic) between the different subsystems;
(ii) the energy flow is between the oscillator groups having resonant frequencies in the frequency bands of interest;
(iii) the oscillators are excited by broadband random excitations with uncorrelated forces (i.e. not point excitation) which are statistically independent – hence there is modal incoherency, and this allows for a linear summation of energies;
(iv) there is equipartition of energy between all the resonant modes within a given frequency band in a given subsystem;
(v) the principle of reciprocity applies between the different subsystems;
(vi) the flow of energy between any two subsystems is proportional to the actual energy difference between the coupled subsystems whilst oscillating – i.e. the flow of energy is proportional to the difference between the average coupled modal energies.

6.3.1 Basic energy flow concepts

The preceding list of general assumptions relates to S.E.A. as it is widely known and applied. Recent research has extended the application of S.E.A. to non-conservatively coupled subsystems, and this aspect is discussed in section 6.8. Also, there has been some debate in the research literature over the assumption that the energy flow is proportional to the average coupled modal energies of the subsystems – this point is discussed shortly.

An individual oscillator driven in the steady-state condition at a single frequency has potential and kinetic energy stored within it. In the steady-state, the input power, Π_{in}, has to balance with the power dissipated, Π_{d}. The power dissipated is related to the energy stored in the oscillator via the damping. From chapter 1

$$\Pi_{\text{d}} = c_{\text{v}}\dot{x}^2 = 2\zeta\omega_n m\dot{x}^2 = 2\zeta\omega_n E = \frac{\omega_n E}{Q} = \omega_n E\eta, \qquad (6.1)$$

where c_v is the viscous-damping coefficient, ζ is the damping ratio (damping/critical damping), ω_n is the radian natural frequency, m is the oscillator mass, E is the stored energy, Q is the quality factor, and η is the loss factor. The power dissipation concepts for a single oscillator can be extended to a collection of oscillators in specified frequency bands (generally octave bands one-third-octave bands or narrower bands with constant bandwidths). Here,

$$\Pi_d = \frac{\omega E}{Q} = \omega E \eta, \tag{6.2}$$

where ω is the geometric mean centre frequency of the band, and η is now the mean loss factor of all the modes within the band.

In the original development of S.E.A., Lyon[6.1] and others considered the flow of energy between two oscillators coupled linearly via stiffness coupling, inertial coupling, and gyrostatic coupling. A good example of gyrostatic coupling is the acoustic coupling between a fluid and a structure. Both oscillators were excited by statistically independent forces with the same broadband spectra. It was shown that the time-averaged energy flow between the two oscillators is given by

$$\langle \Pi_{12} \rangle = \beta' \{ \langle E_1' \rangle - \langle E_2' \rangle \}, \tag{6.3}$$

where β' is a constant of proportionality which is independent of the excitation source strength and is only a function of the oscillator parameters, and the time-averaged energies $\langle E_1' \rangle$ and $\langle E_2' \rangle$ are the blocked energies of the individual oscillators; the blocked energy of an oscillator being the sum of its kinetic and potential energy whilst it is coupled but with the other oscillator held motionless. Subsequent to that original analysis, Lyon[6.1] also showed that the energy flow between the two oscillators is also proportional to the difference between the actual total vibrational energies of the respective coupled oscillators. Thus,

$$\langle \Pi_{12} \rangle = \beta \{ \langle E_1 \rangle - \langle E_2 \rangle \}, \tag{6.4}$$

where $\langle E_1 \rangle$ and $\langle E_2 \rangle$ are now the actual time-averaged energies of the respective coupled oscillators, and β is another constant of proportionality. This equation is the fundamental basis of S.E.A.

There are four important comments to be made in relation to equations (6.3) and (6.4). They are (i) energy flows from an oscillator of higher to lower energy – this is analogous to the previous thermal example; (ii) energy flow is proportional to the time-averaged energy difference; (iii) the constants, β' and β, are related to the blocked natural frequencies and the associated oscillator parameters; and (iv) both equations are exactly correct for energy flow between two linearly coupled oscillators. The reader is referred to Lyon[6.1] for a detailed analysis of the derivation of equations (6.3) and (6.4), and an associated discussion on the energy flow between two linearly coupled oscillators.

6.3.2 Some general comments

Conceptual problems arise when attempting to extend equation (6.4) to the more general case of coupled groups of oscillators. Equation (6.4) is a statement about energy flow between two individual modes, but it is used in S.E.A. to describe the average energy flow between two structures or between a structure and an acoustic volume. The S.E.A. assumption that energy flow is proportional to the difference in average coupled modal energies is thus simply an extension of the two oscillator result to multimodal systems.

Hodges and Woodhouse[6.3] discuss the scope of this S.E.A. assumption in some considerable detail. In their review paper they show that, provided the modal forces are incoherent, the total energy flow between a subsystem and the rest of the system is a sum over differences of uncoupled modal energies. The uncoupled modal energies are defined as the energies of vibration of the individual subsystems whilst vibrating by themselves but being driven by the same external forces that would otherwise have been applied – if a subsystem has no external force applied to it, its uncoupled modal energy would be zero. Hodges and Woodhouse also show that the energy flow is a linear combination of the actual energies of the blocked resonant modes of the various subsystems whilst in a coupled state. Both these types of energy flow models are more rigorous than the S.E.A. assumption (assumption (vi)) and allow for the presence of indirect coupling terms. Indirect coupling indicates that the energy flow between two groups of oscillators is influenced by other oscillator groups in the overall system – i.e. energy difference terms between blocked-mode oscillators which are not directly coupled have to be accounted for. These energy difference terms can only be accounted for in practice if sufficient information is available about the various blocked natural frequencies and the various interaction forces. This information is generally not readily available, hence the S.E.A. assumption that the energy flow between coupled subsystems is proportional to the difference between average coupled modal energies. This S.E.A. assumption is analogous to the heat flow model and it does not therefore allow for the presence of indirect coupling. Therefore, in general terms, S.E.A. is most suitably applied to subsystems which are lightly coupled. When this is the case, there is no indirect coupling and the coupled and uncoupled modal energies are approximately equal. In his review paper, Fahy[6.2] also discusses these subtle differences between coupled and uncoupled modal energies.

The qualitative discussion in the preceding paragraph is hopefully not meant to confuse the reader! It is intended to highlight the fact that S.E.A. is a very powerful engineering tool provided that it is used correctly, there being certain circumstances when its usage (in the form as it is generally known) is inappropriate. The issues raised in the preceding paragraph will become more evident as one progresses through the chapter. At this stage it is sufficient to note that S.E.A. is most successful (i) when there is weak coupling between subsystems; (ii) when the exciting forces are broadband in nature; (iii) when the modal densities of the respective subsystems are high; and (iv) when the assumptions outlined at the beginning of this section are fulfilled. S.E.A. procedures can still be used in practice if any of the preceding criteria are not strictly met. The results are generally not as reliable as they otherwise might be, but they can provide a qualitative assessment of the problem – the strongly coupled vibrational subsystems discussed in section 6.2 (Figure 6.1) is a case in point. Research is currently in progress to reduce the number of assumptions required for, and to broaden the formulation of S.E.A. Some of this work is discussed in sections 6.8 and 6.9.

6.3.3 The two subsystem model

Returning now to S.E.A., in the form as it is generally known, one is now in a position to extend equation (6.4) to cover the more general case of two groups of lightly coupled oscillators with modal densities n_1 and n_2, respectively. The average energy flow, $\langle\Pi_{12}\rangle$, between the two groups of oscillators can be expressed by

$$\langle\Pi_{12}\rangle = \gamma\{\langle E_1\rangle/n_1 - \langle E_2\rangle/n_2\}, \tag{6.5}$$

where γ is another constant of proportionality which is only a function of the oscillator parameters. It should be noted that $\langle E_1\rangle$ and $\langle E_2\rangle$ are the total energies of the respective subsystems; $\langle E_1\rangle/n_1$, etc. are the modal energies.

Equation (6.5) states that the energy flow is proportional to the difference between average coupled modal energies. It can be transformed into a more convenient form by introducing the concept of coupling loss factors which describe the flow of energy between subsystems. The coupling loss factor, η_{ij}, relates to energy flow from subsystem i to subsystem j, and is a function of the modal density, n_i, of subsystem i, the constant of proportionality, γ, and the centre frequency, ω, of the band – it is just a form of the loss factor described in equation (6.1). Equation (6.5) can thus be expressed in power dissipation terms – the nett energy flow from subsystem 1 to subsystem 2 is the difference between the power dissipated during the flow of energy from subsystem 1 to subsystem 2 and the power dissipated during the flow of energy from subsystem 2 back to subsystem 1. Hence, using the power dissipation concepts developed in equation (6.1),

$$\langle\Pi_{12}\rangle = \omega\langle E_1\rangle\eta_{12} - \omega\langle E_2\rangle\eta_{21}, \tag{6.6}$$

where η_{12} and η_{21}, are the coupled loss factors between subsystems 1 and 2, and 2 and 1, respectively.

By inspection of equations (6.5) and (6.6),

$$\frac{\gamma}{n_1} = \omega\eta_{12}, \qquad \text{and } \frac{\gamma}{\eta_2} = \omega\eta_{21}. \tag{6.7}$$

Thus,

$$n_1\eta_2 = n_2\eta_{21}. \tag{6.8}$$

Equation (6.8) is the reciprocity relationship between the two subsystems, and it is sometimes referred to as the consistency relationship. Reciprocity was discussed in section 3.2, chapter 3 in relation to fluid–structure interactions. Hence, by substituting the reciprocity relationship into equation (6.6),

$$\langle \Pi_{12} \rangle = \omega \eta_{12} \left\{ \langle E_1 \rangle - \frac{n_1}{n_2} \langle E_2 \rangle \right\}. \tag{6.9}$$

Now consider a two subsystem model (numerous modes in each subsystem) where one subsystem is driven directly by external forces and the other subsystem is driven only through the coupling. The model is illustrated in Figure 6.5, where Π_1 is the power input to subsystem 1; $\Pi_2 = 0$ is the power input into subsystem 2; n_1 and n_2 are the modal densities, and η_1 and η_2 are the internal loss factors of subsystems 1 and 2, respectively; η_{12} and η_{21} are the coupling loss factors associated with energy flow from 1 to 2 and from 2 to 1; and E_1 and E_2 are the vibrational energies associated with subsystems 1 and 2. All fluctuating terms such as Π or E are assumed to be both time- and space-averaged, and the brackets and overbars have been removed for convenience. Quite often when one is conducting model tests under laboratory conditions, it is convenient to excite a structure at a single point with some sort of an electro-mechanical exciter arrangement. When this is the case, space-averaging is essential; it has been shown by Bies and Hamid[6.5] that single point excitation at several points randomly chosen does satisfy the assumption of statistical independence.

The steady-state power balance equations for the two groups of oscillators are

$$\Pi_1 = \omega E_1 \eta_1 + \omega E_1 \eta_{12} - \omega E_2 \eta_{21}, \tag{6.10}$$

and

$$0 = \omega E_2 \eta_2 + \omega E_2 \eta_{21} - \omega E_1 \eta_{12}. \tag{6.11}$$

Fig. 6.5. A two subsystem S.E.A. model.

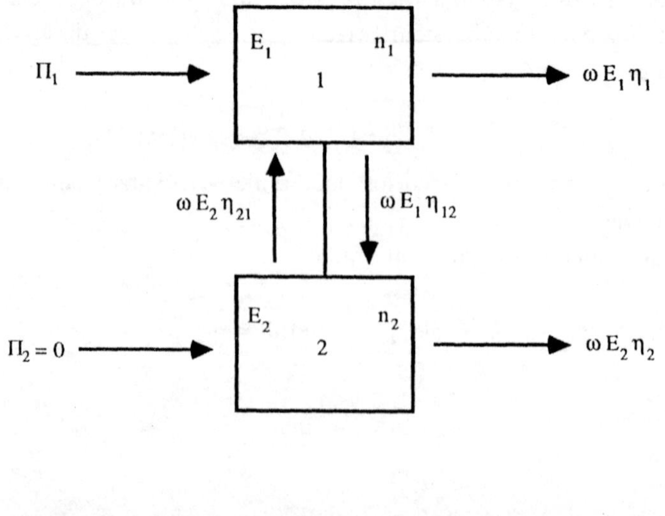

The steady-state energy ratio between the two groups of oscillators can be obtained from equation (6.11). It is

$$\frac{E_2}{E_1} = \frac{\eta_{12}}{\eta_2 + \eta_{21}}. \tag{6.12}$$

Equation (6.12) is a very important conceptual equation. It illustrates how energy ratios between coupled groups of oscillators can be obtained from the internal loss and coupling loss factors. Furthermore, if the input energy to subsystem 1 is known, the output energy from subsystem 2 can be readily estimated. By substituting equation (6.8) into equation (6.12), one gets

$$\frac{E_2^*}{E_1^*} = \frac{\eta_{21}}{\eta_2 + \eta_{21}}, \tag{6.13}$$

where $E_1^* = E_1/n_1$ and $E_2^* = E_2/n_2$. For the special case of two coupled oscillators, rather than two coupled groups of oscillators, the modal densities n_1 and n_2 are both equal to unity, hence $E_1^* = E_1$ and $E_2^* = E_2$.

Two important points, which draw an analogy with the thermal example discussed at the beginning of the chapter, can be made in relation to equation (6.13). Firstly, if $\eta_2 \ll \eta_{21}$ then $E_2^*/E_1^* \to 1$. This suggests that additional damping to subsystem 2 will be ineffective unless η_2 can be brought up to the same level as η_{21}. Secondly, E_2^* is always less than E_1^* since η_{21} has to be positive. When $E_2^*/E_1^* \to 1$ there is equipartition of energy between the two groups of oscillators.

6.3.4 *In-situ estimation procedures*

The energy flow relationships that have just been developed illustrate the basic principles of S.E.A. With a knowledge of the modal densities and internal loss factors of two different subsystems, and the coupling loss factors between the subsystems, one can readily estimate the energy flow ratios. Alternatively, information could be obtained about the internal loss and coupling loss factors from the total energies of vibration and the modal densities.

Equations (6.13) can be re-written in terms of the total energies of vibration, E_1 and E_2, of the two groups of oscillators. Using the consistency relationship (equation 6.8),

$$\frac{E_2}{E_1} = \frac{n_2\eta_{12}}{n_2\eta_2 + n_1\eta_{12}}, \tag{6.14}$$

and thus

$$\frac{\eta_{12}}{\eta_2} = \frac{n_2 E_2}{n_2 E_1 - n_1 E_2}. \tag{6.15}$$

Equations (6.10)–(6.15) are only valid for direct excitation of subsystem 1, with subsystem 2 being excited indirectly via the coupling joint. If the experiment is reversed and subsystem 2 is directly excited with subsystem 1 being excited indirectly via the coupling joint, then

$$\frac{\eta_{12}}{\eta_1} = \frac{n_2 E_1}{n_1 E_2 - n_2 E_1}. \tag{6.16}$$

Equations (6.15) and (6.16) allow one to set up experiments to measure the coupling loss factors between two subsystems provided that one has prior information about the modal densities and the internal loss factors of the individual subsystems. Firstly, subsystem 1 is excited at a single point and the time- and space-averaged energies of vibration of the coupled subsystems are obtained with an accelerometer. The point excitation is repeated at several points, randomly chosen, to satisfy the assumption of statistical independence. Equation (6.15) is then used to obtain the coupling loss factor η_{12} and equation (6.8) is subsequently used to obtain the coupling loss factor η_{21}. Alternatively, if subsystem 2 was excited, equation (6.16) could have been used. Information about the modal densities is obtained separately either experimentally or from theoretical relationships. Information about the internal loss factors is generally obtained from experiments. Procedures for the estimation of modal densities and internal loss factors are discussed later on in this chapter. At this point it is sufficient to note that theoretical estimates of modal densities are readily available for acoustic volumes and for a wide range of basic structural elements: modal densities of composite structural elements generally have to be obtained experimentally. Very little theoretical information is available about internal loss factors; most available information is empirical and is based upon experimental data.

Sometimes, independent information is not available about the internal loss factors η_1 and η_2. When this is the case, equations (6.15) and (6.16) cannot be solved because there are two equations and three unknowns (η_1, η_2, and η_{12} or η_{21}). A third equation is required (without uncoupling the subsystems) to solve for η_1, η_2 and η_{12}. This third equation can be obtained either from a steady-state or from a transient analysis of the coupled subsystems.

Firstly, consider a steady-state analysis. From equations (6.8), (6.10) and (6.12)

$$\frac{\Pi_1}{\omega E_1} = \eta_1 + \frac{\dfrac{n_2}{n_1}\eta_2 \eta_{21}}{\eta_2 + \eta_{21}} = \eta_{\mathrm{TS}1}, \tag{6.17}$$

where $\eta_{\mathrm{TS}1}$ is the total steady-state loss factor of subsystem 1 whilst coupled to subsystem 2 – it is always greater than η_1 because it is a function of the internal loss factor of the second subsystem and the coupling loss factors. $\eta_{\mathrm{TS}1}$ can be measured experimentally by measuring the power input into subsystem 1 and also measuring its vibrational energy. Equation (6.17) provides the third equation which is necessary to solve for the three unknowns (η_1, η_2, and η_{12} or η_{21}).

Alternatively, a transient analysis can be considered by abruptly switching off the power to subsystem 1. The general power balance equation for subsystem 1 (equation 6.10) strictly speaking is given by

$$\Pi_1 = \mathrm{d}E_1/\mathrm{d}t + \omega E_1 \eta_1 + \omega E_1 \eta_{12} - \omega E_2 \eta_{21}. \tag{6.18}$$

For steady-state excitation, $\mathrm{d}E_1/\mathrm{d}t = 0$, and for transient excitation, $\Pi_1 = 0$. Sun *et al.*[6.6] solve the transient equations to obtain a total transient loss factor, η_{TT1}, for a subsystem which is coupled to another one. It is

$$\eta_{TT1} = 0.5[(\eta_1 + \eta_2 + \eta_{12} + \eta_{21}) - \{(\eta_1 + \eta_{12} - \eta_2 - \eta_{21})^2 + 4\eta_{12}\eta_{21}\}^{1/2}]. \tag{6.19}$$

Equation (6.19) also provides the third equation which is necessary to solve for the three unknowns (η_1, η_2, and η_{12} or η_{21}). It is important to note that $\eta_{TS} \neq \eta_{TT}$. The total steady-state loss factor of a subsystem is always larger than its own internal loss factor; this is not necessarily the case for total transient loss factors. Total loss factors are discussed again in section 6.9.

6.3.5 *Multiple subsystems*

The preceding discussions relating to two groups of oscillators can be extended to multiple groups. In the general case, N groups of oscillators yield N simultaneous energy balance equations which can be written in matrix form. The loss factor matrix is symmetric because of the reciprocity relationship as expressed by equation (6.8). The steady-state energy balance matrix is

$$\omega[A]\begin{bmatrix} E_1/n_1 \\ E_2/n_2 \\ \cdot \\ E_N/n_N \end{bmatrix} = \begin{bmatrix} \Pi_1 \\ \Pi_2 \\ \cdot \\ \Pi_N \end{bmatrix}, \tag{6.20}$$

where

$$[A] = \begin{bmatrix} \left(\eta_1 + \sum_{i \neq 1}^{N} \eta_{1i}\right) n_1 & -\eta_{12} n_1 & \cdot & -\eta_{1N} n_1 \\ -\eta_{21} n_2 & \left(\eta_2 + \sum_{i \neq 2}^{N} \eta_{2i}\right) n_2 & \cdot & -\eta_{2N} n_2 \\ \cdot & \cdot & \cdot & \cdot \\ -\eta_{N1} n_N & \cdot & \cdot & \left(\eta_N + \sum_{i \neq N}^{N} \eta_{Ni}\right) n_N \end{bmatrix}. \tag{6.21}$$

The equations in the above matrix are linear equations and they allow for a systematic analysis of the interactions between coupled groups of oscillators and a parametric study of the variables. Any such parametric analysis of a noise or vibrational system using S.E.A. is thus dependent upon a knowledge of the modal densities, the internal loss factors and the coupling loss factors associated with the various subsystems. Because of this, the choice of suitable subsystems is an important factor. As a rule of thumb, it is always appropriate to choose subsystems such that there is weak coupling between them ($\eta_{ij} \ll \eta_i$ and η_j). In S.E.A., it is also assumed that most of the energy flow is between resonant modes of the various subsystems and that the modal density of the resonant modes within a coupled subsystem is equal to the modal density of the uncoupled modes. This assumption is not unreasonable at frequencies where the modal density is high. Thus, a suitable boundary (between two coupled subsystems) is one where there is a large impedance mismatch – i.e. where waves are substantially reflected. Many types of vibrational modes (bending, torsional, shear, longitudinal, etc.) are, however, possible on complex solid bodies and the degree of reflection is very much dependent upon the type of wave that is incident upon the boundary. Thus, a situation could arise where different boundaries have to be defined on the same structure for different wave types. It should also be noted that modal densities of different wave types on the same structure can have very different values. Most vibrations that are associated with noise radiation are, however, associated with bending (flexural) waves and this is especially true for combinations of plates, shells and cylinders. Thus, the application of S.E.A. is generally restricted to the analysis of bending waves generated by flexural or by in-plane (longitudinal and/or shear) wave transmission across joints, and the subsystems are generally selected by their geometric boundaries. Some typical examples are illustrated in Figure 6.6. The first example comprises two coupled plate elements; the second example comprises several coupled plate elements and two coupled volume elements; the third example comprises two coupled cylindrical shell elements and a volume element.

It now remains to discuss the various procedures (experimental and analytical) that are available for the estimation of modal densities, internal loss factors, and coupling loss factors for a variety of common subsystems. An accurate estimation of these parameters is essential for any successful S.E.A. prediction model. Modal densities and loss factors are also of general engineering interest and have applications beyond S.E.A. When conducting experiments to evaluate the three S.E.A. parameters, caution has got to be exercised as one is often dealing with very small numbers and their differences. There have been numerous research papers dealing with the subject over the last twenty years, and Fahy[6.2] provides an extensive bibliography. In general, most of the available publications in the literature on the experimental aspects of S.E.A. are primarily concerned with the application of specialist techniques for a range of specific experimental conditions. The experimentalist must therefore have a comprehensive understanding of the various sources of experimental error, and the suitability or otherwise of the particular technique in relation to the test structure. Of particular concern are the methods of excitation of the structure, the selection of suitable transducers, and the minimisation of feedback and bias errors.

Modal densities, internal loss factors and coupling loss factors are discussed in some detail in the next three sections.

6.4 Modal densities

The vibrational and acoustical response of structural elements, and the acoustical response of volume elements to random excitations, is often dominated by the resonant response of contiguous structural and acoustic modes. It is worth reminding the reader that, when a structure is excited by some form of broadband structural excitation, the dominant structural response is resonant; when a structure is acoustically excited, the dominant response is generally forced although it can also be resonant (chapter 3); and, when a reverberant acoustic volume is excited by some broadband noise source, the dominant response is resonant. It is the energy flow between resonant groups of modes that is of primary concern here. The modal density (number of modes per unit frequency) is therefore a very important parameter for establishing the resonant response of a system to a given forcing function.

Fig. 6.6. Some typical examples of S.E.A. subsystems.

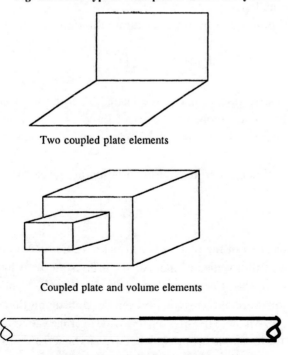

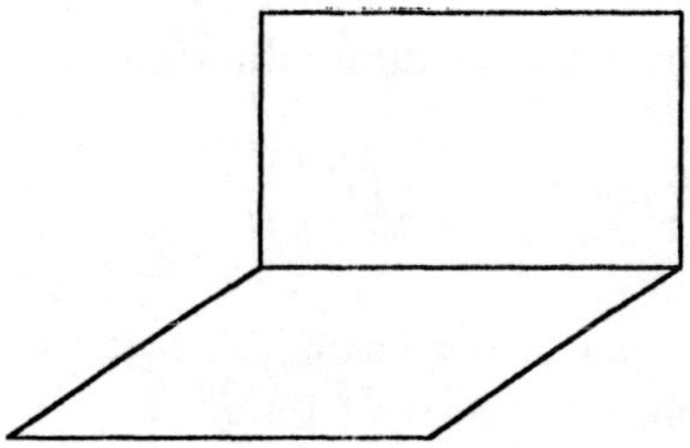

Two coupled plate elements

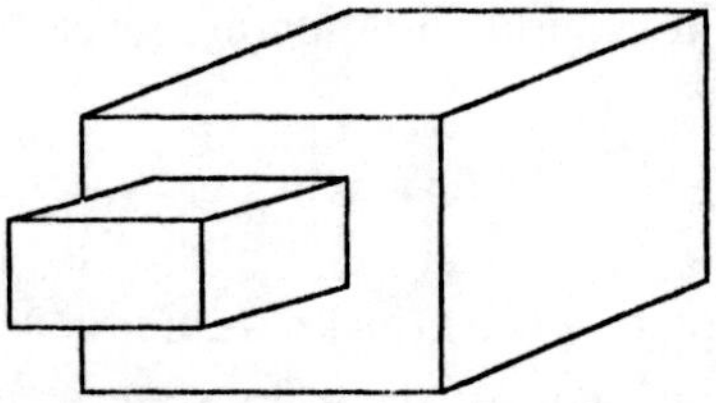

Coupled plate and volume elements

Coupled cylindrical shell and volume elements

Asymptotic modal density formulae are available in the literature[6.1,6.2,6.7–6.10] for a range of idealised subsystems such as bars, beams, flat plates, thin-walled cylinders, acoustic volumes, etc. Theoretical estimates are not readily available for non-ideal subsystems, and under these circumstances experimental techniques[6.11,6.12] are more suitable. Recently, Clarkson and Ranky[6.13] and Ferguson and Clarkson[6.14] have developed theoretical modal density relationships for honeycomb type structures (plates and shells) similar to those used in the aero-space industry.

6.4.1 Modal densities of structural elements

Some simple formulae for estimating the modal densities of some commonly used structural elements are presented in this sub-section. The modal density, $n(f)$, is defined as the number of modes per unit frequency (Hz). It is also sometimes defined in the literature as the number of modes per unit radian frequency – i.e. as $n(\omega)$. Thus,

$$n(f) = 2\pi n(\omega). \tag{6.22}$$

Modal densities of uniform bars in longitudinal vibration are given by[6.7,6.8]

$$n(f) = \frac{2L}{c_{\mathrm{L}}}, \qquad \text{where } c_{\mathrm{L}} = \left(\frac{E}{\rho}\right)^{1/2}, \tag{6.23}$$

and E is Young's modulus of elasticity, ρ is the density (mass per unit volume), and L is the length of the bar.

Modal densities of uniform beams in flexural vibration are given by[6.7,6.8]

$$n(f) = \frac{L}{(2\pi f)^{1/2}}\left(\frac{\rho A}{EI}\right)^{1/4}, \tag{6.24}$$

where A is the cross-sectional area of the beam, EI is the flexural stiffness of the beam, and I is the second moment of area of the cross-section about the neutral plane axis. It is useful to note that the modal density decreases with increasing frequency.

Modal densities of flat plates in flexural vibration are given by[6.7,6.8]

$$n(f) = \frac{S\sqrt{12}}{2c_{\mathrm{L}}t}, \qquad \text{where } c_{\mathrm{L}} = \left\{\frac{E}{\rho(1-\nu^2)}\right\}^{1/2}, \tag{6.25}$$

and S is the surface area of the plate, t is its thickness, and ν is Poisson's ratio.

Modal densities of thin-walled cylindrical shells are somewhat harder to estimate because, not only are they frequency dependent, but this frequency dependency is not a linear function. Several theories have been developed and they are summarised by Hart and Shah[6.9]. All these theories only provide average values of modal densities and do not account for mode groupings that are characteristic of cylindrical shells at frequencies below the ring frequency[6.11]. The ring frequency, f_{r}, of a cylindrical shell is that frequency at which the cylinder vibrates uniformly in the breathing mode, and it is given by

$$f_r = \frac{c_L}{2\pi a_m} = \frac{1}{2\pi a_m}\left\{\frac{E}{\rho(1-\nu^2)}\right\}^{1/2}, \tag{6.26}$$

where a_m is the mean shell radius. Above the ring frequency, the structural wavelengths are such that a cylinder would tend to behave like a flat plate; below the ring frequency, the modal density varies because of selective grouping of structural modes of differing circumferential mode orders. Clarkson and Pope[6.10] utilised approximations developed by Szechenyi[6.15] for estimating average modal densities of cylindrical shells below and above the ring frequency. The relationships are semi-empirical and are based on earlier work reviewed by Hart and Shah[6.9]. They are:

(i) for $f/f_r \leqslant 0.48$

$$n(f) = \frac{5S}{\pi c_L t}\left(\frac{f}{f_r}\right)^{1/2}, \tag{6.27}$$

(ii) for $0.48 < f/f_r \leqslant 0.83$

$$n(f) = \frac{7.2S}{\pi c_L t}\left(\frac{f}{f_r}\right), \tag{6.28}$$

(iii) for $f/f_r > 0.83$

$$n(f) = \frac{2S}{\pi c_L t}\left[2 + \frac{0.596}{F - 1/F}\left\{F\cos\left(\frac{1.745 f_r^2}{F^2 f^2}\right) - \frac{1}{F}\cos\left(\frac{1.745 F^2 f_r^2}{f^2}\right)\right\}\right] \tag{6.29}$$

where S is the surface area of the cylinder, t is its wall thickness, and F is a bandwidth factor ($\{$upper frequency/lower frequency$\}^{1/2}$). For one-third-octave bands, $F = 1.122$, and, for octave bands, $F = 1.414$.

The modal density estimates provided by equations (6.27), (6.28) and (6.29) do not account for the grouping of circumferential modes in cylindrical shells at frequencies below the ring frequency. Hence they do not adequately describe the modal density fluctuations due to the cut-on of these modes. These fluctuations can be very large, particularly for long, thin cylindrical shells. Keswick and Norton[6.11] have recently developed a modal density computer prediction model which accounts for these fluctuations. The computer algorithm is based upon well known strain relationships for thin-walled cylindrical shells.

Honeycomb structures with a deep core are used extensively in the aero-space industry for weight reduction purposes. Recently Clarkson and Ranky[6.13] and Ferguson and Clarkson[6.14] have developed theoretical relationships for a range of honeycomb type structural elements. The bending stiffness can be neglected for honeycomb panels with thin face plates. In this instance, the modal density is given by

$$n(f)=\frac{\pi abm}{gB}f[1+\{m\omega^2+2g^2B(1-\nu^2)\}\{m^2\omega^4+4m\omega^2g^2B(1-m^2)\}^{-1/2}], \quad (6.30)$$

where a and b are the panel dimensions, m is the total mass per unit area, g is the core stiffness parameter, B is the faceplate longitudinal stiffness parameter, f is the frequency in hertz, ω is the radian frequency, and ν is Poisson's ratio. The core stiffness parameter, g, is defined as

$$g=\frac{(G_xG_y)^{1/2}}{h_2}\left(\frac{1}{E_1h_1}+\frac{1}{E_3h_3}\right), \quad (6.31)$$

where h_1 and h_3 are the two faceplate thicknesses, h_2 is the core thickness, G_x and G_y are the shear moduli of the core material in the x- and y-directions, respectively, and E_1 and E_3 are the faceplate moduli of elasticity. The faceplate longitudinal stiffness parameter, B, is given by

$$B=d^2\frac{E_1h_1E_3h_3}{E_1h_1+E_3h_3}, \quad (6.32)$$

where $d=h_2+(h_1+h_3)/2$.

Clarkson *et al.*[6.12–6.14] also discuss the effects of distributed masses, stiffeners, edge closures, attachment members, and corrugation. The general conclusion is that, when the subsystems to be modelled are not ideal structural elements, experimental techniques are more appropriate than a theoretical analysis. These experimental techniques are described in sub-section 6.4.3.

6.4.2 Modal densities of acoustic volumes

The modal density of an acoustic volume varies depending on whether the volume is one-dimensional (a cylindrical tube), two-dimensional (a shallow cavity), or three-dimensional (an enclosure). Lyon[6.1] and Fahy[6.2] provide some semi-empirical relationships for the three cases.

For a one-dimensional acoustic volume (e.g. a long, slender tube), where the wavelength of sound is greater than any of the cross-dimensions,

$$n(f)=\frac{2L}{c}, \quad (6.33)$$

where L is the length of the volume, and c is the speed of sound.

For a two-dimensional shallow acoustic cavity, where the wavelength of sound is at least twice the depth of the cavity,

$$n(f)=\frac{\pi fA}{c^2}+\frac{P}{c}, \quad (6.34)$$

where A is the total surface area of the cavity, and P is its perimeter.

For a three-dimensional volume enclosure,

$$n(f)=\frac{4\pi f^2 V}{c^3}+\frac{\pi f A}{2c^2}+\frac{P}{8c}, \tag{6.35}$$

where V is the volume of the enclosure, A is the total surface area, and P is the total edge length. The modal density of large acoustic volumes (e.g. semi-reverberant rooms) is generally approximated by the first term of equation (6.35).

6.4.3 Modal density measurement techniques

Modal densities of acoustic volumes can be readily obtained from the relationships provided in the previous subsection. However, as far as structural elements are concerned, the relevant subsystems for an S.E.A. analysis are often far from ideal from a geometrical viewpoint. Because of this, theoretical estimates are not readily available, and under these circumstances experimental techniques are more suitable.

Until recently, the structural mode count technique using a sine sweep or an impact hammer has been the only available procedure for estimating modal densities of structures. Whilst these techniques have their applications, they are very cumbersome when having to deal with large numbers of structural modes, and are prone to errors at high frequencies. This is especially true when there is a significant amount of modal overlap. Structural mode count techniques are therefore not suitable for S.E.A. applications where rapid data acquisition is desirable.

Modal densities of structural elements can be reliably obtained via the measurement of the spatially averaged point mobility frequency response function. The point mobility technique originates from some theoretical work by Cremer *et al.*[6.7] and has been successfully used by Clarkson and Pope[6.10], Keswick and Norton[6.11], and others. Mobility is a complex frequency response function (commonly referred to as a transfer function) of an output velocity and an input force. It is defined by

$$\mathbf{Y}(\omega)=\frac{\mathbf{V}(\omega)}{\mathbf{F}(\omega)}, \tag{6.36}$$

where the bold lettering denotes that the quantities are complex. Point mobility is the ratio of velocity to force at a specific point on a structure.

As in the previous chapters, frequency response functions and spectral densities are represented in this chapter as a function of the radian frequency, ω. They can also be represented as functions of frequency, f, and this is the parameter which is used by all modern signal analysis equipment. Both parameters are completely consistent with each other but caution must be exercised when transforming equations. It is useful to note that

$$\frac{\text{quantity}}{\text{Hz}}=\frac{2\pi\times\text{quantity}}{\text{radian frequency}}, \tag{6.37}$$

since $\omega = 2\pi f$. Thus,

$$\mathbf{Y}(f) = 2\pi \mathbf{Y}(\omega), \tag{6.38}$$

and

$$n(f) = 2\pi n(\omega). \qquad \text{(equation 6.22)}$$

It should also be noted that $\mathrm{d}\omega = 2\pi \mathrm{d}f$. As an example, the Fourier transform pair (equation 1.119) can be re-written in terms of f rather than ω. By making the appropriate substitutions (i.e. using equation 6.37 and noting that $\mathrm{d}\omega = 2\pi \mathrm{d}f$),

$$\mathbf{X}(f) = \int_{-\infty}^{\infty} x(t)\, \mathrm{e}^{-\mathrm{i}2\pi f t}\, \mathrm{d}t,$$

and

$$x(t) = \int_{-\infty}^{\infty} \mathbf{X}(f)\, \mathrm{e}^{\mathrm{i}2\pi f t}\, \mathrm{d}f. \qquad \text{(equation 1.119)}$$

Returning now to the discussion on mobilities, the real part of the point mobility, Re $[\mathbf{Y}(\omega)]$, when space-averaged and integrated over the frequency band of interest, is a function of the modal density of the structure. One is concerned with the real part because it represents the mean energy flow which can be dissipated. The imaginary part represents reactive energy exchange in the region of the coupling point. From power balance considerations for point excitation of a finite structure the modal density is given by[6.7,6.12]

$$n(\omega) = 4S\rho_S \overline{\mathrm{Re}\,[\mathbf{Y}(\omega)]}, \tag{6.39}$$

and the band-averaged modal density is given by

$$n(\omega) = \frac{1}{\Delta\omega} \int_{\omega_1}^{\omega_2} 4S\rho_S \overline{\mathrm{Re}\,[\mathbf{Y}(\omega)]}\, \mathrm{d}\omega, \tag{6.40}$$

where S is the surface area of the test structure, ρ_S is the surface mass (mass per unit area), $\Delta\omega$ is the frequency bandwidth, and the overbar represents space-averaging. Hence, the modal density can be obtained experimentally by integrating the real part of the point mobility frequency response function over the frequency band of interest. In principle, equation (6.40) is only applicable to structures with a uniform mass distribution. Numerous experimental results[6.11,6.12] have shown that the equation can be successfully applied to non-uniform structures with varying mass distributions provided that the $S\rho_S$ term is replaced by the total mass.

It now remains to discuss how one obtains a reliable estimate of Re $[\mathbf{Y}(\omega)]$. Two separate issues have to be addressed. Firstly, any external (pre- or post-processing) noise and any subsequent feedback has to be accounted for. Secondly, bias errors associated with the measurement of the mobility frequency response function also have to be accounted for – i.e. errors in the measurement of force and velocity due to the mass and stiffness properties of the transducer.

For an ideal system with no external noise or feedback, as illustrated in Figure 6.7, the point mobility is given by the cross-spectrum of force and velocity and the auto-spectrum of the input force, where

$$\mathbf{Y}(\omega) = \frac{\mathbf{G}_{\mathbf{fv}}(\omega)}{G_{ff}(\omega)}. \tag{6.41}$$

Equation (6.41) neglects the frequency response function of the power amplifier and the exciter system and any feedback due to exciter–structure interactions. The equation also assumes that the gain of the measuring amplifier is such that external noise is reduced to a minimum. These simplifying assumptions produce bias errors which increase as $G_{ff}(\omega) \to 0$, i.e. near a resonance. This can sometimes result in negative peaks.

The feedback noise due to exciter–structure interactions can be minimised via a three channel technique developed by Brown[6.16] that incorporates the signal, $x(t)$, used to drive the system power amplifier in addition to the force and velocity signals. The frequency response function associated with this model is illustrated in Figure 6.8. Here, $\mathbf{H}(\omega)$ is the frequency response function of the power amplifier and the exciter system, $\mathbf{I}(\omega)$ is the feedback frequency response function describing electro-dynamic shaker–structure interactions, $n(t)$ is some external noise at the output stage, $x(t)$ is the original test signal used to drive the power amplifier (most commonly

Fig. 6.7. Idealised frequency response function for point mobility.

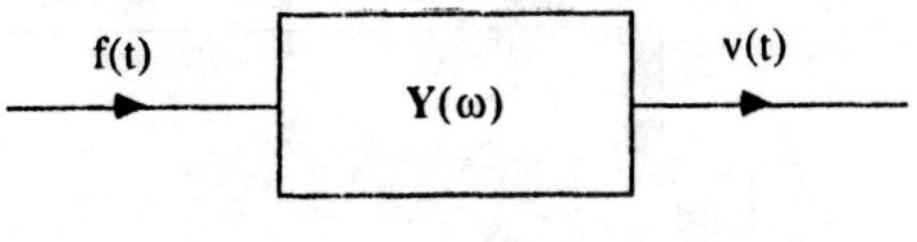

Fig. 6.8. Three channel frequency response function for point mobility. (Adapted from Brown[6.16].)

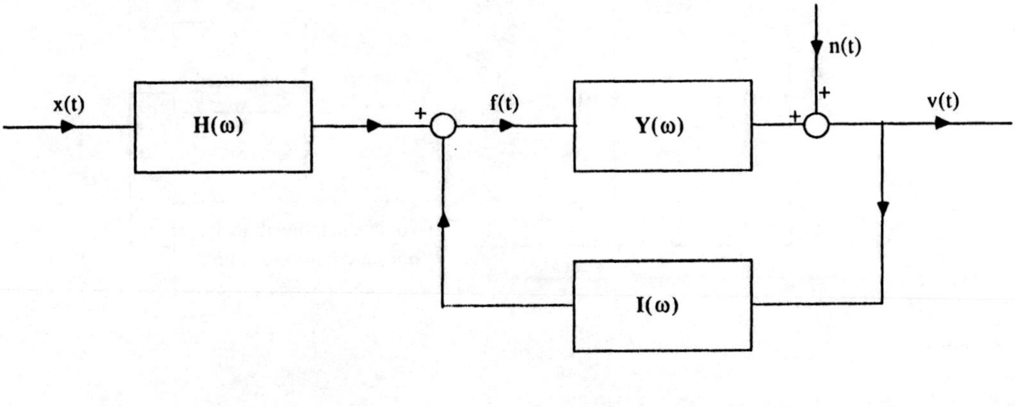

broadband random noise), $f(t)$ is the measured force signal, and $v(t)$ is the measured velocity signal. The point mobility is now given by

$$\mathbf{Y}(\omega)=\frac{\mathbf{G}_{\mathbf{xv}}(\omega)}{\mathbf{G}_{\mathbf{xf}}(\omega)}, \tag{6.42}$$

where $\mathbf{G}_{\mathbf{xv}}(\omega)$ and $\mathbf{G}_{\mathbf{xf}}(\omega)$ are the cross-spectra between the original test signal and the measured velocity signal, and the original test signal and the measured force signal, respectively. The modal density is obtained in the usual manner from the real part of the point mobility (equation 6.40). A typical experimental set-up for the measurement of modal density via the point mobility technique is illustrated in Figure 6.9.

Feedback noise due to exciter–structure interaction is a function of the method of excitation of the test structure and the nature of the vibration induced in the test structure. For modal density measurements, this noise can be reduced by separating the drive coil and the electromagnet from the measurement transducer and the test structure and providing sufficient stiffness in the direction of power flow. The drive rod connector must be stiff in the direction of excitation but flexible in all other directions. Keswick and Norton[6.11] and Brown and Norton[6.17] report on various modal density measurement techniques and provide comparisons between the two

Fig. 6.9. Instrumentation for the measurement of point mobility.

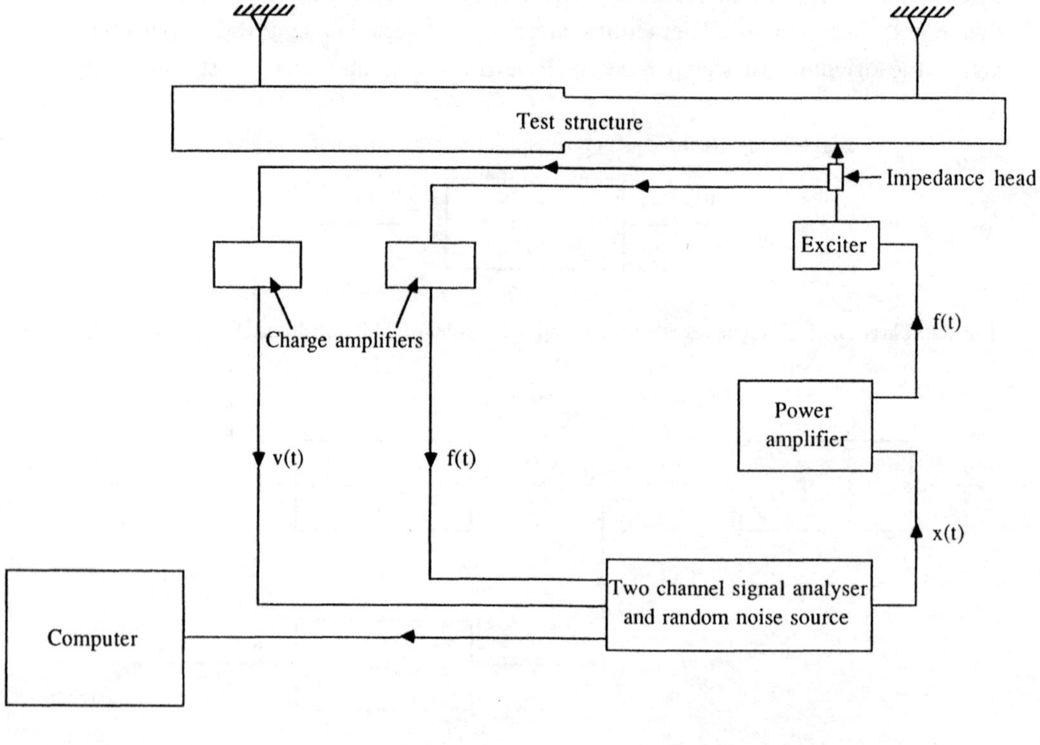

and three channel techniques for a range of different forms of excitation of the test structure. Three different types of commonly used excitation arrangements are illustrated in Figure 6.10.

Mass and stiffness corrections must also be considered when making any frequency response measurements on a structure. Such measurements at a single point on a structure are obtained with an impedance head – a single transducer that combines an accelerometer with a force transducer. There is always some added mass between the force transducer of the impedance head and the measurement point, because (i) some form of attachment is required between the impedance head and the structure, and (ii) a certain proportion of the mass of the impedance head itself (the mass above the force transducer) acts between the sensing element and the driven structure. Also, the accelerometer piezoelectric crystals and the force transducer piezoelectric crystals have certain stiffness characteristics which have to be accounted for.

The mass loading that results from the added mass that appears between the measurement transducer and the structure, to achieve suitable point contact, can affect the point mobility frequency response function. The force and velocity measured by the force transducer and the accelerometer, respectively, in the impedance head are different from those at the point of contact because of this added mass and any associated stiffness effects. A dynamic analysis (see chapter 4, sub-section 4.13.3) of the forces involved on the transducer and the test structure leads to the relationships

$$\frac{\mathbf{F_I}}{\mathbf{F_X}} = 1 + \frac{\mathbf{Y_X}}{\mathbf{Y_M}} \approx 1 + \frac{\mathbf{Y_I}}{\mathbf{Y_M}}, \tag{6.43}$$

and

$$\frac{\mathbf{V_X}}{\mathbf{V_I}} = 1 - \frac{\mathbf{Y_K}}{\mathbf{Y_I}}, \tag{6.44}$$

Fig. 6.10. Commonly used excitation arrangements: (a) spherical point contact; (b) flexible wire drive-rod; (c) free-floating magnetic exciter.

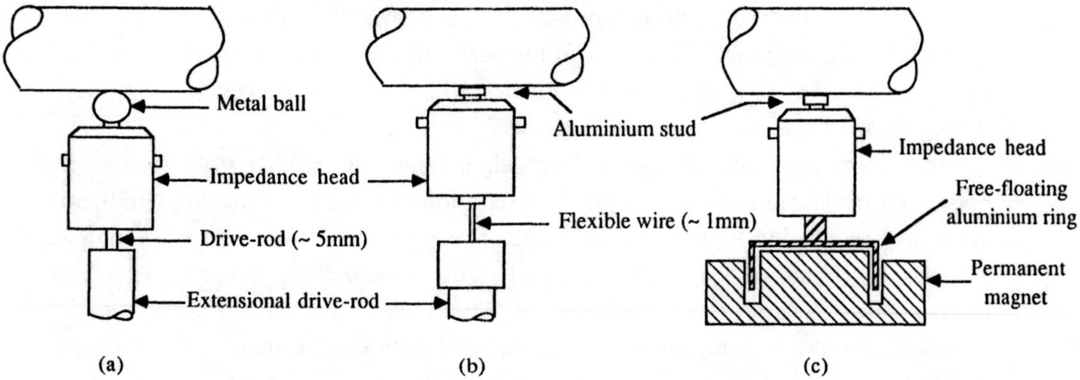

where $\mathbf{Y_X} = \mathbf{V_X}/\mathbf{F_X}$, $\mathbf{Y_I} = \mathbf{V_I}/\mathbf{F_I}$, $\mathbf{Y_M} = 1/i\omega M$, and $\mathbf{Y_K} = i\omega/K_s$. Here, $\mathbf{V_X}$ is the velocity of the structure at the point of excitation, $\mathbf{F_X}$ is the force applied to the structure, $\mathbf{V_I}$ is the velocity measured by the impedance head, $\mathbf{F_I}$ is the force measured by the impedance head, M is the added mass that appears between the force transducer and the structure, and K_s is the associated stiffness between the accelerometer and the structure. Equation (6.43) shows that contamination due to mass loading would result if the mobility of the added mass were small compared to the point mobility of the test structure; i.e. the added mass itself should be small relative to the generalised modal masses in the frequency band of interest. Equation (6.44) shows that the mobility of the stiffness itself should be small; i.e. the stiffness itself should be large. $\mathbf{Y_M}$ and $\mathbf{Y_K}$ are also frequency dependent, and as the frequency increases the mobility of the mass becomes smaller and the mobility of the stiffness becomes larger. This limits the useful frequency range of an impedance head.

When an impedance head is used on a structure that has a low mobility (high impedance), it is the stiffness effects that dominate the bias errors. Likewise, when an impedance head is used on a structure that has a high mobility (low impedance), it is the force variations due to the added mass effects that dominate the bias errors. Noise radiation from structures is generally controlled by flexural (bending) waves. It was shown in chapter 1, sub-section 1.9.5, that bending waves are high mobility (low impedance waves) because their wavespeeds are relatively slow. On the other hand, quasi-longitudinal waves for instance are low mobility (high impedance) waves because their wave speeds are very fast. So, when an impedance head is used whilst measuring mobilities/impedances associated with bending waves, the stiffness errors can be neglected since $\mathbf{Y_K} \ll \mathbf{Y_I}$ and $\mathbf{V_X} \approx \mathbf{V_I}$, and it is the mass loading which is of primary concern.

Mass loading can be accounted for by incorporating mass cancellation in the procedures for estimating the mobility. Keswick and Norton[6.11] and Hakansson and Carlsson[6.18] discuss these procedures in some detail and the interested reader is referred to those publications for the specific details. It is, however, important to note that care must be exercised when implementing mass correction. It is recommended[6.11,6.18] that a spectral approach be adopted rather than the traditional post-processing approach. The spectral approach involves measuring the inertance of the added mass as a function of frequency whilst the post-processing approach simply involves using a constant added mass.

Some typical experimental results[6.11] which demonstrate the effects of (i) feedback noise, and (ii) added mass effects for the excitation of bending waves in cylindrical shells are presented in Figures 6.11 and 6.12. The experimental results are compared with the average modal density theory (equation 6.27) and the computer algorithm[6.11] based on strain relationships developed by Arnold and Warburton[6.19]. The experimental results clearly demonstrate that (i) the elimination of feedback noise, and (ii) suitable mass correction are essential to obtain reliable experimental results.

6.5 Internal loss factors

The internal loss factor, η, is a parameter of primary interest in the prediction of the vibrational response of structures both by S.E.A. and by other more conventional techniques. Whilst analytical expressions of modal densities are available in the literature for a range of geometries, analytical expressions are not generally available for the internal loss factors of structural components and acoustic volumes. The matter is further complicated as the internal loss factor often varies from mode to

Fig. 6.11. Modal density for a clamped pipe with no mass correction: —×—, exact modal density (Arnold and Warburton); ——, average modal density (equation 6.27); —■—, spherical point contact with a 5 mm drive-rod; —□—, 0.35 mm flexible wire drive-rod; —▲—, 1.0 mm flexible wire drive-rod; —△—, 3.0 mm flexible wire drive-rod; —◇—; free-floating magnetic exciter.

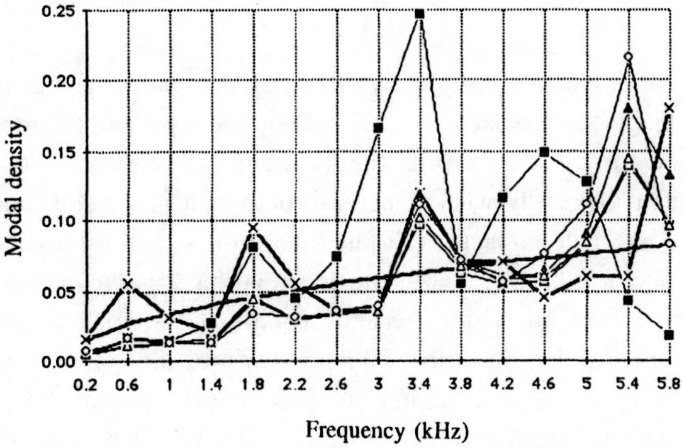

Fig. 6.12. Modal density for a clamped pipe using the three channel method with spectral mass correction: —×—, exact modal density (Arnold and Warburton); ——, average modal density (equation 6.27); —■—, spherical point contact with a 5 mm drive-rod; —□—, 0.35 mm flexible wire drive-rod; —▲—, 1.0 mm flexible wire drive-rod; —△—, 3.0 mm flexible wire drive-rod; —◇—, free-floating magnetic exciter.

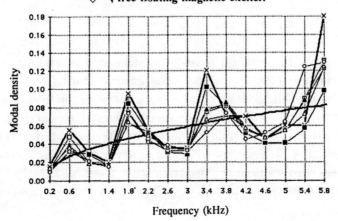

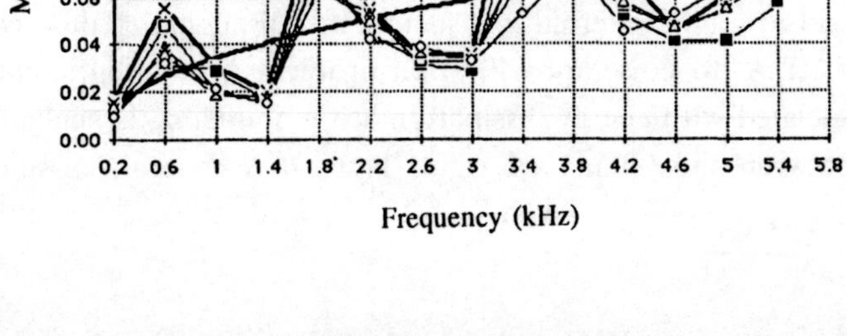

mode, and it is widely recognised that it is the major source of uncertainty in the estimation of the dynamic response of systems.

Internal loss factors incorporate several different damping or energy loss mechanisms, some linear and some non-linear. The two most commonly accepted forms of linear damping are (i) structural (hysteretic or viscoelastic) damping which is a function of the properties of the materials making up the structure, and (ii) acoustic radiation damping which is associated with radiation losses from the surface of the structure into the surrounding fluid medium. In practice, additional non-linear damping mechanisms are sometimes present at the structural boundaries of built-up structures. These include gas pumping at joints or squeeze-film damping, and frictional forces.

The three damping mechanisms act independently of each other, hence the internal loss factor of a structural element which is part of a built-up structure is the linear sum of the three forms of damping. It is given by

$$\eta = \eta_s + \eta_{rad} + \eta_j, \tag{6.45}$$

where η_s is the loss factor associated with energy dissipation within the structural element itself, η_{rad} is the loss factor associated with acoustic radiation damping, and η_j is the loss factor associated with energy dissipation at the boundaries of the structural element. Generally, when structural components are rigidly joined together, $\eta_j < \eta_s$, and the internal loss factor is a function of the structural loss factor and the acoustic radiation loss factor. The acoustic radiation loss factor can become the dominant term in the internal loss factor equation, particularly for lightweight structures with high radiation ratios. This point will be illustrated shortly.

The reader should note that the internal loss factor (equation 6.45) is not to be confused with the total loss factor (equation 6.17) – the total loss factor of a subsystem is a function of its internal loss factor, the internal loss factors of any coupled subsystems, and the associated coupling loss factors. Hence, for two coupled subsystems, the total loss factor of subsystem 1 is given by

$$\eta_{TS1} = \eta_{s1} + \eta_{rad1} + \eta_{j1} + \frac{\frac{n_2}{n_1}\eta_2\eta_{21}}{\eta_2 + \eta_{21}}, \tag{6.46}$$

where the subscript 2 refers to the second subsystem and the first three terms are the various components of the internal loss factor of subsystem 1.

Loss factors of structural elements, η_s, acoustic radiation loss factors, η_{rad}, internal loss factors of acoustic volumes, and various experimental techniques for measuring modal and band-averaged internal loss factors are discussed in this section. In the application of S.E.A. to noise and vibration problems it is often assumed that the loss factor associated with energy dissipation at the joints, η_j, is negligible when the connections between subsystems are rigid. Thus, it is generally assumed that the internal loss factor, η, generally refers to $\eta_s + \eta_{rad}$. When the connections between subsystems are not rigid, η_j becomes significant. These effects, sometimes referred to as coupling damping, are discussed in section 6.8.

6.5.1 Loss factors of structural elements

Internal loss factors of structural elements are generally obtained experimentally by separately measuring the energy dissipation in each of the uncoupled elements. Here, η_j is zero and thus

$$\eta = \eta_s + \eta_{rad}. \tag{6.47}$$

The major practical difficulty in obtaining reliable values of the structural loss factor, η_s, is that most experiments to measure the loss factor of a structural element have to be carried out in air. Hence by necessity, the quantity that is measured is in fact a combination of η_s and η_{rad} as per equation (6.47) above. Accurate measurements of η_s can only be obtained in a vacuum – measurements conducted in an anechoic chamber or under free-field conditions are a linear combination of η_s and η_{rad}. However, provided that the structure is not lightweight, it is reasonable to assume that $\eta_{rad} < \eta_s$ and that the internal loss factor is dominated by the structural damping. This is the assumption that has been made by numerous researchers who have experimentally measured internal loss factors of a variety of structural elements. For lightweight structures (aluminium panels, honeycomb structures, thin-walled cylindrical shells, etc.), however, there is clear evidence that the acoustic radiation loss factor is at least equal to if not greater than the structural loss factor. Rennison and Bull[6.20] and Clarkson and Brown[6.21] have identified and measured acoustic radiation loss factors for lightweight shells and plates, respectively.

It is also very important to note that, if the experimental measurements to obtain the internal loss factors of an individual structural element are conducted in a reverberant (or a semi-reverberant) room, then another subsystem, namely the acoustic volume, inadvertently enters the S.E.A. power balance equation. The loss factor which is now measured is in fact the total loss factor (see equation 6.17 or 6.46) of the structure–acoustic volume system, and the acoustic radiation loss factor becomes a coupling loss factor – i.e. $\eta_{rad} = \eta_{12}$. Hence, the reader should note that, when using internal loss factor data for lightweight structures which has been obtained under reverberant or semi-reverberant conditions, the data include the structural loss factor, η_s, the acoustic radiation loss/coupling loss factor, $\eta_{rad} = \eta_{12}$, and the room volume internal loss factor, η_2. The data are therefore only valid for a specfic set of experimental conditions. The error introduced by using these experimentally obtained loss factors in other situations will result in an overestimation of η_s. This information is only generally representative of η_s if the surface mass of the structure is sufficiently large such that $\eta_{rad} < \eta_s$, and if there is very little energy flow back into the structure from the acoustic volume.

Very little consistent information is readily available about the internal loss factors of structural elements. Most of the data presented in the handbook literature are empirical and it is not at all clear as to whether the tests were conducted in free or in reverberant space. Ungar[6.22] was amongst the first to recognise the various different contributions to the internal loss factor, and provides a detailed discussion on the various damping mechanisms together with typical values of structural loss factors for a range of structural materials. More recently, Richards and Lenzi[6.23] have presented a review of structural damping in machinery. The various non-linear damping mechanisms at structural boundaries (gas pumping, frictional losses, etc.) are discussed in detail and a large range of typical damping values for a wide variety of industrial machinery components is presented. Whilst the data are largely empirical, they are invaluable for obtaining engineering estimates of noise and vibration levels, etc. Ranky and Clarkson[6.24] present detailed band-averaged internal loss factors for aluminium plates and shells, and Norton and Greenhalgh[6.25] present a wide range of modal and band-averaged internal loss factors for steel cylinders. The results presented by Ranky and Clarkson[6.24] and Norton and Greenhalgh[6.25] include both structural and acoustic radiation loss factors. Some typical values of structural loss factors for some common materials are presented in Table 6.1.

6.5.2 *Acoustic radiation loss factors*

Returning to equation (6.45) for a moment, one can clearly see by now that the internal loss factor of a structural element can be dominated by any one of three parameters. If the surface mass of the structure is significant and the losses at the joints are negligible, then $\eta_{\mathrm{rad}} < \eta_{\mathrm{s}}$ and the internal structural damping is the dominant term. If the joints are not rigid, gas pumping mechanisms and frictional losses are the dominant mechanisms and η_{j} is the dominant term. Acoustic radiation damping

Table 6.1. *Structural loss factors for some common materials.*

Material	Structural loss factor, η_{s}
Aluminium	1.0×10^{-4}
Brick, concrete	1.5×10^{-2}
Cast iron	1.0×10^{-3}
Copper	2.0×10^{-3}
Glass	1.0×10^{-3}
Plaster	5.0×10^{-3}
Plywood	1.5×10^{-2}
PVC	0.3×10^{0}
Sand (dry)	0.02–0.2
Steel	$1\text{–}6 \times 10^{-4}$
Tin	2.0×10^{-3}

plays a very important part in the dissipation of energy from lightweight structures when there is very little energy dissipation at the joints.

The acoustic radiation loss factor of a structural element is given by

$$\eta_{\text{rad}} = \frac{\rho_0 c \sigma}{\omega \rho_{\text{S}}}, \tag{6.48}$$

where σ is the radiation ratio of the structure, ρ_{S} is its surface mass (mass per unit area), ρ_0 is the fluid density, c is the speed of sound, and ω is the centre frequency of the band. Equation (6.48) is derived very simply from the radiated sound power using equation (6.2) and equation (3.30) in chapter 3. For a given structural element, σ is generally small at very low frequencies, hence the acoustic radiation loss factor is also small. As ω increases, σ increases rapidly to a value of unity (see chapter 3 for a discussion on radiation ratios), and the acoustic radiation loss factor can dominate the internal loss factor provided that ρ_{S} is small. In this frequency range, σ increases at a faster rate than ω. As one goes up yet higher in frequency (i.e. above the critical frequency for a plate, or above the ring frequency for a cylinder), the radiation ratio remains at unity but the radian frequency term in the denominator in equation (6.48) continues to increase. Hence, the acoustic radiation loss factor, η_{rad}, starts to decrease and a point is reached where the structural loss factor, η_{s}, once again becomes the dominant term in the internal loss factor equation. Clarkson and Brown[6.21] have measured the structural loss factors of aluminium and honeycomb plates in a vacuum, and compared them with the corresponding internal loss factors measured in air. The effects of acoustic radiation damping on the internal loss factors are very evident.

Keswick and Norton[6.26] have measured the internal loss factors ($\eta_{\text{s}} + \eta_{\text{rad}}$) of three steel cylindrical shell arrangements with diameters of 65 mm, 206 mm and 311 mm, and wall thicknesses of 1 mm, 6.5 mm and 6.5 mm, respectively. Internal loss factors were measured up to 1.8 times the ring frequency (equation 6.26) of the largest cylinder in an attempt to separate the structural loss factor effects from the acoustic radiation loss factor effects. The measurements were performed in air and in a room which was relatively 'dead' acoustically. From equation (6.48), one would expect η_{rad} to decrease with increasing frequency. Some typical results are presented in Figure 6.13. The results suggest (i) that structural damping dominates the internal loss factor at very low frequencies, (ii) that acoustic radiation damping dominates the internal loss factor at low and mid frequencies, and (iii) that structural damping again dominates the internal loss factor at high frequencies. The low frequency internal loss factor peaks associated with the smallest cylinder are due to its higher modal density which is associated with its thinner wall – i.e. there are more oscillators present to absorb energy. The results also demonstrate that the structural loss factor, η_{s}, is a lower limit for the internal loss factor.

Internal loss factors of acoustic volumes can be obtained from the reverberation time, T_{60}, of the volume, the reverberation time being the time that the energy level in the volume takes to decay to 1/60 of its original value (i.e. $E/10^6$ or 60 dB). The internal loss factor of an acoustic volume is given by

$$\eta = \frac{\log_e 10^6}{\omega T_{60}} = \frac{13.82}{\omega T_{60}}. \qquad (6.49)$$

6.5.3 Internal loss factor measurement techniques

Numerous techniques are available for the experimental measurements of internal loss factors and some of these have been reviewed in the literature[6.24,6.25]. Sometimes one is interested in modal internal loss factors, but more generally one is interested in band-averaged values.

The two most common techniques for obtaining modal internal loss factors are the half-power bandwidth technique and the envelope decay technique. The half-power bandwidth technique utilises the standard half-power bandwidth relationship which is associated with a 3 dB drop in response from the associated steady-state frequency response function (e.g. receptance) peak. Alternatively, the internal loss factor can be obtained from the vector loci plot of the receptance of a single mode. The envelope decay (reverberation) technique is based on a logarithmic decrement of the transient structural response subsequent to gating of the excitation source. Both techniques have limited applicability and are prone to producing erroneous results particularly on lightly damped structures[6.25]. Furthermore, modal internal loss factors are of limited use for S.E.A. applications and it is the band-averaged values that are of primary interest. The two most common techniques for obtaining band-averaged internal loss factors are the steady-state energy flow technique[6.5] and the random noise burst reverberation decay technique[6.25].

Fig. 6.13. Internal loss factors of cylindrical shells: —●—, 65 mm diameter, 1 mm wall thickness; —○—, 206 mm diameter, 6.5 mm wall thickness; —■—, 311 mm diameter, 6.5 mm wall thickness.

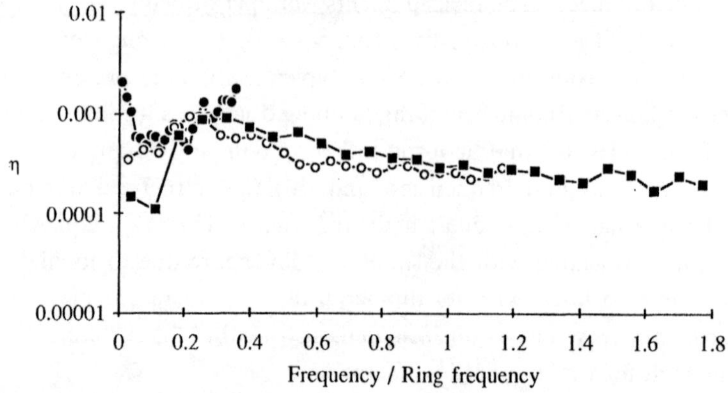

The steady-state energy flow technique has been widely accepted as the most suitable technique for the measurement of internal loss factors, and has often been used in preference to the conventional reverberation decay technique. The procedure requires an evaluation of the steady-state input power, Π_{in}, into the structure, where

$$\begin{aligned}\Pi_{in} &= R_{fv}(\tau = 0) = \int_{-\infty}^{\infty} \mathbf{S_{fv}}(\omega)\,d\omega \\ &= \langle f(t)v(t)\rangle = \text{Re}\,[\mathbf{FV^*}]/2 \\ &= \langle f^2(t)\rangle\,\text{Re}\,[\mathbf{Y}] = |\mathbf{F}|^2\,\text{Re}\,[\mathbf{Y}]/2 \\ &= \langle v^2(t)\rangle\,\text{Re}\,[\mathbf{Z}] = |\mathbf{V}|^2\,\text{Re}\,[\mathbf{Z}]/2. \end{aligned} \tag{6.50}$$

R_{fv} and $\mathbf{S_{fv}}$ are the cross-correlation and cross-spectra between the force and velocity signals, $f(t)$ and $v(t)$ are the time histories of the force and velocity signals, **F** and **V** are the respective Fourier transforms of $f(t)$ and $v(t)$, $\mathbf{V^*}$ is the complex conjugate of **V**, Re [**Y**] is the real part of the point mobility and Re [**Z**] is the real part of the point impedance. The internal loss factor is subsequently obtained from

$$\eta = \frac{\Pi_{in}}{\omega E}, \tag{6.51}$$

where ω is the centre frequency of the band and E is the space-averaged energy of vibration. The instrumentation required for measuring internal loss factors via the steady-state energy flow technique is illustrated in Figure 6.14.

The internal loss factors as obtained by the steady-state energy flow technique require an accurate estimation of the input power. Any experimental errors in the measurement of force and velocity at the points of excitation will be reflected in the internal loss factor estimates. When using continuous, stationary, broadband random noise as an excitation source, the cross-spectrum provides a much more reliable estimate of the input power. This is because the cross-spectrum generates a time-average of the product of force and velocity. This is not the case when one uses the real part of the impedance, and furthermore, since the impedance is very small at a structural resonance, large errors in the estimation of the internal loss factor can result. It has recently been demonstrated by Brown and Clarkson[6.27] that the real part of the impedance can be used to estimate both the input power and the internal loss factor provided that the structure is excited via a deterministic transient excitation where time-averaging is not required, rather than random noise. The deterministic excitation used, that of a rapid swept sine wave, also has the advantage of generating extremely good noise free data. In addition to accurate measurements of the input power, the steady-state energy flow technique also requires an accurate measurement of the spatially averaged mean-square velocity of the structure. It has also been recently found by Norton and Greenhalgh[6.25] that additional errors sometimes exist

Fig. 6.14. Instrumentation for the measurement of internal loss factors via the steady-state energy flow technique.

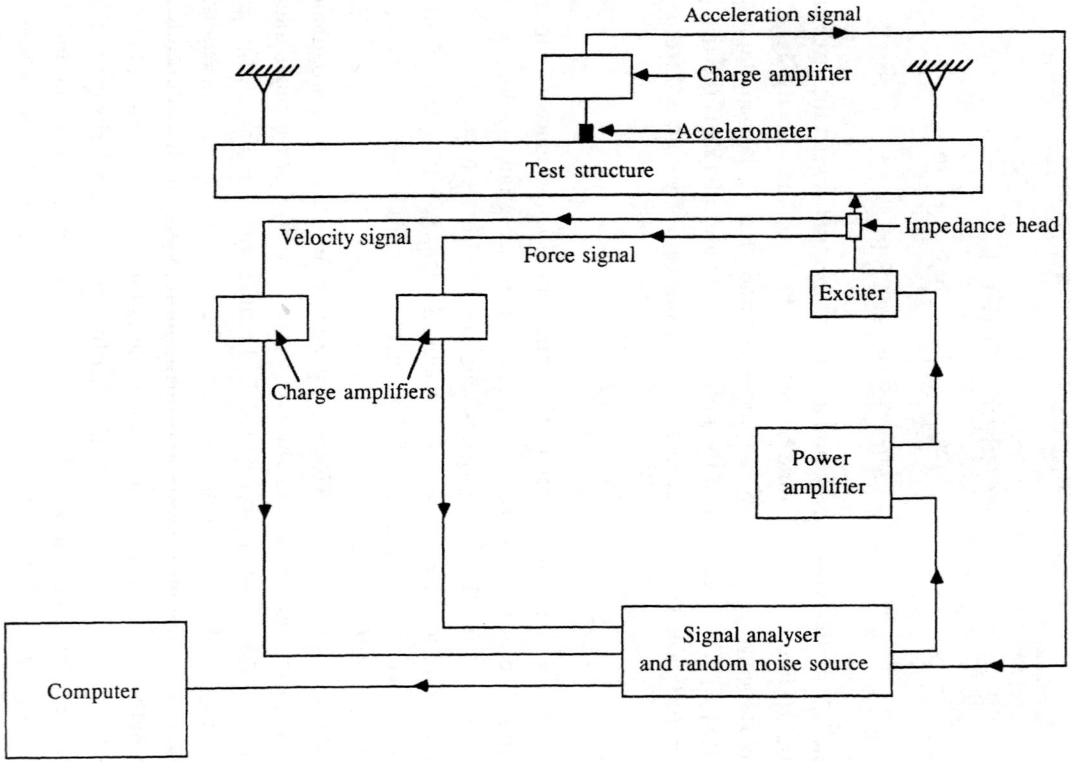

due to contact damping at the excitation point. Contact damping due to losses within the excitation system are significant for very lightly damped structures and are largely dependent upon the type of excitation used. Referring back to Figure 6.10, for instance, excitation arrangemens (a) and (b) would produce more contact damping than excitation arrangement (c).

The random noise burst reverberation decay technique[6.25] allows for a rapid estimation of band-averaged internal loss factors of structures (and acoustic volumes). The method involves the usage of a constant bandwidth random noise burst to excite the structure via a non-contacting electromagnet. The decaying response signal is subsequently averaged and digitally filtered, and the internal loss factor is obtained from equation (6.49). The method provides a very fast way of collecting data with multiple averaging to reduce statistical uncertainty. Generally, the excitation is set up to be a selected percentage (~20%) of the time record length via the usage of a transient capture facility. The instrument required for measuring internal loss factors via the random noise burst reverberation decay technique is illustrated in Figure 6.15. Typical band-averaged reverberation decay time histories for high modal density regions in cylindrical shells are illustrated in Figure 6.16.

Some typical, experimentally obtained, band-averaged internal loss factors[6.17,6.25] for a selection of mild steel cylinders are presented in Figure 6.17. The frequency range investigated is well below the ring frequencies of the cylinders. β is the non-

Fig. 6.15. Instrumentation for the measurement of internal loss factors via the random noise burst reverberation decay technique.

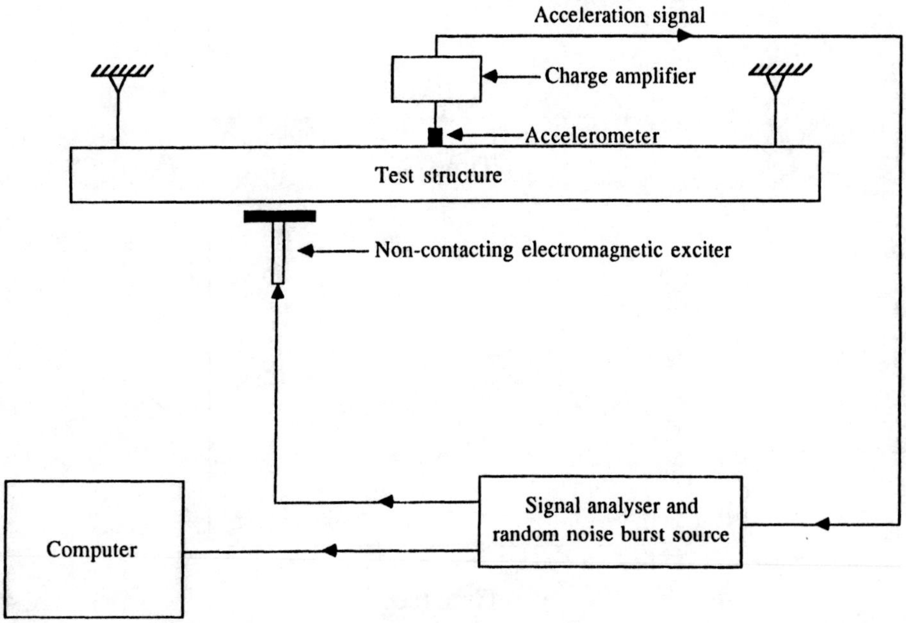

dimensional cylinder wall thickness parameter, and Δ is the ratio of length to radius ($\beta = h/(2\sqrt{3}a_m)$; $\Delta = L/a_m$; h is the cylinder wall thickness; L is the cylinder length; and a_m is the mean cylinder radius). The reverberation decay results suggest that the internal loss factors increase with increasing cylinder wall thickness, but are essentially independent of length. The steady-state results exhibit more variation with frequency and are larger than the random noise burst results. Large variations in internal loss factor estimates are found between those obtained via the cross-spectral estimate of the input power and those obtained via the real part of the impedance (see equation 6.50). The multiplication is performed on band-averaged values of $|v^2(t)|$ and Re $[\mathbf{Z}]$ in order to demonstrate the magnitude of the error obtained, when measuring input power via the real part of the impedance.

Two main conclusions result from the preceding discussions. Firstly, the steady-state energy flow technique for internal loss factor estimation is critically dependent upon the accurate measurement of input power. With continuous, broadband random

Fig. 6.16. Typical band-averaged reverberation decay time histories with random noise burst excitation for high modal density regions in cylindrical shells.

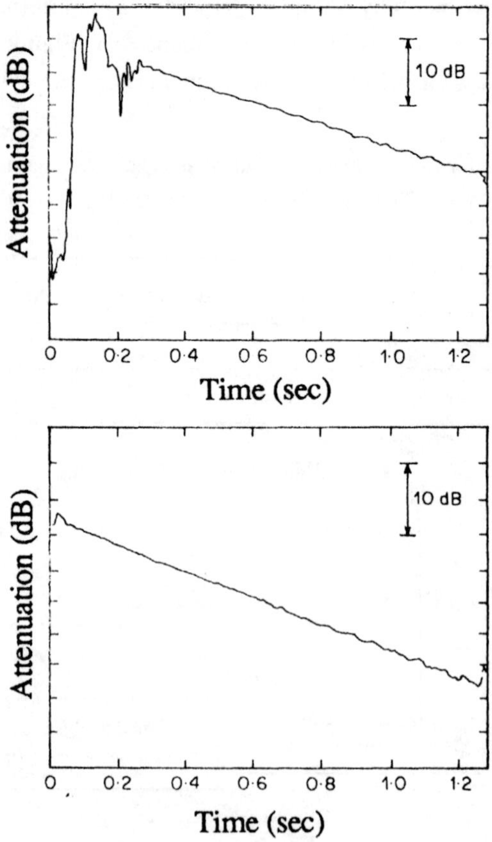

excitation, the cross-spectral technique is more appropriate; with deterministic, transient excitation, such as the swept sine, the impedance technique is recommended. The technique necessitates contact between the structure and the excitation source and therefore particular care has also got to be paid to the excitation arrangement. Secondly, the random noise burst reverberation decay technique with a non-contacting electromagnetic excitation source and appropriate digital filtering and averaging is recommended for internal loss factor measurements on very lightly damped structures.

An alternative digital procedure for estimating modal or band-averaged internal loss factors in lightly damped systems has been recently suggested by Norton and Greenhalgh[6.25]. The technique, referred to as amplitude tracking, is to divide the composite time record of the decay into smaller time-limited signals and to track the attenuation of particular spectral lines of the subsequently transformed time-limited signals – i.e. the attenuation of the amplitude of each resonance in the frequency domain is monitored at specific time intervals after removal of the excitation source.

Fig. 6.17. Typical experimentally obtained band-averaged internal loss factors for cylindrical shells: ●, $\beta = 0.012$, $\Delta = 44.4$, steady-state, impedance; +, $\beta = 0.012$, $\Delta = 44.4$, steady-state cross-spectrum; ×, $\beta = 0.009$, $\Delta = 47.3$, random noise burst; ▲, $\beta = 0.009$, $\Delta = 94.6$, random noise burst; ○, $\beta = 0.026$, $\Delta = 45.9$, random noise burst.

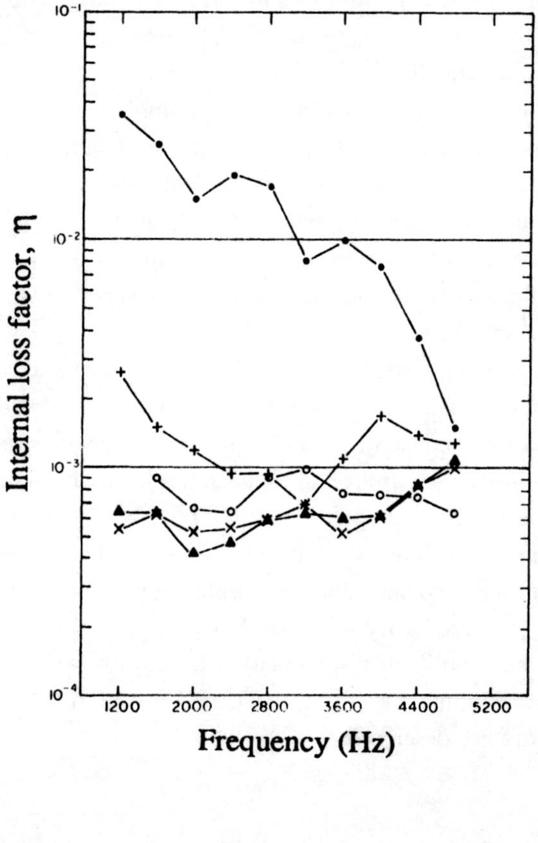

Caution has got to be exercised, however, as digital signal analysis techniques introduce certain limitations which can reduce their measurement flexibility. The usefulness of the fast Fourier transform, for instance, can be compromised since the time record length, frequency step size and the frequency bandwidth are inter-related via the transform algorithm which only calculates amplitude and phase at particular freqencies. The time record length establishes the minimum frequency difference between spectral lines, whilst the sampling interval establishes the maximum frequency. An insufficient time record length within the subdivided time-limited signal will result in loss of resolution. It has been shown that amplitude tracking can provide information about possible coupling between groups of modes within subsystems, and the subsequent energy transfer between them; the experiments show that, when a resonant mode is capable of energy exchange, its internal loss factor varies depending on how it is excited.

6.6 Coupling loss factors

The coupling loss factor, η_{ij}, is unique to S.E.A. and it is the link between two coupled subsystems i and j – i.e. it determines the degree of coupling between the two. If $\eta_{ij} < \eta_i$ or η_j then the subsystems are described as being weakly coupled. In S.E.A. applications it is always desirable to select the subsystems such that they are weakly coupled. There is no single way of evaluating the coupling loss factor both experimentally or analytically.

Theoretical expressions are available for couplings between structural elements (e.g. line junctions between plates, plate–cantilever beam junctions, beam–beam couplings, etc.), couplings between structural elements and acoustic volumes (e.g. plate–acoustic volume couplings, cylindrical shell–acoustic volume couplings, etc.) and acoustic volume–acoustic volume couplings. Couplings between different structural elements are the hardest to define because different types of wave motions can be generated at a discontinuity.

Wave transmission analysis is by far the most successful way of developing theoretical coupling loss factors – the coupling loss factor, η_{ij}, is derived directly from the wave transmission coefficient, τ_{ij}. The transmission coefficients can be evaluated in terms of wave impedance and/or mobilities. Transmission coefficients and wave impedances are discussed in chapter 1 and chapter 3 (also see chapter 4) for some elementary systems. Cremer *et al.*[6.7] provide a detailed coverage of various wave attenuation/transmission coefficients for a wide range of structural discontinuities. The reader is also referred to Lyon[6.1] and Fahy [6.28] for further details.

In this section, the coupling loss factors associated with some of the more common coupling joints are summarised, and the experimental techniques for measuring coupling loss factors are described.

6.6.1 Structure–structure coupling loss factors

The most commonly encountered structure–structure coupling is a line junction between two structures. The coupling loss factor for a line junction has been evaluated by Lyon[6.1] and Cremer *et al.*[6.7], and it is conveniently given in terms of the wave transmission coefficient for a line junction. It is

$$\eta_{12} = \frac{2c_B L\tau_{12}}{\pi\omega S_1}, \tag{6.52}$$

where c_B is the bending wave velocity (or phase velocity) of flexural waves in the first plate (equation 1.322), L is the length of the line, τ_{12} is the wave transmission coefficient of the line junction from subsystem 1 to subsystem 2, ω is the centre frequency of the band of interest, and S_1 is the surface area of the first subsystem. It is useful to note that the group velocity, c_g, of the bending waves (equation 1.4) is twice the phase velocity[6.7] – sometimes equation (6.52) is presented in terms of the group velocity.

Equation (6.52) is a very useful relationship since it allows for the evaluation of the coupling loss factor to be reduced to an evaluation of the wave transmission coefficient. Bies and Hamid[6.5] have used the relationship for comparisons with experimental measurements on coupled flat plates at right angles to each other, and Wöhle *et al.*[6.29], for instance, have evaluated coupling loss factors for rectangular structural slab joints. The wave transmission coefficient for a coupling can be obtained in terms of wave impedances from a wave transmission analysis.

The normal incidence transmission coefficient for two coupled flat plates at right angles to each other is given by[6.5,6.7]

$$\tau_{12}(0) = 2\{\psi^{1/2} + \psi^{-1/2}\}^{-2}, \tag{6.53}$$

where

$$\psi = \frac{\rho_1 c_{L1}^{3/2} t_1^{5/2}}{\rho_2 c_{L2}^{3/2} t_2^{5/2}}, \tag{6.54}$$

and ρ is the density, c_L is the longitudinal wave velocity, t is the thickness, and the subscripts 1 and 2 refer to the subsystems. The random incidence transmission coefficient, τ_{12} is approximated by [6.5]

$$\tau_{12} = \tau_{12}(0)\,\frac{2.754X}{1 + 3.24X}, \tag{6.55}$$

where $X = t_1/t_2$.

The coupling loss factors for two homogeneous plates coupled by point connections (e.g. bolts) rather than a line connection (e.g. a weld) is approximated by[6.30]

$$\eta_{12} = \frac{4Nt_1 c_{L1}(\rho_{S1}^2 t_1^2 c_{L1}^2)(\rho_{S2}^2 t_2^2 c_{L2}^2)}{\sqrt{3}\omega S_1(\rho_{S1}^2 t_1^2 c_{L1}^2 + \rho_{S2}^2 t_2^2 c_{L2}^2)^2}, \tag{6.56}$$

where N is the number of bolts, t is the plate thickness, and the subscripts 1 and 2 refer to the respective subsystems. For riveted or bolted plates, when the bending wavelength in the plates is less than L, equation (6.56) should be used; when the bending wavelength in the plate is greater than L, equation (6.52) is more appropriate.

If a beam is cantilevered to a plate, the coupling loss factor can be expressed in terms of a junction moment impedance[6.1]. The coupling loss factor is given by

$$\eta_{bp} = \frac{(2\rho_b c_{Lb} \kappa_b A_b)^2}{\omega M_b} \operatorname{Re}[\mathbf{Z}_p^{-1}] |\mathbf{Z}_p/(\mathbf{Z}_p + \mathbf{Z}_b)|^2, \tag{6.57}$$

where the subscript b refers to the beam, and the subscript p refers to the plate. κ_b is the radius of gyration of the beam, c_{Lb} is the longitudinal wave velocity of the beam, M_b is the mass of the beam, A_b is the cross-sectional area of the beam, and $\mathbf{Z}$ is the moment impedance. The moment impedances for beams and plates are derived by Cremer *et al.*[6.7]. The moment impedance for a semi-infinite beam which is predominantly in flexure is approximated by

$$\mathbf{Z}_b \approx \frac{0.03\rho_b A_b (c_{Lb}^3 t_b^3)^{1/2}(1-\mathrm{i})}{(\omega/2\pi)^{1/2}}, \tag{6.58}$$

where t_b is the thickness of the beam in the direction of flexure. The moment impedance of an infinite flat plate in flexure is approximated by

$$\mathbf{Z}_p \approx \frac{c_{Lp}^2 \rho_p t_p^3}{\left\{1 - \dfrac{4\mathrm{i}\ln(0.9ka)}{\pi}\right\}\left(\dfrac{2.4\omega}{\pi}\right)}, \tag{6.59}$$

where a is the moment arm of the applied force. In this instance, $a = t_b/2$.

Hopefully, the preceding discussion illustrates the importance of wave transmission analyses and the subsequent evaluation of wave impedances at points, junctions, etc. for the evaluation of structure–structure coupling loss factors. A variety of different combinations is available in the literature[6.7] for plates and beams. The situation is somewhat more complex for coupled structural elements with curvature (e.g. shells), and it is a topic of current ongoing research. When theoretical estimates for coupling loss factors are not available, one generally turns to experimental measurement techniques.

6.6.2 Structure–acoustic volume coupling loss factors

The coupling loss factor for a structure–acoustic volume coupling is somewhat easier to evaluate. It was shown in subsection 6.5.1 that the acoustic radiation loss factor for a structure becomes a coupling loss factor when the structure couples to an acoustic volume. Thus,

$$\eta_{SV} = \frac{\rho_0 c \sigma}{\omega \rho_S}, \tag{6.60}$$

where the subscripts S and V refer to the structure and the volume, respectively. From the reciprocity relationship (equation 6.8), the coupling loss factor between the volume and the structure is

$$\eta_{\mathrm{VS}} = \frac{\rho_0 c \sigma n_{\mathrm{S}}}{\omega \rho_{\mathrm{S}} n_{\mathrm{V}}}, \tag{6.61}$$

where n_{S} is the modal density of the structure and n_{V} is the modal density of the room volume. The problem of structure–acoustic volume coupling thus reduces to one of evaluating radiation ratios.

It was pointed out at the beginning of this chapter that S.E.A. is generally about energy flows between different groups of resonant oscillators. When dealing with structure–structure systems this is generally the case although there are some instances where the energy flow is non-resonant. With sound energy flow through walls, it has been shown in chapter 4 that at frequencies below the critical frequency the dominant sound transmission is non-resonant; it is mass controlled. The coupling loss factor for non-resonant, mass-law, sound transmission through a panel from a source room is given by

$$\eta_{\mathrm{rp}} = \frac{cS}{4\omega V_{\mathrm{r}}} \tau_{\mathrm{rp}}, \tag{6.62}$$

where c is the speed of sound, S is surface area of the panel, V_{r} is the volume of the source room, and τ_{rp} is the sound intensity transmission coefficient (ratio of transmitted to incident sound intensities) from the source room through the panel. The transmission coefficient is given by[6.2]

$$\tau_{\mathrm{rp}}^{-1} = \frac{\pi^9 \rho_{\mathrm{S}}^2}{2^{13} \rho_0^2 S} \left\{1 - \left(\frac{10\omega}{\omega_{\mathrm{C}}}\right)^2\right\} + \left(\frac{\omega \rho_{\mathrm{S}}}{2\rho_0 c}\right)^2, \tag{6.63}$$

for $\omega_0 < \omega < \omega_{\mathrm{C}}/10$, and

$$\tau_{\mathrm{rp}}^{-1} = \left(\frac{\omega \rho_{\mathrm{S}}}{2\rho_0 c}\right)^2, \tag{6.64}$$

for $\omega > \omega_{\mathrm{C}}/10$. In the above equations, ω_{C} is the critical frequency of the panel, ω_0 is the panel fundamental natural frequency, S is the surface area of the panel, ρ_{S} is the surface mass, ρ_0 is the density of the fluid medium (air), and c is the speed of sound. Equations (6.63) and (6.64) are of a similar form to equation (3.99) in chapter 3, which is for an unbounded flexible partition.

6.6.3 *Acoustic volume–acoustic volume coupling loss factors*

The coupling loss factor between two acoustical volumes/cavities (e.g. two connecting rooms with an open door) is identical to the coupling loss factor for non-resonant sound transmission through a panel (equation 6.62). Hence

$$\eta_{12} = \frac{cS}{4\pi V_1} \tau_{12}, \tag{6.65}$$

where S is the area of the coupling aperture, and $\tau_{12} = 1$ for an open window.

6.6.4 Coupling loss factor measurement techniques

The coupling loss factor can be experimentally measured either by setting up a series of controlled experiments under laboratory conditions, or *in situ*. In general, three experimental techniques are available. Great care has always got to be taken because coupling loss factors are at least an order of magnitude below the corresponding internal loss factors when subsystems are lightly coupled, and one is therefore often dealing with very small numbers ($\sim 10^{-4}$) and the differences between them.

The usual laboratory technique for the measurement of coupling loss factors is to restrict the number of subsystems to two, excite one subsystem and subsequently measure the space and time-averaged vibrational energies of both subsystems. Care has got to be taken to ensure that the coupling losses at the boundaries of the coupled structure are negligible and that all the energy flow is only between the two coupled subsystems. For coupled plates and shells it is acceptable to use fine wire point supports. Alternatively, the ends of the coupled structure could be supported on foam rubber pads to simulate free–free end conditions. Equations (6.15) and (6.16), derived in subsection 6.3.4 from the steady-state power balance equations (equation 6.10 and 6.11), are then used to evaluate the coupling loss factors. Additional information is required about the modal densities and the internal loss factors of the respective subsystems. This information is generally obtained by performing separate experiments on the decoupled subsystems, or theoretically as in the case of modal densities. Now, if subsystem 1 is excited, from equation (6.15)

$$\eta_{12} = \frac{\eta_2 n_2 E_2}{n_2 E_1 - n_1 E_2}. \tag{6.66}$$

In this instance, the coupling loss factor η_{12} can be obtained by measuring E_1, E_2, η_2 and either measuring or computing n_1 and n_2. The coupling loss factor η_{21} can be obtained from the reciprocity relationship (equation 6.8). It is good experimental practice to repeat the experiment by exciting the second subsystem and measuring the space- and time-averaged vibrational energies of both subsystems. In this instance equation (6.16) is used and

$$\eta_{12} = \frac{\eta_1 n_2 E_1}{n_1 E_2 - n_2 E_1}. \tag{6.67}$$

The above procedure assumes that the loss factor associated with energy dissipation at the boundaries, η_j, is negligible. If η_j is not negligible, then the above procedure will introduce errors because the uncoupled values of the internal loss factors (obtained from separate experiments) will not be equal to the coupled internal loss factors (note that E_1 and E_2 take on different values for the reverse experiment).

If it is felt that coupling damping is significant, then equations (6.15) and (6.16) can be used in conjunction with a third equation to solve for η_1, η_2, and η_{12} or η_{21}, remembering that η_{12} and η_{21} are related via the reciprocity relationship (equation 6.8). The third equation that is required is the total loss factor (see subsection 6.3.4) and it can be obtained either from a steady-state experiment or from a transient, reverberation decay experiment. The steady-state total loss factor is given by equation (6.17), and the transient total loss factor is given by equation (6.19). The steady-state total loss factor requires the measurement of the input power to the structure; the transient total loss factor only requires a measurement of the reverberation decay of the coupled subsystem. With equations (6.15), (6.16), and (6.17) or (6.19) one has three equations and three unknowns.

The third experimental technique for evaluating *in-situ* coupling loss factors involves the measurement of input power to the coupled system. This technique is referred to as the power injection method[6.5] and it has also been successfully used by Clarkson and Ranky[6.31] and Norton and Keswick[6.32]. For a two subsystem S.E.A. model, the steady-state power balance equations for excitation of subsystem 1 are given by equations (6.10) and (6.11) – i.e.

$$\Pi_1 = \omega E_1\eta_1 + \omega E_1\eta_{12} - \omega E_2\eta_{21}, \qquad \text{(equation 6.10)}$$

and

$$0 = \omega E_2\eta_2 + \omega E_2\eta_{21} - \omega E_1\eta_{12}. \qquad \text{(equation 6.11)}$$

If the experiment is reversed and the second subsystem is excited, then

$$\Pi_2 = \omega E_2\eta_2 + \omega E_2\eta_{21} - \omega E_1\eta_{12}, \qquad (6.68)$$

and

$$0 = \omega E_1\eta_1 + \omega E_1\eta_{12} - \omega E_2\eta_{21}. \qquad (6.69)$$

There are now four equations and four unknowns and one can thus solve them to obtain η_1, η_2, η_{12}, and η_{21} (note that E_1 and E_2 take on different values for the reverse experiment). Furthermore, information about the modal densities is not required any more. This method also has the advantage that the effects of coupling damping are accounted for, and that it can be extended to more than two coupled subsystems via matrix inversion techniques[6.5,6.31] utilising equations (6.20) and (6.21). However because one is now dealing with the measurement of input power, point contact excitation is required and great care has got to be exercised in the experimental techniques. The errors associated with mass loading, contact damping, exciter–structure feedback interaction, etc. have already been discussed in this chapter – they all have to be accounted for.

Some typical coupling loss factors for coupled steel cylindrical shells[6.32], obtained using the power injection technique, are presented in Figure 6.18. The coupling loss factors are for 65 mm diameter, 1 mm wall thickness shells, each 1.5 m long, and coupled by a flanged joint approximately 23 mm thick and 109 mm in diameter. The corresponding internal loss factors (uncoupled and *in situ*) are presented in Figure 6.19 – the differences are associated with coupling damping. The important observation is that the coupling loss factors are generally at least an order of magnitude smaller than the internal loss factors.

Fig. 6.18. Typical coupling loss factors for two cylindrical shells coupled via a flanged joint: —●—, 65 mm diameter, 1 mm wall thickness, air-gap (gas pumping) joint; —○—, 65 mm diameter, 1 mm wall thickness, rubber gasket joint.

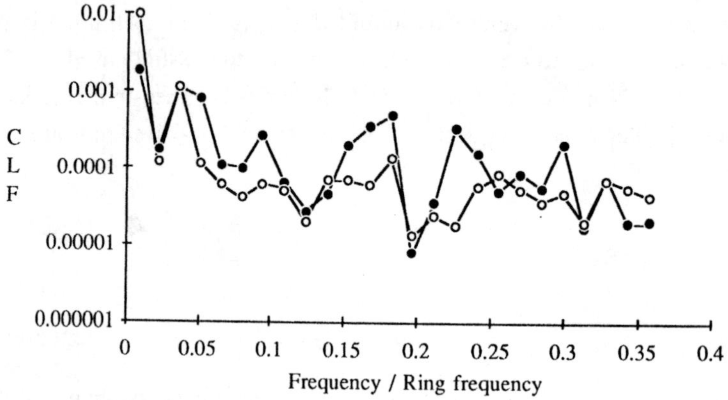

Fig. 6.19. Internal loss factors (coupled and *in situ*) for a 65 mm diameter, 1 mm wall thickness cylindrical shell: —●—, *in situ* – i.e. coupled via an air-gap (gas pumping) joint; —○—, uncoupled.

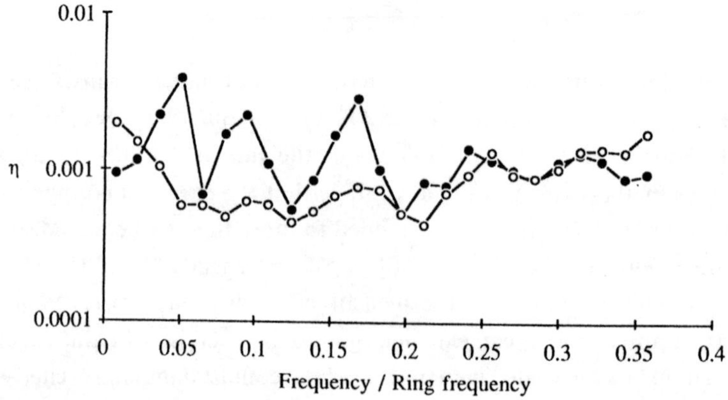

6.7 Examples of the application of S.E.A. to coupled systems

The best way of illustrating the practical significance of S.E.A. is to consider some specific examples. Two commonly encountered examples will be considered here. Firstly, a three subsystem S.E.A. model comprising a beam–plate–room combination will be considered. Secondly, a three subsystem S.E.A. model comprising two rooms coupled by a partition will be considered. The first example is derived from some earlier work on the random vibrations of connected structures by Lyon and Eichler[6.33], and the second example is derived from some earlier work on the sound transmission through panels by Crocker and Kessler[6.34].

6.7.1 A beam–plate–room volume coupled system

As an introduction to the application of S.E.A., consider a system that comprises a large plate-type structure (e.g. a large radiating surface of a machine cover) in a reverberant room. The plate is excited mechanically by a beam-type element (e.g. a directly driven machine element) that is cantilevered to it. Hence, the three S.E.A. subsystems are the beam, the plate, and the room volume. The beam is defined as being subsystem 1, the plate as being subsystem 2, and the room volume as being subsystem 3, and a relationship between the mean-square beam velocity, the mean-square plate velocity, and the resultant mean-square sound pressure level in the room is required.

For the purposes of this example, it is also assumed that the beam is vibrating in flexure – i.e. only the transmission of bending (flexural) waves between the coupled structural elements is considered. If longitudinal, torsional and bending waves are uncoupled from each other, the total coupling loss factor is a linear combination of the three[6.1]. For this particular example, since the axis of the beam is perpendicular to the plate, it can be assumed that the three wave-types are decoupled; furthermore, since the beam is only vibrating in flexure, bending waves will be the dominant source of vibrational energy. It is also assumed for the purposes of this example (i) that the two structural elements are strongly coupled, such that the coupling loss factors are greater than the internal loss factors of the beam and plate, and (ii) that there is no coupling between the beam and the room. Also, because the sound field in the room is not totally diffuse, the analysis provides an upper bound estimate.

Under the conditions described above ($\eta_{12}, \eta_{21} \gg \eta_1, \eta_2$), there is equipartition o modal energy between the beam and the plate, and

$$\frac{E_1}{n_1} = \frac{E_2}{n_2}, \qquad (6.70$$

where E represents the spatially averaged mean-square vibrational energy, n represent the modal density, subscript 1 represents the beam, and subscript 2 represents th plate. If the total masses of the beam and the plate are M_1 and M_2, respectively, then equation (6.70) can be re-written in terms of the time- and space-averaged mean-square velocities of both structures. Hence,

$$\frac{\langle \overline{v_2^2} \rangle}{\langle \overline{v_1^2} \rangle} = \frac{M_1 n_2}{M_2 n_1}. \tag{6.71}$$

Theoretical expressions are readily available for the modal densities of beams, flat plates and room volumes. The modal density for flexural vibrations of a uniform beam of length L is given by equation (6.24). Using equations (1.259) and (6.22) is can be re-expressed as

$$n_1(\omega) = \frac{L}{3.38(c_{L1} t\omega)^{1/2}}, \tag{6.72}$$

where c_{L1} is the longitudinal (compressional) wave velocity of the beam, and t is the thickness of the beam in the direction of transverse vibration excitation. The modal density for flexural vibrations of a flat plate is given by equation (6.25). Using equation (6.22), it can be re-expressed as

$$n_2(\omega) = \frac{S}{3.6 c_{L2} h}, \tag{6.73}$$

where h is the thickness of the plate, S is its surface area, and c_{L2} is the longitudinal wave velocity of the plate. It is worth pointing out that, if the beam and the plate are made of the same material, $c_{L1} \neq c_{L2}$ because of the Poisson contraction effect which is neglected in beam analysis (c_{L1} is given by equation 1.221, and c_{L2} is given by equation 1.321).

The mass of the beam is $M_1 = \rho_1 Lbt$, where b is the other cross-sectional dimension of the beam, and the mass of the plate is $M_2 = \rho_2 Sh$ (ρ_1 and ρ_2 are the respective material densities). By substituting the relevant parameters into equation (6.71)

$$\frac{\langle \overline{v_2^2} \rangle}{\langle \overline{v_1^2} \rangle} = \frac{0.94 \rho_1 bt (c_{L1} t\omega)^{1/2}}{\rho_2 h^2 c_{L2}}. \tag{6.74}$$

Equation (6.74) is a very useful relationship between the vibrational velocities of the beam and the plate.

Now consider the plate–room volume system. The mean-square vibrational energy of the plate can be given in terms of its surface mass, ρ_S – i.e.

$$E_2 = \langle \overline{v_2^2} \rangle \rho_S S, \tag{6.75}$$

where $\rho_S = \rho_2 h$. The mean-square energy level in the reverberant room is given by equation (4.63) – i.e.

$$E_3 = \frac{\langle p \rangle V}{\rho_0 c^2}, \tag{6.76}$$

where V is the volume of the room and $\langle \overline{p^2} \rangle$ is the mean-square sound pressure (the overbar denotes space-averaging). In theory, no space-averaging is required for a reverberant volume, but in practice the volume is often only semi-reverberant. Hence space-averaging is desirable.

The ratio of the energy levels in subsystems 2 and 3 can be obtained from equation (6.12) (note that subsystem 2 is driving subsystem 3 through the coupling link). Thus,

$$\frac{E_3}{E_2} = \frac{\eta_{23}}{\eta_3 + \frac{n_2}{n_3}\eta_{23}}, \tag{6.77}$$

where n_3 is the modal density of the room volume, and η_{23} is the coupling loss factor associated with energy flow from the plate into the room.

The energy flow (radiated sound power) from one side of a plate is given by equation (3.30) – i.e.

$$\Pi_{\text{rad}} = \rho_0 c S \langle \overline{v_2^2} \rangle \sigma, \tag{6.78}$$

where σ is the radiation ratio for the plate for the frequency band centred on ω; σ is a function of frequency. From Figure 6.5 (which adequately models the interaction between the plate and the room),

$$\omega E_2 \eta_{23} = \omega \langle \overline{v_2^2} \rangle \rho_S S \eta_{23} = 2\rho_0 c S \langle \overline{v_2^2} \rangle \sigma = \Pi_{\text{rad}}, \tag{6.79}$$

and thus

$$\eta_{23} = \frac{2\rho_0 c\sigma}{\omega\rho_S}. \tag{6.80}$$

The factor of two is present because it is assumed that the plate structure radiates from both sides into the room.

The modal density of the reverberant room volume can be obtained from equations (6.22) and (6.35). It is approximated by

$$n_3(\omega) = \frac{\omega^2 V}{2\pi^2 c^3}. \tag{6.81}$$

Finally, the internal loss factor of the room is given by equation (6.49). It is

$$\eta_3 = \frac{13.82}{\omega T_{60}}. \tag{6.82}$$

Substituting the relevant parameters into equation (6.77) yields the required relationship between the mean-square plate velocity and the mean-square sound pressure level in the room. Thus,

$$\langle \overline{p^2} \rangle = \frac{\langle \overline{v_2^2} \rangle \rho_S \rho_0 S c^2}{V} \cdot \left(\frac{\dfrac{2\rho_0 c\sigma}{\omega\rho_S}}{\eta_3 + \dfrac{4\pi^2 \rho_0 c^4 \sigma S}{3.6 c_{L2} h V \rho_S \omega^3}} \right). \tag{6.83}$$

The above equation gives the mean-square sound pressure in the room due to radiation from the vibrating plate. It is a useful practical relationship for estimating an upper limit sound pressure level in a room due to a vibrating plate type structure.

Alternatively, if the plate were excited by a diffuse sound field within the room, then

$$\frac{E_2}{E_3} = \frac{\frac{n_2}{n_3}\eta_{23}}{\eta_2 + \eta_{23}}, \tag{6.84}$$

and with the appropriate substitutions

$$\langle \overline{v_2^2} \rangle = \langle \overline{p^2} \rangle \left(\frac{5.45c^2}{\rho_0 chc_{L2}\rho_S\omega^2} \right) \left[\frac{1}{1 + \{(\rho_S\omega\eta_2)/(2\rho_0 c\sigma)\}} \right]. \tag{6.85}$$

An inspection of equation (6.85) shows that, if the energy dissipated in the plate (i.e. η_2) is smaller than the sound power radiated by the plate, then the second term in brackets on the right hand side approximates to unity.

For the limiting case of a lightly damped structure and a reverberant room (i.e. η_2 and η_3 are very small), equations (6.83) and (6.85) are identical because of reciprocity. They both reduce to

$$\langle \overline{v_2^2} \rangle = \langle \overline{p^2} \rangle \left(\frac{5.45c^2}{\rho_0 chc_{L2}\rho_S\omega^2} \right). \tag{6.86}$$

It is important to remember that the equations derived in this subsection only apply to the excitation of resonant subsystem modes and not forced 'mass law' transmission through a panel. Non-resonant transmission through a panel is considered in the next example.

6.7.2 *Two rooms coupled by a partition*

A good example of the application of S.E.A. to both resonant and non-resonant energy flow between subsystems is the transmission of sound through a partition which divides two rooms. This problem has already been treated in section 4.9, chapter 4. In that case, however, the transmission coefficient, τ, was left in a general form to be evaluated either empirically or using manufacturer's data. With S.E.A., the transmission coefficient can be defined specifically in terms of the resonant and non-resonant transmission components.

Consider a three subsystem S.E.A. model as illustrated in Figure 6.20. Subsystem 1 is the source room, subsystem 2 is the partition, and subsystem 3 is the receiving room. The steady-state power balance equations for the three subsystems are

$$\Pi_1 = \omega E_1\eta_1 + \{\omega E_1\eta_{12} - \omega E_2\eta_{21}\} + \{\omega E_1\eta_{13} - \omega E_3\eta_{31}\}, \tag{6.87}$$

$$0 = \omega E_2\eta_2 + \{\omega E_2\eta_{21} - \omega E_1\eta_{12}\} + \{\omega E_2\eta_{23} - \omega E_3\eta_{32}\}, \tag{6.88}$$

and

$$0 = \omega E_3\eta_3 + \{\omega E_3\eta_{31} - \omega E_1\eta_{13}\} + \{\omega E_3\eta_{32} - \omega E_2\eta_{23}\}. \tag{6.89}$$

The associated reciprocity relationships between the three subsystems are

$$n_1\eta_{12} = n_2\eta_{21}, \tag{6.90}$$

$$n_2\eta_{23} = n_3\eta_{32}, \tag{6.91}$$

and

$$n_1\eta_{13} = n_3\eta_{31}. \tag{6.92}$$

The steady-state power balance equations can be re-expressed as

$$\Pi_1 = \omega E_1\eta_1 + \omega\eta_{12}n_1\left\{\frac{E_1}{n_1} - \frac{E_2}{n_2}\right\} + \omega\eta_{13}n_1\left\{\frac{E_1}{n_1} - \frac{E_3}{n_3}\right\}, \tag{6.93}$$

$$0 = \omega E_2\eta_2 - \omega\eta_{12}n_1\left\{\frac{E_1}{n_1} - \frac{E_2}{n_2}\right\} + \omega\eta_{23}n_2\left\{\frac{E_2}{n_2} - \frac{E_3}{n_3}\right\}, \tag{6.94}$$

Fig. 6.20. A three subsystem S.E.A. model.

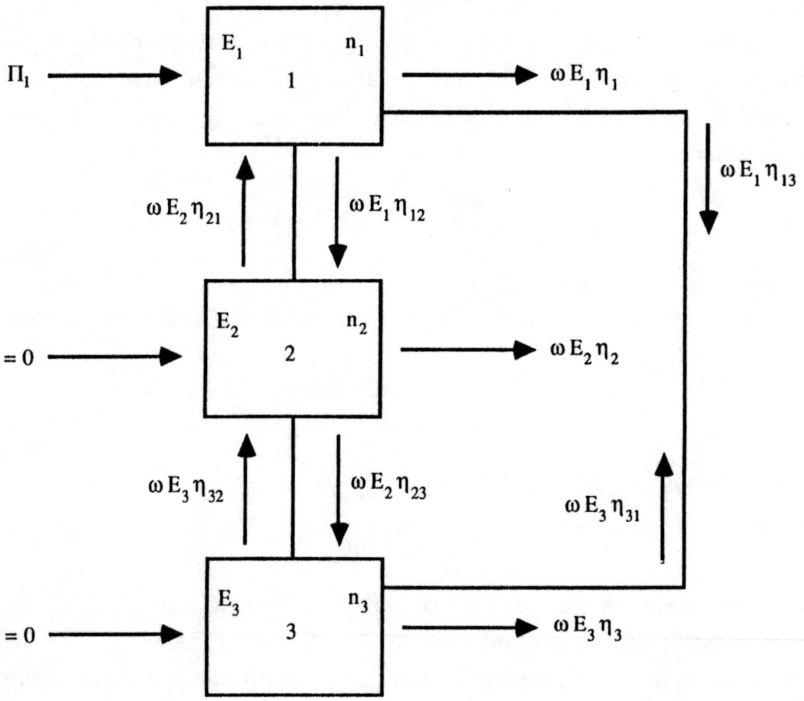

and

$$0 = \omega E_3 \eta_3 - \omega \eta_{13} n_1 \left\{ \frac{E_1}{n_1} - \frac{E_3}{n_3} \right\} - \omega \eta_{23} n_2 \left\{ \frac{E_2}{n_2} - \frac{E_3}{n_3} \right\}. \tag{6.95}$$

From equation (6.94) and the appropiate reciprocity relationships

$$\frac{E_2}{n_2} = \frac{\eta_{21} \frac{E_1}{n_1} + \eta_{23} \frac{E_3}{n_3}}{\eta_2 + \eta_{21} + \eta_{23}}. \tag{6.96}$$

Now, since the sound pressure levels in the source room are significantly greater than those in the receiving room, $E_1/n_1 \gg E_3/n_3$. Also, because the systems are coupled, $\eta_{21} = \eta_{23} = \eta_{\text{rad}}$, and $\eta_2 = \eta_s$, where η_{rad} is the acoustic radiation loss factor of the partition and η_s is the structural loss factor of the partition. Thus,

$$\frac{E_2}{n_2} = \frac{E_1}{n_1} \left\{ \frac{\eta_{\text{rad}}}{\eta_s + 2\eta_{\text{rad}}} \right\}. \tag{6.97}$$

From equation (6.95) and the appropriate reciprocity relationships

$$E_3 = \frac{\eta_{13} E_1 + \eta_{23} E_2}{\eta_3 + \eta_{31} + \eta_{32}}. \tag{6.98}$$

In the above equation, $\eta_{13} E_1$ is the non-resonant mass law transmission component, and $\eta_{23} E_2$ is the resonant transmission component.

The noise reduction, NR, between the source and receiver rooms is given by the difference in the sound pressure levels between the two rooms (i.e. $L_{p1} - L_{p2}$) with the partition in place. This is equivalent to the energy density ratio between the two rooms. Hence,

$$NR = 10 \log_{10} \frac{E_1/V_1}{E_3/V_3}, \tag{6.99}$$

where V_1 and V_3 are the respective room volumes. The ratio E_1/E_3 can be obtained by substituting equation (6.97) into equation (6.98) and using the appropriate reciprocity relationships. It is

$$\frac{E_1}{E_3} = \frac{\eta_3 + \frac{n_1}{n_3} \eta_{13} + \frac{n_2}{n_3} \eta_{\text{rad}}}{\frac{\frac{n_2}{n_1} \eta_{\text{rad}}^2}{\eta_s + 2\eta_{\text{rad}}} + \eta_{13}}. \tag{6.100}$$

The noise reduction can thus be obtained by substituting equation (6.100) into equation (6.99). It is important to note that this noise reduction is associated with both resonant and non-resonant sound transmission through the partition.

The partition transmission loss ($TL = 10\log_{10}1/\tau$) can now be obtained by substituting equations (6.99) and (6.100) into equation (4.101) and solving for τ. The three equations yield a quadratic equation in τ – i.e.

$$\tau^2 S_2 + \tau S_3 \alpha_{3\,\mathrm{avg}} - S_3 \alpha_{3\,\mathrm{avg}} \frac{E_3/V_3}{E_1/V_1} = 0, \tag{6.101}$$

where S_2 is the surface area of the partition, S_3 is the total surface area of the receiving room, and $\alpha_{3\,\mathrm{avg}}$ is the average absorption coefficient of the receiving room. The total absorption of the receiving room can be approximated by $S_3\alpha_{3\,\mathrm{avg}}$. In this instance equation (4.101) becomes

$$NR = TL - 10\log_{10}(1/\alpha_{3\,\mathrm{avg}}), \tag{6.102}$$

where $\alpha_{3\,\mathrm{avg}}$ is given by equation (4.70) – i.e.

$$\alpha_{3\,\mathrm{avg}} = \frac{60V_3}{1.086cS_3T_{60}}, \tag{6.103}$$

where T_{60} is the reverberation time of the receiving room. Thus,

$$TL = 10\log_{10}(1/\tau) = NR + 10\log_{10}\left\{\frac{1.086cS_3T_{60}}{60V_3}\right\}. \tag{6.104}$$

When evaluating the noise reduction from equations (6.99) and (6.100), the modal densities of the rooms (n_1 and n_3) can be evaluated from equation (6.35), and the modal density of the partition, n_2, can be evaluated from equation (6.25). The acoustic radiation damping, η_{rad}, of the partition can be evaluated from equation (6.48), and the non-resonant coupling loss factor, η_{13}, can be evaluated from equations (6.63) and (6.64).

6.8 Non-conservative coupling – coupling damping

The effects of coupling damping, η_j, (i.e. damping at joints in coupled subsystems) have been qualitatively discussed earlier on in this chapter (section 6.5 and subsection 6.6.4), the internal loss factor being defined as the sum of the internal structural damping, η_s, the acoustic radiation damping, η_{rad}, and the coupling damping, η_j.

The quantitative effects of coupling damping on energy flow between coupled structures have been recently investigated by Fahy and Yao[6.35,6.36] and by Sun and Ming[6.37]. In essence, the work is an extension of the original work by Lyon[6.1] and others on the energy flow between two linearly coupled oscillators. In the original work, it was shown that the energy flow between two oscillators is proportional to the difference between the actual total vibrational energies of the respective coupled oscillators (equation 6.4). With groups of oscillators, the energy flow is proportional to the difference in modal energies (equation 6.5).

When coupling damping is included between two oscillators, after considerable algebraic manipulation[6.35–6.37], it can be shown that

$$\langle \Pi_{12} \rangle = \beta\{\langle E_1 \rangle - \langle E_2 \rangle\} + \psi\langle E_1 \rangle + \varphi\langle E_2 \rangle, , \tag{6.105}$$

and that

$$\langle \Pi_{21} \rangle = \beta\{\langle E_2 \rangle - \langle E_1 \rangle\} + \psi'\langle E_2 \rangle + \varphi'\langle E_1 \rangle, \tag{6.106}$$

where β, ψ, φ, ψ' and φ' are constants of proportionality which are functions of the oscillator and coupling parameters (ψ and φ have opposite signs, and so do ψ' and φ'). The physical significance of equations (6.105) and (6.106) is that the energy flow between two non-conservatively coupled oscillators is related (i) to the difference between oscillator energies and (ii) to their respective absolute energies. The equations also show that the energy flow in different directions is not equal anymore.

When the coupling damping is very small compared with the other internal loss factor components (structural and acoustic radiation damping), the constants ψ, φ, ψ' and φ' are $\ll \beta$, and equations (6.105) and (6.106) reduce to equation (6.4). When the coupling damping is of the same order of magnitude as (or larger than) the structural and acoustic radiation damping, it has the nett effect of increasing the internal loss factors of the coupled oscillators (i.e. equations 6.45 and 6.46).

Fahy and Yao[6.36] and Sun and Ming[6.37] also show in their analysis that the energy flow between coupled oscillators depends on the coupling damping – coupling damping reduces the energy of the indirectly driven subsystem, with maximum benefit being attained when the coupling damping is similar in magnitude to the larger of the two damping coefficients. Also, for the case of two coupled oscillators, equipartition of energy does not form an upper limit on oscillator energy ratios when there is coupling damping – the ratio of the energy of the indirectly driven oscillator to that of the directly driven oscillator energy can exceed unity, unlike the conservatively coupled case[6.36]. Nevertheless, for practical S.E.A. applications it is valid to account for the effects of coupling damping as per equation 6.45 since non-conservative coupling only has the nett effect of increasing the internal loss factor of the individual subsystems, the exception being when the coupling damping is very large (~ 1) – i.e. a flexible coupling. Here, the coupling loss factor is also affected by the coupling damping[6.37].

6.9 The estimation of sound radiation from coupled structures using total loss factor concepts

The concept of total loss factors of coupled subsystems was introduced in subsection 6.3.4. These concepts, together with energy accountancy techniques, have been used recently by Stimpson *et al.*[6.38] to predict sound power radiation from built-up structures. Richards[6.39] is largely responsible for extending S.E.A. procedures to include energy accountancy for the optimisation of machinery noise control.

The procedures discussed in this section are derived from the work of Stimpson *et al.*[6.38] and are limited to coupled subsystems where only one subsystem is excited by some external input power – the remaining subsystems are excited via transmission of vibrational energy through the coupling links. Such systems where the main vibrational excitation comes from a single subsystem are relatively common in industry. For such a system, the total radiated sound power is given by[6.38]

$$\Pi_{\text{rad}} = \frac{\rho_0 c \sigma_i E_i}{\rho_i t_i}\left(1 + \sum_{\substack{k=1 \\ k \neq i}}^{N} \frac{E_k \sigma_k t_i \rho_i}{E_i \sigma_i t_k \rho_k}\right), \tag{6.107}$$

where the subscript i refers to the subsystem which is directly excited, and the subscript k refers to all the other subsystems which are coupled to it. The E's represent the space- and time-averaged vibrational energies, the σ's represent the respective radiation ratios, the t's are the respective thicknesses of the subsystems, and the ρ's are the material densities (note that ρ_0 is the fluid density). Stimpson *et al.*[6.38] derive the above equation from the definition of the radiation ratio (i.e. equation 3.30).

When the material density is constant for all subsystems, the above equation simplifies to

$$\Pi_{\text{rad}} = \Pi_{i\,\text{rad}}\left(1 + \sum_{\substack{k=1 \\ k \neq i}}^{N} \frac{E_k \sigma_k t_i}{E_i \sigma_i t_k}\right). \tag{6.108}$$

The sound power radiated by individual subsystems, other than the subsystem which is directly excited, is obtained from the summation term within the brackets – i.e.

$$\Pi_{k\,\text{rad}} = \Pi_{i\,\text{rad}} \frac{E_k \sigma_k t_i}{E_i \sigma_i t_k}, \tag{6.109}$$

where $\Pi_{i\,\text{rad}}$ is the sound power radiated by the subsystem which is directly excited.

Equations (6.108) and (6.109) are very useful for optimising machinery noise control – i.e. they allow for the effects of variations in coupling between subsystems, material shape and size, damping, etc. on radiated sound to be evaluated.

Stimpson *et al.*[6.38] assess the effects of damping by introducing a noise reduction factor, N, where

$$N = 10 \log_{10} \frac{\Pi_{\text{rad (damped)}}}{\Pi_{\text{rad (undamped)}}}. \tag{6.110}$$

The parameter N can be evaluated by considering the total loss factors of the various coupled subsystems. The total steady-state loss factor for a subsystem coupled to another subsystem is given by equation (6.17). This equation can be extended to a subsystem which is coupled to several subsystems. The total steady-state loss factor is now given by

$$\eta_{TSi} = \eta_i + \sum_{\substack{k=1\\k\neq i}}^{N} \eta_{ik}\left(1 - \frac{n_i E_k}{n_k E_i}\right), \tag{6.111}$$

where η_i is the internal loss factor which is given by equation (6.45). In their analysis, Stimpson *et al.*[6.38] use a slightly modified version of equation (6.111). They define the internal loss factor in terms of the structural loss factor, η_s, and the coupling damping loss factor, η_j, and include η_{rad} as a separate term. Their total steady-state loss factor is denoted by η^*_{TS} in this book. The end result is the same and the applied power equals the radiated sound power from the structure plus the power which is dissipated in the structure. Using this energy balance, it can be shown that the noise reduction factor, N, is given by

$$N = 10\log_{10}\frac{\dfrac{\rho_0 c\sigma_i}{\rho t_i}\left(1 + \sum\limits_{\substack{k=1\\k\neq i}}^{N}\dfrac{E'_k\sigma_k t_i}{E'_i\sigma_i t_k}\right) + \omega\eta^{*\prime}_{TSi}\left(1 + \sum\limits_{\substack{k=1\\k\neq i}}^{N}\dfrac{E_k\sigma_k t_i}{E_i\sigma_i t_k}\right)}{\dfrac{\rho_0 c\sigma_i}{\rho t_i}\left(1 + \sum\limits_{\substack{k=1\\k\neq i}}^{N}\dfrac{E_k\sigma_k t_i}{E_i\sigma_i t_k}\right) + \omega\eta^{*}_{TSi}\left(1 + \sum\limits_{\substack{k=1\\k\neq i}}^{N}\dfrac{E'_k\sigma_k t_i}{E'_i\sigma_i t_k}\right)}, \tag{6.112}$$

where E'_1 and E'_k are the vibrational energies after damping treatment has been applied, and $\eta^{*\prime}_{TSi}$ is the new damped total loss factor (i.e. $\eta'_s + \eta'_j$).

Now, most machine structures, unlike lightweight shells and plates, are reasonably heavily damped such that $(\eta_s + \eta_j) \gg \eta_{rad}$. If this is the case, then equation (6.112) can be simplified to

$$N = 10\log_{10}\frac{\eta^{*\prime}_{TSi}\left(1 + \sum\limits_{\substack{k=1\\k\neq i}}^{N}\dfrac{E_k\sigma_k t_i}{E_i\sigma_i t_k}\right)}{\eta^{*}_{TSi}\left(1 + \sum\limits_{\substack{k=1\\k\neq i}}^{N}\dfrac{E'_k\sigma_k t_i}{E'_i\sigma_i t_k}\right)}. \tag{6.113}$$

The above equation illustrates two important points. Firstly, noise reduction will result if the total loss factor of the excited substructure is increased. Secondly, noise reduction will also result if the energy ratio with additional damping treatment, E'_k/E'_i, is greater than the original energy ratio, E_k/E_i.

6.10 Relationships between dynamic stress and strain and structural vibration levels

S.E.A. facilitates the rapid evaluation of mean-square vibrational response levels of coupled structures. For any useful prediction of service life as a result of possible fatigue or failure, these vibrational response levels must be converted into stress levels. If it were possible to correctly predict the dynamic stress levels in a structure directly from its vibrational response levels, then the S.E.A. technique would be usefully extended and become a very powerful monitoring tool – acceleration level measurements are significantly easier to obtain in the field than strain level measurements.

Dynamic stresses and strains result directly from structural vibrations, and Hunt[6.40] and Ungar[6.41] were amongst the first to develop relationships between kinetic energy, velocity and dynamic strain in plates and beams. Lyon[6.1] summarises their work in his book. Fahy[6.42] and Stearn[6.43–6.45] subsequently developed a theoretical analysis, based upon the concept of bending waves in a reverberant field, for the prediction of the spatial variation of dynamic stress, dynamic strain and acceleration in structures subject to multimode frequency excitation. Practical engineering type relationships have evolved from this early theoretical work, and recently Norton and Fahy[6.46] have conducted a series of experiments to establish the correlation of dynamic stress and strain with cylindrical shell wall vibrations. In this work, they paid particular attention to the statistics of the distributions of the ratios of strain with vibrational velocity, and comparisons were made between predicted and actual r.m.s. strains.

Theoretical analyses[6.40–6.45] backed up by experimental evidence[6.43,6.46] have thus allowed for the development of very simple relationships between mean-square vibrational velocities, dynamic stress and dynamic strain for homogeneous structures such as beams, plates, shells, etc. Recent experimental evidence[6.46] also suggests that the relationships can be used in a modified form for regions of stress concentration. From the theoretical analyses, it can be shown that for a variety of mechanical motions (flexure, torsion, compression, etc.) the mean-square dynamic stress is directly related to the mean-square vibrational velocity by the following simple relationship:

$$\frac{\langle\overline{\sigma^2}\rangle}{\langle\overline{v^2}\rangle} = 1.61Kc_{\mathrm{L}}^2\rho^2, \tag{6.114}$$

where σ is the dynamic stress, v is the vibrational velocity, the brackets represent a time-average, the overbar represents a space-average, $c_{\mathrm{L}}^2 = E/\rho$, Poisson's ratio is assumed to be 0.3, and K is a constant which depends upon the type of motion and the system geometry. When Poisson's ratio is not 0.3, a more general form of the equation has to be used[6.43,6.46].

The corresponding relationship between mean-square dynamic strain and mean-square vibrational velocity, once again assuming that Poisson's ratio is 0.3, is

$$\frac{\langle\overline{\xi^2}\rangle}{\langle\overline{v^2}\rangle} = \frac{K}{c_{\mathrm{L}}^2}. \tag{6.115}$$

where ξ is the dynamic strain and K is the same constant as in equation (6.114).

The constant K varies over a small range near unity, except for situations involving stress concentration. For estimation purposes, it is acceptable to set $K = 1$ in regions where there is no stress concentration. Experimental results on cylindrical shells[6.46] suggest that K varies between 3 and 20 in regions where there is some stress concentration.

Some typical experimental results for cylindrical shells are presented in Figures 6.21 and 6.22. The experiments were conducted on a small diameter, fluid-filled cylindrical shell – a steel pipe 6.3 m long with a mean diameter of 160 mm, a 5 mm

Fig. 6.21. Typical time- and space-averaged ratios of velocity to strain for various circumferential excitations of an unconstrained cylindrical shell: ●, $n = 2$ circumferential excitation; ■, $n = 3$ circumferential excitation; ○, multimode circumferential excitation. (n is the number of full waves around the circumference.)

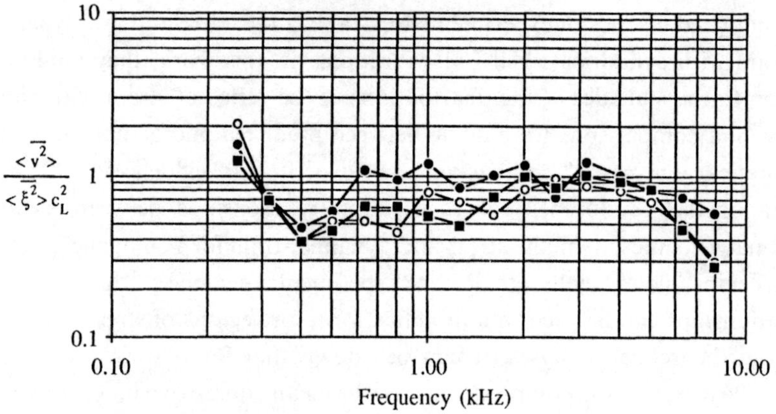

Fig. 6.22. Typical time- and space-averaged ratios of velocity to strain for various circumferential excitations of a constrained cylindrical shell: ○, $n = 2$ circumferential excitation; ●, multimode circumferential excitation. (n is the number of full waves around the circumference.)

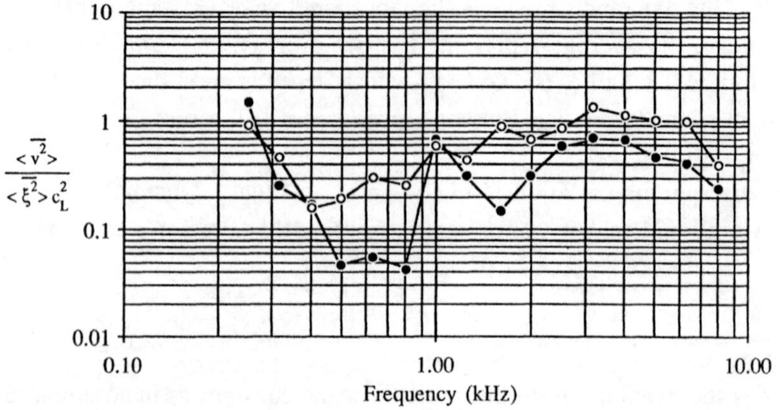

wall thickness, and filled with oil. The pipe was excited with an array of twelve circumferentially spaced exciters. The exciters were spaced such that different circumferential wave patterns could be preferentially selected by controlling the phase and the number of exciters that could be excited at any one time. The pipe wall response was averaged over six randomly chosen positions, and the corresponding dynamic strain was measured at the same locations with semi-conductor strain gauges. Stress concentration was induced in the pipe by clamping a heavy cylindrical ring around its circumference. The experimental results in Figure 6.21 are for the unconstrained (i.e. no stress concentration) pipe, whilst, those in Figure 6.22 are for the pipe constrained with the cylindrical ring clamped around its circumference. The results show that the constant $K \approx 1$ when there is no stress concentration and that it increases when stress concentration is present. Also, the results in Figure 6.22 suggest that the stress concentration is frequency dependent. The higher frequency roll-off is due to signal processing limitations (i.e. strain gauge amplifier frequency response limitations and accelerometer mass loading effects).

Whilst the strict theoretical basis for equations (6.114) and (6.115) requires that a diffuse wave field is present, the experimental evidence[6.46] suggests that the equations can be applied to unconstrained structures in regions of both high and low modal density. In the former case, time-averaged measurements at a single location appear to be adequate, whereas both time and spatial averaging is required in the latter case. When stress concentration is present, the conclusions are not so clear cut; with stress concentration, the constant $K > 1$. The actual value of K would depend upon the specific situation. Further research is required to establish confidence levels for K for a range of different situations.

In conclusion, it should be noted that equations (6.114) and (6.115) are very useful relationships for the prediction of dynamic stress and strain, and can be incorporated into any S.E.A. modelling procedure.

References

6.1 Lyon, R. H. 1975. *Statistical energy analysis of dynamical systems: theory and applications*, M.I.T. Press.

6.2 Fahy, F. J. 1982. 'Statistical energy analysis', chapter 7 in *Noise and vibration*, edited by R. G. White and J. G. Walker, Ellis Horwood.

6.3 Hodges, C. H. and Woodhouse, J. 1986. 'Theories of noise and vibration in complex structures', *Reports on Progress in Physics* **49**, 107–70.

6.4 Woodhouse, J. 1981. 'An introduction to statistical energy analysis of structural vibrations', *Applied Acoustics* **14**, 455–69.

6.5 Bies, D. A. and Hamid, S. 1980. '*In-situ* determination of loss and coupling loss factors by the power injection method', *Journal of Sound and Vibration* **70**(2), 187–204.

6.6 Sun, H. B., Sun, J. C. and Richards, E. J. 1986. 'Prediction of total loss factors of structures, part iii: effective loss factors in quasi-transient conditions', *Journal of Sound and Vibration* **106**(3), 465–79.
6.7 Cremer, L., Heckl, M. and Ungar, E. E. 1973. *Structure-borne sound*, Springer-Verlag.
6.8 Ver, I. L. and Holmer, C. I. 1971. 'Interaction of sound waves with solid structures', chapter 11 in *Noise and vibration control*, edited by L. L. Beranek, McGraw-Hill.
6.9 Hart, F. D. and Shah, K. C. 1971. *Compendium of modal densities for structures*, NASA Contractor Report, CR-1773.
6.10 Clarkson, B. L. and Pope, R. J. 1981. 'Experimental determination of modal densities and loss factors of flat plates and cylinders', *Journal of Sound and Vibration* **77**(4), 535–49.
6.11 Keswick, P. R. and Norton, M. P. 1987. 'A comparison of modal density measurement techniques', *Applied Acoustics* **20**, 137–53.
6.12 Clarkson, B. L. 1986. 'Experimental determination of modal density', chapter 5 in *Random vibration – status and recent developments*, edited by I. Elishakoff and R. H. Lyon, Elsevier.
6.13 Clarkson, B. L. and Ranky, M. F. 1983. 'Modal density of honeycomb plates', *Journal of Sound and Vibration* **91**(1), 103–18.
6.14 Ferguson, N. S. and Clarkson, B. L. 1986. 'The modal density of honeycomb shells', *Journal of Vibration, Acoustics, Stress, and Reliability in Design* **108**, 399–404.
6.15 Szechenyi, E. 1971. 'Modal densities and radiation efficiencies of unstiffened cylinders using statistical methods', *Journal of Sound and Vibration* **19**(1), 65–81.
6.16 Brown, K. T. 1984. 'Measurement of modal density: an improved technique for use on lightly damped structures', *Journal of Sound and Vibration* **96**(1), 127–32.
6.17 Brown, K. T. and Norton, M. P. 1985. 'Some comments on the experimental determination of modal densities and loss factors for statistical energy analysis applications', *Journal of Sound and Vibration* **102**(4), 588–94.
6.18 Hakansson, B. and Carlsson, P. 1987. 'Bias errors in mechanical impedance data obtained with impedance heads', *Journal of Sound and Vibration* **113**(1), 173–83.
6.19 Arnold, R. N. and Warburton, G. B. 1949. 'The flexural vibrations of thin cylinders', *Proceedings of the Royal Society* (*London*) **197A**, 238–56.
6.20 Rennison, D. C. and Bull, M. K. 1977. 'On the modal density and damping of cylindrical pipes', *Journal of Sound and Vibration* **54**(1), 39–53.
6.21 Clarkson, B. L. and Brown, K. T. 1985. 'Acoustic radiation damping', *Journal of Vibration, Acoustics, Stress, and Reliability in Design* **107**, 357–60.
6.22 Ungar, E. E. 1971. 'Damping of panels', chapter 14 in *Noise and vibration control*, edited by L. L. Beranek, McGraw-Hill.
6.23 Richards, E. J. and Lenzi, A. 1984. 'On the prediction of impact noise IV: the structural damping of machinery', *Journal of Sound and Vibration* **97**(4), 549–86.
6.24 Ranky, M. F. and Clarkson, B. L. 1983. 'Frequency average loss factors of plates and shells', *Journal of Sound and Vibration* **89**(3), 309–23.
6.25 Norton, M. P. and Greenhalgh, R. 1986. 'On the estimation of loss factors in lightly damped pipeline systems: some measurement techniques and their limitations', *Journal of Sound and Vibration* **105**(3), 397–423.
6.26 Keswick, P. R. and Norton, M. P. 1987. *Coupling damping estimates of non-conservatively coupled cylindrical shells*, A.S.M.E. Winter Meeting on Statistical Energy Analysis, Boston, pp. 19–24.
6.27 Brown, K. T. and Clarkson, B. L. 1984. *Average loss factors for use in statistical energy analysis*, Vibration Damping Workshop, Wright-Patterson Air Force Base, Ohio, U.S.A.
6.28 Fahy, F. J. 1985. *Sound and structural vibration: radiation, transmission and response*, Academic Press.

6.29 Wöhle, W., Beckmann, Th. and Schreckenbach, H. 1981. 'Coupling loss factors for statistical energy analysis of sound transmission at rectangular structural slab joints part I and II', *Journal of Sound and Vibration* **77**(3), 323–44.
6.30 Wilby, J. P. and Sharton, T. D. 1974. *Acoustic transmission through a fuselage side wall*, Bolt, Beranek, and Newman Report 2742.
6.31 Clarkson, B. L. and Ranky, M. F. 1984. 'On the measurement of coupling loss factors of structural connections', *Journal of Sound and Vibration* **94**(2), 249–61.
6.32 Norton, M. P. and Keswick, P. R. 1987. *Loss and coupling loss factors and coupling damping in non-conservatively coupled cylindrical shells*, Proceedings Inter-Noise '87, Beijing, China, pp. 651–4.
6.33 Lyon, R. H. and Eichler, E. E. 1964. 'Random vibrations of connected structures', *Journal of the Acoustical Society of America* **36**, 1344–54.
6.34 Crocker, M. J. and Kessler, F. M. 1982. *Noise and noise control, volume II*, C.R.C. Press.
6.35 Fahy, F. J. and Yao, D. 1986. *Power flow between non-conservatively coupled oscillators*, Proceedings 12th International Congress of Acoustics, Toronto, Paper D6-2.
6.36 Fahy, F. J. and Yao, D. 1987. 'Power flow between non-conservatively coupled oscillators', *Journal of Sound and Vibration* **114**(1), 1–11.
6.37 Sun, J. C. and Ming, R. S. 1988. *Distributive relationships of dissipated energy by coupling damping in non-conservatively coupled structures*, Proceedings Inter-Noise '88, Avignon, France, pp. 323–6.
6.38 Stimpson, G. J., Sun, J. C and Richards, E. J. 1986. 'Predicting sound power radiation from built-up structures using statistical energy analysis', *Journal of Sound and Vibration* **107**(1), 107–20.
6.39 Richards, E. J. 1981. 'On the prediction of impact noise, III: energy accountancy in industrial machines', *Journal of Sound and Vibration* **76**(2), 187–232.
6.40 Hunt, F. V. 1960. 'Stress and strain limits on the attainable velocity in mechanical vibrations, *Journal of the Acoustical Society of America* **32**(9), 1123–8.
6.41 Ungar, E. E. 1962. 'Maximum stresses in beams and plates vibrating at resonance', *Journal of Engineering for Industry* **84**(1), 149–55.
6.42 Fahy, F. J. 1971. *Statistics of acoustically induced vibration*, 7th International Congress on Acoustics, Budapest, pp. 561–4.
6.43 Stearn, S. M. 1970. *Stress distribution in randomly excited structures*, Ph.D. Thesis, Southampton University.
6.44 Stearn, S. M. 1970. 'Spatial variation of stress, strain and acceleration in structures subject to broad frequency band excitation', *Journal of Sound and Vibration* **12**(1), 85–97.
6.45 Stearn, S. M. 1971. 'The concentration of dynamic stress in a plate at a sharp change of section', *Journal of Sound and Vibration* **15**(3), 353–65.
6.46 Norton, M. P. and Fahy, F. J. 1988. 'Experiments on the correlation of dynamic stress and strain with pipe wall vibrations for statistical energy analysis applications', *Noise Control Engineering* **30**(3), 107–11.

Nomenclature

a	panel dimension, moment arm of an applied force
a_m	mean shell radius
A	cross-sectional area, surface area of a cavity

A_b	cross-sectional area of a beam
b	panel dimension
B	faceplate longitudinal stiffness parameter of a honeycomb panel
c	speed of sound
c_B	bending wave velocity
c_L, c_{L1}, c_{L2}, etc.	quasi-longitudinal wave velocities
c_{LB}	quasi-longitudinal wave velocity of a beam
c_{Lp}	quasi-longitudinal wave velocity of a plate
c_v	viscous-damping coefficient
d	parameter relating to honeycomb panel thickness dimensions (see equation 6.32)
E, E_1, E_2, etc.	stored energies in oscillators or subsystems (groups of oscillators), Young's modulus of elasticity
E_1^*, E_2^*, etc.	modal energies (E/n)
E_i'	vibrational energy of a subsystem after damping treatment (see equations 6.112, 6.113)
$\langle E_1 \rangle, \langle E_2 \rangle$	actual time-averaged energies of respective coupled oscillators or subsystems (groups of oscillators)
$\langle E_1' \rangle, \langle E_2' \rangle$	blocked time-averaged energies of respective coupled oscillators
f	frequency
f_r	ring frequency of a cylindrical shell
$f(t)$	measured input force signal to a linear system
F	bandwidth factor
$\mathbf{F}$	Fourier transform of force (complex function)
$\mathbf{F}_I$	complex force measured by an impedance head transducer
$\mathbf{F}_X$	complex force applied to a structure
$\mathbf{F}(\omega)$	input force to a linear system (complex function)
g	core stiffness parameter of a honeycomb panel
$G_{ff}(\omega)$	one-sided auto-spectral density function of an input force

$\mathbf{G_{fv}}(\omega)$	one-sided cross-spectral density function of functions $f(t)$ and $v(t)$ (complex function)
$\mathbf{G_{xf}}(\omega)$	one-sided cross-spectral density function of functions $x(t)$ and $f(t)$ (complex function)
$\mathbf{G_{xv}}(\omega)$	one-sided cross-spectral density function of functions $x(t)$ and $v(t)$ (complex function)
G_x, G_y	shear moduli of honeycomb panel core material in x- and y-direction, respectively
h	thickness
h_1, h_3	honeycomb panel faceplate thicknesses
h_2	honeycomb panel core thickness
$\mathbf{H}(\omega)$	frequency response function of a power amplifier and exciter system (complex function)
i	integer
I	second moment of area of a cross-section about the neutral plane axis
$\mathbf{I}(\omega)$	feedback frequency response function for shaker–structure interactions (complex function)
j	integer
k	wavenumber, integer
K	constant relating stress and strain to vibrational velocity
K_s	stiffness of a structural element
L	length
m	oscillator mass, total mass per unit area of a honeycomb panel
M, M_1, M_2	mass of structural elements
M_b	beam mass
n	integer
n_1, n_2, etc.	modal densities
n_S	modal density of a structural element
n_V	modal density of an acoustic volume
$n(f)$	modal density as a function of frequency
$n(t)$	noise signal at the output stage
$n(\omega)$	modal density as a function of radian frequency
N	integer, noise reduction factor (see equations 6.112, 6.113)

NR	noise reduction
$\langle \overline{p^2} \rangle$	mean-square sound pressure (space- and time-averaged)
P	perimeter, total edge length
Q	quality factor
$R_{fv}(\tau)$	cross-correlation function of functions $f(t)$ and $v(t)$
S, S_1, S_2, etc.	surface areas
$\mathbf{S}_{\mathbf{fv}}(\omega)$	two-sided cross-spectral density function of functions $f(t)$ and $v(t)$ (complex function)
t, t_1, t_2, etc.	thicknesses
t_b	beam thickness
t_p	plate thickness
T_{60}	reverberation time for a 60 dB decay
TL	transmission loss
$v(t)$	measured output velocity signal from a linear system
$\langle \overline{v_1^2} \rangle$, $\langle \overline{v_2^2} \rangle$	mean-square vibrational velocities (space- and time-averaged)
V, V_r	volumes
$\mathbf{V}$	Fourier transform of velocity (complex function)
$\mathbf{V}^*$	complex conjugate of $\mathbf{V}$
$\mathbf{V}_\mathbf{I}$	complex velocity measured by an impedance head transducer
$\mathbf{V}_\mathbf{X}$	complex velocity of a structure at the point of excitation
$\mathbf{V}(\omega)$	output velocity from a linear system (complex function)
x	distance
$\dot{x}$	velocity
$x(t)$	arbitrary time function, input function to a linear system, original test signal used to drive a power amplifier
X	thickness ratio (t_1/t_2)
$\mathbf{X}(\omega)$	Fourier transform of a function $x(t)$ (complex function)
$\mathbf{Y}$, $\mathbf{Y}(\omega)$	point mobility (complex function)
$\mathbf{Y}_\mathbf{I}$	complex point mobility of an impedance head transducer

$\mathbf{Y_K}$	complex point mobility associated with stiffness
$\mathbf{Y_M}$	complex point mobility associated with mass
$\mathbf{Y_X}$	complex point mobility of a structure
$\mathbf{Z}$	complex point impedance
$\mathbf{Z_b}$	moment impedance of a beam
$\mathbf{Z_p}$	moment impedance of a plate
α_{avg}	space-average sound absorption coefficient
β	constant of proportionality
β'	constant of proportionality
γ	constant of proportionality
ζ	damping coefficient
$\eta, \eta_1, \eta_2, \eta_i, \eta_j$,etc.	loss factors
$\eta_{12}, \eta_{21}, \eta_{ij}$, etc.	coupling loss factors
η_{bp}	coupling loss factor between a beam and a plate
η_j	loss factor associated with energy dissipation at the boundaries of structural elements
η_{rad}	loss factor associated with acoustic damping
η_{rp}	coupling loss factor for non-resonant sound transmission through a panel from a source room
η_s	loss factor associated with energy dissipation within a structural element
η_{SV}	coupling loss factor between a structure and an acoustic volume
η_{VS}	coupling loss factor between an acoustic volume and a structure
η_{TS1}, etc.	total steady-state loss factor of a subsystem (group of oscillators)
η_{TT1}, etc.	total transient loss factor of a subsystem (group of oscillators)
η^*_{TS}	total steady-state loss factor excluding η_{rad} (see equations 6.112, 6.113)
$\eta^{*\prime}_{TS}$	damped total steady-state loss factor excluding η_{rad} (see equations 6.112, 6.113)
κ_b	radius of gyration of a beam

ξ	dynamic strain
$\langle\overline{\xi^2}\rangle$	mean-square dynamic strain (space- and time-averaged)
ν	Poisson's ratio
Π, Π_1, Π_2, etc.	power, input power to oscillators or subsystems (group of oscillators)
Π_d	dissipated power
Π_{in}	input power
Π_{rad}	radiated sound power
$\langle\Pi_{12}\rangle$	time-averaged energy flow between two oscillators or two subsystems (groups of oscillators)
π	3.14 . . .
ρ, ρ_1, ρ_2, etc.	densities
ρ_0	mean fluid density
ρ_b	beam density
ρ_p	plate density
ρ_S, ρ_{S1}, ρ_{S2}, etc.	masses per unit area (surface masses)
σ	radiation ratio, dynamic stress
$\langle\overline{\sigma^2}\rangle$	mean-square dynamic stress (space- and time-averaged)
τ	sound transmission coefficient (wave transmission coefficient)
τ_{12}	wave transmission coefficient
τ_{rp}	wave transmission coefficient through a panel from a source room
ψ	parameter associated with wave transmission coefficients (see equation 6.54), constant of proportionality)
ψ'	constant of proportionality
φ, φ'	constants of proportionality
ω	radian (circular) frequency, geometric mean centre radian frequency of a band
ω_0	panel fundamental natural frequency
ω_C	radian (circular) critical frequency of a panel
ω_n	natural radian (circular) frequency
$\langle\ \rangle$	time-average of a signal
$\overline{}$	space-average of a signal (overbar)

7
Pipe flow noise and vibration: a case study

7.1 Introduction

At the very beginning of this book, the concept of wave–mode duality was emphasised. Its importance to engineering noise and vibration analysis will be illustrated in this chapter via a specific case study relating to pipe flow noise.

The general subject of flow-induced noise and vibrations is a large and complex one. The subject includes: (i) internal axial pipe flows – the transmission of large volume flows of gases, liquids or two phase mixtures across high pressure drops through complex piping systems comprising bends, valves, tee-junctions, orifice plates, expansions, contractions, etc.; (ii) internal cross-flows in heat exchangers, etc. with the associated vortex shedding, acoustic resonances and fluid-elastic instabilities; (iii) external axial and cross-flows – e.g. vortex shedding from chimney stacks; (iv) cavitation; and (v) structure-borne sound associated with some initial aerodynamic type excitation. The reader is referred to Naudascher and Rockwell[7.1], a recent BHRA (British Hydromechanics Research Association) conference publication[7.2] and Blake[7.3] for discussions on a wide range of practical experiences with flow-induced noise and vibrations.

This chapter is, in the main, only concerned with the study of noise and vibration from steel pipelines with internal gas flows[7.4–7.8] – these noise and vibrations are flow-induced and are of considerable interest to the process industries. There are many instances of situations where flow-induced noise and vibration in cylindrical shells have caused catastrophic failures. The mechanisms of the generation of the vibrational response of and the external sound radiation from pipes due to internal flow disturbances are discussed in this chapter. Particular attention is paid to the role of coincidence between structural modes and higher order acoustic modes inside the pipe; the term coincidence, as used in this chapter, relates to the matching of structural wavelengths in the pipe wall with acoustic wavelengths in the contained fluid. Other pipe flow noise sources such as vortex shedding and cavity resonances are also discussed. Semi-empirical prediction schemes are discussed and some general design guidelines are provided. Finally, the usage of a vibration damper for the reduction of pipe flow noise and vibration is discussed.

Pipe flow noise and vibration serves as a good case study because it utilises many of the topics and concepts discussed in the earlier chapters of this book. These include frequency response functions, vibration of continuous systems, solutions to the acoustic wave equation, aerodynamic noise, interactions between sound waves and solid structures, spectral analysis, statistical energy analysis, and dynamic absorption. Furthermore, noise and vibration from cylindrical shells is different to that from flat plates because (i) the effects of curvature of the walls have to be accounted for and (ii) the aerodynamically generated sound field is contained within a 'waveguide'. Also, besides industrial piping systems, the theoretical analyses and the experimental data presented in this chapter have applications in heat exchangers, exhaust systems of internal combustion engines, and nuclear reactors.

In a fully developed turbulent pipe flow (gas phase) through a straight length of pipe with no flow discontinuities or pipe fittings, the vibration of the pipe wall and the associated radiation of sound are due to the random fluctuating pressures along the inside wall of the pipe which are associated with the turbulent flow. This random wall pressure field is statistically uniform both circumferentially and axially, and extends over complete piping lengths and cannot be removed from the flow; it represents a minimum excitation level always present inside the pipe[7.9,7.10]. The situation is somewhat more complex when internal flow disturbances associated with pipe fittings are present in the system. They generate intense internal sound waves which propagate essentially unattenuated through the piping system. The wall pressure fluctuations associated with these sound waves, and the wall pressure fluctuations associated with the local effects of the disturbance itself (e.g. flow separation and increased turbulence levels), contribute significantly to the pipe wall vibration and the external sound radiation. These wall pressure fluctuations are generally only statistically uniform (circumferentially and axially) at large distances from the flow disturbance; at regions in close proximity to the flow disturbance, there are significant circumferential and axial variations.

Thus, in principal, pipe flow noise and vibration can be generated by one or more of the following: (i) the random fluctuating internal wall pressure field associated with fully developed turbulent pipe flow; (ii) the random fluctuating internal wall pressure field resulting from local flow disturbances such as those produced by valves, bends, junctions, and other pipe fittings; (iii) the internal sound pressure field generated by the turbulent pipe flow; (iv) the internal sound pressure field generated by the flow disturbances; and (v) the transmission of mechanical vibrations from pipe fittings which have themselves been excited by the various internal wall pressure fluctuations. In practice, the dominant pipe flow noise and vibration mechanisms tend to be items (ii), (iv) and (v).

7.2 General description of the effects of flow disturbances on pipeline noise and vibration

It is clear from the introduction that noise and vibration is generated in a pipe carrying an internal fully developed turbulent pipe flow even when such a flow is not subjected to any additional disturbances. However, as a result of internal disturbances caused by pipe fittings, the intensity of the pipe wall vibration and the subsequent external noise radiation can be greatly increased. Bull and Norton[7.5–7.6] and Norton and Bull[7.8] have studied the effects of flow disturbances on pipeline noise and vibration in some detail, and a large part of this chapter is based on their work and on the work of others (comprehensive reference lists are provided in references 7.3 and 7.8).

Noise and vibration generation in pipelines involves a sequence of events: disturbance of the flow, generation of internal hydrodynamic or acoustic pressure fluctuations or both by the disturbed flow, excitation of pipe wall vibration by the fluctuating internal wall pressure field, and finally generation of external noise radiation by the vibrating pipe wall. Hence, when the turbulent gas flow inside a pipeline is disturbed by a flow discontinuity such as a bend, a valve, a junction, an orifice plate, or some other form of internal blockage, the statistically uniform fluctuating internal wall pressure field which is characteristic of the undisturbed flow that one would expect in straight runs of pipe, and the associated noise and vibration response, is significantly modified.

The sequence of events that occurs can be described in the following way[7.4–7.8]:

(i) An intense fluctuating non-propagating pressure field is generated in the immediate vicinity of the disturbance. The frequency spectrum of these fluctuations is different from that of the undisturbed flow.

(ii) This fluctuating pressure field decays exponentially with distance from the disturbance, falling off to an essentially constant asymptotic state within a distance of about ten pipe diameters. The fluctuating pressure levels associated with this asymptotic state are still above those of the undisturbed flow, and persist for very large distances downstream of the disturbance.

(iii) At the same time as the fluctuating pressure field decays, the distribution of mean flow velocity over the pipe cross-section returns to its undisturbed state, indicating that the turbulence in the flow also returns to the state characteristic of undisturbed flow.

(iv) The difference between the fluctuating pressure levels of the flow in this 'recovered' state and those of the original undisturbed flow is due to the presence of a superimposed sound field, generated by the disturbance and radiated away from it inside the pipe.

(v) The superimposed sound field consists of plane waves and higher order acoustic modes. The plane waves can, in principle, propagate at all frequencies whilst the higher order acoustic modes can only propagate at frequencies above their cut-off frequencies. These cut-off frequencies are associated with wavelengths that are equal to or smaller than the internal pipe diameter. The cut-off frequency associated with the first higher order acoustic mode of this type is approximately given by

$$f \approx \frac{0.29 c_e}{a_i}, \tag{7.1}$$

where f is the frequency in hertz, c_e is the speed of sound in the external fluid (air), and a_i is the internal pipe radius. This equation is only valid if the temperature inside the pipe is close to atmospheric temperature, and it neglects the effects of flow. The sound field inside the pipe thus consists of plane waves and higher order acoustic modes at frequencies higher than this, but of only plane waves at lower frequencies.

(vi) The mean-square wall pressure fluctuations, $\langle p^2 \rangle$, and the power spectral density, G_{pp}, of undisturbed turbulent wall pressure fluctuations scale as U_0^4 and U_0^3, respectively, at a given Strouhal number $\Omega = \omega a_i / U_0$, where U_0 is the mean velocity, a_i is the internal pipe radius, and ω is the radian frequency. When the flow is disturbed by pipe fittings, this is generally no longer the case, even well downstream of the disturbance where non-propagating components of the disturbance have died out, and the fluctuating wall pressure field comprises a propagating sound field superimposed on the fluctuating pressure field characteristic of the undisturbed turbulent pipe flow. Here, the increment in G_{pp} due to the superimposed sound field scales as U_0^3 at frequencies below the cut-off frequency of the first higher order acoustic mode and as U_0^5 at frequencies above it. The overall mean-square pressure, however, scales as U_0^4 (as is the case for the undisturbed flow) – for very severe disturbances there is some evidence that it scales as a fractionally higher power of flow speed.

(vii) The increased wall pressure fluctuations associated with the flow disturbance caused by a pipe fitting give rise to an increased vibrational response of the pipe wall. This occurs: (a) in the immediate vicinity of the pipe fitting concerned, due to the increased pressure fluctuation levels in the local non-propagating pressure field and also due to propagating sound waves (plane waves and higher order acoustic modes); and (b) over large runs of piping, due to propagating plane waves and higher order acoustic modes. The increased pipe wall vibrational response is accompanied by a corresponding increase in sound radiation into the external medium (air) surrounding the pipe.

In addition to the sequence of events described in (i)–(vii), vibrations in the vicinity of the fittings due to the non-propagating wall pressure fluctuations can be transmitted along the pipe wall to other locations. The extent of this transmission is dependent upon the specific details such as the proximity of the fittings to flanged joints, damping, vibration isolation, etc. However, the whole of a piping system is subjected to increased vibrations due to the propagating sound waves. The effectiveness of the three types of internal wall pressure fluctuations in exciting the pipe wall into vibration increases in the order of non-propagating fluctuating wall pressures, propagating plane waves, and propagating higher order acoustic modes. The effectiveness of the non-propagating pressures in generating pipe wall vibration is of the same order as turbulent boundary layer pressure fluctuations (in the absence of choking of the flow and shock wave generation). In principle, plane waves are not efficient exciters of structural vibrations. This is because, if the duct walls are uniform, a forced peristaltic motion occurs; however, in practice small departures from uniformity allow resonant vibrational modes of the pipe wall to be further excited and the effectiveness of the plane wave as a vibrational exciter is increased. Higher order acoustic modes are the most efficient and effective vibration exciters in gas flows in pipeline systems. This is so because of the possibility of the occurrence of coincidence of these modes with resonant structural vibrational modes of the pipe wall. A 90° mitred bend is a typical example of a flow disturbance which generates significant noise and vibration in pipelines. The reader is referred to Figure 2.1 in chapter 2 for a schematic illustration of the mechanisms of aerodynamic noise generation in pipes and the subsequent external noise radiation.

A series of controlled experiments were conducted on a range of pipe fittings[7.4,7.5,7.7] using air as the gas medium. Attention was concentrated on noise and vibration from pipes in regions where the local effects of a disturbance had died out, and the pipe wall excitation was due to fully developed turbulent pipe flow with a superimposed propagating sound field. Some typical results[7.5] for noise and vibration spectra for a range of pipe fittings are presented in Figure 7.1. The data are presented in non-dimensional form as pipe wall acceleration spectra, Φ_a, and sound power radiation spectra, Φ_π, versus frequency υ where $\Phi_a = G_{aa}/\omega_r^3 a_m^2$, $\Phi_\pi = G_{\pi\pi}/\rho_e c_e^2 S a_m$, $\upsilon = \omega/\omega_r$, $\omega_r = c_L/a_m$, and $c_L = \{E/\rho(1-\nu^2)\}^{1/2}$. E, ρ and ν are, respectively, Young's modulus of elasticity, density, and Poisson's ratio of the pipe material, ρ_e and c_e are the density and speed of sound in the fluid outside the pipe, S is the surface area of the test section, a_m is the mean pipe radius, ω_r is the ring frequency, $G_{\pi\pi}$ is the spectral density of the sound power radiation, and G_{aa} is the spectral density of the pipe wall acceleration.

The large increases in pipe wall vibration and the corresponding external sound radiation are due to various propagating higher order acoustic modes whose cut-off frequencies are illustrated on the diagram (the sound field inside a cylindrical shell is discussed in some detail in the next section). The spectral measurements show the effects of the various pipe fittings in relation to straight pipe flow. The effects, which are quite dramatic in the case of the stronger disturbances ($\sim$30 dB), result primarily from coincidence of higher order acoustic modes and resonant vibrational modes of the pipe wall. The concept of coincidence was introduced in chapter 1 (Figure 1.4) – it allows for a very efficient interaction between the sound waves and the structural waves, and it is discussed in some detail in section 7.5. The effects of coincidence are in general greatest at and essentially confined to frequencies close to the cut-off frequencies of the higher order modes. This proximity of coincidence frequencies and higher order acoustic mode cut-off frequencies plays a significant part in an overall understanding of flow-induced noise and vibration in pipeline systems.

Fig. 7.1. Non-dimensional spectral density of (a) the pipe wall acceleration, and (b) the sound power radiation for $M_0 \sim 0.40$: ○, butterfly valve; ■, 90° mitred bend; +, 45° mitred bend; ▲, gate valve; ◇, 90° radiused bend ($R/a = 6.4$), ∗, 90° degree radiused bend ($R/a = 3.0$); ●, straight pipe; R/a is the average radius ratio. Cut-off frequencies of higher order acoustic modes are also shown. One-third-octave band data.

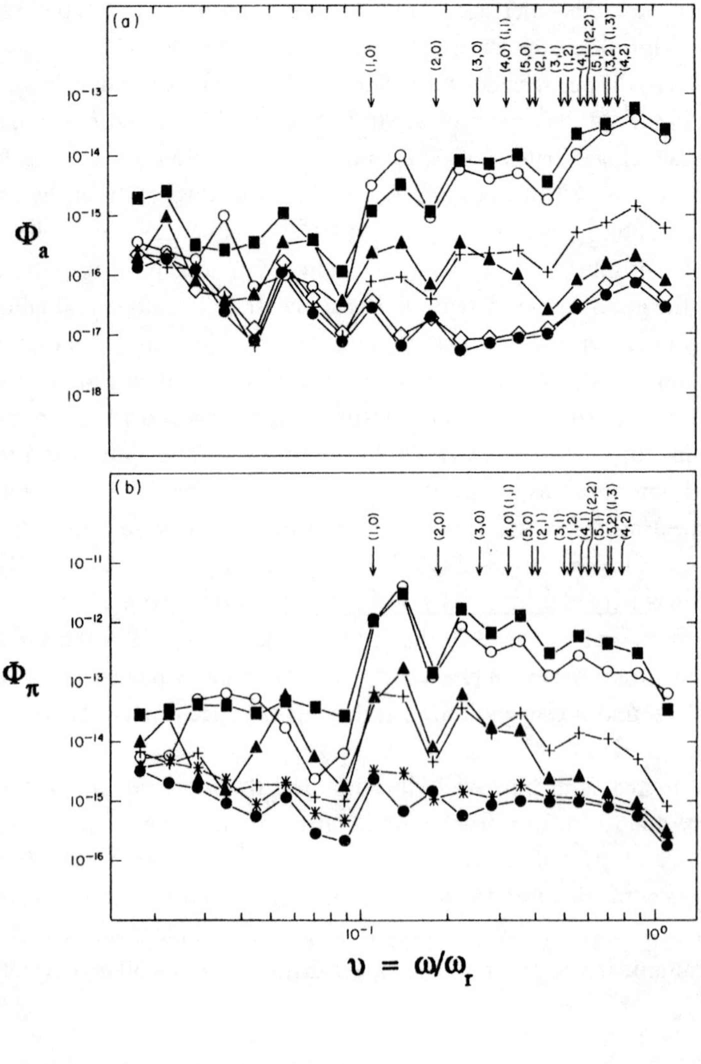

7.3 The sound field inside a cylindrical shell

If one wished to precisely describe the source and sound field inside a cylinder due to some internal flow disturbance, one would have to use the inhomogeneous wave equation, and this would require detailed information about the nature of the acoustic source. In any practical situation this is all but impossible, and the conventional approach to analysing the problem is to obtain statistical information (spectral densities, etc.) about the internal pressure fluctuations, and to attempt to obtain a generalised non-dimensional collapse of the data. This approach has proved to be very successful for turbulent boundary layer pressure fluctuation studies[7.9]. Boundary layer pressure fluctuations are always distributed over the entire surface of a structure which is exposed to fluid flow; hence, the sources of boundary layer noise are distributed. With internal flow disturbances in pipes, the primary sound sources (bends, valves, etc.) are localised. Furthermore, these sound sources tend to dominate over any boundary layer noise. Because these dominant sound sources are localised, one can study the characteristics of pipe flow noise within cylindrical shells (in regions external to the bends, valves, etc.) by first considering the ideal case of sound wave propagation in a cylinder without any superimposed flow conditions. The flow has a convective effect on the propagating sound waves and this can be readily accounted for. Also, because one is dealing with the sound field rather than the source field, the homogeneous wave equation can be used. If an analysis of regions including the internal flow disturbances was required, the inhomogeneous wave equation would have to be used together with suitable estimates of the source strengths, etc.

When sound waves propagate within the confined spaces of a duct, the wave propagation can be either parallel to the duct walls or at some oblique angle to them. The former type of wave propagation is the well known plane wave propagation. The latter type of wave propagation is referred to as higher order acoustic mode or cross mode wave propagation. With plane waves, the acoustic pressure is constant across a given duct cross-section. With higher order acoustic modes, the acoustic pressure is not constant across a given duct cross-section; it is a function of distance across the duct and angular position.

It should be noted that, when studying the interactions between sound waves within a cylindrical shell and the shell itself, it is convenient to assume rigid duct walls for the purposes of describing the sound field within the shell. When dealing with metallic structures (e.g. steel or aluminium shells) the assumptions are justified and are adequate for vibrational response and external noise radiation predictions[7.5–7.8]. In principle, however, three specific cases are possible. They are: (i) shells with perfectly rigid walls, (ii) shells with infinitely flexible walls, and (iii) shells with finitely flexible walls and an assumed elasto-acoustic coupling between the fluid and the shell in which it is contained. In this chapter, it will be assumed that the walls of the cylindrical shells are perfectly rigid for the purposes of describing the contained sound field. In physical terms, this means that the sound waves reflect off the walls and that the vibrational motion of the walls does not affect the internal sound wave pattern. Lin and Morgan[7.11] and El-Rahib[7.12] discuss sound wave propagation in elastic cyinders.

For a rigid cylinder with radial, angular and axial co-ordinates r, θ and x, the solution to the homogeneous wave equation (for propagation in the positive x-direction) for the pressure associated with acoustic propagation in a stationary internal fluid has the following form:

$$\mathbf{p}(r, \theta, x) = \sum_p \sum_q (A_{pq} \cos p\theta + B_{pq} \sin p\theta) J_p(\kappa_{pq} r)\, \mathrm{e}^{\mathrm{i}(k_x x - \omega t)}, \tag{7.2}$$

where

$$\kappa_{pq}^2 + k_x^2 = k^2 = (\omega/c_\mathrm{i})^2, \tag{7.3}$$

and ω is the radian frequency, k_x is the axial acoustic wavenumber, c_i is the speed of sound in the internal fluid, and J_p is the Bessel function of the first kind of order p. The (p, q)th wave or mode has p plane diametral nodal surfaces and q cylindrical nodal surfaces concentric with the cylinder axis, and it can propagate only at frequencies above its cut-off frequency, $(\omega_{\mathrm{co}})_{pq}$, where

$$(\omega_{\mathrm{co}})_{pq} = \kappa_{pq} c_\mathrm{i}. \tag{7.4}$$

Now,

$$\kappa_{pq} = \frac{\pi\alpha_{pq}}{a_\mathrm{i}}, \tag{7.5}$$

where a_i is the internal pipe radius and the $\pi\alpha_{pq}$'s are determined from the eigenvalues satisfying the rigid wall boundary condition $J'_p(\kappa_{pq} a_\mathrm{i}) = 0$, where J' is the first derivative of the Bessel function with respect to r. Equation (7.4) provides the radian frequency, and equation (7.5) provides the wavenumber above which a given higher order acoustic mode can exist. Plane waves can exist at all frequencies in a duct, thus $\kappa_{00} = 0$. The α_{pq}'s for various combinations of p and q are well documented in the literature[7.13] and the values for the first twelve higher order acoustic modes are given in Table 7.1.

Table 7.1. *Solutions to* $J'_p(\kappa_{pq} a_\mathrm{i}) = 0$.

p	q	$\pi\alpha_{pq}$	p	q	$\pi\alpha_{pq}$
1	0	1.8412	5	0	6.4156
2	0	3.0542	2	1	6.7061
0	1	3.8317	0	2	7.0156
3	0	4.2012	6	0	7.5013
4	0	5.3175	3	1	8.0152
1	1	5.3314	1	2	8.5363

Sound waves can thus propagate in a cylindrical shell only as plane waves ($p = q = 0$) if $ka_i < 1.8412$, where the wavenumber k is given by equation (7.3), and as both plane waves and higher order acoustic modes if $ka_i \geqslant 1.8412$. The internal acoustic modes that can be sustained within the pipe can be classified as plane waves ($p = q = 0$), symmetric higher order modes ($p = 0, q \geqslant 1$), and asymmetric higher order 'spinning' modes ($p \geqslant 1, q \geqslant 1$). The duct cross-sectional pressure distributions for a plane wave and the first nine higher order acoustic modes are illustrated in Figure 7.2. The cut-off frequency for a particular acoustic mode is thus

$$(f_{co})_{pq} = \frac{\pi \alpha_{pq} c_i}{2\pi a_i}. \tag{7.6}$$

Fig. 7.2. Internal acoustic modes inside a cylindrical shell (non-dimensional cut-off frequencies are for no flow).

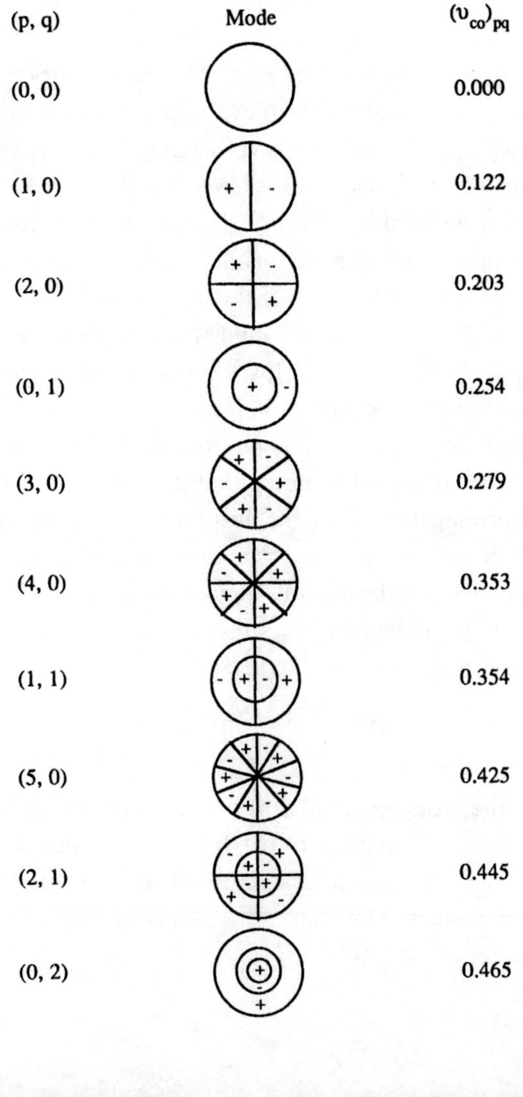
(p, q)
Mode
$(\upsilon_{co})_{pq}$
(0, 0)
0.000
(1, 0)
0.122
(2, 0)
0.203
(0, 1)
0.254
(3, 0)
0.279
(4, 0)
0.353
(1, 1)
0.354
(5, 0)
0.425
(2, 1)
0.445
(0, 2)
0.465

Equation (7.6) is only valid for the case where there is no flow in the pipe (i.e. the fluid is stationary with the exception of the acoustic pressure fluctuations).

In the presence of an idealised uniform flow with velocity U and Mach number $M = U/c_i$ parallel to the pipe axis at all (r, θ), the frequency, as seen by a stationary observer, of a wave with an axial wavenumber component k_x is given by

$$\omega/c_i = (\kappa_{pq}^2 + k_x^2)^{1/2} + Mk_x, \tag{7.7}$$

instead of equation (7.3). The additional term represents the Doppler frequency shift due to the presence of the uniform flow. The cut-off frequency is now reduced to

$$(f_{co})_{pq} = \frac{\pi\alpha_{pq}c_i(1 - M^2)^{1/2}}{2\pi a_i}, \tag{7.8}$$

and it occurs at an axial wavenumber of

$$k_x = -\frac{M\kappa_{pq}}{(1 - M^2)^{1/2}}, \tag{7.9}$$

instead of at $k_x = 0$ as in the no flow case. The dispersion curve (the variation of axial wavenumber with frequency) for these waves, which is symmetrical about the frequency axis in the case of a stationary internal fluid, therefore becomes asymmetrical due to the influence of flow – dispersion curves are discussed in detail in section 7.5. When the flow is not uniform but has a turbulent profile instead, replacement of M by M_0, the centre-line Mach number of the turbulent flow, provides an adequate representation of the convective effect of flow.

It is useful to also briefly consider sound wave propagation in rigid rectangular cross-section ducts. As is the case for circular cross-section ducts, the three-dimensional homogeneous wave equation is used to solve for the pressure associated with acoustic propagation in a stationary internal fluid[7.13], with cartesian co-ordinates being used in this case. It should be noted that no boundary conditions are assumed in the direction of propagation along the duct for both circular and rectangular ducts – the boundary conditions are two-dimensional and are related to the containing walls. The cut-off frequencies for the various higher order acoustic modes that can be sustained in a rectangular duct can be subsequently obtained from the solution to the wave equation and are given by

$$(f_{co})_{pq} = \frac{c}{2\pi}\left\{\left(\frac{p\pi}{a}\right)^2 + \left(\frac{q\pi}{b}\right)^2\right\}^{1/2}, \tag{7.10}$$

where a and b are the cross-sectional dimensions of the rectangular duct, and p and q are the mode orders. Equation (7.10) is useful because it allows for an easy identification of the cut-off frequencies associated with different higher order acoustic modes in rectangular ducts. The convective effects of flow can be accounted for by incorporating a factor of $(1 - M^2)^{1/2}$.

7.4 Response of a cylindrical shell to internal flow

To estimate the vibrational response of the pipe wall and the subsequent external sound radiation due to internal flow, several parameters are required. Firstly, the frequency response function of the cylindrical shell is required. Secondly, information is required about the various natural frequencies of the shell. Thirdly, information is required about the forcing function – i.e. the input to the system. For a cylindrical shell responding to internal flow, the forcing function is the fluctuating internal wall pressure field. This fluctuating wall pressure field comprises turbulent boundary layer pressure fluctuations and acoustic pressure fluctuations. Fourthly, information is required about the degree of spatial coupling between the wall pressure field and the modal structural response – this spatial coupling is referred to as the joint acceptance (or the cross-joint acceptance) and it describes how a distributed input couples to a continuous structure over its length[7.14]. Finally, information is required about the efficiency of sound radiation from the structure – i.e. the radiation ratio of the shell.

7.4.1 General formalism of the vibrational response and sound radiation

Equations derived from a general solution to the dynamic response of a thin-walled cylindrical shell to a propagating sound pressure field dominated by higher order acoustic modes lead to an estimation of the sound radiation from and the vibrational response of pipes with various internal flow disturbances. This procedure will be outlined in section 7.7. To commence, however, the dynamic response of a thin-walled cylindrical shell to an arbitrary random fluctuating wall pressure field, G_{pp}, is required.

For the purposes of analysis, the pipe is modelled as a thin cylindrical shell with simply supported ends, and the calculation of the statistical properties of the vibrational response is based upon the normal mode method of generalised harmonic analysis (sub-section 1.9.6, chapter 1). As a rule of thumb, a cylinder is assumed to be thin-walled if its wall thickness, h, is less than one-tenth of its mean radius, a_m. This is the case for most industrial type pipelines. The pipe structure is considered to be homogeneous over its surface area, and the resonant structural modes which make up the total response of the pipe structure are assumed to be lightly damped. Modal coupling terms are thus neglected in the analysis. Experimental evidence has shown that this assumption, whilst not strictly correct, is acceptable for the prediction of upper and lower bound pipe wall vibration levels and the subsequent sound power radiation.

Consider a section of pipe of length L between supports, with a wall thickness h, which is not too large in relation to the mean radius, a_m ($h/a_m \leqslant 0.1$). In practice, such lengths will be those between flanges or large pipe fittings which can be regarded as supports or end conditions for various sections of pipeline. Each such pipe length constitutes a resonant system with its own set of discrete natural frequencies, and, as far as the external sound radiation is concerned, it is the resonant flexural modes of the pipe wall which are of primary interest.

The power spectral density of the radial displacement response, G_{rr}, averaged over the surface area of the vibrating cylinder, to an arbitrary random fluctuating wall pressure field of power spectral density G_{pp}, can be expressed by [7.4,7.10]

$$\langle\overline{G_{rr}(\omega)}\rangle = \langle\overline{G_{pp}(\omega)}\rangle S^2 \sum_{\alpha} \frac{j_{\alpha\alpha}^2(\omega)\overline{\phi_\alpha^2(\vec{r})}}{|\mathbf{H}_\alpha(\omega)|^2}, \tag{7.11}$$

where $\langle\ \rangle$ represents a time-average, the overbar represents a space-average, $\phi_\alpha(\vec{r})$ defines the shape of the orthogonal normal modes, $j_{\alpha\alpha}^2(\omega)$ is the joint acceptance function for the αth resonant structural mode and the applied pressure field (it is a function which expresses the degree of spatial correlation between the pressure excitation and the structural modes), S is the surface area of the cylinder, and $\mathbf{H}_\alpha(\omega)$ is a modal frequency response function. For this particular case, it is the complex dynamic stiffness (inverse of the receptance) – i.e. $\mathbf{H}_\alpha(\omega) = \mathbf{F}_\alpha/\mathbf{X}_\alpha$.

Now, for a simple one-degree-of-freedom system,

$$\frac{\mathbf{F}}{\mathbf{X}} = k_s - m\omega^2 + \mathrm{i}c_v\omega = m\left(\omega_n^2 - \omega^2 + \mathrm{i}\frac{c_v\omega}{m}\right). \tag{7.12}$$

The viscous damping coefficient can be replaced by the internal loss factor by utilising equations (1.83) and (1.86). From those equations, in general,

$$c_v = \frac{\omega_n^2\eta m}{\omega}. \tag{7.13}$$

However, when the system is lightly damped, equation (7.13) can be approximated by using the definitions preceding equation (1.36), and equation (1.90). Thus,

$$c_v \approx \frac{\omega_n m}{Q}, \tag{7.14}$$

where ω_n is the natural frequency. (For the cylinder with numerous natural frequencies, the subscript n is replaced with the subscript α.)

The modal frequency response function of the cylinder is thus given by

$$\frac{\mathbf{F}_\alpha}{\mathbf{X}_\alpha} = M_\alpha\omega_\alpha^2\left\{1 - \frac{\omega^2}{\omega_\alpha^2} + \mathrm{i}\frac{\omega}{\omega_\alpha Q_\alpha}\right\}, \tag{7.15}$$

and hence

$$|\mathbf{H}_\alpha(\omega)|^2 = M_\alpha^2\omega_\alpha^4\left\{\left(1 - \frac{\omega^2}{\omega_\alpha^2}\right)^2 + \left(\frac{\omega}{\omega_\alpha Q_\alpha}\right)^2\right\}, \tag{7.16}$$

where M_α, ω_α and Q_α are, respectively, the generalised mass, natural frequency and quality factor of the αth mode. Q_α comprises structural damping, acoustic radiating damping and coupling damping at the joints.

The spectral density of the pipe wall acceleration response, G_{aa}, averaged over the surface of the cylinder is related to the spectral density of the radial displacement response, G_{rr}, by

$$\langle\overline{G_{aa}(\omega)}\rangle = \omega^4 \langle\overline{G_{rr}(\omega)}\rangle. \tag{7.17}$$

Thus,

$$\frac{\langle\overline{G_{aa}(\omega)}\rangle}{\langle\overline{G_{pp}(\omega)}\rangle} = \omega^4 S^2 \sum_\alpha \frac{j^2_{\alpha\alpha}(\omega)\overline{\phi^2_\alpha(\vec{r})}}{M^2_\alpha \omega^4_\alpha \left\{\left(1-\frac{\omega^2}{\omega^2_\alpha}\right)^2 + \left(\frac{\omega}{\omega_\alpha Q_\alpha}\right)^2\right\}}. \tag{7.18}$$

Equation (7.18) is a general equation for the acceleration response of a continuous structure to a random pressure field. It can be simplified by recognising that, for homogeneous cylinders and mode shapes corresponding to simply supported ends, $\phi_\alpha(\vec{r})$ and M_α are independent of α; also, $\phi^2_\alpha(\vec{r}) = 1/4$, and $M_\alpha = \rho h S/4$ for all modes. Furthermore, for each natural frequency of a cylindrical shell, there are two modes because of the degeneracy of modes in cylindrical shells – the mode shapes are represented by degenerate mode pairs because of the non-preferential directions available for the mode shapes due to the structural axisymmetry[7.15]. Thus, both sets of modes, as given by the separable functions for the mode shapes of a simply supported cylinder, must be considered for the calculation of a homogeneous vibration response to a statistically homogeneous excitation. Hence,

$$\frac{\langle\overline{G_{aa}(\omega)}\rangle}{\langle\overline{G_{pp}(\omega)}\rangle} = \frac{8}{\rho^2 h^2} \sum_\alpha \frac{\omega^4 j^2_{\alpha\alpha}(\omega)}{\omega^4_\alpha \left\{\left(1-\frac{\omega^2}{\omega^2_\alpha}\right)^2 + \left(\frac{\omega}{\omega_\alpha Q_\alpha}\right)^2\right\}}. \tag{7.19}$$

Equation (7.19) can be generalised by non-dimensionalising the spectral densities of the pipe wall acceleration and the wall pressure fluctuations, respectively. The non-dimensional spectral density of the pipe wall acceleration, averaged over the pipe surface, is $\Phi_a = G_{aa}/\omega^3_r a^2_m$, and the non-dimensional spectral density of the wall pressure fluctuations is $\Phi_p = G_{pp} U_0 / q^2_0 a_i$, where $q_0 = \rho_i U^2_0/2$ and ρ_i is the density of the internal fluid. Thus,

$$\frac{\Phi_a(\omega)}{\Phi_p(\omega)} = \frac{\rho^2_{is} M^3_0 a_i}{6\beta^2 M^3_{LP} a_m} \sum_\alpha \frac{\omega^4 j^2_{\alpha\alpha}(\omega)}{\omega^4_\alpha \left\{\left(1-\frac{\omega^2}{\omega^2_\alpha}\right)^2 + \left(\frac{\omega}{\omega_\alpha Q_\alpha}\right)^2\right\}}, \tag{7.20}$$

where $\rho_{is} = \rho_i/\rho$, ρ_i is the fluid density inside the pipe, ρ is the pipe material density, $M_{LP} = c_L/c_i$, $M_0 = U_0/c_i$, c_i is the speed of sound inside the pipe, and $\beta = h/(2\sqrt{3}a_m)$ is the non-dimensional pipe wall thickness parameter.

The spectral density of the sound power radiated from the pipe, $G_{\pi\pi}$, is defined as

$$G_{\pi\pi}(\omega) = \langle \overline{G_{rr}(\omega)} \rangle \omega^2 \rho_e c_e \sigma_\alpha S, \tag{7.21}$$

where the subscript e refers to the fluid medium outside the pipe, and σ_α is the radiation ratio of the αth mode. The non-dimensional spectral density of the sound power radiated from the pipe, Φ_π, is defined as $\Phi_\pi = G_{\pi\pi}/\rho_e c_e^2 S a_m$; hence,

$$\Phi_\pi(\omega) = \frac{\Phi_a(\omega)\sigma_\alpha M_{LP} c_i}{\upsilon^2 c_e}, \tag{7.22}$$

where $\upsilon = \omega/\omega_r$, $\omega_r = c_L/a_m$, c_i is the speed of sound inside the pipe and c_e is the speed of sound in the external fluid. Thus,

$$\frac{\Phi_\pi(\omega)}{\Phi_p(\omega)} = \frac{\rho_{is}^2 M_0^3 a_i c_i}{6\beta^2 \upsilon^2 M_{LP}^2 a_m c_e} \sum_\alpha \frac{\omega^4 j_{\alpha\alpha}^4(\omega)\sigma_\alpha}{\omega_\alpha^4 \left\{ \left(1 - \frac{\omega^2}{\omega_\alpha^2}\right)^2 + \left(\frac{\omega}{\omega_\alpha Q_\alpha}\right)^2 \right\}}, \tag{7.23}$$

Equations (7.20) and (7.23) are the general formalisms of the vibration response of and the radiated sound power from a thin-walled cylindrical shell which is subjected to a random internal wall pressure field. They can be used to predict vibration and noise from pipelines provided that information is known about the natural frequencies, the internal wall pressure field, the joint acceptance function and the radiation ratios. Even if quantitative information is not readily available, the equations can be used for parametric studies – they clearly illustrate the parametric dependence on flow speed, M_0, pipe wall thickness, β, etc. The prediction of the vibration response of and the sound radiation characteristics from straight sections of pipeline downstream (or upstream) of different types of internal flow disturbances is discussed in sections 7.7 and 7.8.

7.4.2. Natural frequencies of cylindrical shells

When considering the natural frequencies of cylindrical shells, one has to use both wave and modal concepts. It is convenient to describe the natural frequencies associated with flexural wave propagation in terms of axial and circmferential wavenumbers; the natural frequencies of flat plates were described in terms of x and y wavenumbers in chapter 3.

There is a large body of work in the research literature (e.g. Soedel[7.15], Leissa[7.16], Arnold and Warburton[7.17], and Greenspon[7.18]) on the natural frequencies of cylindrical shells. In the main, these theories are exact, are based upon strain relationships, and require extensive computational analysis. Heckl[7.19] has derived a relatively simple relationship for the estimation of the natural frequencies of thin-walled cylindrical shells based upon axial and circumferential wavenumber variations. The work is not dissimilar to that of Szechenyi[6.15] on the modal densities of cylindrical shells. Fahy[7.21] summarises Heckl's work and provides a general discussion on flexural wave propagation in cylindrical shells.

From a modified form of Heckl's results, the natural frequency of the (m, n)th flexural mode of a thin cylindrical shell with wall thickness h is given approximately by[7.8,7.20]

$$\upsilon_{mn}^2 = \beta^2 K^4 + \frac{(1-\nu^2)K_m^4}{K^4}, \tag{7.24}$$

where $\upsilon_{mn} = \omega_{mn}/\omega_r$, $\beta = h/(2\sqrt{3}a_m)$, $K^2 = K_m^2 + K_n^2$, $K_m = k_m a_m = m\pi a_m L$, $K_n = k_n a_m = n$, L is the length of the cylinder, ν is Poisson's ratio, m is the number of half structural waves in the axial direction, and n is the number of full structural waves in the circumferential direction. Thus, for each circumferential mode order (i.e. $n=$ 1, 2, 3, . . . etc.), there are large numbers of axial mode orders and, in general, there are hundreds if not thousands of natural frequencies that can be excited into resonance – the modal density of lightly damped cylindrical shells is generally very high.

Equation (7.24) is applicable to thin cylindrical shells with simply supported ends and its limitations are discussed in detail by Rennison and Bull[7.20] and Fahy[7.21]. A comparison of exact theory (e.g. Arnold and Warburton[7.17]) with equation (7.24) indicates that the latter produces underestimates of the natural frequencies by $\sim 50\%$ for low K_m values for the $n = 1$ and $n = 2$ circumferential modes. For all other values, equation (7.24) gives a good approximation which is quite acceptable for statistical estimates of the vibrational response of and the sound radiation from cylinders. Equation (7.24) can be expressed in dimensional form as

$$f_{mn}^2 = \frac{c_L^2}{4\pi^2 a_m^2}\left\{\beta^2 K^4 + \frac{(1-\nu^2)K_m^4}{K^4}\right\}. \tag{7.25}$$

The first term within the brackets is associated with flexural strain energy in the walls, and the second term is associated with membrane strain energy.

7.4.3 *The internal wall pressure field*

The fluctuating internal wall pressure field is the forcing function to which a cylinder responds. Therefore, any prediction scheme for the estimation of pipe flow noise and vibration requires information about the power spectral density of this forcing function (see equations 7.20 and 7.23).

For the case of noise and vibration generated only by turbulent boundary layer flow, the statistical properties of the wall pressure fluctuations are fairly well defined[7.9] and the variation of the non-dimensional spectral density, Φ_p, with Strouhal number, $\Omega = \omega a_i/U_0$, is similar for flat plates and waveguides (e.g. cylinders)[7.6] – i.e. a universal datum exists. Unfortunately, unlike turbulent boundary layer flow, there is no universal datum for the internal wall pressure field associated with internal flow disturbances in waveguides such as pipelines. If a universal datum was available, then equations (7.20) and (7.23) could be used for estimation purposes without the requirement for experimental measurements – i.e. the internal wall pressure field could be scaled from the universal datum.

When an internal flow disturbance is present in a pipe, additional wall pressure fluctuations are generated due to the internal sound field. At regions in close proximity to these internal flow disturbances, the wall pressure fluctuations are very severe. As mentioned in section 7.2, at regions away from the internal flow disturbances the flow velocity returns to a steady-state characteristic of the undisturbed flow but with a fluctuating pressure level which is higher than that of boundary layer pressure fluctuations. These additional fluctuating pressures are generally due to the superimposed propagating sound field. Experimental data are available for a range of internal flow disturbances for subsonic air flow ($0.2 \leqslant M_0 \leqslant 0.6$) in steel pipelines[7.4–7.8]. Over the flow range investigated, the variation of non-dimensional spectral density, Φ_p, with Strouhal number, Ω, is approximately constant for each of the flow disturbances at locations sufficiently remote from the disturbances. There are, however, variations in mean spectral levels between the different disturbances themselves.

Some typical non-dimensional mean wall pressure spectra, Φ_p, are presented in Figure 7.3 for a range of internal flow disturbances as a function of Strouhal number, Ω. It has to be made very clear to the reader that the spectra are at positions along a straight section of pipe which is well downstream of the disturbances themselves – i.e. they are representative of the wall pressure fluctuations in a fully developed

Fig. 7.3. Some typical non-dimensional mean wall pressure spectra for a range of internal flow disturbances, at positions along a straight section of pipe well downstream (~53 pipe diameters) of the disturbances, themselves, for $0.22 \leqslant M_0 \leqslant 0.51$: —·—, 90° radiused bend ($R/a = 3.0$); —···—, 45° mitred bend; —··—, 90° mitred bend; -----, fully open butterfly valve; --·--, fully open gate valve; ——, undisturbed straight pipe flow; R/a is the average radius ratio. One-third-octave band data.

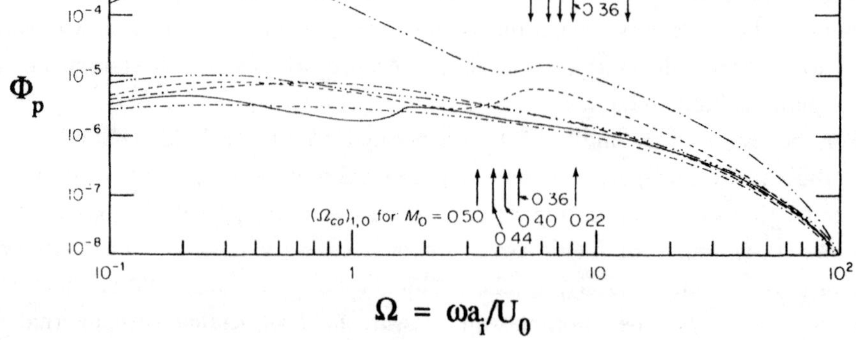

turbulent straight pipe flow with a superimposed propagating sound field due to some upstream (or downstream) flow disturbance. It is clear from the figure that there are large increases over turbulent boundary layer flow for certain types of internal flow disturbances, particularly 90° mitred bends. From the experimental evidence, it could be argued that the wall pressure spectra for the 90° mitred bend and the undisturbed turbulent pipe flow represent upper and lower limits, respectively. The detailed wall pressure spectra data for these two cases are presented in Figure 7.4 – the detailed variations with flow velocity can now be observed (the mean levels of Figure 7.3 are derived from Figure 7.4).

At regions in proximity to internal flow disturbances, the wall pressure fluctuations are much more severe. Whilst of interest from a fundamental and from an aerodynamic noise generation viewpoint, these wall pressure fluctuations are not directly relevant to the prediction of noise and vibration from straight runs of pipeline; these noise sources are localised and can therefore be isolated, boxed in, etc. In addition to being more severe, the wall pressure fluctuations at regions in proximity to a disturbance can be circumferentially non-uniform. In these regions there are non-propagating sound waves (evanescent modes) and increased turbulence levels due to separation, etc., in addition to the propagating plane waves and propagating higher order acoustic modes. Some typical results[7.6] for the wall pressure fluctuations along the inner wall of a 90° mitred bend are presented in Figure 7.5. These results only serve to illustrate the complexity of the problem in the vicinity of a flow disturbance. Once again, the reader is referred to Figure 2.1 for a schematic illustration of the mechanisms of aerodynamic noise generation in pipes.

In summary, information is required about the internal wall pressure field if one wishes to predict pipe flow noise and vibration levels. The prediction of noise at regions in proximity to internal flow disturbances is relatively difficult because of the unique nature of each type of disturbance (each disturbance will have a unique frequency response function) and because the local wall pressure fluctuations are very complex. The prediction of noise and vibration along straight runs of pipeline is somewhat easier since the frequency response function of a cylindrical shell is readily obtained (equation 7.16) and lower and upper limits are available for the internal wall pressure field. It should be pointed out that numerous valve manufacturers, etc., have empirical prediction schemes, generally based upon dimensional analysis, for noise emanating directly from disturbances such as valves, etc. Some of these schemes are discussed in section 7.8.

7.4.4 *The joint acceptance function*

The vibrational response of a pipe wall, in any one of its natural modes of vibration, to excitation by a particular wall pressure field is determined by the joint acceptance function, which expresses the degree of spatial coupling that exists between the

Fig. 7.4. Detailed variations in non-dimensional wall pressure spectra for a 90° mitred bend (~53 diameters downstream) and undisturbed straight pipe flow. Symbols with corresponding values of M_0 are, for the 90° mitred bend, ▼, 0.22, ∗, 0.36; ◒, 0.40; □, 0.44, ⌓, 0.50; and, for the straight pipe, +, 0.22; △, 0.36; ●, 0.41; ○, 0.45; ×, 0.52. One-third-octave band data.

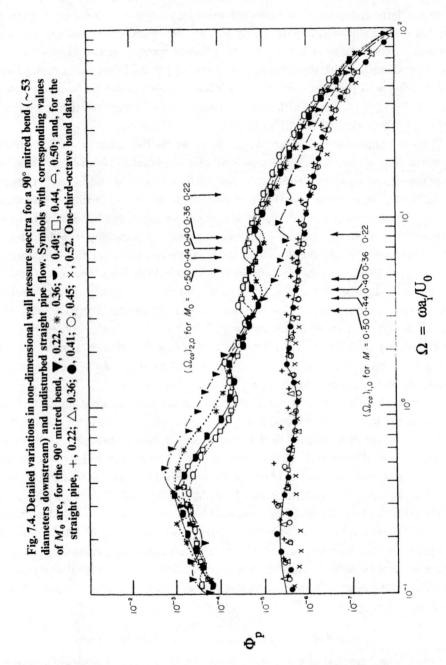

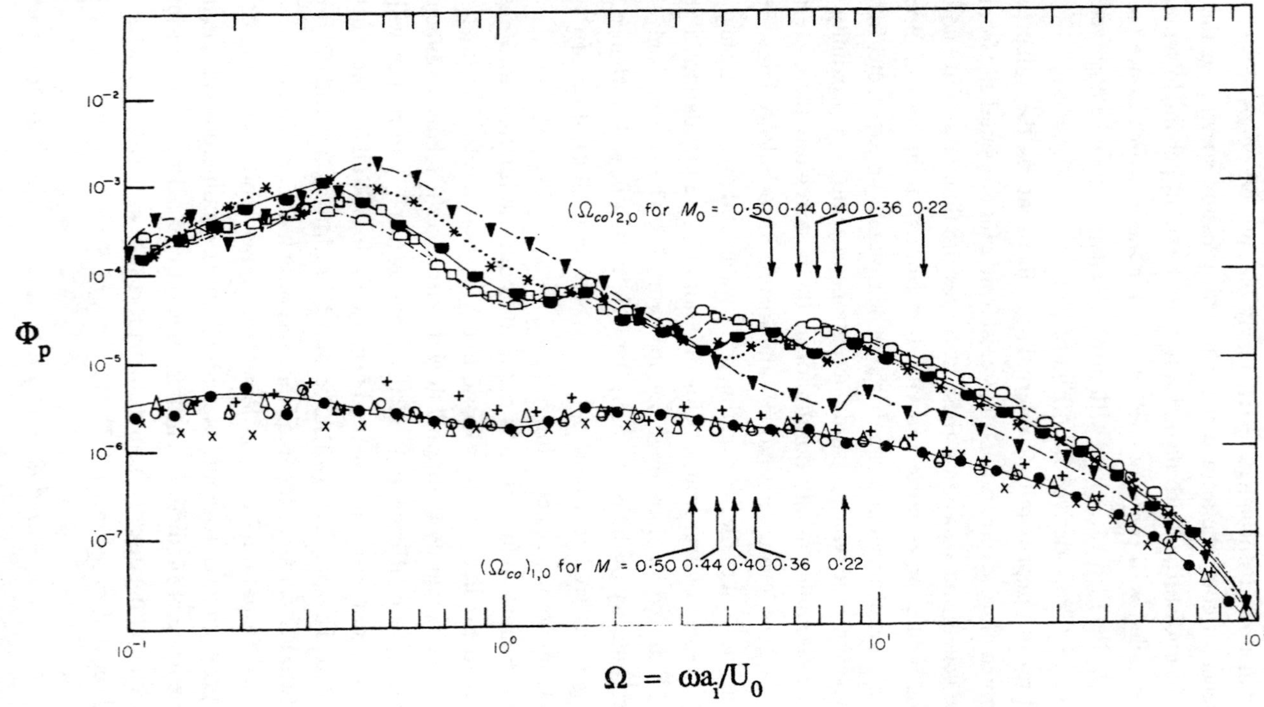

pressure excitation and the structural mode, and by the frequency response function (receptance) of the structural mode – i.e. the vibrational response is proportional to $j^2_{\alpha\alpha}$ and $1/|\mathbf{H}_\alpha(\omega)|^2$. The receptance function is the inverse of the dynamic stiffness, $\mathbf{H}_\alpha(\omega)$, and it is well known. It has a sharp maximum at the resonance frequency for any given structural mode, and the overall response function is proportional to the product of the joint acceptance and the receptance (equations 7.20 and 7.23).

Fig. 7.5. Non-dimensional wall pressure fluctuations along the inner wall of a 90° mitred bend. (a) $M_0 = 0.22$, (b) $M_0 = 0.40$, (c) $M_0 = 0.50$, (d) undisturbed straight pipe flow at $M_0 = 0.40$; X is the number of pipe diameters downstream of the disturbance. One-third octave band data.

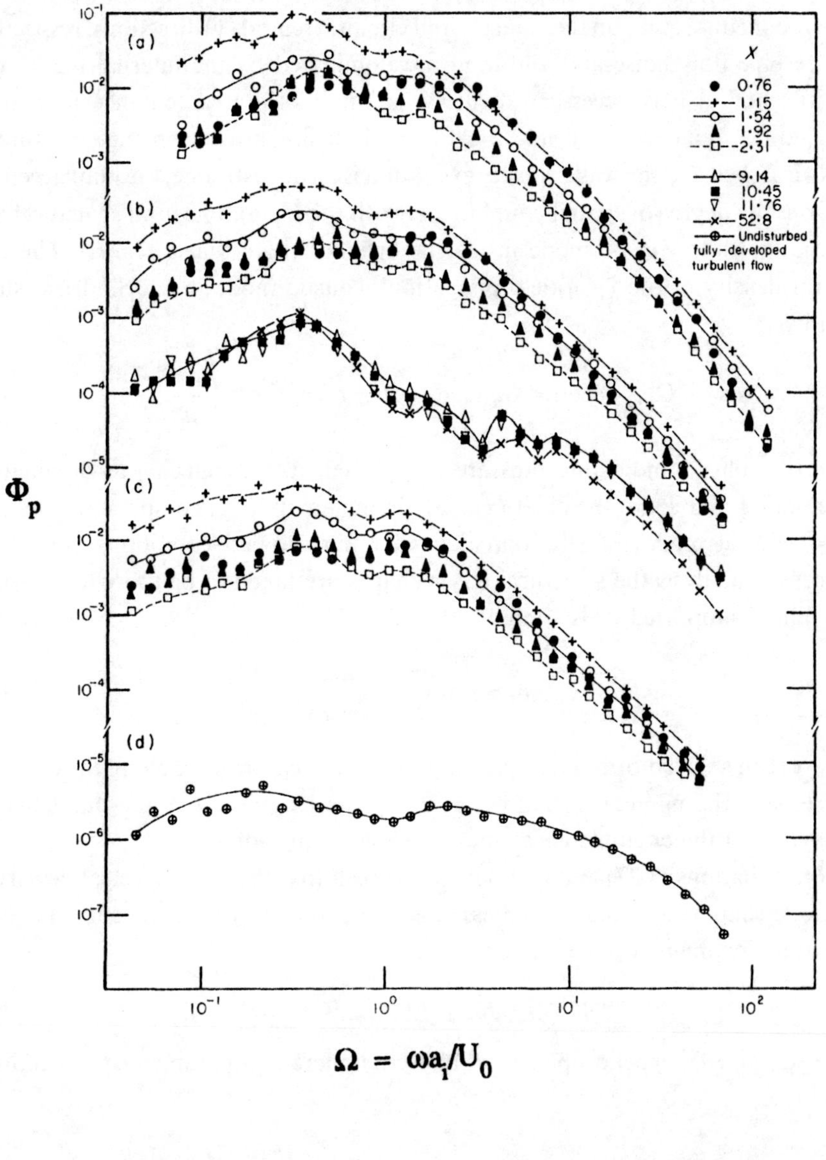

In general terms, the joint acceptance is a function which expresses the degree of spatial coupling/correlation between a distributed input excitation and a structure. For a stationary random input, it is defined as

$$j_{\alpha\alpha}^2(\omega)=\frac{1}{G_{pp}S^2}\int_S\int_S \phi_\alpha(\vec{r})\phi_\alpha(\vec{r}')\mathbf{G}_{\mathbf{p1p2}}(\vec{\varepsilon},\omega)\,\mathrm{d}S(\vec{r})\,\mathrm{d}S(\vec{r}'), \tag{7.26}$$

where G_{pp} is a reference auto-spectral density (generally the auto-spectral density of the stationary random fluctuating wall pressures), S is the surface area of the structure, the vector $\vec{r}$ represents a point on the structure, $\mathbf{G}_{\mathbf{p1p2}}(\vec{\varepsilon},\omega)$ is the cross-spectral density of the wall pressure field, $\vec{\varepsilon}=\vec{r}'-\vec{r}$, and ϕ_α is the mode shape of the αth mode. It is useful to recognise that $G_{pp}=\mathbf{G}_{\mathbf{p1p2}}(0,\omega)$. Also, for a cylindrical shell, $\vec{\varepsilon}=\vec{r}'-\vec{r}$ has components ξ and ψ in the axial (x) and circumferential (y) directions, respectively.

Since pipe flow noise and vibration are dominated by the internal higher order acoustic modes, it is necessary to derive a suitable joint acceptance function for propagating sound waves inside a cylindrical shell. Thus, when the structure is a cylindrical shell and the wall pressure excitation is an acoustic one, the joint acceptance expresses the degree of spatial coupling between the (m,n)th flexural structural mode and the (p,q)th acoustic mode inside the cylinder (note that $\alpha=m,n$). The cross-spectral density of the (p,q)th propagating acoustic mode in a cylindrical shell is given by[7.4]

$$\mathbf{G}_{\mathbf{p1p2}}(\vec{\varepsilon},\omega)=\mathbf{G}_{\mathbf{pq}}(\vec{\varepsilon},\omega)=G_{pq}(\omega)\,\mathrm{e}^{\mathrm{i}k_x\xi}\cos\left(\frac{p\psi}{a_\mathrm{i}}\right), \tag{7.27}$$

where the suffix p indicating pressure is replaced by the suffix pq to designate the mode, and ξ and ψ are the axial (x) and circumferential (y) components of $\vec{\varepsilon}$.

The natural modes of vibration depend upon the end conditions, and for the purposes of analysis the structural mode shapes are taken to be those for a cylinder with simply supported ends. Here

$$\phi_{mm}(\vec{r})=\sin\frac{m\pi x}{L}\begin{bmatrix}\sin ny/a_\mathrm{m}\\ \cos ny/a_\mathrm{m}\end{bmatrix}, \tag{7.28}$$

where x is an axial co-ordinate, y is a circumferential co-ordinate along the cylindrical surface, m is the number of half-waves along the length L, n is the number of full waves around the circumference, and a_m is the mean radius.

From equations (7.27) and (7.28) it can be seen that the cross-spectral density and the mode shapes can each be expressed as the product of independent functions of axial and circumferential parameters. Thus,

$$j_{\alpha\alpha}^2(\omega)=j_{mnmn}^2(\omega)=j_{mm}^2(\omega)\mathrm{j}_{nn}^2(\omega), \tag{7.29}$$

where j_{mm}^2 and j_{nn}^2 depend upon axial and circumferential parameters, respectively.

The joint acceptance of the (m, n)th structural mode excited by the (p, q)th acoustic mode inside a cylindrical shell can be evaluated from equations (7.26)–(7.29). It is given by

$$j_{mm}^2(\omega) = \frac{2K_m^2(1 - \cos \Delta K_m \cos \Delta K_x)}{\Delta^2(K_m^2 - K_x^2)^2}, \tag{7.30}$$

and

$$\begin{aligned} j_{nn}^2(\omega) &= 1/4 \quad \text{for } n = p \\ &= \;\; 0 \quad \text{for } n \neq p, \end{aligned} \tag{7.31}$$

where $\Delta = L/a_m$, $K_x = k_x a_i$, $K_m = m\pi a_m/L$, and k_x is the axial wavenumber of the propagating acoustic mode inside the cylinder. Equations (7.30) and (7.31) show that the joint acceptance will have its maximum value for the condition in which there is spatial or wavenumber matching of the structural and sound waves at the internal pipe wall in both the circumferential and axial directions. The maximum value of j_{mm}^2 occurs when $K_m = K_x$, except for very low m values ($m = 1, 2$), and this maximum value is 1/4. $j_{\alpha\alpha}^2$ thus has a maximum value of 1/16 when $K_m = K_x$.

The interested reader is referred to Bull and Norton[7.7] for a detailed discussion on the properties of the joint acceptance function for cylindrical shells.

7.4.5 *Radiation ratios*

In order to estimate the external sound power radiation from cylindrical shells due to internal flow, the radiation ratios of the shells are required (see equation 7.23). The concepts of radiation ratios were introduced in chapter 3, and radiation ratios of cylindrical shells and other structural elements were discussed in section 3.7.

In principle, there are three types of radiation ratios for cylindrical shells – radiation ratios for uniformly radiating (pulsating) cylinders, radiation ratios for forced peristaltic motion of the pipe wall, and radiation ratios for resonant structural modes[7.8]. Some typical radiation ratios for all three types of shell motions were presented in chapter 3 (Figures 3.17 and 3.18). The radiation ratios of supersonic structural waves (i.e. bending wave speeds, $c_s >$ wave speed in the external medium, c_e) approximate to unity for all three types of shell motions (pulsating cylinders, forced peristaltic motion, and resonant structural modes). The radiation ratios of all types of subsonic structural waves on the other hand are always less than unity.

Pipe flow noise due to internal flow disturbances is dominated by the response of the various resonant modes rather than a forced peristaltic motion or a uniform pulsation of the cylindrical shell (for a detailed discussion on radiation ratios of pipes with internal flows the reader is referred to Norton and Bull[7.8]). Hence, only the radiation ratios of the resonant modes are included in equations (7.21)–(7.23). These radiation ratios, σ_α's, for resonant pipe modes for which the structural wave speed is either subsonic of supersonic with respect to the external fluid medium are given by[7.8]

$$\sigma_\alpha = \frac{16\Delta}{\pi^4 m^2} \int_0^{\pi/2} \frac{\begin{matrix}\cos^2 \\ \sin^2\end{matrix} \{(\Delta K_e/2)\cos\theta\}\, d\theta}{\sin\theta\{1-(K_e/K_m)^2\cos^2\theta\}^2 |H_n^{(1)\prime}(K_e \sin\theta)|^2}, \quad (7.32)$$

where $K_e = k_e a_e$, $k_e = \omega/c_e$, a_e is the external radius of the pipe, c_e is the speed of sound in the external fluid, m is the number of axial half waves, $\cos^2$ is to be used for m odd, and $\sin^2$ is to be used for m even. $H_n^{(1)\prime}(\alpha)$ is the derivative with respect to α of $H_n^{(1)}$, the nth-order Hankel function of the first kind, where n is the number of full waves around the circumference. If the bending wave speed in the pipe wall is supersonic with respect to the external fluid, the equation is greatly simplified and $\sigma_\alpha \approx 1$ for all m and n. The bending wavespeed in a pipe wall can be calculated from equation (3.73).

7.5 Coincidence – vibrational response and sound radiation due to higher order acoustic modes

So far, it has been established in this chapter that a severe disturbance to fully developed turbulent pipe flow in a cylindrical shell results in the generation of intense broadband internal sound waves which can propagate through a piping system. It has also been established that the vibration response of the pipe wall to this excitation, and hence the externally radiated sound power also, are predominantly determined by coincidence of higher order acoustic modes inside the shell and resonant flexural modes of the pipe wall in both the circumferential and axial directions. Finally, it has also been established that higher order acoustic modes, unlike plane waves, are dispersive (see equation 7.3) – i.e. their phase speeds vary with frequency, whereas plane waves propagate at a constant speed.

A propagating sound wave inside a straight section of a rigid cylindrical shell is a travelling wave and, as was shown in section 7.3, it can be modelled as a wave that exists at all frequencies above its cut-off frequency. It therefore exhibits continuous variation of axial wavenumber with frequency. Its circumferential wavenumber component will be fixed, however, because of the boundary conditions imposed upon it. Similarly, standing structural waves in the circumferential direction will also have constant circumferential wavenumber components. Structural waves in the axial direction will be travelling waves only for an infinitely long pipe – any finite section of pipe (such as a straight run of pipeline between support sections) will have standing axial structural waves with discrete values of axial wavenumber components.

The term coincidence refers to matching in both wavelength (wavenumber) and frequency between the modes of the propagating internal sound waves and the resonant flexural modes of the pipe wall. In principle, this matching has to occur in both the axial and circumferential directions; i.e. there has to be exact spatial and frequency coupling. This is not, however, the case in practice because only the sound wave exhibits continuous variation of axial wavenumber. Hence, in general, whilst there is exact spatial and frequency coupling in the circumferential direction, there is spatial but not frequency coupling in the axial direction because of the discrete nature of the structural waves (i.e. they are standing waves or modes). The acoustically determined frequency for spatial (wavenumber) matching will be slightly different from the resonant structural natural frequency. This condition in which the structural and sound waves have equal wavenumbers ($k_x = k_m$) at the pipe wall (but at slightly different frequencies) is termed wavenumber coincidence. Complete coincidence is defined as wavenumber coincidence with, in addition, equality of frequency between the modes of the propagating internal sound waves and the resonant flexural modes of the pipe wall.

In general, because a cylinder has a set of discrete natural frequencies and not a continuum of natural frequencies, only wavenumber coincidence will occur. This is illustrated by the typical wavenumber–frequency dispersion relationships for structural and acoustic modes in Figure 7.6. The acoustic dispersion curves are obtained from a non-dimensional form of equation (7.3) (i.e. $\omega \leftrightarrow \upsilon$, and $k_x \leftrightarrow K_x$). The structural dispersion curves are obtained from equation (7.24).

Figure 7.6 relates specifically to the no flow case (i.e. travelling higher order acoustic modes inside a cylindrical shell with no internal flow) and serves only to illustrate the phenomenon. There has to be circumferential matching ($n = p$) of both wave-types for coincidence to occur. Hence, coincidence can occur between the (m, n) structural

Fig. 7.6. Coincidence of structural pipe modes and propagating internal higher order acoustic modes (no flow): $\beta = 0.007$, $\Delta = 79.4$

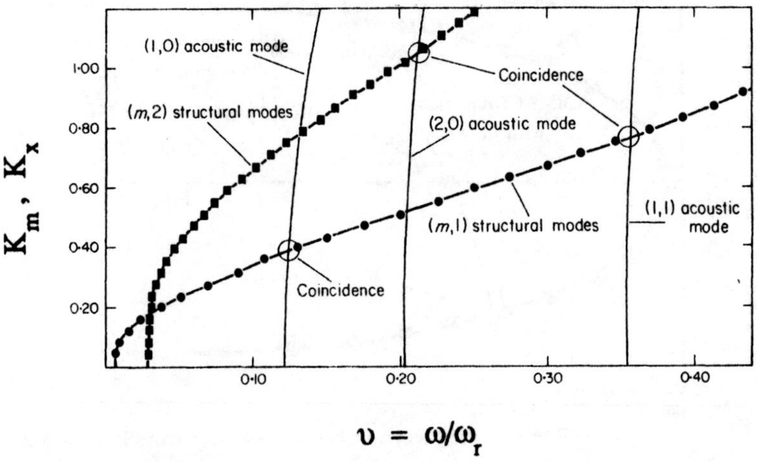

modes and the (n, q) acoustic modes, where $m = 1, 2, 3$, etc., $q = 1, 2, 3$, etc., and $n = p = 1, 2, 3$, etc. Coincidence between the $(m, 1)$ structural modes and the $(1, 0)$ and $(1, 1)$ higher order acoustic modes, and coincidence between the $(m, 2)$ structural modes and the $(2, 0)$ higher order acoustic mode is illustrated in Figure 7.6. It is clear from Figure 7.6 that complete coincidence does not occur because, whilst there is wavenumber matching, frequency matching does not occur.

When there is flow in a pipe, coincidence can occur at both positive and negative values of axial wavenumber. This is because (i) the standing structural waves support both positive and negative wavenumbers due to the degeneracy of modes in cylindrical shell (see paragraph preceding equation 7.19), and (ii) the axial acoustic wavenumber at cut-off, when there is flow in the pipe, is negative (see equation 7.9). In the no flow case, the axial acoustic wavenumber at cut-off is zero, resulting in coincidences at only positive wavenumbers. The flow thus has a significant effect on the acoustic dispersion curves, as already mentioned in section 7.3. The acoustic dispersion relationship (equation 7.7) can be represented in non-dimensional form as

$$\upsilon = \frac{[\{(\kappa_{pq}a_{\mathrm{m}})^2 + K_x^2\}^{1/2} + M_0 K_x]}{M_{\mathrm{LP}}}. \tag{7.33}$$

A typical wavenumber–frequency dispersion relationship for structural and acoustic modes in the presence of flow is illustrated in Figure 7.7. Besides showing that coincidence can occur at both positive and negative values of axial wavenumber, Figure 7.7 shows that, because of the asymmetry of the acoustic mode curve about the frequency axis due to the presence of flow in the pipe, the frequencies of the

Fig. 7.7. The effects of flow on the coincidence of structural pipe modes and propagating internal higher order acoustic modes.

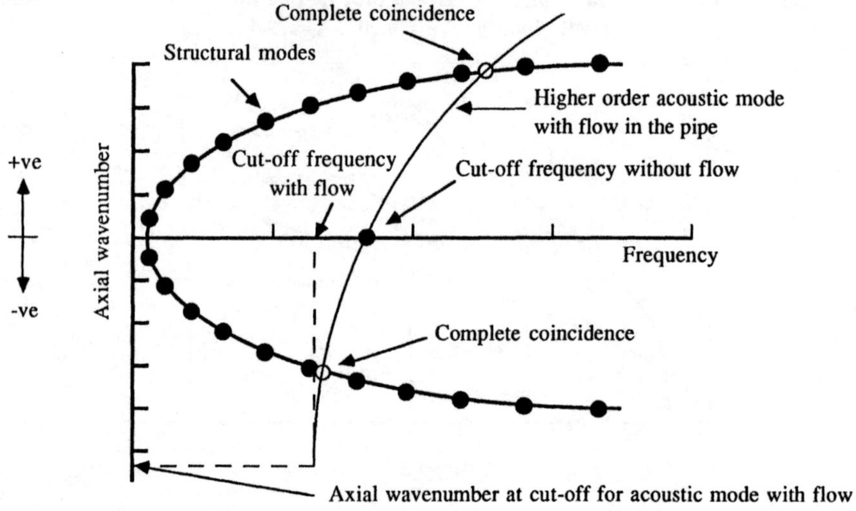

positive and negative wavenumber coincidences will not be the same. The pipe wall response will thus be dominated by four principal structural modes (two positive and two negative) for any given acoustic mode. This is because the acoustic dispersion curve essentially drives two structural resonances in both the positive and negative wavenumber domains. The wavenumber coincidences identified in this way are referred to as principal wavenumber coincidences[7.7,7.8]. Principal coincidences and the subsequent form of the pipe wall response are discussed in detail by Bull and Norton[7.7]. It is sufficient to mention here that, whilst the preceding considerations lead to identification of the structural modes associated with the principal wavenumber coincidences, the maximum structural response in these modes will not in all cases occur precisely at the condition of wavenumber coincidence but at a frequency very close to it. This maximum structural response is critically dependent on the frequency difference between the maximum response of the modal frequency response function, $\mathbf{H}_\alpha(\omega)$, and the maximum response of the modal joint acceptance, $j^2_{\alpha\alpha}(\omega)$ – see equation (7.20). This important fundamental feature of coincidence is illustrated schematically in Figure 7.8.

Figure 7.8(a) illustrates the response for complete coincidence where there is both wavenumber and frequency matching. Damping would reduce the structural response in this situation because the coincidence frequency corresponds with the structural resonance. Figure 7.8(b) illustrates wavenumber coincidence where the frequency difference is large enough to produce two peaks in the structural response, only one of which is damping controlled. Figure 7.8(c) illustrates wavenumber coincidence where the frequency difference is small: damping would still control the structural response in a situation such as this. Thus, when there is both wavelength and frequency matching, the pipe shell is driven at or near a resonance condition, and hence damping has a large effect in reducing the Q factor. When there is poor frequency matching, the shell response is forced by the high response of the contained sound field near the cut-off frequencies, and damping has little effect.

A typical experimentally determined pipe wall acceleration spectrum for turbulent internal pipe flow downstream of a 90° mitred bend[7.7] is presented in Figure 7.9 for a flow speed of Mach number 0.22. Structural resonance frequencies for principal wavenumber coincidence of three particular higher order internal acoustic modes are shown, and the increases in pipe wall response due to the four wavenumber coincidences for each acoustic mode can be clearly seen. It should be noted that the maximum structural responses, due to coincidence, do not necessarily occur precisely at the condition of wavenumber coincidence.

It is useful to note that the radiation ratio, σ, of coincident structural modes is always ~ 1. This is because the bending wave speed in the pipe wall at coincidence is equal to the wavespeed of the surface pressure wave associated with the higher order acoustic mode – at frequencies above the cut-off frequency, the internal surface pressure wave due to a propagating higher order acoustic mode is always supersonic with respect to the contained fluid[7.8]. Hence, if $c_e \leqslant c_i$ the structural wave will be supersonic with respect to the external medium and will therefore radiate very efficiently wth $\sigma \approx 1$.

For metal pipes, the phenomenon of coincidence occurs in close proximity to the cut-off frequencies of the various possible higher order acoustic modes. This is illustrated in Figure 7.10 for a 90° mitred bend. The cut-off frequencies for some of

Fig. 7.8. Schematic illustration of structural response at coincidence (f_α is the structural resonance frequency, and f_c is the coincidence frequency).

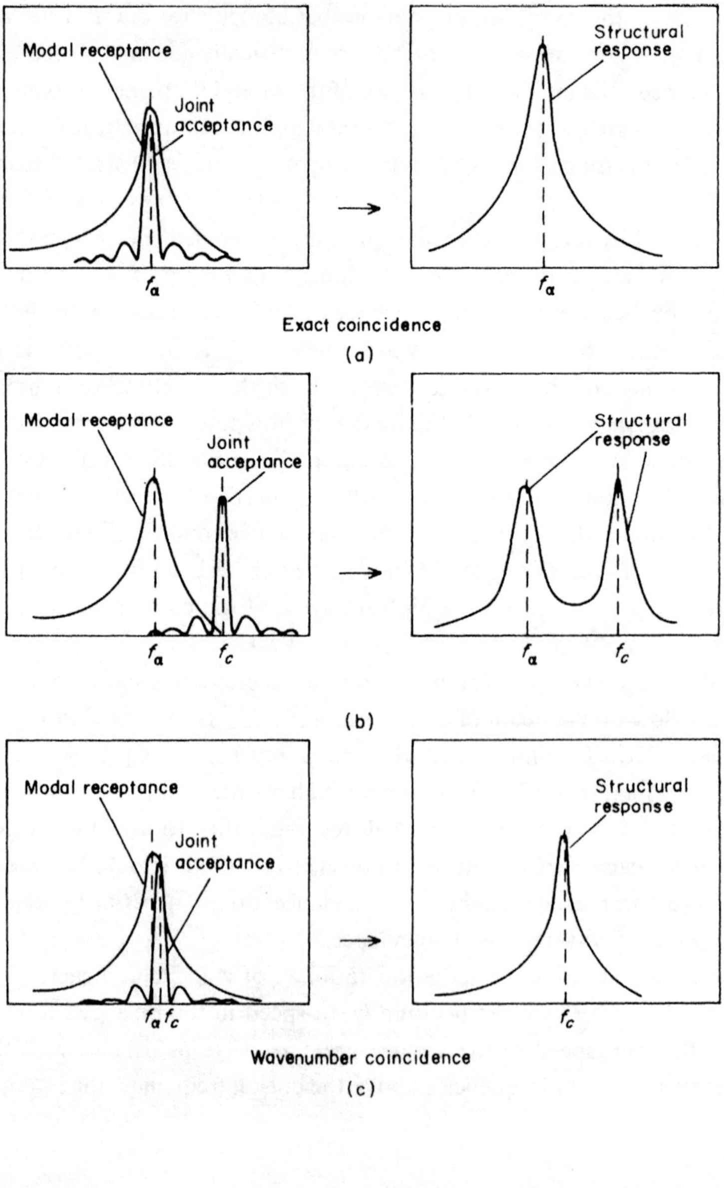

the higher order acoustic modes are illustrated on the figure. These cut-off frequencies can be estimated for the no flow case by simultaneously solving equations (7.24) and (7.33) with $M_0 = 0$. For $M_0 = 0$, equation (7.33) becomes

$$\upsilon^2 = (\upsilon_{co})^2_{pq} + (K_x/M_{LP})^2. \tag{7.34}$$

Thus, assuming that there is a continuum of K_m values for any given K_n in equation (7.24), solving equations (7.24) and (7.34) with $\upsilon = \upsilon_{mn} = \upsilon_c$, $n = p$ and $K_x = K_m$ yields the frequency for complete coincidence, υ_c, and the corresponding value of K_m. If it is assumed that $\beta n^2 \ll \upsilon_{co}$, then an approximate solution for thin cylindrical shells is[7.5]

$$\upsilon_c \approx \upsilon_{co} + \frac{n^2}{2(1-\nu^2)^{1/2} M_{LP}^2}, \tag{7.35}$$

and

$$K_x = K_m \approx \frac{n\upsilon_{co}^{1/2}}{(1-\upsilon^2)^{1/4}}, \tag{7.36}$$

where υ_c is the coincidence frequency and υ_{co} is the cut-off frequency.

An improved approximation (using Newton's method of successive approximations) which accounts for the pipe wall thickness is[7.5]

$$\upsilon_c = \upsilon_{co} + \tfrac{1}{2}\Delta\upsilon_1\{1 + (1+\bar{\upsilon}_{co})^3(\Delta\upsilon_1/\upsilon_{co})\}\{(1-\bar{\upsilon}_{co}) + (1+\bar{\upsilon}_{co})^3 - (\beta^2 n^4/\upsilon_{co})(1+\bar{\upsilon}_{co})^5\}, \tag{7.37}$$

Fig. 7.9. Typical pipe wall acceleration spectrum for Mach number 0.22 for turbulent pipe flow downstream of a 90° mitred bend. Structural frequencies for principal wavenumber coincidences are marked with arrows – the large peaks in proximity to the marked structural response frequencies are the associated coincident responses. 100 Hz bandwidth narrowband data.

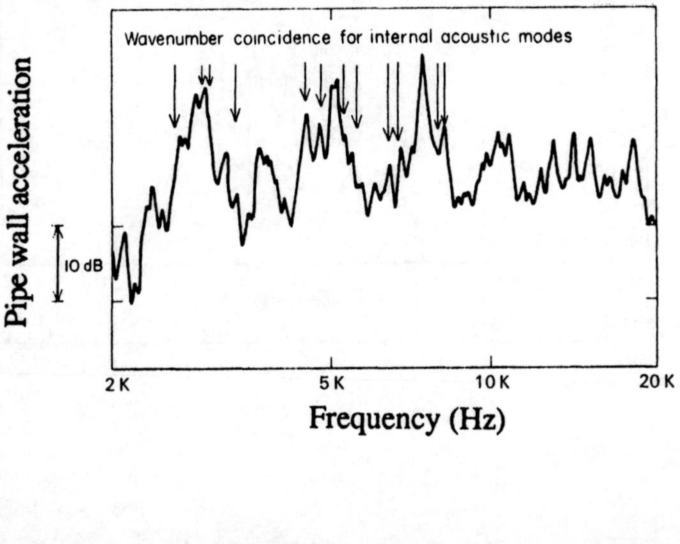

where

$$\Delta\upsilon_1 = \frac{n^2}{2(1-\nu^2)^{1/2}M_{LP}^2}, \tag{7.38}$$

and

$$\bar{\upsilon}_{co} = \frac{\upsilon_{co}}{(1-\nu^2)^{1/2}}. \tag{7.39}$$

Equations (7.35) and (7.37) verify that, for metal pipes, the coincidence frequencies are very close (typically within a few per cent) to the cut-off frequencies of the higher

Fig. 7.10. Non-dimensional sound power radiation from a section of straight pipe well downstream of a 90° mitred bend. (a) $M_0 = 0.22$; (b) $M_0 = 0.40$; (c) $M_0 = 0.50$. 10 Hz bandwidth narrowband data.

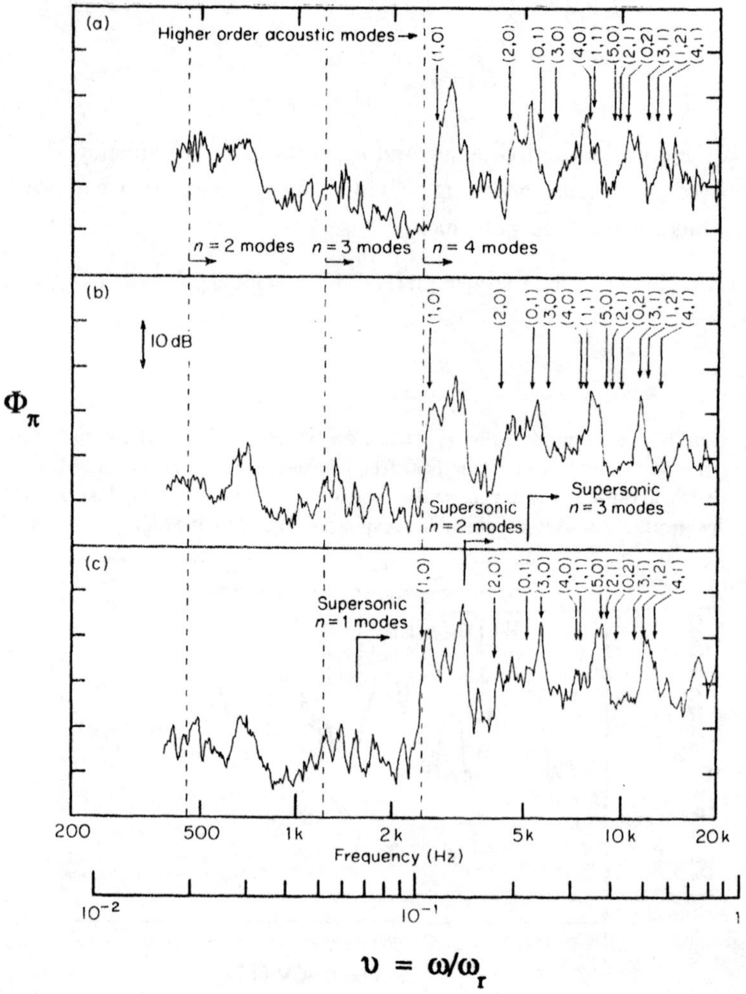

order acoustic modes. The effect of flow is accounted for by replacing the cut-off frequency υ_{co} by $\upsilon_{co}(1 - M_0^2)^{1/2}$.

7.6 Other pipe flow noise sources

Whilst the main thrust of this chapter has been on the effects of higher order acoustic modes on pipe wall noise and vibration, some mention should be made of other possible aerodynamic noise sources. Some typical industrial aerodynamic noise generators in pipes and ducting systems include diffusers, flow spoilers flow through grilles, jets, and cavity resonances[7.8,7.22].

The pressure fluctuations associated with the flow-induced excitations in these cases are generally broadband in nature (shaped like a haystack) and peak at a characteristic frequency. The peak level of the spectrum is proportional to the dynamic head, q, where

$$q = \frac{\rho_i U^2}{2}, \tag{7.40}$$

and ρ_i is the fluid density, and U is the characteristic velocity. The characteristic frequency associated with this peak level is proportional to the Strouhal number S, where

$$S = \frac{\Omega}{2\pi}, \tag{7.41}$$

and Ω is the Strouhal number associated with the radian frequency. S is a function of frequency, characteristic velocity and a characteristic length scale.

The characteristic velocity, U, can be the mean flow velocity as in the case of grilles or diffusers, some constricted flow velocity as in the case of flow spoilers, an exit velocity as in the case of jets, or the speed of sound as in the case of valve noise; pressure ratios across valves are generally such that the flow is sonic at the exit. For gas flows in ducts and piping systems, S is typically $\sim$0.20, although it can be higher in some special case such as choked flows or very high pressure differentials ($\sim$40 kPa) across a spoiler.

When one considers the case of two flows with the same mean velocity, if one flow has a peak at a lower frequency in the frequency spectrum of the pressure fluctuations than the other, it can be concluded that its turbulence is of a larger scale – i.e. its turbulent boundary layer is thicker. The 'haystack' pressure spectrum is generally associated with the shedding of turbulence by an obstruction in the flow (this phenomenon is commonly known as vortex shedding), or with the impingement of a fluid jet flow onto a solid surface. A typical frequency spectrum associated with such flows is illustrated schematically in Figure 7.11.

It should be appreciated by the reader at this stage that the flow disturbances discussed earlier on in this chapter in relation to higher order acoustic mode propagation also have wall pressure frequency spectra of the shape shown in Figure 7.11. The type of flow-induced noise discussed here is also present, but in those instances the higher order acoustic modes dominate the structural response and the external sound power radiation, even though their energy levels (inside the pipe) are sometimes well below the generalised wall pressure spectrum. This is a very important observation.

For gas jets, the Strouhal number is given by

$$S = \frac{f_p \phi}{U_e} = 0.2, \tag{7.42}$$

where f_p is the peak frequency in the far-field, ϕ is the nozzle diameter, and U_e is the exit velocity.

For spoilers (splitter plates, etc.) in ducted flows, the Strouhal number is given by

$$S = \frac{f_p t}{U_c} = 0.2 \quad \text{for } \Delta P,\ 4\,\text{kPa}$$

$$= 0.5 \quad \text{for } \Delta P,\ 40\,\text{kPa}, \tag{7.43}$$

where f_p is the frequency associated with the spectral peak, t is the projected thickness of the spoiler, P is the total pressure, and U_c is the constricted flow speed. For $4\,\text{kPa} < \Delta P < 40\,\text{kPa}$, S can be obtained by interpolation.

For grilles in ducted flows, the Strouhal number is given by

$$S = \frac{f_p \phi_r}{U_0} = 0.20, \tag{7.44}$$

where f_p is the vortex shedding frequency, ϕ_r is the diameter of a typical rod element in the grille, and U_0 is the mean flow speed. The main difference between a grille

Fig. 7.11. Generalised broadband spectrum for in-pipe flow generated noise.

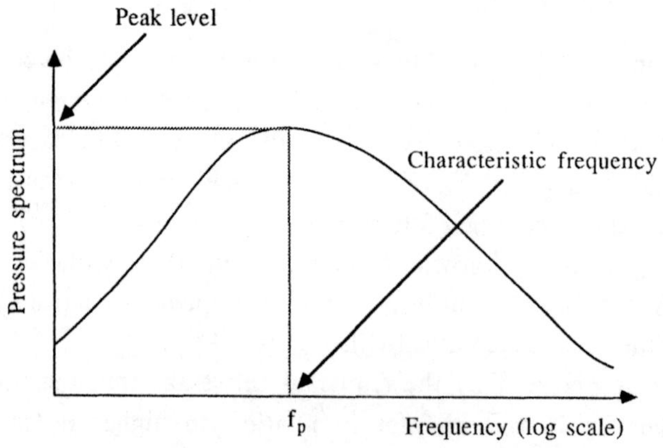

and a spoiler in a ducted system is that in the former it is assumed that the duct cross-sectional area is sizeable ($>0.2\ m^2$) and that the flow velocity is low ($\leqslant 60\ m\ s^{-1}$) such that the jet related noise is insignificant. Grille noise is thus the result of interactions between the flow and the rigid bodies – the periodic vortices generate lift-force fluctuations on the individual rods in the grille. It should be noted that, for the case of spoiler noise, the frequency peak, f_p, can also be regarded as a vortex shedding frequency in some instances (flow/rigid body interactions) rather than a turbulent mixing process excitation.

For air valves controlling the flow of gas through a pipe, it is not unreasonable to assume that the flow is choked – i.e. the Mach number at the valve exit is unity. There are two noise generation mechanisms associated with choked flows. They are (i) turbulent mixing in the vicinity of the valve, and (ii) shock noise downstream of the valve. For valve pressure ratios <3 both mechanisms must be considered, and for valve pressure ratios >3 shock noise is predominant. For both mechanisms, the sound power spectrum has the characteristic 'haystack' shape with

$$S = \frac{f_p D}{c}, \tag{7.45}$$

where c is the speed of sound in the gas in the valve, D is the narrowest cross-sectional dimension of the valve opening, and f_p is the frequency associated with the spectral peak. This is illustrated schematically in Figure 7.12. The peak Strouhal number, S, for turbulent mixing is 0.20. For shock noise, the peak Strouhal number can be obtained from Table 7.2. It is a function of the pressure ratio across the valve.

Fig. 7.12. Schematic illustration of valve opening.

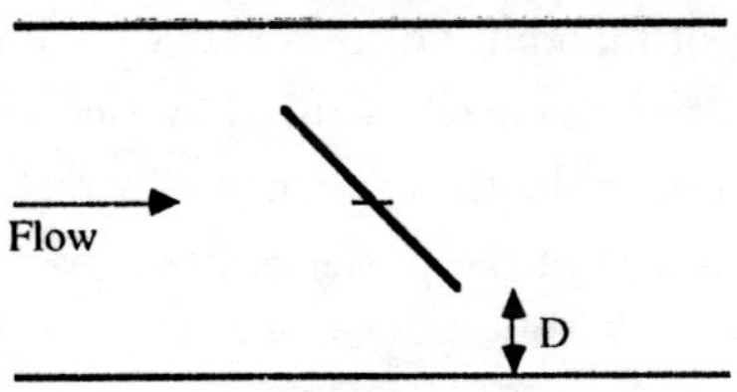

Table 7.2. *Peak Strouhal number for shock noise mechanisms (choked air valves).*

Pressure ratio across valve	Peak Strouhal number
2.0	0.65
2.5	0.29
3.0	0.20
4.0	0.16
5.0	0.13
6.0	0.11
7.0	0.07
8.0	0.05

The discussions in this section are compatible with the previous discussions on higher order acoustic mode generation. If the peak frequency associated with the broadband 'haystack' spectrum is below the cut-off frequency of the first higher order acoustic mode, then vortex shedding and/or plane waves and/or boundary layer turbulence are the dominant sources of noise and vibration at these frequencies. In addition, there will also be some noise and vibration generated by coincident higher order acoustic modes at the higher frequencies. The relative contributions of the low frequency components (vortex shedding, plane waves and turbulence) and the high frequency components (coincident higher order acoustic modes) will depend upon the type of internal flow disturbance. If the peak frequency associated with the 'haystack' spectrum is above the cut-off frequency of the first higher order acoustic mode, then wavenumber coincidence is the dominant source of noise and vibration, although there will be some secondary high frequency contribution from vortex shedding, plane waves and turbulence. In this instance, the low frequency noise and vibration will only be due to plane waves and turbulence, and it will not have a vortex shedding component.

If the spectral density peaks of the noise from a pipe/duct flow are due to vortex shedding, etc. then one would expect all the spectral peaks to be at the same Strouhal number, S, for all flow speeds. Sometimes, however, in duct flow situations flow independent acoustic resonances are encountered. Such phenomena are typically found in gate valves.

For the particular case of a gate valve, the spectral maxima result from an acoustic response, such as a cavity resonance, to vortex shedding from the edges of the cavity (see sub-section 2.4.4, chapter 2). The maximum pressure response of the cavity is determined by the degree of matching of these two phenomena, and the frequency for maximum response is also influenced by this flow–acoustic interaction. The lowest frequency at which the acoustic depth resonance of a rectangular cavity of depth D and streamwise length L occurs in the presence of a low Mach number flow can be expressed as

$$\frac{fD}{c_{\mathrm{i}}} = \frac{0.25}{1 + 0.8L/D}. \tag{7.46}$$

The frequency of vortex shedding from the upstream edge of such a cavity can be estimated from

$$\frac{fL}{U_0} = \frac{0.75}{M_0 + 1/k_{\mathrm{v}}}, \tag{7.47}$$

where $k_{\mathrm{v}}U_0$ is the convection velocity of the vortex (k_{v} is typically 0.57). The excitation of a cavity resonance by vortex shedding becomes more effective as the two frequencies merge together.

7.7 Prediction of vibrational response and sound radiation characteristics

The prediction of absolute vibrational response levels of and absolute sound radiation levels from pipes with internal gas flows is not an easy task. As is evident from the discussions so far in this chapter, there are numerous source mechanisms which complicate the issue. Because of this, it is far more appropriate to adopt a parametric type study – i.e. to analyse the effects of changes in pipe wall thickness, pipe material, pipe dimensions, fluid properties, flow speed, etc., on pipe wall vibration and subsequent sound radiation. It is the relative reductions in noise and vibration due to the effects of varying these parameters which is of direct concern to pipeline designers, etc.

Several authors have postulated different procedures for the estimation of noise and vibration transmission through pipe walls. Some of these include Bull and Norton[7.7] (or Norton and Bull[7.8]), White and Sawley[7.23], Fagerlund and Chou[7.24], Holmer and Heymann[7.25], and Pinder[7.26]. Pinder's[7.26] report, in particular, is an excellent critical review of the available procedures. For the purposes of this book, it is sufficient to summarise some of these procedures. It is also useful to remind the reader that this chapter is primarily concerned with metallic (steel, aluminium, etc.) pipes, hence it is assumed that the shell walls are perfectly rigid as far as the internal propagating sound waves are concerned.

Equations (7.20) and (7.23) can be used to estimate the vibrational response and the sound power radiation, but they require considerable detailed knowledge of the internal sound field and of the vibrational characteristics of the pipe. It is generally convenient to perform the analysis in one-third-octave bands, and information is required about the modal quality factors, Q_α, the wall pressure spectra, G_{pp}, the joint acceptance, $j^2_{\alpha\alpha}$, and the frequency response function. Whilst the prediction of the absolute levels requires extensive computations, a study of the governing equations provides sufficient information about the relevant parameters required for an estimation of the relative increases or decreases in pipe wall vibrational response and subsequent sound radiation. From a detailed inspection of equations (7.20) and (7.23), the following observations can be made.

(i) The response is a function of the spectral density, G_{pp}, of the internal wall pressure field. This parameter is a function of the flow speed and the geometry of the flow disturbance (e.g. 45° mitred bend, 90° mitred bend, gate valve, etc.). Larger values of G_{pp} are associated with severe disturbances. Detailed experimental studies of a range of flow disturbances have provided a data base in non-dimensional form[7.5,7.6,7.8]. Some of these experimentally obtained wall pressure fluctuations have been presented in this chapter (Figures 7.3–7.5). The studies show that the wall pressure spectra, G_{pp}, are affected by flow: turbulent wall pressure spectra scale as U_0^3; wall pressure spectra associated with plane waves scale as U_0^3; and wall pressure spectra associated with higher order acoustic modes scale as U_0^5.

(ii) The response is a function of the non-dimensional pipe wall thickness parameter, β – i.e. it is a function of the ratio of pipe wall thickness to pipe diameter. There is a twofold effect. Firstly, there is a direct effect which is inversely proportional to β^2. Secondly, there is an additional effect due to the variations of the modal responses within the summation sign. This additional effect is particularly important when considering the response at frequencies where coincidence is possible – i.e. at frequencies above the cut-off frequency of the first higher order acoustic mode. Here, variations in β produce significant variations in the possible number of wavenumber coincidences.

A numerical procedure has been developed[7.7] to evaluate the total possible number of wavenumber coincidences for a given pipe wall thickness, internal diameter and upper limiting frequency. It has been shown that the number of coincidences, N_c, with frequencies below a given limiting frequency, υ_L, is of the form

$$N_c = N_c(\upsilon_L, M_{LP}, \beta, M_0), \tag{7.48}$$

and is essentially independent of non-dimensional length, Δ. It was found that, except at the smallest value of β, changing the flow speed produced no significant change in the number of coincidences. The variations of N_c with υ_L and β for $M_0 = 0$, and $M_0 = 0.5$ are presented in Figure 7.13 and Figure 7.14, respectively. The variations of N_c with β for a particular limiting frequency, υ_L, (namely $\upsilon_L = 0.88$, which corresponds to $f = 20$ kHz and 18.5 kHz for the experimental test pipes with $h = 0.89$ mm and 6.36 mm, and $a_m = 36$ mm) are presented in Figure 7.15.

If in practice pipe wall vibration and sound radiation at frequencies up to a limiting dimensional frequency, f_L (e.g. the limit of the audio-frequency range) are the main

Fig. 7.13. Variation of N_c with υ_L and β for $M_0 = 0$.

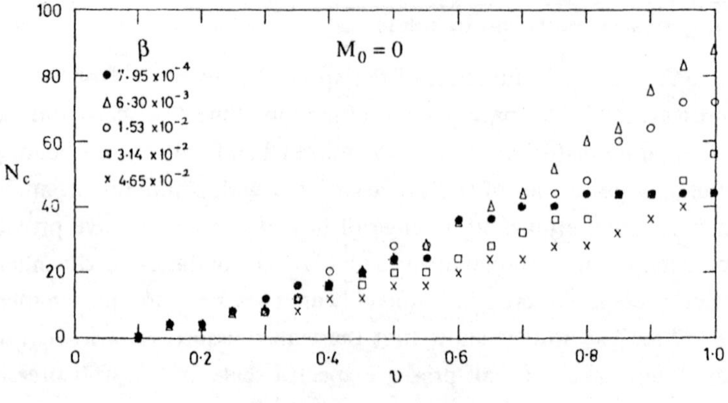

consideration, the limiting non-dimensional frequency, $\upsilon_L = \omega_L/\omega_r = 2\pi f_L a_m / c_i M_{LP}$, will increase with increasing pipe radius. For $f_L = 20$ kHz, say, $\upsilon_L \sim 0.7$ for a pipe with $a_m = 30$ mm, and Figures 7.13 and 7.14 indicate that the number of coincidences in the range of interest would be typically about thirty-five. However, for a pipe with $a_m = 300$ mm, $\upsilon_L \sim 7.0$ and the number of coincidences would be correspondingly larger.

It is clear that, for pipes of practical interest, the number of coincidences in the audio-frequency range which will contribute to the pipe wall vibrational response due to broadband internal acoustic excitation (such as that associated with internal flow disturbances due to pipe fittings) will be large. A correct selection of β will, however, allow for an avoidance of the maximum coincidence situation arising (see Figure 7.15).

White and Sawley[7.23] have produced expressions for energy sharing between the pipe wall and the contained fluid for frequencies below and above the cut-off frequency of the first higher order acoustic mode. Their expressions are based upon S.E.A.

Fig. 7.14. Variation of N_c with υ_L and β for $M_0 = 0.5$.

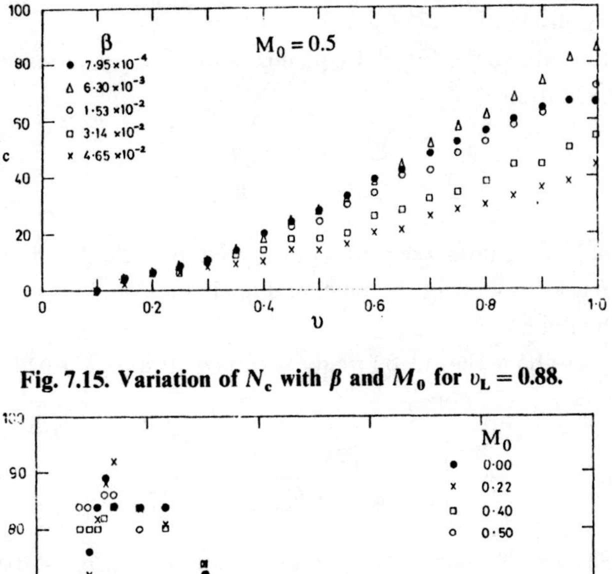

Fig. 7.15. Variation of N_c with β and M_0 for $\upsilon_L = 0.88$.

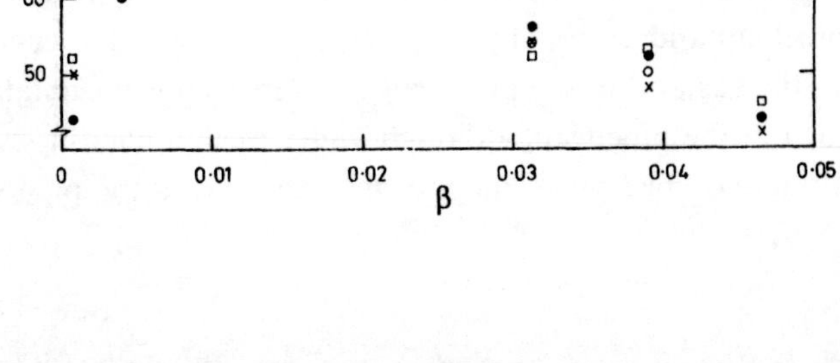

procedures and relate to (i) fluid excitation of the coupled systems, and (ii) mechanical excitation of the coupled systems.

For frequencies below the cut-off frequency of the first higher order acoustic mode and for fluid excitation

$$\frac{E_f}{E_p} = \frac{hE}{\rho_i c_i^2 D\{1 + (f/f_r)^2\}}, \tag{7.49}$$

where E_f is the energy in the contained fluid, E_p is the energy in the pipe wall, E is Young's modulus, D is the mean pipe diameter, and the other parameters are as defined previously.

For frequencies below the cut-off frequency of the first higher order acoustic mode and for mechanical excitation

$$\frac{E_f}{E_p} = \frac{M_f}{2M_p}, \tag{7.50}$$

where M_f and M_p are the fluid and pipe wall masses, respectively, for a unit length of pipe. For flow excitation, a larger percentage of the energy is in the fluid, particularly if it is a gas. For mechanical excitation, on the other hand, most of the energy will be in the pipe wall.

For frequencies above the cut-off frequency of the first higher order acoustic mode and for fluid excitation

$$\frac{E_f}{E_p} = \frac{n_f}{n_p}\left(1 + \frac{\eta_p}{\eta_{pf}}\right), \tag{7.51}$$

where n_f and n_p are the modal densities of the contained fluid and pipe, respectively, η_p is the internal loss factor of the pipe, and η_{pf} is the coupling loss factor from the pipe wall to the fluid.

For frequencies above the cut-off frequency of the first higher order acoustic mode and for mechanical excitation

$$\frac{E_f}{E_p} = \frac{n_f}{n_p}\left\{\frac{1}{1 + (\eta_f/\eta_{pf})(n_f/n_p)}\right\}. \tag{7.52}$$

White and Sawley[7.23] suggest that $\eta_p/\eta_{pf} \sim 1.0$ and that $\eta_f/\eta_{pf} \sim 10.0$. Thus for fluid excitation the energy ratio is dependent upon the modal density ratio, and for mechanical excitation it is a function of both the modal density ratio and η_f/η_{pf}. With gas flows in pipelines and excited by an acoustic excitation, the energy is generally carried both in the gas and in the pipe wall; with mechanical excitation, the energy is usually carried in the pipe wall, although under certain conditions the gas might dominate. With liquid filled pipes, most of the energy is in the pipe wall.

Fagerlund and Chou[7.24] derive expressions for sound transmission through pipe walls based upon S.E.A. procedures. They provide a useful relationship between the mean-square sound pressure inside the pipe and the mean-square sound pressure in the far-field outside the pipe. It is

$$\frac{p_e^2}{p_i^2} = \frac{5\rho_e^2 c_e^2 c_i^2 (2a_i + 2h)(\Delta k_x) G(M) \sigma_i \sigma_e}{18 \rho a_i r h \omega^2 \Delta\omega (\rho_i c_i \sigma_i + \rho_e c_e \sigma_e + h \rho \omega \eta_p)}, \tag{7.53}$$

where the subscript e refers to the external fluid, the subscript i refers to the internal fluid, r is the radial distance from the cylinder axis, ρ is the density of the pipe material, $G(M)$ is a velocity correlation factor, a is the pipe radius, h is the pipe wall thickness, σ is the radiation ratio, η_p is the average internal loss factor of the pipe in the frequency band $\Delta\omega$ with centre frequency ω, and Δk_x is the change in axial structural wavenumber. This change in axial structural wavenumber is proportional to the number of modes within the frequency band. The velocity correlation factor is a function of Mach number and varies linearly from 1.0 to ~3.5 for Mach number variations from 0 to 0.5. Equation (7.53) is critically dependent upon variations in Δk_x and the radiation ratios.

Holmer and Heymann[7.25] also provide expressions for sound transmission through pipe walls in the presence of flow. They define a transmission loss, *TL*, and a transmission coefficient, τ, for a length of cylinder of radiating area equal to the cross-sectional area of the pipe. These are the definitions of *TL* and τ which are used here since the transmission loss will only be a unique value if a reference length is used. Hence, the transmission loss, TL_{pipe}, for a length of cylinder of radiating area equal to the cross-sectional area of the pipe, is given by

$$TL_{pipe} = 10 \log_{10}\Pi_i - 10 \log_{10}\Pi_e + 10 \log_{10}(4L/D), \tag{7.54}$$

where the subscript i refers to the inside of the pipe, the subscript e refers to the outside of the pipe, Π is the sound power, L is the length, and D is the mean diameter.

With this definition of sound transmission through pipe walls, both Holmer and Heymann[7.25] and Pinder[7.26] provide several semi-empirical relationships for predicting *TL*. Pinder[7.26] in particular provides a relationship for frequencies below the cut-off frequency of the first higher order acoustic mode based on work by Kuhn and Morfey[7.27]. It is

$$TL_{pipe} = 10 \log_{10}\left[\left\{\frac{\rho^2 c_L}{\rho_i^2 c_i}\right\}(h/2a_m)^2(\omega_r/\omega)^3\right] - 24 \text{ dB}, \tag{7.55}$$

where all the terms are as defined previously. Experimental evidence suggests that the theory is conservative in that it underestimates the *TL* by at least 20 dB, especially for small diameter pipes.

For frequencies above the cut-off frequency of the first higher order acoustic mode but below the ring frequency, f_r, the transmission loss through the pipe wall is dominated by coincidence. Pinder[7.26] derives a parametric dependence for the transmission coefficient, τ, based on the work of Bull and Norton[7.7]. It is given by

$$\tau \propto \frac{\rho_i^2 c_i^2 (2a_m)^2}{h^2 \rho^2 c_L^2 \eta_p v^2}, \tag{7.56}$$

and it leads to a transmission loss, TL_{pipe}, which is dependent upon $(h/2a_m)^2$. Equation (7.56) only allows for the evaluation of relative effects and does not allow for the prediction of absolute pipe transmission losses; absolute pipe transmission losses can only be obtained via extensive computations using either equations (7.20) and (7.23), or equation (7.53).

Very little experimental data are available for frequencies above the ring frequency, f_r. Coincidence is still the primary mechanism by which the pipe responds to the internal flow, and TL varies with $(h/2a_m)^2$. At these high frequencies, the number of coincidences is significantly increased (see Figures 7.13 and 7.14). Pinder[7.26] proposes the following empirical procedure which is a function of h^2. Firstly, TL_{pipe}, is calculated at $4f_r$ from

$$TL_{pipe} = 60 \log_{10} f + 20 \log_{10} h - 150 \text{ dB}, \tag{7.57}$$

and then it is blended to the transmission loss at f_r by assuming a zero gradient at f_r and an 18 dB per octave slope at $4f_r$. This procedure fits the test results of Holmer and Heymann[7.25].

7.8 Some general design guidelines

Based on the discussions in the preceding sections, several general design guidelines can be drawn up to assist designers and plant engineers in analysing any potential problems associated with flow-induced noise and vibration in pipelines. It should be clear by now that the primary sources of pipe flow noise and vibration are (i) coincidence, (ii) Strouhal number dependent vortex shedding phenomena, (iii) Strouhal number independent cavity resonances, (iv) boundary layer turbulence, (v) propagating plane waves, (vi) flow separation and increased turbulence at discontinuities, and (vii) mechanical excitation.

In general terms, pipe flow noise and vibration levels are controlled by (i) the geometry of the disturbance, (ii) the flow speed of the gas, and (iii) the pipe wall thickness. There are no explicit mathematical relationships for the parametric dependence on (i) – predicted results are qualitative and have to be obtained from an experimental data bank. Some typical spectra of internal fluctuating wall pressures (in a section of piping well downstream of the disturbances) were presented in Figures 7.3 and 7.4. The data for the 90° mitred bend and straight pipe flow represent upper and lower limits, respectively, for the range of pipe fittings tested. The internal wall pressure fluctuations are substantially more severe and not circumferentially uniform at regions in proximity to a disturbance. This point was illustrated in Figure 7.5.

For straight runs of pipeline downstream of bends, tee-junctions, valves, etc., coincidence is generally the dominant source of noise and vibration. The low frequency noise (i.e. noise below the cut-off frequency of the first higher order acoustic mode) is due to plane waves or Strouhal number dependent vortex shedding phenomena or Strouhal number independent cavity resonances. Boundary layer turbulence is generally not a major noise or vibration problem. At regions in close proximity to pipe fittings, the major sources of noise and vibration are flow separation and increased turbulence, and mechanical excitation.

Any design or trouble shooting exercise should commence with an attempt to identify the frequencies associated with the various possible mechanisms. If the installation already exists, the acquisition of noise and vibration spectra greatly facilitates the noise source identification procedures.

To begin with, the cut-off frequencies of the various higher order acoustic modes need to be established for a given pipe diameter, wall thickness and flow speed. It is only necessary to establish the cut-off frequencies of the first few higher order acoustic modes; at higher frequencies they are very hard to identify as they tend to merge together. The various cut-off frequencies can be obtained from equation (7.8) and Table 7.1. The corresponding coincidence frequencies can be evaluated from equations (7.35), (7.36) and (7.37) (note that the effect of flow is accounted for by replacing the cut-off frequency f_{co} by $f_{co}(1 - M_0^2)^{1/2}$). For steel pipes, the coincidence frequencies are in close proximity to the cut-off frequencies themselves. One also needs to ensure that the pipe wall thickness parameter β is not such that it allows for a maximum coincidence situation to arise. This question can be addressed by reference to Figure 7.15.

Strouhal number dependent vortex shedding phenomena and Strouhal number independent cavity resonances can be identified from equations (7.42)–(7.47) depending upon the type of pipe fitting.

Boundary layer turbulence and propagating plane waves produce broadband wall pressure spectra, which in turn produce a broadband pipe wall vibrational response. The response due to these mechanisms is generally of a lower level than a coincident response, or a response due to vortex shedding and/or cavity resonances. Because of the broadband nature of these excitation types, it is not generally possible to associate them with any dominant spectral peaks, the exception being when a single frequency plane wave excitation such as a pulsation from a pump, etc. is present.

Noise and vibration, due to flow separation and increased turbulence at discontinuities, and any associated mechanical excitation are harder to quantify. The noise and vibration spectra are generally unique to the type of pipe fitting. Several empirical prediction schemes are available from manufacturers for control valves such as globe valves, ball valves and butterfly valves. All these prediction schemes are a function of the pressure ratio across the valve. The predominant noise generation mechanisms in control valves are flow separation and shock wave/increased turbulence interactions downstream of the throttling elements. Localised shock noise is generated by interactions between shock waves (due to the pressure ratio) and increased turbulence due to separation, etc. At large distances from the valves, it is the propagating higher order acoustic modes that are the dominant noise sources.

Allen[7.28], Reethof and Ward[7.29] and Ng[7.30] all discuss semi-empirical procedures for estimating valve noise. It is useful to note that the valve noise spectra take on the broadband 'haystack' shape with a spectral peak frequency given by equation (7.45). The sound pressure level at some distance r from a valve can be approximated by[7.30]

$$L_{\mathrm{valve}} \approx 158.5 + 10 \log_{10}\left(\frac{D^2\chi^4}{r^2}\right), \tag{7.58}$$

where D is the narrowest cross-sectional dimension of the valve opening, r is the distance to the observer, and χ is a parameter which is related to the fully expanded jet Mach number. It is given by[7.30]

$$\chi = \left[\frac{2}{\gamma - 1}\left\{\left(\frac{P_{1t}}{P_2}\right)^{(\gamma-1)/\gamma} - \frac{(\gamma+1)}{2}\right\}\right]^{1/2}, \tag{7.59}$$

where γ is the specific heat ratio, P_{1t} is the upstream total pressure, and P_2 is the downstream static pressure. Equations (7.58) and (7.59) represent only one of several available procedures for the estimation of valve noise. Jenvey[7.31] presents fundamental parametric relationships, based on dimensional analysis, for the estimation of radiated sound power from valves for both subsonic and choked flow. For subsonic flow

$$\Pi \propto \left(\frac{\Delta P_t}{P_{1t}}\right)^2 \Delta P_t^2 A_j^{2.4}, \tag{7.60}$$

and for choked flow

$$\Pi \propto \left(\frac{\Delta P_t}{P_{1t}}\right) \Delta P_t^2 A_j^{2.4}, \tag{7.61}$$

where ΔP_t is the total pressure ratio across the valve, P_{1t} is the upstream total pressure, and A_j is the cross-sectional area of the jet (i.e. the orifice cross-sectional area). Equations (7.60) and (7.61) are fairly useful in that they relate simply and directly to the primary valve parameters – i.e. total pressure ratio across the valve, upstream total pressure and cross-sectional area.

7.9 A vibration damper for the reduction of pipe flow noise and vibration

The fundamental mechanism for the generation of pipe wall vibration and subsequent external acoustic radiation from straight runs of pipelines, downstream of internal flow disturbances, is the coincidence of internal higher order acoustic modes with resonant flexural modes in the pipe wall. The form of the structural response at coincidence was discussed in section 7.5 and illustrated schematically in Figure 7.8. When the coincidence frequency is close to a structural resonance frequency (see Figure 7.8a and c), the structural response will be damping controlled. When the coincident structural response is damping controlled, a vibration damper can be utilised to reduce the noise and vibration.

Howard *et al.*[7.32] discuss such a coincidence damper. The damper consists of a rigid metal ring attached to the pipe by several discrete rubber inserts. A theoretical model of the pipe with the attached damper is developed by using the receptance technique where the ring is modelled as a rigid mass and the rubber inserts are modelled as complex massless springs by using a complex stiffness. The model enables a prediction of the optimum reduction in structural response. By reducing the receptance of the structural modes in the region of the various coincidence frequencies, the enhanced structural response caused by coincidence matching is reduced.

It is important to re-emphasise that the coincidence damper specifically addresses the problem of reducing the structural responses resulting from wavenumber coincidences. As already mentioned, the form of coincidence, and the subsequent effectiveness of any coincidence damper, are critically dependent on the proximity of the modal receptance to the joint acceptance. A coincidence damper will significantly increase the damping of structural modes that are associated with coincidence; hence, if the frequency difference between the structural resonance frequency and the coincidence frequency is small, it would be expected that the damper would reduce the coincident structural response. On the other hand, if the structural modal density is low, then the frequency difference would be greater and the effectiveness of the damper would be substantially reduced.

In their analysis, Howard *et al.*[7.32] developed receptances (frequency response functions relating displacements to force) for cylindrical shells to radial forces, and direct and cross-receptances for the rubber mounted rigid rings. A schematic representation of the rigid ring with the rubber inserts is illustrated in Figure 7.16. The ring is subsequently mounted on a cylindrical pipe with Allen screws as illustrated in Figure 7.17. The two subsystems (pipe and rubber mounted ring) are thus coupled via the coupling points; the force and displacement relationships which are known to occur at the coupling points between the two subsystems are the link between the cylindrical shell receptances and the rubber mounted ring receptances. The number of equations which result depends upon the number of coupling points N which are used to connect the two subsystems. The number of rubber inserts which are used to dampen out the specific structural modes of the pipe are dependent upon the particular circumferential order mode which is being reduced.

For the purpose of the experiments[7.32], the $(m, 2)$ structural modes were chosen – i.e. modes with two full waves around the circumference. Also, a ring damper having three equispaced rubber inserts was used such that the $(m, 2)$ modes could not orient themselves so as to have all the circumferential nodes occurring at the insert points.

The response of the complete system is obtained by coupling the receptances of the various subsystems[7.32], and the subsequent reduction in response of resonant

Fig. 7.16. Schematic illustration of the rigid ring with the rubber inserts.

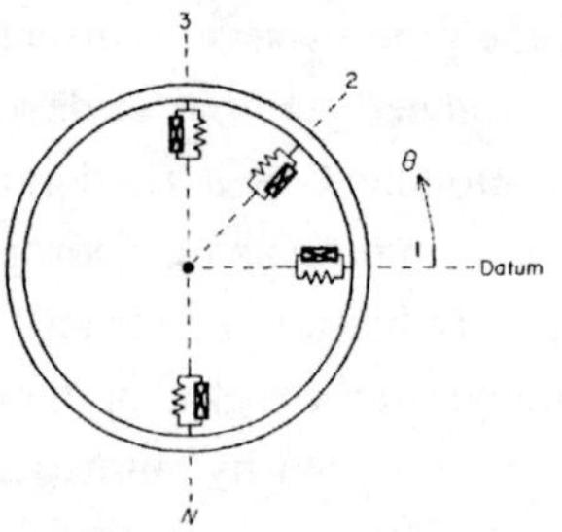

Fig. 7.17. Rubber mounted ring, showing method used to locate rubber inserts between outer pipe wall and inner diameter of ring.

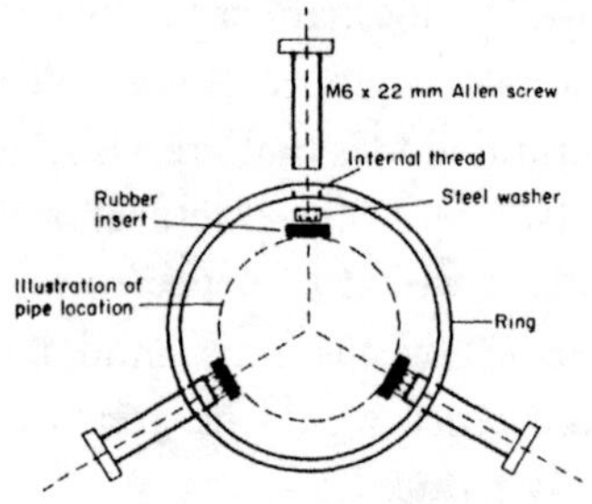

Fig. 7.18. Experimentally measured receptance of the simply supported pipe at mid-span (relative units, linear scale).

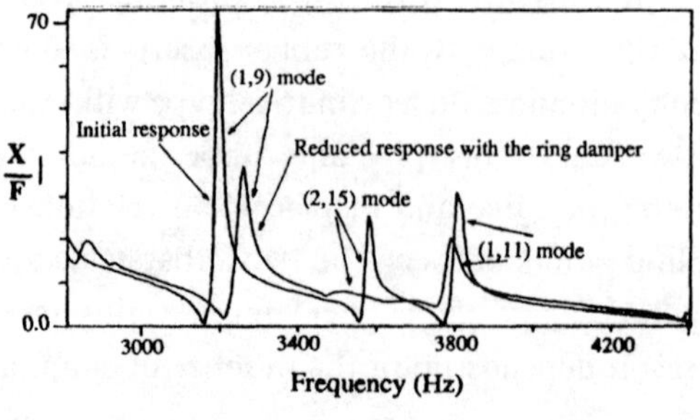

modes depends upon such factors as the rubber stiffness, the level of hysteretic damping, the ring mass, and the axial location of the ring along a section of pipeline. The reduction in response can be optimised by tuning these parameters. The reader is referred to Howard *et al.*[7.32] for details of the theoretical model and the optimisation procedure. Some typical experimental results are presented in Figure 7.18 for a rubber mounted ring with three inserts (butyl rubber). It is interesting to note that the ring not only reduced the $(m, 2)$ mode in the frequency band of interest, but also the $(m, 1)$ modes. Also, and more importantly, the measured quality factors (Q's) were significantly reduced – i.e. from $\sim$730 to $\sim$20. This significant increase in damping due to the insertion of the ring damper is an important finding since it provides a simple and effective way of selectively reducing the response of different circumferential order pipe modes. If these pipe modes are in close proximity to the corresponding coincidence frequencies then the damper will be effective in reducing the coincidence structural response.

References

7.1 Naudascher, E. and Rockwell, D. 1979. *Practical experiences with flow-induced vibrations*, Springer-Verlag.

7.2 BHRA Fluid Engineering, 1987. *Flow-induced vibrations*, International Conference, Bowness-on-Windermere, England, Conference Proceedings.

7.3 Blake, W. K. 1986. *Mechanics of flow-induced sound and vibration*, Academic Press.

7.4 Norton, M. P. 1979. *The effects of internal flow disturbances on the vibration response of and the acoustic radiation from pipes*, Ph.D. Thesis, University of Adelaide.

7.5 Bull, M. K. and Norton, M. P. 1980. 'The proximity of coincidence and acoustic cut-off frequencies in relation to acoustic radiation from pipes with disturbed internal turbulent flow', *Journal of Sound and Vibration* **69**(1), 1–11.

7.6 Bull, M. K. and Norton, M. P. 1981. 'On the hydrodynamic and acoustic wall pressure fluctuations in turbulent pipe flow due to a 90° mitred bend', *Journal of Sound and Vibration* **76**(4), 561–86.

7.7 Bull, M. K. and Norton, M. P. 1982. *On coincidence in relation to prediction of pipe wall vibration and noise radiation due to turbulent pipe flow disturbed by pipe fittings*, Proceedings of BHRA International Conference on Flow-Induced Vibrations in Fluid Engineering, Reading, England, pp. 347–68.

7.8 Norton, M. P. and Bull, M. K. 1984. 'Mechanisms of the generation of external acoustic radiation from pipes due to internal flow disturbances', *Journal of Sound and Vibration* **94**(1), 105–46.

7.9 Bull, M. K. 1967. 'Wall pressure fluctuations associated with subsonic turbulent boundary layer flow', *Journal of Fluid Mechanics* **28**(4), 719–54.

7.10 Rennison, D. C. 1976. *The vibrational response of and the acoustic radiation from thin-walled pipes excited by random fluctuating pressure fields*, Ph.D. Thesis, University of Adelaide.

7.11 Lin, T. C. and Morgan, G. W. 1956. 'Wave propagation through fluid contained in a cylindrical elastic shell', *Journal of the Acoustical Society of America* **28**(4), 1165.

7.12 El-Rahib, M. 1982. 'Acoustic propagation in finite length elastic cylinders: parts I and II', *Journal of the Acoustical Society of America* **71**(2), 296–317.
7.13 Morse, P. M. and Ingard, K. U. 1968. *Theoretical acoustics*, McGraw-Hill.
7.14 Bendat, J. S. and Piersol, A. G. 1980. *Engineering applications of correlation and spectral analysis*, John Wiley & Sons.
7.15 Soedel, W. 1981. *Vibrations of shells and plates*, Marcel Dekker.
7.16 Leissa, A. W. 1973, *Vibration of shells*, NASA Special Report SP-288.
7.17 Arnold, R. N. and Warburton, G. B. 1949. 'The flexural vibration of thin cylinders', *Proceedings of the Royal Society (London)* **197A**, 238–56.
7.18 Greenspon, J. E. 1960. 'Vibrations of a thick-walled cylindrical shell – comparison of exact theory with approximate theories', *Journal of the Acoustical Society of America* **32**(2), 571–8.
7.19 Heckl, M. 1962. 'Vibration of point-driven cylindrical shells', *Journal of the Acoustical Society of America* **34**(5), 1553–7.
7.20 Rennison, D. C. and Bull, M. K. 1977. 'On the modal density and damping of cylindrical pipes', *Journal of Sound and Vibration* **54**(1), 39–53.
7.21 Fahy, F. J. 1985. *Sound and structural vibration: radiation, transmission and response*, Academic Press.
7.22 Heller, H. H. and Franken, P. A. 'Noise of gas flows', chapter 16 in *Noise and vibration control*, edited by L. L. Beranek, McGraw-Hill.
7.23 White, P. H. and Sawley, R. J. 1972, 'Energy transmission in piping systems and its relation to noise control', *Journal of Engineering for Industry (ASME Transactions)*, May, pp. 746–51.
7.24 Fagerlund, A. C. and Chou, D. C. 1981. 'Sound transmission through a cylindrical pipe wall', *Journal of Engineering for Industry (ASME Transactions)* **103**, 355–60.
7.25 Holmer, C. I. and Heymann, F. J. 1980. 'Transmission of sound through pipe walls in the presence of flow', *Journal of Sound and Vibration* **70**(2), 275–301.
7.26 Pinder, N. J. 1984. *The study of noise from steel pipelines*, CONCAWE Report No 84/55 (The Oil Companies' European Organisation for Environmental and Health Protection).
7.27 Kuhn, G. F. and Morfey, C. L. 1976. 'Transmission of low frequency internal sound through pipe walls', *Journal of Sound and Vibration* **47**(1), 147–61.
7.28 Allen, E. E. 1976. 'Fluid piping system noise', chapter 11 in *Handbook of industrial noise control*, edited by L. L. Faulkner, Industrial Press.
7.29 Reethof, G. and Ward, W. C. 1986. 'A theoretically based valve noise prediction method for compressible fluids', *Journal of Vibration, Acoustics, Stress, and Reliability in Design* **108**, 329–38.
7.30 Ng, K. W. 1980. *Aerodynamic noise generation in control valves*, paper presented at ASME Winter Annual Meeting of Noise Control and Acoustics National Group (Chicago).
7.31 Jenvey, P. L. 1975. 'Gas pressure reducing valve noise', *Journal of Sound and Vibration* **41**(1), 506–9.
7.32 Howard, I. M., Norton, M. P. and Stone, B. J. 1987. 'A coincidence damper for reducing pipe wall vibrations in piping systems with disturbed internal turbulent flow', *Journal of Sound and Vibration* **113**(2), 377–93.

Nomenclature

a	cross-sectional dimension of a rectangular duct
a_i	internal pipe radius
a_m	mean pipe radius
A_j	cross-sectional area of a jet
A_{pq}	constant associated with diametral and cylindrical nodal surfaces of the (p, q)th higher order acoustic mode inside a cylindrical shell
B_{pq}	constant associated with diametral and cylindrical nodal surfaces of the (p, q)th higher order acoustic mode inside a cylindrical shell
c	speed of sound
c_e	speed of sound in the fluid outside a pipe
c_i	speed of sound in the fluid inside a pipe
c_L	quasi-longitudinal wave velocity in a pipe wall material
c_s	bending wave velocity in a cylindrical shell
c_v	viscous-damping coefficient
D	narrowest cross-sectional dimension of a valve opening, cavity depth, mean pipe diameter
E	Young's modulus of elasticity
E_f	energy in a contained fluid
E_p	energy in a pipe wall
f	frequency
f_c	complete coincidence frequency
f_{co}, $(f_{co})_{pq}$	cut-off frequency of the (p, q)th higher order acoustic mode
f_L	limiting frequency associated with the number of coincidences
f_{mn}	natural frequency of the (m, n)th pipe structural mode
f_p	peak frequency in the far-field, frequency associated with a spectral peak, vortex shedding frequency
f_r	ring frequency of a cylindrical shell

$\mathbf{F}$	complex force
$\mathbf{F}_\alpha$	complex modal force input
G_{aa}	one-sided auto-spectral density function of pipe wall acceleration levels
G_{pp}, G_{pq}	one-sided auto-spectral density function of internal pipe wall pressure fluctuations
$\mathbf{G_{p1p2}}(\vec{\varepsilon}, \omega)$, $\mathbf{G_{pq}}(\vec{\varepsilon}, \omega)$	one-sided cross-spectral density function of internal wall pressure fluctuations (complex function)
G_{rr}	one-sided auto-spectral density function of pipe wall displacement levels
$G_{\pi\pi}$	one-sided auto-spectral density function of external sound power radiation from a pipe
$G(M)$	velocity correlation factor
$\langle \overline{G_{pp}(\omega)} \rangle$	one-sided auto-spectral density function of internal pipe wall pressure fluctuations (space- and time-averaged)
$\langle \overline{\mathrm{G}_{rr}(\omega)} \rangle$	one-sided auto-spectral density function of pipe wall displacement levels (space- and time-averaged)
$\langle \overline{\mathrm{G}_{\pi\pi}(\omega)} \rangle$	one-sided auto-spectral density function of external sound power radiation from a pipe (space- and time-averaged)
h	pipe wall thickness
$H_n^{(1)}$	first-order Hankel function of the first kind
$H_n^{(1)'}$	derivative of a first-order Hankel function of the first kind
$\mathbf{H}_\alpha(\omega)$	modal frequency response function of the αth pipe structural mode (complex function)
$j_{mm}^2(\omega)$	axial joint acceptance function for the αth resonance pipe structural mode and the applied pressure field
$j_{nn}^2(\omega)$	circumferential joint acceptance function for the αth resonant pipe structural mode and the applied pressure field
$j_{\alpha\alpha}^2(\omega)$, $\mathrm{j}_{mnmn}^2(\omega)$	joint acceptance function for the αth resonant pipe structural mode and the applied pressure field
J_p	Bessel function of the first kind of order p

J'_p	first derivative of the Bessel function of the first kind of order p
k	acoustic wavenumber inside a pipe
k_e	acoustic wavenumber outside a pipe (ω/c_e)
k_m	axial structural wavenumber
k_n	circumferential structural wavenumber
k_s	spring stiffness
k_v	constant associated with a vortex convection velocity
k_x	axial acoustic wavenumber
K	non-dimensional structural wavenumber
K_e	$k_e a_e$
K_m	non-dimensional axial structural wavenumber
K_n	non-dimensional circumferential structural wavenumber
L	length
L_{valve}	sound pressure level at some distance r from a valve
m	mass, number of half structural waves in the axial direction
M	Mach number
M_0	mean Mach number
M_f	contained fluid mass for a unit length of pipe
M_{LP}	c_L/c_i
M_p	pipe wall mass for a unit length of pipe
M_α	generalised mass of the αth pipe structural mode
n	number of full structural waves in the circumferential direction
n_f	modal density of the contained fluid in a pipe
n_p	modal density of a cylindrical shell
N	number of coupling points on a ring damper
N_c	number of coincidences
p	number of diametral nodal surfaces on a cylindrical shell, mode order of a rectangular duct
p_e	acoustic pressure fluctuations external to a pipe

p_i	acoustic pressure fluctuations inside a pipe
$\mathbf{p}(r, \theta, x)$	pressure associated with acoustic propagation in a stationary fluid inside a cylindrical shell
$\langle p^2 \rangle$	mean-square wall pressure fluctuations
P_{1t}	total pressure upstream of a valve
P_2	static pressure downstream of a valve
q	number of cylindrical nodal surfaces concentric with the axis of a cylindrical shell, mode order of a rectangular duct, dynamic head
q_0	dynamic head
Q	quality factor
Q_α	quality factor of the αth pipe structural mode
r	radial distance
$\vec{r}, \vec{r}'$	positions on the pipe surface (vector quantity)
S	surface area of a pipe test section, Strouhal number associated with frequency
t	time, thickness
TL, TL_{pipe}	transmission loss
U	uniform flow velocity inside a pipe, characteristic velocity
U_0	mean flow velocity inside a pipe
U_c	constricted flow speed
U_e	jet exit velocity
x	axial distance
$\mathbf{X}$	complex displacement
$\mathbf{X}_\alpha$	complex modal displacement
α	pipe structural mode ($\alpha = m, n$)
α_{pq}	constant associated with eigenvalues satisfying the rigid pipe wall boundary conditions for the (p, q)th higher order acoustic mode inside a cylindrical shell
β	non-dimensional pipe wall thickness parameter
γ	specific heat ratio
Δ	L/a_m
Δk_x	change in axial structural wavenumber

ΔP_t	total pressure ratio across a valve
$\Delta\omega$	radian (circular) frequency band
$\vec{\varepsilon}$	$\vec{r}' - \vec{r}$ (vector quantity)
η	loss factor
η_p	loss factor for a pipe
η_{pf}	coupling loss factor from a pipe wall to the fluid
θ	angle
κ_{pq}	acoustic wavenumber associated with the (p, q)th higher order acoustic mode inside a cylindrical shell
ν	Poisson's ratio
ξ	axial component of $\vec{\varepsilon}$
π	3.14 . . .
Π_e	radiated sound power outside a pipe
Π_i	radiated sound power inside a pipe
ρ	pipe material density
ρ_e	density of the fluid outside a pipe
ρ_i	density of the fluid inside a pipe
ρ_{is}	ρ_i/ρ
σ_e	radiation ratio in the fluid external to a pipe
σ_i	radiation ratio in the fluid inside a pipe
σ_α	radiation ratio of the αth pipe structural mode
τ	sound transmission coefficient (wave transmission coefficient)
υ	non-dimensional frequency (ω/ω_r)
υ_c	non-dimensional complete coincidence frequency
υ_{co}, $(\upsilon_{co})_{pq}$	non-dimensional cut-off frequency of the (p, q)th higher order acoustic mode
υ_L	limiting non-dimensional frequency associated with the number of coincidences
υ_{mn}	non-dimensional natural frequency of (m, n)th pipe structural mode
υ_r	non-dimensional ring frequency of a cylindrical shell
ϕ	nozzle diameter
ϕ_r	diameter of a grille rod element

$\phi_\alpha(\vec{r})$	mode shape of the αth orthogonal normal mode
Φ_a	non-dimensional pipe wall acceleration auto-spectral density function
Φ_p	non-dimensional auto-spectral density function of internal pipe wall pressure fluctuations
Φ_π	non-dimensional sound power radiation auto-spectral density function
χ	parameter related to a fully expanded jet Mach number (see equation 7.59)
ψ	circumferential component of $\vec{\varepsilon}$
ω	radian (circular) frequency
ω_{co}, $(\omega_{\text{co}})_{pq}$	natural radian (circular) cut-off frequency of the (p, q)th higher order acoustic mode
ω_{L}	limiting radian (circular) frequency associated with the number of coincidences
ω_{mn}, ω_α	natural radian (circular) frequency of the (m, n)th or αth pipe structural mode
ω_n	natural radian (circular) frequency
ω_{r}	radian (circular) ring frequency of a cylindrical shell
Ω	Strouhal number associated with radian frequency
$\langle\ \rangle$	time-average of a signal
$\overline{}$	space-average of a signal (overbar)

8
Noise and vibration as a diagnostic tool

8.1 Introduction

It is becoming increasingly apparent to engineers that condition monitoring of machinery reduces operational and maintenance costs, and provides a significant improvement in plant economy. Condition monitoring involves the continuous or periodic assessment of the condition of a plant or a machine component whilst it is running, or a structural component whilst it is in service. It allows for fault detection and prediction of any anticipated failure, and it has significant benefits including (i) decreased maintenance costs, (ii) increased availability of machinery, (iii) reduced spare part stock holdings, and (iv) improved safety.

With condition monitoring, the maintenance interval is determined by the condition of a machine. This is quite different to a scheduled maintenance programme where a machine is serviced at a specific period in time irrespective of its condition, and a breakdown maintenance programme where a machine is run until it fails. Noise and vibration analysis is but one of several condition monitoring techniques. Other techniques include temperature monitoring, current and voltage monitoring, metallurgical failure analysis, and wear debris analysis. Current spectral analysis is sometimes used for condition monitoring of electrical drives such as generators and large induction motors. Wear debris analysis (e.g. ferrography, atomic absorption, atomic emission, etc.) is often used to supplement noise and vibration as a diagnostic tool.

This chapter is concerned with the usage of noise and vibration as a diagnostic tool. Firstly, available signal analysis techniques (most of which were introduced in chapter 5) are reviewed. Secondly, procedures for source identification and fault detection from a variety of different noise and vibration signals are developed. Thirdly, and finally, some specific test cases are discussed.

Noise signals measured at regions in proximity to, and vibration signals measured on, the external surfaces of machines can contain vital information about the internal processes, and can provide valuable information about a machine's running condition. When machines are in a good condition, their noise and vibration frequency spectra have characteristic shapes. As faults begin to develop, the frequency spectra change. This is the fundamental basis for using noise and vibration measurements and analysis in condition monitoring. Of course, sometimes the signal which is to be monitored is submerged within some other signal and it cannot be detected by a straightforward time history or spectral analysis. When this is the case, specialised signal processing techniques have to be utilised.

8.2 Some general comments on noise and vibration as a diagnostic tool

Single overall (broadband) noise and vibration measurements are useful for evaluating a machine's general condition. However, a detailed frequency analysis is generally necessary for diagnostic purposes. Various frequency components in the noise or vibration frequency spectrum can often be related to certain rotational or reciprocating motions such as shaft rotational speeds, gear tooth meshing frequencies, bearing rotational frequencies, piston reciprocating motions, etc. These signals change in amplitude and/or frequency as a result of wear, eccentricity, unbalanced masses, etc. and can be readily monitored.

As a general rule, machines do not break down without some form of warning – pending machine troubles are characterised by an increase in noise and/or vibration levels, and this is used as an indicator. The well known 'bathtub' curve, illustrated in Figure 8.1, is a plot of noise or vibration versus time for a machine. The level decreases during the running-in period, increases very slowly during the normal operational duration as normal wear occurs, and finally increases very rapidly as it approaches a possible breakdown. If normal preventative maintenance repairs were performed, repairs would be carried out at specified fixed intervals irrespective of whether or not they were required. By delaying the repair until the monitored noise or vibration levels indicate a significant increase, unnecessary maintenance and

Fig. 8.1. Typical trends of overall noise or vibration levels of a machine during normal operation.

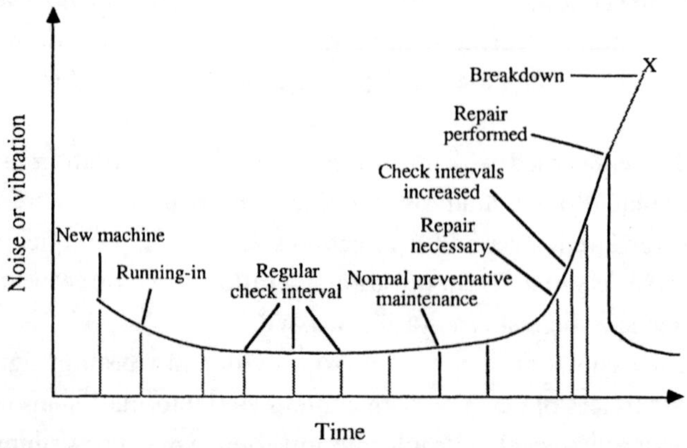

'strip-down' can be avoided. Not only does this minimise delays in production, but unnecessary errors during 'strip-down' which could produce further faults are avoided.

This technique of continuous 'on-condition' monitoring has three primary advantages. Firstly, it avoids catastrophic failures by shutting down a machine when noise or vibration levels reach a pre-determined level; secondly, there are significant economic advantages as a result of increasing the running time between shut-downs; and, thirdly, since the frequency spectrum of a machine in the normal running condition can be used as a reference signal for the machine, subsequent signals when compared with this signal allow for an identification of the source of the fault.

The third advantage mentioned above is very important and needs some amplification. It was discussed briefly in chapter 5 (Figure 5.3) and it is worth re-emphasising it again here. As previously mentioned, although overall noise and/or vibration measurements provide a good starting point for fault detection, frequency analysis provides much more information. In addition to diagnosing the fault, it gives an earlier indication of the development of the fault than an overall vibration measurement does. This very important point is illustrated in Figure 8.2 where early fault detection

Fig. 8.2. Schematic illustration of early fault detection via spectrum analysis.

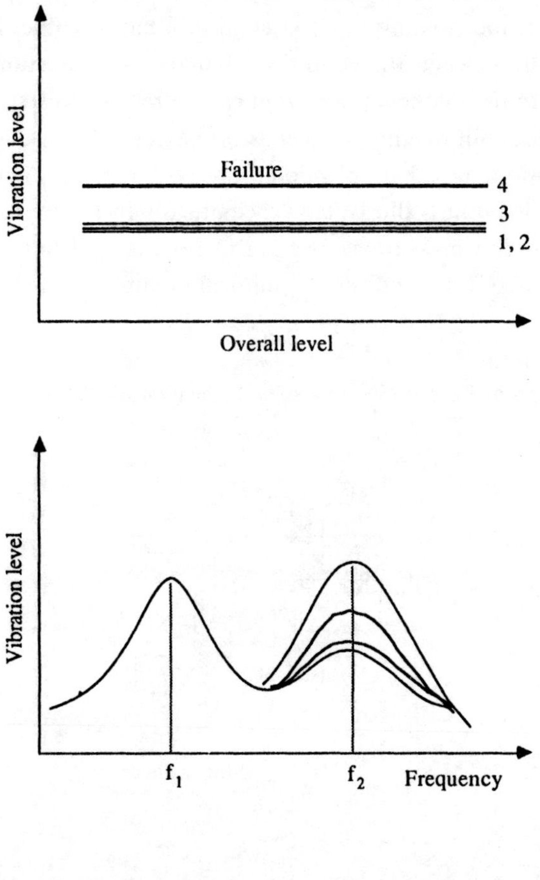

via the spectrum analysis results in an early warning. The gradual increase in the noise and/or vibration level at frequency f_2 would not have been detected in the overall noise and/or vibration level until it was actually greater than the signal at frequency f_1.

The choice of a suitable location for the measurement transducer is also very important. This is especially so for vibration measurement transducers (accelerometers). As an illustrative example, consider the bearing housing in Figure 8.3. The acceleration measurements are used to monitor the running condition of the shaft and bearing since wear usually occurs at the connection between rotating parts and the stationary support frame, i.e. at the bearings. The accelerometers must be placed such as to obtain as direct a path for vibration as is possible. If this is not the case then the measured signal will be 'contaminated' by the frequency response characteristics of the path and will not be a true representation of the source signal. Accelerometers A and C are positioned in a more direct path than B or D. Accelerometer C feels the vibration from the bearing more than vibration from other parts of the machine. Accelerometer D would receive a confusion of signals from the bearing and other machine parts. Likewise, accelerometer A is positioned in a more direct path for axial vibrations than is accelerometer B.

In summary, noise and vibration signals are utilised for condition monitoring because: (i) a machine running in good condition has a stable noise and vibration frequency spectrum – when the condition changes, the spectrum changes; and (ii) each component in the frequency spectrum can be related to a specific source within the machine. Thus, fault diagnosis depends on having a knowledge of the particular machine in question, i.e. shaft rotational speeds, number of gear teeth, bearing geometry, etc. This point is illustrated schematically in Figure 8.4.

Noise and vibration measurements have to provide a definite cost saving before they should be used for condition monitoring/maintenance/diagnostic purposes.

Fig. 8.3. Illustration of the selection of a suitable location for the measurement transducer.

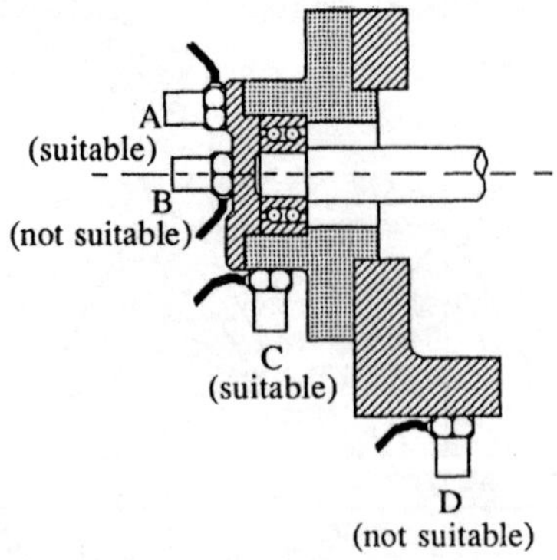

Several general questions arise.

(1) Do noise and vibration measurements suit the particular maintenance system and the machines being used?

(2) What instrumentation is required to provide the most economical system?

(3) Are specialised personnel essential or can personnel already available perform the task?

(4) Can the usage of noise and vibration measurements reduce operation or maintenance costs to give an improvement in plant economy?

Several other general factors have to be considered in any decision to set up a condition monitoring programme.

(1) If a condition monitoring system is to be used to shut down a machine in response to a sudden change, then permanent monitoring is recommended. If, however, the signals are only being used to obtain an early warning of a developing fault, then intermittent monitoring is recommended.

(2) Where damage to a machine itself is of prime economic importance, permanent monitoring is required. If production loss is of prime economic importance, intermittent monitoring is more suitable.

(3) A permanent monitoring system must be substantially more reliable than an intermittent manually operated system. It must have a rugged environ-

Fig. 8.4. Schematic illustration of identification of various frequency components.

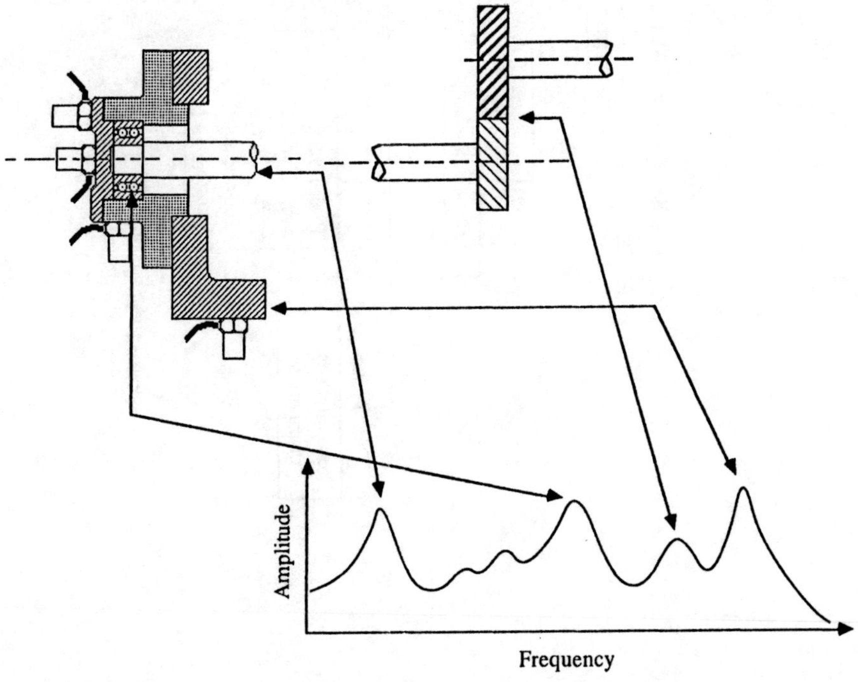

Fig. 8.5. Conventional magnitude and time domain analysis techniques.

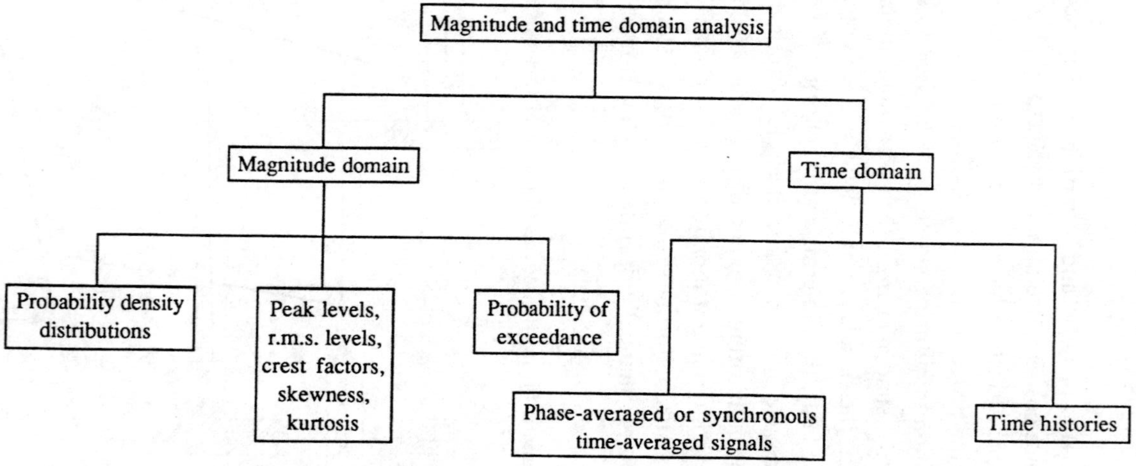

mental casing and must be insensitive to both mechanical and electrical transients.

(4) Regardless of the type of analysis to be done, it is important to choose an appropriate number of measurement points on the machines to be monitored and to develop a readily accessible data base.

8.3 Review of available signal analysis techniques

When using noise and vibration as a diagnostic tool, the type of signal analysis technique required depends very much upon the level of sophistication that is required to diagnose the problem. Numerous analysis techniques are available for the condition monitoring of machinery or structural components with noise and vibration signals.

The commonly used signal analysis techniques (magnitude analysis, time domain analysis and frequency domain analysis) were discussed in chapter 5, and the reader is referred to Figure 5.7 for a quick overview. In this section, available signal analysis techniques are reviewed with particular emphasis being placed upon their usage as a diagnostic tool. In condition monitoring, it is common to group magnitude and time domain analysis procedures together. This is essentially because the magnitude parameters (r.m.s. values, peak values, skewness, etc.) are generally trended over an extended period of time. Thus, magnitude and time domain analysis techniques, as discussed in chapter 5, are grouped together here. In addition to conventional time and frequency domain analysis techniques, advanced techniques like cepstrum analysis, sound intensity techniques for sound source location, envelope spectrum analysis, recovery of source signals, and propagation path identification are reviewed in this section.

8.3.1 Conventional magnitude and time domain analysis techniques

Numerous magnitude and time domain techniques are available for noise and vibration diagnostics. They are summarised in Figure 8.5.

The analysis of individual time histories of noise or vibration signals is in itself a very useful diagnostic procedure. Quite often, a significant amount of information can be extracted from a simple time history which can be obtained by playing back a tape recorded signal onto a storage oscilloscope, an *x-y* plotter, or a digital signal analyser. For a start, the nature of the signal can be clearly identified – i.e. is it transient (impulsive), random or periodic. Furthermore, peak levels of noise or vibration can be detected. Also, as noise and/or vibration levels start to increase due to a deterioration of the condition of the equipment being monitored, the time history changes.

A classical example of the usage of time histories for diagnostic purposes is the acceleration time history of a bearing supporting a rotating shaft. When a bearing is in good condition, the vibration signal from the bearing housing is broadband and random, and significantly lower than when it is not in a good condition. When a discrete defect is introduced, the bearing is subjected to an impulse, and this is reflected in the time history. This point is illustrated in Figures 8.6(a) and (b) which are the acceleration time histories of the vibrations on a bearing housing of a large skip drum winder for a mine cage. The drum winder has a diameter of ~6 m, rotates at ~0.5 Hz (i.e. ~10 m s^{-1}), and is mounted on a shaft with a diameter of ~0.9 m. Figure 8.6(a) represents the acceleration time history for normal operating conditions. The signal is continuous, random and broadband. The periodicity of the impulsive time history observed in Figure 8.6(b) corresponds to a once per revolution excitation. The acceleration signal level has also increased in comparison to Figure 8.6(a).

Time histories can also be used (i) to analyse start-up transients in electrical motors, (ii) to identify the severity of electrical vibrations (by observing the change in time history after the electrical power is switched off), and (iii) to distinguish between unbalance and discrete once per revolution excitations. Sometimes, when the dominant excitation source is a discrete frequency, it is necessary to phase-average

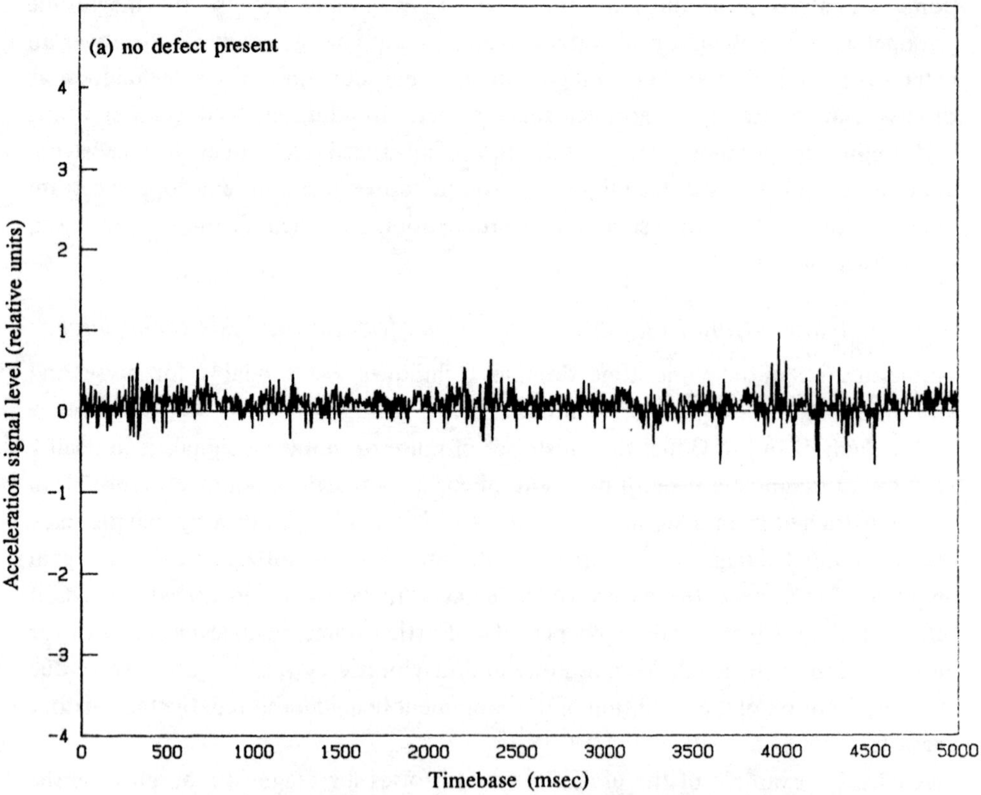

Fig. 8.6. Acceleration time histories of the vibrations on a bearing housing of a large skip drum winder for a mine cage: (a) no defect present; (b) discrete defect present.

a noise and/or vibration time history. This is achieved by synchronising the signal to be measured with the excitation signal, i.e. the measurement is triggered at a specific point (usually a zero crossing) in the excitation cycle. This allows for the removal of unwanted random and periodic signal components. Signals which are synchronous with the trigger will average to their mean value whilst noise or non-synchronous signals will average to zero. Phase-averaging in the time domain is sometimes referred to as synchronous time-averaging. Figure 8.7(a) is a time history of a signal from a bearing on an electric motor. The corresponding linear spectrum (Fourier transform of the time history) is presented in Figure 8.7(b). Information about the time history of the dominant frequency at 1400 Hz (which incidentally is associated with an electrical fault in the motor) can be obtained by phase-averaging the time history of the bearing signal whilst synchronised to the 1400 Hz frequency. The phase-averaged time history is presented in Figure 8.8(a), and the corresponding linear spectrum is presented in Figure 8.8(b). The non-synchronous signal components have been removed.

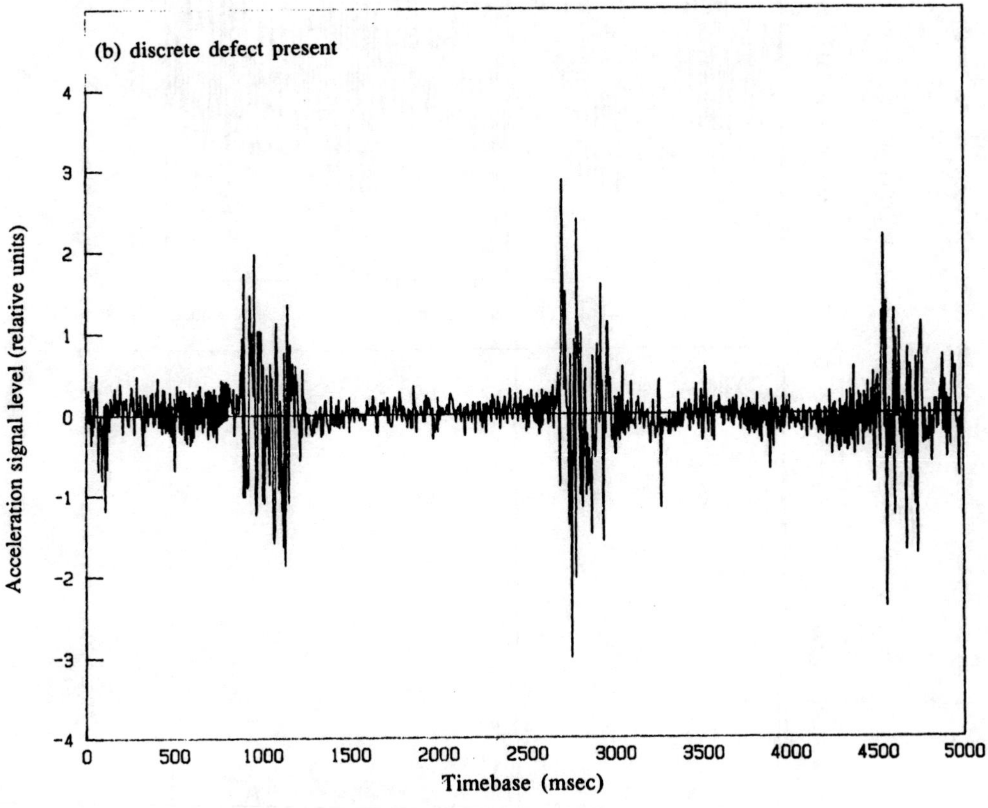

Several magnitude parameters can be extracted from the time history of a noise or vibration signal. They are (i) the peak level, (ii) the r.m.s. level, and (iii) the crest factor. The crest factor is the ratio of the peak level to the r.m.s. level, and it is given by

$$\text{crest factor} = \frac{\text{peak level}}{\text{root-mean-square level}}. \tag{8.1}$$

The crest factor is a measure of the impulsiveness of a noise or vibration signal. It

Fig. 8.7. Acceleration time history and corresponding linear spectrum of a bearing signal from an electric motor.

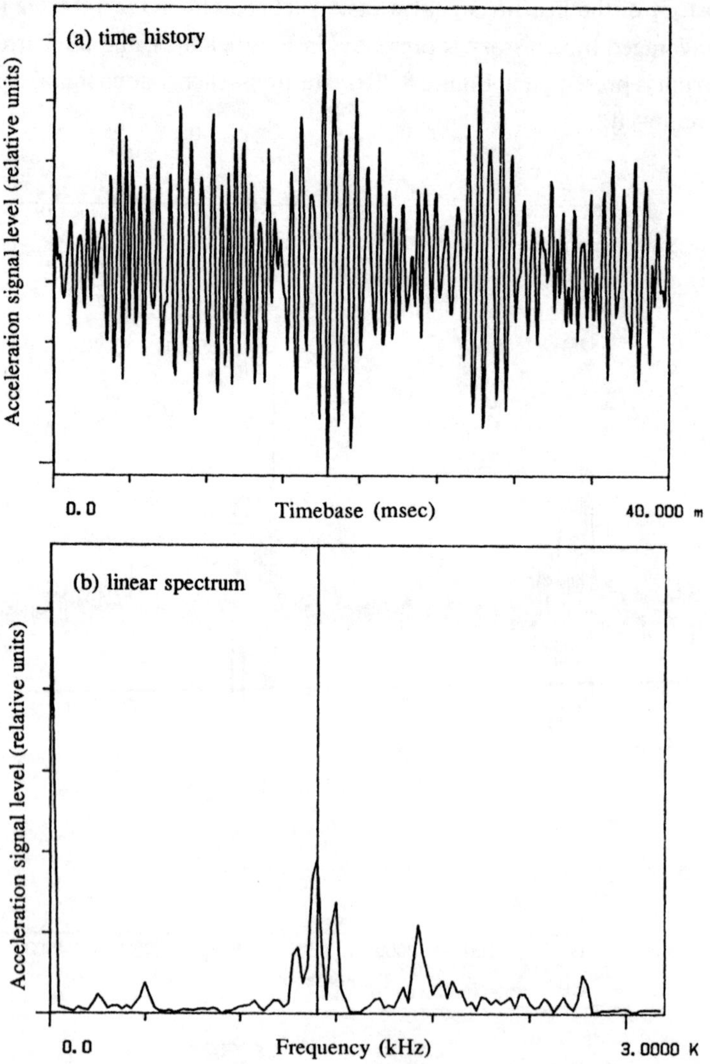

is often used when dealing with shocks, impulsive noise and short events. The crest factor for a sine wave is 1.414, and the crest factor for a truly random noise signal is generally less than 3. The crest factor is commonly used to detect impulsive vibrations produced by damaged bearings. As a rule of thumb, good bearings have vibration crest factors of ~2.5–3.5, and damaged bearings have crest factors >3.5. Values as high as ~7 can sometimes be recorded prior to failure. The pros and cons of the crest factor diagnostic technique for bearings and gears are discussed in section 8.4.

Fig. 8.8. Phase-averaged acceleration time history and corresponding phase-averaged linear spectrum of a bearing signal from a electric motor.

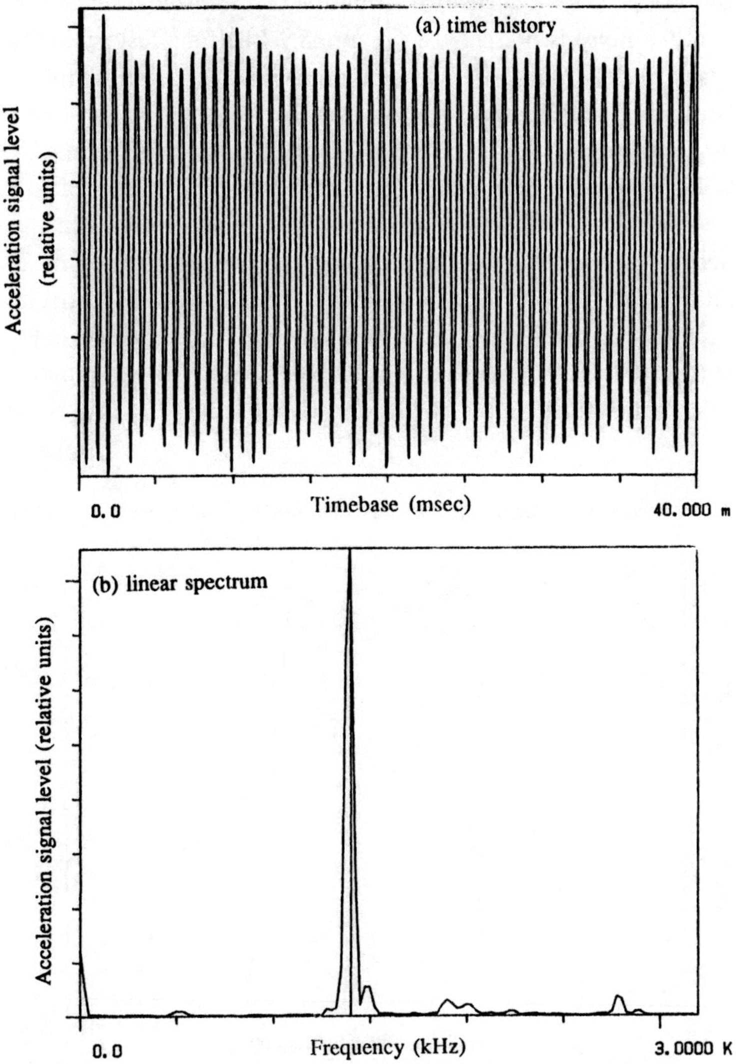

In addition to peak levels, r.m.s. levels and crest factors, various other statistical parameters can be extracted from the time histories of noise and vibration signals. These include (i) probability density distributions, (ii) second-, third- and fourth-order statistical moments, and probability of exceedance relationships.

Probability density distributions of noise levels or vibration amplitudes are often trended over time. Probability density distributions were discussed in chapter 5. They are simply continuous histograms of signal amplitudes – i.e. the well known individual cells of a histogram are reduced to zero width, a continuous curve is fitted to the data points, and the abscissa is normalised such that the total area under the curve is unity. Most modern digital signal analysers have the facility of generating histograms and subsequently converting the information into a probability density distribution. The probability density distribution of a sinusoidal signal superimposed on some random noise is presented in Figure 8.9 (the time history in Figure 8.7(a) corresponds to this distribution). A true random signal has a bell shaped probability density distribution (see Figure 5.9), and a true sine wave has a U-shaped probability density distribution (also see Figure 5.9). As a monitored noise and/or vibration signal level increases with time, due to some increasing defect, its probability density distribution changes in both shape and amplitude. These changes in the condition of the machine can be identified by trending the probability density distribution curves with time. This allows for a comparison of the spread (or distribution) of the monitored signal level with time. This point is illustrated schematically in Figure 8.10.

The first two statistical moments of a probability density distribution are the mean value and the mean-square value. The reader should be familiar with the significance

Fig. 8.9. Probability density distribution of a sinusoidal signal superimposed on some random noise.

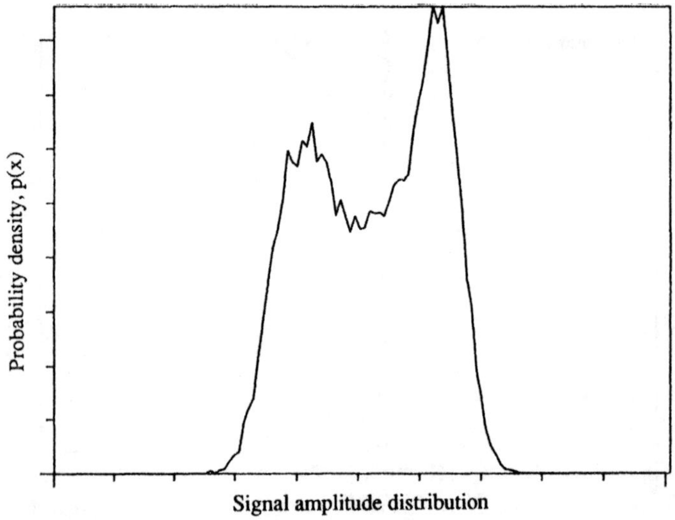

of these two statistical parameters by now. The third statistical moment is the skewness of a distribution, and this parameter was introduced in chapter 5. It is a measure of the symmetry of the probability density function. The fourth statistical moment is widely used in machinery diagnostics, particularly for rolling element bearings. It is called kurtosis, and it is given by

$$\text{kurtosis} = \frac{E[x^4]}{\sigma^4} = \frac{1}{\sigma^4}\int_{-\infty}^{\infty} x^4 p(x)\,\mathrm{d}x = \frac{1}{\sigma^4 T}\int_0^T x^4\,\mathrm{d}t. \tag{8.2}$$

Because the fourth power is involved, the value of kurtosis is weighted towards the values in the tails of the probability density distributions – i.e. it is related to the spread in the distribution. As a general rule, odd statistical moments provide information about the disposition of the peak relative to the median value, and even statistical moments provide information about the shape of the probability distribution curve. The value of kurtosis for a Gaussian distribution is 3. A higher kurtosis value indicates that there is a larger spread of higher signal values than would generally be the case for a Gaussian distribution.

The kurtosis of a signal is very useful for detecting the presence of an impulse within the signal. It is widely used for detecting discrete, impulsive faults in rolling element bearings. Good bearings tend to have a kurtosis value of ~ 3, and bearings with impulsive defects tend to have higher values (generally >4). The usage of kurtosis is limited because, as the damage to a bearing becomes distributed, the impulsive content of the signal decreases with a subsequent decrease in the kurtosis value. This point is illustrated in Figure 8.11 where the kurtosis is trended with time. The usage of kurtosis for diagnosing the condition of bearings is discussed in section 8.4.

Fig. 8.10. Schematic illustration of the trending of probability distribution curves with time.

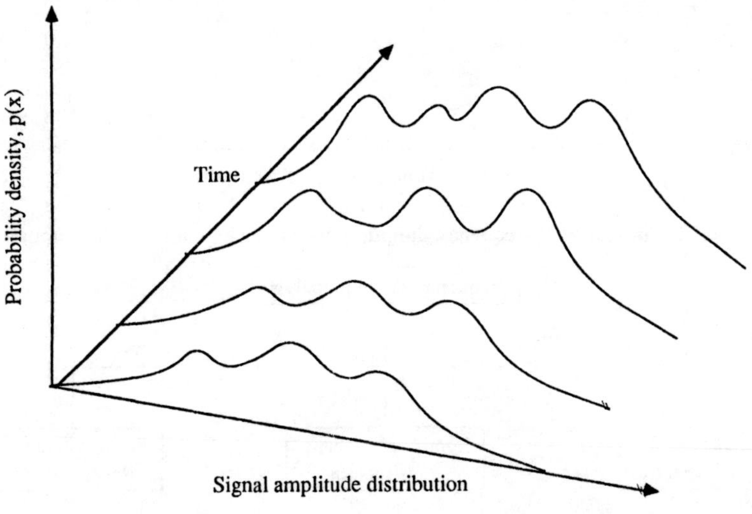

Probability of exceedance relationships, such as Weibull distributions of peaks and Gumble logarithmic relationships, are sometimes used to predict the probability that an instantaneous signal amplitude exceeds a particular value. These relationships were briefly discussed in chapter 5 (section 5.3.1), and appropriate references were provided. They are not particularly suited as field diagnostic tools, but are particularly useful for correlating past results with future outcomes.

8.3.2 Conventional frequency domain analysis techniques

Whilst numerous advances have been made in recent years in the usage of the frequency domain as a noise and vibration diagnostic tool, only the conventional, single channel, frequency domain analysis techniques will be reviewed in this section. These techniques are summarised in Figure 8.12.

The baseband auto-spectral density (0 Hz to upper frequency limit of the instrumentation) is the most common form of frequency domain analysis for noise and vibration diagnostics. In most instances, significant diagnostic information can be obtained from the auto-spectral density, which is generated by Fourier transforming

Fig. 8.11. The trending of kurtosis with time.

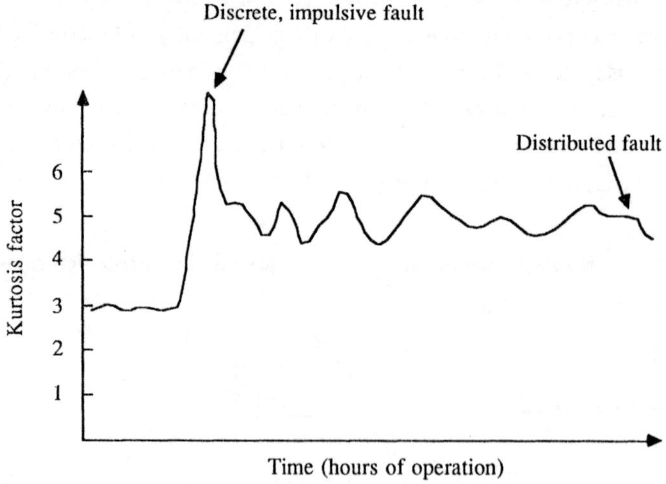

Fig. 8.12. Conventional frequency domain analysis techniques (single channel).

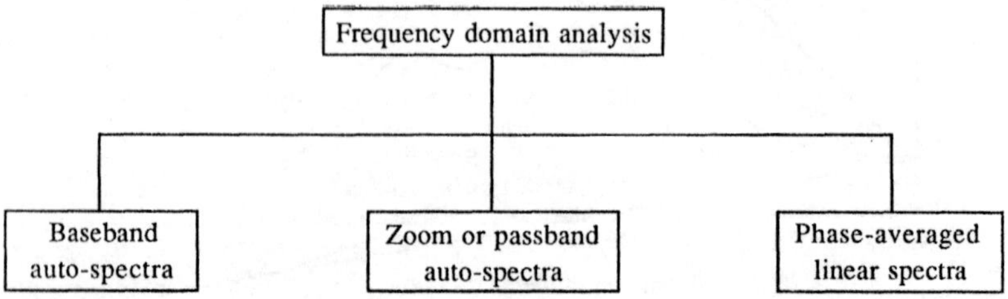

the time history and multiplying it by its complex conjugate (i.e. see equation 5.15). Quite often, the baseband auto-spectral density is trended over an extended period of time and the results presented in a cascade (or waterfall) plot. In this way, variations in different frequency components with time can be observed. A typical waterfall plot is illustrated in Figure 8.13. The identification of various frequency components associated with different items (e.g. shaft rotational speeds, bearings, gears, etc.) is discussed in section 8.4.

The zoom or passband auto-spectral density is often used to provide detailed information within a specified frequency band. Typical baseband and passband auto-spectral densities of a bearing vibration signal from an electric motor with a dominant electrical fault at 1400 Hz are presented in Figures 8.14(a) and (b). The sidebands on both sides of the dominant peak (which are typical of certain electrical faults) are clearly evident from the passband spectrum. It is a fairly routine procedure to programme a spectrum analyser to provide a passband frequency analysis. Commercial units are available for specific passband frequency analysis applications. The shock pulse meter for monitoring ultrasonic frequency components of high speed rolling element bearings is a typical example.

Sometimes, rather than using the auto-spectral density, it is more useful to analyse the linear frequency spectrum. The linear frequency spectrum is the Fourier transform of the time history (the auto-spectrum is the linear spectrum multiplied by its complex conjugate) and it gives both magnitude and absolute phase information at each frequency in the analysis band. Because of this, it requires a trigger condition for averaging and, as with phase-averaging of signals in the time domain, any non-synchronous signal will average to zero. Linear frequency spectrum averaging is

Fig. 8.13. Typical waterfall plot of frequency and amplitude versus time.

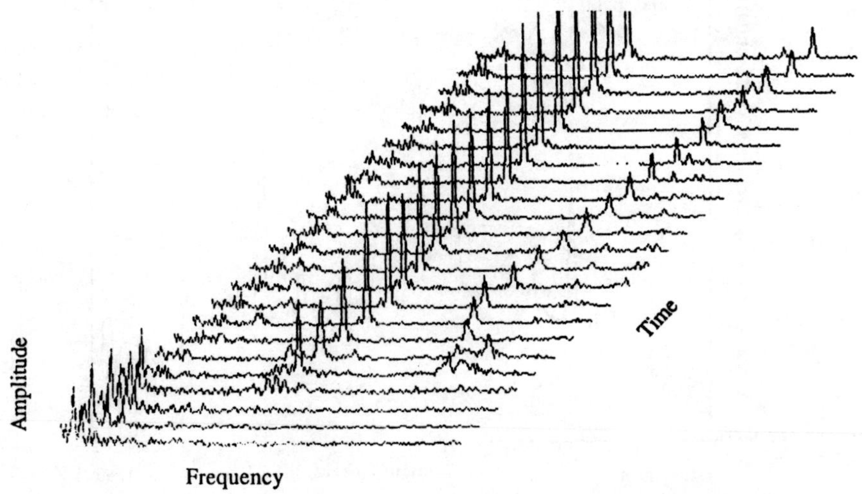

particularly useful when the background noise level is high, and the required frequency components cannot be readily identified from an auto-spectrum – this is often the case with rotating machinery. A typical example of an auto-spectrum and the corresponding linear spectrum (both averaged 400 times) is illustrated in Figure 8.15. The mean value of the non-synchronous linear spectrum components is zero, whereas, with the auto-spectrum, the noise averages to its mean-square value.

Fig. 8.14. Typical baseband and passband acceleration auto-spectral densities of a bearing vibration signal from an electric motor.

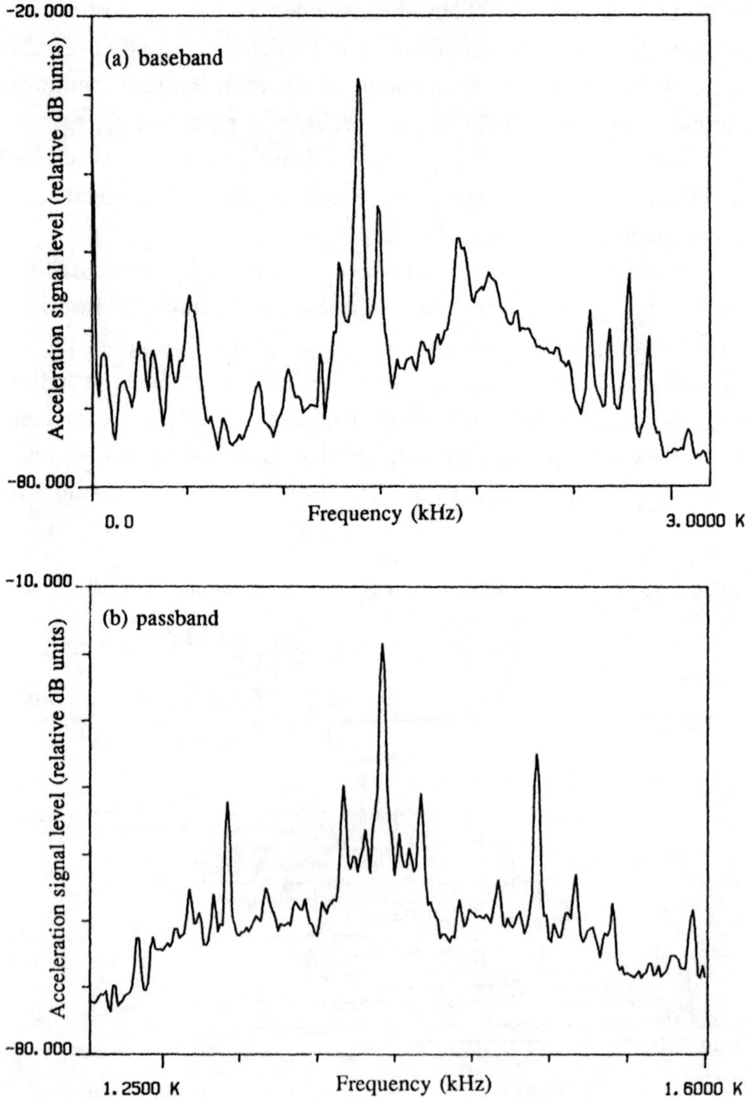

8.3.3. Cepstrum analysis techniques

The concepts of cepstrum analysis were discussed in some detail in chapter 5 (section 5.3.3) and will therefore only be briefly reviewed here. Practical examples relating to rolling element bearings are discussed in section 8.5.

Two types of cepstra exist – the power cepstrum and the complex cepstrum. Both types of cepstra are real-valued functions. The power cepstrum is the inverse Fourier transform of the logarithm of the power spectrum (or the square of the modulus of the forward Fourier transform of the logarithm of the power spectrum) of a time signal, both definitions being consistent with each other. The complex cepstrum is the inverse Fourier transform of the logarithm of the forward Fourier transform of a time signal.

The power cepstrum is used to identify periodicity in the frequency spectrum, just like the frequency spectrum is used to identify periodicity in the time history of a signal. It is also used for echo detection and removal, for speech analysis, and for the measurement of properties of reflecting surfaces.

The complex cepstrum is used when one wishes to edit (deconvolute) a signal in the quefrency domain and subsequently return to the time domain. This procedure is possible because the complex cepstrum contains both magnitude and phase information.

Cepstrum analysis is generally used as a complementary technique to spectral analysis. It is seldom used on its own as a diagnostic tool as it tends to suppress information about the global shape of the spectrum. Furthermore, the derivation of the cepstrum is not a routine signal analysis procedure, and one needs to exercise a certain amount of care.

Fig. 8.15. An example of an auto-spectrum and the corresponding linear spectrum (both averaged 400 times).

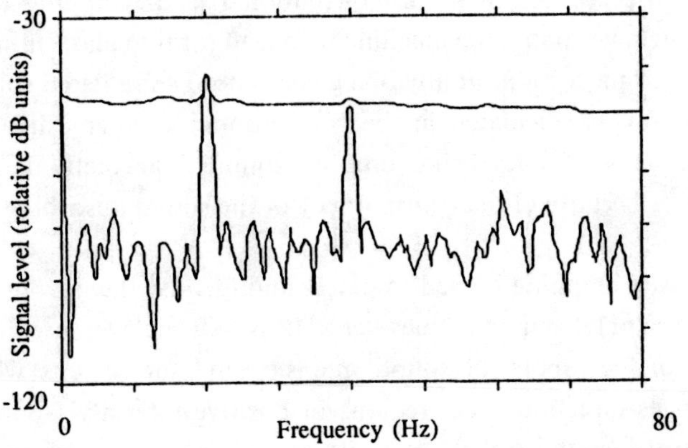

8.3.4 Sound intensity analysis techniques

Sound intensity is the flux of sound energy in a given direction – it is a vector quantity and therefore has both magnitude and direction. Procedures for measuring sound intensity, and techniques for the measurement of sound power, utilising sound intensity measurements, were discussed in sub-section 4.7.4, chapter 4.

In addition to the measurement of sound power, sound intensity measurements can be used for a variety of noise control engineering applications. These include sound field visualisation, sound source location and identification, transmission loss measurements, determination of sound absorption coefficients, and the detection of acoustic enclosure cover leakages. Sound source location, in particular, is a major diagnostic application of sound intensity measurements. It is usually divided into two groups – sound source ranking and sound intensity mapping.

Sound source ranking involves the measurement of sound intensity at numerous regions which are close to the source (e.g. a machine), the objective being to evaluate the sound power radiated from different parts of the machine by subdividing the selected measurement surface area around the complete machine into smaller control surfaces. Both the overall sound power level and the sound power radiated in different frequency bands can thus be evaluated for each of the different control surfaces (using equation 4.87). The sound intensity of each of the smaller control surfaces can be obtained either via a continuous sweep of the microphone pair, or by subdividing the smaller control surface into a series of grid points and measuring the normal component of the intensity vector at each grid point. With this procedure, a rank-ordering of the different sound sources on the complete machine can be obtained. Sound source ranking is often used to compare and identify the sound power radiated by various components of an engine. It has also been used to identify sound sources associated with vacuum cleaners, industrial looms, diamond drilling equipment, etc. The usage of sound intensity measurements greatly simplifies sound source identification procedures. It is significantly less time consuming (by a ratio of $\sim$1:15), and significantly more reliable than the conventional lead wrapping technique. Lead wrapping involves wrapping the machine with lead (or fibreglass, mineral, wool, etc.) and exposing certain sections at any one given time. The radiated sound power from the open 'window' is calculated in the conventional manner using sound pressure level measurements. Besides being time consuming, the method is prone to low frequency errors because of the transparency of the sound absorbing material at low frequencies.

Sound intensity mapping is used to detect and identify the flow of sound intensity (which is a vector) from machines, etc. It is particularly useful for the rapid identification of 'hot spots' of sound intensity and for regions where the vector quantity changes direction – i.e. regions of positive intensity (sound sources) and negative intensity (sound sinks). Intensity mapping can be performed in real time (i.e. with a continuous sweep of the microphone pair) if the appropriate signal processing instrumentation is available. Alternatively, it can be performed by breaking up the area of interest into a grid and measuring the normal component of the intensity vector at each grid point. There are numerous ways of presenting sound intensity mapping data, including vector field plots, contour plots and three-dimensional waterfall type plots. With three-dimensional plots, the 'hills' represent regions of positive intensity, the 'valleys' represent regions of negative intensity, and the other two co-ordinates represent spatial locations. In each of the three conventional ways of presenting sound intensity, the data can be presented as (i) single frequency data, (ii) frequency interval data, and (iii) overall levels. A typical sound intensity vector field plot is illustrated schematically in Figure 8.16.

Figures 8.17(a) and (b) are plots of sound pressure level spectra and sound intensity level spectra for an electric motor drive unit located near a broadband noise source. The sound pressure level spectra is a scalar quantity and provides the overall pressure fluctuations at a given point in space. The sound intensity spectra (obtained from equation 4.90), on the other hand, clearly shows the direction of flow of sound energy; the 100 Hz noise from the motor can be clearly identified. Figures 8.17(a) and (b) are only presented to illustrate positive and negative intensity at a single position. A vector plot, contour plot, etc., would be required for sound intensity mapping.

Sound intensity techniques are also used for a variety of advanced signal analysis procedures including vibration (structure-borne) intensity using two accelerometers, sound intensity in fluids using two hydrophones, and gated intensity for the analysis of synchronous signals using a trigger function. The interested reader is referred to the Proceedings of the 2nd International Congress on Acoustic Intensity[8.1] for a series of papers on the latest advances in sound intensity measurement procedures. Gade[8.2] and Maling[8.3] also provide useful summaries of the applications of sound

Fig. 8.16. Schematic illustration of a sound intensity vector field for two sources and a sink.

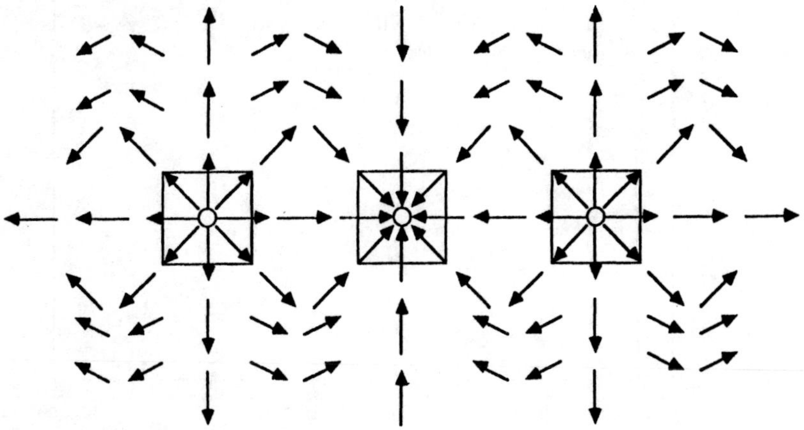

intensity measurements in noise control engineering, together with numerous additional references.

8.3.5 Other advanced signal analysis techniques

There are numerous other advanced noise and vibration signal analysis techniques that are available for diagnostic purposes. Some of these include the analysis of envelope spectra, propagation path identification using causality correlation tech-

Fig. 8.17. An example of the sound pressure level spectra and the corresponding sound intensity spectra for an electric motor drive unit located near a broadband noise source.

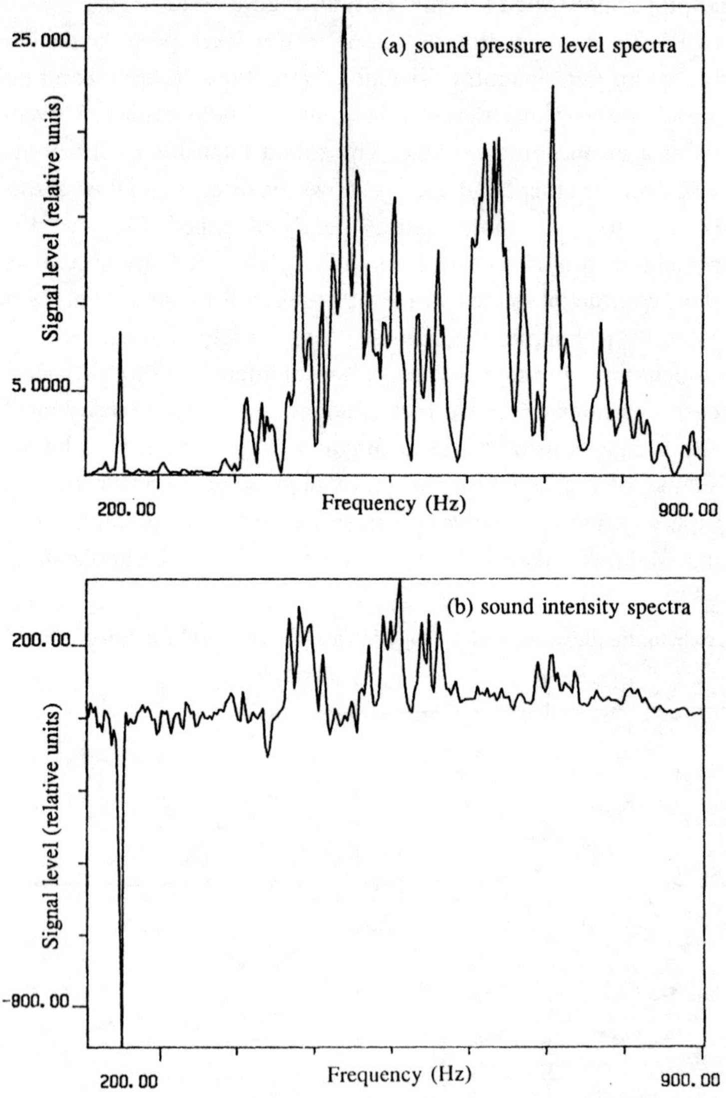

niques, frequency response functions (transfer functions), and the recovery of temporal waveforms of source signals.

The analysis of envelope spectra involves spectral analysing the envelope or amplitude modulation component of a time history. It is particularly useful for providing diagnostic information concerning early damage to rolling element bearings. In the early stages of a developing bearing fault, the impulses produced by the fault are very short in duration and the energy associated with the impulse is usually distributed over a very wide frequency range (often well into the ultrasonic region). Because of this, frequency analysis in the range of the fundamental bearing frequencies will often not reveal any developing bearing faults. The envelope of the time history, however, contains information about (i) the impact rate and (ii) the amplitude modulation. Discrete faults along the inner and outer races of a rolling element bearing generate impulses at a rate which corresponds to the contact with the rolling elements. Discrete faults along the inner race rotate in and out of the loaded zone, generating the amplitude modulation. These faults can be detected by spectrum analysing the envelope spectrum – peaks which are concealed in the spectrum of the original time history (due to the high frequency content of the impulses) can now be identified. The envelope signal is usually generated by the following procedure. Firstly, the bearing vibration time history is octave bandpass filtered around a bearing resonance to reduce components which are unrelated to the bearing. Secondly, the signal is enveloped (amplitude demodulated) by full-wave rectification and low-pass filtration or via Hilbert transformation[8.4]. Finally, the envelope signal is spectrum analysed. The process is illustrated schematically in Figure 8.18. The envelope spectrum is only useful for detecting early bearing damage. As the damage spreads and becomes randomly distributed along the bearing races, the peaks in the envelope

Fig. 8.18. Procedures required to generate the envelope power spectrum.

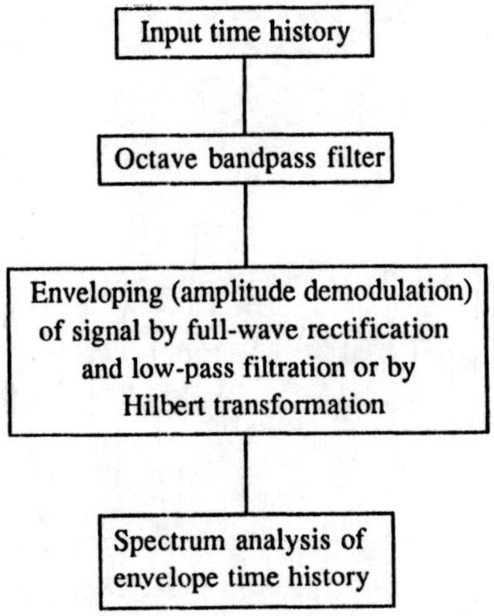

spectrum smear out and the spectra becomes broadband. At this stage, a simple r.m.s. vibration level would indicate that the vibration levels are excessive. Source identification and fault detection in bearings are discussed in section 8.4.

Propagation path identification using causality correlation techniques involves placing microphones at each of the various source locations and at the receiver location, and evaluating the cross-correlation coefficients between the various signals. The effects of discrete reflections can be readily accounted for by careful examination of the time delays associated with the respective cross-correlation peaks. Sometimes, it is more appropriate to evaluate the impulse response function instead, the impulse response being the time domain representation of the frequency response function between the two signals. As a general rule, when the input signal is dispersive, the impulse response is a more sensitive indicator of time delays between signals. Sometimes it is more appropriate to evaluate the time delay from the phase angle associated with the cross-spectrum between the signals, rather than from the cross-correlation between the signals. The cross-spectral method minimises output noise problems. The basic concepts of propagation path identification were discussed in chapter 5, and the reader is referred to Figures 5.12 and 5.13 in particular. Bendat and Piersol[8.5] provide detailed information on the three different procedures (cross-correlations, impulse responses and cross-spectra) for source location. The main problem that is encountered in practice for all three procedures is that most practical noise sources have finite dimensions.

A typical example of the usage of a cross-correlation for propagation path identification is illustrated in Figure 8.19. A simple experiment was conducted in a

Fig. 8.19. Typical example of cross-correlation for propagation path identification.

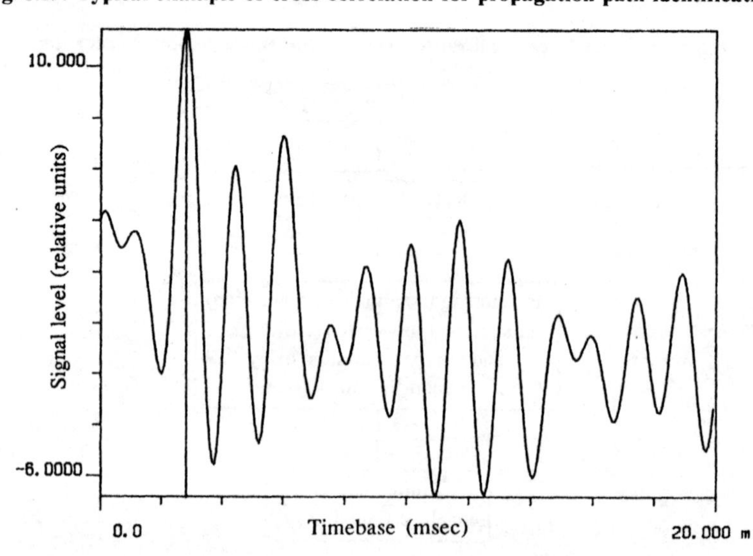

laboratory whereby two microphones were located ~1 m apart. A sound source was located near one of the microphones. With the distance between the two microphones known, the speed of sound is readily evaluated from the appropriate time delay. With reference to Figure 8.19, the major peak at 2.8133 ms corresponds to the time it takes for the sound waves to travel from the first to the second microphone. The speed of sound is thus evaluated to be ~355 m s^{-1}. Whilst this is not a terribly reliable estimate of the speed of sound, it demonstrates the relative ease with which it can be evaluated. The secondary peaks in the cross-correlation can be related to reflections from various walls, objects, etc., in the laboratory.

Sometimes, the propagating waves are dispersive. Typical examples of dispersive waves are bending waves in structures and higher order acoustic modes in a duct. Because the waves are dispersive, their wave velocities vary with frequency (e.g. see equation 3.11 in chapter 3 and equation 7.3 in chapter 7). When propagating waves are dispersive, the apparent propagation speed of the waves at a given frequency is the group velocity, c_g. It represents the speed of propagation of a wave packet. It is useful to remember that the group velocity of bending waves is twice the bending wave velocity (or phase velocity), c_B. This point was mentioned in chapter 6 (sub-section 6.6.1). A typical example of a cross-correlation between two accelerometers on a long beam is shown in Figure 8.20. The dispersive nature of the bending waves is clearly evident. The reader is once again referred to Bendat and Piersol[8.5] for further discussions on dispersive propagation path identification.

Frequency response functions (transfer functions) are also widely used in noise and vibration signal analysis. A wide range of frequency response functions are

Fig. 8.20. Typical example of dispersive wave propagation in a beam.

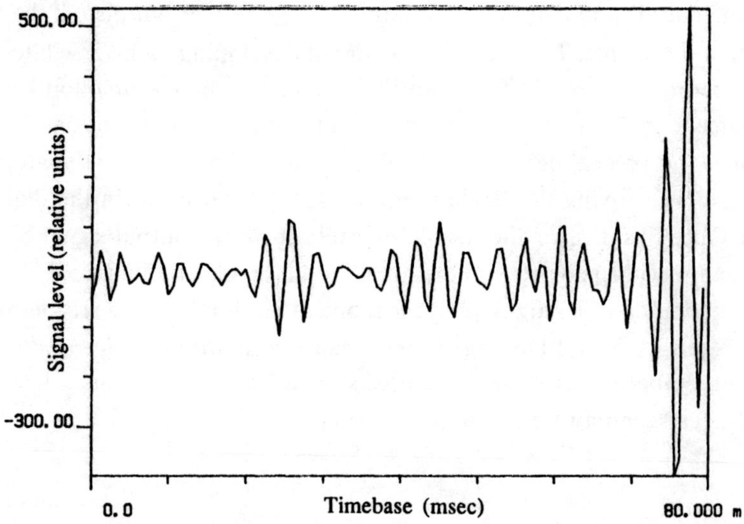

available including receptances, mobilities, impedances, etc. Most of these have been covered in significant detail in all the chapters in this book. The three important parameters with any frequency response function are magnitude, phase and coherence. Frequency response functions are very useful for the rapid identification of natural frequencies of structures. The natural frequencies can be identified either from the magnitude or the phase of the frequency response. It is important, however, to note that good coherence is essential for the identification of natural frequencies. Some typical experimental results of acceleration output/force input for a small sheet metal box type structure are presented in Figures 8.21(a) and (b). The results were obtained with a calibrated impact hammer; the force transducer was mounted on the hammer head, and the accelerometer was mounted on the structure.

The recovery of temporal waveforms of source signals from measured vibration signals is another advanced signal analysis technique that can be used for machinery diagnostics. The main impetus for this work has come from Lyon and DeJong[8.6,8.7]. It has particular application in internal combustion engines, particularly for combustion pressure recovery, piston slap, and valve/valve seat impact. In each of these instances, it is very difficult to continuously monitor the source signal, and it is significantly easier to monitor an acceleration signal on the casing/housing. The acceleration signal is, however, contaminated by the frequency response characteristics of the path, and needs to be 'manipulated' in order to reconstruct the temporal waveform of the source signal.

The process of 'manipulating' the output signal to reconstruct the source signal involves a process called inverse filtering. It involves generating an inverse filter which has the negative of the magnitude and the negative of the phase of the measured frequency response function. When this filter is placed in sequence with the measured frequency response function, the overall system frequency response function has uniform magnitude and constant phase. In this way, it does not distort the temporal waveform of the source. The actual procedure of developing an inverse filter requires that an experiment be set up to obtain the frequency response function between the input source signal and the vibration output with special attention be given to maintaining the precise details of the phase and its unwrapping (unwrapping the phase involves removing the random jumps of $\pm 2\pi$ that occur in the digital signal analysis). Once this inverse filter has been established, the output signal can be used in conjunction with it for continuous or periodic diagnostic purposes. The precise details of these techniques are beyond the scope of this book and the reader is referred to Lyon[8.6] and Lyon and DeJong[8.7] for further information. The procedures for the recovery of temporal waveforms of source signals have been outlined here in order to give the reader an awareness of their availability.

8.4 Source identification and fault detection from noise and vibration signals

Source identification and fault detection from noise and vibration signals associated with items which involve rotational motion such as gears, rotors and shafts, rolling element bearings, journal bearings, flexible couplings, and electrical machines depends upon several factors. Some of these factors are (i) the rotational speed of the item,

Fig. 8.21. Magnitude, phase and coherence for the frequency response function (acceleration/force) of a simple test structure.

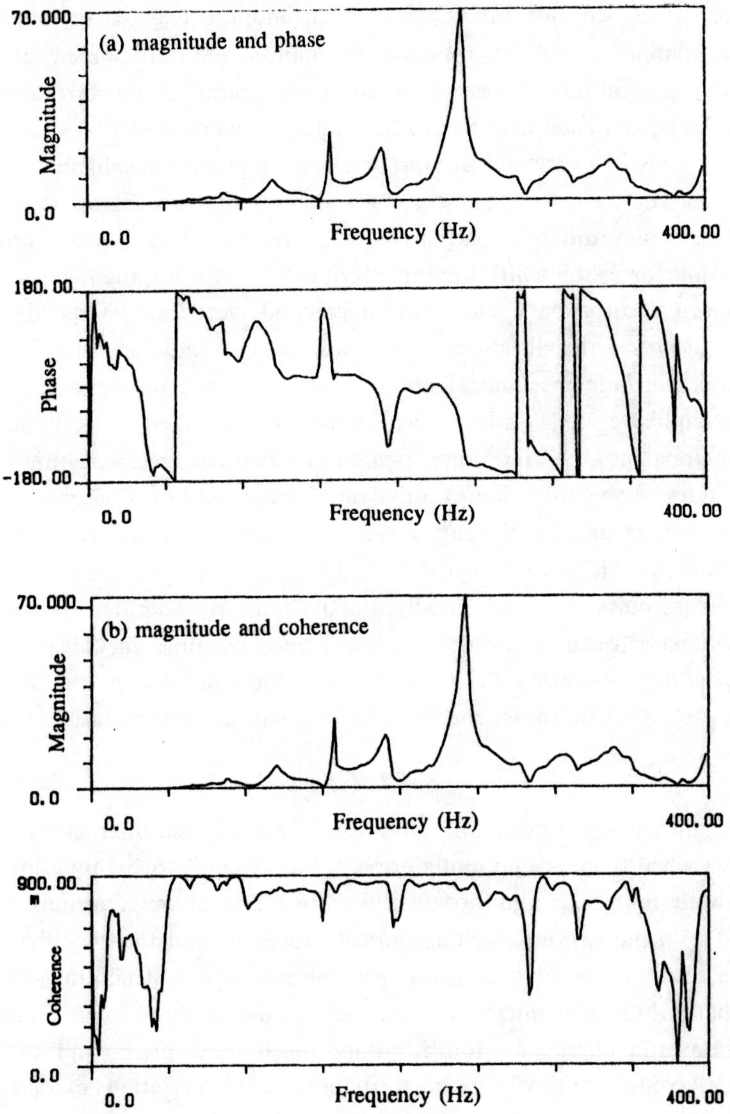

(ii) the background noise and/or vibration level, (iii) the location of the monitoring transducer, (iv) the load sharing characteristics of the item, and (v) the dynamic interaction between the item and other items in contact with it. Stewart[8.8] reviews the application of signal processing techniques to machine health monitoring with particular emphasis on gears, rotors and bearings. He demonstrates that the main factors for gears are (i), (iii) and (v); the main factors for rotors are (i), (iv) and (v); and for bearings the main factors are (i), (ii) and (iii). The factor which is common to all items involving rotational motion is simply the rotational motion itself – the dominant noise and/or vibration frequency is always related in some manner to it. Source identification and fault detection from noise and vibration signals associated with items which do not involve rotational motion (e.g. casing and support resonances, piping resonances, torsional resonances, etc.) are somewhat harder to quantify in a general sense – i.e. a common denominator such as a rotational speed is not available, and each item has to be treated on its own merits.

The main causes of mechanical vibration are unbalance, misalignment, looseness and distortion, defective bearings, gearing and coupling inaccuracies, critical speeds, various forms of resonance, bad drive belts, reciprocating forces, aerodynamic/hydrodynamic forces, oil whirl, friction whirl, rotor/stator misalignments, bent rotor shafts, defective rotor bars, etc. Mechanical and electrical defects also manifest themselves as noise – the vibrations are transformed into radiated noise. The common noise sources include mechanical noise, electrical noise, aerodynamic noise, and impactive/impulsive noise. Mechanical noise is associated with items such as fan/motor unbalance, bearing noise, structural vibrations, reciprocating forces, etc. Electrical noise is generally due to unbalanced magnetic forces associated with flux density variations and/or air gap geometry, brush noise, electrical arcing, etc. Aerodynamic noise is related to vortex shedding, turbulence, acoustic modes inside ducts, pressure pulsations, etc. Finally, impact noise is generated by sharp, short, forceful contact between two or more bodies. Typical examples include punch presses, tooth impact during gearing, drills, etc. Some of the more common faults or defects that can be detected using noise and/or vibration analysis are summarised in Table 8.1.

8.4.1 Gears

The dominant source of noise and vibration in gears is the interaction of the gear teeth. Even when there are no faults present, the dynamic forces that are generated produce both impulsive and broadband noise. The discrete, impulsive noise is associated with the various meshing impact processes, and the broadband noise is associated with friction, fluid flow, and general gear system structural vibration and noise radiation. Gear geometry factors, such as the pressure angle, contact ratio, tooth face width, alignment, tooth surface finish, gear pitch, and tooth profile accuracy, all contribute to vibration and radiated noise. Variations of load and speed also contribute to gear noise. Finally, the expulsion of fluid (air and/or lubricant) from meshing gear teeth can sometimes produce shock waves, particularly in high speed gears.

Gear faults generally fall into one or more of three categories. They are: (i) discrete gear tooth irregularities – localised faults; (ii) uniform wear around the whole gear – distributed faults; and (iii) tooth deflections under high external dynamic loads.

The main frequency at which gearing induced vibrations will be generated is the gear meshing or toothpassing frequency, f_m. It is given by

$$f_m = \frac{N \times \text{r.p.m.}}{60}, \tag{8.3}$$

Table 8.1. *Some typical faults and defects that can be detected with noise and vibration analysis.*

Item	Fault
Gears	Tooth meshing faults Misalignment Cracked and/or worn teeth Eccentric gears
Rotors and shafts	Unbalance Bent shafts Misalignment Eccentric journals Loose components Rubs Critical speeds Cracked shafts Blade loss Blade resonance
Rolling element bearings	Pitting of race and ball/roller Spalling Other rolling element defects
Journal bearings	Oil whirl Oval or barrelled journals Journal/bearing rub
Flexible couplings	Misalignment Unbalance
Electrical machines	Unbalanced magnetic pulls Broken/damaged rotor bars Air gap geometry variations
Miscellaneous	Structural and foundation faults Structural resonances Piping resonances Vortex shedding

where N is the number of teeth, and the r.p.m. is the rotational speed of the gear. It is useful to note that several gear meshing frequencies are present in a complex gear train. Also, because of the periodic nature of gear meshing, integer harmonics are also present. The direction of the vibrations can be either radial or axial, and increases in vibration levels at the gear meshing frequency and its associated harmonics are typical criteria for fault detection. This point is illustrated schematically in Figure 8.22. The increases at the gear meshing frequency and the various harmonics are associated with wear. As a general rule of thumb, the higher harmonics generally have lower amplitudes, even when the gear is 'worn'. A higher harmonic with a large amplitude generally indicates the presence of a gear wheel resonance – i.e. the higher harmonic coincides with a natural frequency of the gear wheel or some other structural component within the gearbox system.

When gear meshing frequencies cannot be readily identified from a noise and/or vibration spectrum due to (i) the presence of several gears in a complex train, and (ii) high background random noise and vibration, techniques such as synchronous signal-averaging (phase-averaging) or cepstrum analysis can be used to detect the various periodic components and any associated damage. The interested reader is referred to Stewart[8.8] and to Randall[8.9] for precise details. Stewart[8.8] discusses various advanced techniques for the analysis of gearbox signals in the time domain which are either synchronous or asynchronous with gear rotation. These time domain techniques generally require a detailed assessment of the amplitude modulation characteristics of the gearbox vibration signals. Two distinct forms of amplitude modulation are generally analysed. They are (i) the overall modulation of the envelope of the time signal, and (ii) internal modulation of specific frequency components. Randall[8.9] discusses cepstrum analysis techniques for separating excitation and

Fig. 8.22. Schematic illustration of elementary gearbox vibration spectra.

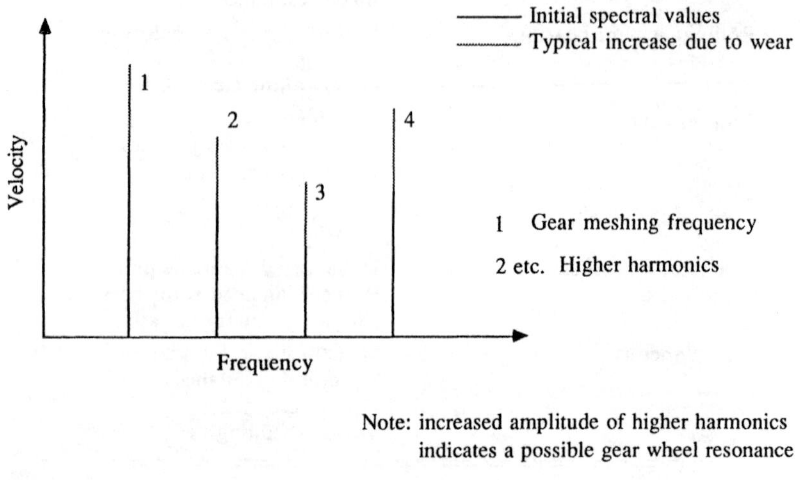

structural response effects in gearboxes. The reader is referred to Figure 5.10 in chapter 5 for a typical example.

The most common gear fault is a discrete gear tooth irregularity such as a broken or chipped tooth. With a single discrete fault, high noise and vibration levels can be expected at the shaft rotational frequency, f_s, and its associated harmonics. These narrowband peaks are in addition to the various gear meshing frequencies and their associated harmonics, which are also present. This point is illustrated in Figure 8.23. Also, discrete faults tend to produce low level, flat, sideband spectra (at $\pm$ the shaft rotational speed and its associated harmonics) around the various gear meshing frequency harmonics.

Distributed faults such as uniform wear around a whole gear tend to produce high level sidebands (at $\pm$ the shaft rotational speed and its associated harmonics) in narrow groups around the gear meshing frequencies. This point is illustrated in Figure 8.24. When the high level sidebands are restricted to the fundamental gear meshing frequency, the gear meshing noise and vibration are being modulated periodically at

Fig. 8.23. Frequency spectra associated with a discrete gear tooth irregularity.

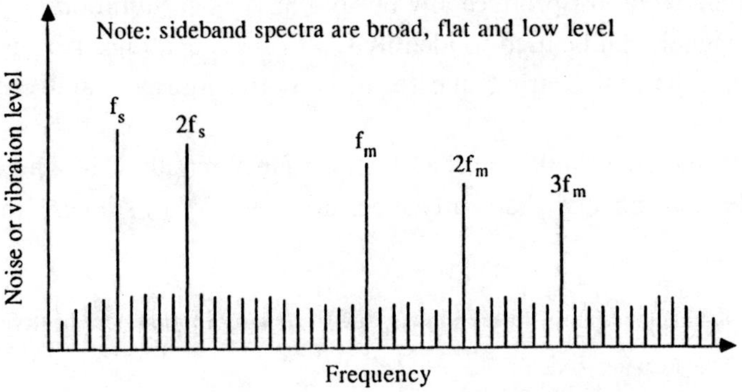

Fig. 8.24. Frequency spectra associated with uniform wear around a whole gear.

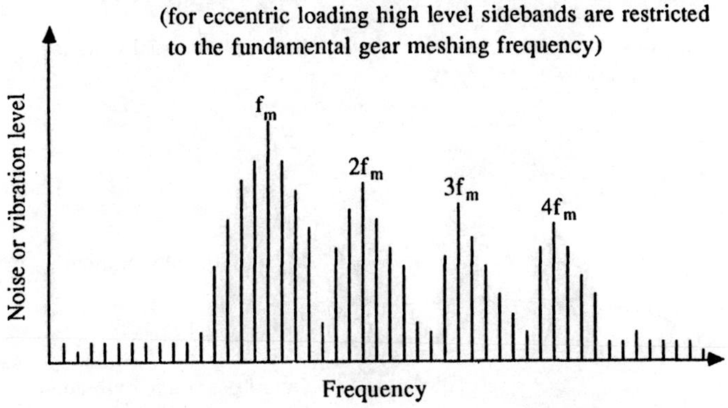

a frequency which corresponds to the shaft rotational speed and its associated harmonics. This generally occurs when the gear is eccentric, or if the shaft is misaligned and there is a high fluctuating dynamic load on the teeth.

8.4.2 Rotors and shafts

The two most common faults associated with rotating shafts are misalignment and unbalance. With misalignment, the vibration is both radial and axial, and the increase in vibration is at the rotational frequency and the first few harmonics. With unbalance, the vibration is generally radial and the increase in vibration is at the rotational frequency.

Phase measurements allow one to distinguish between rocking motion and a bent shaft during rotation. If the radial vibrations on the two bearings are out of phase, then the motion is a rocking one; if they are in phase, then the shaft is bent. This point is illustrated in Figure 8.25. Misalignment can also be detected by an out of phase axial vibration. For the case of a force unbalance, there will be no phase difference at the rotational frequency, whereas for a couple unbalance it is about 180°. This point is illustrated in Figure 8.26. It is useful to remember that force and couple unbalances do not produce any out of phase axial vibrations.

Vibration signals can be used to identify shaft rubs (i.e. a once per revolution rub or impact in a journal bearing due to an eccentric journal), and critical speeds (whirling).

Most rotor and shaft faults can be identified fairly readily. The one exception is a cracked shaft. A cracked shaft introduces non-linearity of stiffness and damping,

Fig. 8.25. Detection of misalignment due to rocking motion or a bent shaft.

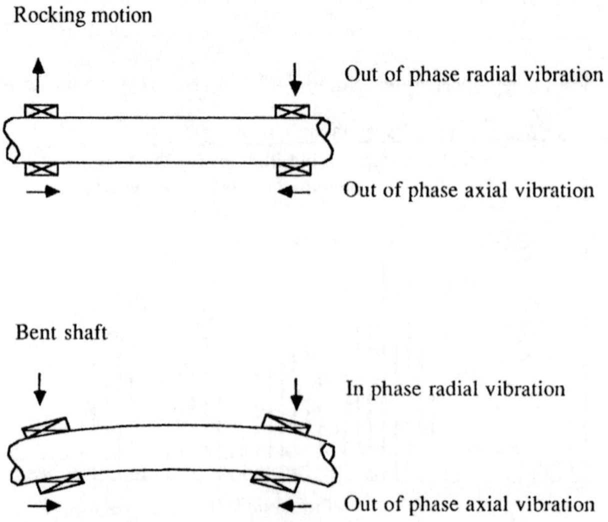

and several specialist procedures are available[8.8,8.10]. In essence, they all predict that the growth of a crack can be detected from the vibrational response at the main and subcritical speeds of the rotor. Care has got to be exercised not to confuse balancing and bending problems with cracks. When a rotor is run at a constant operational speed, the shaft takes on a deflected rotating shape and the crack continuously opens and closes. In order to identify a crack and separate it from balancing and bending problems, it has to be 'exercised' by varying the rotational speed of the shaft. Adams *et al.*[8.11] have developed a unique vibration technique for non-destructively assessing the integrity of structures. Their method of damage location depends upon measuring the natural frequencies at two or more stages of damage growth, one of which may be the undamaged condition. It also depends upon the nature of the vibration response (i.e. the mode shapes) remaining unchanged. Thus, their technique has particular application to the early detection of rotor cracks. The technique involves recognising that the crack introduces an additional flexibility into the rotor and modelling the rotor and the stiffness using receptance analysis techniques. This additional flexibility causes the natural frequencies to change, and the damage location is identified as being the position(s) where the magnitude of the flexibility that would cause the change in natural frequency is the same for several modes. To successfully locate a crack, it is necessary to evaluate the flexibility variation curves for the first three or four natural frequencies of the rotor.

8.4.3 Bearings

In machine condition monitoring, most attention is generally given to the monitoring of bearing conditions because (i) it is the most common component; (ii) it possesses a finite lifespan and fails through fatigue; and (iii) it is often subjected to abuse and

Fig. 8.26. Detection of force and couple unbalance.

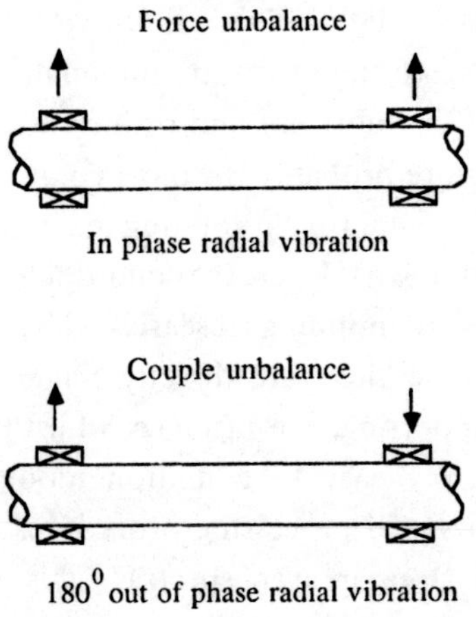

fails more frequently than other components. Two types of bearings are used. They are rolling-contact bearings and sliding-contact bearings.

Rolling-contact bearings can have either point contact or line contact with the bearing race. Furthermore, the forces can be sustained either in the radial direction (radial ball or roller bearings) with no axial load, or in both the radial and axial directions (angular ball or roller bearings). In this instance, the bearings are capable of sustaining an axial (thrust) load. The elements for rolling-contact bearings can be spherical, cylindrical, tapered or barrel-shaped.

Sliding-contact bearings can be journal, thrust or guide bearings. Journal bearings require fluid lubrication and are cylindrical in shape. Thrust bearings prevent motion along the axis of a shaft. Guide bearings are commonly used to guide the motion of a machine component along its length without rotation.

As a general rule of thumb, sliding-contact bearings are quieter than rolling-contact bearings. The primary noise and vibration mechanisms for rolling-contact bearings is the impact process between the rolling elements and the bearing races. The primary noise and vibration mechanisms for sliding-contact bearings is the friction and rubbing that occurs when there is inadequate or improper lubrication.

Noise and vibration can be used as a diagnostic tool for sliding-contact bearings, and in particular journal bearings, to identify conditions such as journal/bearing rub, and oval or barrelled journals. With inadequate or improper lubrication, the journal film can break down producing a 'stick-slip' excitation of the shaft and other connected machine components. Stick-slip involves short durations of metal-to-metal contact. Another common source of noise and vibration in journal bearings is oil whirl. Oil whirl can be identified as a noise or vibration at a frequency which is approximately half the shaft rotational speed. It occurs at half the shaft rotational speed because the oil film next to the shaft rotates at the shaft speed and the oil film next to the bearing is stationary; hence, the average oil velocity is half the shaft speed. This is the frequency at which the shaft in the bearing is excited by the oil surrounding it, particularly if the bearings are lightly loaded. Oil whirl is most common in lightly loaded shafts because the restoring forces are minimal. Oil whirl can be minimised by varying the viscosity of the lubricant and/or increasing the oil pressure.

Rolling-contact bearings are probably the most common type of bearings that are used in industry. They play a vital role in rotating machinery, and their failure results in the machinery being shut down. Hence, the condition monitoring of rolling-contact bearings is the subject of continuing research. The problems associated with monitoring rolling-contact bearings are directly related to the complexity of the machine which they are supporting. For instance, whilst turbo-generator and electric motor bearings are relatively easy to condition monitor, aero-engine mainshaft bearings require advanced signal processing procedures such as zoom or passband spectral analysis and envelope power spectra. This is essentially because the background noise and vibration levels are generally very high for the latter case, and simple time domain techniques such as overall or r.m.s. level detection, crest factors and kurtosis are not suitable.

The vibration level measured from the housing of a rolling-contact bearing comes from four main sources. They are (i) bearing element rotations, (ii) resonance of the bearing elements and attached structural supports, (iii) acoustic emission, and (iv) intrusive vibrations. Bearing element rotations generate vibrational excitation at a series of discrete frequencies which are a function of the bearing geometry and the rotational speed. These are the frequencies which provide information about the condition of the inner race, outer race and rolling elements of a bearing. The bearing elements and the various structural components that support the bearing housing all have natural frequencies. More often than not, these natural frequencies are excited into resonance and they appear in the vibration signature from the bearing housing. It is often desirable to identify these frequencies via impact tests, etc., when the machine is not running and to establish whether or not they coincide with the various bearing element rotation frequencies. Acoustic emission is associated with short, impulsive, stress waves generated by very small scale plastic deformation, crack propagation or other atomic scale movements in high stress regions within the bearing. The vibration signals associated with acoustic emission tend to be high frequency (kHz to MHz) in nature and can be used to detect both early and advanced damage. Finally, intrusive vibrations relate to the transmission of vibrations from other parts of the machine to the bearing housing. These external vibrations can be due to a variety of causes. As a result of all these 'additional' noises and vibrations, it is not always easy to identify the discrete frequencies associated with the bearing element rotations.

Several discrete frequencies (and their associated harmonics) can be expected from rolling-contact bearings. As already mentioned, they are a function of bearing geometry and the rotational speed. They are summarised by Shahan and Kamperman[8.12] and are reproduced here. They are:

(i) the shaft rotational frequency, f_s, where

$$f_s = N/60; \tag{8.4}$$

(ii) the rotational frequency of the ball cage with a stationary outer race, f_{bcsor}, where

$$f_{bcsor} = \{f_s/2\}\{1 - (d/D)\cos\phi\}; \tag{8.5}$$

(iii) the rotational frequency of the ball cage with a stationary inner race, f_{bcsir}, where

$$f_{bcsir} = \{f_s/2\}\{1 + (d/D)\cos\phi\}; \tag{8.6}$$

(iv) the rotational frequency of a rolling element, f_{re}, where

$$f_{re} = \{f_s/2\}\{D/d\}\{1 - (d/D)^2 \cos^2 \phi\}; \quad (8.7)$$

(v) the rolling element pass frequency on a stationary outer race, f_{repfo}, where

$$f_{repfo} = \{Zf_s/2\}\{1 - (d/D) \cos \phi\}; \quad (8.8)$$

(vi) the rolling element pass frequency on a stationary inner race, f_{repfi}, where

$$f_{repfi} = \{Zf_s/2\}\{1 + (d/D) \cos \phi\}; \quad (8.9)$$

(vii) the rolling element spin frequency, f_{resf} (contact frequency between a fixed point on a rolling element with the inner and outer races), where

$$f_{resf} = f_s\{D/d\}\{1 - (d/D)^2 \cos^2 \phi\}; \quad (8.10)$$

(viii) the frequency of relative rotation between the cage and the rotating inner race with a stationary outer race, f_{rciso}, where

$$f_{rciso} = f_s\{1 - 0.5\{1 - (d/D) \cos \phi\}\}; \quad (8.11)$$

(ix) the frequency of relative rotation between the cage and the rotating outer race with a stationary inner race, f_{rcosi}, where

$$f_{rcosi} = f_s\{1 - 0.5\{1 + (d/D) \cos \phi\}\}; \quad (8.12)$$

(x) the frequency at which a rolling element contacts a fixed point on a rotating inner race with a stationary outer race, f_{recri}, where

$$f_{recri} = Zf_s\{1 - 0.5\{1 - (d/D) \cos \phi\}\}; \quad (8.13)$$

(xi) the frequency at which a rolling element contacts a fixed point on a rotating outer race with a stationary inner race, f_{recro}, where

$$f_{recro} = Zf_s\{1 - 0.5\{1 + (d/D) \cos \phi\}\}; \quad (8.14)$$

and N is the shaft rotational speed in r.p.m., d is the roller diameter, D is the pitch diameter of the bearing, ϕ is the contact angle beween the rolling element and the raceway in degrees, Z is the number of rolling elements, and $\phi = 0°$ for a radial ball bearing.

The above series of eleven equations (f_s, f_{re}, f_{resf}, and four pairs since equations 8.5 and 8.12, 8.6 and 8.11, 8.8 and 8.14, and 8.9 and 8.13 are identical) defines all the possible discrete frequencies that can be expected. In addition to these frequencies their harmonics will also be excited. Hence, cepstrum analysis is particularly useful in identifying the various periodic families. Of the eleven, there are three major frequencies that are commonly identified and associated with defective bearings. They are: (i) the rolling element pass frequency on the outer race, f_{repfo}, which is associated with outer race defects; (ii) the rolling element pass frequency on the inner race, f_{repfi}, which is associated with inner race defects; and (iii) the rolling element spin frequency, f_{resf}, which is associated with ball or ball cage defects. All these defects initially manifest themselves as narrowband spikes at the respective frequencies. Sometimes, when there is excessive internal clearance or if a bearing turns on a shaft, narrowband spikes will be detected at several multiples of the shaft rotational frequency, f_s. As the size of a bearing defect increases, the bandwidth of the narrowband spike increases and it eventually becomes broadband and the overall vibrational energy associated with the defect increases.

When the discrete bearing frequencies cannot be identified because of high background noise and/or widespread damage, advanced signal analysis techniques have to be employed. There is no single best technique for the condition monitoring of bearings. Some useful guidelines are provided here.

(i) Crest factors are reliable only in the presence of significant impulsiveness. Typical values of crest factors for bearings in a good condition range from 2.5 to 3.5, and values for bearings with impulsive defects are higher, ranging up to ~ 11. Generally speaking, crest factors higher than 3.5 are indicative of damage. Crest factor values of a vibration signal are relatively insensitive to operating speed and bearing load, provided that sufficient speed is maintained to generate a bearing vibration which is above the background noise level, and sufficient load is applied to maintain full contact. At higher operating speeds, both the peak and the r.m.s. values increase proportionally, giving a relatively constant crest factor. In the absence of significant impulsiveness, the reliability of the crest factor technique to detect bearing damage breaks down. Some typical examples include bearings with shallow defects which have no significant edge, bearings with advanced wear damage, and bearings with a large number of defects or widespread damage.

(ii) The kurtosis technique is also only reliable in the presence of significant impulsiveness. It is based on detecting changes in the fourth statistical moment as impulsive faults develop. Typical values of the kurtosis of a signal range from 3 to 45, depending upon the condition of the bearing. As a general rule, variations in kurtosis closely follow variations in the crest factor, the only difference being the variations in numerical magnitude – i.e. kurtosis provides a much wider 'dynamic range'. Generally, kurtosis values higher than ~ 4 are indicative of damage. Like the crest factor, the kurtosis of a bearing vibration signal is unaffected by changes in speed and loading. Because the kurtosis is based upon detecting impulsiveness, it is subject to the same limitations as crest factors.

(iii) Spectral analysis of bearing signals is the most useful diagnostic and fault detection technique, but it requires details about the bearing geometry and the operating conditions. Provided that the bearing vibration signals are not submerged in background noise, the various discrete frequencies can be readily identified. Defects on the outer race tend to dominate because the vibrations which are generated here have the shortest path to the measurement transducer. Also, vibration levels increase with defect size.

(iv) Cepstrum analysis is an invaluable complementary technique to spectral analysis – it allows for an identification of all the different harmonic components and any associated sidebands. Cepstrum analysis also separates the internal vibration of the bearing from the transfer function of the path to the measurement transducer[8.9]. Cepstrum analysis is seldom used on its own because it tends to suppress information about the global shape of the spectrum which may contain diagnostic information of its own.

(v) The envelope power spectrum (sometimes known as the high frequency resonance technique) is very useful in the presence of a high background noise level. The vibration peaks of a bearing in good condition are always very low in amplitude compared to those obtained from defective bearings, and the frequency distribution of the peaks is random. The absence of significant non-harmonic peaks in the envelope spectrum suggests either a non-defective bearing or one in which there is widespread damage. The vibration levels for the latter case are significantly higher than for the case of undamaged bearings. The envelope power spectrum technique requires information about the bearing and its resonances, and it also requires the suitable selection of filter bandwidths and centre frequencies.

(vi) It is often appropriate to consider a combination of the above mentioned techniques to improve both diagnosis and fault detection.

A specific test case relating to the identification of rolling-contact bearing damage is presented in sub-section 8.5.3.

8.4.4 *Fans and blowers*

Fans and blowers are used in industry for air movements and product handling requirements. Fans are generally used to move large volumes of air (generally for ventilation), and blowers are used for conveying products and materials. The two main types of fans are centrifugal and axial. Centrifugal fans can be backward-curved, forward-curved or radial. Axial fans can be vane-axial or tube-axial. As a general rule, axial fans are noisier than centrifugal fans because they require higher pressures. The two main types of blowers are (i) rotary positive displacement blowers, and (ii) high speed radial centrifugal fans.

Noise from rotating fans and blowers can be classified as (i) self-noise, and (ii) interaction noise. Self-noise is associated with the generation of sound by fluid (i.e. air) flow over the blades, and interaction noise is associated with the reaction of the blades with disturbances which are moving in the same reference frame as the blade itself. Depending on the type of rotating unit, there are several possible causes for both self-noise and interaction noise. Examples of self-noise include sound from steady loading (rotational noise due to thrust and torque), the continuous passage of boundary layer turbulence past the trailing edges, rotational monopole type noise due to the finite thickness of the blades, and vortex development in the trailing wake. Examples of interaction noise include the interaction of ducted rotor blades with annular boundary layers, unsteady loading, blade–blade tip vortex interactions, rotor–stator interactions, and inlet flow disturbances. The reader is referred to Blake[8.13] and to Glegg[8.14] for a comprehensive review of the mechanisms of noise generation in rotating machinery. As a qualitative overview, it is useful to note that both self-noise and interaction noise manifest themselves as (i) discrete tonal components at the blade passing frequency and its associated harmonics, (ii) discrete tonal components at the shaft rotational speed and its associated harmonics, and (iii) broadband aerodynamic noise. The blade passing tones tend to dominate certain narrow frequency bands, and the overall sound levels are often dominated by the broadband aerodynamic noise. Specialist techniques[8.13,8.14] are required to separate the various self-noise and interaction noise components. Interaction noise tends to be more significant in axial flow fans than in centrifugal fans.

The main sources of discrete noise in centrifugal fans are (i) the pressure fluctuations that are generated as the blades pass a fixed point in space, and (ii) the pressure fluctations that are generated as the blades pass the scroll cut-off point. These sources are illustrated in Figure 8.27. They generate a family of discrete tones with the blade passing frequency being the fundamental. This blade passing frequency (and its associated harmonics) is given by

$$f_b = \frac{\text{r.p.m.} \times N \times n}{60}, \tag{8.15}$$

Fig. 8.27. Noise generation in a centrifugal fan.

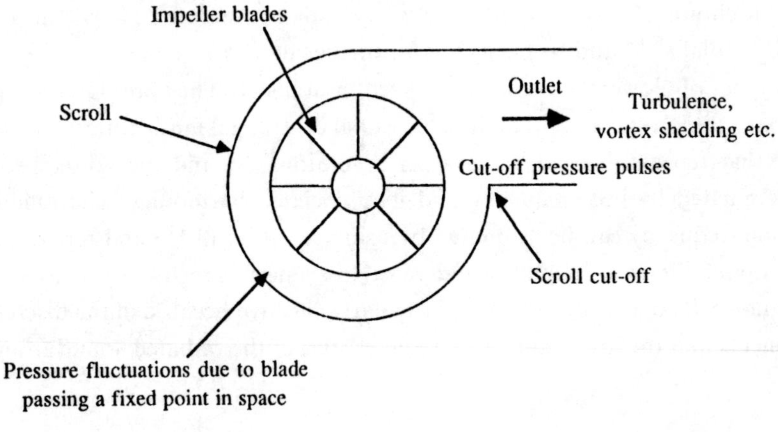

where N is the number of blades, and $n = 1, 2, 3$, etc. In addition to this harmonic family, there is significant broadband aerodynamic noise associated with vortex shedding, turbulence, etc. Furthermore, acoustic resonances in the scroll casing also often influence the sound spectrum. Small fans in particular can have a Helmholtz resonance in the audible frequency range. In addition, higher order acoustic modes within the duct casing can also be a problem. Because centrifugal fans are low pressure and high volume devices, reactive (absorptive) duct silencers are an efficient means of noise control[8.15].

Axial flow fans generate more noise than centrifugal fans. In addition to the discrete blade noise and broadband aerodynamic noise, there are several other mechanisms which result from non-linear interactions between the blades and the fluid, and the blade and the wake. The dominant interaction noise components are associated with rotating unsteady pressure patterns, the resultant noise of which is also at the blade passing frequency and its associated harmonics. These rotating pressure patterns are set up with pressure lobes at each blade. Axial flow fans are usually contained within ducts and, when this is the case, boundary conditions are imposed on the sound field. Higher order acoustic duct modes (as discussed in chapter 7) can thus be set up if a suitable combination of blades and vanes is available to set up rotating pressure patterns that correspond to their pressure distributions. The number of lobes, m_L, of the interaction pressure pattern is given by

$$m_L = nN \pm kV, \tag{8.16}$$

where N is the number of blades, $n = 1, 2, 3$, etc., V is the number of vanes, and $k = \pm 1, \pm 2, \pm 3$, etc. The rotational speed associated with the m_L lobed interaction pattern is given by

$$M_L = \frac{n \times N \times \text{r.p.m.}}{m_L}. \tag{8.17}$$

As the speed of the interaction pattern increases, so does the radiated sound power. A careful choice of N and V minimises the radiated sound power. The reader is referred to Blake[8.13] and to Glegg[8.14] for further details.

Two types of blowers are commonly used in industry. They are (i) rotary positive displacement blowers, and (ii) high speed radial centrifugal fans. Rotary blowers have a cycle that repeats itself four times per revolution[8.15], and the noise spectrum is thus dominated by this frequency and its associated harmonics. The fundamental excitation frequency can be estimated by using equation (8.15) and replacing r.p.m. by 4 × r.p.m. Flexible couplings and reactive silencers are usually used to reduce blower noise. Reactive silencers are particularly effective because of the discrete tonal components and the low frequency characteristics of the radiated sound from rotary blowers.

8.4.5 Furnaces and burners

The noise levels associated with furnaces, burners and other combustion processes arise from complex fluid interactions due to turbulent mixing, etc. Combustion noise can be classified into four groups. They are: (i) combustion roar, (ii) combustion driven oscillations, (iii) unstable combustion noise, and (iv) combustion amplification of periodic flow phenomena. Putnam[8.16] provides an excellent comprehensive review of combustion and furnace noise in the industrial environment, and this sub-section is, in the main, a brief summary of that work.

Combustion roar is a broadband noise with a smooth frequency spectrum which is dependent on the level of turbulence. The basic noise frequency spectrum is very similar to a jet noise frequency spectrum. Low and high frequency components of combustion roar tend to be modified by the characteristics of the environment. Room response effects tend to modify the low frequencies, and burner-tiles (protective shield around the flame) amplify the radiated noise at their natural frequencies. Additional noise from valves, bends, etc., in fuel and air supply lines also tends to amplify the high frequency components of the combustion process. These effects are illustrated schematically in Figure 8.28 together with the four different types of combustion noise. As a general rule, combustion roar increases with the square of turbulence intensity – the turbulent fluctuations are about 5 to 20% of the characteristic flow velocity in the burner. Other parameters that affect combustion roar include the firing rate of single burners, burner size, flame size, fuel consumption and fuel type. Combustion roar can be reduced by installing inlet and exhaust mufflers and by incorporating Helmholtz resonators into the shape of the burner-tiles around the flame.

Fig. 8.28. Schematic illustration of combustion noise. (Adapted from Putnam[8.16].)

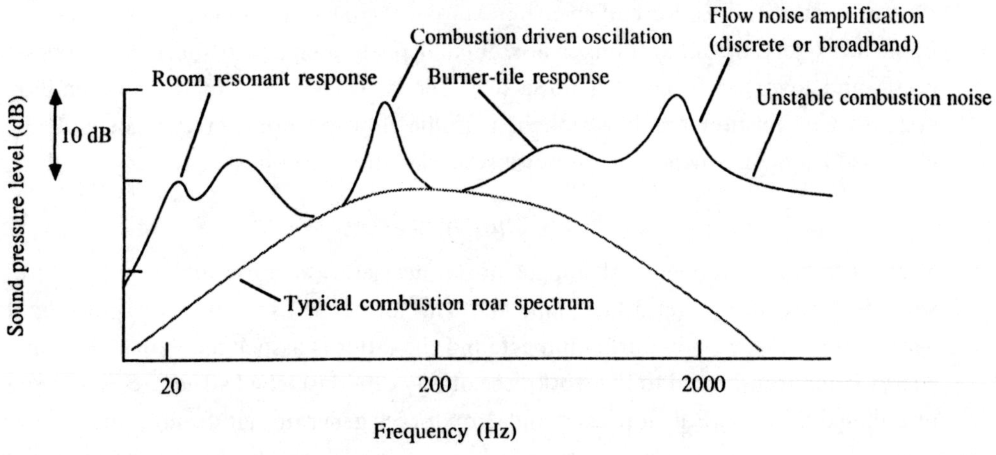

Combustion driven oscillations occur at discrete frequencies and involve a feedback cycle where a change in the heat release sets up an acoustic oscillation, and the acoustic oscillation alters the heat release rate. Four of the most common types of combustion driven oscillations are (i) singing flames (ii) fuel-oil combustors, (iii) tunnel burners, and (iv) vortex shedding. As an example, diffusion flames emit a periodic sound (singing flame) when fuel supply lines are inserted sufficiently far into a combustion tube – periodic changes in pressure at regions close to the flame and at a natural frequency of the combustion tube produce a periodic change in the fuel supply rate which in turn causes a periodic change in the heat release rate. Vortex shedding can also set up a feedback cycle. Sometimes, vortices are generated near the end of a burner. When the vortex shedding frequency is close to the natural acoustic frequency of the furnace, the acoustic oscillation triggers the vortex shedding which in turn results in a periodic change in flame surface area. This periodic change in flame surface area results in a periodic heat release which in turn maintains the acoustic oscillation. Combustion driven oscillations can be eliminated or suppressed by clearly identifying the feedback cycle. For instance, with fuel supply lines in combustion tubes, the burner can be moved away from regions of sound pressure maxima inside the combustion tube, or the regions of maximum sound pressure can be relieved by adding small ports to the tube. Vortex shedding can be minimised by removing the flame front from regions of high vorticity.

Unstable combustion noise occurs at the high end of the combustion roar frequency spectrum. It is due to the increase in the effective turbulence level at regions close to the limit of flame stability (blow-off or flash back) and its frequency spectrum characteristics are smooth although amplification can occur at the natural frequencies of the enclosure. Increases of ~ 10 dB can be expected when the mixture ratio or the flow rate is changed sufficiently to place a flame in an unstable regime.

Combustion amplification of periodic flow phenomena generally occurs at discrete frequencies although some amplification of broadband noise can occur. This particular type of combustion noise is driven by external signal sources including ultrasonic frequencies, jets, Strouhal number flow related phenomena, swirl-burner precession noise, and upstream generated noise (e.g. bends, valves, etc., in the air handling system). Combustion amplification of periodic flow phenomena can generally be identified by the frequency of the noise source without combustion.

8.4.6 Punch presses

The two main components of impact noise (acceleration noise and ringing noise) were discussed in section 3.11, chapter 3. The former is associated with the rapid decelerations of the body during impact, and the latter is associated with vibrational energy being transmitted to the workpiece or any other attached structures and being re-radiated as noise. Punch presses and drop forges generate significant impact noise and, in addition to this impact noise, there are two further noise sources. They are (i) billet expansion noise and (ii) air expulsion noise. Billet expansion noise is generated when the ram impacts it and causes sudden deformation and outward radial movement – it is highly impulsive and only lasts for a few microseconds. Air expulsion noise is generated due to high velocity air being ejected between the dies immediately prior to impact. Acceleration, billet expansion and air expulsion noise all last for a very short duration but can generate intense peak sound levels of ~140–150 dB(A). As a general rule, billet expansion noise and air expulsion noise attenuate very rapidly with distance and do not pose a noise problem to the operator. Thus, the two dominant sources of punch press noise are acceleration and ringing noise. Acceleration noise is restricted to the die space area and ringing noise is associated with structural radiation from the press structure, press equipment and controls, ground radiation and material handling.

Acceleration noise from a punch press or a drop forge is a function of ram velocity, ram volume and the duration of the contact time with the workpiece. It can be reduced by (i) reducing the ram velocity, (ii) reducing the ram volume, and (iii) increasing the contact time. The reader is referred to section 3.11 in chapter 3.

Ringing noise from a punch press or a drop forge is associated with a variety of different items. The relative importance of each of these items is very much dependent upon the individual situation. As a general rule, most of the ringing noise is associated with regions in proximity to the workpiece area. Studies by Halliwell and Richards[8.17] indicate that up to ~60% of the radiated sound comes from the workpiece area (ram, dies, bolster and sowblock) which can be considered as a mass–spring system, the stiffness of which is controlled by the type and tightness of the keying system used. Tight die keys produce significantly less ringing noise than loose die keys (~5 dB). Also, distancing the keying system from the die edge reduces the ringing noise by ~10 dB[8.17]. Additional noise reduction can be attempted by re-design of the workpiece area, selective damping, modal de-coupling, etc.

8.4.7. Pumps

Hydraulic pumps are widely used in industry. Typical examples include centrifugal, reciprocating, screw and gear pumps. The most common vibration problems associated with pumps are unbalance, misalignment, defective bearings and resonance. These topics have already been covered in this book in a general sense. As far as pumps are concerned, they manifest themselves as a result of (i) hydraulic forces, (ii) cavitation, and (iii) recirculation.

Hydraulic forces manifest themselves as discrete frequency noise and/or vibration at frequencies corresponding to the total number of compression or pumping events per revolution multiplied by the shaft r.p.m. and its associated harmonics – i.e.

$$f_p = (n \times \text{r.p.m.} \times N)/60, \tag{8.18}$$

where N is the number of compression or pumping events per revolution, and $n = 1$, 2, 3, etc. For instance, the total number of pumping events associated with centrifugal fans is related to the number of impeller vanes – the hydraulic forces are associated with pressure pulsations within the pump which are generated as an impeller vane passes a stationary diffuser. Provided that the impeller is centrally located and suitably aligned with the pump diffusers, the hydraulic pulsations will be balanced and minimal.

In addition to the above mentioned discrete frequency noise and vibration, pumps also display broadband noise characteristics due to turbulence, cavitation and recirculation. Cavitation is a somewhat common problem with centrifugal pumps in particular and produces significant wear and erosion. Cavitation generally occurs when a centrifugal pump is being operated with an inadequate suction pressure or below its design capacity. In this instance, the fluid entering the pump is insufficient to fill the volume and this generates small, microscopic, unstable vacuum cavities which collapse (implode) very rapidly. This continuing series of implosions is due to high frequency alternating pressures within the liquid – the bubbles grow to a particular size and then collapse causing very high instantaneous pressures and temperatures. This continuous implosion process produces significant damage to the impeller and pump housing – the associated noise and vibration is random and broadband. A typical bearing vibration spectrum from a cavitating centrifugal pump driven by an induction motor is illustrated in Figure 8.29.

Recirculation (reverse flow) also produces broadband noise and vibration. It is generally not as severe as cavitation but it can also cause damage to seals, bearings

Fig. 8.29. A typical bearing vibration spectrum from a cavitating centrifugal pump.

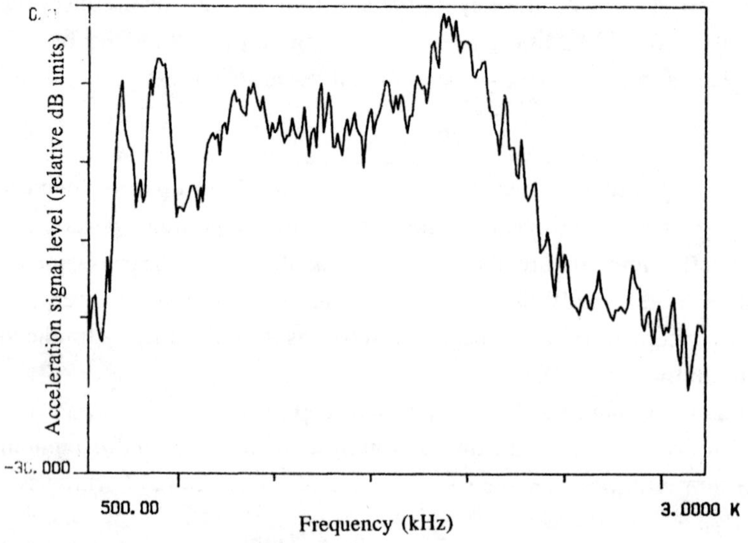

and other related pump components. It normally occurs in centrifugal pumps which have a large impeller eye diameter to outlet diameter ratio.

8.4.8 Electrical equipment

Most types of electrical equipment are sources of noise and vibration. Some typical examples include transformers, electric motors, generators and alternators. The noise and vibration from electric motors, generators and alternators is particularly useful as a diagnostic tool.

Transformer noise is generally associated with the vibrations of the core and the windings due to magnetostriction or magnetomotive forces. The noise associated with these vibrations is a discrete frequency hum at twice the supply frequency and at its associated harmonics. As a general rule, the larger the transformer, the louder are the low frequency harmonics; the smaller the transformer, the louder are the high frequency harmonics.

Noise and vibration sources in electric motors can be (i) mechanical, (ii) aerodynamic, or (iii) electromagnetic. Moreland[8.18], Bloch and Geitner[8.19], Hargis *et al.*[8.20], Cameron *et al.*[8.21], and Tavner *et al.*[8.22] all provide reviews of the different sources of electric motor noise and vibration. These sources are summarised in Figure 8.30.

Mechanical problems are generally associated with defective bearings, unbalance, looseness, misalignments, end winding damage due to mechanical shock, impact or fretting, etc. Procedures for diagnosing these types of faults have already been discussed. Aerodynamic problems are generally associated with ventilation fans and

Fig. 8.30. Major sources of electric motor noise and vibration.

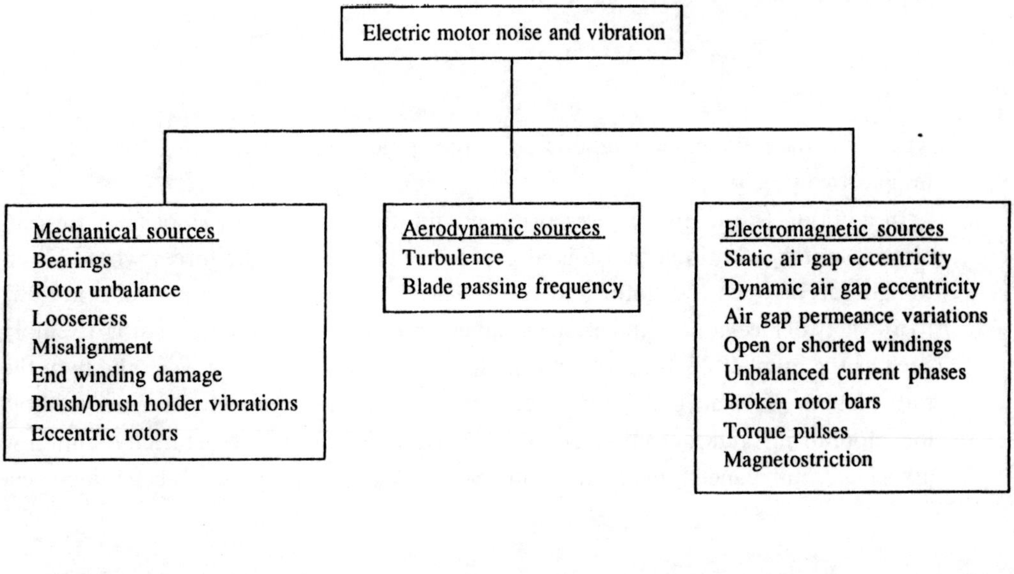

include such items as discrete blade passing frequencies, resonant volume excitations within the motor housing, broadband turbulence, etc.

Noise and vibration associated with electrical problems are generally due to unequal electromagnetic forces acing on the stator or rotor. Some typical causes of these unequal magnetic forces include broken rotor bars, static and/or dynamic air gap eccentricity, uneven air gap flux distribution, open or shorted rotor and stator windings and other inter-turn winding faults, unbalanced current phases, torque oscillations or pulses, and magnetostriction. A very simple test to establish whether or not a noise or vibration signal from an electric motor is due to an electromagnetic fault is to switch the machine off. If the noise or vibration signal disappears instantly, then the source is electromagnetic. If it does not, then it is either mechanical or aerodynamic.

Some useful basic relationships for electric motor bearing vibration signals are: (i) 1 × shaft rotational frequency and associated harmonics – mechanical unbalance; (ii) 2 × shaft rotational frequency and associated harmonics – misalignment between the motor and the driven load; (iii) 2 × electrical supply frequency and associated harmonics – misalignment between bearing centres resulting in a non-uniform air gap (this produces an unbalanced magnetic pull), torque pulses, and/or a range of other specific electrical faults associated with the armature and/or the stator. These include broken rotor bars, open or shorted rotor windings, open or shorted stator windings, inter-turn winding faults, and unbalanced electrical phases. It should be noted that any vibration signal at the electrical supply frequency itself is due to magnetic interference and is therefore not a true vibration signal.

The preceding three discrete frequency components will always be present to some degree in an electric motor bearing vibration signal; the diagnostic comments relate to trended increases in bearing vibration levels over a period of time. The electrical supply frequency, f_e, is related to the shaft rotational frequency, f_s, by

$$f_e = \frac{f_s p}{2}, \tag{8.19}$$

where p is the number of magnetic poles (not pole-pairs) – e.g. four magnetic poles implies two pole-pairs.

In addition to the above three primary vibration frequencies associated with an electric motor, additional noise is also produced from periodic forces which are in the air gap between the stator and rotor. These noise and vibration signals are due to a variety of mechanical and electromagnetic properties of the stator–rotor assembly such as the number of rotor and stator slots and the difference between the two, the radial length of the air gap, permeance variations in the air gap, etc., and are suitable for monitoring and detecting static and dynamic air gap eccentricity and any associated unbalanced magnetic pull. Static eccentricity is caused by incorrect positioning of the rotor or stator, or stator core ovality. It generally does not change provided that the rotor-shaft assembly is sufficiently stiff. Dynamic eccentricity, on the other hand, is caused by the centre of the rotor not being at the centre of rotation. Bent shafts, unbalance, cracked rotor bars, etc., generate dynamic eccentricity. High levels of static eccentricity can generate significant unbalanced magnetic pull which in turn generates dynamic eccentricity. Of particular importance in the detection of rotor defects (such as broken rotor bars) which produce static and dynamic eccentricity in induction motors are the slot harmonic frequencies, f_{sh}, associated with stator core vibrations. They arise from the interaction of the fundamental magnetic flux wave with its harmonics and with the rotor-slot components, and are given by

$$f_{sh} = f_e\left\{\frac{2R}{p}(1-s) \pm 2(n-1)\right\}, \tag{8.20}$$

where R is the number of rotor slots, p is the number of magnetic poles, s is the unit slip (in practice it ranges in value typically between 0.02 and 0.05) between the rotating speed of the magnetic field and the rotating speed of the armature, and $n = 1, 2, 3$, etc. Slot harmonic frequencies are commonly detected on the bearing vibrations even when the stator core vibrations are normal. Increased static eccentricity can be identified by changes in the vibration slot harmonic freqencies[8.21]. Electromagnetic irregularities associated with dynamic eccentricity in a rotor produce modulation of the dominant slot harmonic frequency at ± the rotational frequency, and at ± the slip frequency – i.e. an irreguarity, associated with dynamic eccentricity, manifests itself as a family of sidebands around the dominant slot harmonic frequency. The function is strong in harmonics in the case of a discontinuity such as a broken rotor bar because of the high flux density around the bar[8.20,8.21]. A typical example for a bearing vibration signal from an electric induction motor (fifty-two rotor bars, sixty stator slots, four magnetic poles) with an electromagnetic irregularity is illustrated in Figure 8.14. The slot harmonic sidebands are very evident from the passband spectra.

8.4.9 Source ranking in complex machinery

Noise source identification and ranking in complex machinery is a fundamental requirement for the implementation of effective noise control measures. It requires the usage of a variety of different techniques. The traditional methods of identifying and ranking various noise sources in complex machinery include (i) subjective assessment of the different types of sounds, (ii) selective operation of different parts of the machine, and (iii) wrapping/enclosing the complete machine with lead, fibreglass, mineral wool, etc., and selectively unwrapping different parts of the machine. In recent years several new techniques which are dependent upon advanced signal processing procedures have emerged. They include surface velocity measurement techniques, sound intensity measurement techniques, and coherence and/or cross-correlation measurement techniques. The various available techniques are summarised in Figure 8.31. Each of these techniques, with the exception of the surface and vibration intensity measurement technique, has already been discussed in this book, and will therefore only be briefly reviewed here.

Selectively unwrapping different parts of an enclosed/wrapped machine is a common way of attempting to rank the various noise sources within the machine. The technique is very time consuming, and one has to be careful that the machine does not overheat. The technique is also not suitable for low frequencies (<300 Hz) because the transmission loss of the wrapping material (lead, fibreglass, mineral wool, etc.) is generally inadequate at these frequencies. Also, the accuracy of the measured sound power depends on the difference in level between the sound power of the completely wrapped machine and the sound power emanating from the unwrapped section.

Surface velocity measurement techniques are suited to noise source identification and the ranking of machines where the dominant radiated noise is structure-borne. In essence, the technique utilises equation (3.30) in chapter 3, where

$$\Pi = \sigma\rho_0 c S\langle\overline{v^2}\rangle, \qquad \text{(equation 3.30)}$$

and Π is the radiated sound power, σ is the radiation ratio, and v^2 is the mean-square surface vibrational velocity ($\langle\ \rangle$ is a time-average and $\overline{}$ is a space-average). The accuracy of the technique is critically dependent upon the availability of information about the radiation ratio. Hence, for flat plate-type structures, the technique is more reliable at frequencies above the critical frequency since $\sigma \approx 1$. The surface velocity method is not particularly suited to rotating machinery or to very hot surfaces.

Fig. 8.31. Various noise source identification techniques.

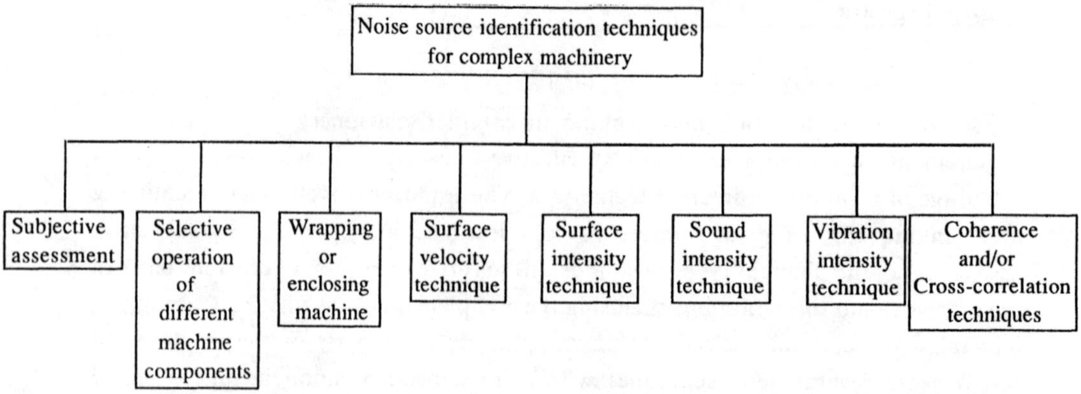

The measurement of intensity is a useful way of establishing a source ranking in complex machinery. Three types of intensity measurement techniques are available, the most common of which is the sound intensity technique. The sound intensity technique has already been described in sub-section 4.7.4, chapter 4, and in section 8.3.4. The other two are the surface intensity technique and the vibration intensity technique. Surface intensity measurement techniques involve the usage of one microphone and one accelerometer, and vibration intensity measurement techniques involve the usage of two accelerometers, as illustrated in Figure 8.32. In practice, care has got to be exercised with all three intensity techniques to ensure that phase errors are minimised. With the sound intensity technique, several procedures have been developed to reduce any phase errors between the two microphones. These include (i) the usage of phase-matched microphones, (ii) microphone switching procedures, which, in principle, eliminate the phase mismatch, and (iii) a frequency response function procedure which corrects the intensity data with the measured phase mismatch. Reinhart and Crocker[8.23] review these various procedures.

With the surface intensity technique, an accelerometer is mounted on the surface of the vibrating structure and a pressure microphone is held in close proximity to it. It is assumed that the velocity of the vibrating surface is equal to the acoustic particle velocity and that the magnitude of the pressure does not vary significantly from the vibrating surface to the microphone. The resultant surface intensity normal to the surface (sound intensity at the surface of the structure) is given by Crocker and Zockel[8.24] as

$$I_x = \frac{1}{2\pi}\int_0^f \frac{Q_{pa}\cos\phi + C_{pa}\sin\phi}{f}\,df, \tag{8.21}$$

Fig. 8.32. Sound, surface and vibration intensity measurement techniques.

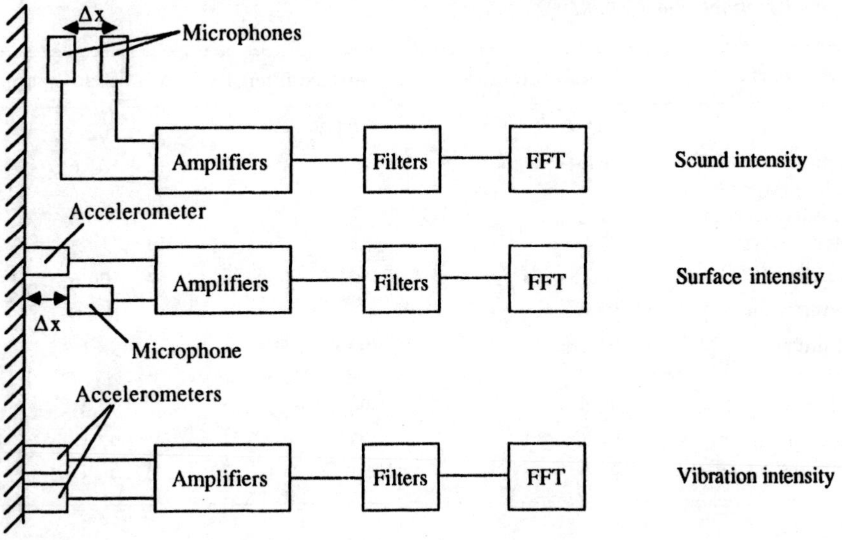

where Q_{pa} is the quadrature spectrum (imaginary part of the one-sided cross-spectral density), C_{pa} is the coincident spectrum (real part of the one-sided cross-spectral density), and ϕ is the phase shift between the two signals due to the instrumentation and due to the separation distance between the microphone and the accelerometer. This phase shift has to be accounted for in the analysis. The time lag phase shift can be evaluated for each measurement point, and the instrumentation phase shift has to be evaluated during the calibration procedure. A detailed discussion on the phase shift errors associated with surface intensity measurements is given by McGary and Crocker[8.25]. The total sound power radiated from the vibrating surface can be evaluated from[8.24]

$$\Pi = \sum_{j=1}^{n} \left\{ \sum_{i=1}^{N} (Q_{\mathrm{pa}_{ij}} \cos \phi_j + C_{\mathrm{pa}_{ij}} \sin \phi_j) A_i \right\} \frac{\Delta f}{2\pi f_j}, \tag{8.22}$$

where n is the number of data points in the frequency domain, N is the number of area increments (A_i) on the whole surface, and Δf is the frequency resolution (i.e. $\Delta f = f/n$). The main advantage of the surface intensity technique is that information is not required about the radiation ratio of a vibrating surface. It is also useful in highly reverberant spaces where a reverberant field exists very close to the surface of a machine. Its main disadvantage is that the phase difference between the microphone and the accelerometer has to be accurately accounted for.

Reinhart and Crocker[8.23] performed a series of experiments to compare the sound power radiated by different components of a diesel engine using the sound intensity technique, the surface intensity technique and the traditional lead wrapping technique. Their results are summarised in Table 8.2. The advantages of the sound intensity and

Table 8.2. *Sound power levels (dB re 10^{-12} W) of different diesel engine components (from Reinhart and Crocker[8.23]).*

Engine part	Sound intensity	Surface intensity	Lead wrapping
Oil pan	102.7	103.3	102.6
Engine manifold Turbocharger Cylinder head Valve covers	101.4	–	101.6
Aftercooler	100.8	101.9	100.6
Engine front	95.0	–	100.0
Oil filter Cooler	91.1	93.4	98.1
Left block wall	97.4	94.6	97.3
Right block wall	94.8	93.3	97.3
Fuel and oil pumps	91.5	–	96.3

the surface intensity techniques over the lead wrapping is apparent – in several instances, the lead wrapping technique overestimated the actual radiated sound power from components such as the engine front, the fuel and oil pumps, and the right block wall because of 'background noise' caused by the leakage of sound from other dominant sources through the wrapping.

The vibration intensity measurement technique is used to identify free-field energy flow due to bending waves in a solid. It can be shown[8.2] that the vibration intensity, I_v, in a given direction is

$$I_v = \frac{(B\rho_S)^{1/2}}{2\pi f \Delta x} \int_0^T \left\{ \frac{(a_1 + a_2)}{2} \int_0^t (a_2 - a_1)\,d\tau \right\} dt, \tag{8.23}$$

where B is the bending stiffness of the structure, ρ_S is the mass per unit area, Δx is the separation between the two accelerometers, and a_1 and a_2 are the two accelerometer signals. It should be noted that the scaling factor is frequency dependent.

Coherence and/or cross-correlation measurement techniques[8.5] are sometimes used for propagation path identification and for noise source identificaton. The technique is highly specialised and has to be used with caution. The general procedures for propagation path identification have been discussed in sub-section 8.3.5. For noise source identification, the usual procedure is to utilise a multiple input–single output model with accelerometers as the inputs and a single microphone as the output. This procedure has been successfully used on punch presses and diesel engines by evaluating the coherent output power for each of the different inputs. For instance, Crocker[8.26] discusses the usage of the coherence function for noise source identification in engines and vehicles. Signal processing problems arise when the various inputs are coherent amongst themselves (i.e. the procedure assumes that the various noise sources are incoherent). Data handling and computational errors including random, bias and time delay errors are also present and have to be quantified.

8.4.10 Structural components

Noise and vibration signals can be used for a variety of applications relating to quantification of the response characteristics of different types of structural components. Most of these applications have already been discussed in this book, and are therefore only briefly reviewed here. They are summarised in Figure 8.33.

Natural frequencies of structural components can be readily obtained by using frequency response functions. Either steady-state or impact techniques can be used. In the field, it is generally more convenient to use an impact hammer and an accelerometer and to obtain a frequency response function between the force input and the acceleration, velocity, or displacement output. In principle, by carefully monitoring the output response characteristics (at the identified natural frequencies) at different points on a structure, one can obtain the mode shapes for the first few natural frequencies – in practice this technique is limited to simple structural geometries such as vibrating beams, plates, etc.

Radiation ratio characteristics of different structural elements can be obtained from equation (3.30) by measuring the mean-square velocity response of the structure and the radiated sound power. The sound or surface intensity technique is the most appropriate means of obtaining the radiated sound power from an individual structural component of a complex piece of machinery.

Modal densities of structural components can be obtained from point mobility (frequency response function of velocity on force) measurements. These procedures have been discussed in chapter 6. Modal density information is very important for statistical energy analysis (S.E.A.) applications. It is also useful for obtaining an overview of the resonant response characteristics of different structural components. In addition to mobility, numerous other frequency response measurements are sometimes obtained including impedances, receptances, etc.

Noise and vibration signal analysis techniques can also be used to measure structural, acoustic radiation and coupling loss factors, and the energy flow between coupled structures. Once again, these procedures have been discussed in some detail in chapter 6. Like modal densities, these parameters are very important for S.E.A. applications. Furthermore, damping characteristics of structural components are also of general interest in structural dynamics. Damping characteristics of individual structural modes can be obtained directly from frequency response functions (i.e. the 3 dB points). When the damping characteristics of groups of closely spaced modes are required, reverberation decay or steady-state energy flow techniques are recommended. These techniques provide information about band-averaged values.

An additional procedure which, as yet, has not been discussed is the identification of impulse excitation spectra. The force spectrum of an impulse process can be used

Fig. 8.33. The usage of noise and vibration signals to quantify the response characteristics of structural components.

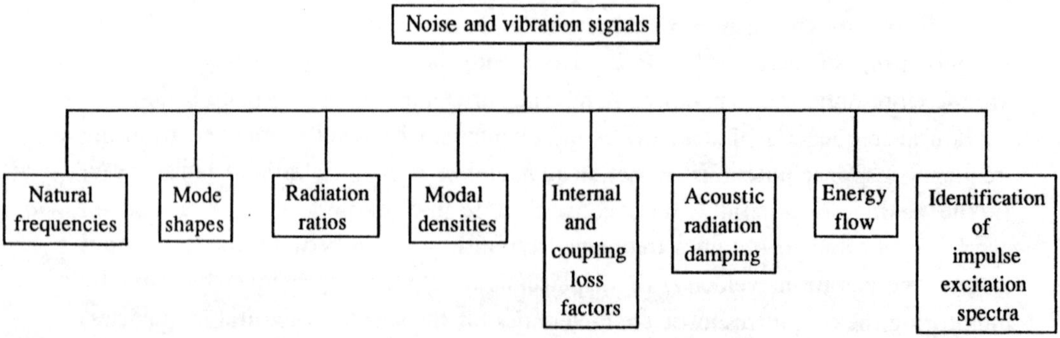

to identify the characteristics of the impulse rise time and to quantify the type of excitation force. An ideal 'impact' has a continuous broadband frequency spectrum. A real 'impulse' on the other hand, has a finite rise time and this manifests itself as a high frequency drop off in the energy spectrum. This point is illustrated in Figure 8.34. It can be shown[8.6] that (i) if an impulse rise time is very abrupt (i.e. a step force input), the high frequency content decays at 6 dB per octave. (ii) if the rise time (force onset) increases linearly, the high frequency content decays at 12 dB per octave, and (iii) if the rise time is quadratic, the high frequency content decays at 18 dB per octave. The transition between the low and high frequency regions is established by the time constant of the interaction. The time constant is approximated by c_v/m, where c_v is a viscous-damping representation of the dissipated vibrational energy, and m is the mass of the impacting body[8.6]. Thus, the high frequency content of a rigid plate impact would decay at 12 dB per octave, and the high frequency content of a flexible plate impact would decay at 6 dB per octave. The force spectrum of a compliant sphere impacting an elastic plate is illustrated schematically in Figure 8.35. The initial 6 dB per octave decay is due to the abrupt increase in force on the flexible plate. The final 12 dB per octave roll off is associated with contact stiffness.

Fig. 8.34. Schematic illustration of ideal 'impact' and real 'impulse' spectra.

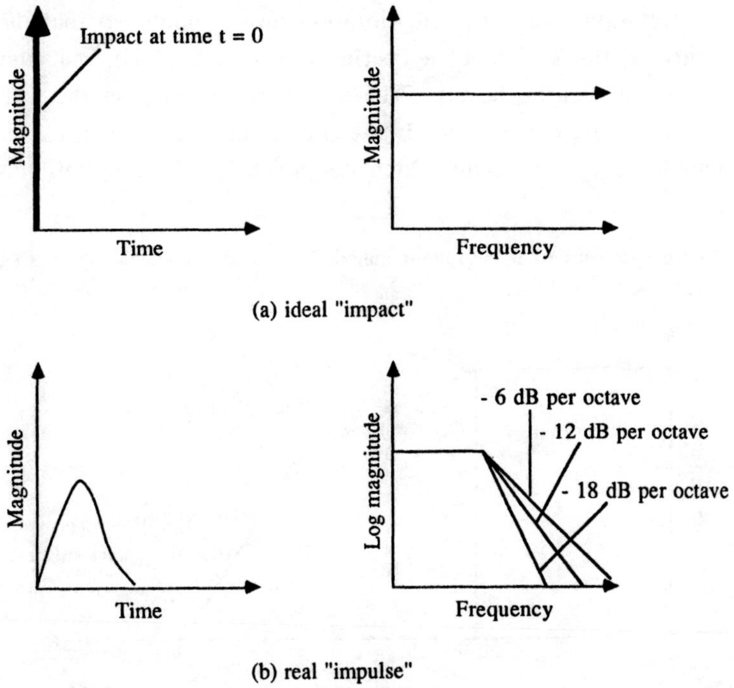

8.4.11 Vibration severity guides

Several vibration severity guides are available for assessing the severity of machinery vibrations. The major standards currently in use include the I.S.O. standards (International Standards Organisation), the V.D.I. standards (German National Standards), and the B.S. standards (British National Standards). The most widely used vibration severity criteria, which are common to I.S.O., V.D.I. and B.S. (I.S.O. 2372, V.D.I. 2056 and B.S. 4675) are based on broadband vibrational velocity levels between 10 Hz and 1000 Hz. Practical experience has, however, illustrated that quite often significant frequency components are present at higher frequencies. When this is the case, the standards have to be interpreted with caution, and it is often best to carefully trend increases in vibrational levels instead. The Canadian Government Specification, CDA/MS/NVSH 107: Vibration Limits for Maintenance, is a table of criteria for bearing vibration measurements from 10 Hz up to 10 000 Hz, thus allowing for better detection at higher frequencies.

The I.S.O. 2372, V.D.I. 2056 and B.S. 4675 vibration severity criteria for bearings of rotating machinery (10 Hz to 1000 Hz) are presented in Figure 8.36, and the Canadian Government Specification, CDA/MS/NVSH 107 vibration severity criteria (10 Hz to 10 000 Hz) for bearing vibration maintenance limits are presented in Table 8.3. It is very important to note that Figure 8.36 relates to broadband vibrational velocities between 10 Hz and 1000 Hz and that Table 8.3 relates to broadband vibrational velocities between 10 Hz and 10 000 Hz.

A major limitation of all absolute vibration severity guides is that the vibration measurements on the surface of a bearing also contains the frequency response characteristics of the bearing housing. The further away one is from the 'true' vibration signal to be monitored, the larger is the probability that the measured signal is 'contaminated'. This point was illustrated in Figure 8.3, and the limitation is overcome

Fig. 8.35. Force spectrum of a compliant sphere impacting an elastic plate. (Adapted from Lyon[8.6].)

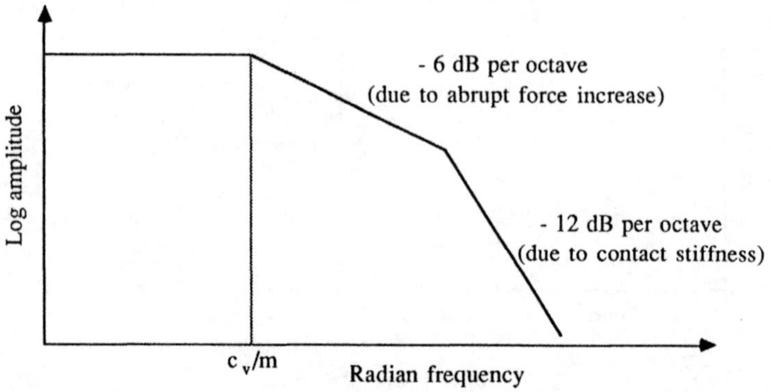

in practice by trending the data at specific measurement points over lengthy periods of time in order to observe relative increases from a predefined baseline spectrum.

As a general rule, at frequencies up to 1000 Hz, increases in vibrational velocity of factors of up to 2.5 are considered to be significant, and increases of factors of 10 and more require immediate attention. At frequencies above 4000 Hz, the above factors are increased to 6 and 100, respectively. Quite often, these guidelines provide a very reliable and useful indication of a machine's general condition.

Eshleman[8.27] provides a comprehensive discussion on a range of vibration standards including ships, aircraft, structures, etc., together with an extensive list of references including numerous national standards.

8.5 Some specific test cases

This final section of the book is devoted to some specific test cases. Five cases are discussed and they have been picked at random. Only relevant parts of the test cases, pertaining to the usage of noise and vibration as a diagnostic tool, are discussed. The five test cases include: (i) cabin noise source identification on a load–haul–dump vehicle; (ii) noise and vibration source identification on a large induction motor; (iii) identification of rolling-contact bearing damage; (iv) flow-induced noise and vibration

Fig. 8.36. Vibration severity criteria for 10 Hz to 1000 Hz (I.S.O. 2372, V.D.I. 2056, and B.S. 4675.)

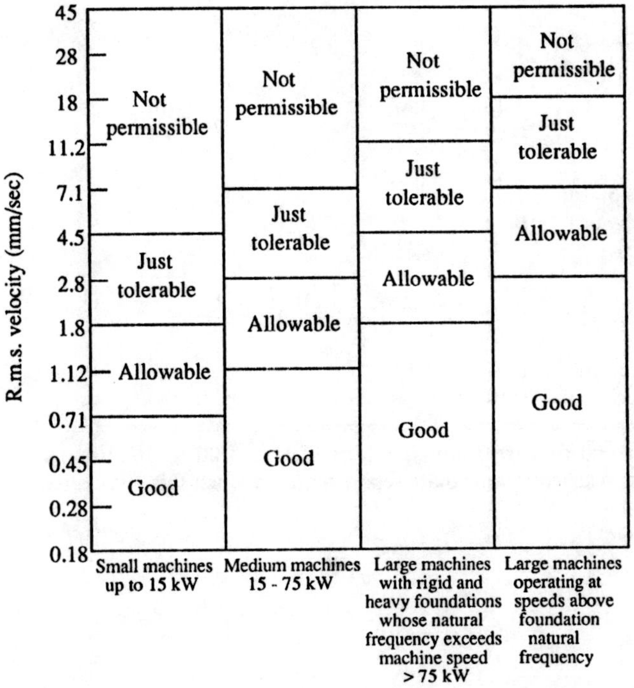

Table 8.3. *Canadian Government Specification CDA/MS/NVSH 107 for bearing vibration limits (10 Hz to 10 000 Hz) for maintenance.*

	For new machines		For worn machines	
	Long life[a] (mm s^{-1})	Short life[b] (mm s^{-1})	Check level[c] (mm s^{-1})	Recondition[d] (mm s^{-1})
Gas turbines				
(over 20 kHP)	7.9	18	18	32
(6 to 20 kHP)	2.5	56	10	18
(up to 5 kHP)	0.79	3.2	5.6	10
Steam turbines				
(over 20 kHP)	1.8	18	18	32
(6 to 20 kHP)	1.0	5.6	18	32
(up to 5 kHP)	0.56	3.2	10	18
Compressors				
(free piston)	10	32	32	56
(HP air, air cond)	4.5	10	10	18
(LP air)	1.4	5.6	10	18
(refridge)	0.56	5.6	10	18
Diesel generators	1.4	10	18	32
Centrifuges	1.4	10	18	32
Gear boxes				
(over 10 kHP)	1.0	10	18	32
(10 HP to 10 kHP)	0.56	5.6	18	32
(up to 10 HP)	0.32	3.2	10	18
Boilers	1.0	3.2	5.6	10
Motor generator sets	1.0	3.2	5.6	10
Pumps				
(over 5 HP)	1.4	5.6	10	18
(up to 5 HP)	0.79	3.2	5.6	10
Fans				
(below 1800 r.p.m.)	1.0	3.2	5.6	10
(above 1800 r.p.m.)	0.56	3.2	5.6	10
Electric motors				
(over 5 HP)	0.25	1.8	3.2	5.6
(up to 5 HP)	0.14	1.8	3.2	5.6
Transformers				
(over 1 kVA)	0.14	–	0.56	1.0
(up to 1 kVA)	0.10	–	0.32	0.56

([a] long life is ~1000 to 10000 hours, [b] short life is ~100 to 1000 hours, [c] service required when this level is reached, [d] immediate repair required when this level is reached in any octave band) associated with a gas pipeline; and (v) flow-induced noise and vibration associated with an aluminium hulled racing sloop (yacht).

8.5.1 Cabin noise source identification on a load–haul–dump vehicle

A company that manufactures/assembles large load–haul–dump vehicles was concerned with excessive noise levels in the operator cabin. Some preliminary work had suggested that the main noise source was gearbox noise, and it had to be established whether or not the quietening of the gearbox (by redesign or by acoustic treatment) would in fact reduce the noise levels to an acceptable level.

A noise and vibration survey of the operator cabin was subsequently conducted. All sound pressure level measurements were obtained at the position of the driver's head, and all vibration measurements were obtained on the gearbox cover, which was located below the driver cabin.

The linear sound pressure levels and the A-weighted sound levels at the driver position are presented in tabular form in Table 8.4 for several operating conditions (which are described and labelled as tests 1 to 9). The average A-weighted sound level at the driver position during the various modes of operation were well in excess

Table 8.4. *Overall sound pressure level (driver position) and acceleration level (gearbox cover) measurements*

Test	Overall sound pressure level (driver position)		Acceleration level (gearbox cover)	
	(dB(Lin))	(dB(A))	(dB(Lin)) re 1 g r.m.s.	(dB(A)) re 1 g r.m.s.
(1) Low idle with no hydraulics	93	84	27.4	25.5
(2) Maximum idle with no hydraulics (full throttle)	105.4	99.4	40.8	40.6
(3) Full throttle, hydraulic stall	105.8	99.5	41.0	40.6
(4) Full throttle, converter stall	105.2	96.4	25.6	22.2
(5) Moving forward unloaded in first gear	104.2	100.2	42.4	42.0
(6) Moving forward unloaded in second gear	106.6	101.5	43.5	43.6
(7) Pushing, digging, etc. in pit	107.0	103.0	45.8	42.8
(8) (a) Moving forward loaded in first gear	104.2	99.5	42.0	41.5
(b) Reversing loaded in first gear	105.0	100.0	42.5	41.6
(9) (a) Moving forward loaded in second gear	105.2	100.8	43.2	43.0
(b) Reversing loaded in second gear	106.8	104.2	45.6	45.5

of the 90 dBA limit. Vibrational acceleration measurements on the gearbox cover showed levels of up to 45 dBA re 1 g.

The data were also narrowband spectral analysed (12.5 Hz resolution) in an attempt to identify the relative magnitudes of the major noise sources. A-weighted spectra for the sound levels at the operator position, whilst the loader was stationary at idle speed (with no hydraulics) and at full throttle (with no hydraulics) are shown in Figure 8.37. Full throttle with converter stall can be considered to be a datum for the overall engine noise at the operator position because, in that state, the transmission is stationary. The spectra for full throttle hydraulic stall, in third gear, are also presented in Figure 8.37. The spectra for maximum idle with no hydraulics and full throttle with hydraulic stall are similar, and this is reflected in the overall sound levels (tests 2 and 3). The A-weighted spectra for the full throttle converter stall are somewhat lower than those for hydraulic stall or maximum idle, the difference ($\sim$4 dB(A)) being due to gearbox transmission noise.

Sound pressure levels at the operator position whilst the loader was moving forward unloaded in first and second gears are presented in Figure 8.38. In addition to the low frequency noise (< 500 Hz), which can be associated directly with the engine, there are spectral increases (cf. with Figure 8.37, test 2) centred approximately on

Fig. 8.37. Spectrum of A-weighted sound levels at the driver position (tests 1–4).

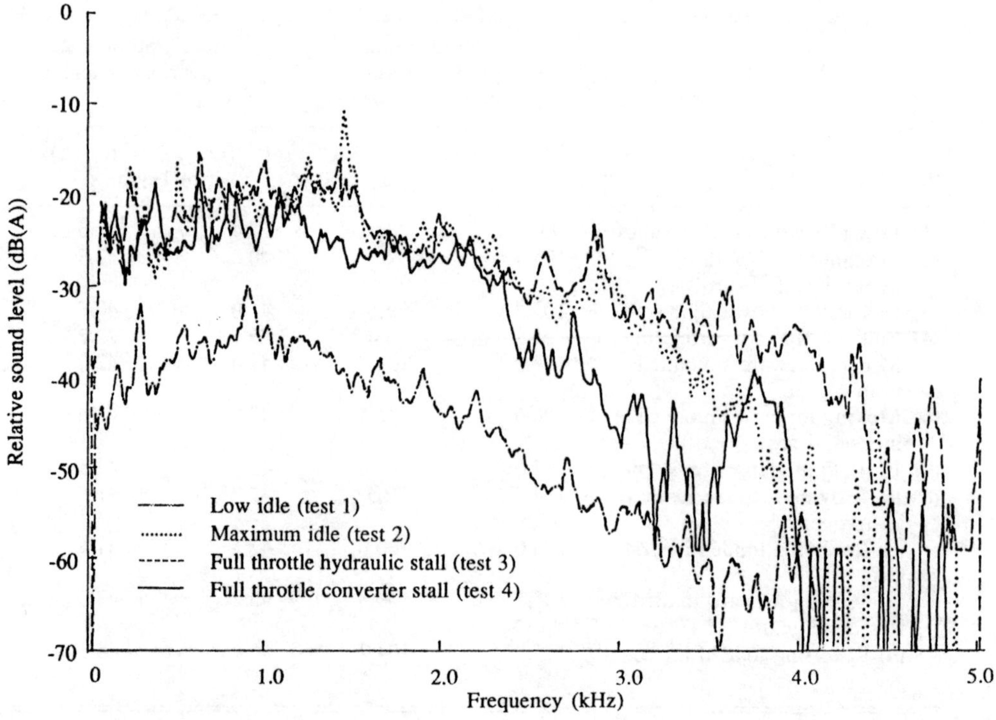

1250 Hz, 2250 Hz and 2700 Hz. There are corresponding increases at those frequencies in the acceleration level spectra on the gearbox cover, as illustrated in Figure 8.39. Similar trends are observed in Figure 8.40 and Figure 8.41 for the loader moving forward and reversing, whilst loaded in the first gear, and in Figure 8.42 and Figure 8.43 for second gear. The identical characteristics of the spectral peaks of the vibrational acceleration measurements on the gearbox cover and the noise measurements at the driver position indicate that the various peaks are associated with gear meshing frequencies and corresponding harmonics. The sharp peak centred on ~2250 Hz during the reversal motion is associated with the gear meshing frequency of the reverse gearing mechanism. Further quantitative confirmation is obtained from an inspection of Figure 8.37 (test 4) for which the vehicle was stationary and the gears not meshing – the spectral increases that are present in Figures 8.38–8.43 are absent here.

Estimates from the available data suggest that the elimination of the vibration and noise due to the gear meshing would only reduce the sound levels by ~3 to 6 dB(A) at the very most. The engine noise alone is of the order of 96 dB(A) (test 4), and no further reduction could be expected without treatment of that noise source.

The main conclusion resulting from the investigation was that, whilst gear meshing

Fig. 8.38. Spectrum of A-weighted sound levels at the driver position (tests 5 and 6).

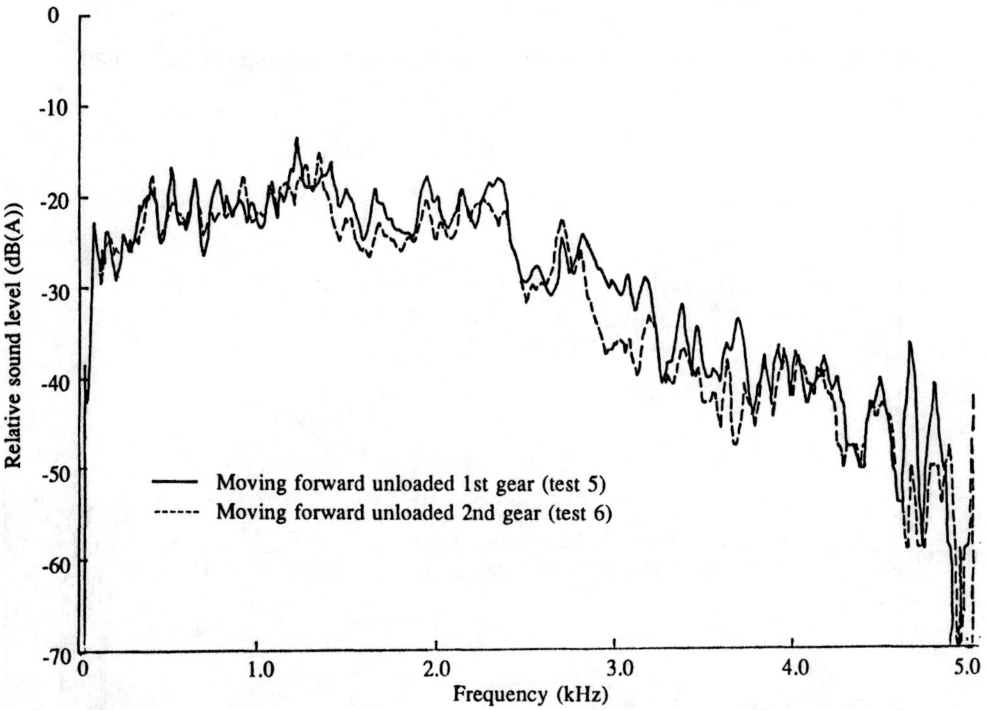

Fig. 8.39. Vibrational acceleration spectrum on the gearbox cover (tests 5 and 6).

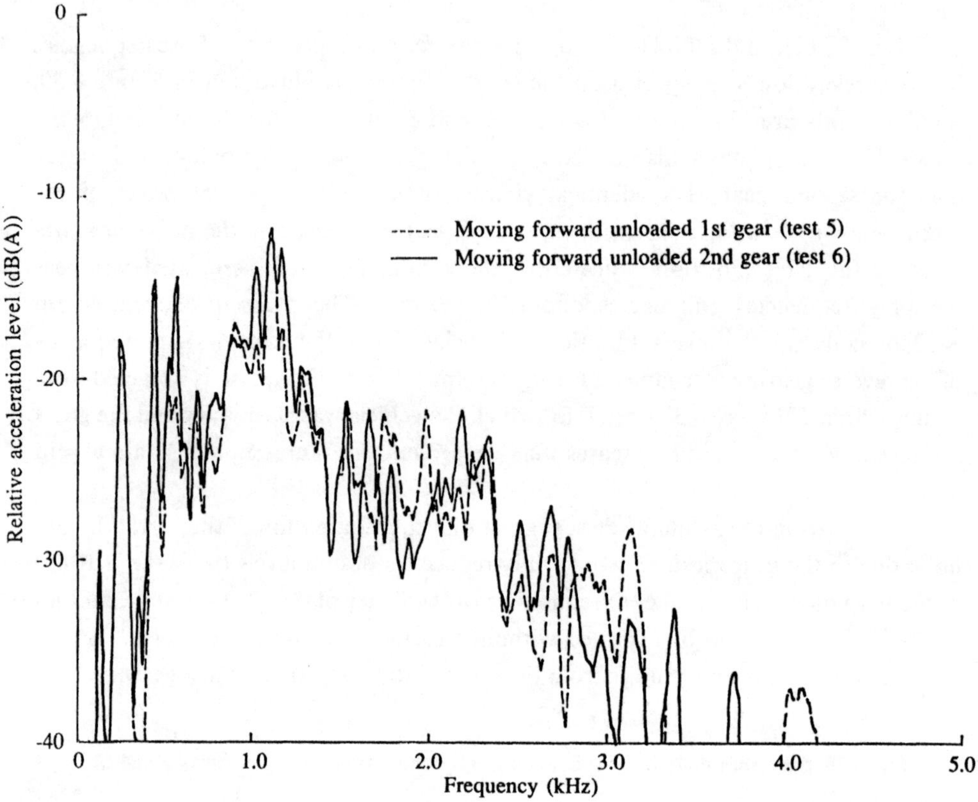

Fig. 8.40. Spectrum of A-weighted sound levels at the driver position (tests 8(a) and 8(b)).

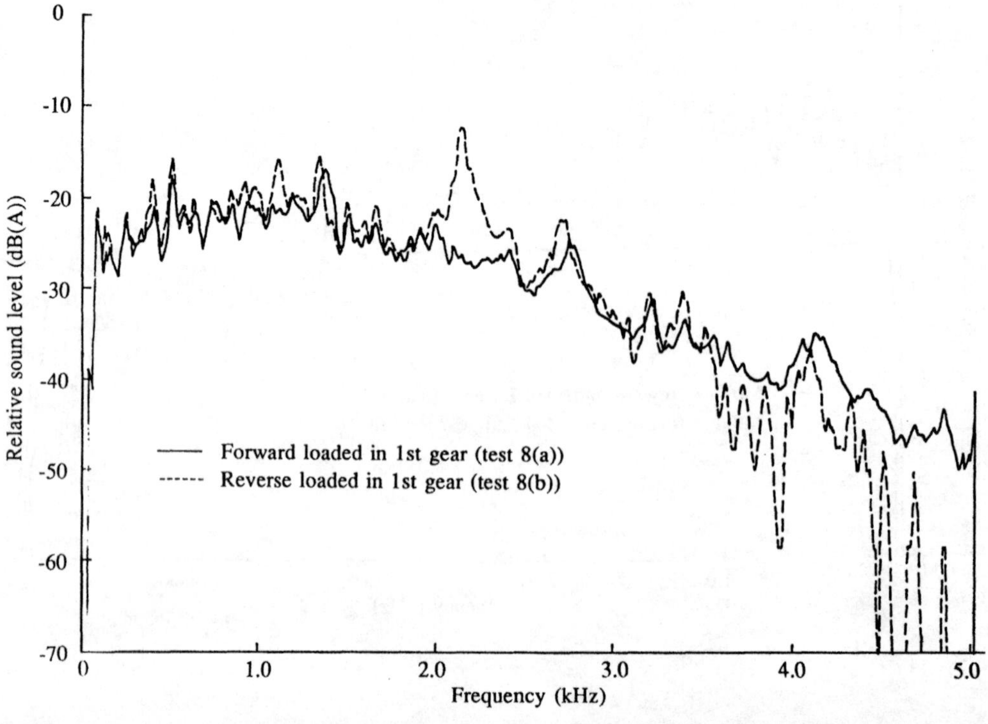

Fig. 8.41. Vibrational acceleration spectrum on the gearbox cover (tests 8(a) and 8(b)).

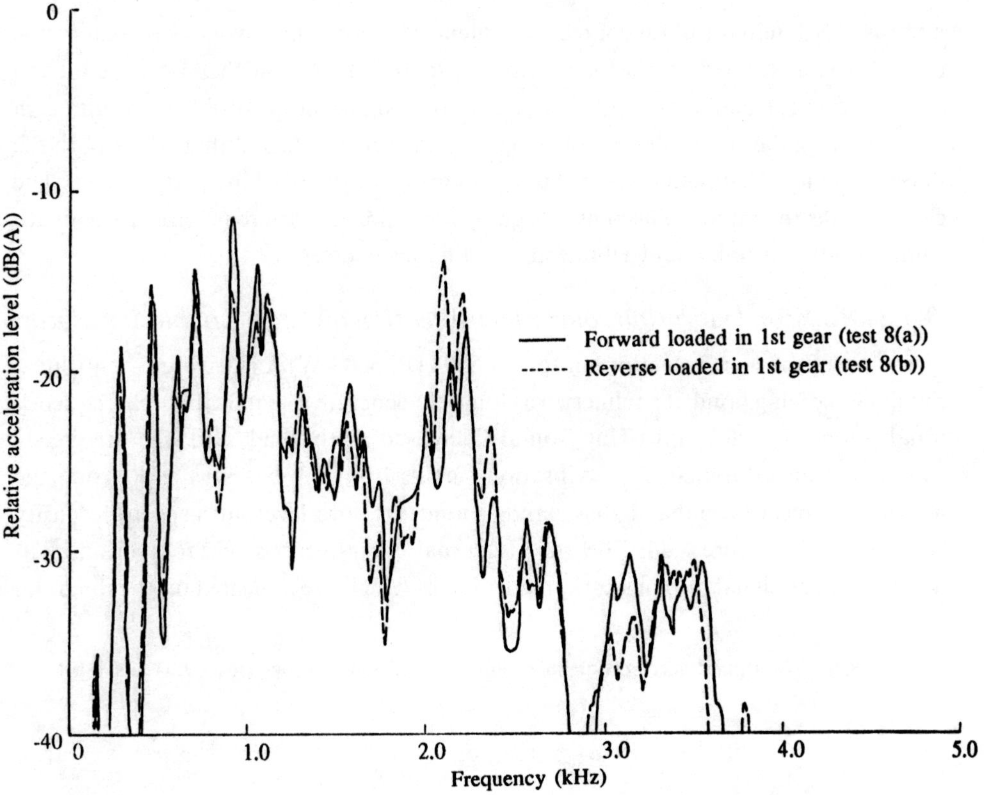

Fig. 8.42. Spectrum of A-weighted sound levels at the driver position (tests 9(a) and 9(b)).

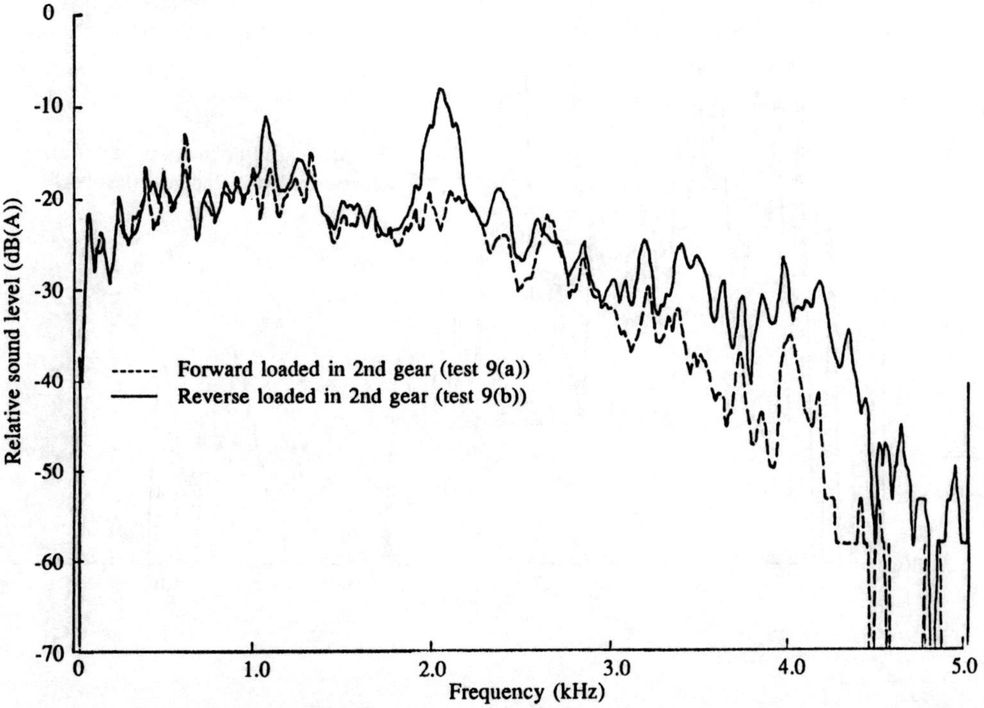

noise is the dominant noise source, treatment of the gearbox would not reduce the noise levels at the driver position below ~96 dB(A). Any further noise reduction would require a detailed investigation into the engine noise problem, resulting in either altering the engine design, or designing an effective shroud that meets both the noise reduction requirements and the requirements of the machine performance. The gearbox noise could be reduced by designing an enclosure with removable panels with sound absorbent lining and vibration isolating fasteners.

8.5.2 Noise and vibration source identification on a large induction motor

A large squirrel cage induction motor (~650 HP, 550 kW) for driving a centrifugal pump in a cooling pond at a refinery was found to generate a significant high frequency tonal sound (~1400 Hz). This sound increased with load and the frequency component also dominated the vibration signals from the bearings supporting the motor. A typical broadband (baseband) sound pressure level auto-spectral density is presented in Figure 8.44. The 1400 Hz tonal component is by far the dominant frequency. The signal also has certain harmonics which are indicated on the diagram.

Fig. 8.43. Vibrational acceleration spectrum on the gearbox cover (tests 9(a) and 9(b)).

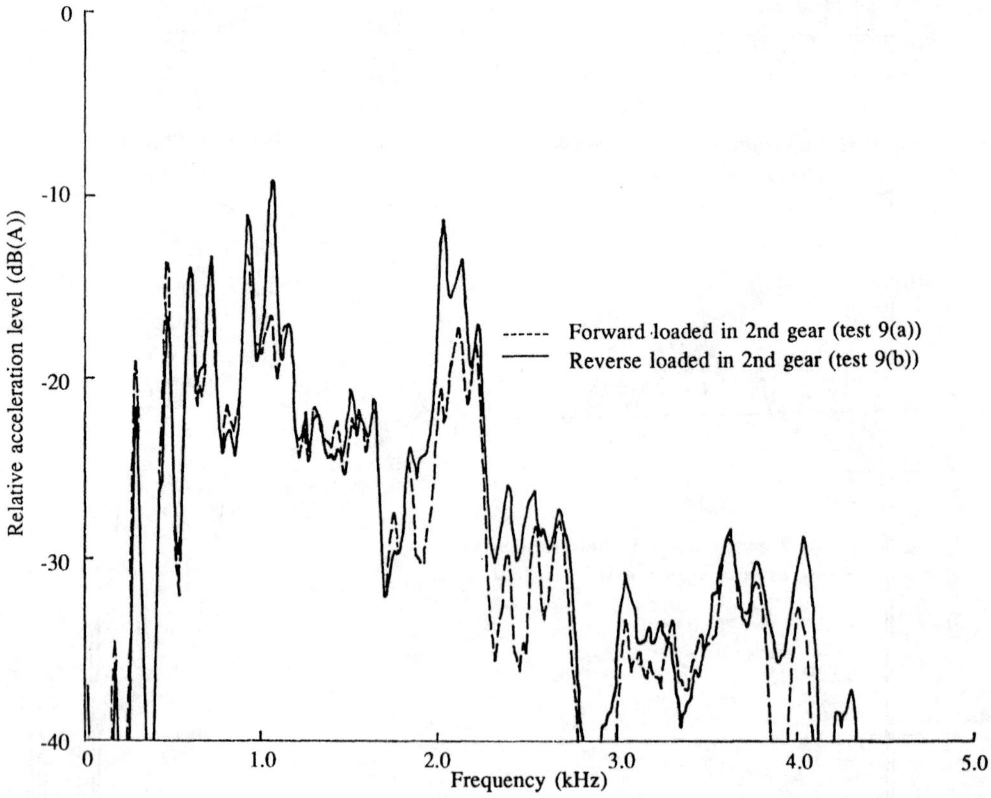

A typical bearing vibration acceleration signal is presented in Figure 8.45. Once again, the 1400 Hz tonal component is by far the dominant frequency, and the harmonic content is clearly identified.

A cepstrum analysis was conducted to confirm the periodicity of the signal. A typical power cepstrum for a vibration signal from the bearings supporting the motor is presented in Figure 8.46. The dominant quefrency peak at ~0.72 ms corresponds

Fig. 8.44. Typical baseband sound pressure level auto-spectral density from an electric motor.

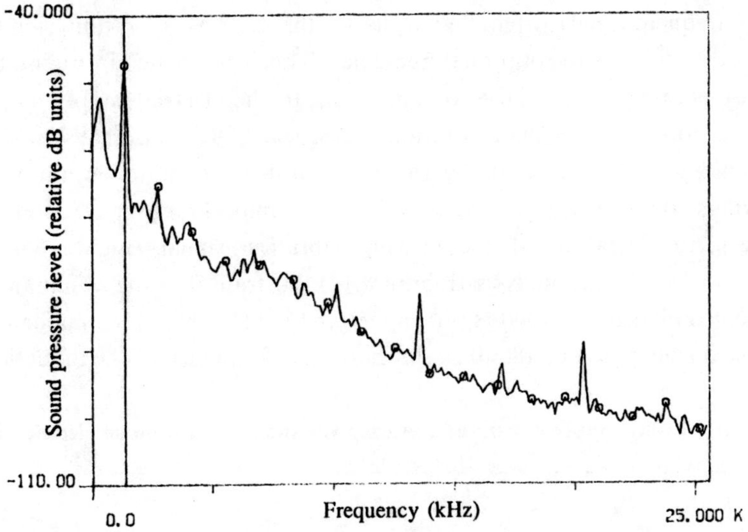

Fig. 8.45. Typical baseband acceleration auto-spectral density of a bearing vibration signal from an electric motor.

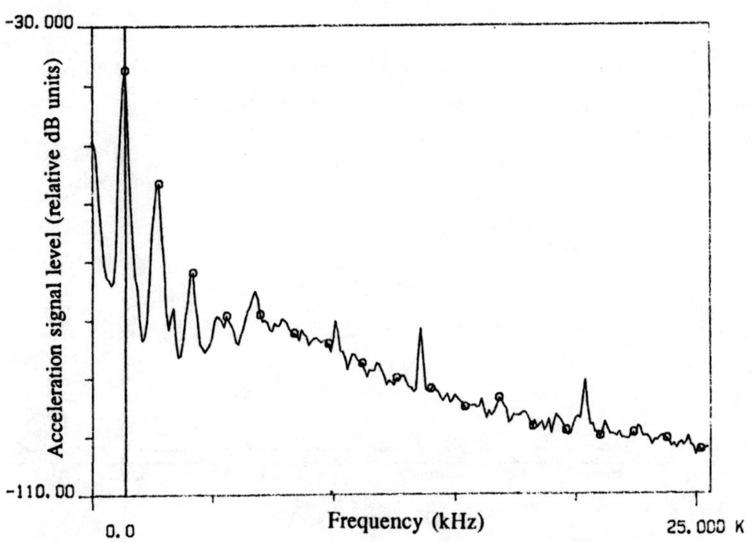

to a periodic signal at ~1389 Hz. Figure 8.46 clearly illustrates how the power cepstrum brings out the periodicity in a signal.

From the discussions in sub-section 8.4.8, noise and vibration sources in electric motors can be (i) mechanical, (ii) aerodynamic, and (iii) electromagnetic. It is very easy to quantify the electrómagnetic content of a noise or vibration signal from an electric motor – if the signal disappears instantaneously when the machine is switched off, then the source is electromagnetic. This was in fact the case for the dominant spectral component in Figures 8.44 and 8.45 – i.e. the forces and resultant vibrations were only generated whilst the electrical power was being applied.

A low frequency narrowband analysis of the bearing vibration signal clearly identified the 25 Hz shaft rotational frequency. This is illustrated in Figure 8.47. The 50 Hz discrete frequency, whilst corresponding to the electrial supply frequency, is also a harmonic of the shaft rotational frequency. The discrete low frequency component (~3 Hz) is associated with base motion – the induction motor unit and the centrifugal pump were mounted on a steel, mezzanine floor over the cooling pond.

The cepstrum analysis of the bearing vibration signal (Figure 8.46) and a narrowband (passband) analysis (Figure 8.14) illustrate that the dominant spectral component (nominally ~1400 Hz) is in fact at 1389 Hz. The three major peaks can be identified from equation (8.20) as slot harmonic frequencies. The induction motor

Fig. 8.46. Typical power cepstra of a bearing vibration signal from an electric motor.

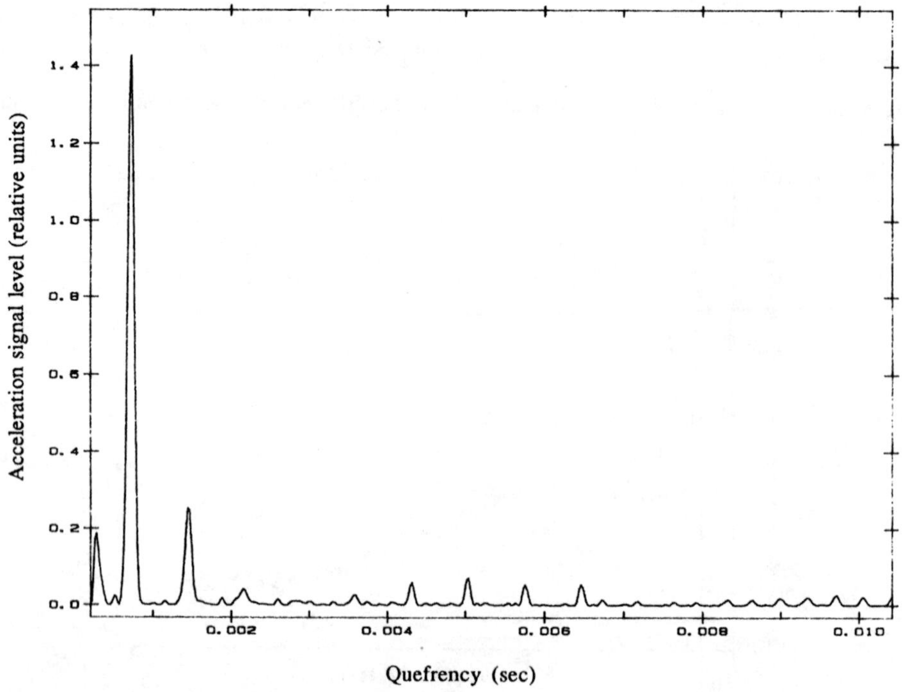

has fifty-two rotor bars, sixty stator slots, and four magnetic poles, and without any slip this corresponds to slot harmonic frequencies at 1300 Hz, 1400 Hz and 1500 Hz, etc. The dominant peak at 1389 Hz is thus associated with the second harmonic of the slot harmonic frequency, or the arithmetic average of the number of rotor bars and stator slots. The 11 Hz difference between the estimated 1400 Hz and the observed 1389 Hz is associated with slip between the rotating speed of the magnetic field and the rotating speed of the armature. In sub-section 8.4.8 it was mentioned that electromagnetic irregularities associated with dynamic eccentricity in a rotor produce modulation of the dominant slot harmonic frequency at $\pm$ the rotational frequency, and at $\pm$ the slip frequency. This phenomenon is clearly observed in Figure 8.14, indicating that the dominant noise and vibration signal is (a) electromagnetic, and (b) associated with dynamic air gap eccentricity.

8.5.3 Identification of rolling-contact bearing damage

Some laboratory controlled experimental results are presented in this sub-section on the identification of rolling-contact bearing damage. Single row tapered roller bearings (NSK 30204) with a 20 mm bore (pitch diameter, $D = 34$ mm; roller diameter, $d = 6$ mm; contact angle, $\phi = 12.96°$; number of rolling elements, $Z = 15$) were tested at different rotational speeds. Tapered roller bearings were chosen because they are easily dismantled in order to introduce artificial defects. It should be pointed out to the reader that all the spectra presented in this sub-section have been obtained from laboratory test records, and have been digitised manually on a graphics tablet with

Fig. 8.47. Low frequency passband auto-spectral density of a bearing vibration signal from an electric motor.

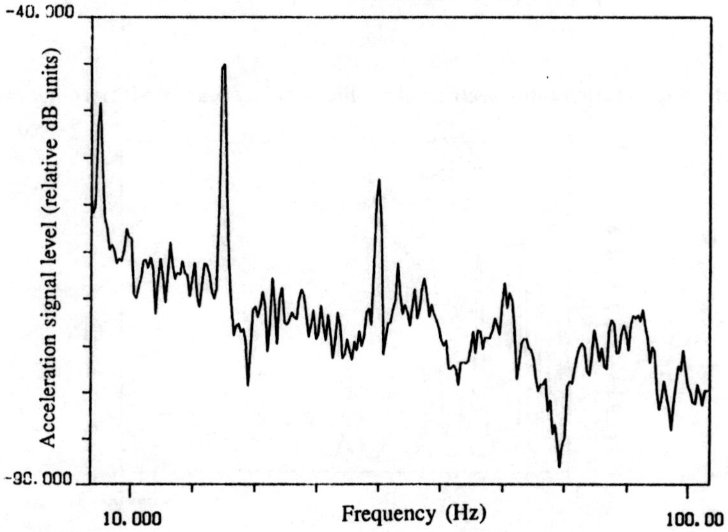

a wand for presentation in this book. Hence, the peaks, troughs, frequency scale, dips and rises in the spectra are not perfect representations of the actual spectra in the laboratory records. They are, however, adequate for illustrative purposes and quantitative discussions on the identification of rolling-contact bearing damage.

A typical bearing vibration (acceleration) auto-spectrum of an undamaged bearing is presented in Figure 8.48. The shaft rotational frequency and its associated harmonics are clearly identified. The important point about this spectrum is that all the distinct peaks observed correspond to the fundamental frequency and its harmonics.

A typical bearing vibration (acceleration) auto-spectrum of a bearing with a discrete defect on the outer race is presented in Figure 8.49. The dominant signal is at the rolling element pass frequency on the outer race, f_{repfo}, which is given by equation (8.8), and its associated harmonics. Discrete outer race defects are easily detected since the transducer can often be mounted on the outer race itself, minimising any frequency response function effects of the transmission path.

Fig. 8.48. Bearing vibration auto-spectrum of a rolling-contact bearing in good condition.

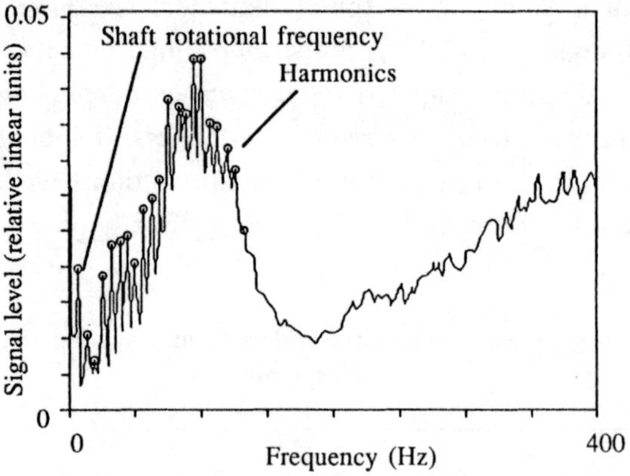

Fig. 8.49. Bearing vibration auto-spectrum of a rolling-contact bearing with an outer race defect.

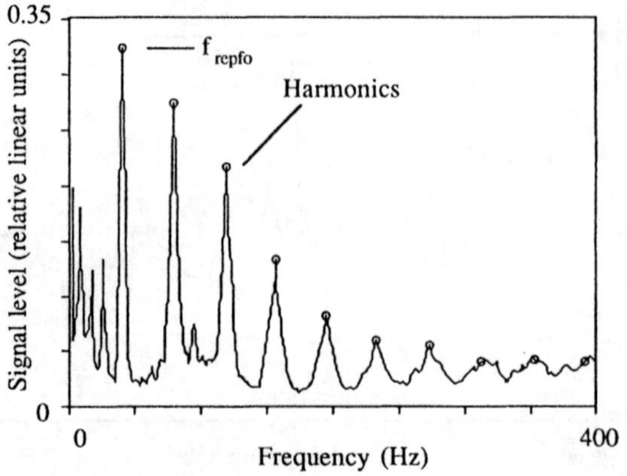

A typical bearing vibration (acceleration) auto-spectrum of a bearing with a discrete roller defect is presented in Figure 8.50. The rolling element spin frequency (equation 8.10) is not readily identified in this instance, although its harmonics are. Furthermore, if both outer race and roller defects are present, the outer race defects tend to dominate the spectra because of their proximity to the transducer and the roller defects tend to be submerged. Under these conditions, spectral analysis by itself is generally not adequate and one has to turn to more sophisticated condition monitoring techniques. The power cepstrum is a particularly useful technique for identifying and separating different periodic families. The power cepstrum corresponding to the bearing vibration signal associated with Figure 8.50 is presented in Figure 8.51. The quefrency peak at ~17.5 ms corresponds to the rolling element spin frequency, f_{resf}.

When the bearing defects are shallow (early damage), it is not always easy to detect bearing damage from a frequency analysis in the range of the fundamental bearing

Fig. 8.50. Bearing vibration auto-spectrum of a rolling-contact bearing with a roller defect.

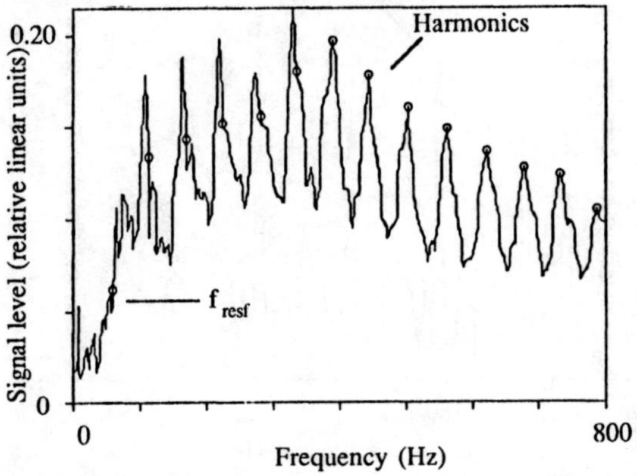

Fig. 8.51. Bearing vibration power cepstrum of a rolling-contact bearing with a roller defect.

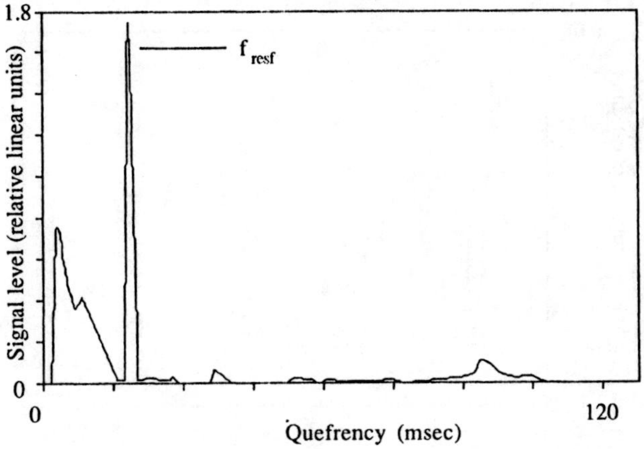

frequencies. As previously mentioned (see sub-section 8.3.5), in the early stages of a developing fault the impulses produced by the fault are very short in duration and the energy associated with the impulse is usually distributed over a wide frequency range. Under these circumstances, the envelope power spectrum is a useful means of identifying the different harmonic components associated with the defect. The bearing vibration (acceleration) envelope power spectrum of a rolling-contact bearing with a shallow outer race defect is illustrated in Figure 8.52. The corresponding auto-spectrum is illustrated in Figure 8.53. Because the outer race defect is shallow, the harmonic content of the rotating element pass frequency on the outer race is not as clearly identifiable as in Figure 8.49 which relates to a larger discrete defect. In fact, the spectrum in Figure 8.53 is not dissimilar to the spectrum in Figure 8.48 which

Fig. 8.52. Bearing vibration envelope power spectrum of a rolling-contact bearing with a shallow outer race defect.

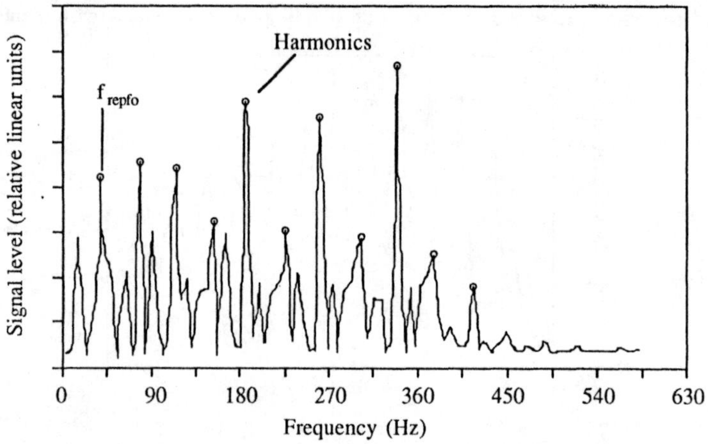

Fig. 8.53. Bearing vibration auto-spectrum of a rolling-contact bearing with a shallow outer race defect.

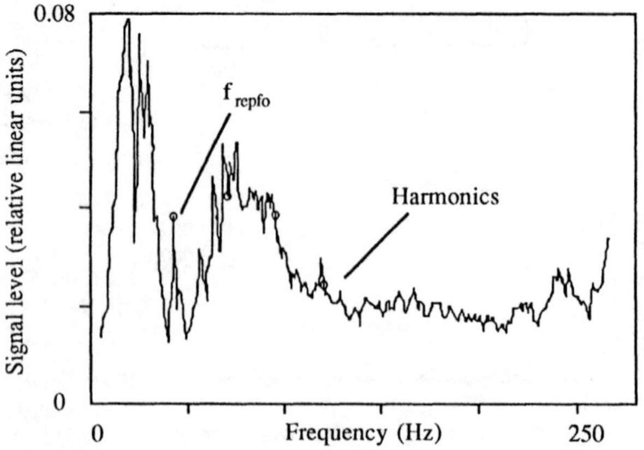

is for a rolling element bearing in good condition. However, the envelope power spectrum (Figure 8.52) contains information about the impact rate and the amplitude modulation and therefore allows for the identification of the rolling element pass frequency on the outer race and the various harmonics.

The bearing vibration (acceleration) envelope power spectrum of a rolling-contact bearing with a roller defect is illustrated in Figure 8.54. Once again, the rolling element spin frequency and its various harmonics are clearly identified. The envelope power spectrum is thus useful for separating the harmonics of different discrete defects.

Envelope power spectra are only useful for providing diagnostic information about early damage because they highlight the various impulsive peaks and their associated harmonics. When a bearing is in good condition, the envelope power spectrum is generally broadband. Any peaks which are present tend to be very low in amplitude compared to those obtained from defective bearings. Also, the frequency distribution tends to be random. When the bearing defects are distributed (late damage) the envelope power spectrum is also broadband with a random frequency distribution. The main difference between this condition and the 'good' condition is that the envelope power spectrum levels associated with late damage are significantly higher. A typical bearing vibration (acceleration) envelope power spectrum of a rolling-contact bearing in good condition is presented in Figure 8.55.

8.5.4 Flow-induced noise and vibration associated with a gas pipeline

In chapter 7 it was illustrated that coincidence between higher order acoustic modes inside a cylindrical shell and pipe wall structural modes is a dominant mechanism

Fig. 8.54. Bearing vibration envelope power spectrum of a rolling-contact bearing with a roller defect.

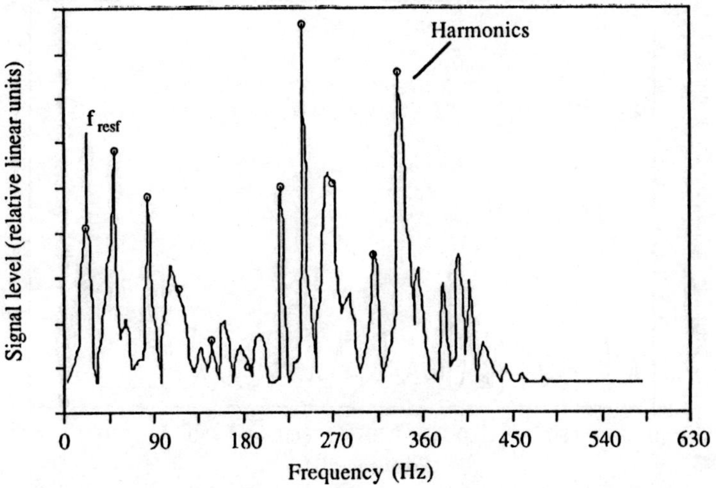

for the generation of flow-induced noise and vibration in pipelines. Some field data relating to a gas pipeline installation where this is the case is presented here.

Excessive noise and vibration levels were experienced at a gas pipeline installation. It was observed that the noise and vibration originated from various tee-junction intersections, and that it propagated for large distances along straight runs of pipeline. A typical pipe wall acceleration spectrum is presented in Figure 8.56, and the corresponding externally radiating sound pressure level spectrum at some appropriate radial distance from the gas pipeline is presented in Figure 8.57. The internal diameter of the steel pipeline is 0.914 m, the speed of sound in the internal gas is ~ 385 m s^{-1}, and the mean flow Mach number is ~ 0.1. The precise details of the measurement locations are not directly relevant to this test case as one is only concerned with identifying the mechanism of noise and vibration generation.

From Figures 8.56 and 8.57, it is very evident that there are large increases in pipe wall vibration and externally radiated noise at certain discrete frequencies. Also, there is a one to one correlation between the vibration discrete frequencies and the radiated noise discrete frequencies. A close examination of Figures 8.56 and 8.57 reveals that the first three dominant peaks are at 275 Hz, 461 Hz and 676 Hz.

Having studied the nature of the noise and vibration spectra, it is evident from the discussions in chapter 7 that coincidence is a possible source of the dominant peaks at 275 Hz, 461 Hz and 676 Hz. The cut-off frequencies of the various higher order acoustic modes can be readily evaluated from equation (7.8) and Table 7.1. The cut-off frequencies are given by

$$(f_{co})_{pq} = \frac{\pi\alpha_{pq}c_i(1-M^2)^{1/2}}{2\pi a_i}, \qquad \text{(equation 7.8)}$$

Fig. 8.55. Bearing vibration envelope power spectrum of a rolling-contact bearing in good condition.

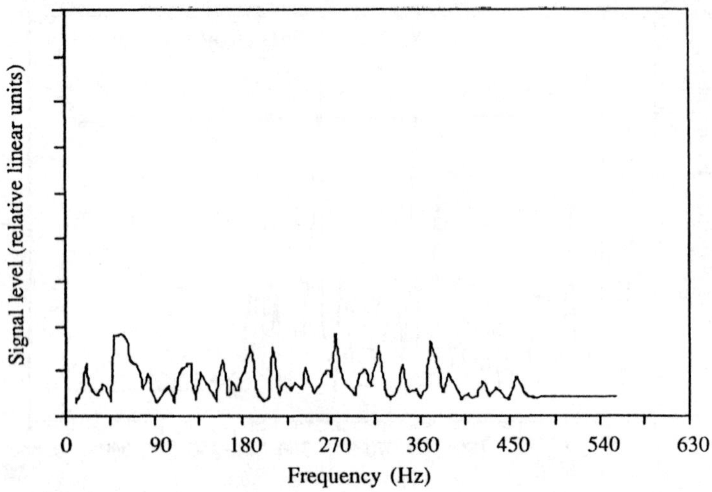

where c_i is the speed of sound inside the pipe, M is the mean flow Mach number, a_i is the internal pipe radius, and the $\pi\alpha_{pq}$'s are obtained from Table 7.1. In chapter 7 (section 7.5) it was shown that the coincidence frequency, f_c, is usually higher than the cut-off frequency, f_{co}, of the relevant higher order acoustic mode. This important point is clearly illustrated in Figure 7.7 (the complete coincidence frequencies will

Fig. 8.56. Typical pipe wall acceleration spectrum for a gas pipeline (speed of sound $\sim$385 m s^{-1}; mean flow Mach number $\sim$0.1; internal diameter $\sim$0.914 m).

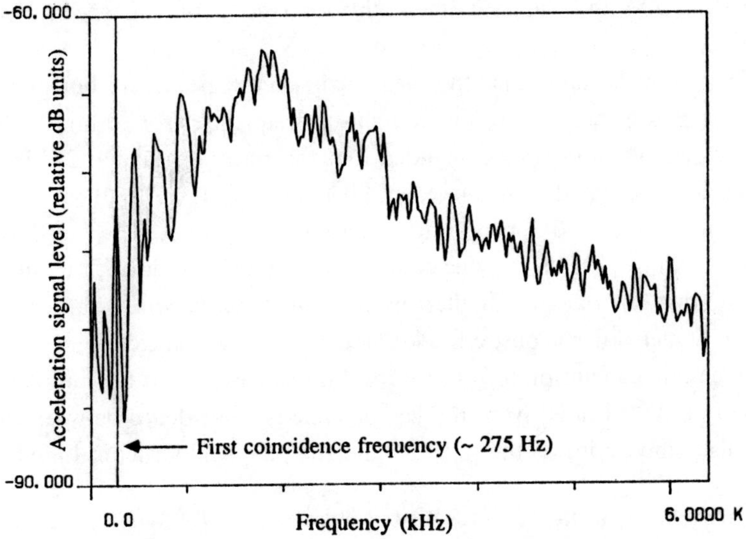

Fig. 8.57. Typical externally radiated sound pressure levels for a gas pipeline (speed of sound $\sim$385 m s^{-1}; mean flow Mach number $\sim$0.1; internal diameter $\sim$0.914 m).

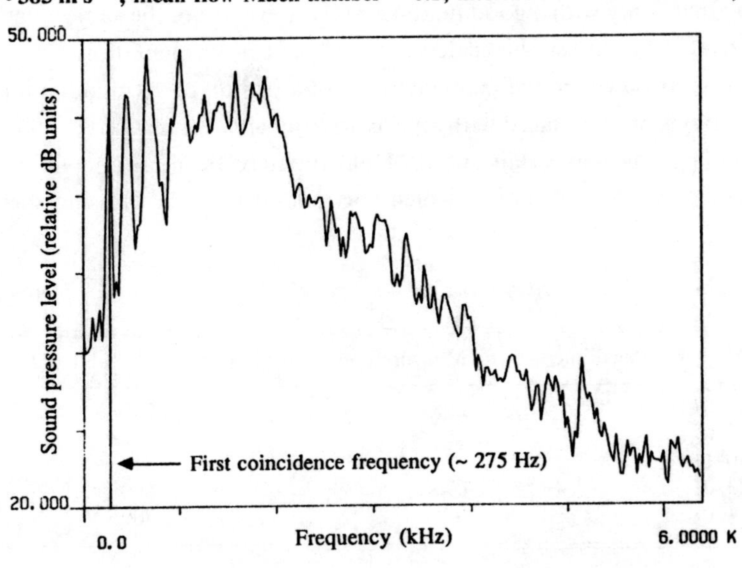

always be higher than the associated higher order acoustic mode cut-off frequencies; this is not always the case for wavenumber coincidence, as one of the structural modes with a negative axial wavenumber could have a frequency less than the cut-off frequency). Hence, if coincidence is the dominant mechanism for the vibration and noise spectra in Figures 8.56 and 8.57, respectively, then one would expect that the measured coincident peaks are in close proximity to, and generally higher than, the respective cut-off frequencies of the appropriate higher order acoustic modes. The calculated cut-off frequencies for the relevant higher order acoustic modes (as obtained from equation 7.8) and the experimentally observed dominant spectral peaks are presented in Table 8.5.

It is clear from the table that the measured spectral peaks are both higher than and in close proximity to the calculated cut-off frequencies for certain higher order acoustic modes. It can thus be concluded that the spectral peaks at 275 Hz, 461 Hz and 676 Hz are associated with coincidence between the (1, 0), (2, 0) and (3, 0) higher order acoustic modes and appropriate structural pipe modes. The fourth dominant measured peak at ~1022 Hz is due to a higher acoustic mode such as the (5, 0) or the (2, 1) modes – as one goes higher up in frequency it becomes harder to identify the specific higher order acoustic modes associated with coincidence.

Thus, the main conclusion to result from this test case is that the dominant source of externally radiated noise from the gas pipeline is coincidence between the higher order acoustic modes inside the pipeline and the pipe wall structural modes.

8.5.5 Flow-induced noise and vibration associated with a racing sloop (yacht)

This final test case illustrates how one can sometimes identify sources of noise and/or vibration armed only with a good fundamental knowledge of the subject and without the requirement for any sophisticated measurement instrumentation!

A situation arose where the manufacturers of a prototype 10 m aluminium hulled racing sloop (yacht) were faced with an excessive low frequency (~100–200 Hz) noise and vibration problem at certain speeds. Unfortunately, the noise and vibration levels peaked at a speed of ~10 knots, which corresponded to the cruising speed of the yacht.

Table 8.5. *Measured spectral peaks and calculated cut-off frequencies.*

Higher order mode	Measured spectral peaks	Calculated f_{co}
(1, 0)	275	245
(2, 0)	461	407
(0, 1)	–	511
(3, 0)	676	560
(4, 0)	–	709

Because the problem was speed related, it was immediately recognised that the noise and vibration was probably flow-induced and in particular probably associated with vortex shedding – i.e. it was highly probable that the vortex shedding frequency coincided with a major structural resonance. Furthermore, since the yacht was constructed out of aluminium, which is very lightly damped, the problem was amplified.

Some simple field trials allowed for a rapid identification of the source of the problem. The possible primary sources of the vortex shedding were the mast and the keel. Towing the yacht forwards (at the 10 knot cruising speed) without the mast did not eliminate the noise and vibration. This suggested that the keel was the probable source. Towing the yacht backwards (!!) at 10 knots resulted in the complete elimination of the offending noise and vibration! This proved beyond any doubt that the source of the problem was vortex shedding from the keel – the flow pattern around the keel altered when the yacht was towed backwards and this eliminated/reduced the vortex shedding. The source of the problem was confirmed yet again by temporarily attaching some diagonal boundary layer trip wires to the keel and sailing the yacht in its normal manner at its cruising speed. As expected, the boundary layer trip wires uncorrelated the vortex shedding pattern and this in turn eliminated the noise and vibration. Vortices only produce excessive structural vibrations when they are well correlated along the length of a structure – they oscillate in sympathy with each other and this sets up a tonal excitation. When they are uncorrelated they do not sustain the tonal excitation.

Thus, the source of the excessive noise and vibration on the yacht was identified without the need to obtain any quantitative measurement.

References

8.1 Proceedings of 2nd International Congress on Acoustic Intensity, CETIM, Senlis, France, 1985.

8.2 Gade, S. 1985. 'Sound intensity and its application in noise control', *Sound and vibration*, **3/85**, 14–26.

8.3 Maling, G. C. 1986. *Progress in the application of sound intensity techniques to noise control engineering*, Proceedings Inter-Noise '86, Cambridge, U.S.A., pp. 41–74.

8.4 Bell, D. 1984. *An envelope technique for the detection and diagnosis of incipient faults in rolling element bearings*, Brüel & Kjaer.

8.5 Bendat, J. S. and Piersol, A. G. 1980. *Engineering applications of correlation and spectral analysis*, John Wiley & Sons.

8.6 Lyon, R. H. 1987. *Machinery noise and diagnostics*, Butterworths.

8.7 Lyon, R. H. and DeJong, R. G. 1984. 'Design of a high-level diagnostic system', *Journal of Vibration, Acoustics, Stress, and Reliability in Design* **106**, 17–21.

8.8 Stewart, R. M. 1982. 'Application of signal processing techniques to machinery health monitoring', chapter 23 in *Noise and vibration*, edited by R. G. White and J. G. Walker, Ellis Horwood.

8.9 Randall, R. B. 1984. *Separating excitation and structural response effects in gearboxes*, Proceedings of the Third International Conference on Vibrations in Rotating Machinery (I.Mech.E.), York, England, pp. 101–8.
8.10 Henry, T. A. and Okah-Avae, B. E. 1976. *Vibrations in cracked shafts*, Proceedings of the First International Conference on Vibrations in Rotating Machinery (I.Mech.E), Cambridge, England, pp. 15–20.
8.11 Adams, R. D., Cawley, P., Pye, C. J. and Stone, B. J. 1978. 'A vibration technique for non-destructively assessing the integrity of structures', *Journal of Mechanical Engineering Science* (*I. Mech. E.*) **20**(2), 93–100.
8.12 Shahan, J. E. and Kamperman, G. 1976. 'Machine element noise', chapter 8 in *Handbook of industrial noise control*, edited by L. L. Faulkner, Industrial Press.
8.13 Blake, W. K. 1986. *Mechanics of flow-induced sound and vibration*, Academic Press.
8.14 Glegg, S. 1982. 'Fan noise', chapter 19 in *Noise and vibration*, edited by R. G. White and J. G. Walker, Ellis Horwood.
8.15 Bell, L. H. 1982. *Industrial noise control*, Marcel Dekker.
8.16 Putnam, A. A. 1976. 'Combustion and furnace noise', chapter 10 in *Handbook of industrial noise control*, edited by L. L. Faulkner, Industrial Press.
8.17 Halliwell, N. A. and Richards, E. J. 1980. *Acoustical study of a forging drop hammer*, Satellite Symposium on Engineering for Noise Control, 10th International Congress on Acoustics, Adelaide, pp. D1–D8.
8.18 Moreland, J. B. 1979. 'Electrical equipment', chapter 25 in *Handbook of noise control*, edited by C. M. Harris, McGraw-Hill.
8.19 Bloch, H. P. and Geitner, F. K. 1986. *Machinery failure analysis and troubleshooting*, Gulf Publishing.
8.20 Hargis, C., Gaydon, B. G. and Kamash, K. 1982. *The detection of rotor defects in induction motors*, I.E.E. Conference Publication no. 213, 216-220.
8.21 Cameron, J. R., Thomson, W. T. and Dow, A. B. 1986. 'Vibration and current monitoring for detecting air gap eccentricity in large induction motors', *I.E.E. Proceedings B* **133**(3), 155–63.
8.22 Tavner, P. J., Gaydon, B. G. and Ward, D. M. 1986. 'Monitoring generators and large motors', *I.E.E. Proceedings B* **133**(3), 169–80.
8.23 Reinhart, T. E. and Crocker, M. J. 1982. 'Source identification on a diesel engine using acoustic intensity measurements, *Noise Control Engineering* **18**(3), 84–92.
8.24 Crocker, M. J. and Zockel, M. 1980. *Techniques for noise source identification in complex machinery*, Satellite Symposium on Engineering for Noise Control, 10th International Congress on Acoustics, Adelaide, pp. B1–B10.
8.25 McGary, M. C. and Crocker, M. J. 1982. 'Phase shift errors in the theory and practice of surface intensity measurements', *Journal of Sound and Vibration* **82**(2), 275–88.
8.26 Crocker, M. J. 1979. *Identifying sources of noise in engines and vehicles*, Proceedings Inter-Noise '79, Warsaw, Poland, pp. 347–56.
8.27 Eshleman, R. L. 1976. 'Vibration standards', chapter 19 in *Shock and vibration handbook*, edited by C. M. Harris and C. E. Crede, McGraw-Hill.

Nomenclature

a_1, a_2	accelerometer signals (acceleration)
a_i	internal pipe radius
A	surface area
B	bending stiffness

c	speed of sound
c_i	speed of sound in the fluid inside a pipe
C_{pa}	coincident spectrum (real part of the one-sided cross-spectral density)
d	roller diamter
D	pitch diameter of a rolling element bearing
$E[x^4]$	fourth statistical moment (kurtosis) of a function
f, f_1, f_2, etc.	frequencies
f_b	blade passing frequency
f_{bcsir}	rotational frequency of a ball cage with a stationary inner race
f_{bcsor}	rotational frequency of a ball cage with a stationary outer race
$(f_{co})_{pq}$	cut-off frequency of the (p, q)th higher order acoustic mode
f_e	electrical supply frequency
f_m	gear meshing or toothpassing frequency
f_p	discrete frequency associated with hydraulic forces in pumps
f_{rciso}	frequency of relative rotation between the cage and the rotating inner race with a stationary outer race
f_{rcosi}	frequency of relative rotation between the cage and the rotating outer race with a stationary inner race
f_{re}	rotational frequency of a rolling element
f_{recri}	frequency at which a rolling element contacts a fixed point on a rotating inner race with a fixed outer race
f_{recro}	frequency at which a rolling element contacts a fixed point on a rotating outer race with a fixed inner race
f_{repfi}	rolling element pass frequency on the inner race
f_{repfo}	rolling element pass frequency on the outer race
f_{resf}	rolling element spin frequency
f_s	shaft rotational frequency
f_{sh}	slot harmonic frequencies

i	integer
I_v	vibration intensity
I_x	sound intensity in the x-direction
j	integer
k	integer
m_L	number of interaction pressure pattern lobes
M	Mach number
M_L	rotational speed associated with the number of interaction pressure pattern lobes
n	integer
N	number of gear teeth, shaft rotational speed, number of blades, number of compression or pumping events per revolution
p	number of magnetic poles
$p(x)$	probability density function
Q_{pa}	quadrature spectrum (imaginary part of the one-sided cross-spectral density)
R	number of rotor slots
s	unit slip
S	surface area
t	time
T	time
$\langle \overline{v^2} \rangle$	mean-square vibrational velocity (space- and time-averaged)
V	number of vanes
x	input signal, random variable
Z	number of rolling elements
α_{pq}	constant associated with eigenvalue satisfying the rigid pipe wall boundary conditions for the (p, q)th higher order acoustic mode inside a cylindrical shell
Δf	frequency resolution
Δx	separation between two accelerometers
π	3.14 . . .
Π	radiated sound power
ρ_0	mean fluid density
ρ_S	mass per unit area (surface mass)
σ	standard deviation, radiation ratio
τ	time

ϕ	contact angle between a rolling element and the raceway, phase shift between two signals
ω	radian (circular) frequency
$\langle\ \rangle$	time-average of a signal
$\overline{}$	space-average of a signal (overbar)

Problems

Chapter 1

1.1 A sound wave propagating through a medium has period of 0.02 s and a wavelength of 6.86 m. Evaluate the phase velocity and the group velocity of the wave.

1.2 A travelling wave is described by the following expression:

$z(x, t) = 6.25\, e^{i(471.2t - 1.37x)}$.

Evaluate (i) the wave velocity, (ii) the wavelength, and (iii) the period of the wave (note that x is in metres).

1.3 Derive the natural frequencies for both the spring–mass sytems in Figure P1.3.

Fig. P1.3.

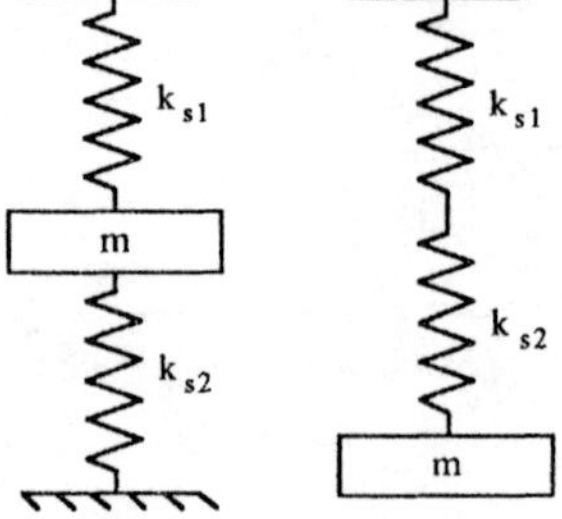

1.4 Estimate the frequency of vertical oscillation of a right cylindrical buoy which floats upright in salt water (density $\sim 1026\ \mathrm{kg\ m^{-3}}$). The buoy has a mass of 1250 kg and a diameter of 1.35 m.

1.5 Assuming that the hinged horizontal bar in Figure P1.5 is rigid and of negligible mass, derive an expression for the natural frequency of the system.

Fig. P1.5.

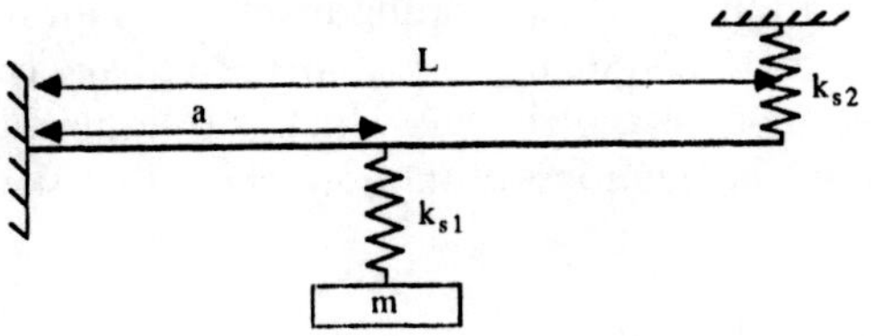

1.6 Why would if be almost impossible to play any musical instrument, e.g. a piano, in tune if $F \neq k_s s$?

1.7 A vibrating industrial machine with a mass of 7100 kg is mounted on four rubber mounts each with a static deflection of ~5.6 mm. Estimate the stiffness of each of the rubber mounts, and the natural frequency of the mounted system. If the r.m.s. vibrational velocity is ~48 mm s^{-1}, estimate the vibrational energy.

1.8 A lecture theatre/auditorium door is to be designed such that the return swing occurs in the shortest possible time without oscillating. This can be achieved with a viscous damper and a torsional spring arrangement. If the door is 2.2 m high, 1.2 m wide, 65 mm thick, and weighs 85 kg, estimate the viscous-damping coefficient required to achieve the design criteria. Also, estimate how long it will take for the door to come within 3° of closing if it is initially opened to 80° (assume a torsional spring stiffness of 22 Nm rad^{-1}).

1.9 A thin rectangular plate of mass M and surface area S (per side) is suspended in a viscous fluid of kinematic (shear) viscosity ν via a spring of stiffness k_s. Derive an expression for the damped natural frequency of vibration of the plate in terms of the undamped natural frequency, ω_n, the surface area, the mass, and the viscosity.

1.10 Consider a mass–spring system with viscous damping. The mass is displaced by an amount x_0 from its static equilibrium position and then released with zero initial velocity. Show that the logarithmic decrement $\delta = \ln(x_1/x_2)$ of any two consecutive amplitudes is

$$\delta = (2\pi\zeta)/(1-\zeta^2)^{1/2}.$$

Note: the logarithmic decrement is a measure of the damping factor and it is a convenient way of measuring the damping of a system.

Show that the logarithmic decrement for N amplitudes is

$$\delta = (1/N)\ln x_0/x_n,$$

where δ is defined as above.

The following data are given for a system with viscous damping: mass 6 kg; spring constant 15 kN m^{-1}; and the amplitude decreases to 0.20 of the initial value after seven consecutive cycles. Evaluate the damping coefficient of the damper.

1.11 Consider a spring–mass system constrained to move in the vertical direction and excited by an unbalanced rotating machine as illustrated in Figure P1.11. The total mass of the system is M, and the unbalance is represented by an eccentric mass m with eccentricity e, which is rotating with an angular velocity ω. Starting from the equation of motion, show that the steady-state response of the system is

$$\frac{MX}{me} = \frac{(\omega/\omega_n)^2}{[\{1-(\omega/\omega_n)^2\}^2 + \{2\zeta\omega/\omega_n\}^2]^{1/2}},$$

where X is the amplitude of the displacement of the non-rotating mass from the static equilibrium position. Sketch the form of MX/me for a range of damping ratios $0 < \zeta < 1$.

A counter-rotating eccentric mass is used to produce forced oscillations of a spring–damper supported mass as illustrated in Figure P1.11. By varying the speed of rotation, a resonant amplitude of 16 mm is recorded. When the speed of rotation is increased considerably beyond the resonant frequency, the amplitude approaches a constant value of 1.52 mm. Estimate the damping ratio of the system.

Fig. P1.11.

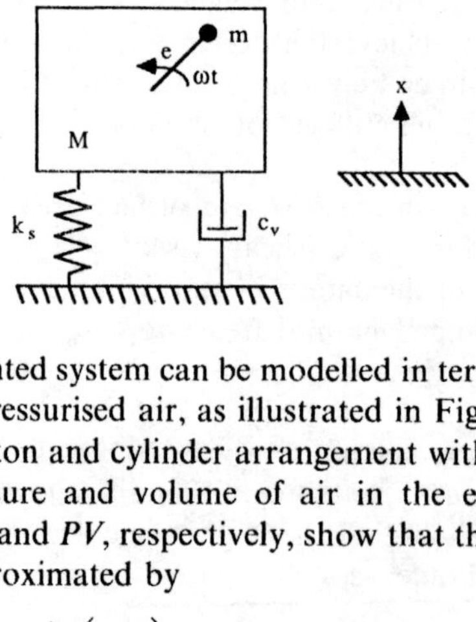

1.12 An air-spring mounted system can be modelled in terms of a mass M, a bellows and a volume of pressurised air, as illustrated in Figure P1.12. Assuming that the bellows is a piston and cylinder arrangement with a cross-sectional area A, and that the pressure and volume of air in the equilibrium and displaced positions are P_0V_0 and PV, respectively, show that the natural frequency of the system can be approximated by

$$f_n = \frac{1}{2\pi}\left(\frac{\gamma P_0 A^2}{V_0 M}\right)^{1/2} = \frac{1}{2\pi}\left(\frac{\gamma A g}{V_0}\right),$$

where γ is the ratio of the specific heats (assume that the process is adiabatic and that the stiffness characteristics of the air spring are linear).

Fig. P1.12.

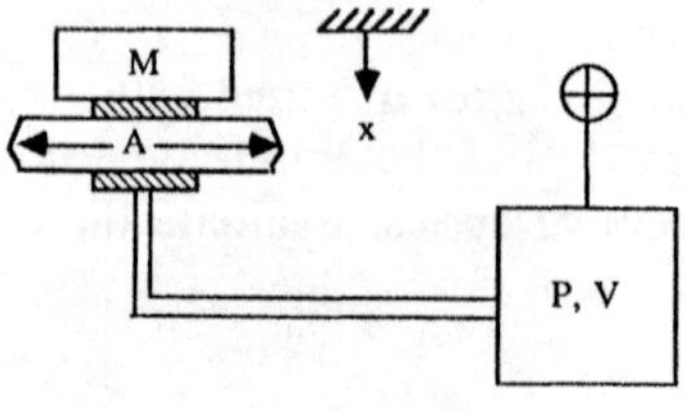

1.13 A centrifuge unit in a salt wash plant can be modelled as a large rectangular unit of total mass 7100 kg, with a handling capacity of 250 000 kg of salt per hour. It has two functional vibrations – firstly, a 300 r.p.m. rotational speed for a bucket which rotates in the vertical plane on a horizontal axis, and secondly a 25 Hz shaking motion in the horizontal plane along the same horizontal axis. The unit is mounted on 24 rubber isolator pads (six in each corner) each of which are 55 mm thick prior to compression. The average compressed thickness of each pad is 49.40 mm. The individual spring stiffness of each of the rubber pads is 5.9×10^5 N m^{-1}. Perform the necessary calculations to show that the existing rubber pads are not adequate for vibration isolation, and estimate the amplitude of the vibrations that exist. Assuming that the same type of rubber pads are used, establish the number of pads that are required to achieve suitable vibration isolation. If air springs were recommended, as an alternative, to provide a high degree of vibration isolation and the unit was mounted on four air springs (one at each corner), estimate the required height of the air springs. Also estimate the required equilibrium air pressure for a 250 mm diameter air spring.

1.14 Derive expressions for the mobility and impedance of a mass element, a spring element, a viscous-damping element, and a mass–spring–damper element.

1.15 An air compressor of 750 kg mass operates at a constant speed of 2750 r.p.m. The rotating parts are well balanced, the reciprocating parts have a mass of 15 kg, and the crank radius is 150 mm. If it is desired that only 20% of the unbalance force is transmitted to the foundation, determine the overall spring stiffness required (assume $\zeta = 0.15$). Also determine the amplitude of the transmitted force.

1.16 A machine weighing 20 kg and supported on springs of total stiffness 100 kN m^{-1} has an unbalanced rotating element which generates a 8000 N disturbing force at 3000 r.p.m. Assuming a damping factor of $\zeta = 0.20$, determine (i) the amplitude of motion due to the unbalance, (ii) the transmissibility, and (iii) the transmitted force.

1.17 A light, hollow box, as illustrated in Figure P1.17, is subjected to forced vibrations in a fluid of low viscosity such as air. Because the mass of the body is small, and its volume is large, the damping influence of the fluid resistance is significant. The resisting force due to the fluid can be approximated by

$$F_D = \pm 0.5 \rho v^2 C_D A_p,$$

where ρ is the fluid density, $v = dx/dt$ is the velocity of the body, C_D is the drag coefficient, and A_p is the area of the body projected on a plane perpendicular to the direction of motion. Determine (i) the equivalent viscous-damping coefficient, (ii) the amplitude of the steady-state response, and (iii) the amplitude at resonance.

Fig. P1.17.

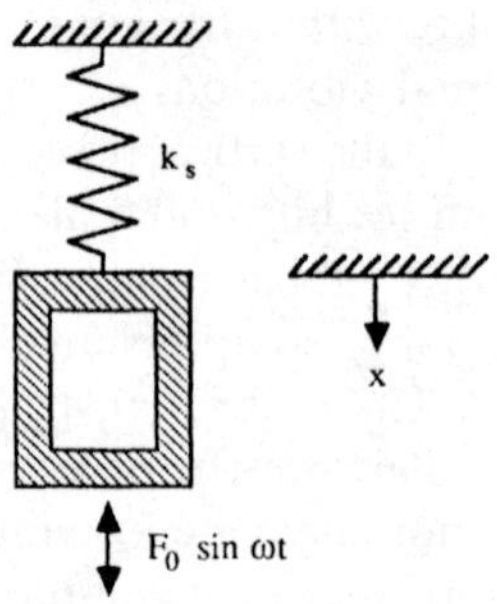

1.18 An undamped single-degree-of-freedom system is subjected to a step excitation ($f(t) = F_0$ for $t > 0$). Using the convolution integral, evaluate the system response, $x(t)$, to the input force. What is the peak response in relation to the static deflection of the system?

1.19 A 40 kg mass is mounted on a spring of stiffness 220 kN m^{-1}, and the spring is mounted within a large box as illustrated in Figure P1.19. The box is dropped from a height of 10 m. Derive an expression for the maximum displacement of the mass and estimate the maximum force transmitted to it. Assume that the box remains in contact with the floor upon striking it.

Fig. P1.19.

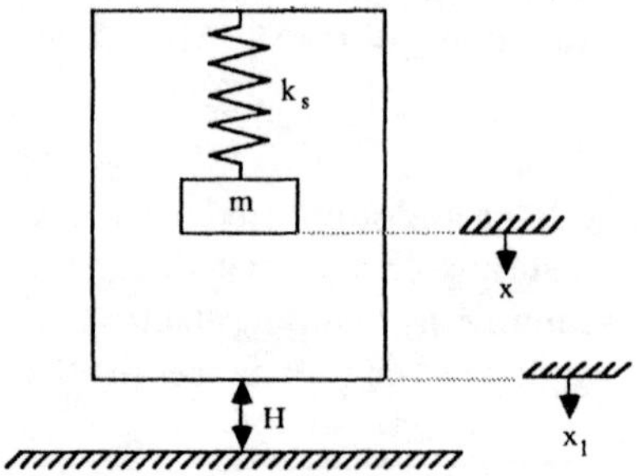

1.20 The deflection at some point x on a beam due to a load $W_1(t)$ at some specified point is $y_1(t)$. The deflection at the same point on the beam due to a second load $W_2(t)$ at a different point to the first is $y_2(t)$, as illustrated in Figure P1.20. Given that the deflection due to both loads is $y(t) = y_1(t) + y_2(t)$, determine the auto-correlation function of $y(t)$ in terms of the correlation functions associated with $y_1(t)$ and $y_2(t)$.

1.21 Show how one can obtain the impulse response function of a linear system via the cross-correlation between a white noise input and the output response signal.

Fig. P1.20.

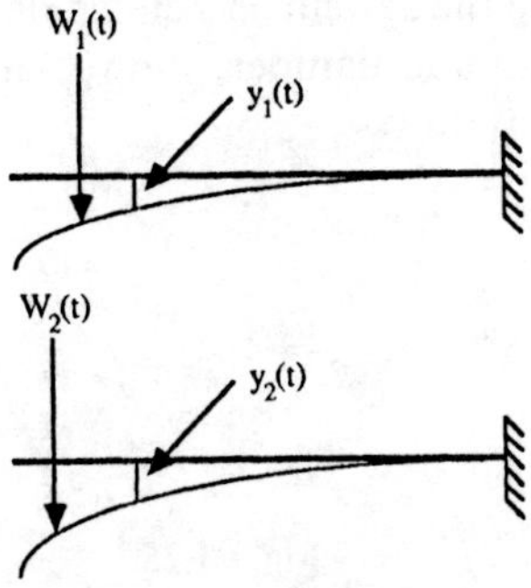

1.22 A stationary random process has an auto-correlation function, $R_{xx}(\tau)$, given by

$$R_{xx}(\tau) = X\,e^{-a|\tau|},$$

where $a > 0$. Determine (i) the mean value, (ii) the mean-square value, and (iii) the power spectral density of the process.

1.23 The single-sided spectral density of the deflection $y(t)$ of an electric motor bearing is

$$G_{yy}(\omega) = 0.0462\ \text{mm}^2\ \text{Hz}^{-1},$$

over a frequency band 0 to 1000 Hz, and is essentially zero for all other frequencies. Determine the mean-square deflection, and obtain an expression for the auto-correlation function, $R_{yy}(\tau)$, for the $y(t)$ process.

1.24 A machine component weighing 42 kg is mounted on four rubber pads. The average static deflection is 3.5 mm. The component is excited by a random force whose single-sided spectral density is given in Figure P1.24. Assuming that the damping ratio of the pads is 0.20, derive an expression for the spectral density of the output vibrational displacement of the machine component, and estimate the r.m.s. displacement amplitude.

Fig. P1.24.

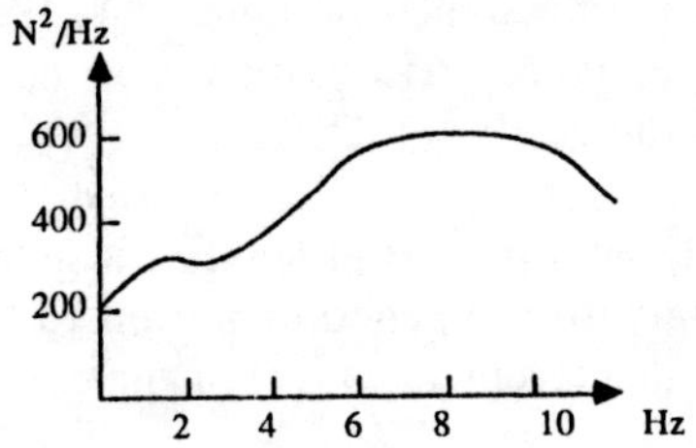

1.25 A machine component at its static equilibrium position is represented by a uniform slender bar of mass m and length L, a spring, k_s, and a damper, c_v, as illustrated in Figure P1.25(a). The tip of the bar is subjected to a rectangular force impulse $f(t)$ with an auto-correlation function as illustrated in Figure P1.25(b). By modelling the system as a single-degree-of-freedom system with an equivalent mass, spring and damper, show that the output spectral density of the displacement of the tip is

$$S_{yy}(\omega) = \frac{\frac{aT}{2\pi}\left(\frac{\sin \omega T/2}{\omega T/2}\right)^2}{\left(k_s - \frac{7}{27}m\omega^2\right)^2 + c_v^2\omega^2}.$$

Fig. P1.25.

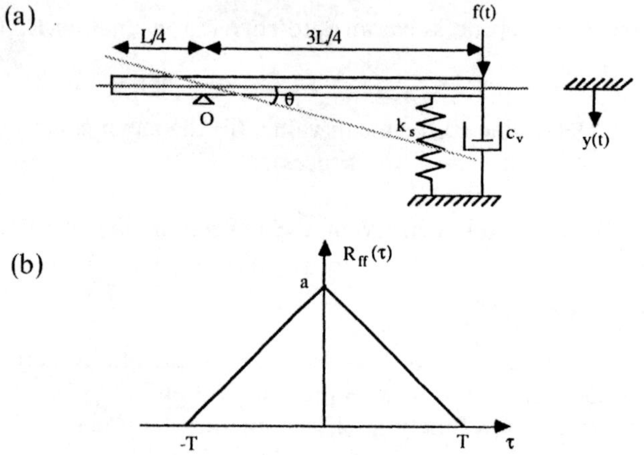

1.26 A 1.5 m long steel cord is fixed (clamped) at its ends. It has a diameter of 1.5 mm, a density of 7800 kg m^{-3}, and a fundamental natural frequency of 125 Hz. What is the tension of the cord?

1.27 The cord in problem 1.26 is forced at one end at 85 Hz. Assuming that the tension remains the same and that the cord is undamped, evaluate the drive-point mechanical impedance (assume that the clamped ends have an infinite mechanical impedance). Also estimate the nett energy transfer between the driving force and the cord.

1.28 A clamped, tensioned, flexible cord of length l is given an initial velocity V at its mid-point. Starting with the general solution to the wave equation, derive an expression for the displacement $u(x, t)$ of the cord.

1.29 A 3.4 kg mass is suspended by a 1 mm diameter steel wire which is 0.8 m long. What is the natural frequency of the system? Qualitatively describe the vibrational characteristics of the system if the wire mass is significantly larger than the tip mass.

1.30 Two aluminium bars and a steel bar are press fitted together to form a step-discontinuity as illustrated in Figure P1.30. The aluminium bars each have a cross-sectional area of 10^4 mm^2, and the steel bar has a cross-sectional area of 4×10^4 mm^2. The aluminium bars are each 400 mm long, and the steel bar is 300 mm long. A harmonic, incident longitudinal wave with a displacement amplitude of 0.25 mm and a frequency of 80 Hz is applied to one end of the system. Calculate (i) the amplitude of the reflected wave at the point of excitation which is due to the first discontinuity; (ii) the amplitude of the reflected wave at the point of excitation which is due to the second discontinuity; and (iii) the times at which each of the reflections will be detected at the excitation point. Will the amplitude of the resultant wave at the excitation point be the sum of the amplitudes of the incident and reflected waves?

Calculate the amplitude of the first transmitted wave and the corresponding transmission coefficient (i.e. the first transmitted wave is the first complete wave train to pass through the discontinuity). Qualitatively discuss the steady-state transmission coefficient due to multiple reflections at the discontinuity.

Fig. P1.30.

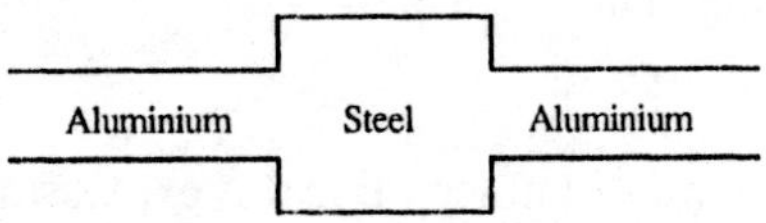

1.31 A concentrated point force, $P_0 f(t)$, is applied to the centre of a simply supported uniform beam of length l, modulus of elasticity E and second moment of area I. Starting from the general solution to the transverse beam equation, evaluate (i) the mode shapes normalised with respect to the mass per unit length, (ii) the natural frequencies, ω_n's, and (iii) the mode participation factor, H_n.

Show that the deflection at the centre of the beam is given by

$$u(x, t) = \frac{2P_0 l^3}{EI} \sum_{n=1}^{\infty} \frac{\sin\frac{n\pi}{2} \sin\frac{n\pi x}{l}}{(n\pi)^4} D_n(t),$$

for $n = 1, 3, 5$, etc., where $D_n(t)$ is the dynamic load factor. If the time variation is a unit step function between $t = 0$ and $t = t$, evaluate the dynamic load factor.

1.32 Show that the mode shapes (eigenfunctions) for the transverse vibration of a beam which is rigidly clamped at both ends are orthogonal.

1.33 A uniform circular shaft of length l and fixed at one end has an external torque $T \sin \omega t$ applied to its free end. Using generalised co-ordinates and the method of normal modes, show that the steady-state torsional shaft vibration can be expressed as

$$\theta(x, t) = \frac{2T}{\rho I_p l} \sum_{n=1,3,5}^{\infty} \left\{ (-1)^{(n-1)/2} \sin\frac{n\pi x}{2l} \frac{\sin \omega t}{\omega_n^2 - \omega^2} \right\},$$

where ρ is the density of the shaft material, and I_p is the polar second moment of area (i.e. the torsional stiffness of the shaft is I_pG, where G is the shear modulus, and the mass moment of inertia is ρI_p).

1.34 A 3.0 m long uniform steel beam with rectangular cross-sectional dimensions of 4 mm and 5 mm is simply supported at its ends. A zero mean stationary random point force with an auto-correlation function approximated by

$$R_{FF}(\tau) = F^2 \, \mathrm{e}^{-a|\tau|},$$

with $F = 750$ N and $a = 1000^{-1}$ s is applied at a third of the distance from a support. Estimate the r.m.s. displacement at the centre of the beam for the $n = 5$ mode. Assume that the density of steel is 7700 kg m^{-3} and that the structural loss factor, η, is 5×10^{-4}.

Chapter 2

2.1 A plane wave travelling in nitrogen in the +ve x-direction can be described by

$$\mathbf{p}(\vec{x}, t) = \mathbf{A} \, \mathrm{e}^{\mathrm{i}(\omega t - kx)},$$

where **A** and **p** are complex numbers. Use this expression to derive a relationship for the mean sound intensity. If the temperature of the gas is 75°C and the absolute pressure is 3 atm, evaluate (i) the mean sound intensity, (ii) the mean kinetic energy per unit volume, and (iii) the mean energy density for a peak acoustic pressure fluctuation of 1.125 N m^{-2}.

2.2 Assuming an inviscid fluid in the absence of sources of mass or body forces, use the continuity equation and the equation of conservation of momentum to derive expressions for the particle velocity amplitude and the particle displacement amplitude for (a) a positive plane wave, and (b) a negative plane wave. Describe the phase relationships between the acoustic pressure, p, the particle velocity, u, and the particle displacement, ξ, in each case.

A plane sound wave in air, of 100 Hz frequency, has a peak acoustic pressure amplitude of 2 N m^{-2}. What is its intensity level, its particle displacement amplitude, its particle velocity amplitude, its r.m.s. pressure, and its sound pressure level?

2.3 Consider the instantaneous pressure of two harmonic sound waves at some fixed point in space ($x = 0$). Assuming that there is a phase difference between the two waves, evaluate the total time-averaged r.m.s. pressure when (i) $\omega_1 \neq \omega_2$, and (ii) when $\omega_1 = \omega_2$.

Two sound sources are radiating sound waves at different discrete frequencies. If their individual sound pressure levels, recorded at a position, are 75 and 80 dB, respectively, find the total sound pressure level due to the sources together. What is the total sound pressure level if they are radiating at the same discrete frequency?

2.4 What is the speed of sound for an incompressible fluid?

2.5 An underwater sonar beam delivers 120 W of sound power at 25 kHz, in the form of a plane wave. Estimate the amplitude of the particle velocity and the particle displacement of the plane wave beam. Assume that the beam has a diameter of 0.4 m.

2.6 A 0.3 m diameter sound source oscillates at a frequency of 2000 Hz with a peak source strength of $6.97 \times 10^{-2}\ m^3\ s^{-1}$. Calculate (i) the transition distance between the near and far fields, (ii) the mean-square sound pressure at a distance of 3 m from the source, and (iii) the sound power of the sound source.

2.7 The sound pressure level at the exit plane of a 20 m high steam exhaust stack with a 0.5 m diameter is 125 dB. The dominant frequency in the duct is a 200 Hz tone. Estimate the sound pressure level to be expected on the ground plane at a distance of 30 m from the stack.

2.8 A rotating fan blade in a duct generates 123 dB of sound power at 300 Hz. Estimate the r.m.s. fluctuating pressure forces associated with the aerodynamic source within the duct.

2.9 A monopole sound source of source strength $8 \times 10^{-2}\ \mathrm{m^3\ s^{-1}}$ and radius 30 mm, radiates at 700 Hz in air. If a similar monopole was placed 10 mm from it, and the sources radiated 180° out of phase with each other, estimate the ratio of the sound power radiated by the combined source to that by either of the monopoles by themselves.

2.10 A sound source has a dipole source strength given by

$$\mathbf{Q}_d(t - r/c) = Q_{dp}\, \mathrm{e}^{i\omega(t - r/c)}.$$

Starting with the dipole velocity potential (equation 2.111), derive an expression for the acoustic pressure fluctuations associated with the source. Also derive expressions for the radial and tangential components of the acoustic particle velocity. What is the specific acoustic impedance of the sound source at a large distance away in the far-field? What are the units of $\mathbf{Q}_d$?

2.11 A spherical sound source with an effective radius of 0.10 m and a r.m.s. vibrational velocity of $0.004\ \mathrm{m\ s^{-1}}$ is mounted on a concrete floor. Estimate the radiated sound power if the source radiates at (i) 300 Hz, and (ii) 30 kHz.

2.12 A flat piston of radius 0.25 m radiates into water on one side of an infinite baffle. The piston radiates 120 W of sound power at 25 kHz. Estimate (i) the velocity amplitude of the piston, and (ii) the radiation mass loading. If the piston has a mass of 0.2 kg, a stiffness of $1200\ \mathrm{N\ m^{-1}}$, and a damping ratio, ζ, of 0.1, estimate the applied force required to produce the velocity amplitude calculated in (i).

2.13 A very long straight run of gas pipeline with a nominal 0.5 m diameter has a uniform pulsating harmonic surface velocity amplitude of 0.067 m s^{-1} at 140 Hz. Estimate the sound power radiated per unit length.

2.14 A spherical sound source of radius 0.05 m has a harmonic normal surface velocity of 0.02 m s^{-1}. Estimate the source strength and the mass flux per unit time. Assume that the source radiates into air. If the density fluctuations, ρ', are also harmonic (i.e. $\rho=\rho_0+\rho' e^{i\omega t}$), derive an expression for the rate of change of mass flux.

2.15 The monofrequency spherical wave $G(t-r/c)/r$ is a solution to the homogeneous wave equation everywhere except at $r=0$. By considering the homogeneous wave equation in the region of a singularity ($r=0$), incorporates a point source into it and thus show that

$$\left(\frac{1}{c^2}\frac{\partial^2}{\partial t^2}-\nabla^2\right)\frac{G(t-r/c)}{r}=4\pi G(t)\,\delta(\vec{x}).$$

Note: the relationship

$$\int_V \nabla^2\frac{G(t-r/c)}{r}\,dV=\int_S\frac{\partial}{\partial n}\frac{G(t-r/c)}{r}\,dS$$

should be used to solve the above problem.

2.16 Illustrate with a simple example how information obtained in the wave field (from the homogeneous wave equation) will not provide any information about the source distribution.

2.17 Derive expressions for the ratio of the fluctuating pressure amplitude at $r=\delta$ ($\ll c/\omega$) to the fluctuating pressure amplitude at $r=\Delta$ ($\gg c/\omega$) for (i) a point harmonic monopole, and (ii) a point harmonic dipole. Both sources only radiate sound at a frequency ω.

2.18 Consider the unsteady addition of heat to a perfect gas. The gas density is a function of both the pressure and the heat supplied – i.e. $\rho=\rho(P,h)$, where P is the total pressure, and h is the enthalpy per unit mass. Using the conventional thermodynamic perfect gas relationships ($\rho=P/RT$, and $c_p=dh/dT=\gamma R/(\gamma-1)$), show that the wave equation describing the motion of a perfect gas with unsteady heat addition is

$$\frac{1}{c^2}\frac{\partial^2 p}{\partial t^2}-\nabla^2 p=\frac{\rho_0(\gamma-1)}{c^2}\frac{\partial^2 h}{\partial t^2}.$$

Also show that the solution to this equation is

$$p(\vec{x},t)=\frac{\rho_0(\gamma-1)}{4\pi c^2}\frac{\partial^2}{\partial t^2}\int_V\frac{h\left(\vec{y},t-\frac{|\vec{x}-\vec{y}|}{c}\right)}{|\vec{x}-\vec{y}|}\,d^3\vec{y},$$

and qualitatively describe the characteristics of the source of combustion noise.

2.19 Show how variations in the Lighthill stress tensor, T_{ij}, vanish for linear, inviscid flow, resulting in the linear, homogeneous wave equation.

2.20 Qualitatively discuss the effects of a rigid, reflecting ground plane on the sound power of a monopole sound source. Explain why the source behaves like a dipole at very large distances. What does the sound intensity scale as at these large distances?

2.21 Using dimensional scaling parameters (U for velocity, D for dimension, λ for wavelength, c for the speed of sound, ρ for density) show that the ratio of the sound power radiated by a dipole to that radiated by a monopole is

$$\frac{\Pi_D}{\Pi_M} = \left(\frac{D}{\lambda}\right)^2,$$

and that the ratio of the sound power radiated by a quadrupole to that radiated by a monopole is

$$\frac{\Pi_Q}{\Pi_M} = \left(\frac{D}{\lambda}\right)^4.$$

Estimate the reduction in radiated sound pressure levels to be expected from a free jet if the jet exit velocity is reduced by 30%.

2.22 The radiated sound pressure level at a distance of 30 m from a steam exhaust stack with a nominal 0.3 m diameter is 97 dB. Estimate the mean flow velocity out of the duct.

2.23 Two aircraft jet engines operate under identical conditions and both provide equal thrust. The only difference between the two engines is that the diameter of the first engine is 65% of the diameter of the second engine. Estimate the difference in radiated noise levels between the two.

2.24 If one wished to achieve a doubling of thrust in an aircraft jet engine, would it be better to increase the nozzle area or the exhaust speed from a noise control viewpoint?

2.25 A small, rigid flow spoiler in a large duct has an effective exposed diameter of 0.4 m, and the 98 dB of sound generated at a distance of 4 m has a dominant 200 Hz frequency. Estimate the order of magnitude of the fluctuating forces associated with the radiated noise.

2.26 Consider a supersonic Mach number flow (e.g. a supersonic jet) such that the source region is not acoustically compact (i.e. $\lambda < D$). Use the solution to Lighthill's equation for the radiated sound pressure in an unbounded region of space together with suitable scaling parameters to show that the far-field radiated sound pressure scales as

$$p^2 \sim \rho_0^2 c^4 \frac{M^2 D^2}{r^2}.$$

Hint: because the source region is not acoustically compact, the retardation time variations over the source region have to be accounted for.

Chapter 3

3.1 Evaluate (a) the bending wavespeed at 1000 Hz, and (b) the critical frequency for 2 mm flat plates made of (i) aluminium, (ii) brass, (iii) glass, (iv) lead, and (v) steel.

3.2 If one were to model a very large flat steel plate as an undamped, infinite plate which is mechanically driven, estimate the peak radiated sound pressure level due to the vibrating plate at (i) 800 Hz, and (ii) 10 000 Hz. The plate is 1.5 mm thick, and its peak surface velocity is 8 mm s^{-1} at both frequencies.

3.3 A small machine component can be modelled as a spherical source with a typical radial dimension of 50 mm. If it has an r.m.s. surface vibrational velocity of 6.5 mm s^{-1} in the 500 Hz octave band, estimate the radiated sound power.

3.4 What is the radiation ratio of a spherical sound source whose circumference is equal to half the wavelength of the radiated sound.

3.5 A clamped, flat, steel plate with dimensions 2.5 m × 1.5 m × 4 mm is driven mechanically in the 1000 Hz octave band (707–1414 Hz). Evaluate the necessary parameters to sketch the form of the wavenumber diagram, and qualitatively describe the sound radiation characteristics of the plate. How many resonant modes are present in the octave band? What is the radiation ratio of the plate in the frequency band of interest? If the plate has an r.m.s. surface vibrational velocity of 5.2 mm s^{-1}, estimate the radiated sound power assoociated with all the resonant modes.

3.6 If the plate in problem 3.5 has a loss factor of 4.2×10^{-4}, estimate the drive-point sound power, Π_{dp}, and hence estimate the total sound power radiated by the plate.

3.7 Consider a 3.5 m × 2.5 m × 5 mm clamped, damped, aluminium flat plate which is mechanically excited by (i) a point source, and (ii) a line source. The plate has a structural loss factor of 3×10^{-2}. Compute (i) the ratio of the drive-point sound power to the radiated resonant (reverberant) sound power, and (ii) the ratio of the drive-line sound power to the radiated resonant (reverberant) sound power for all the octave bands below the critical frequency. Comment on the ratios at frequencies above the critical frequency.

3.8 Starting with equations (3.57) and (3.67), derive equation (3.68).

3.9 A large diesel engine can be approximated as a cube with 1.2 m sides. The far-field sound radiation is dominated by the 500 Hz octave. Estimate the radiation ratio characteristics of the engine, and the sound power radiated if the r.m.s. vibration levels are 12.2 mm s^{-1}.

3.10 A 3 mm thick steel panel is suspended between two rooms and acoustically excited in the 1000 Hz octave. Estimate the magnitude of the ratio of transmitted to incident waves, and the mechanical impedance per unit area in that octave band.

3.11 A room is to be partitioned by an 8 m × 3 m solid brick wall. The wall is 110 mm thick and has a surface density of 2.1 kg m^{-2} per mm wall thickness. Using the plateau method, estimate the field incidence transmission loss (*TL*) characteristics of the wall as a function of octave band frequencies.

3.12 A building material panel has a critical frequency of 4240 Hz, and the field-incidence sound transmission loss in the 2000 Hz octave is 27 dB. Estimate the bending stiffness (per unit width) of the material.

3.13 Starting with the definition of the bending wave velocity in a plate (i.e. $c_B = \{1.8c_L tf\}^{1/2}$), derive an expression for the coincidence frequencies for glass panels. What is the lowest frequency at which coincidence can occur for a 6 mm thick glass panel?

3.14 A 20 mm thick particle board panel which is 6 m × 6 m forms a partition between two rooms. It is subjected to an incident sound field (in one of the rooms) which is not diffuse. The panel has a structural loss factor of $\sim 1.5 \times 10^{-2}$. Estimate the octave band transmission loss characteristics of the panel from 31.5 Hz to 16 000 Hz. Compare these values with the transmission loss characteristics of the panel if it were subjected to a diffuse sound field.

3.15 A double-leaf pyrex glass pane is selected for the windows of a high rise building development. The panel comprises two 6 mm glass panes separated by a 12 mm air-gap. Estimate the double-leaf panel resonance frequency. What would one do if one wanted to improve the transmission loss performance at the double-leaf panel resonance? Is this requirement compatible with optimum high frequency performance?

3.16 A simply supported flat aluminium plate is submerged in water. If the plate is 1.4 m × 1.4 m × 7 mm, estimate the effect of fluid loading on the natural frequency of the (1, 1) mode and the (12, 12) mode.

3.17 A cylindrical drop forge hammer has a 200 mm diameter, and is 300 mm long. It is dropped from a height of 10 m onto a metal workpiece. Assuming that the contact time is very small, estimate the peak sound pressure level associated with the impact at a distance of 10 m from the forge.

Chapter 4

4.1 At a particular position in a workshop, three machines produce individual sound pressure levels of 90, 93 and 95 dB re 2×10^{-5} N m^{-2}. Determine the total sound pressure level if all the machines are running simultaneously.

4.2 What will be the total sound pressure level of two typewriters each producing a sound pressure level of 83 dB at a particular measuring position? What will be the total sound pressure level if a third typewriter was introduced?

4.3 Determine the sound pressure level at a certain point due to a machine running alone, if measurements at that point with the machine 'on' and 'off' give sound pressure levels of 94 dB and 90 dB, respectively.

4.4 What are the upper and lower frequency limits for a one-quarter octave band centred on 4000 Hz? If the sound pressure level in the band is 96.4 dB, estimate the sound pressure spectrum level for a given sub-band with a 2 Hz width.

4.5 An omni-directional noise source is located at the centre of an anechoic chamber. The sound pressure level measured at 1 m from the acoustical centre of the source is 90 dB in a particular low frequency band. Calculate the sound power level of the source in the frequency band and the sound pressure level 5 m from the source. If the same omni-directional source is placed on a hard ground surface outdoors, calculate the sound pressure level at a distance of 5 m.

4.6 The 250 Hz octave band sound pressure levels measured at a radius of 4.6 m from an exhaust stack are 96, 89, 95, 95, 92, 94, 91 and 88 dB. Calculate the directivity index and the directivity factor for each of the eight microphone locations, and hence determine the sound power level. What is the actual sound power emitted from the source? Determine the corresponding sound pressure levels at similar microphone locations 9.2 m from the source.

4.7 A speedway complex poses a possible community noise problem. Average background noise levels (due to the wind, external traffic noise, etc.) at the geometrical centre of the circuit are estimated at 70 dB(A). Average noise levels at the same location with a 'full house' are 88 dB(A). With the 'full house', the average noise level at a distance of 8 m from a single speedway motor cycle is 95 dB(A). Six such machines are generally in use for a race at any given time. Estimate (i) the overall noise levels and (ii) the noise levels due to all the machines only at distances of 1 km and 2 km from the six motorcycles. Treat the motorcycles as a group of uncorrelated noise sources in close proximity to each other. Does the sound source type (constant power, constant volume, etc.) have any bearing on the estimated noise levels in this particular case?

4.8 Consider a stream of traffic flow on a major highway as comprising a row of point sources, each of sound power $L_{\Pi} = 104$ dB under free-field conditions. The point sources are 5 m apart. Estimate the sound pressure level at a position 12 m away from the road. What would be the decay rate of the sound pressure level at this distance from the traffic flow? What would be the sound pressure level at a position 24 m away from the road? If a 2 m brick wall were erected at a distance of 12 m from the road, what would the sound pressure level be at the second position (i.e. 24 m away from the source)? Assume that the dominant frequencies are in the 500 Hz octave band.

4.9 A person at a rifle range is shooting at a fixed target which is located y metres away. A large lake separates the target and the rifleman, hence it is not convenient to measure y directly. It is proposed to use a sound level meter to assist in the measurement of the distance y. After averaging over several shots, a pressure measurement of 2.3 Pa is obtained at the position of the target. The distance from the rifleman is increased by a further 25 m, and, after appropriate averaging, a sound pressure of 1.3 Pa is obtained. Estimate the distance between the marksman and the target.

4.10 An electric motor for a swimming pool pump has a sound power rating of 88 dB. This rating relates to tests conducted by the manufacturer in an anechoic chamber. If the motor is mounted on a brick paved ground plane against the back wall of a residential dwelling (i.e. the intersection of two large flat surfaces), estimate the upper and lower limits of the sound pressure level to be expected at the fence of the bounding property which is 8 m away. Assume that the acoustic centre of the motor is 150 mm from the ground plane and that the fence is 2.5 m high. If the fence is a brick wall, estimate the upper and lower limits of the sound pressure level at a receiver who is 12 m away from the wall on the other side. Assume that the receiver is 1.9 m tall, and that the dominant sound is at 1 kHz.

4.11 A machine is placed on a concrete floor in a shop area. The average sound pressure levels in various octave bands at a distance of 1.5 m from the source are tabulated below together with the averge sound pressure levels at 3.0 m. The room is 8 m × 8 m × 3 m. Determine (i) the sound power level in each octave band, (ii) the overall sound pressure levels, (iii) the overall sound power level, (iv) the room constant in each octave band, (v) the average absorption coefficient in each octave band, and (vi) the overall room constant and absorption coefficient. Comment on whether the room is live or dead.

Octave band (Hz)	63	125	250	500	1000	2000	4000	8000
L_p (1.5 m) (dB)	90	95	100	93	82	75	70	70
L_p (3.0 m) (dB)	87	92	99	91	80	72	68	67

4.12 A workshop has dimensions as sketched in Figure P4.12. The floor is concrete, the walls are brick, and the ceiling is of suspended panels, the sound absorption coefficients of which are listed below. There are two fully open windows, each

3 × 1.5 m, in one wall. A machine with an effective radius of 0.3 m at position A has sound power output figures (as determined by its manufacturer in an anechoic chamber) as tabulated below and has, in itself, no pronounced directivity. Estimate the overall sound pressure level, and overall A-weighted sound level experienced by operators at locations B and C which are, respectively, 3 m and 11 m from the machine. Are they in the direct or reverberant field?

Octave band (Hz)	125	250	500	1000	2000	4000
Sound absorption coefficients						
Brick	0.03	0.03	0.03	0.04	0.05	0.07
Concrete	0.01	0.01	0.015	0.02	0.02	0.02
Ceiling	0.2	0.15	0.20	0.20	0.30	0.30
Sound power						
Source L_Π	85	98	92	97	96	95

Fig. P4.12.

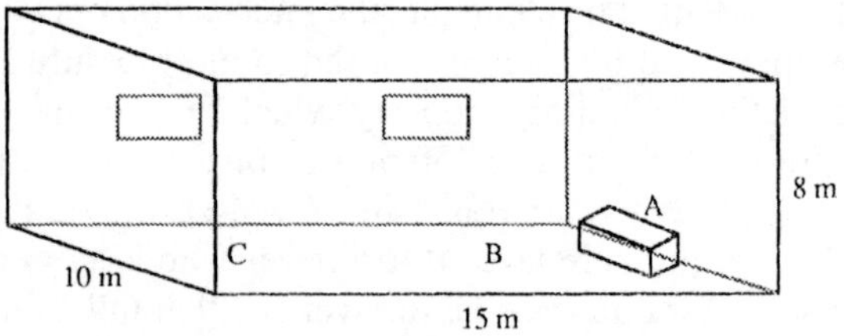

4.13 A machine which radiates isotropically produces a free-field sound power level of 135 dB in the 250 Hz band, and 128 dB in the 1000 Hz band. These two octave bands are the dominant frequencies of interest. The machine, which has dimensions of 1.5 m × 1 m × 0.8 m, is situated in a room which is 7 m × 12 m × 3 m with an average absorption coefficient $\alpha_{avg} = 0.1$ at 250 Hz and $\alpha_{avg} = 0.13$ at 1000 Hz. The machine is enclosed by a 2.5 m × 2 m × 2 m hood which has an average absorption coefficient $\alpha_{avg} = 0.65$ at 250 Hz and $\alpha_{avg} = 0.70$ at 1000 Hz, and transmission coefficients $\tau = 0.0012$ at 250 Hz and $\tau = 0.0010$ at 1000 Hz. Assuming that the machine is located in the centre of the room, estimate the overall sound pressure level in the room with the hood (enclosure) in place. Is the hood adequate from an industrial hearing conservation point of view? Briefly discuss what could be done to reduce the noise levels in the room even further.

4.14 Plane waves and spherical waves are commonly used as sound source models for noise and vibration analysis. Explain why the specific acoustic impedance of a plane wave is resistive whereas the specific acoustic impedance of a spherical sound wave has both a resistive and a reactive component.

4.15 A room that is 4 m × 8 m × 3 m has an average absorption coefficient of 0.1 in the 1 kHz octave band. A machine (1 m × 1 m × 1 m), enclosed by a 2 m × 2 m × 2 m hood is located in the room. The machine, which radiates isotropically, produces a sound power level of 120 dB in an anechoic chamber in the 1 kHz band. The hood has an average absorption coefficient and a transmission coefficient of 0.7 and 0.001, respectively, in this band. With the hood in place, estimate the sound pressure level in the room.

4.16 A water-cooled refrigeration compressor is installed on a concrete floor in a room producing a reverberant sound pressure level, L_{p1}, which is tabulated below. The physical shape of the machine is effectively a rectangular cube 1.5 m × 2.5 m × 1.5 m. A technician has to operate a control panel in the room and as such the noise level is unacceptable. An enclosure is to be designed for the compressor with sufficient space left within the enclosure for normal maintenance on all sides of the machine. The recommended enclosure dimensions are 2.5 m × 3.5 m × 2.5 m, and the interior surfaces of the enclosure are to be lined with 50 mm thick mineral wool blanket having the absorption characteristics shown in the table. The machine surface can be assumed to have the absorptive properties of concrete.

Calculate the transmission loss required for the enclosure walls and roof if the reverberant sound pressure level in the room with the machine enclosed, L_{p2}, is not to exceed the NC-45 Noise Criteria rating provided. Comment on the noise reduction problems that might be encountered if a close-fitting acoustic enclosure was used instead.

Octave band (Hz)	63	125	250	500	1000	2000	4000	8000
Concrete (α)	0.01	.01	0.01	0.02	0.02	0.02	0.03	0.03
Mineral wool (α)	0.10	0.20	0.45	0.65	0.75	0.80	0.80	0.80
L_{p1} (dB)	72	79	81	84	83	81	80	75
L_{p2} (dB) (NC-45)	67	60	54	49	46	44	43	41

4.17 A centrifugal air compressor produces 110 dB in the 1000 Hz octave band at a distance of 1 m from its nearest major surface. The operator of another machine is 7 m from the compressor. An enclosure is to be installed over the compressor to reduce the sound pressure level in the 1000 Hz band to 85 dB at the operator's position. The compressor is 2 m long, 2 m wide, and 1.2 m high, and it is located in a room 25 m long, 20 m wide and 8.3 m high. The enclosure is 3 m × 3 m × 2.2 m. Assuming that the floor and ceiling have an absorption coefficient of 0.02, and that the walls have an absorption coefficient of 0.29, determine the required transmission loss of the enclosure for (i) an enclosure internal absorption coefficient of 0.1, and (ii) an enclosure internal absorption coefficient of 0.75.

4.18 The main noise source in a plant room is the blower system for air distribution to the rest of the building. The dimensions of the room are 8 m × 10 m × 3 m and the blower system is located on the ground along the middle of the 8 m wall as illustrated in Figure P4.18. The dimensions of the blower are 1 m × 2 m × 1 m. The sound power levels (in free space) associated with the blower are provided in tabular form. The ceiling of the plant room is covered with sound absorbing material with absorption coefficients α_C, the floor has absorption coefficients α_F, and the walls have absorption coefficients α_W as indicated in the table. Adjacent to the plant room is an operator room which is 5 m × 5 m × 3 m. The wall separating the two rooms has transmission loss characteristics TL which are also given in tabular form together with the absorption coefficients α'_F of the carpeted floor in the operator room. The walls and ceiling of the operator room have the same sound absorbing characteristics as the plant room. Provide a conservative estimate of the octave band sound pressure levels to be expected in the operator room. What are the corresponding A-weighted octave sound levels?

Octave band (Hz)	125	250	500	1000	2000	4000
L_Π (dB)	105	103	98	108	107	109
α_C	0.07	0.20	0.40	0.52	0.60	0.67
α_F	0.01	0.01	0.015	0.02	0.02	0.02
α_W	0.03	0.03	0.03	0.04	0.05	0.07
TL (dB)	39	42	50	58	63	67
α'_F	0.08	0.24	0.57	0.69	0.71	0.73

Fig. P4.18.

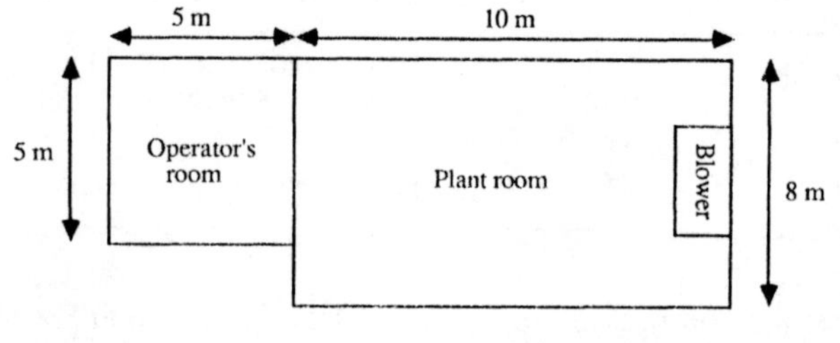

4.19 Determine the overall sound pressure level at the centre of a room 12 m × 10 m × 3 m, with a 20 mW sound source located in the centre of one of the 10 m walls, at the intersection between the floor and the wall. Assume that the walls have an absorption coefficient $\alpha_W = 0.02$, the floor $\alpha_F = 0.08$, and the celing $\alpha_C = 0.24$. Is the room live or dead?

4.20 The room in problem 4.19 is to be divided into two equal spaces by a partition wall 10 m wide and 3 m high. Assuming that the edges of the partition wall are clamped, investigate the possibility of using the following materials: plywood, particle board, lead sheet, and plasterboard. Attention should be given to the possible effects of resonance and coincidence, making any assumptions that you consider appropriate. The required transmission loss (TL) over the frequency range 125 Hz to 4000 Hz is specified below, together with the thickness, density and longitudinal (compressional) wave velocity of the four materials.

One-third-octave band (Hz)	125	160	200	250	315	400	500	600
Transmission loss (*TL*)	8	12	16	20	24	28	30	32
One-third-octave band (Hz)	800	1000	1250	1600	2000	2500	3150	4000
Transmission loss (*TL*)	34	36	38	39	39	39	39	39

Material	Thickness (m)	Density (kg m^{-3})	Compressional wave velocity (m s^{-1})
Plywood	0.038	600	3080
Particle board	0.019	750	669
Lead sheet	1.70×10^{-3}	11 340	1235
Plasterboard	0.013	650	6800

4.21 Compare the octave band noise reduction (*NR*) performance of (a) two 13 mm gypsum wallboards separated by a 64 mm air-gap, (b) a 125 mm plastered brick wall, and (c) a double brick wall (50 mm cavity, 100 mm plastered bricks) in relation to the sound transmission from one room to another. The octave band sound transmission loss characteristics are provided below in tabular form. The receiving room is 8 m wide, 9 m long and 3 m high, and the average octave band sound absorption coefficients of the walls, floor and ceiling are provided. Comment on the factors that could cause deterioration in the calculated noise reduction performance.

Octave band (Hz)	125	250	5000	1000	2000	4000
			Transmission loss characteristics			
Two 13 mm wallboards separated by a 64 mm air-gap	18	27	37	45	43	39
125 mm plastered brick wall	36	36	40	46	54	57
Double brick wall (50 mm cavity, 100 mm plastered bricks)	37	41	48	60	61	61
			Sound absorption coefficients			
Walls	0.04	0.04	0.09	0.15	0.17	0.23
Floor	0.02	0.06	0.14	0.37	0.60	0.66
Ceiling	0.30	0.20	0.15	0.05	0.05	0.05

4.22 Evaluate the effects of air absorption on the average sound absorption coefficient in a 25 m × 20 m × 10 m room at 20°C with a 50% relative humidity.

4.23 The surface of an acoustic tile has a normal specific acoustic impedance of $\mathbf{Z}_s = 650 - i1450$ rayls. Evaluate the absorption coefficient of the tile for normal incident sound waves in air.

4.24 A three metre high solid brick wall is built around the speedway complex in problem 4.7, at a radial distance of 70 m from the geometrical centre, in an attempt to reduce the radiated noise. Evaluate the effects of the barrier at 70 m, 1000 m and 2000 m (the maximum noise level occurs at about 250 Hz).

4.25 A centrifugal unit in a salt wash plant can be modelled as a large rectangular unit with dimensions of 1.4 m, 2.4 m and 1.2 m in the x, y and z planes, respectively. The unit has a total effective mass of 7170 kg and its centre of gravity is located at its geometrical centre. The unit is mounted on twenty-four rubber isolators (six in each corner), each of which has a spring stiffness of 5.9×10^5 N m^{-1}. Assuming that the horizontal stiffness of the isolators is 50% of the vertical stiffness, evaluate the six natural frequencies of the rigid body motions.

4.26 A sonar transducer on a submarine hull is protected by a streamlined, stainless steel, spherical dome. The transducer radiates sound waves at 24 kHz. Assuming that the dome is free-flooding (i.e. the cavity surrounding the transducer inside the dome is filled with sea water), estimate the greatest thickness of sheet steel that can be used such that the radiated signal is to be attenuated by no more than 6 dB on passing through the dome.

4.27 A 300 kg mass is mounted on six rubber isolators each with a stiffness of 5.9×10^5 N m^{-1}, a mass of 0.33 kg, and a damping ratio of 0.15. The mass of the foundation slab upon which the isolated mass is mounted is 4000 kg. As a first approximation, neglect the stiffness and damping in the mass and the foundation and estimate the transmissibility of the isolated system at 200 rad s^{-1}.

4.28 A spring–mass–damper system is subjected to a steady-state abutment excitation $x_0(t) = X_0 \sin \omega t$ and it is required to reduce the steady-state response $x_1(t)$ of the mass m_1 to zero. As a solution, a dynamic absorber is added, as shown in Figure P4.28, with the result that $x_1(t) = 0$ in the steady-state. If $m_2 = 0.1m_1$ and $\omega = 1.2(k_{s1}/m_1)^{1/2}$ what is the required value of k_{s2} in terms of k_{s1}? Also, what is the amplitude of the motion of the mass m_2 in terms of k_{s1}, c_{v1}, ω and X_0?

Fig. P4.28.

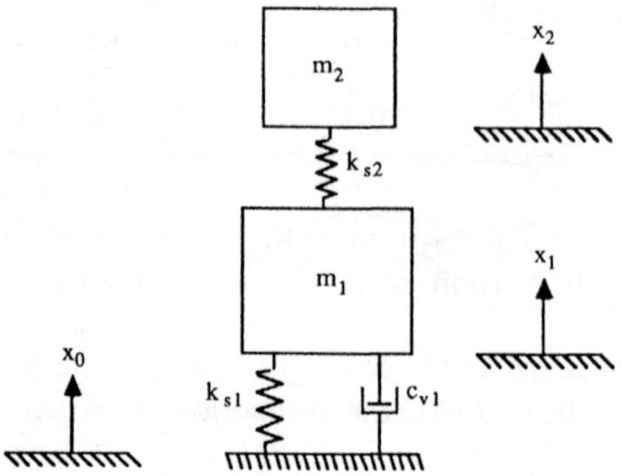

Chapter 5

5.1 A random variable, x, has the following distribution:

$$p(x) = 0.25 \quad \text{for } 0 < x \leqslant 0.25$$
$$= 0.75 \quad \text{for } 0.25 < x \leqslant 1$$
$$= 0.25 \quad \text{for } 1 < x \leqslant x_{max},$$

where x_{max} represents the maximum amplitude of the signal. Estimate x_{max}, the mean value, $E[x]$, the mean-square value, $E[x^2]$, and the standard deviation, σ.

5.2 Evaluate the skewness and the kurtosis of the random variable in problem 5.1.

5.3 A time history of an engine casing vibration with a signal to noise ratio of at least 40 dB is required. Estimate the number of time records that have to be averaged to achieve this.

5.4 The coherence of a frequency response measurement is 0.92. Assuming that the system is linear, what is the signal to noise ratio? If a signal to noise ratio of 40 dB was required, what would the coherence have to be?

5.5 Determine the auto-correlation of a cosine wave $x(t) = A \cos t$, and plot it against τ.

5.6 What are the values of (a) an auto-correlation coefficient, and (b) a normalised auto-covariance for (i) $\tau = 0$ and (ii) $\tau = \infty$?

Fig. P5.7.

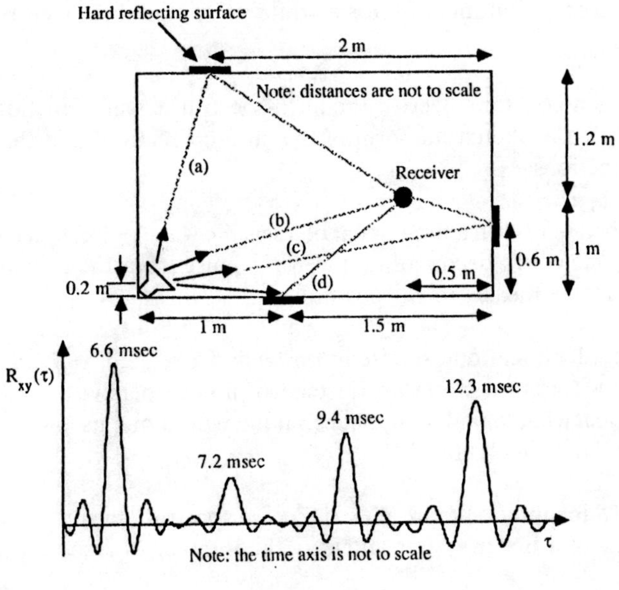

5.7 Consider the acoustic propagation problem outlined in Figure P5.7. Broadband sound covering a frequency range from 25 Hz to 5000 Hz is applied to the speaker, and a cross-correlation is obtained between the input and output signals. Assuming that the speed of sound in air is 340 m s^{-1}, identify the major peaks in the cross-correlation function. Which propagation paths have the largest and the smallest contributions to the overall noise level at the receiver? What would happen if the above experiment was repeated with a reduced frequency bandwidth?

5.8 A non-dispersive propagation path can be modelled as a linear system with a constant frequency response function (i.e. $\mathbf{H}(\omega) = H$). Consider such a system as illustrated in Figure P5.8. Given that $x(t)$ is the input signal, $y(t)$ is the measured output response signal, and $n(t)$ is an output noise signal which is statistically independent from the input signal, show that

$$R_{xy}(\tau) = HR_{xx}(\tau - d/c),$$

where d is the propagation distance, and c is the propagation velocity which is constant for a non-dispersive medium. If the medium is a fluid of density 1026 kg m^{-3}, the wave propagation speed in the medium is 1500 m s^{-1}, and the propagation distance is 20 m, at what time delay would one expect the cross-correlaton function to peak?

Fig. P5.8.

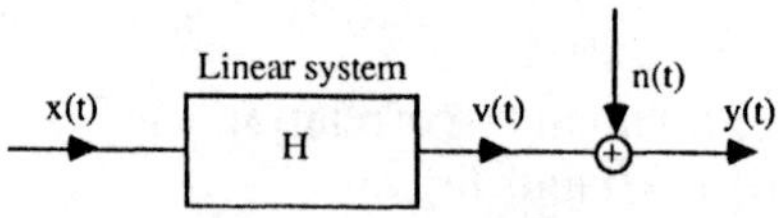

5.9 The auto-correlation function for a white noise signal is given by

$$R_{xx}(\tau) = 2\pi S_0 \delta(\tau),$$

where S_0 is a constant. Derive the auto-spectral density function of the white noise signal, and sketch the form of the auto-correlation and the auto-spectral density functions.

5.10 If one wished to obtain a time signal of a seismic wave pulse, describe a technique by which one could deconvolute the wave pulse from the impulse response of the earth at the measurement position.

5.11 If the path identification exercise in problem 5.7 related to the identification of dispersive waves in a structure (instead of non-dispersive sound waves in the atmosphere), what signal analysis technique would one use in preference to the cross-correlation technique? Why?

5.12 Figure 5.15 (chapter 5) represents the cross-spectral density between the input and output of a linear system. Identify the number of resonant modes.

5.13 Whilst measuring the frequency response function of a linear system with noticeable extraneous noise at the output stage, a coherence value of 0.79 was obtained at the dominant frequency. What percentage of the true output signal is associated with the extraneous noise?

5.14 How much faster is a 10 000 point fast Fourier transform than a similar discrete Fourier transform?

5.15 A time record is digitised on a signal analyser into a sequence of 1024 equally spaced sample values. The frequency resolution of the corresponding auto-spectra is 12.5 Hz. Evaluate (i) the digitising rate, (ii) the Nyquist cut-off frequency, and (iii) the normalised random error for a spectral average over 100 time records. How many time averages would one require to achieve a normalised random error of 0.01?

5.16 The following specifications are required for a digital spectral analysis: (i) $\varepsilon_r \leqslant 0.05$, (ii) $B_e = 1$ Hz; (iii) $f_c = 20\,000$ Hz. Evaluate (i) the number of spectral averages required, and (ii) the number of spectral lines required per transform. How many calculations are required using the FFT algorithm?

5.17 Show that the spectral window of a rectangular lag window with $w(\tau) = 1$, and with $-T \leqslant \tau \leqslant T$ is given by

$$W(\omega) = \frac{T}{\pi}\left\{\frac{\sin \omega T}{\omega T}\right\},$$

and by

$$W(f) = 2T\left\{\frac{\sin 2\pi f T}{2\pi f T}\right\}.$$

5.18 Show that the spectral window of a triangular lag window with

$$w(\tau) = 1 - \frac{|\tau|}{T} \quad \text{for } 0 \leqslant |\tau| \leqslant T$$

$$= 0 \quad \text{otherwise}$$

is given by

$$W(\omega) = \frac{T}{2\pi}\left\{\frac{\sin(\omega T/2)}{\omega T/2}\right\}^2,$$

and by

$$W(f) = T\left\{\frac{\sin \pi f T}{\pi f T}\right\}^2.$$

5.19 Why is time record averaging essential in digital signal analysis?

5.20 Derive an expression for the measured frequency response function, $\mathbf{H}_{xy}(\omega)$, in terms of the true frequency response function, $\mathbf{H}(\omega)$, and the associated auto- and cross-spectral densities, for a single input–output linear system with extraneous noise, $m(t)$, at the input stage which passes through the system. Also derive a similar expression for the measured frequency response function, $\mathbf{H}_{xy}(\omega)$, for a single input–output system with extraneous noise, $m(t)$, at the input stage which does not pass through the system (see Figure 5.30 in chapter 5). Comment on the measured frequency response function for the case where (i) the extraneous noise is correlated with the measured input signal, and (ii) the extraneous noise is not correlated with the measured input signal.

5.21 Evaluate the Fourier transform of the function in Figure P5.21. When $t = T$, the function becomes an odd weighting function. What is the magnitude of the corresponding spectral window?

Fig. P5.21.

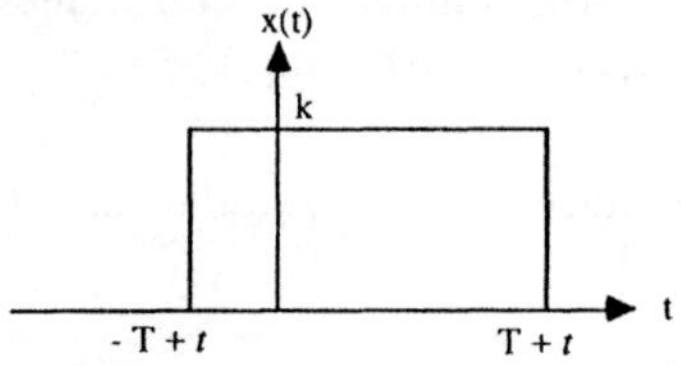

Chapter 6

6.1 A spring-mounted rigid body with a 150 kg mass can be modelled as a single oscillator with a stiffness of 6.5×10^6 N m^{-1}. A steady-state applied force of 100 N produces a velocity of 0.2 m s^{-1}. Estimate the damping ratio, the loss factor and the quality factor of the system.

6.2 Consider two coupled groups of oscillators with similar modal densities, in which only the first group is directly driven in the steady-state. Using the steady-state power balance equations, show that

$$\frac{E_2}{E_1} = \frac{\eta_{21}}{\eta_2 + \eta_{21}}.$$

Now, assuming that the oscillators are strongly coupled, the first group is lightly damped, the second group is heavily damped, and that one wishes to minimise the vibrational levels transmitted to the second group, what should one do?

6.3 Show that

$$\frac{\Pi_1}{\omega E_1} = \eta_1 + \frac{\dfrac{n_2}{n_1}\eta_2\eta_{21}}{\eta_2 + \eta_{21}}$$

for two coupled groups of oscillators in steady-state vibration.

6.4 Consider a two subsystem S.E.A. model in which steady-state power is injected directly into both subsystems. If the power injected into subsystem 2 is one-quarter of the power injected into subsystem 1, derive an expression for the modal energy ratio in terms of the modal densities, the loss factors, and the coupling loss factor between subsystems 1 and 2 (η_{12}). If there is equipartition of modal energy between the two subsystems, derive an expression for the loss factor of the second subsystem in terms of the loss factor of the first subsystem, and the respective modal densities.

6.5 Consider a three subsystem S.E.A. model in which steady-state power is injected directly into subsystems 1 and 3. Subsystem 2 is directly coupled to both subsystems 1 and 3, but subsystems 1 and 3 are not directly coupled themselves. Also, the power injected into subsystem 3 is twice the power that is injected into subsystem 1. Show that the time- and space-averaged mean-square vibrational energy of subsystem 2 in terms of the time- and space-averaged mean-square vibrational energy of subsystem 1, the total masses M_1 and M_2 of the two subsystems, and the relevant S.E.A. parameters is given by

$$\langle\overline{v_2^2}\rangle = \frac{M_1}{M_2}\langle\overline{v_1^2}\rangle\frac{\eta_{12}}{(\eta_2+\eta_{21}+\eta_{23})} + \frac{M_1}{M_2}\langle\overline{v_1^2}\rangle\frac{\eta_{32}}{(\eta_2+\eta_{21}+\eta_{23})}\cdot \left\{\frac{2\eta_1\eta_2 + 2\eta_1\eta_{21} + 2\eta_1\eta_{23} + 2\eta_2\eta_{12} + 3\eta_{12}\eta_{23}}{\eta_2\eta_3 + \eta_3\eta_{21} + \eta_3\eta_{23} + \eta_2\eta_{32} + 3\eta_{21}\eta_{32}}\right\}.$$

6.6 Consider two coupled oscillators where only one is directly driven by external forces and the other is driven only through the coupling. Derive expressions for the total vibrational energies of each of the oscillators in terms of the input power, the loss factors, the coupling loss factors, and the natural frequencies of the oscillators. If (a) $\eta_{21} \gg \eta_1$ and $\eta_{21} \gg \eta_2$, (b) $\eta_{21} \ll \eta_1$ and $\eta_{21} \ll \eta_2$, (c) $\eta_2 \ll \eta_{21} \ll \eta_1$, and (d) $\eta_1 \ll \eta_{21} \ll \eta_2$ what parameters govern the vibrational responses of each of the two oscillators?

6.7 Evaluate (a) the modal density and (b) the number of modes in each of the octave bands from 500 Hz to 8000 Hz for a 10 m long steel bar (with cross-sectional dimensions 100 mm × 100 mm) for (a) longitudinal, and (b) flexural vibrations.

6.8 If one wished to reduce the modal density of flat plate elements (e.g. those used for machine covers, etc.) what practical options are available?

6.9 A large factory space is approximately 25 m × 30 m × 10 m. Evaluate the modal density and the number of modes in each of the octave bands from 500 Hz to 8000 Hz.

6.10 As a first approximation, a satellite structure can be modelled as a large flat aluminium platform which is coupled to a large aluminium cylinder, as illustrated in Figure P6.10. The aluminium plate is 5 mm thick and is 3.5 m × 3 m. The cylinder is 2 m long, has a mean diameter of 1.5 m and has a 3 mm wall thickness. The following information is available about the structure in the 500 Hz octave band: the platform is directly driven and the cylinder is only driven via the coupling joints; the internal loss factor of the platform, η_1, is 4.4×10^{-3}, the internal loss factor of the cylinder, η_2, is 2.4×10^{-3}; the platform r.m.s. vibrational velocity is 27.2 mm s^{-1}; and the cylinder r.m.s. vibrational velocity is 13.2 mm s^{-1}. Estimate the coupling loss factors, η_{12} and η_{21}, and the input power.

Fig. P6.10.

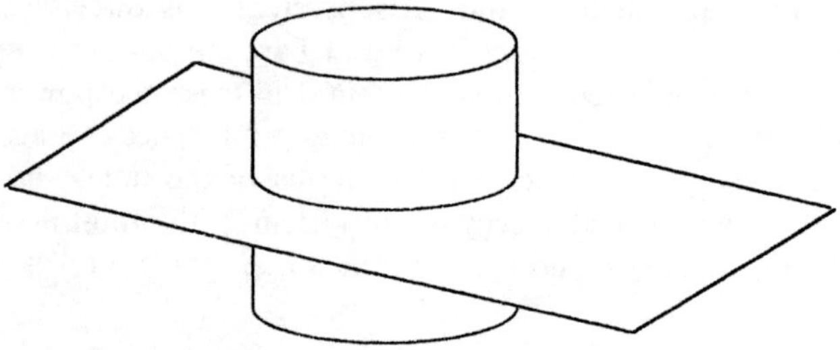

6.11 Explain why bias errors associated with stiffness effects can be neglected when an impedance head is used to measure the mobility/impedance associated with bending waves in a structure. What dominates the bias errors in this instance?

6.12 Consider the clamped aluminium flat plate in problem 3.7. Given that the structural loss factor for aluminium is 1.0×10^{-4}, evaluate the internal loss factors in each of the octave bands from 31.5 Hz to 2000 Hz (assume that there is no energy dissipation at the boundaries). What numerical value do the structural loss factors asymptotically approach at high frequencies (i.e. frequencies well in excess of the plate critical frequency)?

6.13 Consider two flat aluminium plates which are coupled at right angles to each other. The first plate is 3 mm thick and is 2.5 m × 1.2 m, and the second plate is 5.5 mm thick and is 2.0 m × 1.2 m. Evaluate the coupling loss factors in all the octave bands from 125 Hz to 2000 Hz for (a) a welded joint along the 1.2 m edge, and (b) a bolted joint with twelve bolts along the 1.2 m edge.

6.14 From Figure 6.19, estimate the coupling damping associated with gas pumping at the coupling (flanged joint) between the two 65 mm diameter, 1 mm wall thickness cylindrical shells at 1290 Hz.

6.15 For the case of the beam–plate–room coupled system in sub-section 6.7.1, show that equations (6.83) and (6.85) are identical when the plate is lightly damped and the room is highly reverberant.

6.16 Two volume spaces separated by a partition which is mechanically excited with broadband noise can be modelled as a three subsystem S.E.A. model. Derive an expression for the difference in sound pressure levels between the two volume regions in terms of the relevant S.E.A. parameters associated with each of the subsystems.

6.17 Consider a machine structure where $\eta_s + \eta_j \gg \eta_{rad}$. Assuming that (i) E represents space- and time-averaged vibrational energies, (ii) the subscript i refers to the system which is directly excited, and (iii) the subscript k refers to all other subsystems which are coupled to subsystem i, would noise reduction be achieved if the ratios E_k/E_i increase with the addition of damping treatment to the structure?

6.18 Peak overall r.m.s. vibrational velocity levels of $\sim$0.5 m s^{-1} are recorded on a section of steel pipeline at a gas refinery installation. The average overall levels are $\sim$0.15 m s^{-1}. Estimate the peak and the average overall dynamic stress levels.

Chapter 7

7.1 Evaluate the cut-off frequencies and the associated axial wavenumbers for the first six higher order acoustic modes in a 254 mm diameter circular duct containing steam for (a) the no flow case, and (b) a mean flow velocity of 200 m s^{-1}.

7.2 Evaluate the first three cut-off frequencies for higher order acoustic modes in a rectangular air conditioning duct with dimensions 0.65 m × 0.4 m, and with a mean air flow of 15 m s^{-1}.

7.3 Starting with equations (7.11) and (7.12), work through all the relevant equations in sub-section 7.4.1 to derive equations (7.20) and (7.21). Check both equations for dimensional consistency.

7.4 Determine the non-dimensional pipe wall thickness parameter, β, required for the ring frequency of a steel pipe to equal its critical frequency in air at (i) 15°C and (ii) 150°C. What are the corresponding diameter to pipe wall thickness ratios?

7.5 Consider a 2.92 m long steel cylinder pipe with a mean pipe radius of 36.72 mm, and a pipe wall thickness of 0.89 mm. Draw the wavenumber dispersion relationships for the $(m, 1)$ and $(m, 2)$ structural modes and the $(1, 0)$, $(1, 1)$ and $(2, 0)$ higher order acoustic modes for the no flow case (equation 7.24 can be used to identify the structural modes). Identify the various coincidence regions on the dispersion plots. Which specific structural modes are coincident with the higher order acoustic modes? Are the coincidences complete coincidences or wavenumber coincidences?

Assuming that there is a continuum of K_m values for any given K_n value, and using the thin shell approximations, estimate the complete coincidence frequencies. Are the complete coincidence frequencies greater than or less than the associated higher order acoustic mode cut-off frequencies? Would this necessarily still be the case if there is flow in the pipe? What about the wavenumber coincidence frequencies?

7.6 A pressure relief valve in a gas pipeline has an inlet pressure of 5800 kPa, and an outlet pressure of 730 kPa. The narrowest cross-sectional dimension of the valve-opening is 30 mm. The ratio of specific heats of the gas in the pipeline is 1.29, and the speed of sound is 396 m s^{-1} at 5°C. Estimate the dominant frequency of the noise generated by the valve at a gas temperature of 5°C and at 25°C. What is the dominant mechanism associated with the valve noise?

7.7 A pressure relief valve and the associated piping in a gas pipeline installation are illustrated schematically in Figure P7.7. The valve can be modelled as a free-floating piston arrangement with a pilot controlled valve which ensures equality of pressure on both sides of the piston until such time that it is desired that the valve be opened. When the pilot valve is switched off (either manually or automatically), the pressure builds up on the underside, the piston is pushed upwards, and a relief path is established for the gas. The piston mass is 45 kg, the maximum allowable inlet pressure is 5800 kPa, the outlet pressure is 730 kPa, and the ratio of specific heats is 1.29. The relevant dimensions are presented in Figure P7.7

Fig. P7.7

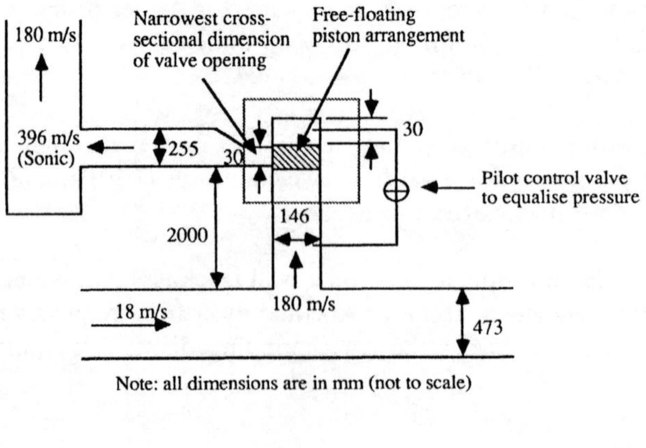

Note: all dimensions are in mm (not to scale)

There are three primary sources of possible vibration and noise in the valve/inlet piping/outlet piping arrangement. Identify these three sources and evaluate the dominant frequencies associated with each of them. Take the speed of sound in the gas to be 396 m s^{-1}. Also, the longitudinal natural frequencies in a pipe which is closed at one end are approximately given by $(2n-1)c/4L$, where n is an integer number, c is the speed of sound in the pipe, and L is the pipe length.

7.8 At what wall thickness will a nominal 0.5 m steel pipe have the maximum possible number of coincidences at frequencies below the ring frequency?

7.9 Estimate the sound pressure level due to the valve noise (not the inlet/outlet piping noise) associated with the pressure relief valve in problem 7.7, at a distance of 10 m from the valve.

7.10 Given that the nominal diameter of a gas pipeline is fixed and that the flow parameters (density, temperature, flow velocity, etc.) are also fixed, what parameter controls the sound transmission loss through the pipe wall at (a) frequencies below the cut-off frequencies of the first higher order acoustic mode, (b) frequencies above the cut-off frequency of the first higher order acoustic mode but below the ring frequency, and (c) frequencies above the ring frequency? Would damping improve the transmission loss performance in any of these three regions?

Chapter 8

8.1 What is the crest factor of (i) a sine wave, and (ii) a broadband random noise signal? As a rule of thumb, what is the typical range of crest factors for good and for damaged bearings? Why is the crest factor not suitable as a diagnostic tool to detect advanced/widespread bearing damage?

8.2 What is the typical range of kurtosis values for good and damaged bearings? What are the limitations of kurtosis as a diagnostic tool?

8.3 Define and briefly discuss the application of each of the following signal analysis terms: (i) kurtosis, (ii) impulse response, (iii) coherence, (iv) power cepstrum, (v) complex cepstrum, (vi) sound intensity, (vii) synchronous time-averaging, (viii) phase-averaging, (ix) crest factor, (x) skewness, (xi) envelope power spectrum, (xii) inverse filtering, (xiii) surface intensity, (xiv) vibration intensity.

8.4 If one had access to a small portable exciter system, a single accelerometer, a phase meter and a r.m.s. vibration meter with a trigger facility, briefly describe how one would go about obtaining the first few vibrational mode shapes of the chassis of a vehicle.

8.5 Identify the various discrete vibrational frequencies associated with a rolling-contact bearing with a rotating inner race and a stationary outer race. The bearing has fifteen rollers, a pitch diameter of 34 mm, a roller diameter of 6 mm, a 12.96° contact angle, and the shaft rotates at 2000 r.p.m.

8.6 A six blade, four vane axial fan in a circular duct with a 0.572 m internal radius, rotates at 3500 r.p.m. Estimate (i) the blade passing frequency and (ii) the frequencies associated with the first set of lobed interaction pressure patterns. Is it possible for any higher order acoustic duct modes to be set up? Assume that the mean flow speed in the duct is 30 m s^{-1} and that the speed of sound is 343 m s^{-1}.

8.7 An electrical induction motor has sixty rotor bars, sixty-eight stator slots and six magnetic poles. The shaft which it drives rotates at 3600 r.p.m. Identify the three discrete frequency components that would be present in a vibration signal from one of the motor bearings. Also, identify the primary vibrational frequency associated with interactions of the fundamental magnetic flux wave with its harmonics and with the rotor-slot components (assume zero slip). If electro-magnetic irregularities associated with dynamic eccentricity are present, how could they be identified?

8.8 Consider a length of straight pipe with a fully developed internal turbulent gas flow. In general, the fluctuating wall pressure field (on the internal pipe wall) in the piping system will be the sum of a fluctuating turbulence pressure field, p_{T}, and a fluctuating acoustic pressure field comprising plane acoustic waves, p_{P}, and higher order acoustic modes, p_{H}. The resultant instantaneous pressure at a point on the wall at a particular circumferential position is therefore given by

$$p(x,t) = p_{\mathrm{T}}(x,t) + p_{\mathrm{P}}(x,t) + p_{\mathrm{H}}(x,t).$$

Assuming that each fluctuating component of the wall pressure field is a random function of space and time, and is stationary, show that the cross-correlation of the fluctuating wall pressure between two points at the same circumferential position is

$$R_{pp}(\xi,\tau) = R_{\mathrm{TT}}(\xi,\tau) + R_{\mathrm{PP}}(\xi,\tau) + R_{\mathrm{HH}}(\xi,\tau) + R_{\mathrm{TP}}(\xi,\tau) + R_{\mathrm{PT}}(\xi,\tau) + R_{\mathrm{TH}}(\xi,\tau) + R_{\mathrm{HT}}(\xi,\tau) + R_{\mathrm{PH}}(\xi,\tau) + R_{\mathrm{HP}}(\xi,\tau),$$

where ξ is the longitudinal separation distance between the two measuring positions at the same circumferential position, and τ is the corresponding time delay. The subscripts T, P and H refer to turbulence, plane acoustic waves, and higher order acoustic modes, respectively.

Assuming zero mean values, show that the above expresion for the cross-correlation when expressed in terms of correlation coefficients is

$$\begin{aligned}\langle p^2\rangle\rho_{pp} = {} & \langle p_{\mathrm{T}}^2\rangle\rho_{\mathrm{TT}} + \langle p_{\mathrm{P}}^2\rangle\rho_{\mathrm{PP}} + \langle p_{\mathrm{H}}^2\rangle\rho_{\mathrm{HH}} + \langle p_{\mathrm{T}}^2\rangle^{1/2}\langle p_{\mathrm{P}}^2\rangle^{1/2}\rho_{\mathrm{TP}} \\ & + \langle p_{\mathrm{T}}^2\rangle^{1/2}\langle p_{\mathrm{P}}^2\rangle^{1/2}\rho_{\mathrm{PT}} + \langle p_{\mathrm{T}}^2\rangle^{1/2}\langle p_{\mathrm{H}}^2\rangle^{1/2}\rho_{\mathrm{TH}} + \langle p_{\mathrm{T}}^2\rangle^{1/2}\langle p_{\mathrm{H}}^2\rangle^{1/2}\rho_{\mathrm{HT}} \\ & + \langle p_{\mathrm{P}}^2\rangle^{1/2}\langle p_{\mathrm{H}}^2\rangle^{1/2}\rho_{\mathrm{PH}} + \langle p_{\mathrm{P}}^2\rangle^{1/2}\langle p_{\mathrm{H}}^2\rangle^{1/2}\rho_{\mathrm{HP}},\end{aligned}$$

where the ρ's are the correlation coefficients, and $\langle\ \rangle$ represents mean-square values. What is the equivalent expression if all three components of the wall pressure field are uncorrelated?

Appendix 1
Relevant engineering noise and vibration control journals

A list of several international journals that publish research and development articles related to various aspects of engineering noise and vibration control is presented below.

Acustica – S. Hirzel Verlag
Applied Acoustics – Elsevier Applied Science
Current Awareness Abstracts – Vibration Institute
Journal of the Acoustical Society of America – Acoustical Society of America
Journal of Fluid Mechanics –Cambridge University Press
Journal of Fluids and Structures – Academic Press
Journal of Sound and Vibration – Academic Press
Journal of Vibration, Acoustics, Stress, and Reliability in Design – American Society of Mechanical Engineers
Mechanical Systems and Signal Processing – Academic Press
Noise and Vibration in Industry – Multi-Science
Noise Control Engineering Journal – Institute of Noise Control Engineers
Shock and Vibration Digest – Vibration Institute
Sound and Vibration – Acoustical Publications, Inc.

Appendix 2

Typical sound transmission loss values and sound absorption coefficients for some common building materials

A Typical sound transmission loss (*TL*) values

Description	Sound transmission loss (dB) (octave bands)					
	125	250	500	1000	2000	4000
Single panels						
1 mm aluminium sheet (stiffened)	11	10	10	18	23	25
125 mm thick plastered brick	36	36	40	46	54	57
360 mm thick plastered brick	44	43	49	57	66	70
150 mm hollow concrete (painted)	36	36	42	50	55	6(
75 mm solid concrete	35	40	44	52	59	6
150 mm plastered solid concrete	40	43	50	58	64	67
Chipboard (~20 mm) on a wooden frame	17	18	25	30	26	32
6 mm monolithic glass	24	26	31	34	30	37
12 mm monolithic glass	27	32	36	33	40	49
Hardwood panels (~50 mm)	19	23	25	30	37	42
1.5 mm lead sheet	28	32	33	32	32	33
3.0 mm lead sheet	30	31	27	38	44	33
Loaded vinyl sheet (~3 mm)	12	15	21	27	31	37
Loaded vinyl sheet (~6 mm)	21	23	30	35	40	49
Plasterboard (~10 mm) on a wooden frame	15	20	24	29	32	35
Plywood (~5 mm) on a wooden frame	9	13	16	21	27	29
1 mm galvanised steel sheet	8	14	20	26	32	38
1.6 mm galvanised steel sheet	14	21	27	32	37	43
Sandwich panels						
Laminated glass (3 mm × 0.75 mm × 3 mm)	26	29	32	35	35	42
Laminated glass (6 mm × 1.5 mm × 6 mm)	28	32	36	38	41	51
1.5 mm lead between two sheets of 5 mm plywood	26	30	34	38	42	44

Table A (*cont.*)

Description	Sound transmission loss (dB) (octave bands)					
	125	250	500	1000	2000	4000
Double-leaf panels						
Double brick wall (50 mm cavity, 100 mm plastered bricks)	37	41	48	60	61	61
Double brick wall (150 mm cavity, 100 mm plastered bricks)	51	54	58	63	69	74
Two 6 mm glass panes (separated by a 12 mm air gap)	26	23	32	38	37	52
Two 6 mm glass panes (separated by a 25 mm air gap)	23	28	35	41	38	51
Two 6 mm laminated glass panes (separated by a 12 mm air gap)	25	31	39	44	46	56
Two 13 mm gypsum wallboards (separated by a 64 mm air gap)	18	27	37	45	43	39
Two 16 mm gypsum wallboards (separated by a 64 mm air gap)	19	29	40	46	37	44
Two 16 mm gypsum wallboards (separated by a 64 mm air gap filled with fibreglass)	26	36	45	50	41	46
Six 16 mm gypsum wallboards (separated by a 100 mm central air gap filled with fibreglass; three coupled wallboards on each side)	42	46	54	63	62	66

Note: the above sound transmission loss values only represent typical laboratory values for the types of materials described. Manufacturers of different types of building materials (walls, ceilings, floors, doors, windows, etc.) generally provide sound transmission loss data sheets which have been derived from certified one-third-octave band laboratory tests. It is important to note that manufacturers' data generally relate to laboratory type measurements. Field installed transmission loss values are generally lower due to flanking transmission, leakage, etc.

B Typical sound absorption coefficients

Description	Sound absorption coefficient (octave bands)					
	125	250	500	1000	2000	4000
Exposed brick	0.05	0.04	0.02	0.04	0.05	0.05
Brick (painted)	0.01	0.01	0.02	0.02	0.02	0.03
Normal carpet	0.02	0.06	0.14	0.37	0.60	0.66
Thick pile carpet	0.15	0.25	0.50	0.60	0.70	0.70
Concrete (solid)	0.01	0.01	0.02	0.02	0.02	0.02
Porous concrete block (painted)	0.10	0.05	0.06	0.07	0.09	0.08
Curtains (heavy draped)	0.07	0.31	0.49	0.75	0.70	0.60
Fibrous glass wool (25 mm)	0.07	0.23	0.48	0.83	0.88	0.80
Fibrous glass wool (100 mm)	0.39	0.91	0.99	0.97	0.94	0.89
Plate glass	0.25	0.25	0.18	0.12	0.07	0.05
Hardboard	0.10	0.10	0.15	0.15	0.10	0.10
Person in a wood or padded seat	0.15	0.25	0.40	0.40	0.45	0.40
Person in a fully upholstered seat	0.20	0.40	0.45	0.45	0.50	0.45
Plasterboard	0.30	0.20	0.15	0.05	0.05	0.05
Fibrous plaster	0.04	0.05	0.06	0.08	0.04	0.06
Plasterboard ceiling	0.20	0.20	0.15	0.10	0.05	0.05
Open cell polyurethane acoustic foam (25 mm)	0.14	0.30	0.63	0.91	0.98	0.91
Open cell polyurethane acoustic foam (50 mm)	0.35	0.51	0.82	0.98	0.97	0.95
Unoccupied wood or padded seat	0.03	0.05	0.05	0.10	0.15	0.10
Unoccupied fully upholstered seat	0.10	0.20	0.30	0.30	0.30	0.35
Textile faced acoustic foam (25 mm)	0.10	0.25	0.59	0.98	0.92	0.98
Terrazzo flooring	0.01	0.01	0.01	0.01	0.01	0.01
Vinyl faced acoustic foam (25 mm)	0.14	0.25	0.63	0.92	0.82	0.65
Wood	0.15	0.11	0.10	0.07	0.09	0.03

Note: the data provided in this appendix has been collated from several sources including references 1.4, 2.6 and 2.7.

Appendix 3
Units and conversion factors

A Some common units used in engineering noise and vibration control

Primary units

length metre (m)
mass kilogramme (kg)
quantity of a substance mol
temperature kelvin (K)
time second (s)

Secondary units

energy joule (J ≡ Nm ≡ Ws)
force newton ($N \equiv kg\ m\ s^{-2}$)
frequency hertz (s^{-1})
molecular weight mol^{-1}
power watt ($W \equiv J\ s^{-1} \equiv Nm\ s^{-1}$)
pressure Pascal ($Pa \equiv N\ m^{-2}$)
radian frequency rad s^{-1}

Derived units

acceleration $m\ s^{-2}$
acoustic (radiation) impedance*
(force per unit area/volume velocity per unit area) Ns m^{-1}
acoustic intensity $W\ m^{-2}$
auto-spectral density (power spectral density) $units^2\ Hz^{-1}$
energy density $J\ m^{-3}$
energy spectral density $units^2\ s\ Hz^{-1}$
entropy $J\ kg^{-1}\ K^{-1}$
gas constant $J\ kg^{-1}\ K^{-1}$
mechanical impedance
(force/velocity) $N\ s\ m^{-1}$
mechanical stiffness $N\ m^{-1}$
mobility $m\ N^{-1}\ s^{-1}$
modulus of elasticity, adiabatic bulk modulus $N\ m^{-2}$
quefrency s
specific acoustic impedance
(pressure/velocity) $N\ s\ m^{-3}$
surface density $kg\ m^{-2}$
universal gas constant $J\ mol^{-1}\ K^{-1}$
velocity $m\ s^{-1}$
viscosity $N\ s\ m^{-2}$
viscous damping $N\ s\ m^{-1}$
volume density $kg\ m^{-3}$
volume velocity $m^3\ s^{-1}$

* Acoustic impedance is sometimes defined as pressure/volume velocity

B Conversion factors

Length

1 ft = 0.3048 m
1 in = 25.4 mm
1 mile = 1.609344 km
1 mph = 1.6093 km h^{-1} = 0.44704 m s^{-1}
1 nautical mile = 1.852 km
1 knot = 1 nm h^{-1} = 0.5144 m s^{-1}

Area

1 ft^2 = 0.09290304 m^2
1 in^2 = 0.00064516 m^2
1 acre = 4046.86 m^2
1 hectare = 10^4 m^2

Volume

1 ft^3 = 28.3168 litre
1 litre = 10^{-3} m^3
1 U.K. gal = 4.54609 litre
1 U.K. pint = 0.568261 litre
1 U.S. gal = 3.7853 litre

Mass

1 lb = 0.45359237 kg
1 oz = 28.3495 g
1 ton = 1.01605 tonne
1 lb ft^{-3} = 16.0185 kg m^{-3}

Force (N, kg m s^{-2})

1 lbf = 4.44822 N
1 kgf = 1 kp = 9.80665 N
1 bar = 14.50 psi = 10^6 dyne/cm^{-1} = 10^5 Pa = 10^5 N m^{-2}
1 psi = 6.89476 kPa
1 mm H_2O = 9.80665 Pa
1 mm Hg = 133.322 Pa
1 atm = 101.325 kPa

Energy (J, Nm, Ws)

1 ft-lbf = 1.355818 J
1 Btu = 1055.06 J
1 kWh = 3.6 MJ
1 kcal = 4.1868 kJ

Power (W, J s^{-1}, N m s^{-1})

1 ft-lbs s^{-1} = 1.355818 W
1 hp = 745.7 W
1 kcal h^{-1} = 1.163 W

Temperature

a °C = b K − 273.15
a °C = (b F − 32)/1.8
b °F = (1.8 × a °C) + 32

Appendix 4
Physical properties of some common substances

A Solids

Solid	Density, ρ_0 (kg m^{-3})	Young's modulus, E (Pa)	Poisson's ratio, ν	Wavespeed, c_L Bar (m s^{-1})	Bulk	Product of critical frequency (f_c) and bar thickness* (m s^{-1})
Aluminium	2700	7.1×10^{10}	0.33	5150	6300	12.7
Brass	8500	10.4×10^{10}	0.37	3500	4700	18.7
Concrete (dense)	2600	$\sim 2.5 \times 10^{10}$	–	–	3100	21.1
Copper	8900	12.2×10^{10}	0.35	3700	5000	17.7
Cork	250	6.2×10^{10}	–	–	500	130.7
Cast iron	7700	10.5×10^{10}	0.28	3700	4350	17.7
Glass (Pyrex)	2300	6.2×10^{10}	0.24	5200	5600	12.6
Gypsum (plasterboard)	650	–	–	–	6800	9.61
Lead	11 300	1.65×10^{10}	0.44	1200	2050	54.5
Nickel	8800	21×10^{10}	0.31	4900	5850	13.3
Particle board	750	–	–	–	669	97.7
Polyurethane	72	1.9×10^{7}	–	–	513	127.4
Polystyrene	42	1.1×10^{7}	–	–	512	127.6
PVC	66	5.5×10^{7}	–	–	913	71.6
Plywood	600	–	–	–	3080	21.2
Rubber (hard)	1100	2.3×10^{9}	0.4	1450	2400	45.1
Rubber (soft)	950	5×10^{6}	–	–	1050	62.2
Silver	10 500	7.8×10^{10}	0.37	2700	3700	24.2
Steel	7700	19.5×10^{10}	0.28	5050	6100	12.9
Tin	7300	4.5×10^{10}	0.33	2500	–	26.1
Wood (hard)	650	1.2×10^{10}	–	4300	–	15.2

* Where a bar thickness does not apply, the thickness of the bulk material is used.

B Liquids

Liquid	Density, ρ_0 (kg m^{-3})	Temperature (°C)	Specific heat ratio, γ	Wavespeed, c (m s^{-1})
Castor oil	950	20	–	1540
Ethyl alcohol	790	20	–	1150
Fresh water	998	20	1.004	1483
Fresh water	998	13	1.004	1441
Glycerin	1260	20	–	1980
Mercury	13 600	20	1.13	1450
Petrol	680	20	–	1390
Sea water	1026	13	1.01	1500
Turpentine	870	20	1.27	1250

C Gases

Gas	Density, ρ_0 (kg m^{-3})	Temperature (°C)	Specific heat ratio, γ	Wavespeed, c (m s^{-1})
Air	1.293	0	1.402	332
Air	1.21	20	1.402	343
Carbon dioxide	1.84	20	1.40	267
Hydrogen	0.084	0	1.41	1270
Hydrogen	0.084	20	1.41	1330
Nitrogen	1.17	20	1.40	349
Oxygen	1.43	0	1.40	317
Oxygen	1.43	20	1.40	326
Steam	0.6	100	1.324	405

Note: the data provided in this appendix has been collated from several sources including references 1.3, 1.4, 2.6 and 2.7.

Answers to problems

Chapter 1

1.1 343 m s^{-1}; 343 m s^{-1}.

1.2 344 m s^{-1}; 4.58 m; 0.0133 s.

$$\frac{1}{2\pi}\left(\frac{k_{s1}+k_{s2}}{m}\right)^{1/2}; \qquad \frac{1}{2\pi}\left\{\frac{k_{s1}k_{s2}}{m(k_{s1}+k_{s2})}\right\}^{1/2}.$$

1.4 0.54 Hz.

1.5

$$\frac{1}{2\pi}\left\{\frac{L^2k_{s1}k_{s2}}{m(a^2k_{s1}+L^2k_{s2})}\right\}^{1/2}.$$

1.6 As long as $F = k_s x$, the frequency of oscillation is independent of amplitude – i.e. f is proportional to k_s and m. If the system is non-linear, i.e. $F \neq k_s x$, the frequency of each note would depend upon how hard one strikes the keys.

1.7 3.11×10^6 N m^{-1}; 6.66 Hz; 16.36 J.

1.8 59.9 N m s rad^{-1}; 6.9 s.

1.9

$$f_d = \frac{\omega_n\sqrt{1-\left(\frac{Sv}{M\omega_n}\right)^2}}{2\pi}.$$

1.10 21.96 N s m^{-1}.

1.11 0.0475.

1.13 The existing vibration amplification is ~ 1.8. The effects of the rotational unbalance forces can be minimised by (i) increasing the frequency ratio to >3, or by (ii) by reducing the frequency ratio to <0.5. Increasing the frequency ratio means reducing the effective spring stiffness, and decreasing the frequency ratio means increasing the effective spring stiffness. Increasing the number of pads to forty-eight would reduce the effects of the unbalanced forces by about 60% as compared with the existing arrangement. Decreasing the number of pads to four would reduce the effects of the unbalanced forces by about 10%, but would generate instability problems.

56 mm; 3.58×10^5 N m^{-2}.

1.14

	Mobility	Impedance
Mass	$1/(\mathrm{i}m\omega)$	$\mathrm{i}m\omega$
Spring	$\mathrm{i}\omega/k_s$	$k_s/(\mathrm{i}\omega)$
Damper	$1/c_v$	c_v
Mass–spring–damper	$[c_v + \mathrm{i}(m\omega - k_s/\omega)]^{-1}$	$[c_v + \mathrm{i}(m\omega - k_s/\omega)]$

1.15 8411 kN m^{-1}; 37.3 kN.

1.16 4.26 mm; 0.1084; 867.3 N.

1.17

$$\frac{8C_F X\omega}{3\pi}; \quad X = \frac{F_0/k_s}{\{(1-\omega^2/\omega_n^2)^2 + (8C_F X\omega^2/3\pi k_s)^2\}^{1/2}};$$

$$X_{res} = \left(\frac{3\pi F_0}{8C_F\omega_n^2}\right)^{1/2}; \quad \text{(Note: } C_F = \rho C_D A_p/2\text{)}.$$

1.18 $(F_0/k_s)(1 - \cos\omega_n t)$; the peak response is twice the static deflection.

1.19 $(g/\omega_n^2)\{(1-\cos\omega_n t_0)^2 + (\omega_n t_0 - \sin\omega_n t_0)^2\}^{1/2}$

(t_0 is the time taken for the box to strike the floor); 41.8 kN.

1.20 $R_{yy}(\tau) = R_{y1y1}(\tau) + R_{y1y2}(\tau) + R_{y2y1}(\tau) + R_{y2y2}(\tau)$. Note the auto-correlation of a deflection at a given point due to separate loads cannot be determined by adding the auto-correlations resulting from each load acting separately.

1.21 $h(t) = (R_{xy}(\tau)/2\pi S_0)$.

1.22 0; X; $\{aX/\pi(a^2+\omega^2)\}$.

1.23 46.20 mm^2;

$$46.2\left(\frac{\sin 2000\pi\tau}{2000\pi\tau}\right)\text{ mm}^2.$$

1.24

$$G_{yy}(\omega)=\frac{G_{FF}(\omega)}{(k_s-m\omega^2)^2+c_v^2\omega^2};\ 4.24\ \text{mm}.$$

1.26 1938 N.

1.27 $-\text{i}13.28$ Ns m^{-1} (the impedance is reactive); the nett energy transfer is zero.

1.28

$$u(x,t)=\frac{8Vl}{\pi^3 c_s}\sum_{n=0,1,3}^{\infty}\frac{(-1)^{(n-1)/2}}{n^3}\sin\frac{n\pi x}{l}\sin\frac{n\pi c_s t}{l}.$$

1.29 37.8 Hz; when $\rho AL \gg M$ the system behaves like a bar fixed at one end and free at the other.

1.30 0.209 mm: 6.30×10^{-2} mm; 0.155 ms; 0.274 ms; not a straight algebraic sum because of phase differences; 7.54×10^{-2} mm; 0.302; evaluation of the steady-state transmission coefficient due to multiple reflections requires computation of multiple reflections and transmissions at the discontinuities – the amplitude ratio will eventually reach a constant steady-state value.

1.31

$$\phi_n(x)=\left(\frac{2}{l}\right)^{1/2}\sin\frac{n\pi x}{l};\qquad \omega_n=\left(\frac{n^4\pi^4}{l^4}\frac{EI}{\rho_L}\right)^{1/2};$$

$$H_i=\left(\frac{2}{l}\right)^{1/2}\sin\frac{n\pi}{2};\qquad D_n(t)=1-\cos\omega_n t.$$

1.34 0.200 mm.

Chapter 2

2.1 $I=(\hat{p}^2/2\rho_0 c)=(p_{\text{rms}}^2)/\rho_0 c)$; 5.655×10^{-4} N m^{-1} s^{-1}; 7.435×10^{-7} J m^{-3}; 1.487×10^{-6} J m^{-3}.

2.2 +ve plane wave: $\hat{u}=\hat{p}/\rho_0 c$; $\hat{\xi}=\hat{u}/\text{i}\omega=\hat{p}/\text{i}\omega\rho_0 c$; p is in phase with u; p leads ξ by 90°.
−ve plane wave: $\hat{u}=-\hat{p}/\rho_0 c$; $\hat{\xi}=\text{i}\hat{p}/\omega\rho_0 c$; p is 180° out of phase with u; p lags ξ by 90°; 4.82×10^{-3} N m^{-2} (96.8 dB); 7.67 μm; 4.82 mm s^{-1}; 1.41 N m^{-2}; 97 dB.

2.3 $p_{\text{rms}}=\{\langle p_1^2\rangle+\langle p_2^2\rangle\}^{1/2}$; $p_{\text{rms}}=\{\langle p_1^2\rangle+\langle p_2^2\rangle+2p_{1\text{rms}}p_{2\text{rms}}\cos(\theta_1-\theta_2)\}^{1/2}$, where θ_1 and θ_2 are the respective phases; 81.2 dB; 83.9 dB (assume that the sources are in phase, i.e. $\cos(\theta_1-\theta_2)=1$).

2.4 $c = \infty$.

2.5 $0.0352\ \mathrm{m\,s^{-1}}$; 2.24×10^{-7} m.

2.6 0.274 m; $12.65\ \mathrm{N^2\,m^{-4}}$; 3.45 W.

2.7 77.6 dB.

2.8 10.7 N.

2.9 0.0055.

2.10

$$\mathbf{p}(r, \theta, t) = \frac{\rho_0 \omega \cos\theta}{4\pi}\left(\frac{\omega Q_{\mathrm{dp}}}{cr} - \frac{\mathrm{i}Q_{\mathrm{dp}}}{r^2}\right)\mathrm{e}^{\mathrm{i}\omega(t - r/c)};$$

$$\mathbf{u}_r(r, \theta, t) = \frac{\cos\theta}{4\pi}\mathrm{e}^{\mathrm{i}\omega(t - r/c)}\left(\frac{-\mathrm{i}\omega Q_{\mathrm{dp}}}{cr^2} - \frac{2Q_{\mathrm{dp}}}{r^3}\right)$$

$$+\left(\frac{-\mathrm{i}\omega}{c}\right)\mathrm{e}^{\mathrm{i}\omega(t - r/c)}\frac{\cos\theta}{4\pi}\left(\frac{\mathrm{i}\omega Q_{\mathrm{dp}}}{cr} + \frac{Q_{\mathrm{dp}}}{r^2}\right);$$

$$\mathbf{u}_\theta(r, \theta, t) = = \frac{-\sin\theta}{4\pi}\left(\frac{\mathrm{i}\omega Q_{\mathrm{dp}}}{cr^2} + \frac{Q_{\mathrm{dp}}}{r^3}\right)\mathrm{e}^{\mathrm{i}\omega(t - r/c)}; \qquad \mathbf{Z}_\mathrm{a} = \rho_0 c; \qquad \mathrm{m^4\,s^{-1}}.$$

2.11 5.055×10^{-4} W (87 dB); 8.353×10^{-4} W (89.2 dB).

2.12 $2.855 \times 10^{-2}\ \mathrm{m\,s^{-1}}$; 0.0456 kg; 8586 N.

2.13 1.474 W.

2.14 $6.28 \times 10^{-4}\ \mathrm{m^3\,s^{-1}}$; $7.60 \times 10^{-4}\ \mathrm{kg\,s^{-1}}$; $Q'(t) = \mathrm{Re}\,\{\mathrm{i}4\pi a^2 U_\mathrm{a}\omega\, \mathrm{e}^{\mathrm{i}\omega t}(\rho_0 + 14.8\rho')\}$.

2.17 Δ/δ; $(\Delta/\delta)(c/\omega\delta)$.

2.20 r^{-4}.

2.21 12.4 dB.

2.22 $54.4\ \mathrm{m\,s^{-1}}$.

2.23 The smaller engine is 11.2 dB louder.

2.24 Doubling the nozzle area produces a 3 dB increase in radiated noise (and a doubling of the thrust). Increasing the velocity by $\sqrt{2}$ produces a 12 dB increase in radiated noise (and a doubling of the thrust).

2.25 285 N.

Chapter 3

3.1

	Aluminium	Brass	Glass	Lead	Steel
c_B (m s^{-1})	140.1	116.1	138.9	69.3	137.6
f_C (Hz)	5990	8675	6102	24 461	6213

3.2 No nett sound is radiated at 800 Hz (only a near field is present). At 10 000 Hz, the peak radiated sound pressure level is 112 dB.

3.3 9.55×10^{-5} W (80 dB).

3.4 0.20.

3.5 The dominant modes are the acoustically slow resonant edge modes; thirty-five resonant modes are present in the octave band; $\sigma = 0.0407$; 1.713×10^{-3} W (92.3 dB).

3.6 7.25×10^{-6} W (68.6 dB); 92.3 dB.

3.7

Octave band (Hz)	Π_{dp}/Π_{rad}	Π_{dl}/Π_{rad}
31.5	0.1068	1.4637
63	0.1415	1.3703
125	0.1862	1.2808
250	0.2460	1.1964
500	0.3253	1.1187
1000	0.1878	0.4566

At frequencies above the critical frequency, the ratios start to decrease significantly – i.e. the resonant response dominates.

3.9

$$\sigma = \frac{(1.55 \times 10^{-2} f)^4}{12 + 4(1.55 \times 10^{-2} f)^4} = 0.25 \text{ at 500 Hz; 0.1112 W (110.5 dB).}$$

3.10 0.134; $(5360/\cos\theta)$ N s m^{-3}.

3.11

Octave band (Hz)	Transmission loss (TL)
63	35.8
125	37
250	37
500	42.5
1000	52.5
2000	62.5
4000	72.5

3.12 51.29 Nm.

3.13 $f_{CO} = 12.2/t \sin^2 \theta$; 2033 Hz.

3.14

Octave band (Hz)	Diffuse field TL (dB)	Non-diffuse field TL (dB)
31.5	6.4	0
63	12.2	3
125	18.1	9
250	24.1	15
500	30.1	21
1000	36.1	27
2000	42.1	33
4000	48.1	39
8000	27	27
16000	38.6	38.6

3.15 209 Hz; the TL performance can be improved by increasing the surface mass ratio (i.e. $\rho_{s1} \neq \rho_{s2}$) – this requirement is not compatible with optimum high frequency performance which requires that $\rho_{s1} = \rho_{s2}$.

3.16

	Air	Fluid-loaded
$f_{1,1}$	17.5 Hz	3.53 Hz
$f_{12,12}$	2525 Hz	1466 Hz

3.17 132.4 dB.

Chapter 4

4.1 97.9 dB.

4.2 86 dB; 87.8 dB.

4.3 91.7 dB.

4.4 3668 Hz; 4362 Hz; 71 dB.

4.5 101 dB; 76 dB; 82 dB.

4.6

DI_θ	Q_θ	L_Π	$L_{P\theta}$ at 9.2 m
+2.7	1.86	117.6	90
−4.3	0.37	117.6	83
+1.7	1.48	117.6	89
+1.7	1.48	117.6	89
−1.3	0.74	117.6	86
+0.7	1.17	117.6	88
−2.3	0.59	117.6	85
−5.3	0.30	117.6	82

Sound power emitted is 0.570 W.

4.7 Overall noise levels: 60.9 dB(A); 54.8 dB(A).
Motor cycle noise levels; 59.9 dB(A); 53.8 dB(A).
The sound source type does not have any bearing on the estimated noise levels because one is dealing directly with sound levels and distances from the source rather than sound power.

4.8 83.4 dB; 3 dB per doubling of distance; 80.4 dB; 74.4 dB.

4.9 33 m.

4.10 70.6 dB (constant volume source);
58.6 dB (constant pressure source);
49.4 dB (constant volume source);
37.4 dB (constant pressure source).

4.11

Octave band (Hz)	63	125	250	500	1000	2000	4000	8000
L_Π (dB)	99.7	104.7	105.9	101.4	90.4	84.7	78.4	79.7
R (m^3)	109.8	109.8	21.5	54.2	54.2	109.8	54.2	109.8
α	0.33	0.33	0.088	0.195	0.195	0.33	0.195	0.33

$\bar{L}_p$ (1.5 m) = 102.1 dB; $\bar{L}_p$ (3.0 m) = 100.6 dB; $\bar{L}_\Pi = 109.7$ dB;
$\bar{R} = 38.82$ m^2; $\alpha_{avg} = 0.1477$.
Room is essentially average (neither live nor dead) – i.e. it is semi-reverberant.

4.12 Position B: 96.9 dB; 92.9 dB(A).
Position C: 95.3 dB; 91.0 dB(A).
The room radius (r_C) is <3 m for all octave bands, hence both people are in the reverberant field.

4.13 96 dB. This sound pressure level is still very high. Hence one needs to either alter the internal absorption coefficients or the transmission loss of the enclosure material. Increasing the internal absorption is not efficient since the absorption is already ~0.7. It is more appropriate to choose a different material for the hood with a larger transmission loss.

4.14 For a plane wave, there is no addition or subtraction of mass from the source region. Hence, the impedance is purely resistive. For a spherical wave, there is addition and subtraction of mass from the volume source. Here, the impedance is both resistive and reactive.

4.15 79 dB.

4.16

Octave band (Hz)	63	125	250	500	1000	2000	4000	8000
TL (dB)	14.4	25.2	28.9	34.4	35.2	34.7	34.7	31.6

A close-fitting enclosure would (i) produce a low frequency, double-leaf panel resonance, and (ii) produce higher frequency cavity resonances.

4.17 27.6 dB; 12.7 dB.

4.18

Octave band (Hz)	125	250	500	1000	2000	4000
L_P (dB)	72.3	60.4	41.4	41.0	33.8	30.7
L_{pA} (dB(A))	56.2	51.5	38.2	41.0	35.0	31.5

4.19 92.8 dB (using the constant power model); 98.8 dB (using the constant volume model). The reverberant component dominates the sound field, hence the room is live.

4.20 Plywood and plasterboard have higher fundamental natural frequencies than lead or particle board. They also have significantly lower critical frequencies. Lead meets the required *TL* in all one-third-octave bands. Particle board has slightly lower TL values in the mid frequency range. It is a good compromise since it is considerably lighter and cheaper.

	f_{11} (Hz)	f_c (Hz)
Plywood	57	558
Particle board	6.2	5142
Lead sheet	1.0	31 132
Plasterboard	43.4	739

One-third-octave band (Hz)	125	160	200	250	315	400	500	600
TL (lead)	20.2	22.4	24.2	26.2	28.3	30.3	32.3	34.3
TL (particle board)	17.9	19.8	21.7	23.6	25.6	27.7	29.6	31.7
One-third-octave band (Hz)	800	1000	1250	1600	2000	2500	3150	4000
TL (lead)	36.3	38.3	40.2	42.4	44.3	46.2	48.3	50.3
TL (particle board)	33.7	35.7	37.6	39.7	41.7	43.6	45.6	47.6

4.21

Octave band (Hz)	125	250	500	1000	2000	4000
NR gypsum	18.5	26.8	38.0	47.8	47.3	43.9
NR single brick	36.5	35.8	41.0	48.8	58.3	61.9
NR double brick	37.5	40.8	49.0	62.8	65.3	65.9

Factors that could cause deterioration in the calculated noise reduction performance include flanking transmission via mechanical connections and air leaks.

4.22 At 2000 Hz, α is increased by 0.0253; at 4000 Hz, α is increased by 0.0642; at 6300 Hz, α is increased by 0.1342; at 8000 Hz, α is increased by 0.2263.

4.23 0.333.

4.24 6.8 dB; 6 dB; 6 dB.

4.25 $f_z = 7.07$ Hz, $f_{xy} = 8.66$ Hz; $f_{xzr} = 2.47$ Hz; $f_{xzp} = 6.79$ Hz; $f_{yzr} = 2.64$ Hz; $f_{yzp} = 11.11$ Hz.

4.26 4.58 mm.

4.27 0.45.

4.28 $k_{s2} = 0.144k_{s1}$;

$$X_2 = \frac{X_0\{(k_{s1}^2 + c_{v1}^2\omega^2)\}^{1/2}}{0.144k_{s1}}.$$

Chapter 5

5.1 2.5; 1.016; 1.466; 0.659.

5.2 8.96; 26.39.

5.3 10 000.

5.4 10.6 dB; 0.9999.

5.5 $(A^2/2)\cos\tau$.

5.6 1; 0; 1; 0.

5.7 Propagation path (b) has the largest contribution to the overall noise level at the receiver. Propagation path (d) has the smallest contribution. The ability to identify correlation peaks corresponding to individual paths becomes increasingly difficult as the frequency bandwidth is reduced – i.e. the oscillations within the envelope become larger as the bandwidth is reduced.

5.8 13.33 ms.

5.9 S_0.

5.10 Echo removal utilising the complex cepstrum.

5.11 The impulse response technique.

5.12 2.

5.13 0.266%.

5.14 753 times faster.

5.15 12 800 Hz; 6400 Hz; 0.1; 10 000.

5.16 400; 40 000; 6.12×10^5.

5.19 The normalised random error for a single digital time record is unity.

5.20

$$\mathbf{H}_{\mathbf{xy}}(\omega) = \mathbf{H}(\omega)\left\{1 + \frac{\mathbf{G}_{\mathbf{xm}}(\omega)}{G_{xx}(\omega)}\right\};$$

$$\mathbf{H}_{\mathbf{xy}}(\omega) = \mathbf{H}(\omega)\left\{1 + \frac{G_{mm}(\omega)}{\mathbf{G}_{uu}(\omega)}\right\}^{-1}.$$

When the extraneous noise, $m(t)$ is correlated with the measured input signal, $x(t)$, the first equation is a biased estimate of $\mathbf{H}(\omega)$. If $\mathbf{G}_{\mathbf{xm}}(\omega) = 0$ (i.e. the extraneous noise is uncorrelated), then it is a true estimate. The second equation is always a biased estimate of $\mathbf{H}(\omega)$ and it is a function of the signal to noise ratio.

5.21

$$\mathbf{X}(\omega) = \left(\frac{kT}{\pi}\right)\frac{\sin \omega T}{\omega T}\mathrm{e}^{-i\omega t}; \qquad \frac{T}{\pi}\frac{\sin \omega T}{\omega T}.$$

Chapter 6

6.1 8.00×10^{-3}; 1.60×10^{-2}; 62.40.

6.2 Add damping treatment to the first group and provide vibration isolation between the two subsystems (see section 6.2).

6.4

$$\frac{E_1^*}{E_2^*} = \frac{n_2}{n_1}\left(\frac{4\eta_2 + 5\dfrac{n_1}{n_2}\eta_{12}}{\eta_1 + 5\eta_{12}}\right); \qquad \eta_2 = \left(\frac{n_1}{n_2}\right)\left(\frac{\eta_1}{4}\right).$$

6.6

$$E_1 = \frac{\Pi_1(1 + \eta_{21}/\eta_2)}{\omega_1\eta_1(1 + \eta_{21}/\eta_2) + \omega_2\eta_{21}}; \qquad E_2 = \frac{\Pi_1\eta_{21}/\eta_2}{\omega_1\eta_1(1 + \eta_{21}/\eta_2) + \omega_2\eta_{21}}.$$

	E_1	E_2
(a)	η_1, η_2	η_1, η_2
(b)	η_1	$\eta_1, \eta_2, \eta_{21}$
(c)	η_1	η_1
(d)	η_{21}	η_2

6.7

Octave band (Hz)	Longitudinal		Flexural	
	$n(f)$	$n(f)\Delta f$	$n(f)$	$n(f)\Delta f$
500	0.004	1.40	0.0148	5.24
1000	0.004	2.80	0.0105	7.42
2000	0.004	5.60	0.0074	10.44
4000	0.004	11.20	0.0052	14.69
8000	0.004	22.40	0.0037	20.91

6.8 The practical options available are: (i) decreasing the effective surface area by stiffening the plate, and/or (ii) increasing the thickness of the plate.

6.9

Octave band (Hz)	$n(f)$	$n(f)\Delta f$
500	617	2.18×10^5
1000	2433	1.72×10^6
2000	9662	1.36×10^7
4000	38 508	1.09×10^8
8000	153 750	8.69×10^8

6.10 4.26×10^{-4}; 3.92×10^{-4}; 1.31 W.

6.11 Stiffness effects dominate low mobility waves, whereas added mass effects dominate high mobility waves. Bending waves are high mobility waves because their wavespeeds are relatively low. Hence, any stiffness effects are small compared to the added mass effects which produce significant force variations.

6.12

Octave band (Hz)	Transmission loss (TL)
31.5	8.30×10^{-4}
63	6.50×10^{-4}
125	5.19×10^{-4}
250	4.17×10^{-4}
500	3.40×10^{-4}
1000	5.15×10^{-4}
2000	4.72×10^{-3}

1.0×10^{-4} (at high frequencies, $\sigma \rightarrow 1$, hence η_{rad} decreases with increasing ω).

6.13

Octave band (Hz)	Welded joint η_{12}	Welded joint η_{21}	Bolted joint η_{12}	Bolted joint η_{21}
125	3.15×10^{-3}	5.77×10^{-3}	1.44×10^{-2}	2.64×10^{-2}
250	2.23×10^{-3}	4.09×10^{-3}	7.19×10^{-3}	1.32×10^{-2}
500	1.58×10^{-3}	2.90×10^{-3}	3.60×10^{-3}	6.60×10^{-3}
1000	1.12×10^{-3}	2.05×10^{-3}	1.80×10^{-3}	3.30×10^{-3}
2000	7.89×10^{-4}	1.45×10^{-3}	8.99×10^{-4}	1.65×10^{-3}

6.14 3.33×10^{-3}.

6.16 $L_{p1} - L_{p2} = NR = 10 \log_{10}(E_1/V_1)/(E_3/V_3)$, where V_1 and V_3 are the respective volumes, and

$$\frac{E_1}{E_3} = \frac{\eta_3 + 2\dfrac{n_1}{n_3}\eta_{13} + \dfrac{n_2}{n_3}\eta_{\text{rad}}}{\eta_1 + \eta_{12} + 2\eta_{13}}.$$

6.17 Yes (see equation 6.113).

6.18 25.7 MPa; 7.7 MPa.

Chapter 7

7.1

Mode	No flow $(f_{co})_{pq}$ (Hz)	No flow k_x (m^{-1})	With flow $(f_{co})_{pq}$ (Hz)	With flow k_x (m^{-1})
(1, 0)	934	0	812	−8.23
(2, 0)	1550	0	1348	−13.66
(0, 1)	1945	0	1691	−17.13
(3, 0)	2132	0	1854	−18.79
(4, 0)	2699	0	2347	−23.78
(1, 1)	2706	0	2354	−23.84

7.2 264 Hz; 428 Hz; 503 Hz.

7.4 4.22×10^{-3}; 6.20×10^{-3}; 136.8; 93.1.

7.5 There are three coincidence regions – two with the (m 1) structural modes, and one with the (m, 2) structural modes. The (9, 1) and (10, 1) structural modes are coincident with the (1, 0) acoustic mode; the (19, 1) and (20, 1) structural modes are coincident with the (1, 1) acoustic mode; the (26, 2) and (27, 2) structural modes are coincident with the (2, 0) acoustic mode. All the coincidences are wavenumber coincidences. The complete coincidence frequency associated with the (1, 0) acoustic mode is 0.1238; the complete coincidence frequency associated with the (2, 0) acoustic mode is 0.2105; the complete coincidence frequency associated with the (1, 1) acoustic mode is 0.3542. All three complete coincidence frequencies are very close to (but greater than) the associated higher order acoustic mode cut-off frequencies.

When there is a mean flow superimposed on the propagating acoustic wave, the cut-off acoustic wavenumber becomes negative (see Figure 7.7). The complete coincidence frequencies will always be greater than the associated higher order acoustic mode cut-off frequencies. This is not always the case for wavenumber coincidence, as one of the structural modes with a negative axial wavenumber could have a frequency less than the cut-off frequency.

7.6 660 Hz; 683.3 Hz; shock noise downstream of the valve.

7.7 The three sources of vibration and noise are summarised below. Firstly, low frequency vibration excitation is produced because of interactions between longitudinal sound waves in the inlet pipe (riser) and the spring–mass effect of the piston and the volume of air above it in the relief valve. Secondly, coincidence between internal higher order acoustic modes inside the piping and the pipe wall natural frequencies produces significant vibration and noise in the inlet and outlet piping. Spiralling acoustic waves are set up at the inlet/outlet/tee-junction and at the sonic throat inside the valve. One would expect the inlet tee-junction to be the dominant source of this mechanism because of the large change in flow velocity (i.e. large momentum change) at that point. Thirdly, complex flow–acoustic interactions at regions in proximity to the valve opening produce significant noise and vibration in the valve itself. The frequency spectrum is dependent upon the valve pressure ratio. For pressure ratios <3 turbulent mixing in the vicinity of the valve is the dominant mechanism; for pressure ratios >3 shock noise immediately downstream of the valve is the dominant mechanism. The dominant frequencies associated with the first source are the inlet riser longitudinal natural frequencies and its associated harmonics – i.e. 49.5 Hz, 148.5 Hz, 247.5 Hz, 346.5 Hz, etc. The dominant frequencies associated with the second source are the coincidence frequencies associated with the first few higher order acoustic modes in the inlet riser. The first two higher order acoustic modes occur at 1416 Hz and 2349 Hz. The coincidence frequencies will be close to these frequencies. The dominant frequency associated with the third source (shock noise downstream of the valve) is 660 Hz.

7.8 5.54 mm.

7.9 117.6 dB.

7.10 In all three cases, the sound transmission loss is proportional to the square of the pipe wall thickness (see equations 7.55, 7.56 and 7.57). Additional damping will only be effective in (b).

Chapter 8

8.1 1.414; ~3; 2.5–3.5; 3–10.
Crest factors are reliable only in the presence of significant impulsiveness. The impulsive content is smeared out as the damage becomes widespread.

8.2 ~3; >4.
Because kurtosis is based upon detecting impulsiveness, it is subjected to the same limitations as crest factors – i.e. it is not suitable for the detection of widespread damage.

8.4 Excite the structure at a discrete frequency, vary (sweep) the frequency until a resonance is detected, and synchronise the signal to be measured with the excitation signal, i.e. trigger the measurement at a specific point in the excitation cycle. This procedure removes unwanted signal components – synchronous signals average to their mean value, whilst the non-synchronous signals average to zero. One can thus map the deflections at the various positions on the chassis relative to a specific point in the excitation cycle.

8.5 $f_s = 33.33$ Hz; $f_{bcsor} = 13.80$ Hz; $f_{re} = 91.6$ Hz; $f_{repfo} = 207$ Hz; $f_{resf} = 183.3$ Hz; $f_{rciso} = 19.53$ Hz; $f_{recri} = 293$ Hz.

8.6 350 Hz; 35 Hz; 175 Hz.
The first higher order acoustic mode will be set up because the first rotating pressure pattern ($k = -1$) will correspond to its pressure distribution.

8.7 60 Hz (1 × shaft rotational frequency); 120 Hz (2 × shaft rotational frequency); 360 Hz (2 × electrical supply frequency); 3600 Hz (fundamental slot harmonic frequency).

Electromagnetic irregularities associated with dynamic eccentricity manifest themselves as a family of sidebands around the dominant slot harmonic frequency, at ± the rotational frequency, and at ± the slip frequency.

8.8 $\langle p^2 \rangle \rho_{pp} = \langle p_T^2 \rangle \rho_{TT} + \langle p_P^2 \rangle \rho_{PP} + \langle p_H^2 \rangle \rho_{HH}$.

Index

Printed in the United States
1945